Machining

Machining—Recent Advances, Applications and Challenges

Special Issue Editors

Luis Norberto López de Lacalle
Gorka Urbicain

MDPI • Basel • Beijing • Wuhan • Barcelona • Belgrade

Special Issue Editors
Luis Norberto López de Lacalle
University of the Basque Country
Spain

Gorka Urbicain
University of the Basque Country
Spain

Editorial Office
MDPI
St. Alban-Anlage 66
4052 Basel, Switzerland

This is a reprint of articles from the Special Issue published online in the open access journal *Materials* (ISSN 1996-1944) from 2018 to 2019 (available at: https://www.mdpi.com/journal/materials/special_issues/machining_recent_advances_applications_and_challenges)

For citation purposes, cite each article independently as indicated on the article page online and as indicated below:

LastName, A.A.; LastName, B.B.; LastName, C.C. Article Title. *Journal Name* **Year**, *Article Number*, Page Range.

ISBN 978-3-03921-377-1 (Pbk)
ISBN 978-3-03921-378-8 (PDF)

Cover image courtesy of Luis Norberto López de Lacalle Marcaide.

Contents

About the Special Issue Editors

Luis Norberto López de Lacalle is a full professor at the University of Basque Country, teaching Machine Dynamics, Manufacturing Systems, and Machine-Tools. He is the founder and manager of the CFAA (Centro de Fabricación Avanzada en Aeronáutica or Advanced Manufacturing Research Centre) and responsible for leading international and national projects. His topics of interest include, among others, traditional as nontraditional machining processes, process reliability, fault diagnosis, sustainability and efficiency in manufacturing, vibrations, etc. He is also the leader of the High-Performance Manufacturing Group at the University of the Basque Country and works for different national agencies as a project manager and reviewer. He is the author of a number of publications in high-impact journals (more than 150) and international conferences.

Gorka Urbicain is an assistant professor at the University of Basque Country, teaching Mechanics, Materials Resistance, and Manufacturing Systems. He completed his PhD in 2014, which was focused on stability in turning processes. His main topics of interest include the study of machining processes, cutting dynamics, including forces, vibrations, surface quality models, process monitoring, sustainability and efficiency in manufacturing, modal analysis, etc. He is the author of a number of publications in high-impact journals and international conferences.

Preface to "Machining—Recent Advances, Applications and Challenges"

Machining was the first manufacturing technology of all humankind since the Oldowan (or Mode I) stone tool was discovered, one of the first archaeological artifacts (style) in prehistory. This was a real cutting device that has been improving over the last 2.6 million years to reach the current technology of multitasking and high-speed machine tools. Both then and now, the material removal process has remained basic for people—in ancient times to cut meat and bones, and today to achieve tight tolerances in the metal cutting industry. This book is devoted to the latest advances in machining technology, involving process, machine-tools, cutting tools, and software. Over the last 20 years, productivity, defined as the material removal rate, has doubled, and the order of magnitude of conventional scale machining is around the hundredth of millimeters. Machining using tools with a defined cutting edge is a basic technology for making car components, windmill components, aero engine and fuselage elements, dies and molds, prosthesis, clocks and watches, and many other applications, combining precision and high productivity. Even new approaches such as additive manufacturing are not displacing machining as the only way to achieve precision and productivity. Machinists are realistic people, all the time considering that a good solution is not only for one specimen, but it must provide a sound and reliable solution for many similar pieces. That is why the new approaches related with Industry 4.0 concepts, or digitalization, or just the so-called IoT, can help to control the continuous processes of daily life production. Regarding machine tools, books written only 20 years ago still divided the machine tools for metalworking into lathes and milling machines, all were painted in standard green, and numerical control was defined as a high-tech feature. In 2020, machine tools are multistaking and multiprocess stations, suitable for turning, multiaxis milling, drilling, threading, gear making, and provided with a powerful numerical control. The name CNC is still in use, but hardware and internal control architecture are not an issue today, but the software and function that can obtain all possibilities to machine structures and characteristics and to connect the machine to full smart factories is. The first uses of the new 5G networks are in scientific journals today. Otherwise, current colors, general appearance, and marketing around the world of machine tools transmit the extended functionality of current machine tools.Cutting tools are in a rapid evolution, with new tool materials based on different grades of ceramic inserts, sintered carbide, and extra-hard PCBN and PCD. Some of them include new coatings based on PVD and CVD to reach maximum hardness, strength, and wear resistance. Over the last decade, there was a great step ahead in the novel use of super alloys, titanium alloys, composites, high-added value steels, and other materials. The first question in machining has always been which materials are going to be applied. Emerging alloys always push research to get practical and written data for the target end users. The usual term "machinability" depends not only on hardness but on many other physical and mechanical properties, such as thermal conductivity, tendency to strain hardening or thermal softening, reactivity, etc. All aspects related to final surface integrity and quality are of paramount importance, especially in aeronautics and aerospace applications. Some of the works use laser as a basic tool. Since its discovery and evolution in the 1960s it has become a basic tool for cutting, welding, surface treatment, and additive manufacturing in our workshops. It is nice now to watch again the industrial application in James Bond film Goldfinger (1964) and see how laser has spread in the industry. Both Book editors, Prof López de Lacalle and Ast.Prof Urbikain, are active members of the Basque Country community around manufacturing technologies. The former is the Director of a new center about aeroengine manufacturing and machine tools for this business. The latter is a leading expert in machine dynamics and vibrational problems in manufacturing. All the effervescent activities in this highly industrialized and small region regarding machining were initial ideas for this book. However, contributions in this book came from very different parts of the world, another proof that machining is spread in all our industries, and good ideas can come from very well-developed but emerging countries as well. Even in 2020, creative minds can propose open-minded solutions only applying the principle of "think it differently". Finally, we are grateful to all the support of Basque Government, Council or Province of Biscay, and the partnership of companies of the CFAA for the ideas and promotion of research and knowledge in this key technology: machining and machine tool.

Luis Norberto López de Lacalle, Gorka Urbicain
Special Issue Editors

Review

Thin-Wall Machining of Light Alloys: A Review of Models and Industrial Approaches

Irene Del Sol [1],*, Asuncion Rivero [2], Luis Norberto López de Lacalle [3] and Antonio Juan Gamez [1]

[1]   Department of Mechanical Engineering and Industrial Design, Faculty of Engineering, Universidad de Cádiz, 11519 Puerto Real, Spain; antoniojuan.gamez@uca.es

[2]   Tecnalia Research & Innovation. Scientific and Technological Park of Guipuzkoa, 20009 Donostia-San Sebastian, Spain; asun.rivero@tecnalia.com

[3]   Department of Mechanical Engineering, University of the Basque Country, 48013 Bilbao, Spain; norberto.lzlacalle@ehu.eus

*   Correspondence: irene.delsol@uca.es; Tel.: +34956483513

Received: 28 May 2019; Accepted: 19 June 2019; Published: 23 June 2019

Abstract: Thin-wall parts are common in the aeronautical sector. However, their machining presents serious challenges such as vibrations and part deflections. To deal with these challenges, different approaches have been followed in recent years. This work presents the state of the art of thin-wall light-alloy machining, analyzing the problems related to each type of thin-wall parts, exposing the causes of both instability and deformation through analytical models, summarizing the computational techniques used, and presenting the solutions proposed by different authors from an industrial point of view. Finally, some further research lines are proposed.

Keywords: thin-wall machining; chatter; vibration; deflection; damping; prediction; workholding; fixture; dynamic; stability

---

1. Introduction

A wide range of aeronautical parts such as stringers, ribs, frames, spars, hubs, blisks, turbine blades, shells, bulkheads or skin panels can be classified as thin-wall structures [1]. They are designed to avoid mechanical assembly using bolts or rivets and to keep a uniform behavior all over the part. Thin-wall structures are usually manufactured out of advanced materials widely used in the aeronautical sector such as aluminum and titanium alloys, although some high-performance materials such as Inconel or specific types of stainless steel can also be used. Their good corrosion resistance and strength properties allow reducing the quantity of material needed, obtaining slim parts with a good weight–resistance ratio [2]. Therefore, thin-wall parts are manufactured by removing large quantities of material from the original block through a machining process, typically achieving a 30:1 buy-to-fly ratio [1]. The in-process parts are characterized by their low stiffness and constant change of mechanical properties. Their thickness is at least six times lower than the two other relevant directions, thus being flexible and easy to bend.

This fact produces dynamic and static problems during the machining operation. On the one hand, dynamic instabilities become frequent, and self-excited vibrations or chatter are more likely to appear, increasing the roughness of the final part, the tool wear and the wear of the different machine components [3,4]. Forced vibrations provoked by the dynamic of the machining operation also affect the final part, therefore reducing its quality. On the other hand, from a static point of view, both clamping and cutting forces produce elastic deformation that can affect the final dimension and the roughness of the part [5,6]. Induced residual stressed may also modify the final geometry of the part during the process [7]. Additionally, preforms or skin panels have to be accurately positioned on the machining center to ensure the tight tolerances [8].

If these problems cannot be controlled, companies are forced to include reprocessing steps or discard useless parts, increasing the production cost. Being difficult to manufacture, thin-wall machining costs are justified just by weight reduction and fuel efficiency. For this reason, different organizations such as the European H2020 [9] have funded research projects on the improvement of thin-wall machining. Globally, this research line has been increasing, as can be appreciated by the number of papers focused on this topic published in international scientific databases (Figure 1).

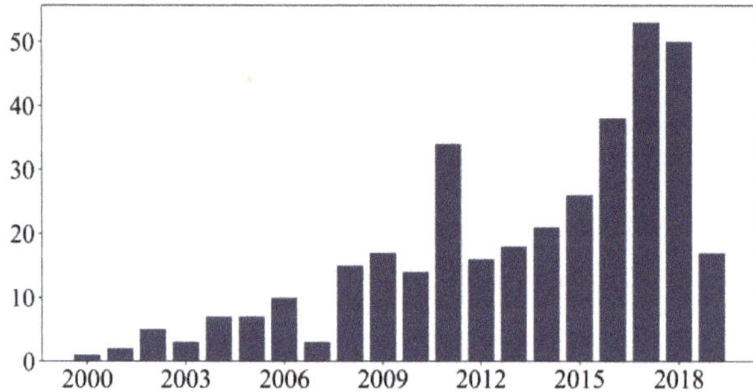

Figure 1. The number papers published on Web of Science related to thin-wall machining.

To ensure thin-wall quality, several authors have proposed analytical or computational dynamic models to predict the behavior of the part and select the best parameters to reduce chatter [10] or to damp the system [2]. Others have studied the deflection of the part due to the force interaction [11] or to the induced residual stress [12].

Alternatively, industry has developed special clamping or monitoring systems combined with adaptive control, such that quality is measured on-line and statistical parameters optimizations are used to monitor periodic changes [13].

The purpose of this work is to review the state of the art of thin-wall machining. Initially, thin-wall parts are identified, classified and associated to the critical problems previously introduced. Then, to understand the dynamic and static behavior of these machining operation, the most common analytics models are exposed. The proposed solutions, focused on models or industrial approaches, are explained in different sections and the reference distribution is summarized in Table 1. Finally, conclusions are drawn and possible future research lines are proposed.

Table 1. Thin-wall machining solutions. Model and industrial approaches.

	Models	
Thin-wall dynamic problems	Chatter and self-exciting aspects	[14–42]
	Resonance and amplification	[33,41,43–60]
	Quasi-static models	[36,49,61–70]
Thin-wall deformation	FEM modeling	[51,61,62,65,71–78]
	Residual Stresses	[79–88]
	Industrial Approach	
Parameter selection	Statistic and machine learning models	[62,89–95]
	Virtual Twins	[66,78,96–99]
Active solutions	Monitoring	[32,41,95,100–111]
	Measurements	[106,112–116]
	Fixtures	[83,116–126]
	Workholding	[19,75,127–131]
	Active damping actuators	[132–135]
	Stiffening devices	[136–140]

2. Type of Parts and Associated Problems

2.1. Thin-Wall Parts: Characteristics and Types

Thin-wall parts can include different types of parts, being their main characteristic their lack of stiffness and final slim factor, which is defined as their height divided by their thickness. Regarding the machining process and their characteristics, parts can be classified into two groups: monolithic blocks and skin panels.

Parts composing the first group have a geometry machined from monolithic blocks (Figure 2a–e). The part is obtained by removing a 90–95% of the initial material volume of the block by machining operations [141,142]. Stringers, ribs or frames are machined to obtain different pocket shapes, keeping their structural strength and reducing their weight. They are usually manufactured using three-axis machining, when the non-rigidity machining problems appear on the last steps for both, thin-floors and thin-walls. Impellers, blisks and blades are also included in this group, but their complex shapes require a constant change of the tool angle, changing the way the cutting forces are applied. Moreover, the cantilever produces a high deflection of the parts, making the control of the real depth of cut more difficult. Researchers use a sample part, represented in Figure 2e, as a simplification of the real cases to test and verify the machining parameters or to develop dynamic models reducing the computational time [143,144]. The behavior of the last roughing steps and end milling operation is then extrapolated to more complex shapes.

The other group, commonly known as skin panels, is mainly composed of shells, wings, fuselage parts (Figure 2g), bulkheads (Figure 2f), doors, satellite parts and frames, presenting a slim factor higher than ten [145]. Their buy-to-fly ratio is generally lower than those of monolithic parts. They are machined to pocket a large area of their surface, reducing the weight of the part. These pockets were traditionally machined using chemical milling, a highly pollutant process that does not induce residual stress and simplifies the clamping system [146]. However, since 2007, and mainly due to environmental reasons, special milling CNC centers have been designed and used for this purpose [147]. They are built with two symmetrical heads, hence the operation is named mirror milling. The machining head works perpendicular to the surface and a second head follows the machining head to ensure the support of the part and to reduce the deflection.

2.2. Dynamic and Static Problems

Generally, the main problem of thin-wall machining is the vibrations associated to their low rigidity. Depending on their cause, vibrations are considered self-induced (chatter) or forced.

Chatter takes place when the natural frequency response (FRF) of the system is excited due to the machining operation [14]. These instabilities are usually related to the tool vibrations produced during the machining but the most important one is the FRF of the part [17,25,27], which is constantly changing due to geometry variations. This cyclical behavior changes the FRF of the system and generates an unstable machining process [11,28]. Forced vibration or amplification takes place when the stiffness of the part is not enough to maintain a constant chip thickness. The workpiece and the tool deflect down to the edge action, producing a vibration at the same frequency as the spindle speed or multiples of it [32]. Both cases modify the contact between part and tool edge, changing the chip width, which affects the real cutting forces. Instabilities usually produce marks on the part that increase the roughness of the final product, affecting its final surface quality [45,60].

The other main problem associated to the low stiffness of the part is the dimensional error produced due to the deflection of the part, a static issue not considered on the machining of rigid parts. Machining of rigid parts usually deals with the flexibility of the cutter system [74], although it is not usually taken into account in traditional milling models. However, it is common to have an influence of the part and the cutting tool flexibility in thin-wall machining.

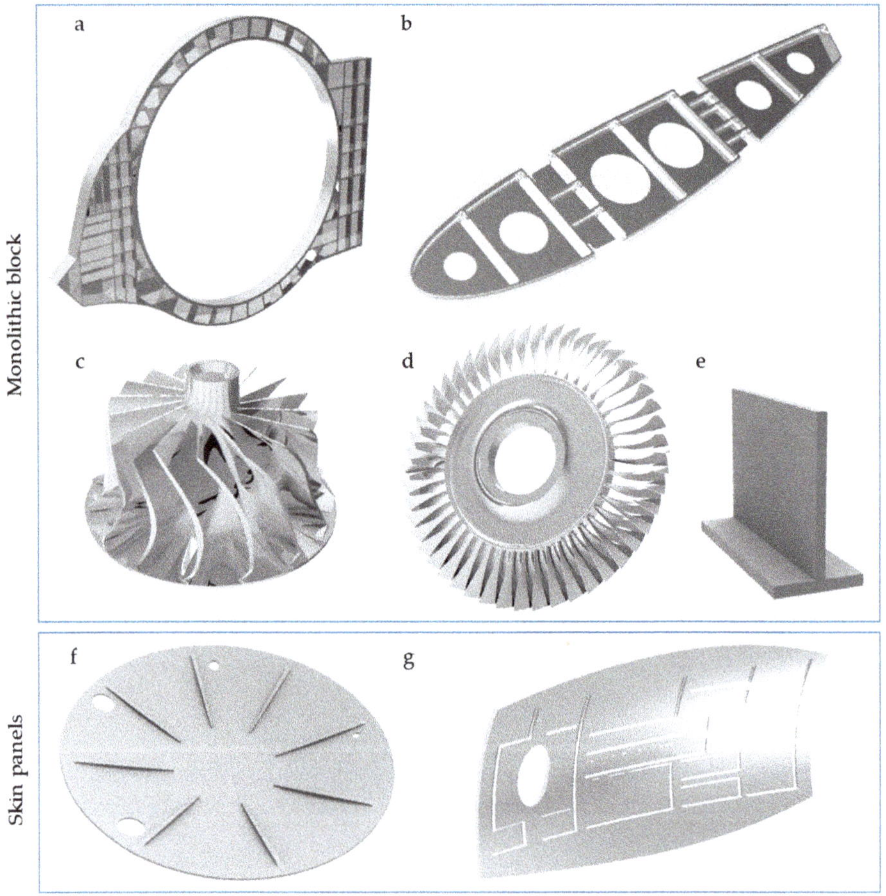

Figure 2. Examples of thin-wall parts: (**a**) frame; (**b**) rib; (**c**) impeller; (**d**) blisk; (**e**) sample parts; (**f**) bulkhead; and (**g**) fuselage skin.

Static deflection can appear due to the interaction of the cutting forces [63]. In this case, the deformation usually depends on the cutting strategy (up-milling or down-milling) and the cutting parameters, which define the cutting forces and, therefore, the deformation of the system [64,66–68,148]. Currently, high speed milling reduces cutting forces [2,10] and induces less residual stresses [7], but sometimes this technology cannot remove the deflection completely. This fact is aggravated in mirror milling due to the real part geometry variations [75,76]. Skins come from double curve processes and their position on the clamping system do not usually match the designed one, producing overcutting areas. Additionally, those parts are larger than the used in monolithic blocks, so the workholdings and fixtures do not usually ensure the tolerances of machining for the final parts and increase the difficulty of the whole machining process.

The different approaches found to predict and solve both issues are summarized in Figure 3, showing a complete work flow for thin-wall machining process. It covers the sequences used in physical and statistical models, commonly applied to improve the machining performance. For design and pre-industrialization stages, computational models and virtual twins allow selecting parameters or toolpath to reduce chatter and part deformation as well as to validate the design of specific work holding and stiffening devices. In industry, different workholding, stiffening devices or adaptive control solutions based on statistics and machine learning models, have been developed following the outlines of industry 4.0.

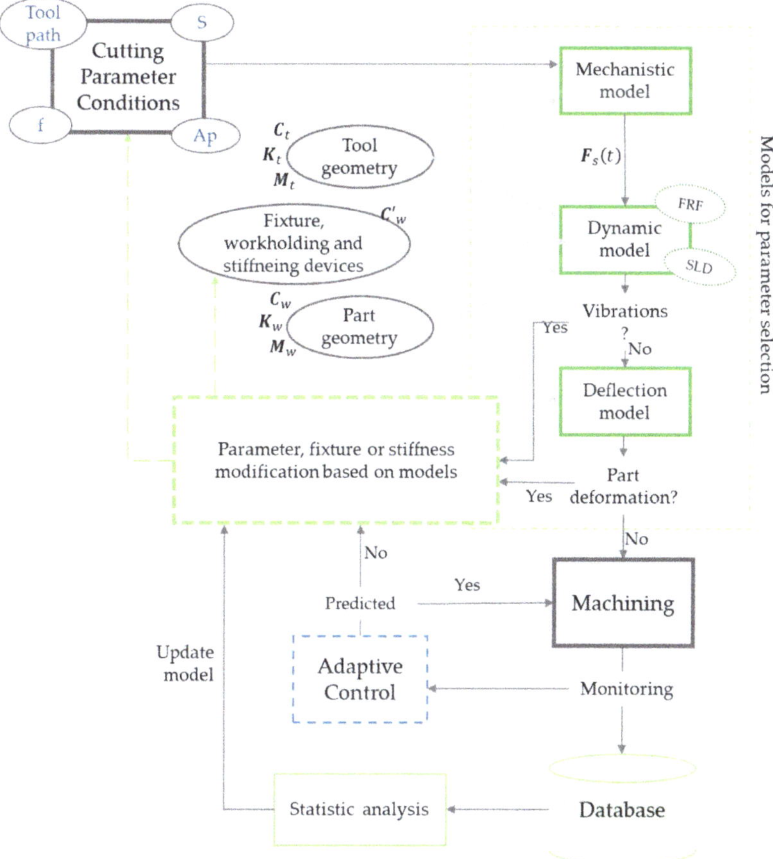

Figure 3. Scheme of the thin-wall machining process work flow.

3. Analytic Models

Dynamic response and machining performance of thin-wall parts were widely studied in the 1990s by Budak et al. [22,23]. Most of the research studies focused on the design of new methods to predict the behavior of the system are based on the FRF [57] and the deformations produced by the cutting forces [67]. Their final objective is the selection of the tool geometry and cutting parameters such that chip thickness is constant and vibrations are reduced. For this purpose, and to understand the physical fundamentals of thin-wall machining, analytic models of the forces, dynamics and deflection are explained in the following sections.

3.1. Cutting Force Prediction

The expected forces can be calculated using a mechanistic model, which is adapted to the machining parameters, the tool, and the material as a function of the force coefficient. Tool geometry is very important, particularly to reduce vibration [1], and mechanistic models may be adapted to it. For this reason, Gradisek et al. [52] presented a generalization of the classical mechanistic model for most commercial tool geometries. Urbikain et al. [43,149] developed a specific model for a barrel tool shape, in which the positioning angle of the tool and runout—eccentricity of the tool due to its positioning on the head spindle—is also considered. Ma et al. [44] performed a similar work including the relative position of the tool for five-axis applications and its effect on dynamic stability. In this section, the mechanistic model is described using a standard bull end mill (Figure 4a) considering also the tool angle position (λ).

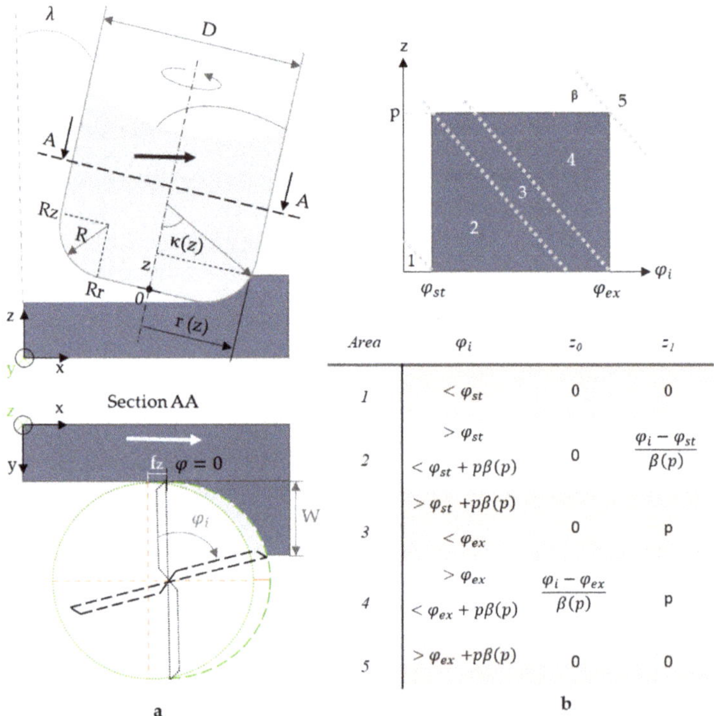

Area	φ_i	z_0	z_1
1	$< \varphi_{st}$	0	0
2	$> \varphi_{st}$, $< \varphi_{st} + p\beta(p)$	0	$\dfrac{\varphi_i - \varphi_{st}}{\beta(p)}$
3	$> \varphi_{st} + p\beta(p)$, $< \varphi_{ex}$	0	p
4	$> \varphi_{ex}$, $< \varphi_{ex} + p\beta(p)$	$\dfrac{\varphi_i - \varphi_{ex}}{\beta(p)}$	p
5	$> \varphi_{ex} + p\beta(p)$	0	0

Figure 4. (a) Tool geometry; and (b) integration limits selection.

The force is obtained by integration of the differential edge elements contributing to the cut and summing the engaged teeth. The differential equation of the forces considering $q = \{t, r, a\}$ as the component tangential, radial or axial, respectively, can be defined as:

$$\partial F_q(\varphi, z) = K_{qe}\partial S + K_{qc}h(\varphi, z)\partial b \tag{1}$$

where K_{qe} comprises the force coefficients related to the friction phenomena and K_{qc} the ones related to the cutting action. ∂b is the differential chip width and ∂S is the differential edge length. h is the chip thickness as a function of the rotated angle (φ) and the axial depth of cut (z).

$$\partial b = \frac{\partial z}{\sin \kappa(z)} \qquad \partial S = \frac{\partial z}{\sqrt{1 - \left(\frac{R_a - z}{R}\right)^2}} \tag{2}$$

The instant rotation angle is defined as:

$$\varphi(\varphi_i, z) = \varphi_i - (j-1)\frac{2\pi}{N} - \beta(z) \tag{3}$$

where j is the teeth number from 0 to N, considering N the total number of teeth and $\beta(z)$ the helix angle as a function of the instant depth of cut, and can be defined as:

$$\beta(z) = \left(1 - \frac{R_a - z}{R}\right)\tan\beta_0 \tag{4}$$

The instant chip thickness is defined as:

$$h(\varphi, z) = f_z \sin\varphi(\varphi_i, z)\sin\kappa(z)\sin\lambda \tag{5}$$

The integration limits are determined by the starting and ending angle of engagement of the tool and the z minimum (z_0) and maximum (z_1) values applied on the instant angle (Figure 4b).

Forces in Cartesian coordinates can be calculated including the position angle of the tool.

$$\frac{\partial F_{x,y,z}}{\partial z} = \begin{bmatrix} \sin\lambda & 0 & -\cos\lambda \\ 0 & 1 & 0 \\ \cos\lambda & 0 & \sin\lambda \end{bmatrix} \begin{bmatrix} -\cos\varphi & -\sin\varphi\sin\kappa & -\sin\varphi\cos\kappa \\ \sin\varphi & -\cos\varphi\sin\kappa & -\cos\varphi\cos\kappa \\ 0 & \cos\kappa & -\sin\kappa \end{bmatrix} \begin{bmatrix} \partial F_t \\ \partial F_r \\ \partial F_a \end{bmatrix} \tag{6}$$

Considering more than a single tooth, the total force applied is defined as:

$$F_{x,y,z}(\varphi_i) = \sum_{j=0}^{N}\left[\int_{z_1}^{z_2}\partial F_{xyz,j}(\varphi_i, z)\right] \tag{7}$$

The accuracy of those models depends on the fitting used for the calculation of the force coefficients (K_{qc} and K_{qe}), and they are commonly predicted through experimental tests. Coefficient values are considered constants for different feed rate per tooth (f_z) while ∂b and ∂S depend on the tool geometry. The differential forces are studied on the Cartesian edges. Average forces for each test are used to calibrate the values of K_{qe} and K_{qc} for each condition of tool angle, material and spindle speed. Using at least two f_z, the six equations system is solved, obtaining the force coefficient. To improve the accuracy of the values, Liu et al. [105] used three spline interpolation. There are alternative methods to calculate K_{qc} and K_{qe} such as that of Du et al. [69], who used the Fourier form to determine the milling forces and then the force coefficient.

3.2. Dynamic Model

Once the cutting forces are calculated, stability of the system can be predicted based on the FRF and the cutting parameters values. To establish it, the degrees of freedom of the system are required. Tool and workpiece flexibility can be considered as independent, providing three different situations (Figure 5). The deflection of the part [10], the cutter system [64] or both [46] may affect the quality of the part in terms of final thickness and the roughness. For this reason, several researchers modeled the dynamic deformation of the system to predict the real quantity of material removed in each step and avoid reprocessing.

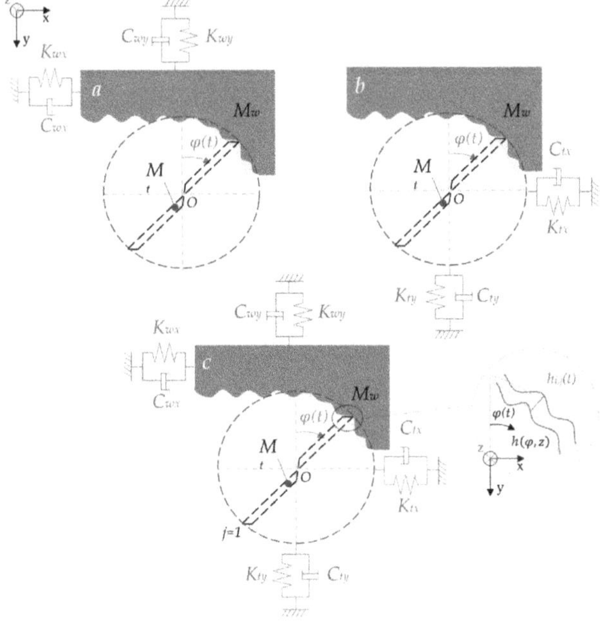

Figure 5. Flexibility of the system categorized as: (**a**) rigid cutter–flexible workpiece system; (**b**) rigid workpiece–flexible cutter system; and (**c**) double flexible system.

Most studies are based on the following assumptions:

- Temperature and other factors related to the machining process do not affect the behavior of the tool and the workpiece during the cutting operation.
- The only force considered is the cutting force, and deformation is only elastic.

Considering a mixed situation and a multiple contact model, the dynamic behavior of the system can be modeled by an equation of the form:

$$M_s \ddot{Q}_s(t) + C_s \dot{Q}_s(t) + K_s Q_s(t) = F_s(t) \qquad (s = w, t) \tag{8}$$

where mass (M_s), damping (C_s) and stiffness (K_s) are matrices with dimension $3n_s \times 3n_s$ on the workpiece (w) and the tool (t). The vibration vector $Q_s(t)$ is defined by the modal displacement ($\Gamma_s(t)$) and the mass normalized mode (U_s). ζ_s is the modal damping ratio matrix and ω_s is the diagonal FRF matrix, both matrices having the dimension of $m_s \times m_s$.

$$\ddot{\Gamma}_s(t) + (2\zeta_s \omega_s)\dot{\Gamma}_s(t) + \omega_s^2 \Gamma_s(t) = U_s^T F_s(t) \tag{9}$$

where

$$\Gamma_s(t) = \left\{ \begin{array}{c} \Gamma_t(t) \\ \Gamma_w(t) \end{array} \right\}, \quad \omega_s = \left[\begin{array}{cc} \omega_t & 0 \\ 0 & \omega_w \end{array} \right], \quad \zeta_s = \left[\begin{array}{cc} \zeta_t & 0 \\ 0 & \zeta_w \end{array} \right], \quad U_s = |U_t - U_w|, \tag{10}$$

$$F_s = F_t = -F_w \tag{11}$$

$F_s(t)$ correspond to the cutting forces and is calculated following the force prediction section but, in this case, chip thickness ($h_{i,j}$) and axial immersion angle (κ) should consider the dynamic interaction:

$$h_{i,j}(t) = \left[x_{i,j}(t) - x_{i,j}(t - T_{j-1}) \right] \sin \varphi(t) + \left[x_{i,j}(t) - x_{i,j}(t - T_{j-1}) \right] \cos \varphi(t) \tag{12}$$

$$\kappa(t) = \cos^{-1}\left(\frac{V \cdot P}{|P|^2 |V|}\right) \tag{13}$$

x and y are the displacement between tool and workpiece during the effect of the j cutting flute at the time interval t. T_{j-1} is the time interval between two following flutes. V is the tool axis vector and P is the relative position vector to the instant t, both of which depend on the instant relative position between the part and the tool (Figure 6).

$$V = \left(V_x, V_y, V_z\right) \qquad P = \left(x_0 - x_{i,j}(t), y_0 - y_{i,j}(t), z_0 - z_{i,j}(t)\right) \tag{14}$$

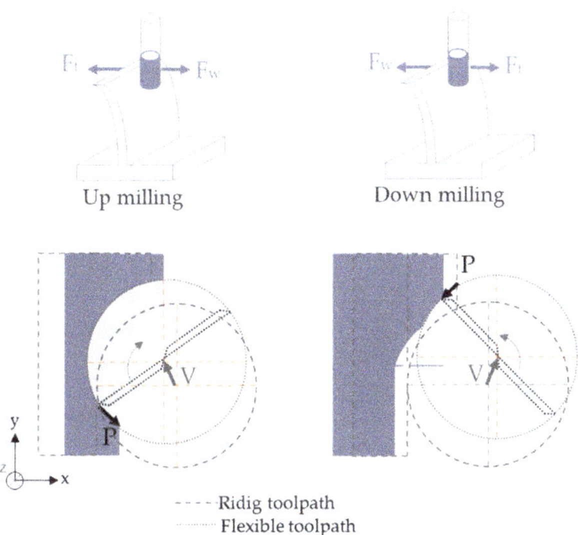

Figure 6. Deflection produced on the parts considering displacement of workpiece and tool.

3.3. Deflection Model

Once the forces are stable and no chatter appears, deflection can still produce over cutting or under cutting. The normal force applied to the part can bend it depending on its stiffness (k), producing a displacement (δ) of the contact point between the workpiece and the tool.

$$F = k\,\delta \tag{15}$$

Fixing the deflection, F can be related to the thickness of the part (Figure 7). The maximum value of the force is determined by the Young's modulus (E), the Poisson ratio (μ), the position of the tool, and the total thickness (w). This thickness should be corrected considering the addendum to the real thickness depending on the location ($\Delta w(u,v)$), based on the tool referenced axis, and considering the residual thickness over the designed surface [66]. The final displacement is obtained based on the predicted forces. This model can be applied not only to cantilever plates but also to more complex geometries.

$$F = \frac{E}{(1 - \mu^2)} f(a, b, c)(w + \Delta w(u, v))^3 \tag{16}$$

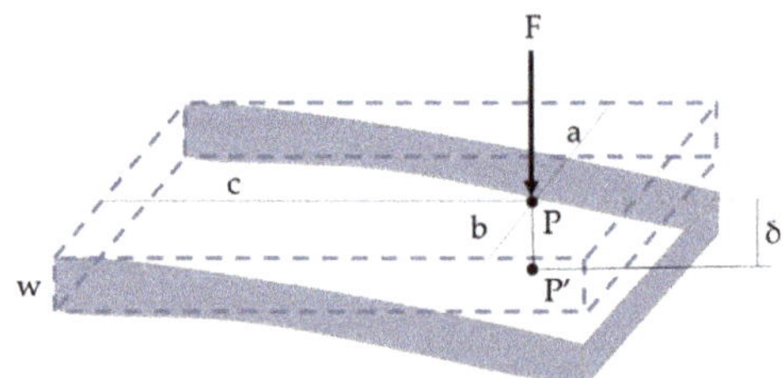

Figure 7. Scheme of the bending of thin-wall produced by the cutting forces.

4. Computational Solutions

Computational analysis is used for the study of the final behavior of the part. It analyzes and predicts the vibration and deflection behavior during thin-wall milling. In both cases, initial forces are calculated using mechanistic models based on experimental data [64] or commercial software such as AdvanEdgeTM [70], VERICUT® [148] or DEFORMTM [77]. They are used as inputs for the initial conditions of the workpiece. Then, self-developed or commercial software such as ANSYSTM [20,24,34,71] or ABAQUS [29,35,53,58,71,150] are used to obtain the FRF of the system, the dynamic behavior or its deflection.

4.1. Vibration Prediction

Different authors tried to establish new dynamic models based on computational experiments in order to predict chatter or forced vibrations during the machining of thin-wall parts. Most of them are based on the study of the FRF by analyzing Stability Lobes Diagrams (SLD) and instant chip engagement to choose the correct cutting parameters in order to improve surface quality of the part and reducing tool wear [21].

4.1.1. Chatter

Chatter prediction starts by calculating the FRF of the workpiece and tool-spindle using impact hammer test [15,26,30,31,36–38]. One point of the tool and different locations of the workpiece are hit, and excitation responses are recorded by accelerometers. The weight of the accelerometers is subtracted from the total weight of the system. The data are treated and filtered to determine the FRF matrix and the modal damping ratio matrix. Damping ratios are usually considered constant during milling and FRF be studied under the most critical situations [3].

Those results lead to a general dynamic model that is dependent on the machining parameters. As is well known, machining parameters directly affect the efficiency of the process and, in this particular case, its stability. To ensure both of them, most researchers study the SLD of the system.

SLD are one of the most common tools used in thin-wall machining to select parameters in order to reduce chatter by just setting the correct machining parameters in terms of efficiency [39–41]. SLD usually represents the stability areas based on the axial depth of cut and the spindle speed (Figure 8). However, typical methods for calculating SLD determine more restrictive areas than they should [40,46]. This is the reason recent works focus on the improvement of its calculation.

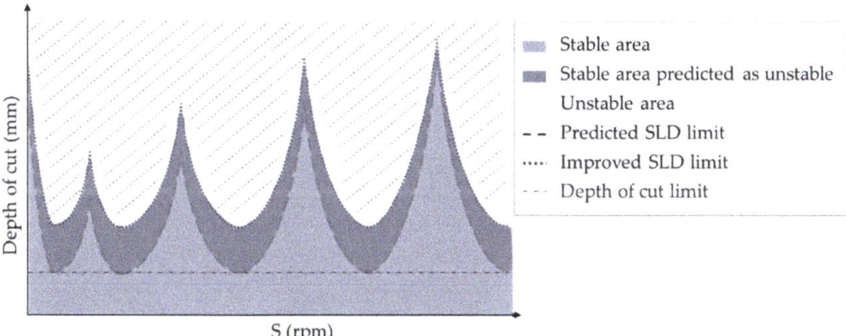

Figure 8. Schematic SLD presenting stable and unstable areas and the possible improvement of the SLD limit curve.

Several authors [25,30,42,150] considered more than one parameter machining condition, generating 3D-SLD. Their studies mainly focus on sample parts to generalize monolithic part machining behavior. For instance, Olvera et al. [39], Germashev et al. [40] and Guo et al. [16] studied the runout and helix angle effect in a SLD, while Urbikain et al. [149] and Jing et al. [20] compared the effect of the tool position. Liu et al. [105] focused their research in the effect of the radial depth of cut and Qu et al. [25] analyzed the feed rate effect. Feng et al. [18] evaluated the effect of velocity-dependent mechanisms to obtain a closer SLD, obtaining better results than the ones using a plowing force coefficient. Finally, some studies are focused on the fixture effect, such as the flexible support [46] or the damping system [19,31,47].

However, few works were found where SLD approach is applied to mirror milling applications or similar test combinations [15,20]. For those cases, bull and ball end mills are used to analyze the effect of the angle or relative position of the tool in the SLD [15].

4.1.2. Amplification

Resonance and amplification can be predicted based on the differential equations of the dynamic behavior of the system, which can be calculated through semi-analytical methods. These models have usually been developed in MATLAB [25,48,49] or C++ [45]. However, this solution is time consuming and has low accuracy. For this reason, one of the most recent approaches is to develop more efficient ways to solve the stability differential equations. Song et al. [50] used the Sherman–Morrison–Woodbury formula to calculate FRF considering the mass loss, whereas Li et al. [64] used a Runge–Kutta method for the same purpose. Feng et al. [18] used Taylor series to linearize the dynamic equations and Olvera et al. [39] solved the model using enhanced multistage homotopy perturbation (EMHP) and Chebyshev method in order to improve the accuracy.

Other authors use computational methods to predict vibrations. In this case, it is important to consider mass loss and tool wear because they can also modify the dynamics of the system and thus the stability of the machining [51]. This consideration implies a constant remeshing and reanalysis, involving a considerable computing time. Some authors tried to include the effect of the material loss. Meshreki et al. [151] proposed the use of 2D multispan plate (MSP). It improved the computational efficiency but it can only be applied to simple geometries. Budak et al. and Yang et al. [21,33] developed Structural Dynamic Modification (SDM) by updating the mass loss by the time domain. Tuysuz and Altintas [57] developed an iterated improved reduction system technique combined with a matrix perturbation technique to use the computational time only once. Yang et al. [29] used component mode synthesis (CMS) and space structural modification to develop a decomposition-condensation model that reduce computational time. Fei et al. [20] solved the dynamic model using a semi-discretization method. Ding et al. [19] established a dynamic model dividing the part and analyzing the FRF on both parts. Li et al. [64] improved roughness by developing a dynamic model for machining of integral impellers blades. Shuang et al. [59] used a coupled Eulerian–Lagrangian model to relate the chip formation to the cutting forces oscillation amplitudes, reducing the surface roughness produced by part deformation. The model used by Tian et al. [26] is presented as a theoretical base for suppressing resonance in the milling process. Ahmadi [55] compared a Finite Strip Model (FSM), FEM analysis and a semi-analytical model for the study of the dynamics of thin-wall machining (Figure 9). Lin et al. [56] studied the FRF of the machining system and related the waviness of the part with the force vibrations and not to the self-exited vibrations.

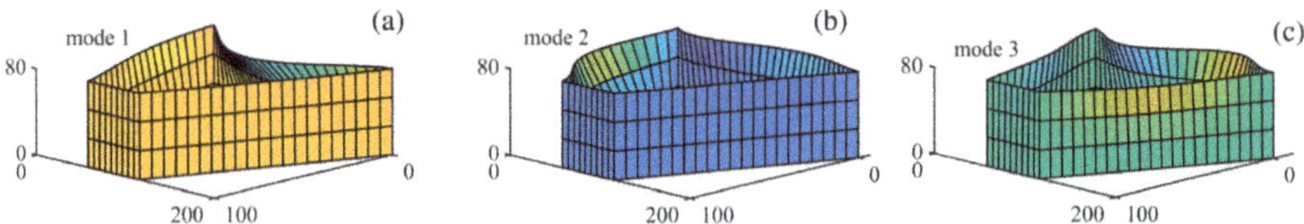

Figure 9. Average deflection obtained for (**a**) the first mode, (**b**) the second mode and (**c**) the third mode of a pocket structure [55].

4.2. Dimensional Error Prediction

Computational methods for part deflection analysis usually involve simulations to estimate deflection of the part and reduce the dimensional tolerances of the process neglecting the dynamic response of the system [61].

Initially, the applied force is calculated based on mechanistic models and considering the instant chip formation. To improve the accuracy of the model, approaches such as the finite difference method are employed only for the contact interference between workpiece and tool [34] or Eulerian-Lagrangian methods to predict the chip formation and the final force applied [74].

Generally, the workpiece material is modeled using the Johnson–Cook law [74,78] and elastoplastic behavior should be considered [5]. Residual stress can be excluded from the material model since the induced residual stresses are more significant than those produced by previous forming steps such as rolling or forging [65]. However, their effect should be taken into account to predict the final part deflection.

The workpiece deflection is predicted through iteration and considering a quasi-static situation [61,70]. The analysis is performed following the toolpath and iteration must be performed for every new tool position due to the change of the workpiece–tool contact and the workpiece stiffness. In fact, the deformation produced by these changes can vary considerably just in one rotation of the spindle, as illustrated in Figure 10. Consequently, for each new position, the part should update the existing material, remeshing the workpiece and considerably increasing the computing time. Izamshah et al. [142] combined FEM and statistical models to reduce it in the simulation of the surface error. Ratchev et al. [62] addressed the solution by using a volume element-based surface generation approach to predict the deflection of the part. Similarly, Si-meng et al. [72] increased the solving speed by changing the simulation method. They considered the material loss using Boolean operations and hexahedral mapping algorithms, including tool and workpiece springback. Their models were validated with an error lower than 15%. Wang and Si [49] discarded the mesh subdivision or mesh adaptive technology because both considerably increased the computational burden, whereas the accuracy was not improved. Meanwhile, they simulated stiffness variation by removing the two elements adjacent to the cutter location, improving deflection accuracy.

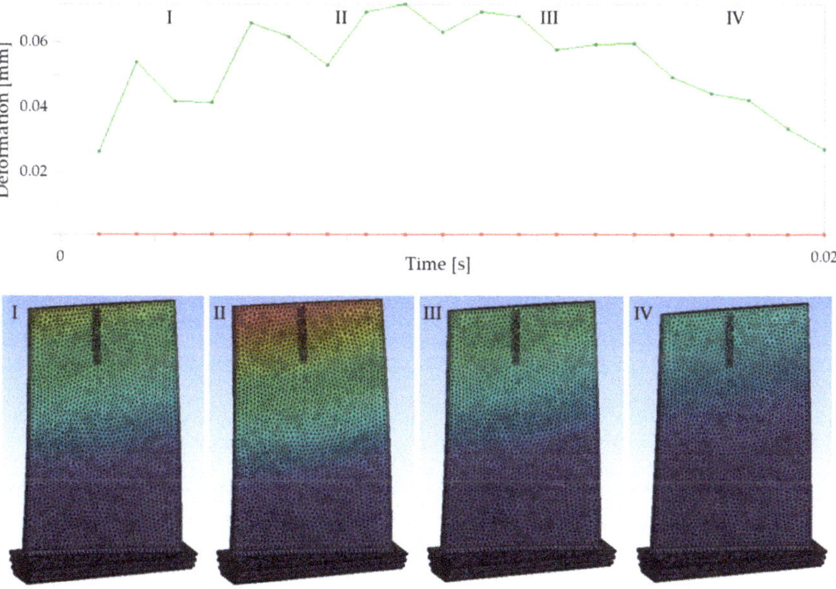

Figure 10. Deformation of a blade during one period rotation of the spindle [78].

Other authors focused on how the parameters affect the deflection of the part. Some of them studied the effect of the wall thickness [67,73], the depth of cut [66] or the feed rate [61,78] on the dimensional error. Others authors considered the tool position [49,70] and the fixture system [5]. Zhang et al. [28] went further by including the effect of the distance to the clamping system.

Another factor to take into account in the final dimensional error is the induced residual stress. It is usually evaluated for thin-walled parts through X-ray [79–82], neutron diffraction [83] or XDR [84] technologies. Residual stress can be affected by the cutting parameters, such as tool geometry [81,82], depth of cut [79], final quality, tool path [65], process temperature [84] and cutting forces [84]. In fact, the selection of the proper parameters with the aim of reducing the induced residual stress can lower the part deflection up to 45% [85]. For instance, Masoudi et al. [84] proved the effect of high-speed machining conditions on the reduction of the distortions produced by residual stresses, considering also that an increase of the depth of cut would increase the internal stress. Gao et al. [86] proposed a semi-analytical method to predict the deformation of thin-wall machining parts based on the effect of the residual stress present on the part. Jiang et al. [81,82] studied the uncut chip thickness effect on the induced residual stresses, relating also the strategy—up or down milling—and the change of tool diameter.

Once the deflection is predicted, many methods apply the results to compensate the tool path or to modify the cutting parameters in order to reduce it [62]. Hao et al. [87] used this approach to correct blade deflections and to geometrically predict the roughness of the final part produced by the separation between the workpiece and the tool. Chen et al. [6] reduced the final error by applying a toolpath compensation strategy. Richter-Trummer et al. [12] presented a simulation method that predicted the distortion produced by residual stresses and allowed managing it, ensuring the dimensional quality of the machining parts. Similarly, Wu et al. [88] used quasi-symmetric machining to reduce the deformation produced by residual stresses. The machining results are considerably more accurate when the compensation is made at the last layer (Figure 11).

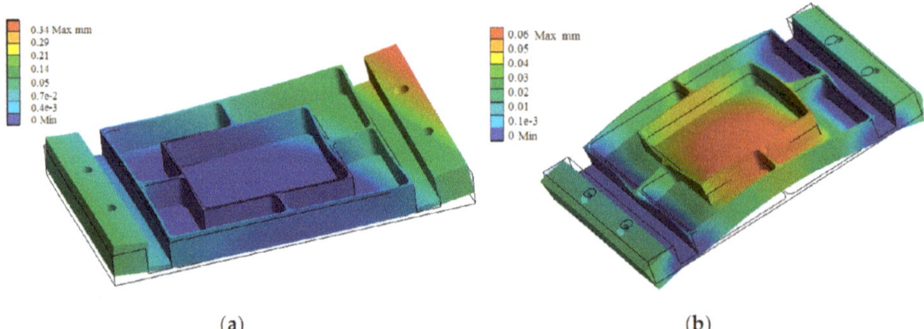

(a)   (b)

Figure 11. Deformation of a thin-wall part: (**a**) not considering the residual stress; and (**b**) following a quasi-symmetric machining reducing the residual stress [88].

5. Industrial Approach

Apart from computational modeling, several solutions following an industrial approach have been developed such as parameter selection, adaptive control, or workholdings and fixture design.

5.1. Parameters Selection

Parameter selection can be established through experimental data-based models, commonly statistics or neural network based models, or through virtual twins studies, in which depth of cut and toolpath are modified to ensure the final thickness of the part.

5.1.1. Database Models

Cutting parameters for thin-wall machining are also studied from an experimental and statistical point of view [13,89–94]. Researchers developed experimental models based on ANOVA results by analyzing how the cutting parameters and machining strategies affect the roughness and the dimensional error. Sonawane et al. [89] used ANOVA analysis to reduce the deflection of a cantilever sheet considering different inclination angles. Qu et al. [90] optimized the experimental models analysis using a neural network analysis type NSGA II. Izamshah et al. [142] obtained a generalized force model based on ANOVA analysis, training the dataset using FEM simulations. Oliveira et al. [91] found the milling strategy (down or up milling) as the most influencing parameter for dimensional error. They also remarked that f_z had an influence on surface roughness, but only when down milling strategies were used. Borojevic et al. [92] optimized the machining time based on the machining strategy and the cutting parameters. Bolar et al. [93] and Jiang et al. [7] detected three different areas of study for roughness when flank milling of thin-wall components was performed. The first area (initial engagement) and the last one (final disengagement) are more unstable than the center of the part. Both surface and residual stresses are increased due to the forced vibrations produced by the tool on those two areas. Yan et al. [66] implemented an experimental method that allow setting the maximum depth of cut as a function of the cutting force, thus its effect does not produce any displacement on the part.

All these models can be used as simplified models to implement on adaptive control, reducing time response and modifying the cutting parameters more quickly. The effect of the cutting parameters on residual stresses, cutting forces, deflection and surface roughness is summarized in Table 2.

Table 2. Cutting parameters effect on residual stress, forces, deflection and roughness. S, Spindle Speed; f, feed rate; Ap, depth of cut; NP, N° of paths; MRR, Material Removal Rate; RS, Residual Stress; F, Forces; Def, Deflection; Rg, Surface Roughness.

	RS	F	Def	Rg
S	↓	↑	↑	↓
f	↑	↑	↑/↓ [1]	↑
Ap	↑	↑	↑	↑
NP	↑		↑	↑
MRR		↑		↑

[1] Down milling strategies increase the deflection of the part while up milling decrease it.

5.1.2. Virtual Twins

Another industrial approach is to develop virtual twins that will ensure the correct selection of parameters and machining strategies. Some CAM commercial software packages have optimization modules to integrate the dynamic error induced by the cutting forces as data but others only integrate the force and toolpath analysis on a FEM software. Rai et al. [5] considered elastoplastic deformation on a 3D virtual environment predicting the nonlinear behavior during machining. Jiang et al. [148] used the module VERICUT optimization to select the best parameters based on the part and tool model. This software uses neural networks to select the optimum set of parameters.

Under a set of conditions, constant thickness and cutting force, different tool paths are evaluated. Yan et al. [66] programed a depth of cut strategy simulating the physical behavior of a blade depending on the tool path generated on UG NX. The suggested variable depth of cut improves the machining error by up to 80% and save a third of the machining time. Rashev et al. [96] included artificial neural network to the CAM to improve the accuracy of the predicted deflection. Wan et al. [97] used deformation simulations to predict the optimum position of the support, evaluating the relative workpiece–part position.

Another application of virtual twins is to simulate the real position of the part. Especially mirror milling, due to their double curve, needs to use premeasuring techniques to redesign the tool-path considering the real position of the part [8,98,99]. Once it is calculated, software determines the tool path transplantation between the nominal surface and the actual one [98].

5.2. Adaptive Control

Adaptive control is a solution based on the on-line monitoring of the machining and an instant intervention on the process to ensure the final quality of the part. Signal data processing and monitoring systems are the base for automatic responses through parameter modification or active damping actuators. These options generally improve the vibration behavior of the system. Meanwhile, on-line measurement systems are used to reduce dimensional errors.

5.2.1. Monitoring

The on-line detection of vibrations, combined with SLD data and dynamic or database models, can lead to adaptive control systems able to improve final quality of the part [106]. For that, a monitoring system able to distinguish the stability or instability of the process is required. Different authors have worked to implement filters and detection systems so parameter changes can be applied.

On the vibration field, Rubeo et al. [107] used the peak-to-peak force diagrams to detect instability. Germashev et al. [40] presented a simple FFT as a tool for prediction analysis and related the fluctuation of the tool with the surface quality. Tian et al. [26] proposed a matrix perturbation method as a time saving way to obtain the natural resonance frequency in thin-wall parts while they are machined. Tian et al. [45] used an eigenvalue sensitivity method to improve machining stability and the final surface finishing. Liu et al. [105] applied different filters to the cutting force in order to analyze the cutting coefficient behavior and its effect on the stability of the system. Liu et al. [108,109] used a Q-factor method to identify the change on the machining operation between the stable and chatter regions. The method was used for flank and mirror milling and quantified the level of chatter based on the force signal. Muhammad et al. [110] designed an active control system based on operational amplifier circuits where they can control the instant vibration, recording the acoustic signal with a microphone. Based on a dynamic model, the damping system changed the applied force so the chatter was reduced. Liu et al. [95] considered the deformation of the tool and the workpiece using an approach based on multisensor fusion and support vector machine (SVM) as a machine learning analysis. The recorded signal was analyzed using wavelet decomposition, and then SVM was applied to signal whether a change on the machining condition was needed. Ma et al. [111] developed a model to be implemented in adaptive control in which the feed rate was modified as a function of the real chip thickness. Feng et al. [32] established a different chatter model based on wavelet analysis of the cutting forces. They also studied its influence on the roughness of the part. Wan et al. [100] proposed a method for the chatter identification on thin-wall machining using a Hilberg–Huan transform and Gao et al. [101] used Complex Morlet Wavelet Transform (CMWT) to detect chatter in thin-wall machining.

Similarly, different techniques have been tested to prevent part deflection. On-line techniques have been developed based on signal treatment or on-line measurements. Wang et al. [102] used the lifting wavelet transform of the cutting forces to identify the bending of the part. Liu et al. [103] employed the cutting forces combined with a dynamic feature model that established the error compensation on real time to avoid the deformation. This solution considerably reduced the thickness error of the final part. Similarly, Han et al. [78] designed a parameter optimization federate control algorithm based on a previous simulation study of the deflection of the part that can be implemented as a control strategy schedule in an open modular architecture CNC system (OMACS). The control strategy schedule was based on the Brent–Dekker algorithm and it was successfully implemented as an adaptive control. Ma et al. [104] discovered the relationship between the induced residual stresses and the cutting power that can be used as a parameter in an on-line measuring method to avoid the deflections caused by residual stress.

5.2.2. Measurements

As explained above, geometrical errors can be reduced by adjusting the depth of cut as a function of the real position of the part surface. However, the instant deformation of the part caused by the machining process is difficult to predict with pre-machining analysis, especially for complex components. Following this principle, some authors have proposed measuring the position of the workpiece on-line in order to ensure its final dimension and its surface quality.

Optical techniques were tested to define the cause of the part deflection [112]. Despite experiments related to the cutting conditions with the mechanical deformation, the acquired geometrical data were not used to reduce dimensional errors.

Touching displacement sensors have also been studied for on-line measurements. Wang et al. [113] adjusted the cutting depth according to the geometrical deviations of the thin-wall, which were measured before the finishing stage on the same milling machine where the rough machining was performed. The measurements allowed calculating the depth compensation value. Similarly, Hao et al. [106] reconstructed the real in-process surface using a displacement sensor. They developed an algorithm to adjust the toolpath and the machining sequence depending on the instant deformation of the part. Their results are shown in Figure 12.

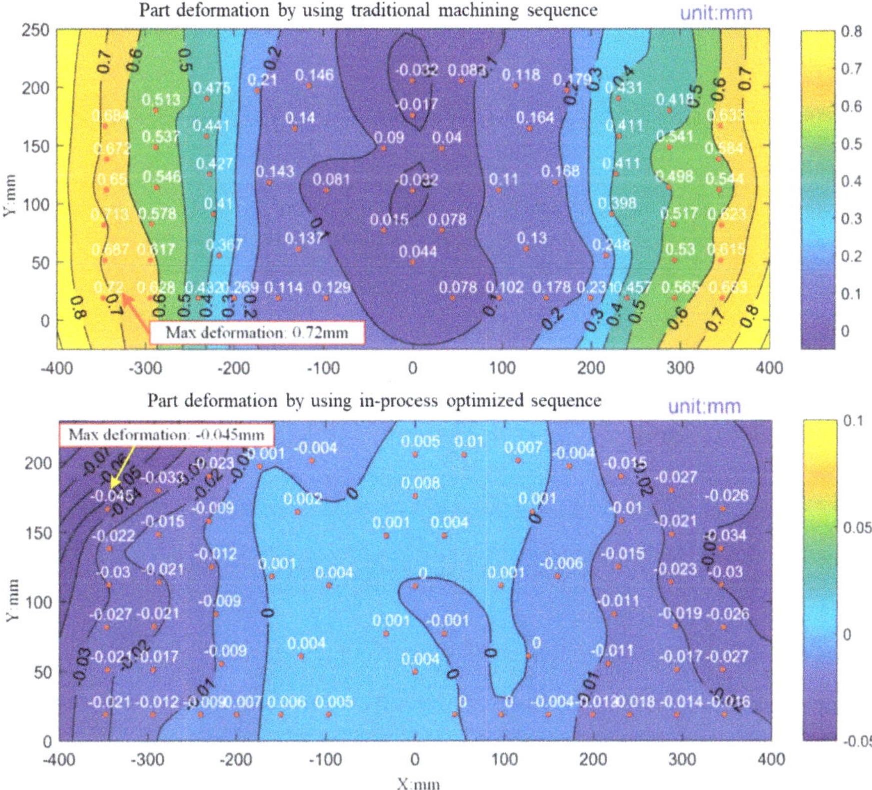

Figure 12. Part deformation obtained with a traditional machining sequence (upper) and with an adaptive machining system selecting an optimized sequence (lower) [106].

Other authors have tested ultrasonic devices for this purpose. Huang et al. [114] automatically recalculated the new tool position combining ultrasonic on-line measures and touching measures of a tank bottom. Rubio et al. [115] designed a flexible clamping system with an ultrasonic premeasure device that automatically adjusted the depth of cut to ensure the final thickness of the skin.

In the same way, Wang et al. [116] used a laser displacement sensor included on the supporting head to measure the in-process displacements of the workpiece. They implemented a forecasting compensatory control system to predict the skin deflection and to adaptively control the width of cut improving the quality of the final part.

5.3. Fixtures, Workholdings and Stiffening Devices

A different strategy to prevent instabilities during low rigidity part machining is to address this problem from the fixing perspective. This strategy could be more efficient than selecting free chatter cutting parameters for complex parts, as FRF is difficult to obtain and it is continuously changing during machining operation.

5.3.1. Fixtures and Workholdings

One of the most common fixture systems for thin-wall parts are vacuum fixtures [83,116–120]. Vacuum fixture applications are based on two main two options: customized vacuum systems or flexible vacuum cups. Their use can reduce vibration and deflection [27]. On the one hand, customized vacuum systems use special vacuum adsorption equipment for each part to be machined [120]. This option is expensive and limited to the number of equal parts produced. Moreover, the vacuum table can produce a tensile stress on the part, affecting the part deformation [12,83]. On the other hand, adaptable vacuum cups or beds increase the flexibility of the clamping fixture. This solution is based on flexible pins combined with vacuum caps or heads that adapt the position to the curves of the part [121]. Some researchers have used the virtual twin concept to select the positions of the additional supports needed for these parts [122–126]. They simulated the system obtaining a minimum part deformation. However, for large skin panels, the support provided by the cups is not enough to reduce deflection or could need a complex optimization of the cup position to reduce it [76,117]. For this reason, Rubio et al. [127] developed a flexible vacuum bed insuring the contact between the part and the workholding, reducing both vibration and deflection.

Impellers, blades and blisks are usually clamped using hydraulic chunks or special jaws that try to reduce the clamping pressure, reducing the possible in-process deformation. These systems can avoid vibrations and deflection for the initial roughing steps but machining performance can be improved using additional workholding. To increase the machining parameters and the operation efficiency, flexible workholdings have been designed for the machining of thin-wall parts. These devices apply a support on the predicted optimal position based on FEM studies [24,124]. Figure 13 shows a commercial example of the company INNOCLAMP® [128]. The workholding is specially designed to compensate the cutting energy all over the part. The position is defined using simulations, and the supports are applied at the most flexible positions. The system usually has embedded sensors, thus it is possible to change the behavior of the workholding, depending on the operation and to register historical data to feed new databases.

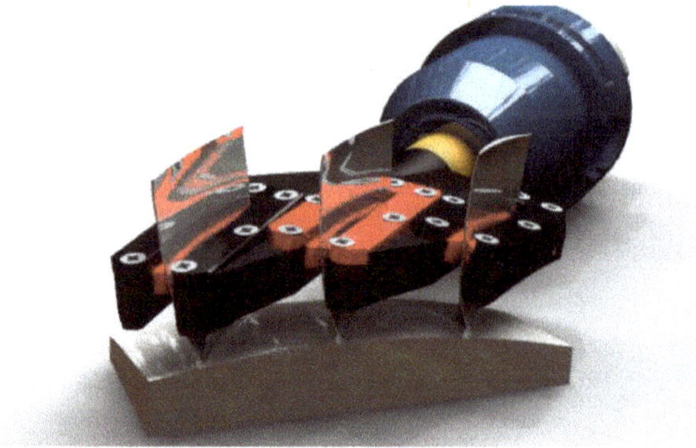

Figure 13. Innoclamp® holding system [128].

Alternatively, moving fixtures are used to maintain stability during low rigidity part machining. Most of the solutions are based on the mirror milling concept, a supporting element that moves synchronously to the tool. For instance, Fei et al. [48] designed a supporting element that moves collinear to the tool-path, acting as support of the cutting force. They compared the results with and without the fixing element, verifying the suppression of the deformation during the machining and the decrease of vibration amplitude, improving consequently the final roughness. Similarly, Liu et al. [129] proposed the use of an air jet to reduce deflection on the part. The jet was synchronized with the machined head following the mirror milling criteria. It impacted on one side of the wall, acting as a support of the cutting forces. Its effect was evaluated through vibration, cutting forces, thickness and final roughness data. Cutting experiments proved that the air jet assistance reduced the vibration of the workpiece up to 47% and both, thickness error and machined surface quality, were improved. However, both systems [48,129] are used for sample parts and the flexibility of it application is questionable, especially for complex geometries.

Mahmud et al. [75] tested a more sophisticated moving fixture to hold a fuselage panel, while skin panel pockets were milled. A magnetic workholding, made up of two sets of magnets, held the fuselage panel. The master set of magnets was placed on the mill tool side, following the tool trajectory provided by the milling machine. The slave set was located on the fuselage panel back and it followed the master module, compensating the cutting forces by the magnetic attraction force. It was experimentally verified that magnetic forces supported the milling thrust force and they overcame the frictional force on the slave unit.

Similar fixing systems were designed by machine tool makers such as Dufieux [130] and M. Torres [131]. They developed a mirror machine center provided with a double-head mechanism. The cutting tool installed on one head is used to remove materials from one side of the skin. Meanwhile, a twin head moves simultaneously providing an auxiliary support while the skin thickness is measured on-line to control the tool position and the depth of cut along the tool trajectory. Even though these solutions have already been implemented at industrial level for fuselage panels milling, they are still limited due to their high investment costs and need to use premeasuring techniques to redesign the tool-path considering the real position of the part.

5.3.2. Active Damping Actuators

Active dampers have the ability to accommodate variable conditions where they are more difficult to implement. It is a specific part of the adaptive control in which the main objective is to avoid vibrations. In this case, the adaptive control makes the decision based on the monitoring system and actuates through piezo-actuator sensors or eddy current damping (EDC). Although active damping is more difficult to implement in real industrial environments, using this approach, a sevenfold improvement in the limiting depth of cut can be obtained. For example, Zhang and Sims [132] reported workpiece chatter avoidance in milling using piezoelectric active damping mounted directly on the workpiece. The research was done on simple geometry parts such as a cantilever plate. Diez et al. [133] used this technology to compensate the deformations of the part, improving final dimensional error. Rashid and Nicolescu [134] proposed an active control of workpiece vibrations in milling through piezo-actuators embedded in workholding systems. Although the tested part had a simple geometry (rectangular blocks) and it was dynamically stiff, the vibration active control improved the dynamics of the workholding system by cancelling the vibration signal generated by the cutting process, succeeding in improving surface quality and tool life.

Yang et al. [135] designed an active and lightweight device based on EDC to attenuate the vibrations produced during the machining of a thin-walled aluminum frame. They attached an ECD to the workpiece, which vibrated synchronously to it. Based on the electromagnetic induction, when a relative motion against the magnet appears, the magnetic flux through a conductor changes and a repulsive force is generated, attenuating the vibration. ECD demonstrated the capacity of keeping the process stable under different cutting parameters (f_z, S, and a_p), achieving a reduction of the machining vibrations of up to 84%.

5.3.3. Stiffening Devices

Stiffening devices, opposite to active damping, add an extra device to the flexible component such as mass compensation system [136], magnetorheological fluids [47,137] or low melting alloys [138]. Their main advantages are the simplicity of their design and their easy implementation. However, they do not consider the stiffness change.

Trying to increase the rigidity of the part, Diaz et al. [137] investigated the use magnetorheological fluids (MRF) to prevent instabilities during thin-floor parts machining. These types of fluids have the ability to change from liquid to a semi-solid state due to the action of a magnetic field. Instead of applying variable cutting conditions to avoid chatter, they proposed the use of a shock absorber based on MRF. The way to assembly the shock absorber to the workpiece, the amount of fluid and how much voltage needed to be applied are issues of vital importance to make the shock absorber work correctly.

Experiments proved it was possible to reach optimum machining speeds in the absence of instabilities during the machining of a thin-floor part. Wang et al. [138] investigated the use of a low-melting point alloy (LMPA) as a phase change material to configure flexible fixtures. This research was focused on complex thin-walled parts, which are difficult to clamp due to the low thickness of the walls. The LMPA was heated up to 70 °C—its melting point—and casted in the gap between the part and a rigid fixture to form a rigid body with fixture among the LMPA, the fixture and the part. The LMPA increased the rigidity of the component during machining. It also reduced the deformation and the vibrations caused by the milling forces, significantly improving the machining accuracy. Furthermore, once the LMPA was melted again, almost no impact was observable on the workpiece.

Other authors have focused their studies on mass compensation systems. Kolluru et al. [139] added a viscoelastic passive dampers to minimize machining vibrations in a ring-type workpiece. A viscoelastic tape (3M® ISD112) was used to place the tuned damper blocks. The viscoelastic tape thickness, its weight and the position of the tuned mass were optimized using FEM. The dynamic response of the workpiece with and without dampers was simulated and the predicted responses were validated by impact hammer tests. The efficacy of dampers blocks was evaluated by machining undamped casing and damped casing and their use provided a significant reduction in vibration in terms of root mean square error. The author affirmed this solution can be rapidly adapted to other workpiece geometries using the FEM model they developed.

A similar approach was reported by Woody et al. [140], who addressed the potential of energy absorbing urethane foams as a passive damping to improve the dynamic behavior of an open-back. Different damping configurations were simulated and tested by FEM and experimental analysis. FRF measurements proved that the energy absorbing foam fabricated from urethane increased the damping performance by sixty times, adding less than 6% of mass compared to the overall structure. Both studies [139,140] agreed that the position of the material was important to define the clamping system, being the highest weight the most relevant factor.

Nevertheless, once a specific step of the machining process is reached, it is possible to detect a decrease on the efficiency of the damper compensating the FRF of the system. For this reason, Yang et al. [136] proposed a passive damper with tunable stiffness to suppress the wide range of vibration frequencies covered on the thin-wall machining. The ratio between the FRF of the passive damper and the workpiece varies with the material removal, especially for the thin-walled part. For this reason, they located a damper inside the workpiece and changed its orientation, modifying the workpiece stiffness. The amplitude of the workpiece FRF at the target mode was reduced by 40%, decreasing the machining vibrations and the surface roughness.

6. Conclusions

This work presents the state of the art of thin-wall machining. This machining usually presents dynamic and static problems that should be controlled to ensure the final quality of the part. Analytic models allow understanding the behavior of the system, defining where vibration or deflection appear. They are also the base of the computational solutions where the most significant advances have been focused on including mass loss and stiffness variation, reducing computational time and increasing prediction accuracy.

Industrial approach have been focused on the following:

- Virtual twins development integrates CAM and simulation to predict the machining behavior, the future real position of the surface to cut, and improve the machining efficiency by selecting the proper cutting parameter, toolpath, or prediction [152].
- Adaptive control, which is used in production to improve the part quality and to feed the database models, detects the process instabilities or deformation through signal analysis and changing on-line the machining parameters.

- Fixture systems design and new approaches try to include adaptive control on the workholding or the stiffening devices to increase the product efficiency, allowing to use more aggressive cutting parameters.

Overall, these studies highlight the need for more accurate simulations and control of the machining process [153]. On the one hand, there remain several aspects of computational modeling about which is little known, particularly computational time reduction is needed to consider the simulation of complex final geometries.

On the other hand, a roughness control method should be designed to detect on-line vibration and deflection with a shorter time response. For example, real-time networks could be implemented using an Internet of Things approach to collect and treat data. The improvement of interfaces and connectivity could lead to avoiding reprocessing steps. Intelligent machine control could also be achieved through machine learning or metaheuristic algorithm analysis of big data. New processing schemes can be investigated for machining time and part deformation reduction, such as simultaneous machining for monolithic parts. Considering all of this evidence, it seems that industry 4.0 outlines are not completely integrated and their integration could make thin-wall machining more profitable.

Author Contributions: Conceptualization, I.D.S. and A.J.G.; formal analysis, I.D.S. and A.R.; investigation, I.D.S.; writing—original draft preparation, I.D.S.; writing—review and editing, A.R., L.N.L.d.L. and A.J.G.; supervision, L.N.L.d.L. and A.R.; project administration, A.J.G.; and funding acquisition, L.N.L.d.L.

Funding: This research was funded by University of Cadiz, UCA/REC01VI/2016.

Acknowledgments: Figure 9 reprinted by permission from RightsLink Permissions: (Springer) (The International Journal of Advanced Manufacturing Technology) (Finite strip modeling of the varying dynamics of thin-walled pocket structures during machining, Keivan Ahmadi), (2016). Figure 10 reprinted by permission from RightsLink Permissions: (Springer) (The International Journal of Advanced Manufacturing Technology) (Cutting deflection control of the blade based on real-time feedrate scheduling in open modular architecture CNC system, Zhenyu Han, Hongyu Jin, Yunzhong Fu et al.), (2016). Figure 11 reprinted by permission of the publisher (Taylor & Francis Ltd., http://www.tandfonline.com) from Dynamic machining process planning incorporating in-process workpiece deformation data for large-size aircraft structural parts, Xiaozhong Hao, Yingguang Li, et al., International Journal of Computer Integrated Manufacturing, 2019.

References

1. Herranz, S.; Campa, F.J.; López de Lacalle, L.N.; Rivero, A.; Lamikiz, A.; Ukar, E.; Sánchez, J.A.; Bravo, U. The milling of airframe components with low rigidity: A general approach to avoid static and dynamic problems. *Proc. Inst. Mech. Eng. Part B J. Eng. Manuf.* **2005**, *219*, 789–801. [CrossRef]

2. Li, X.; Zhao, W.; Li, L.; He, N.; Chi, S. Modeling and application of process damping in milling of thin-walled workpiece made of titanium alloy. *Shock Vib.* **2015**, *2015*. [CrossRef]

3. Bravo, U.; Altuzarra, O.; López de Lacalle, L.N.; Sánchez, J.A.A.; Campa, F.J.J. Stability limits of milling considering the flexibility of the workpiece and the machine. *Int. J. Mach. Tools Manuf.* **2005**, *45*, 1669–1680. [CrossRef]

4. Arnaud, L.; Gonzalo, O.; Seguy, S.; Jauregi, H.; Peigné, G. Simulation of low rigidity part machining applied to thin-walled structures. *Int. J. Adv. Manuf. Technol.* **2010**, *54*, 479–488. [CrossRef]

5. Rai, J.K.; Xirouchakis, P. Finite element method based machining simulation environment for analyzing part errors induced during milling of thin-walled components. *Int. J. Mach. Tools Manuf.* **2008**, *48*, 629–643. [CrossRef]

6. Chen, W.; Xue, J.; Tang, D.; Chen, H.; Qu, S. Deformation prediction and error compensation in multilayer milling processes for thin-walled parts. *Int. J. Mach. Tools Manuf.* **2009**, *49*, 859–864. [CrossRef]

7. Jiang, X.; Zhu, Y.; Zhang, Z.; Guo, M.; Ding, Z. Investigation of residual impact stress and its effects on the precision during milling of the thin-walled part. *Int. J. Adv. Manuf. Technol.* **2018**, *97*, 877–892. [CrossRef]

8. Bi, Q.; Huang, N.; Zhang, S.; Shuai, C.; Wang, Y. Adaptive machining for curved contour on deformed large skin based on on-machine measurement and isometric mapping. *Int. J. Mach. Tools Manuf.* **2019**, *136*, 34–44. [CrossRef]

9. *Amended WP & Budget 4th Call for Proposals (CFP04): Preliminary List and Full Description of Topics*; European Union Funding for Research & Innovation: Brussel, Belgium, 2016; p. 219.

10. Tang, A.; Liu, Z. Three-dimensional stability lobe and maximum material removal rate in end milling of thin-walled plate. *Int. J. Adv. Manuf. Technol.* **2009**, *43*, 33–39. [CrossRef]

11. Ratchev, S.; Liu, S.; Huang, W.; Becker, A.A. A flexible force model for end milling of low-rigidity parts. *J. Mater. Process. Technol.* **2004**, *153–154*, 134–138. [CrossRef]

12. Richter-Trummer, V.; Koch, D.; Witte, A.; Dos Santos, J.F.; De Castro, P.M.S.T. Methodology for prediction of distortion of workpieces manufactured by high speed machining based on an accurate through-the-thickness residual stress determination. *Int. J. Adv. Manuf. Technol.* **2013**, *68*, 2271–2281. [CrossRef]

13. Izamshah, R.A. *Hybrid Deflection Prediction for Machining Thin-Wall Titanium Alloy Aerospace Component*; RMIT: Melbourne, Australia, 2011.

14. Biermann, D.; Kersting, P.; Surmann, T. A general approach to simulating workpiece vibrations during five-axis milling of turbine blades. *CIRP Ann. Manuf. Technol.* **2010**, *59*, 125–128. [CrossRef]

15. Yan, R.; Gong, Y.; Peng, F.; Tang, X.; Li, H.; Li, B. Three degrees of freedom stability analysis in the milling with bull-nosed end mills. *Int. J. Adv. Manuf. Technol.* **2016**, *86*, 71–85. [CrossRef]

16. Guo, Q.; Jiang, Y.; Zhao, B.; Ming, P. Chatter modeling and stability lobes predicting for non-uniform helix tools. *Int. J. Adv. Manuf. Technol.* **2016**, *87*, 251–266. [CrossRef]

17. Kolluru, K.V.; Axinte, D.; Becker, A. A solution for minimising vibrations in milling of thin walled casings by applying dampers to workpiece surface. *CIRP Ann. Manuf. Technol.* **2013**, *62*, 415–418. [CrossRef]

18. Feng, J.; Wan, M.; Gao, T.Q.; Zhang, W.H. Mechanism of process damping in milling of thin-walled workpiece. *Int. J. Mach. Tools Manuf.* **2018**, *134*, 1–19. [CrossRef]

19. Ding, Y.; Zhu, L. Investigation on chatter stability of thin-walled parts considering its flexibility based on finite element analysis. *Int. J. Adv. Manuf. Technol.* **2018**, *94*, 3173–3187. [CrossRef]

20. Fei, J.; Lin, B.; Yan, S.; Zhang, X.; Lan, J.; Dai, S. Chatter prediction for milling of flexible pocket-structure. *Int. J. Adv. Manuf. Technol.* **2017**, *89*, 2721–2730. [CrossRef]

21. Yang, Y.; Zhang, W.H.; Ma, Y.C.; Wan, M. Chatter prediction for the peripheral milling of thin-walled workpieces with curved surfaces. *Int. J. Mach. Tools Manuf.* **2016**, *109*, 36–48. [CrossRef]

22. Budak, E. Mechanics and Dynamics of Milling Thin Walled Structures. Ph.D. Thesis, University of British Columbia, Vancouver, BC, Canada, 1994. [CrossRef]

23. Budak, E.; Altintas, Y.; Armarego, E.J.A. Prediction of Milling Force Coefficients from Orthogonal Cutting Data. *J. Manuf. Sci. Eng.* **1996**, *118*, 216. [CrossRef]

24. Wang, H.; Huang, L.; Yao, C.; Kou, M.; Wang, W.; Huang, B.; Zheng, W. Integrated analysis method of thin-walled turbine blade precise machining. *Int. J. Precis. Eng. Manuf.* **2015**, *16*, 1011–1019. [CrossRef]

25. Qu, S.; Zhao, J.; Wang, T. Three-dimensional stability prediction and chatter analysis in milling of thin-walled plate. *Int. J. Adv. Manuf. Technol.* **2016**, *86*, 2291–2300. [CrossRef]

26. Tian, W.; Ren, J.; Zhou, J.; Wang, D. Dynamic modal prediction and experimental study of thin-walled workpiece removal based on perturbation method. *Int. J. Adv. Manuf. Technol.* **2018**, *94*, 2099–2113. [CrossRef]

27. Campa, F.J.; López de Lacalle, L.N.; Celaya, A. Chatter avoidance in the milling of thin floors with bull-nose end mills: Model and stability diagrams. *Int. J. Mach. Tools Manuf.* **2011**, *51*, 43–53. [CrossRef]

28. Zhang, J.; Lin, B.; Fei, J.; Huang, T.; Xiao, J.; Zhang, X.; Ji, C. Modeling and experimental validation for surface error caused by axial cutting force in end-milling process. *Int. J. Adv. Manuf. Technol.* **2018**, *99*, 327–335. [CrossRef]

29. Yang, Y.; Zhang, W.-H.; Ma, Y.-C.; Wan, M.; Dang, X.-B. An efficient decomposition-condensation method for chatter prediction in milling large-scale thin-walled structures. *Mech. Syst. Signal. Process.* **2019**, *121*, 58–76. [CrossRef]

30. Jin, X.; Sun, Y.; Guo, Q.; Guo, D. 3D stability lobe considering the helix angle effect in thin-wall milling. *Int. J. Adv. Manuf. Technol.* **2016**, *82*, 2123–2136. [CrossRef]

31. Li, Z.; Sun, Y.; Guo, D. Chatter prediction utilizing stability lobes with process damping in finish milling of titanium alloy thin-walled workpiece. *Int. J. Adv. Manuf. Technol.* **2017**, *89*, 2663–2674. [CrossRef]

32. Feng, J.; Sun, Z.; Jiang, Z.; Yang, L. Identification of chatter in milling of Ti-6Al-4V titanium alloy thin-walled workpieces based on cutting force signals and surface topography. *Int. J. Adv. Manuf. Technol.* **2016**, *82*, 1909–1920. [CrossRef]

33. Budak, E.; Tunç, L.T.; Alan, S.; Özgüven, H.N. Prediction of workpiece dynamics and its effects on chatter stability in milling. *CIRP Ann.* **2012**, *61*, 339–342. [CrossRef]

34. Wu, Q.; Li, D.-P.; Ren, L.; Mo, S. Detecting milling deformation in 7075 aluminum alloy thin-walled plates using finite difference method. *Int. J. Adv. Manuf. Technol.* **2016**, *85*, 1291–1302. [CrossRef]

35. Han, D.; Li, P.; An, S.; Shi, P. Multi-frequency weak signal detection based on wavelet transform and parameter compensation band-pass multi-stable stochastic resonance. *Mech. Syst. Signal. Process.* **2016**, *71–71*, 995–1010. [CrossRef]

36. Scippa, A.; Grossi, N.; Campatelli, G. FEM based cutting velocity selection for thin walled part machining. *Procedia CIRP* **2014**, *14*, 287–292. [CrossRef]

37. Campa, F.J.; López de Lacalle, L.N.; Lamikiz, A.; Sánchez, J.A. Selection of cutting conditions for a stable milling of flexible parts with bull-nose end mills. *J. Mater. Process. Technol.* **2007**, *191*, 279–282. [CrossRef]

38. Bolsunovskiy, S.; Vermel, V.; Gubanov, G.; Kacharava, I.; Kudryashov, A. Thin-Walled Part Machining Process Parameters Optimization based on Finite-Element Modeling of Workpiece Vibrations. *Procedia CIRP* **2013**, *8*, 276–280. [CrossRef]

39. Olvera, D.; Urbikain, G.; Elías-Zuñiga, A.; López de Lacalle, L. Improving Stability Prediction in Peripheral Milling of Al7075T6. *Appl. Sci.* **2018**, *8*, 1316. [CrossRef]

40. Germashev, A.; Logominov, V.; Anpilogov, D.; Vnukov, Y.; Khristal, V. Optimal cutting condition determination for milling thin-walled details. *Adv. Manuf.* **2018**, *6*, 280–290. [CrossRef]

41. Urbikain Pelayo, G.; López De La Calle, L. Stability charts with large curve-flute end-mills for thin-walled workpieces. *Mach. Sci. Technol.* **2018**, *22*, 585–603. [CrossRef]

42. Yan, Z.; Liu, Z.; Wang, X.; Liu, B.; Luo, Z.; Wang, D. Stability prediction of thin-walled workpiece made of Al7075 in milling based on shifted Chebyshev polynomials. *Int. J. Adv. Manuf. Technol.* **2016**, *87*, 115–124. [CrossRef]

43. Urbikain, G.; Artetxe, E.; López de Lacalle, L.N. Numerical simulation of milling forces with barrel-shaped tools considering runout and tool inclination angles. *Appl. Math. Model.* **2017**, *47*, 619–636. [CrossRef]

44. Ma, J.; Zhang, D.; Liu, Y.; Wu, B.; Luo, M. Tool posture dependent chatter suppression in five-axis milling of thin-walled workpiece with ball-end cutter. *Int. J. Adv. Manuf. Technol.* **2017**, *91*, 287–299. [CrossRef]

45. Tian, W.; Ren, J.; Wang, D.; Zhang, B. Optimization of non-uniform allowance process of thin-walled parts based on eigenvalue sensitivity. *Int. J. Adv. Manuf. Technol.* **2018**, *96*, 2101–2116. [CrossRef]

46. Xu, C.; Feng, P.; Zhang, J.; Yu, D.; Wu, Z. Milling stability prediction for flexible workpiece using dynamics of coupled machining system. *Int. J. Adv. Manuf. Technol.* **2017**, *90*, 3217–3227. [CrossRef]

47. Ma, J.; Zhang, D.; Wu, B.; Luo, M.; Chen, B. Vibration suppression of thin-walled workpiece machining considering external damping properties based on magnetorheological fluids flexible fixture. *Chin. J. Aeronaut.* **2016**, *29*, 1074–1083. [CrossRef]

48. Fei, J.; Lin, B.; Xiao, J.; Ding, M.; Yan, S.; Zhang, X.; Zhang, J. Investigation of moving fixture on deformation suppression during milling process of thin-walled structures. *J. Manuf. Process.* **2018**, *32*, 403–411. [CrossRef]

49. Wang, L.; Si, H. Machining deformation prediction of thin-walled workpieces in five-axis flank milling. *Int. J. Adv. Manuf. Technol.* **2018**, *97*, 4179–4193. [CrossRef]

50. Song, Q.; Liu, Z.; Wan, Y.; Ju, G.; Shi, J. Application of Sherman-Morrison-Woodbury formulas in instantaneous dynamic of peripheral milling for thin-walled component. *Int. J. Mech. Sci.* **2015**, *96–97*, 79–90. [CrossRef]

51. Ding, H.; Ke, Y. Study on Machining Deformation of Aircraft Monolithic Component by FEM and Experiment. *Chin. J. Aeronaut.* **2006**, *19*, 247–254. [CrossRef]

52. Gradišek, J.; Kalveram, M.; Weinert, K. Mechanistic identification of specific force coefficients for a general end mill. *Int. J. Mach. Tools Manuf.* **2004**, *44*, 401–414. [CrossRef]

53. Kang, Y.-G.; Wang, Z.-Q. Two efficient iterative algorithms for error prediction in peripheral milling of thin-walled workpieces considering the in-cutting chip. *Int. J. Mach. Tools Manuf.* **2013**, *73*, 55–61. [CrossRef]

54. Barbero, B.R.; Ureta, E.S. Comparative study of different digitization techniques and their accuracy. *CAD Comput. Aided Des.* **2011**, *43*, 188–206. [CrossRef]

55. Ahmadi, K. Finite strip modeling of the varying dynamics of thin-walled pocket structures during machining. *Int. J. Adv. Manuf. Technol.* **2017**, *89*, 2691–2699. [CrossRef]

56. Lin, X.; Wu, D.; Yang, B.; Wu, G.; Shan, X.; Xiao, Q.; Hu, L.; Yu, J. Research on the mechanism of milling surface waviness formation in thin-walled blades. *Int. J. Adv. Manuf. Technol.* **2017**, *93*, 2459–2470. [CrossRef]

57. Tuysuz, O.; Altintas, Y. Frequency Domain Updating of Thin-Walled Workpiece Dynamics Using Reduced Order Substructuring Method in Machining. *J. Manuf. Sci. Eng.* **2017**, *139*, 071013. [CrossRef]

58. Ratchev, S.; Liu, S.; Becker, A.A.A. Error compensation strategy in milling flexible thin-wall parts. *J. Mater. Process. Technol.* **2005**, *162–163*, 673–681. [CrossRef]

59. Shuang, F.; Chen, X.; Ma, W. Numerical analysis of chip formation mechanisms in orthogonal cutting of Ti6Al4V alloy based on a CEL model. *Int. J. Mater.* **2018**, *11*, 185–198. [CrossRef]

60. Elbestawi, M.A.; Sagherian, R. Dynamic modeling for the prediction of surface errors in the milling of thin-walled sections. *J. Mater. Process. Technol.* **1991**, *25*, 215–228. [CrossRef]

61. Tsai, J.-S.; Liao, C.-L. Finite-element modeling of static surface errors in the peripheral milling of thin-walled workpieces. *J. Mater. Process. Technol.* **1999**, *94*, 235–246. [CrossRef]

62. Ratchev, S.; Govender, E.; Nikov, S.; Phuah, K.; Tsiklos, G. Force and deflection modelling in milling of low-rigidity complex parts. *J. Mater. Process. Technol.* **2003**, *143–144*, 796–801. [CrossRef]

63. Wan, M.; Zhang, W.H.; Qin, G.H.; Wang, Z.P. Strategies for error prediction and error control in peripheral milling of thin-walled workpiece. *Int. J. Mach. Tools Manuf.* **2008**, *48*, 1366–1374. [CrossRef]

64. Li, M.; Huang, J.; Ding, W.; Liu, X.; Li, L. Dynamic response analysis of a ball-end milling cutter and optimization of the machining parameters for a ruled surface. *Proc. Inst. Mech. Eng. Part B J. Eng. Manuf.* **2019**, *233*, 588–599. [CrossRef]

65. Jiang, X.; Wang, Y.; Ding, Z.; Li, H. An approach to predict the distortion of thin-walled parts affected by residual stress during the milling process. *Int. J. Adv. Manuf. Technol.* **2017**, *93*, 4203–4216. [CrossRef]

66. Yan, Q.; Luo, M.; Tang, K. Multi-axis variable depth-of-cut machining of thin-walled workpieces based on the workpiece deflection constraint. *CAD Comput. Aided Des.* **2018**, *100*, 14–29. [CrossRef]

67. Aijun, T.; Zhanqiang, L. Deformations of thin-walled plate due to static end milling force. *J. Mater. Process. Technol.* **2008**, *206*, 345–351. [CrossRef]

68. Wan, M.; Zhang, W.H. Calculations of chip thickness and cutting forces in flexible end milling. *Int. J. Adv. Manuf. Technol.* **2006**, *29*, 637–647. [CrossRef]

69. Du, Z.; Zhang, D.; Hou, H.; Liang, S.Y. Peripheral milling force induced error compensation using analytical force model and APDL deformation calculation. *Int. J. Adv. Manuf. Technol.* **2017**, *88*, 3405–3417. [CrossRef]

70. Wang, L.; Huang, H.; West, R.W.; Li, H.; Du, J. A model of deformation of thin-wall surface parts during milling machining process. *J. Cent. South. Univ.* **2018**, *25*, 1107–1115. [CrossRef]

71. Ma, J.W.; Liu, Z.; Jia, Z.Y.; Song, D.N.; Gao, Y.Y.; Si, L.K. Stability recognition for high-speed milling of TC4 thin-walled parts with curved surface. *Int. J. Adv. Manuf. Technol.* **2017**, *91*, 1–11. [CrossRef]

72. Liu, S.M.; Shao, X.D.; Ge, X.B.; Wang, D. Simulation of the deformation caused by the machining cutting force on thin-walled deep cavity parts. *Int. J. Adv. Manuf. Technol.* **2017**, *92*, 3503–3517. [CrossRef]

73. Ning, H.; Zhigang, W.; Chengyu, J.; Bing, Z. Finite element method analysis and control stratagem for machining deformation of thin-walled components. *J. Mater. Process. Technol.* **2003**, *139*, 332–336. [CrossRef]

74. Gang, L. Study on deformation of titanium thin-walled part in milling process. *J. Mater. Process. Technol.* **2009**, *209*, 2788–2793. [CrossRef]

75. Mahmud, A.; Mayer, J.R.R.; Baron, L. Magnetic attraction forces between permanent magnet group arrays in a mobile magnetic clamp for pocket machining. *CIRP J. Manuf. Sci. Technol.* **2015**, *11*, 82–88. [CrossRef]

76. Do, M.D.; Son, Y.; Choi, H.J. Optimal workpiece positioning in flexible fixtures for thin-walled components. *CAD Comput. Aided Des.* **2018**, *95*, 14–23. [CrossRef]

77. Wu, Q.; Li, D.-P. Analysis and X-ray measurements of cutting residual stresses in 7075 aluminum alloy in high speed machining. *Int. J. Precis. Eng. Manuf.* **2014**, *15*, 1499–1506. [CrossRef]

78. Han, Z.; Jin, H.; Fu, Y.; Fu, H. Cutting deflection control of the blade based on real-time feedrate scheduling in open modular architecture CNC system. *Int. J. Adv. Manuf. Technol.* **2017**, *90*, 2567–2579. [CrossRef]

79. Li, B.; Jiang, X.; Yang, J.; Liang, S.Y. Effects of depth of cut on the redistribution of residual stress and distortion during the milling of thin-walled part. *J. Mater. Process. Technol.* **2015**, *216*, 223–233. [CrossRef]

80. Ma, Y.; Liu, S.; Feng, P.F.; Yu, D.W. Finite element analysis of residual stresses and thin plate distortion after face milling. In Proceedings of the 2015 12th International Bhurban Conference on Applied Sciences and Technology (IBCAST), Islamabad, Pakistan, 13–17 January 2015; pp. 67–71. [CrossRef]

81. Jiang, X.; Li, B.; Yang, J.; Zuo, X.Y.; Li, K. An approach for analyzing and controlling residual stress generation during high-speed circular milling. *Int. J. Adv. Manuf. Technol.* **2013**, *66*, 1439–1448. [CrossRef]

82. Jiang, X.; Li, B.; Yang, J.; Zuo, X.Y. Effects of tool diameters on the residual stress and distortion induced by milling of thin-walled part. *Int. J. Adv. Manuf. Technol.* **2013**, *68*, 175–186. [CrossRef]

83. Chatelain, J.-F.; Lalonde, J.-F.; Tahan, A.S. Effect of Residual Stresses Embedded within workpieces on the distortion of parts after machining. *Int. J. Mech.* **2012**, *6*, 43–51.

84. Masoudi, S.; Amini, S.; Saeidi, E.; Eslami-Chalander, H. Effect of machining-induced residual stress on the distortion of thin-walled parts. *Int. J. Adv. Manuf. Technol.* **2014**, *76*, 597–608. [CrossRef]

85. Arrazola, P.J.; Özel, T.; Umbrello, D.; Davies, M.; Jawahir, I.S. Recent advances in modelling of metal machining processes. *CIRP Ann. Manuf. Technol.* **2013**, *62*, 695–718. [CrossRef]

86. Gao, H.; Zhang, Y.; Wu, Q.; Li, B. Investigation on influences of initial residual stress on thin-walled part machining deformation based on a semi-analytical model. *J. Mater. Process. Technol.* **2018**, *262*, 437–448. [CrossRef]

87. Hao, Y.; Liu, Y. Analysis of milling surface roughness prediction for thin-walled parts with curved surface. *Int. J. Adv. Manuf. Technol.* **2017**, *93*, 2289–2297. [CrossRef]

88. Wu, Q.; Li, D.-P.; Zhang, Y.-D. Detecting Milling Deformation in 7075 Aluminum Alloy Aeronautical Monolithic Components Using the Quasi-Symmetric Machining Method. *MET Archit.* **2016**, *6*, 80. [CrossRef]

89. Sonawane, H.A.; Joshi, S.S. Modeling of machined surface quality in high-speed ball-end milling of Inconel-718 thin cantilevers. *Int. J. Adv. Manuf. Technol.* **2015**, *78*, 1751–1768. [CrossRef]

90. Qu, S.; Zhao, J.; Wang, T. Experimental study and machining parameter optimization in milling thin-walled plates based on NSGA-II. *Int. J. Adv. Manuf. Technol.* **2017**, *89*, 2399–2409. [CrossRef]

91. de Oliveira, E.L.; de Souza, A.F.; Diniz, A.E. Evaluating the influences of the cutting parameters on the surface roughness and form errors in 4-axis milling of thin-walled free-form parts of AISI H13 steel. *J. Braz. Soc. Mech. Sci. Eng.* **2018**, *40*, 334. [CrossRef]

92. Borojević, S.; Lukić, D.; Milošević, M.; Vukman, J.; Kramar, D. Optimization of process parameters for machining of Al 7075 thin—Walled structures. *Adv. Prod. Manag.* **2018**, *13*, 125–135. [CrossRef]

93. Bolar, G.; Das, A.; Joshi, S.N. Measurement and analysis of cutting force and product surface quality during end-milling of thin-wall components. *Meas. J. Int. Meas. Confed.* **2018**, *121*, 190–204. [CrossRef]

94. Del Sol, I.; Rivero, A.; Salguero, J.; Fernández-Vidal, S.R.; Marcos, M. Tool-path effect on the geometric deviations in the machining of UNS A92024 aeronautic skins. *Procedia Manuf.* **2017**, *13*, 639–646. [CrossRef]

95. Liu, C.; Li, Y.; Zhou, G.; Shen, W. A sensor fusion and support vector machine based approach for recognition of complex machining conditions. *J. Intell. Manuf.* **2018**, *29*, 1739–1752. [CrossRef]

96. Ratchev, S.; Govender, E.; Nikov, S. Towards deflection prediction and compensation in machining of low-rigidity parts. *Proc. Inst. Mech. Eng. Part B J. Eng. Manuf.* **2002**, *216*, 129–134. [CrossRef]

97. Wan, X.-J.; Hua, L.; Wang, X.-F.; Peng, Q.-Z.; Qin, X. An error control approach to tool path adjustment conforming to the deformation of thin-walled workpiece. *Int. J. Mach. Tools Manuf.* **2011**, *51*, 221–229. [CrossRef]

98. Hao, X.; Li, Y.; Deng, T.; Liu, C.; Xiang, B. Tool path transplantation method for adaptive machining of large-sized and thin-walled free form surface parts based on error distribution. *Robot. Comput. Integr. Manuf.* **2019**, *56*, 222–232. [CrossRef]

99. Wang, Y.; Hou, B.; Wang, F.; Ji, Z.; Liang, Z. Research on a thin-walled part manufacturing method based on information-localizing technology. *Proc. Inst. Mech. Eng. Part C J. Mech. Eng. Sci.* **2017**, *231*, 4099–4109. [CrossRef]

100. Wan, S.; Li, X.; Chen, W.; Hong, J. Investigation on milling chatter identification at early stage with variance ratio and Hilbert–Huang transform. *Int. J. Adv. Manuf. Technol.* **2018**, *95*, 3563–3573. [CrossRef]

101. Gao, J.; Song, Q.; Liu, Z. Chatter detection and stability region acquisition in thin-walled workpiece milling based on CMWT. *Int. J. Adv. Manuf. Technol.* **2018**, *98*, 699–713. [CrossRef]

102. Wang, G.; Yang, X.; Wang, Z. On-line deformation monitoring of thin-walled parts based on least square fitting method and lifting wavelet transform. *Int. J. Adv. Manuf. Technol.* **2018**, *94*, 4237–4246. [CrossRef]

103. Liu, C.; Li, Y.; Shen, W. A real time machining error compensation method based on dynamic features for cutting force induced elastic deformation in flank milling. *Mach. Sci. Technol.* **2018**, *22*, 766–786. [CrossRef]

104. Ma, Y.; Feng, P.; Zhang, J.; Wu, Z.; Yu, D. Energy criteria for machining-induced residual stresses in face milling and their relation with cutting power. *Int. J. Adv. Manuf. Technol.* **2015**, *81*, 1023–1032. [CrossRef]

105. Liu, X.; Gao, H.; Yue, C.; Li, R.; Jiang, N.; Yang, L. Investigation of the milling stability based on modified variable cutting force coefficients. *Int. J. Adv. Manuf. Technol.* **2018**, *96*, 2991–3002. [CrossRef]

106. Hao, X.; Li, Y.; Zhao, Z.; Liu, C. Dynamic machining process planning incorporating in-process workpiece deformation data for large-size aircraft structural parts. *Int. J. Comput. Integr. Manuf.* **2019**, *32*, 136–147. [CrossRef]

107. Rubeo, M.A.; Schmitz, T.L. Global stability predictions for flexible workpiece milling using time domain simulation. *J. Manuf. Syst.* **2016**, *40*, 8–14. [CrossRef]

108. Liu, H.; Bo, Q.; Zhang, H.; Wang, Y. Analysis of Q-factor's identification ability for thin-walled part flank and mirror milling chatter. *Int. J. Adv. Manuf. Technol.* **2018**, *99*, 1673–1686. [CrossRef]

109. Wang, Y.; Bo, Q.; Liu, H.; Hu, L.; Zhang, H. Mirror milling chatter identification using Q-factor and SVM. *Int. J. Adv. Manuf. Technol.* **2018**, *98*, 1163–1177. [CrossRef]

110. Muhammad, B.B.; Wan, M.; Liu, Y.; Yuan, H. Active Damping of Milling Vibration Using Operational Amplifier Circuit. *Chin. J. Mech. Eng.* **2018**, *31*, 90. [CrossRef]

111. Ma, J.W.; He, G.Z.; Liu, Z.; Qin, F.Z.; Chen, S.Y.; Zhao, X.X. Instantaneous cutting-amount planning for machining deformation homogenization based on position-dependent rigidity of thin-walled surface parts. *J. Manuf. Process.* **2018**, *34*, 401–411. [CrossRef]

112. Loehe, J.; Zaeh, M.F.; Roesch, O. In-Process Deformation Measurement of Thin-walled Workpieces. *Procedia CIRP* **2012**, *1*, 546–551. [CrossRef]

113. Wang, G.; Li, W.; Tong, G.; Pang, C. Improving the machining accuracy of thin-walled parts by online measuring and allowance compensation. *Int. J. Adv. Manuf. Technol.* **2017**, *92*, 2755–2763. [CrossRef]

114. Huang, N.; Yin, C.; Liang, L.; Hu, J.; Wu, S. Error compensation for machining of large thin-walled part with sculptured surface based on on-machine measurement. *Int. J. Adv. Manuf. Technol.* **2018**, *96*, 4345–4352. [CrossRef]

115. Rubio, A.; Rivero, A.; Del Sol, I.; Ukar, E.; Lamikiz, A. Capacitation of flexibles fixtures for its use in high quality machining processes: An application case of the industry 4.0 paradigm. *Dyna* **2018**, *93*, 608–612.

116. Wang, X.; Bi, Q.; Zhu, L.; Ding, H. Improved forecasting compensatory control to guarantee the remaining wall thickness for pocket milling of a large thin-walled part. *Int. J. Adv. Manuf. Technol.* **2018**, *94*, 1677–1688. [CrossRef]

117. Junbai, L.; Kai, Z. Multi-point location theory, method, and application for flexible tooling system in aircraft manufacturing. *Int. J. Adv. Manuf. Technol.* **2010**, *54*, 729–736. [CrossRef]

118. Hao, M.; Xu, D.; Wei, F.; Li, Q. Quantitative analysis of frictional behavior of cupronickel B10 at the tool-chip interface during dry cutting. *Tribol. Int.* **2018**, *118*, 163–169. [CrossRef]

119. Zhou, G.; Li, Y.; Liu, C.; Hao, X. A feature-based automatic broken surfaces fitting method for complex aircraft skin parts. *Int. J. Adv. Manuf. Technol.* **2016**, *84*, 1001–1011. [CrossRef]

120. Zhou, Z.; Zhang, H.; Xu, M. Research on precision and greenhouse manufacturing technology for large aircraft panels. *Procedia CIRP* **2016**, *56*, 565–568. [CrossRef]

121. Bumgarner, K.; Lebakken, C.; Vando, C.; Reddie, W.; Jacovetti, G. Universal Holding Fixture. U.S. Patent No. 4,684,113, 3 December 2009.

122. Zeng, S.; Wan, X.; Li, W.; Yin, Z.; Xiong, Y. A novel approach to fixture design on suppressing machining vibration of flexible workpiece. *Int. J. Mach. Tools Manuf.* **2012**, *58*, 29–43. [CrossRef]

123. Nee, A.Y.C.; Kurnar, A.S.; Tao, Z.J. An intelligent fixture with a dynamic clamping scheme. *Proc. Inst. Mech. Eng. Part B J. Eng. Manuf.* **2000**, *214*, 183–196. [CrossRef]

124. Wan, X.-J.; Zhang, Y.; Huang, X.-D. Investigation of influence of fixture layout on dynamic response of thin-wall multi-framed work-piece in machining. *Int. J. Mach. Tools Manuf.* **2013**, *75*, 87–99. [CrossRef]

125. Liu, S.G.; Jin, Q.; Wang, P. Effect of additional supports on surface errors in the peripheral milling of a flexible workpiece. *Int. J. Mater. Prod. Technol.* **2008**, *31*, 214–223. [CrossRef]

126. Liu, S.; Zheng, L.; Zhang, Z.H.; Wen, D.H. Optimal fixture design in peripheral milling of thin-walled workpiece. *Int. J. Adv. Manuf. Technol.* **2006**, *28*, 653–658. [CrossRef]

127. Rubio, A.; Calleja, L.; Orive, J.; Mújica, Á.; Rivero, A. Flexible Machining System for an Efficient Skin Machining. *SAE Tech.* **2016**, *1*, 2129. [CrossRef]

128. Innoclamp GmbH Innoclamp. Available online: http://www.innoclamp.de/ (accessed on 6 May 2019).

129. Liu, C.; Sun, J.; Li, Y.; Li, J. Investigation on the milling performance of titanium alloy thin-walled part with air jet assistance. *Int. J. Adv. Manuf. Technol.* **2018**, *95*, 2865–2874. [CrossRef]

130. Dufieux Dufieux Industrie—Modular. Available online: http://www.dufieux-industrie.com/en/mirror-milling-system-mms (accessed on 2 June 2016).

131. Mtorres Surface Milling Machining. Available online: http://www.mtorres.es/en/aeronautics/products/metallic/torres-surface-milling (accessed on 2 June 2016).
132. Zhang, Y.M.; Sims, N.D. Milling workpiece chatter avoidance using piezoelectric active damping: A feasibility study. *Smart Mater. Struct.* **2005**, *14*, N65–N70. [CrossRef]
133. Diez, E.; Perez, H.; Marquez, J.; Vizan, A. Feasibility study of in-process compensation of deformations in flexible milling. *Int. J. Mach. Tools Manuf.* **2015**, *94*, 1–14. [CrossRef]
134. Rashid, A.; Mihai Nicolescu, C. Active vibration control in palletised workholding system for milling. *Int. J. Mach. Tools Manuf.* **2006**, *46*, 1626–1636. [CrossRef]
135. Yang, Y.; Xu, D.; Liu, Q. Milling vibration attenuation by eddy current damping. *Int. J. Adv. Manuf. Technol.* **2015**, *81*, 445–454. [CrossRef]
136. Yang, Y.; Xie, R.; Liu, Q. Design of a passive damper with tunable stiffness and its application in thin-walled part milling. *Int. J. Adv. Manuf. Technol.* **2017**, *89*, 2713–2720. [CrossRef]
137. Díaz-Tena, E.; Marcaide, L.N.L.D.L.; Gómez, F.J.C.; Bocanegra, D.L.C. Use of Magnetorheological Fluids for Vibration Reduction on the Milling of Thin Floor Parts. *Procedia Eng.* **2013**, *63*, 835–842. [CrossRef]
138. Wang, T.; Zha, J.; Jia, Q.; Chen, Y. Application of low-melting alloy in the fixture for machining aeronautical thin-walled component. *Int. J. Adv. Manuf. Technol.* **2016**, *87*, 2797–2807. [CrossRef]
139. Kolluru, K.V.; Axinte, D.A.; Raffles, M.H.; Becker, A.A. Vibration suppression and coupled interaction study in milling of thin wall casings in the presence of tuned mass dampers. *Proc. Inst. Mech. Eng. Part B J. Eng. Manuf.* **2013**, *228*, 826–836. [CrossRef]
140. Woody, S.C.; Smith, S.T. Damping of a thin-walled honeycomb structure using energy absorbing foam. *J. Sound Vib.* **2006**, *291*, 491–502. [CrossRef]
141. Garimella, S.; Ramesh Babu, P. Understanding the challenges in machining thin walled thin floored Avionics components. *Int. J. Appl. Sci. Eng. Res.* **2013**, *2*, 93–100.
142. Izamshah, R.A.; Mo, J.P.T.; Ding, S. Hybrid deflection prediction on machining thin-wall monolithic aerospace components. *Proc. Inst. Mech. Eng. Part B J. Eng. Manuf.* **2011**, *226*, 592–605. [CrossRef]
143. Khandagale, P.; Bhakar, G.; Kartik, V.; Joshi, S.S. Modelling time-domain vibratory deflection response of thin-walled cantilever workpieces during flank milling. *J. Manuf. Process.* **2018**, *33*, 278–290. [CrossRef]
144. Zhang, L.; Gao, W.; Zhang, D.; Tian, Y. Prediction of Dynamic Milling Stability considering Time Variation of Deflection and Dynamic Characteristics in Thin-Walled Component Milling Process. *Shock Vib.* **2016**, *2016*, 3984186. [CrossRef]
145. Scheider, I.; Brocks, W. Residual strength prediction of a complex structure using crack extension analyses. *Eng. Fract. Mech.* **2009**, *76*, 149–163. [CrossRef]
146. Mahmud, A.; Mayer, J.R.R.; Baron, L. Determining the minimum clamping force by cutting force simulation in aerospace fuselage pocket machining. *Int. J. Adv. Manuf. Technol.* **2015**, *80*, 1751–1758. [CrossRef]
147. Panczuk, R. Clean alternative technology to chemical milling: Demonstration of technical, environmental and economic performance of mechanical milling for the machining of complex shaped panels used in the aeronautical and space industries—GAP (Green Advanced Panel). 2007. Available online: https://nanopdf.com/download/gap-clean-alternative-technology-to-chemical-milling_pdf (accessed on 25 June 2018).
148. Jiang, X.; Lu, W.; Zhang, Z. An approach for improving the machining efficiency and quality of aerospace curved thin-walled parts during five-axis NC machining. *Int. J. Adv. Manuf. Technol.* **2018**, *97*, 2477–2488. [CrossRef]
149. Urbikain, G.; Olvera, D.; de Lacalle, L.N.L. Stability contour maps with barrel cutters considering the tool orientation. *Int. J. Adv. Manuf. Technol.* **2017**, *89*, 2491–2501. [CrossRef]
150. Zhang, X.; Yu, T.; Wang, W.; Ehmann, K.F. Three-dimensional process stability prediction of thin-walled workpiece in milling operation. *Mach. Sci. Technol.* **2016**, *20*, 406–424. [CrossRef]
151. Meshreki, M.; Attia, H.; Kövecses, J. Development of a New Model for the Varying Dynamics of Flexible Pocket-Structures During Machining. *J. Manuf. Sci. Eng.* **2011**, *133*, 041002. [CrossRef]

152. Fernández, B.; González, B.; Artola, G.; López de Lacalle, N.; Angulo, C. A Quick Cycle Time Sensitivity Analysis of Boron Steel Hot Stamping. *Metals* **2019**, *9*, 235. [CrossRef]
153. Calleja, A.; Bo, P.; Gonzalez, H.; Bartoň, M.; López de Lacalle, L.N. Highly accurate 5-axis flank CNC machining with conical tools. *Int. J. Adv. Manuf. Technol.* **2018**, *97*, 1605–1615.

Article

On the Cutting Performance of Segmented Diamond Blades when Dry-Cutting Concrete

A. J. Sánchez Egea [1,*], V. Martynenko [2], D. Martínez Krahmer [2], L. N. López de Lacalle [1], A. Benítez [3] and G. Genovese [4]

[1] Department of Mechanical Engineering, Aeronautics Advanced Manufacturing Center (CFAA), Faculty of Engineering of Bilbao, Alameda de Urquijo s/n, 48013 Bilbao, Spain; norberto.lzlacalle@ehu.eus

[2] Centro de Investigación y Desarrollo en Mecánica, Instituto Nacional de Tecnología Industrial INTI, Avenida General Paz 5445, 1650 Miguelete, Provincia de Buenos Aires, Argentina; vmart@inti.gob.ar (V.M.); mkrahmer@inti.gob.ar (D.M.K.)

[3] Centro de Investigación y Desarrollo en Construcciones, Instituto Nacional de Tecnología Industrial, Avenida General Paz 5445, 1650 Miguelete, Provincia de Buenos Aires, Argentina; alemir@inti.gob.ar

[4] All Import SA, Paraná 4532, 1605 Munro, Provincia de Buenos Aires, Argentina; gustavo@e-bremen.com

* Correspondence: antonio.egea@ehu.eus

Received: 18 January 2018; Accepted: 7 February 2018; Published: 9 February 2018

Abstract: The objective of the present study is to analyze and compare the cutting performance of segmented diamond blades when dry-cutting concrete. A cutting criteria is proposed to characterize the wear of the blades by measuring the variation of the external diameter and the weight loss of the blade. The results exhibit the cutting blade SB-A, which has twice the density of diamonds and large contact area, exhibits less wear even though the material removal rate is higher compared with the other two cutting blades. Additionally, the surface topography of the different blades is evaluated to examine the impact of wear depending on the surface profile and the distribution of the diamonds in the blade's matrix. Large number of diamonds pull-out are found in blades type SB-C, which additionally shows the worst wear resistant capability. As a conclusion, the cutting efficiency of the blade is found to be related to the density of embedded diamonds and the type of the surface profile of the cutting blade after reaching the stop criteria.

Keywords: dry-cutting; concrete; segmented diamond blade; topography; diameter variation; weight loss

---

1. Introduction

Diamond blades were designed to cut hard or abrasive materials such as concrete, marble and ceramics. During the last 15 years, blades with diamond segments (diamond particles embedded in a cementing matrix) have allowed substantially increasing the effectiveness of these cutting operations [1]. Generally, researchers have focused on two topics when studying the performance of diamond blades: the first topic deals with the lifespan and tool wear, while the second topic studies the effectiveness in term of energy consumption under different operational configurations, such as cutting forces, vibrations and influence of lubricants. Regarding to the first topic, González et al. [2] described analytical model to predict wear of different metal alloys during a cutting process. They took into account the relative speed of the abrasive material, the applied force, the type of material and the size of the abrasive grains to describe the abrasive mechanism. Consequently, wear coefficients are proposed that behave linearly with the size of the abrasive grain for each metal alloy. Sánchez et al. [3] described a novel dressing method to recovery large-size grinding wheels. In this sense, a specific electrical mechanism to heal the conductive-bond with several electric variables such as electrode size, wheel speed and several current configurations were proposed. They found that the current configurations mainly affect the material removal rate, whereas the other parameters are linked to the process capability. Grinding experiments showed that this novel method can reduce up to 50% of the cutting forces and, ultimately, higher depth of cut can be assessed. Furthermore, Lopez de Lacalle et al. [4] focused on study how coating improves the specific cutting force on high speed steel (HSS) and cemented carbide flying tools while cutting wood. As a result, an increase of the lifespan of 116% and more than 30 km of cutting length was achieved with the HSS tool coated with TiCN or AlCrSiN. Recently, Guerra Rosa and coworkers [5–7] investigated the relationship between wear and the mechanical parameters of different metallic binders utilized to manufacture cutting tools. In particular, they studied the wear behavior of powder metallurgy matrices used in diamond tools and the possible relationship with their mechanical properties. The wear behavior of these matrices was evaluated during the processing of some common polycarbonate and soda-lime glass plates, and also in much harder materials, e.g., granite tiles. The results showed a linear relationship of the wear with the normal force, whereas the wear evolution of these matrices processing granite was about 100 times higher than processing glass; and in glass it was approximately 10 times higher than processing polycarbonate. Even though the mentioned literature presented the wear evolution and the behavior of the lifespan of different cutting tools, there is no standardized test in the industry to evaluate the performance of the cutting discs under common operational configurations.

Much research has been published about the cutting effectiveness using different operational configuration. For example, vibration issues when high-speed cutting stone with large diamond blades are commonly found [8]. The effects of the outer and inner diameters of the blade can be determined in the frequency spectrum, whereas the size of blades is linearly related to the spindle speed, which induces resonance and high vibrations. This issue can be reduced with a proper selection of the operating conditions and by increasing the stiffness and/or the internal strength of the blade. Additionally, Ucun and coworkers [9] investigated the diamond density and the type of matrix material when cutting granite stone. During the experiment the cutting forces and the specific cutting energy were analyzed to enhance the cutting performance. They concluded that using a high ratio of W-Co as a matrix compound enhance the cutting efficiency by reducing the tool wear, the cutting forces and the specific cutting energy. Slight improvement of wear was found when high diamond density was assessed in the segment area. Few years later, they [10] studied the influence of four lubricants when cutting stones with diamond blades to compare the results with respect to the same cutting process but employing water as a lubricant. They found that water mixed with boron oil achieved lower power consumption and higher wear resistance of the diamond blade. Furthermore, Oliveira et al. [11] analyzed several metallic matrices with embedded diamonds during sawing process. In particular, they analyzed the cutting forces and the tool wear when different iron-based powders were used to prevent diamond pull-outs. Consequently, lower tool wear was denoted with these type of powders compared with a standard tool, although higher cutting forces were found due to a higher number of diamonds at the contact surface. Regarding the cutting forces, the influence of the axial cutting force on the tool wear and lifespan of the blade when cutting marble was addressed [12]. The increasing the feed rate and depth of cut are associated with an axial deviation, which can promote circular cracks in the flange region. Moreover, the cutting capabilities of different diamond blades and metallic matrices were investigated by using several working parameters when cutting two types of stones (limestone and marble) [13,14]. During the cutting tests, vertical and horizontal forces, electric energy consumption and vibrations were analyzed with the aim to evaluate and, ultimately, the efficiency of the cutting tool was characterized according to the mechanical properties of the different binders, the size of diamond grit and type of the bounding matrix used to embed the diamonds. Furthermore, Konstanty and Tyrala [15] described the wear mechanism of cutting tool with diamonds embedded under real industrial conditions. This work revealed that the wear rate was major affected due to the diamond density and the iron-base matrix compound exhibit the best cutting performance results against wear. Finally, they described the interaction between the diamonds and the surface to cut, denoting the importance of a matrix to retain the diamond during the cutting process to avoid pull-outs.

In addition, Ucun et al. [16] examined the energy consumption variation according to several cutting parameters when cutting natural stones with diamond blades. During the cutting process, down cutting was performed as the cutting operation, granite (Blue Pearl) was used as the natural stone and water was preferred as the cutting lubricant. Three different cutting parameters, depth of cut, circular velocity and cutting speed, were chosen for the cutting experiments. Specific cutting energy values were obtained through an analytic method using the data recorded from the energy analyzer and the dynamometer. According to these results, the power consumption increased as the cutting parameters increased, but the specific cutting energy decreased. Therefore, the depth of cut was the most sensitive cutting parameter for the rise of power consumption. Accordingly, a new computer algorithm to maximize tool productivity of a sawing process is addressed, taking into account the stone characteristics and the quality required for the end product [17]. This algorithm essentially depends on three variables, namely, the cutting depth, the feed rate and the rotational velocity, as well as how these variables are related with the forces acting on the tool. Oliveira et al. [18] examined the viability of adding different additives to the Fe-Cu matrix to make a binder material with the adequate wear properties needed for manufacturing diamond impregnated tools for cutting stone. After the hot-pressing cycles, the main mechanical properties of the sintered bodies were evaluated. Cutting tests under real conditions with Porriño granite were carried out to compare the performance of the tools. The results from the cutting trials revealed that the tools show quite similar behavior during the cutting operations, thus indicating that replacement of Nb with Co is a promising challenge to be followed in the near future. Recently, Krolczyk and coworkers [19,20] investigated in detail measurement protocols of effective areas of different cutting tools and propose which are the advantages and limitations of the optical measurement devices depending on the type of abrasive tool. Additionally, they stated the differences of dry cutting and cutting with lubricant and the onset wear effects in the effective cutting area such as adherence and cracking. As a result, they found that the surface profile after the cutting process revealed higher tribological irregularities when machining without coolant, subsequently higher plastic flow at the cutting zone. Finally, Hu et al. [21] studied the cutting performance of three kinds of diamond blades with different structure parameters. Sawing force and vibration were measured by cutting several concretes with different strengths with different cutting parameters. Later, the characterization of the sawing forces and vibrations helped to find the optimal structure of diamond blade to properly operate with the following parameters: segment width, sawing velocity and the type of the material to cut. Following this research topic, the aim of this work is to evaluate and compare the cutting performance of segmented diamond blades when dry-cutting concrete, where the stop cutting criterion is fixed based on the total surface of concrete to cut. Likewise, the tool wear is measured by the variation of the external diameter and the weight loss of the blade. The final topography of the blades is evaluated to determine the impact of wear at the surface and, ultimately, the integration of the diamonds in the matrix. In this sense, it is possible to quantify when a blade possesses a superior cutting capability, in terms of the common machining operation without risking the lifespan of the cutting machine.

2. Methodology

A total of 12 universal Segmented Blades (SB-A, SB-B and SB-C, four samples per brand) with embedded diamonds were investigated. The specifications of the blades are: 115 mm of diameter, 13,200 rpm of maximum spindle speed, 80 m/s of maximum cutting velocity and all the blades fulfilled the EN 13236 standards. The three segmented blades SB-A, SB-B and SB-C dispose 9 segments for the first two models and 8 segments for SB-C model. Figure 1 shows two box plots with the geometrical measurements of thickness and height of each of the cutting segments for each type of diamond blade. Lines within the box plots represent the median; squares represent the average; whiskers represent the tenth and ninetieth percentiles; and crosses below and above the box plots represent maximum and minimum sided values.

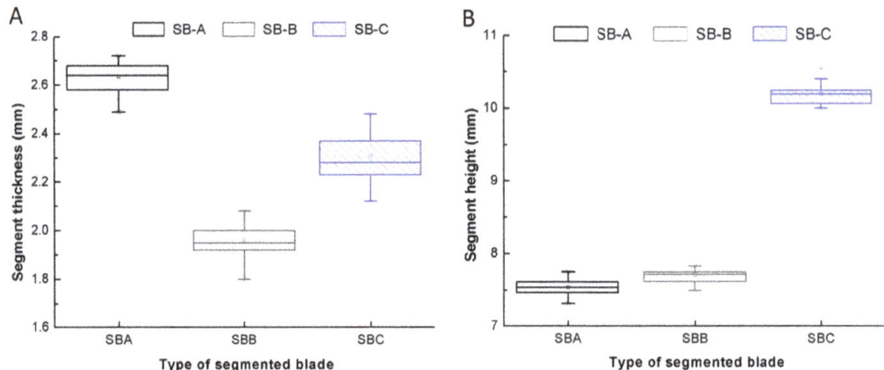

Figure 1. Boxplot of the: thickness (**A**); and height (**B**) of the cutting segments for each type of diamond blades.

A rectangular mortar matrix was designed and developed to fabricate the concrete blocks by following the standard IRAM 1534:2004. A total of 65 prismatic blocks of $290 \times 260 \times 40$ mm^3 were developed by mixing concrete with medium size stones. Particularly, the composition of the concrete was 170 kg/m^3 of water, 350 kg/m^3 of concrete Portland CPC-40, 800 kg/m^3 of siliceous sand, 191 kg/m^3 of granitic sand (dimension up to 6 mm), 875 kg/m^3 of medium size stones (dimension of 6 to 12 mm), 1% and 0.55% of mass of cement of two superplasticizer additives Mira-57 and Daracem-18, respectively. The procedure for preparing the specimens was the following. Firstly, the sand and gravels were combined in the concrete mixer during 30 s. Then, the Portland cement was added and mixed during 30 s. Subsequently, the half part of the water and the additives were added and combined during 90 s. Finally, the rest of water and additives were introduced and mixed in the mixer during 180 s. Following this procedure, a homogeneous blend was obtained. Later, the mechanical characterization of the concrete was performed with compression tests of 8 cylindrical specimens of Φ 150×300 mm^2 with 28 days of curation following the standard IRAM 1546:1992. The average of the ultimate tensile strength of the concrete blocks was found to be 39.1 ± 1.2 MPa, which is the crucial parameter to evaluate the cutting performance of the diamond blades.

All the linear cuts were made on the two opposite faces of the concrete blocks with greater longitudinal area. A total of 28 linear cuts with a lateral displacement of 9 mm were performed in each face. A cutting area of 650,000 mm^2 (0.65 m^2) was established as the total length or the stop criteria to measure the wear of the blades; similar values were used by Oliveira et al. [18]. This area is equivalent to make 56 cuts of 10 mm deep and 290 mm long in 4 different concrete blocks. An in-house adapted bench was utilized to accurately perform all these linear cuts. This bench can move in two axes of the horizontal plane due to a pair of servomotors which are regulated with a programmable control. A manual angular saw (Makita, model: 9557HPG, Anjō, Japan) of 840 W and 11,000 rpm was attached in a structural bridge and, then, allocated above the bench. The depth of cut was manually set with a height adjustment module. To select the proper operational cutting parameters, calibration tests were carried out to define the maximum feed rate for a depth of cut of 10 mm and without exceeding a consumption of 7 A. As a result, the feed rate selected to perform the cutting experiments was 320 mm/min. Figure 2 exhibits a schematic diagram with pictures of the in-house cutting machine of concrete blocks with segmented diamond blades.

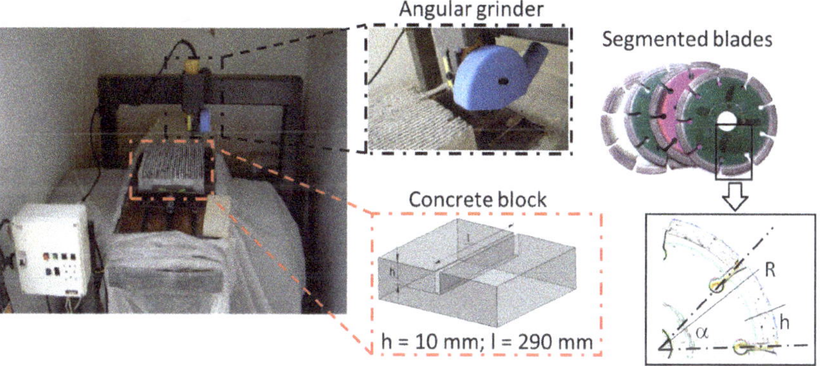

Figure 2. Cutting process of dry-concrete by using different segmented diamond blades in an in-house developed automatic machine.

The thickness and height of the segment area were assessed using a digital caliber (Tesa Technology, model: 00530084, Renens, Switzerland). The same caliber was employed to measure the diameters of blades before and after the cutting experiment. An electronic scale (Moretti, model: OAC 2.4, South Deerfield, MA, USA) with an accuracy of 0.2 g was utilized to record the weights of the as-received blade and after reaching the stop criteria. Additionally, optical microscope (Carl Zeiss, model: Axiotech 100HD-3D, Oberkochen, Germany) and electron microscopy (Philips, model: SEM 505, Amsterdam, The Netherlands) were employed to capture images of the cutting characteristics and the diamond size of each blade. Finally, a 3D profilometer (Leica, model: DCM3D, UPV, Bilbao, Spain) with a HCX PL Fluotar objectives of 10× magnification was used to measure the surface and profile topographies of the segment area. The resolution was 140 nm in-plane and 100 nm in vertical direction by using confocal technique. The experiments were accomplished at the INTI-Mechanics Center in Argentina.

3. Results

3.1. Cutting Tool Characterization

Firstly, the density of diamonds in each segmented blade is measured to determine if significant differences are presented between the studied blades. To do that, the diamonds expose on both sides of the segment area are counted. Table 1 shows the number of diamonds measured on both side of the segment, the average number of diamonds per blade and the error dispersion.

Regarding the blades differences, it is desired to estimate the density of diamonds in each blade. To do that, the effective area of each segmented blade type is calculated as follow:

$$A_s = \frac{\alpha}{360} \cdot [\pi \cdot (R^2 - (R - h)^2)] \tag{1}$$

where A_s is the segment area, R is the radio of the blade, h is the segment height measured in Figure 1, and α is the effective angle of the segments area which depend on the type of blade used. Later, the density of the diamond per segment area is determined by multiplying the effective segment area per the average diamond quantity in a single segment. Table 2 shows the estimated density of diamonds per effective segment area in each blade.

Table 1. Quantity of diamonds per segment of each type of blade.

Parameter/Blade Type	SB-A	SB-B	SB-C
	57	36	42
	54	32	52
Sample 1	63	46	50
	59	40	48
	55	49	56
	61	35	51
Sample 2	59	56	35
	54	24	35
	49	45	42
	50	32	42
Sample 3	75	38	40
	55	50	37
	74	46	48
	48	50	47
Sample 4	40	31	52
	53	36	70
Average	56.63	40.38	46.69
Dispersion	8.93	8.79	8.93

Table 2. Estimated density of diamonds per effective segment area of each blade.

Parameter/Blade Type	SB-A	SB-B	SB-C
Av. segment height (mm)	7.54	7.71	10.19
Number of segments	9	9	8
Effective angle of area (°)	33.50	37.50	53.30
Effective segment area (mm^2)	237.69	265.43	377.33
Av. diamond quantity per segment (n° grains)	56.62	40.37	46.68
Diamond density per segment (n° grains/mm^2)	0.24	0.15	0.12

The density values show that the SB-A blade dispose twice the number of diamonds in a segmented area compared to the SB-C blade. Many diamonds embedded in the blade's matrix increase the cutting efficiency and let us increase the feed rate to cut rough grain structure such as concrete [21]. Therefore, higher diamond density in the segment area is expected to extend the lifespan of blade due to decrease the average cutting force of each diamond, despite that increasing the number of diamond will increase the cost of blade. To characterize the profile of the matrix–diamonds distribution of each blade, a segment area of each blade was examined by electron microscopy. Figure 3 exhibits the matrix–diamond distribution for each type of blade.

Five measurements of the diamond size were taken to determine that the diamond dimension are equivalent among the three blades. In this sense, the diamond size are 0.26 ± 0.11 mm, 0.36 ± 0.08 mm, and 0.27 ± 0.08 mm for the blade type SB-A, SB-B and SB-C, respectively. Figure 3 shows that many diamonds are found on the surface of SB-A compared to the other two types. Therefore, more diamonds will be in contact with the surface of the concrete to cut; thus, a higher cutting efficiency is expected during the process.

Figure 3. Grain size, thickness and matrix–diamond distribution for each type of blade.

Finally, the material removal rate is computed for each diamond blade to compare the tool wear with comparable cutting configurations. Accordingly, the material removal rate is a function of the cutting conditions and of the operation performed. For a sawing operation, the material removal rate (*MRR*) can be determined as follow:

$$MRR = f \cdot A_c = f \cdot \left(\frac{\pi \cdot R^2}{2} + w \cdot (d - r) \right)$$ (2)

where f is the feed rate (mm/min), A_c is the effective cross sectional area, R is the radio of the blade, w is the average segment width measured in Figure 1, and d is the depth of cut set during the cutting trials. As an approximation, the area of the blade's edge was considered as a semicircle of diameter equal to the width. Subsequently, the values of the material removal rate are 22.0 mm^3/min, 19.3 mm^3/min and 19.8 mm^3/min for the blade type SB-A, SB-B and SB-C, respectively. Therefore, SB-A is cutting more concrete in the same time and, consequently, a higher tool wear is expected compared to SB-B and SB-C.

3.2. Tool Wear Measurement

As mentioned above, 0.65 m^2 of cutting area was assessed as the total length/stop criteria to measure the wear of each blade. Two different ways of measuring the tool wear are commonly used: diameter difference and weight loss of the blade. The aim here is to determine if both methods for measuring the tool wear are comparable. Firstly, the tool wear is estimated by measuring the difference of the initial diameter and diameter of blade once the stop criteria is assessed, as listed in Table 3.

Table 3. Tool wear analyses by measuring the diameter difference.

Blade Type/Parameter	Initial Diameter (mm)	Final Diameter (mm)	Difference (mm)	Average (mm)	Dispersion * (mm)
SB-A	118.45	118.25	0.20	0.38	0.21
	118.49	118.25	0.24		
	118.25	117.60	0.65		
	118.61	118.18	0.43		
SB-B	115.07	114.35	0.72	0.79	0.05
	115.29	114.45	0.84		
	115.25	114.48	0.77		
	115.02	114.20	0.82		

Table 3. *Cont.*

Blade Type/Parameter	Initial Diameter (mm)	Final Diameter (mm)	Difference (mm)	Average (mm)	Dispersion * (mm)
SB-C	115.39	114.00	1.39	1.26	0.11
	115.52	114.24	1.28		
	115.32	114.19	1.13		
	115.49	114.26	1.23		

* The error dispersion is estimated with the standard deviation.

On the other hand, the tool wear is also estimated by measuring the difference of the initial weight and weight of blade once the stop criteria is reached, as listed in Table 4.

Table 4. Tool wear analyses by measuring the weight loss difference.

Blade Type/Parameter	Initial Weight (g)	Final Weight (g)	Difference (g)	Average (g)	Dispersion * (g)
SB-A	158.2	157.0	1.2	1.50	0.38
	158.4	157.2	1.2		
	160.2	158.2	2.0		
	160.8	159.2	1.6		
SB-B	96.1	94.2	2.6	2.75	0.19
	96.8	93.8	3.0		
	97.6	94.8	2.8		
	95.2	92.6	2.6		
SB-C	106.4	101.2	5.2	4.30	0.62
	103.6	99.8	3.8		
	105.8	101.8	4.0		
	104.2	100.0	4.2		

* The error dispersion is estimated with the standard deviation.

Boxplots are assessed to address the dispersion of the diameter variation and weight loss and, subsequently, to statistically compare the tool wear differences between the three blades. Figure 4 shows the tool wear in terms of diameter variation and weight loss when reaching the stop criteria of the cutting length.

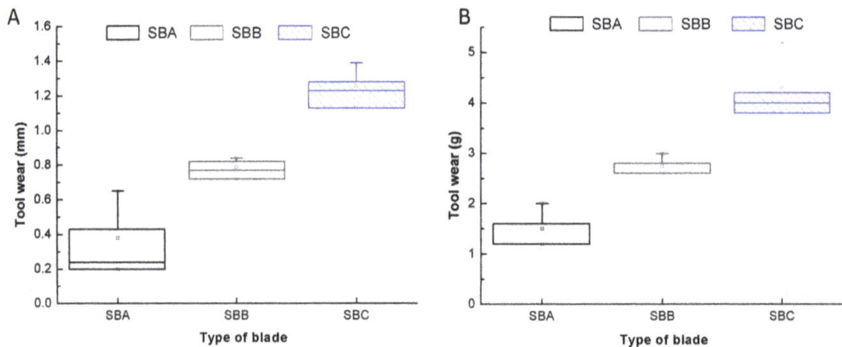

Figure 4. Boxplots of the tool wear in the three studied diamond blades when reaching the stop criteria in terms of: diameter variation (**A**); and weight loss (**B**).

From the boxplot diagrams of diameter variation and weight loss, it is clear that wear differences are found in each blade. Lowest values of wear are found in SB-A compared with SB-B and SB-C. On the contrary, wear values of 1.3 mm of diameter difference and about 4 g of weight loss is recorded in the SB-C type after completing the cutting criteria. SB-A, SB-B and SB-C were staggered in an increasing wear sequence. Consequently, the blade type SB-C shows the worse behavior against wear, exhibiting a diameter change of 1.26 mm and a weight loss of 4.30 g on average. On the other hand, SB-A exhibits the best performance with a diameter variation of 0.38 mm and a weight loss of 1.50 g on average. Consequently, the results show that tool wear can be measured either by recording the weight loss difference or by measuring the diameter variation, as both methods have been found to be equivalent. To study in detail the different tool wear findings, the surface topography was evaluated to analyze the surface profile of each blade. Figure 5 exhibits the surface of the edge of the segment area for the three blades, SB-A, SB-B, and SB-C, after reaching the stop criteria. The peaks found in the contour of the surface topography is noise due to the side-effects from the optical acquisition procedure.

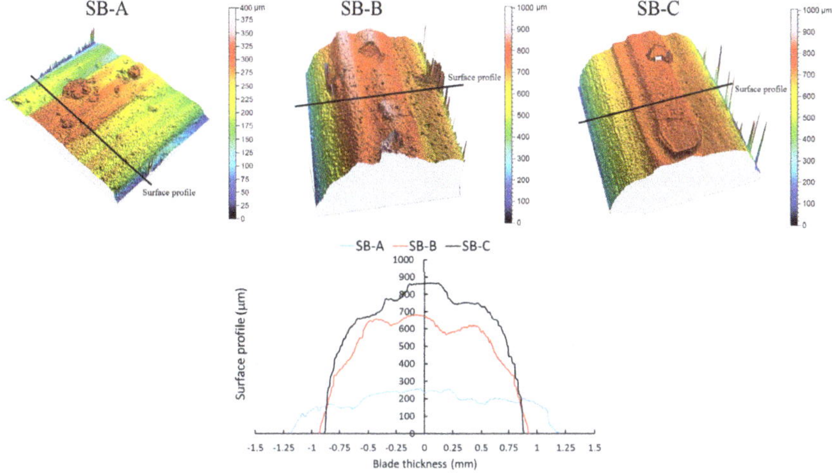

Figure 5. Surface topography of the edge of the segment area for each type of blade.

The results show that the topography of a surface is different for each case, as expected from the wear resistant values found in Figure 4. In particular, the SB-A presents a higher flatness surface and, consequently, more diamonds in the surface compared to the other two blades. This type of surface will bring a higher contact area during the cutting process, which means higher cutting forces due to friction. A higher cutting efficiency is found due to high number of diamonds in contact with the concrete during the dry-cutting process. A similar conclusion was stated by Turchetta [22], where he described that cutting forces and the specific cutting energy increase depending on the shape of the interface between cutting tool and workpiece, as a higher material removal rate is achieved. SB-B and SB-C show similar surface profiles within the range of 700–800 µm of maximum height and about 2 mm of thickness. Both cases exhibit that the diamonds are allocated in the center of the blades thickness and the agglutinant is wearing on both sides of the blade, leaving a prominent semicircle in the cross sectional area of the segment area. During the sawing process, it is crucial the compound matrix of the segments area keep the diamond grains for as long as possible during the cutting process. The blade type SB-C was analyzed using an electronic microscope after finishing the cutting criteria to determine if any evidence can explain the wear differences, in terms of diameter and weight, between SB-B and SB-C. Many holes due to diamond grain pull-out were observed in the SB-C. This loss of diamond grains in the agglutinant clearly affects the cutting capacity of the blade and enhances the tool wear. No diamond detachment occurred in cutting blades SB-A and SB-B. It is possible to minimize the diamond grit pull-outs by using active elements in bounding matrix (Bailey and Collin [23]), such as Cr and Co, which have been found to enhance the diamond holding effectiveness and improve wear performance of the diamond disc [18].

In future work, the cutting forces will be recorded when cutting dry-cutting concrete with the aim of estimating the power consumption and material removal rate for each type of blade. It is expected that, assuming the same depth of cut, the surface topography will modify the cutting forces and, consequently, the power consumption of the saw. Additionally, the wear evolution will be analyzed by measuring the diameter variation and weight loss for different cutting length values until reaching the failure of the cutting tool. Accordingly, both the wear evolution and the lifespan of the segmented diamond blades can be characterized. Finally, several bounding matrices and surfaces coatings, such as TiN and AlTiN, will be characterized to address the adhesion of diamonds to improve the wear resistance of these cutting tools.

4. Conclusions

The present work describes a procedure to evaluate the wear performance of different segmented diamond blades. The following findings can be withdrawn:

- It is necessary to adapt a cutting machine to test several cutting conditions with a previously defined stop criteria and, ultimately, measure the tool wear according to the tool diameter variation and the weight loss of the blade.
- The density of the diamonds in a segmented area is experimentally estimated by dividing the number of diamonds found in the segment surface per the effective segment areas of each type of blade. The wear experiments show that the diamond density exhibits a strong impact on the cutting efficiency of the blade.
- The surface profile seems to influence of the cutting capability, where a higher surface flatness enhances the cutting efficiency, due to higher contact area of the surface to cut with the diamonds grains embedded in the metallic matrix.
- Larger cutting areas and higher cutting feed rate are necessary to evaluate the wear performance of the blades as well as the wear capability of the blades before the failure.
- The limitation of the presented study deals with the bounding matrix composition of each blade, which is unknown. Likely, the diamond pull-out found in SB-C blades is due to not using effective active elements in the bounding matrix. Unfortunately, this information is neither provided by the supplier nor listed in the technical data sheet of the product.

Acknowledgments: All Import SA is acknowledged for their support and contribution to this work. This work is supported by the Ministry of Economy and Competitiveness of Spain (reference project: FJCI-2016-29297), Instituto Nacional de Tecnología Industrial (INTI) of Argentina and the Aeronautics Advanced Manufacturing Center (CFAA) of Bilbao.

Author Contributions: Conceptualization: D. Martínez Krahmer and L. N. López de Lacalle. Data curation: V. Martynenko and A. J. Sánchez Egea. Formal analysis: D. Martínez Krahmer, A. J. Sánchez Egea and L. N. López de Lacalle. Funding acquisition: D. Martínez Krahmer, A. J. G. Genovese and L. N. López de Lacalle. Methodology: V. Martynenko, A. Benítez and D. Martínez Krahmer. Software: V. Martynenko, A. J. Sánchez Egea and D. Martínez Krahmer. Supervision: D. Martínez Krahmer and L. N. López de Lacalle. Validation: V. Martynenko, A. Benítez, G. Genovese and A. J. Sánchez Egea. Writing original draft: V. Martynenko, D. Martínez Krahmer and A. J. Sánchez Egea. Writing, reviewing and editing the manuscript: D. Martínez Krahmer, G. Genovese and L. N. López de Lacalle.

References

1. Fernandes, J.C.; Pinto, S.; Amaral, P.M.; Rosa, L.G. Development of a new testing method for assessing the wear behaviour of circular cutting discs. *Mater. Sci. Forum* **2008**, *587*, 966–970, doi:10.4028/www.scientific.net/MSF.587-588.966.
2. González, H.A.; Sánchez Egea, A.J.; Travieso-Rodríguez, J.A.; Llumà i Fuentes, J.; Jorba Peiró, J. Estimation of the polishing time for different metallic alloys in surface texture removal. *Mach. Sci. Technol.* **2018**, *22*, 1–13, doi:10.1080/10910344.2017.1402931.
3. Sánchez, J.A.; Ortega, N.; Lopez de Lacalle, L.N.; Lamikiz, A.; Marañon, J.A. Analysis of the electro discharge dressing (EDD) process of large-grit size cBN grinding wheels. *Int. J. Adv. Manuf. Technol.* **2006**, *29*, 688–694, doi:10.1007/s00170-005-2579-z.
4. Lopez de Lacalle, L.N.; Larrinoa, J.; Rodriguez, A.; Fernández, A.; López, R.; Azkona, I. On the cutting of wood for joinery applications. *Proc. Inst. Mech. Eng. Part B J. Eng. Manuf.* **2014**, *229*, 1–46, doi:10.1177/0954405414534431.
5. Guerra Rosa, L.; Coelho, A.; Amaral, P.M.; Cruz Fernandes, J. Test Methodology to evaluate the wear performance of PM matrices used in Diamond impregnated tools for cutting hard materials. In *Power Metallurgy for Automotive*, 1st ed.; Ramakrishnan, P., Ed.; New New Age International Ltd.: Delhi, India, 2012; Chapter 22.
6. Amaral, P.M.; Coelho, A.; Anjinho, C.A.; Cruz Fernandes, J.; Guerra Rosa, L. Evaluation of the relationship between diamond tool wear performance and mechanical properties of the individual metallic binders. *Mater. Sci. Forum* **2010**, *636*, 1467–1474, doi:10.4028/www.scientific.net/MSF.636-637.1467.
7. Guerra Rosa, L.; Amaral, P.M.; Fernandes, J.C. Experimental determination of Young's modulus in PM metal matrices used in diamond impregnated tools for cutting hard materials. *Powder Metall.* **2013**, *51*, 46–52, doi:10.1179/174329008X277451.
8. Zhao, M.; Zhang, L.; Wu, B. Study on the dynamic characteristics of the large diamond circular saw blade. *Adv. Mater. Res.* **2014**, *945–949*, 781–784, doi:10.4028/www.scientific.net/AMR.945-949.781.
9. Ucun, I.; Aslantas, K.; Sedat Buyuksagis, I.; Tasgetiren, S. An investigation on the effect of diamond concentration and matrix material composition in the circular sawing process of granites. *Proc. Inst. Mech. Part C J. Mech. Eng. Sci.* **2011**, *225*, 17–27, doi:10.1243/09544062JMES2104.
10. Ucun, I.; Aslantas, K.; Sedat Buyuksagis, I.; Tasgetiren, S. Effect of cooling liquids on cutting process using diamond segmented disc of natural stones. *Proc. Inst. Mech. Part C J. Mech. Eng. Sci.* **2012**, *227*, 2315–2327, doi:10.1177/0954406212473555.
11. Oliveira, F.A.C.; Anjinho, C.A.; Coelho, A.; Amaral, P.M.; Coelho, M. PM materials selection: The key for improved performance of diamond tools. *Met. Powder Rep.* **2017**, *72*, 339–344, doi:10.1016/j.mprp.2016.04.002.
12. Aslantas, K.; Ozbek, O.; Ucun, I.; Sedat Buyuksagis, I. Investigation of the effect of axial cutting force on circular Diamond saw blade used in marble cutting process. *Mater. Manuf. Process.* **2009**, *24*, 1423–1430, doi:10.1080/10426910903344039.
13. Anjinho, C.A.; Sá Coelho, A.M.; Amaral, P.M.; Fernandes, J.C.; Guerra Rosa, L. *New Standard Methodologies of Classify the Efficiency of Diamond Cutting Discs*; Global Stone Congress—AIDICO: Alicante, Spain, 2010.
14. Anjinho, C.A.; Amaral, P.M.; Fernandes, J.C.; Guerra Rosa, L. A new laboratory methodology for assessing the cutting behaviour of gangsawing blades. *Key Eng. Mater.* **2013**, *548*, 72–81, doi:10.4028/www.scientific.net/KEM.548.72.
15. Konstanty, J.S.; Tyrala, D. Wear mechanism of iron-base diamond-impregnated tool composites. *Wear* **2013** *303*, 533–540, doi:10.1016/j.wear.2013.04.016.
16. Ucun, I.; Aslantas, K.; Sedat Buyuksagis, I.; Tasgetiren, S. Determination of specific energy in cutting process using diamond saw blade of natural stone. *Energy Educ. Sci. Technol. Part A Energy Sci. Res.* **2012**, *28*, 641–648.
17. García, J.C.; Amaral, P.M.; Coelho, A.; Tavares, A.; Fernandes, J.C.; Guerra Rosa, L. Optimisation of circular sawing conditions to maximize tool productivity for each class of material. *Key Eng. Mater.* **2013**, *547*, 106–114, doi:10.4028/www.scientific.net/KEM.548.106.
18. Oliveira, H.C.P.; Coelho, A.; Amaral, P.M.; Fernandez, J.C.; Guerra Rosa, L. Comparison between cobalt and niobium as a matrix component for diamond impregnated tools used for stone cutting. *Key Eng. Mater.* **2013**, *547*, 98–105, doi:10.4028/www.scientific.net/KEM.548.98.

19. Krolczyk, G.M.; Nieslony, P.; Maruda, R.W.; Wojciechowsk, S. Dry cutting effect in turning of a duplex stainless steel as a key factor in Clean Production. *J. Clean. Prod.* **2017**, *142*, 3343–3354, doi:10.1016/j.jclepro.2016.10.136.

20. Kaplonek, W.; Nadolny, K.; Krolczyk, G.M. The use of focus-variation microscopy for the assessment of active surfaces of a new generation of coated abrasive tools. *Meas. Sci. Rev.* **2016**, *16*, 42–53, doi:10.1515/msr-2016-0007.

21. Hu, S.; Wang, C.; Chen, B.; Hu, Y. Dry-cutting concrete study of diamond saw blade with different segment width. *Mater. Sci. Forum* **2006**, *532*, 321–324, doi:10.4028/www.scientific.net/MSF.532-533.321.

22. Turchetta, S. Cutting force in stone machining by diamond disk. *Adv. Mater. Sci. Eng.* **2010**, *2010*, 631437, doi:10.1155/2010/631437.

23. Bailey, M.W.; Collin, W.D. De Beers Titanized metal bond diamond grit and related studies on the sawing of stone and concrete. *Ind. Diam. Rev.* **1978**, *1*, 8–13.

Article

Characteristics of the Arcing Plasma Formation Effect in Spark-Assisted Chemical Engraving of Glass, Based on Machine Vision

Chao-Ching Ho [1],* and Dung-Sheng Wu [2]

[1] Graduate Institute of Manufacturing Technology and Department of Mechanical Engineering, National Taipei University of Technology, 1, Sec. 3, Zhongxiao E. Rd., Taipei 10608, Taiwan

[2] Department of Mechanical Engineering, National Yunlin University of Science and Technology, Yunlin 64002, Taiwan; m10211002@yuntech.org.tw

* Correspondence: hochao@mail.ntut.edu.tw; Tel.: +886-2-2771-2171; Fax: +886-2-2731-7191

Received: 6 February 2018; Accepted: 20 March 2018; Published: 22 March 2018

Abstract: Spark-assisted chemical engraving (SACE) is a non-traditional machining technology that is used to machine electrically non-conducting materials including glass, ceramics, and quartz. The processing accuracy, machining efficiency, and reproducibility are the key factors in the SACE process. In the present study, a machine vision method is applied to monitor and estimate the status of a SACE-drilled hole in quartz glass. During the machining of quartz glass, the spring-fed tool electrode was pre-pressured on the quartz glass surface to feed the electrode that was in contact with the machining surface of the quartz glass. In situ image acquisition and analysis of the SACE drilling processes were used to analyze the captured image of the state of the spark discharge at the tip and sidewall of the electrode. The results indicated an association between the accumulative size of the SACE-induced spark area and deepness of the hole. The results indicated that the evaluated depths of the SACE-machined holes were a proportional function of the accumulative spark size with a high degree of correlation. The study proposes an innovative computer vision-based method to estimate the deepness and status of SACE-drilled holes in real time.

Keywords: in situ estimation; SACE-drilled hole depth; spark-assisted chemical engraving; glass machining; computer vision; electrochemical discharge machining

---

1. Introduction

Non-traditional engraving processes have been becoming a promising technology for micro-machining. New groups of unconventional engraving processes such as the sustainable biomachining of metals by using bacteria [1–3] and spark-assisted chemical engraving of non-conductive materials [4] can remove material from a workpiece without affecting the mechanical properties of the material due to the lack of machining forces produced in the process. Spark-assisted chemical engraving (SACE) is also known as electrochemical discharge machining (ECDM) and is a non-traditional machining process that is advantageous for machining non-conducting [5], hard and brittle materials with low residual stress to produce micron-sized holes. In the field of electrochemical discharge machining, it is important to focus on enhancing the processing efficiency, hole quality, and reproducibility. The processing efficiency is improved if the state of the SACE machining is monitored immediately, and this improves the quality of the processing holes and reduces the processing cost. However, the machining of non-conducting materials with the SACE machining process is associated with challenges [6]. A literature survey indicates that several attempts have focused on improving the machining performance of SACE [4–6].

The gas film thickness is the main limiting factor of electrochemical discharge phenomena. Surfactants were added to an electrolyte to increase the wettability of the tool electrode and to thereby reduce the gas film thickness. It was experimentally observed that the critical voltage was significantly reduced due to the increase in the wettability of the tool electrode, and the gas film thickness was thereby reduced by adding surfactants to the electrolyte. In 2007, Zheng et al. [7] changed the electrode geometry to reduce the impact area of the hole discharge and improve the wettability of the electrode to reduce the thickness of the gas film. The combination of the flat sidewall and flat front tool produced a compact gas film attached to the tool electrode. Additionally, the experiment confirmed that this improved the processing accuracy and reduced the entrance reaming problems. In 2010, Cheng et al. [8] suggested that the stability of the gas film formation determined the machining accuracy, surface roughness, and reproducibility of the machined parts and indicated that a correlation existed between the current signal and the gas film quality. An increase in the processing deepness beyond 300 µm made it difficult to maintain the machining efficiency due to the worsening of the electrolyte circulation at the tooltip and the increase of the average current value. In 2011, Jiang et al. [9] established a finite element model to correlate spark energy and the geometry of removed material. However, uncertainties were involved in the discharging activity during ECDM due to the high instability of the gas film. In 2015, Gouda et al. [10] reviewed the parameters that influence the material removal rate for electrochemical discharge machining and concluded that parameters including DC power supply, tool electrode, electrolyte, and work-piece materials affect the machining performance. The parameters also contributed to the stability of the gas film. During the ECDM process, an unstable gas film around the tool electrode reduces machining reproducibility. In 2016, Giandomenico et al. [11] indicated that the disadvantage of the electro-discharge machining process includes relatively poor accuracy due to gas bubbles that increase the current density at the side gap.

Investigations have been conducted to identify the main parameters that influence SACE machining. Thus, there is a paucity of studies that offer a complete understanding of the machining process. During the SACE machining, spark discharges induced inside a gas film surround the tool electrode. Spark discharges occur between the tool electrode and the electrolyte when the voltage between the electrodes exceeds the critical voltage. Previous studies [12,13] indicate that the gas film is a key factor for spark discharge and that the quality of machining performed by spark discharging is essentially controlled by the quality of the gas film during the SACE process. The quality of the gas film is a dominant factor that determines the machining qualities. Although the machining mechanism of SACE is not established to date [14], it is generally agreed that the main machining process is the thermal melting of the workpiece. However, active feedback monitoring of the process is needed to further improve the machining characteristics of SACE.

In this study, we propose an innovative monitoring method for the in-situ measuring of SACE-drilled holes based on computer vision analysis. As a result, a low cost in-situ optical inspection system was developed to characterize the spark discharge during SACE machining. The monitoring of spark discharge is used as the basis for an in situ measuring system for SACE hole drilling. This monitoring arrangement includes image acquisition and pre-processing units, a SACE drilling apparatus, and a computer analysis unit as presented in Figure 1. First, we employed the industrial camera of the image acquisition and pre-processing unit for the in situ acquisition of an image of the SACE drilling region of the surface of the workpiece. The high spark energy caused the melting and evaporation of the material during the SACE drilling and resulted in thermal erosion due to the heat simultaneously generated by the discharges. The sparking action phenomenon involved arcing action, and this is captured by the industrial cameras. The images comprised of light emission observed in the film in which electrical discharges occur between the tool electrode and the surrounding electrolyte. Finally, image data are analyzed at the computer analysis unit to conduct an in situ processing of the relationship between the geometrical forms of the drilling region comprising of the deepness of a blind hole and image information. The experimental arrangement realizes benefits including lower costs, in situ measurement, and improved performance.

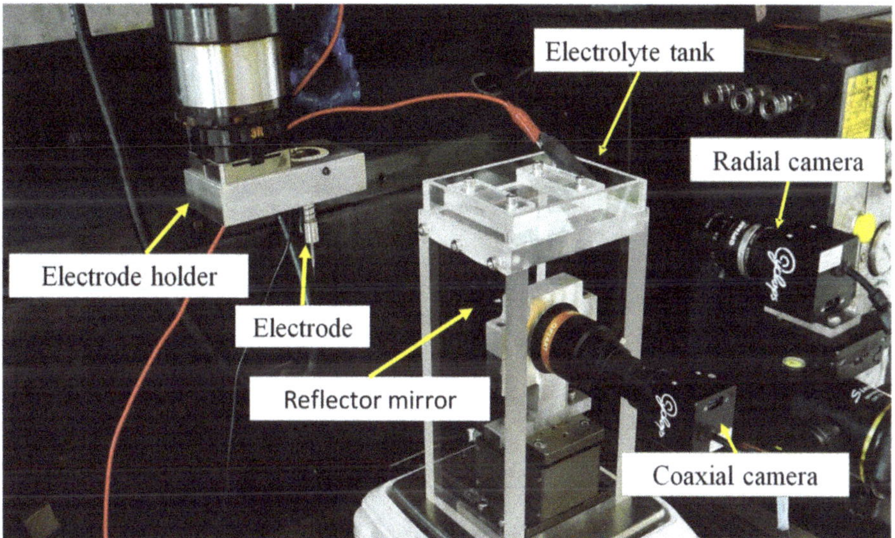

Figure 1. The experimental arrangement for spark assisted chemical engraving drilling.

2. Experimental Setup and Method

2.1. Setup

Figure 1 shows the experimental arrangement comprising of the SACE drilling module, the power supply, the image acquisition and pre-processing unit, and the computer control and analysis system. A schematic diagram of the SACE experimental setup is illustrated in Figure 2. The computer is linked to a programmable intelligent computer (PIC) microcontroller (Microchip Technology Inc., Chandler, AZ, USA) that controls the SACE drilling parameters including the drilling voltage and current to trigger and control the SACE drilling module. The workpiece was dipped in a bath that contained a solution with a 5 M KOH (the electrolyte solution), and a tool electrode was positioned above the workpiece surface (quartz glass, 1.0 mm thick). The electrolyte bath served as the machining chamber, the attachment to supply power to the electrode tool, a spring-fed tool arrangement, and a DC power supply (the programmable DC power analyzer Agilent N6705, Santa Clara, CA, USA). During the machining of quartz glass, the spring-fed tool electrode was pre-pressured on the quartz glass surface to feed the electrode that was in contact with the machining surface of the quartz glass. The spring-fed tool arrangement was designed and fabricated to enable the smooth feeding of the electrode tool and to avoid mechanical breaking action that would break the quartz glass.

The application of a suitable electric potential (i.e., 40 V in our system) across the two electrodes led to the construction of the gas film around the tool electrode, and the arcing plasma induced by the discharge phenomenon that directly followed contributed to the material removal. Figure 3 shows the image of the gas film formation by using a tapered tungsten carbide (WC) electrode with a diameter of 249 μm that is arranged based on the tool immersion depth. The tapered tool electrode design (i.e., complex shaped electrodes [15]) is able to help to drill deep and narrow cavities [16] and stabilize the machining process [17]. A tapered tool electrode was clamped on the chuck and employed in the experiments to increase the consistency of spark generation and to focus the discharges on a concentrated region. The tool and chuck were connected to the power supply during the SACE process. When the DC voltage was applied, a spark discharge occurred primarily at the electrolyte surface as well as at the tooltip and the electrolyte surface exhibited a concave shape due to the surface tension. As shown in Figure 4, the presence of a bubble layer was observed around the electrodes and it coalesced into a gas film, thereby leading to the expected light emission (i.e., the sparking phenomenon).

The arcing plasma image was obtained using the two complementary metal–oxide–semiconductor (CMOS) cameras. One of the cameras was mounted radially 190 mm from the work site with a 35 mm focus lens and a 1.5× close-up lens. The other camera was mounted coaxially at a distance of 100 mm below the electrolyte tank with a 0.7× telocentric lens. The PIC microcontroller unit triggered the SACE power module to produce a single SACE voltage pulse and instantaneously delayed it for a short time period in order to trigger the image acquisition and pre-processing unit. This captured each single arcing plasma image of the sparking phenomenon. The surroundings settings and drilling parameters of the experiment are listed in Table 1.

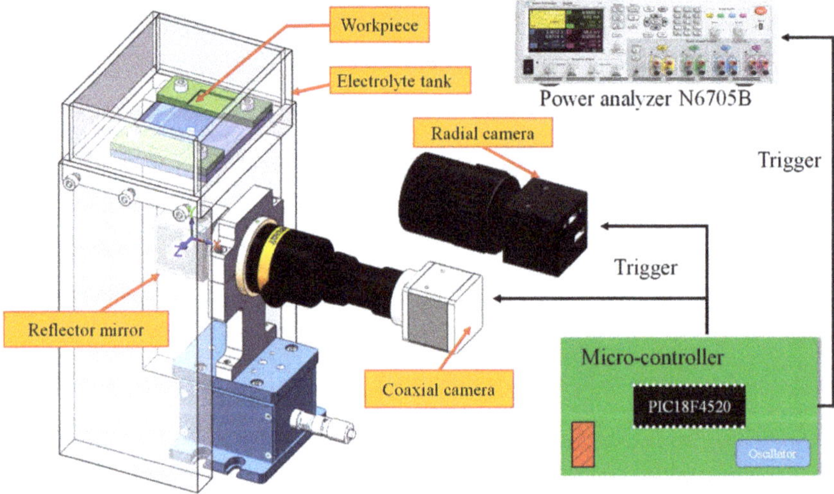

Figure 2. The schematic diagram of the spark-assisted chemical engraving (SACE) experimental setup.

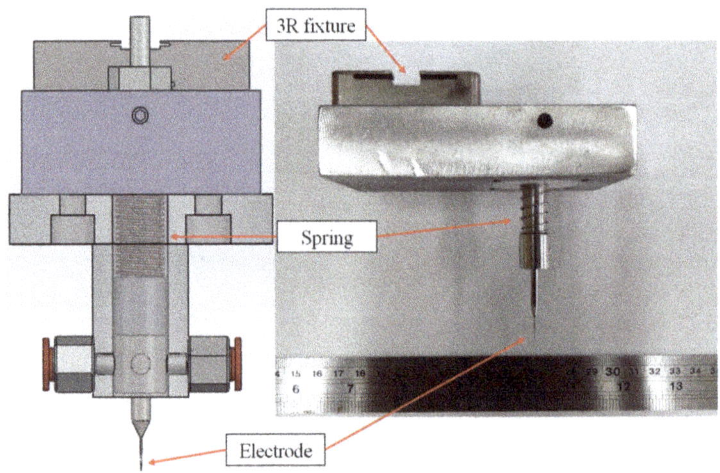

Figure 3. The tapered tool electrodes with a spring-fed mechanism.

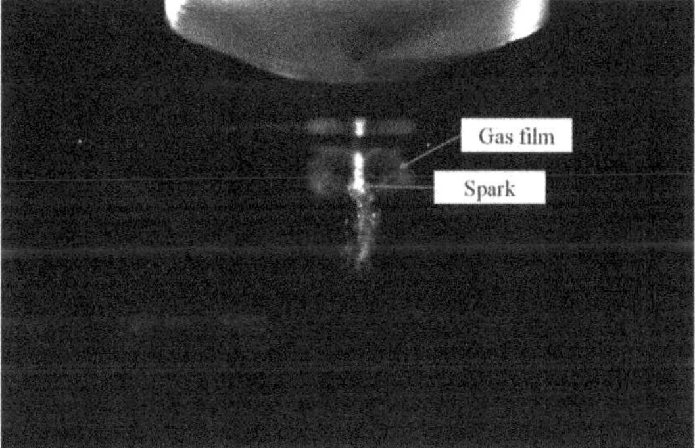

Figure 4. The appearance of the sparking phenomenon is observed when the gas film is formed.

Table 1. The experimental parameters.

Parameter	Value
Voltage	40 V
Peak current	1.5 A
Workpiece material	Quartz glass
Thickness of workpiece	10 mm
Tool electrode material	Tungsten carbide
Electrolyte	KOH

2.2. Estimation Method

The SACE-drilled system comprising of the blind hole deepness analysis were acquired and evaluated using 8-bit encoding and 752 × 480 pixel images. We used the thresholding technique during the image processing to calculate the arcing plasma emission area and compute the entire pixel area of the arcing plasma region. The images of the quartz glass after the thresholding technique are shown in Figure 5 in the radial direction and Figure 6 in the coaxial direction. The machining time is used to estimate the deepness of a SACE-drilled hole in the conventional control method. An increase in the machining time increased the deepness of the hole. The conventional control method used experience-based control strategies to sense the finalization of the drilling process off-line.

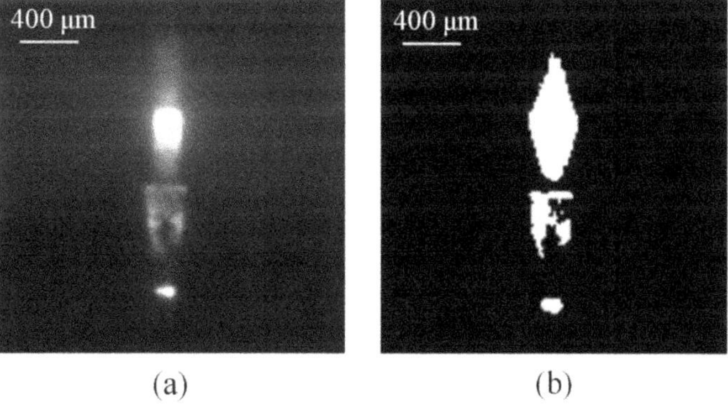

Figure 5. The arcing plasma area of an image of quartz glass in the radial direction (**a**) acquired by an industrial camera and (**b**) after the thresholding technique.

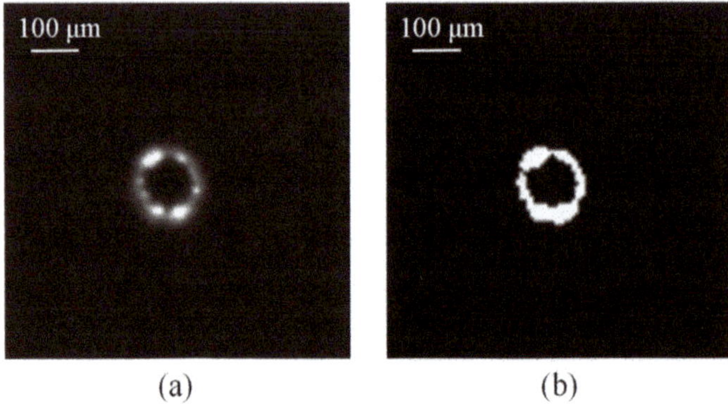

Figure 6. The arcing plasma area of an image of quartz glass in the coaxial direction (**a**) acquired by an industrial camera and (**b**) after the thresholding technique.

The computer vision method was used to determine the correlation between the SACE-induced arcing plasma and the SACE-drilled hole deepness. It is an innovative evaluation mechanism to determine the deepness of a SACE-drilled hole during drilling. During the SACE-drilled machining, a spark discharge is usually induced as a result of the melting and evaporation of the material from the arcing-heated region. The electrical spark discharges occurred through a gas film constructed around the tool electrode and were observed by the setup of the camera in the radial and coaxial directions. The pixels of the SACE-induced plasma area for each individual single image frame from the start to the end of the drilling were calculated using machine vision processing. We subsequently accumulated the entire pixel area for single image frames of the plasma area through SACE drilling and obtained the whole pixel area of the plasma area from the start to end.

3. Results and Discussion

In these experiments, the drilling deepness was accounted for using a load-cell based mechanical scaler to establish the processing depth. The diameter of the holes and the area of the thermal-affected region were investigated using optical microscopy (OM) as shown in Table 2 and Figure 7. The results indicated deviations in the diameter of the machined contour and the thermal-affected region of approximately 3.7% and 2.7%, respectively. The experimental results indicated good agreement with those obtained by extant studies [18]. In reference [18], Wüthrich et al. indicated that deviations of the micro-holes with a mean diameter of a few hundred microns typically corresponded to approximately 20% of perfect circular contour.

Table 2. Hole diameter and diameter of the heat affected zone.

Experimental Dataset	Hole Diameter (μm)	Heat Affected Diameter (μm)
I	275	521
II	257	550
III	259	532

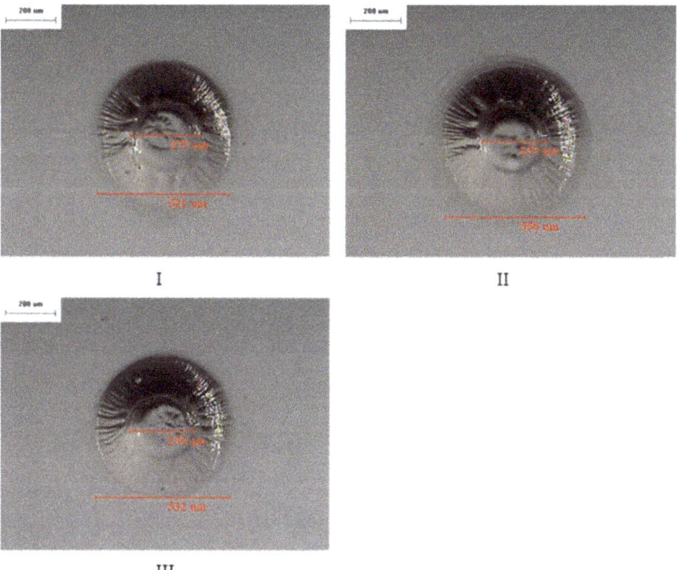

Figure 7. The diameter of the holes and the area of the thermally affected region are investigated by optical microscopy (OM) (**I–III**) 50×.

The resulting light emission images in the radial and coaxial directions for SACE-drilled machining were obtained using a process time of 60 s in which the machining energy of the SACE was 40 V and 1.5 A for quartz glass as shown in Figures 8 and 9, respectively. The results suggested that a certain degree of deviation continued to exist in the light emission pixels. This was caused by the bubbles produced at the front end of the electrode and also around the electrode sidewall [19]. Therefore, a significant spark light emission was observed around the electrode. The unstable gas film fluctuated arbitrarily and randomly, and this resulted in irregular spark behavior [20].

The accumulated light emission pixels of the SACE-drilled holes evidently increased with increases in the drilling time. The correlation between the drilling-hole deepness and the pixel value of the SACE-induced plasma area is presented in the radial direction in Figure 10. The results indicate that the deepness of the SACE-drilled holes was estimated by a monotonically increasing function.

The deepness of the SACE-drilled hole was investigated as a function of the accumulative plasma size to fully monitor the process parameters. Figure 10 shows the deepness as a function of the machining time at which the camera detected the arcing plasma pixel. The results revealed that the accumulative arcing plasma size exhibited a high coefficient of determination (R^2) of 0.946 with respect to the machining depth. The experimental results indicated a good agreement [21] with those obtained by SACE machining, and this was caused mainly by spark discharges that occurred through a gas film constructed around the tool electrode. As shown in Figure 11, the hole deepness is proportionally related to the drilling time and the total number of the pixels of all the frames of the plasma area. Based on the proportional relation, we used the entire accumulative pixel values to evaluate and sense the SACE-drilled hole deepness. In this study, we calculated that the size of the SACE-induced plasma area at the surface of the workpiece can be converted into the value of the pixels. The spark energy conditions determined both the SACE-drilled hole deepness and size of the plasma area relative to the drilling time. Thus, we analyzed and obtained the entire pixel area of the plasma area in situ during processing. Additionally, we realized the in situ monitoring and evaluation of the deepness of a SACE-drilled hole. Figure 11 shows the relationship between pixels and deepness relative to the process time. The linear dependency of the two variables revealed the association between the measured spark pixel and drilled depth. In this case, it is observed that the machined deepness was saturated at a time corresponding to 50 s at an approximate deepness of 0.6 mm. The accumulated spark pixel also stopped increasing at the end of the machining stage. It was observed that the R^2 of the linear regression between the accumulative arcing plasma size and the machining time is 0.978. Furthermore, the R^2 of the linear regression between the machined deepness and the machining time is 0.948 before the saturation depth. The results also indicated that the machined status was estimated using the acquired image pixel.

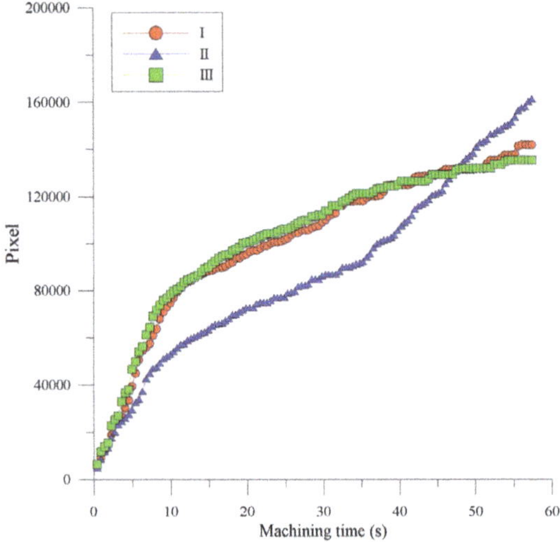

Figure 8. The accumulated light emission pixels of the SACE-drilled holes, acquired by using a coaxially mounted industrial camera.

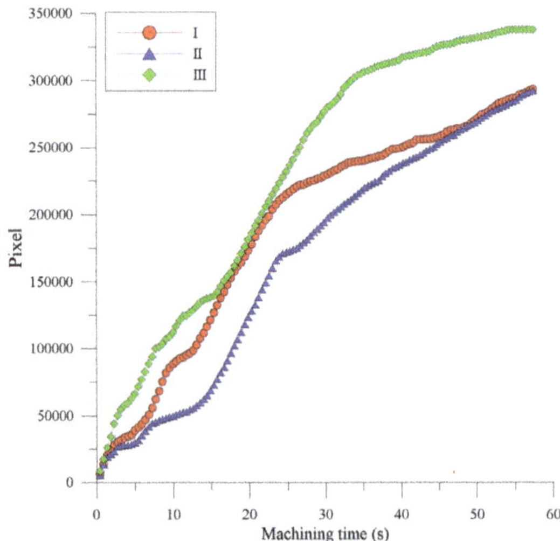

Figure 9. The accumulated light emission pixels of the SACE-drilled holes, acquired by using a radially mounted industrial camera.

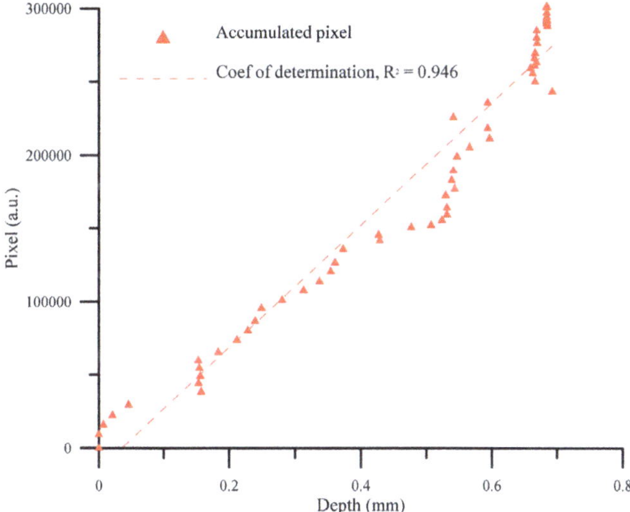

Figure 10. The correlation between the deepness of the SACE-drilled hole and the pixels of the arcing plasma area, acquired by using a radially mounted industrial camera.

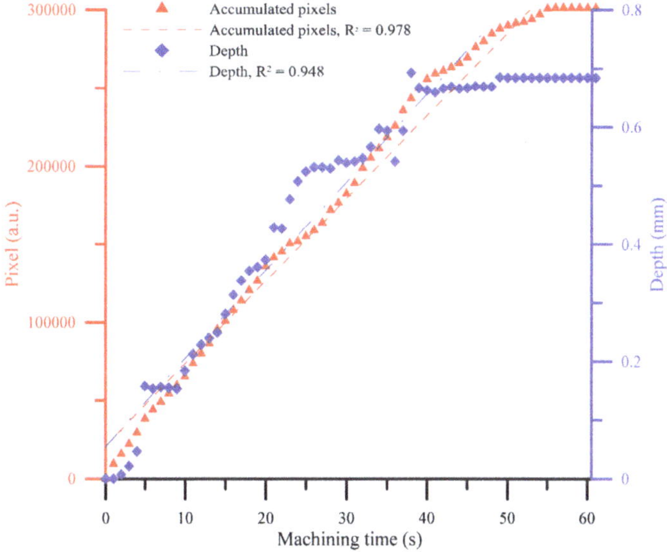

Figure 11. The correlation between pixels and deepness relative to the drilling time.

4. Conclusions

This study examined a novel in situ technique to estimate the depths of SACE-drilled holes. In this study, visual signals that could be used for the on-line monitoring of SACE machining were discussed first. The vision signal was directly accessible for the measurement and was therefore specifically discussed. Based on the SACE-induced arcing plasma area that is converted into a pixel value, we obtained an association between the accumulative pixels and hole depth. The system acquired and analyzed images of SACE drilling in real time and, thus, the hole deepness evaluation was also in real time. The evaluation mechanism based on a machine vision processing method was combined with a deepness control method to perform conventional SACE machining and to realize increasingly accurate deepness control for a SACE-drilled hole. The key deepness information on the machining process was accessible through a relatively simple analysis without requiring additional sensors. As ECDM is a spark-based and highly stochastic micromachining method, future work will focus on further investigation into correlating the spark pixel with the effects of various process parameters such as applied voltage, electrolyte concentration and tool geometry on machining status.

Acknowledgments: The authors thank the anonymous reviewers for their helpful suggestions. We also want to thank Prof. Chia-Lung Kuo for providing the experiment equipment and his assistance in the design of the tool electrode. This research was supported by the Ministry of Science and Technology, Taiwan under Grant no. MOST 106-2221-E-027-051.

Author Contributions: Chao-Ching Ho conceived and designed the experiments; Dung-Sheng Wu performed the experiments; Chao-Ching Ho and Dung-Sheng Wu analyzed the data; Chao-Ching Ho and Dung-Sheng Wu contributed analysis tools; Chao-Ching Ho wrote the paper.

Conflicts of Interest: The authors declare no conflict of interest.

References

1. Díaz-Tena, E.; Rodríguez-Ezquerro, A.; de Lacalle Marcaide, L.L.; Bustinduy, L.G.; Sáenz, A.E. A sustainable process for material removal on pure copper by use of extremophile bacteria. *J. Clean. Prod.* **2014**, *84*, 752–760. [CrossRef]

2. Díaz-Tena, E.; Rojo, N.; Gurtubay, L.; Rodríguez-Ezquerro, A.; López de Lacalle, L.N.; Oyanguren, I.; Barbero, F.; Elías, A. Biomachining: Preservation of acidithiobacillus ferrooxidans and treatment of the liquid residue. *Eng. Life Sci.* **2017**, *17*, 382–391. [CrossRef]

3. Díaz-Tena, E.; Barona, A.; Gallastegui, G.; Rodríguez, A.; López de Lacalle, L.N.; Elías, A. Biomachining: Metal etching via microorganisms. *Crit. Rev. Biotechnol.* **2017**, *37*, 323–332. [CrossRef] [PubMed]

4. Wüthrich, R.; Fascio, V. Machining of non-conducting materials using electrochemical discharge phenomenon—An overview. *Int. J. Mach. Tools Manuf.* **2005**, *45*, 1095–1108. [CrossRef]

5. Wuthrich, R.; Ziki, J.D.A. *Micromachining Using Electrochemical Discharge Phenomenon: Fundamentals and Application of Spark Assisted Chemical Engraving*; William Andrew: New York, NY, USA, 2014.

6. Pachaury, Y.; Tandon, P. An overview of electric discharge machining of ceramics and ceramic based composites. *J. Manuf. Process.* **2017**, *25*, 369–390. [CrossRef]

7. Zheng, Z.-P.; Su, H.-C.; Huang, F.-Y.; Yan, B.-H. The tool geometrical shape and pulse-off time of pulse voltage effects in a Pyrex glass electrochemical discharge microdrilling process. *J. Micromech. Microeng.* **2007**, *17*, 265. [CrossRef]

8. Cheng, C.-P.; Wu, K.-L.; Mai, C.-C.; Yang, C.-K.; Hsu, Y.-S.; Yan, B.-H. Study of gas film quality in electrochemical discharge machining. *Int. J. Mach. Tools Manuf.* **2010**, *50*, 689–697. [CrossRef]

9. Jiang, B.; Lan, S.; Ni, J.; Zhang, Z. Experimental investigation of spark generation in electrochemical discharge machining of non-conducting materials. *J. Mater. Process. Technol.* **2014**, *214*, 892–898. [CrossRef]

10. Goud, M.; Sharma, A.K.; Jawalkar, C. A review on material removal mechanism in electrochemical discharge machining (ECDM) and possibilities to enhance the material removal rate. *Precis. Eng.* **2016**, *45*, 1–17. [CrossRef]

11. Giandomenico, N.; Meylan, O. Development of a new generator for electrochemical micro-machining. *Procedia CIRP* **2016**, *42*, 804–808. [CrossRef]

12. Huang, S.F.; Zhu, D.; Zeng, Y.B.; Wang, W.; Liu, Y. Micro-hole machined by electrochemical discharge machining (ECDM) with high speed rotating cathode. *Adv. Mater. Res.* **2011**, *295*, 1794–1799. [CrossRef]

13. Laio, Y.; Wu, L.; Peng, W. A study to improve drilling quality of electrochemical discharge machining (ECDM) process. *Procedia CIRP* **2013**, *6*, 609–614. [CrossRef]

14. Kulkarni, A.; Sharan, R.; Lal, G. Measurement of temperature transients in the electrochemical discharge machining process. *AIP Conf. Proc.* **2003**, *684*, 1069–1074.

15. Rajurkar, K.; Yu, Z. 3d micro-edm using cad/cam. *CIRP Ann.-Manuf. Technol.* **2000**, *49*, 127–130. [CrossRef]

16. Ayesta, I.; Izquierdo, B.; Sanchez, J.; Ramos, J.; Plaza, S.; Pombo, I.; Ortega, N. Electrode path generation algorithm for complex shape cavities. Electrode design and application for blisk machining. In Proceedings of the 5th Manufacturing Engineering Society International Conference, Zaragoza, Spain, 26–28 June 2013.
17. Flaño, O.; Ayesta, I.; Izquierdo, B.; Sánchez, J.; Zhao, Y.; Kunieda, M. Improvement of EDM performance in high-aspect ratio slot machining using multi-holed electrodes. *Precis. Eng.* **2018**, *51*, 223–231. [CrossRef]
18. Wüthrich, R.; Spaelter, U.; Bleuler, H. The current signal in spark-assisted chemical engraving (SACE): What does it tell us? *J. Micromech. Microeng.* **2006**, *16*, 779. [CrossRef]
19. Tang, W.; Kang, X.; Zhao, W. Enhancement of electrochemical discharge machining accuracy and surface integrity using side-insulated tool electrode with diamond coating. *J. Micromech. Microeng.* **2017**, *27*, 065013. [CrossRef]
20. Han, M.-S.; Min, B.-K.; Lee, S.J. Modeling gas film formation in electrochemical discharge machining processes using a side-insulated electrode. *J. Micromech. Microeng.* **2008**, *18*, 045019. [CrossRef]
21. Wüthrich, R.; Hof, L.A.; Lal, A.; Fujisaki, K.; Bleuler, H.; Mandin, P.; Picard, G. Physical principles and miniaturization of spark assisted chemical engraving (SACE). *J. Micromech. Microeng.* **2005**, *15*, S268. [CrossRef]

 *materials*

Article

The Effect of Weld Reinforcement and Post-Welding Cooling Cycles on Fatigue Strength of Butt-Welded Joints under Cyclic Tensile Loading

Oscar Araque [1,*], Nelson Arzola [2] and Edgar Hernández [2]

[1] Department of Mechanical Engineering, Universidad de Ibagué, Ibagué 730001, Colombia
[2] Research Group in Multidisciplinary Optimal Design, Department of Mechanical and Mechatronics Engineering, Universidad Nacional de Colombia, Bogota 111321, Colombia; narzola@unal.edu.co (N.A.); edhernandezl@unal.edu.co (E.H.)
* Correspondence: oscar.araque@unibague.edu.co; Tel.: +57-8-270-9400

Received: 7 March 2018; Accepted: 7 April 2018; Published: 12 April 2018

Abstract: This research deals with the fatigue behavior of butt-welded joints, by considering the geometry and post-welding cooling cycles, as a result of cooling in quiet air and immersed in water. ASTM A-36 HR structural steel was used as the base metal for the shielded metal arc welding (SMAW) process with welding electrode E6013. The welding reinforcement was 1 mm and 3 mm, respectively; axial fatigue tests were carried out to determine the life and behavior in cracks propagation of the tested welded joints, mechanical characterization tests of properties in welded joints such as microhardness, Charpy impact test and metallographic analysis were carried out. The latter were used as input for the analysis by finite elements which influence the initiation and propagation of cracks and the evaluation of stress intensity factors (SIF). The latter led to obtaining the crack propagation rate and the geometric factor. The tested specimens were analyzed, by taking photographs of the cracks at its beginning in order to make a count of the marks at the origin of the crack. From the results obtained and the marks count, the fatigue crack growth rate and the influence of the cooling media on the life of the welded joint are validated, according to the experimental results. It can be concluded that the welded joints with a higher weld reinforcement have a shorter fatigue life. This is due to the stress concentration that occurs in the vicinity of the weld toe.

Keywords: butt weld joint; fatigue; crack growth rate; weld reinforcement; cooling rate

---

1. Introduction

In comparison with other methods for the joint of materials, welding is one of the most used ones. However, it holds diverse flaws related to the manufacturing process. The most common imperfections are discontinuities, porosities, undercuts, cracks, inclusions, and lack of fusion, among others. These flaws are produced by the increase of speed, change in polarity of the current, and the extension and diameter of the electrode. Additionally, these variables modify the grain size and the areas surrounding the weld bead, such as: the fusion zone (FZ), the heat affected zone (HAZ), and the base metal (BM) [1].

An evaluation of fatigue strength for welded joints consist of the study of crack initiation and propagation phenomena. Due to this, it is necessary to know the mechanisms to determine the various typical factors of fracture mechanics, with the aim to explain the behavior of cracked material when subjected to stresses and the mechanisms that lead to premature failure in load conditions below the yield and breakage limits [2]. The fracture mechanics shows that the parameters tenacity, crack size, and level of strength can be related to predict the possibility of a fracture, such as stress intensity factors (SIF) in butt welds [3–5]. This parameter depends on the geometry and the load and it is a measure of the degree to which a load is amplified at the tip of a crack. Additionally, the fracture mechanics raises other expressions for the evaluation of the crack propagation rate (da/dN), as function of the SIF range (ΔK), where C and m are material constant values. This relationship is known as the Paris law [6].

$$\frac{da}{dN} = C(\Delta K)^{m} \tag{1}$$

Several researches have been developed aimed at determining the areas that make up the welded joint [7–9]. They have generally identified the fusion zone (FZ), which presents a grain growth from the fusion line into the interior of the fusion zone, the heat affected zone (HAZ) characterized by the hardening during the heating. In the cooling, the microstructure can present evidences of liquefaction in edges of grain and the base metal (BM), corresponding to the formation limit of the hardener precipitate. In the research of [10], there are observed microstructural variations in a series of low-alloy steel weld which deposit different carbon concentrations (product from experimental electrodes), getting allotriomorphic ferrite growth is assumed to occur by an equilibrium transformation mechanism; its formation is found to determine the development of both Widmanstätten and acicular ferrite.

The observation of the fault surfaces in welding are useful to estimate the mechanisms of crack propagation, using the mark counting. In the research of [11], it has been observed that the marks are normally formed when the load is intermittent during the service and the marks in a much finer scale. They can show the tip position on the crack after each cycle. The observation of the marks always suggests a fatigue failure, but the absence of this pattern does not rule it out. The crack origin and, in some cases, the marks can be observed through a scanning electron microscope (SEM). In the work of [12], the modeling and implementation of criteria for multiple crack propagation, including interaction and coalescence was made. In this research, the correlation between the number of initiation sites and the applied stress level, showing that crack propagation is related with fatigue strength of weld.

The determination of the mechanical properties of the weld joints such as hardness, fracture toughness, yield strength, and tensile strength are not constant, Bullón et al. [13] indicates that they are affected by the thermodynamics or thermal field (temperature field) due to the high gradient temperature changes that are generated causing microstructural transformations, which affect the mechanical field (field of stresses and strain) and metallographic field (field of microstructural state) causing morphological changes. In addition, there is a mutual influence between this fields represented by the continuous and discontinuous lines.

Other researchers have studied the effect of microstructural changes and mechanical properties on the welded joint. Various researchers [14,15] show that geometric discontinuities affect the fatigue strength of welded joints. It is in the HAZ where the discontinuities can nucleate, which in general are the initiators of fatigue cracks in the welded joints. The fatigue fissure begins in a local flaw of the structure, either internal or external. In fact, the bead geometry already constitutes a geometric discontinuity, which turns out to be a stress concentrator.

In regards to the analysis by fracture mechanics on welded joints, several authors have researched the topic over the years through the selection of a combination of factors, such as material, the model geometry, and load conditions. However, most researchers have opted for the use of computational tools and the finite element method (FEM). A study using this technique by the Berrios' researcher [16], found a remarkable correspondence between the experimental trajectories of crack propagation on high resistance-steel and the propagation trajectories obtained using the software ANSYS APDL. The research carried out by Araque and Arzola [17], developed an experimental-theoretical analysis about the influence of the cooling medium and the geometry of the welding bead profile in fatigue life and the associated parameters with structural integrity of welded joints. The results were a set of analytical expressions for the weld magnification factor Mk, the obtained models of behavior of the weld magnification factor are compared with the results from other researchers with some small differences.

On the other hand, Toribio [18] has identified life fatigue for a pressure vessel under cyclic loads of thermal origin, as well as proposal on a general procedure to calculate the life fatigue on any structure subjected to cyclic load. He has been determined that shorter fracture trajectories during the life service generate leaks prior to break.

On the same field of research, Lewandowski [19] has analyzed the behavior of crack growth in ferritic-perlitic structures subjected to cyclic flexion and has determined that, in all cases, life fatigue on welded specimens is less than on completely solid specimens, due to different mechanical properties to the characteristics of the welded joints (base material, weld bead, HAZ thermal affected area).

About the use of software to analyze fractures, Franc3D has gained popularity among researchers in the last decade. Yang [20], has proposed an algorithm based on the linear elastic fracture mechanics for the simulation of fatigue crack growth under non-proportional loads. He has found that cycles of growth calculated on the basis of stress intensity factors (SIF) correspond to the experimental data for specimens subjected to slightly lower loads. Similarly, Xiao [21] has made a numerical analysis for surface cracks in on a hemispherical hole and has concluded that the geometry of the whole surface is a determining factor in the SIF. By using this software, researcher Chin [22] has estimated how a crack spreads on a support made from an alloy of titanium and aluminum (very common in the aerospace industry) and has found that the Type I (SIF's) is below the fracture toughness to the fracture material. Then, he has concluded this support could tolerate a certain type of cracks.

This article shows an experimental and theoretical study of the fatigue failures on butt welded joints for structural carbon steel ASTM A36 HR that used shielded metal arc welding (SMAW) with electrode E6013. The influence of the weld reinforcement was evaluated, as well as the post weld cooling environment by the uniaxial fatigue tests and the failure mechanism analysis present in the fracture surface. In other tests, the mechanical and fractomechanical properties from the welded joints were analyzed in relation with the crack origin and propagation caused by fatigue.

2. Materials and Methods

This section can be divided by subheadings. It should provide a concise and acute description of the experimental results, their interpretation as well as the experimental conclusions that can be drawn.

2.1. Experimental Design

For the uniaxial fatigue tests, 8 replicates were selected for each treatment accounting for a total of 32 test specimens. Two experimental factors were established: weld bead geometry and the post-welding cooling media. Two levels for each factor the welding reinforcement (1 and 3 mm), and the cooling rate, slow (quiet air) and fast (immersion in water) are evaluated, respectively. For the Charpy impact energy test six replicates for each treatment were carried out, 3 in the (FZ) and 3 in the (HAZ), respectively. Moreover, 3 more replicates were analyzed for the base metal (BM), thus 27 test specimens were obtained for the late type of test. Finally, for the uniaxial tension test three samples were evaluated, a total sampling amount for each treatment, accounting a total of 12 test specimens. To make the data management more feasible a codification system was assigned for each test type and cooling medium, correspondingly. This can be found in Table 1 for the uniaxial fatigue test specimens, Table 2 for Charpy impact energy test specimens and, finally in Table 3 for the tension test specimens, respectively. For example, 1F-A code is the test specimen with 1 mm of welding reinforcement, cooled in water and is the replica A.

Table 1. Coding used for the uniaxial fatigue test specimens.

Code	Experimental Factors		Number of Replicas
	Welding Reinforcement (mm)	Post-Welding Cooling Medium	
1S-(replica)	1	Quiet air	8
1F-(replica)	1	Water	8
3S-(replica)	3	Quiet air	8
3F-(replica)	3	Water	8

Table 2. Coding used for the Charpy impact energy test specimens.

Code	Experimental Factors			Number of Replicas
	Welding Reinforcement (mm)	Post-Welding Cooling Media	Impact Zone Located at	
1S-FZ-(replica)	1	Quiet air	Fusion zone	3
1S-HAZ-(replica)	1	Quiet air	Heat affected zone	3
1F-FZ-(replica)	1	Water	Fusion zone	3
1F-HAZ-(replica)	1	Water	Heat affected zone	3
3S-FZ-(replica)	3	Quiet air	Fusion zone	3
3S-HAZ-(replica)	3	Quiet air	Heat affected zone	3
3F-FZ-(replica)	3	Water	Fusion zone	3
3F-HAZ-(replica)	3	Water	Heat affected zone	3
BM	-	-	Base metal	3

Table 3. Coding used for the uniaxial tension test specimens.

Code	Experimental Factors		Number of Replicas
	Welding Reinforcement (mm)	Post-Welding Cooling Medium	
1ST-(replica)	1	Quiet air	3
1FT-(replica)	1	Water	3
3ST-(replica)	3	Quiet air	3
3FT-(replica)	3	Water	3

2.2. Materials

Rolled carbon structural steel ASTM A36 with a plate thickness of 6 mm was used on the welded samples for the uniaxial fatigue and uniaxial tension. For the Charpy impact energy test a plate thickness of 12 mm was used. The mechanical properties for ASTM A36 are shown in Table 4. The chemical composition of the material was checked by spark spectrometry test UV-VIS (Shimadzu Scientific Instruments, Kyoto, Japan) and the results are presented in Table 5.

Table 4. Mechanical properties for ASTM A36 steel.

Yield Stress (MPa)	Ultimate Tensile Strength (MPa)	Elongation (%)
262	434	38

Table 5. Chemical composition for ASTM A36 steel, (wt. %).

Chemical Composition	Weight Percentage (wt. %)
C	0.202
Mn	0.532
Si	0.030
P	0.028
S	0.011

2.3. Welding Procedure

The E6013 electrode is used as a filler for the welded joints. The shielded metal arc welding process SMAW was used for welding procedure, the plane to top position was utilized with a butt welded joint design following, voltage ranging between 15–42 V and current between 85–140 A DC. The welding was performed perpendicularly to laminar direction of BM. Linear speed between 0.12 and 0.18 m/min. The applied weld passes depended on the over-thickness width, as follows: single pass for 1 mm over-thickness and two passes for 2 and 3 mm over-thickness, respectively. The welding process was carried out following a standard SMAW procedure and executed by a qualified welder [23]. The flaws found in the welded joints were within what was expected in this type of joints [24]. Subsequently the plates were cut using abrasive waterjet (AWJ) method.

2.4. Uniaxial Fatigue Test

The same geometry was used for the fatigue and uniaxial tension test specimen. In Figure 1 a diagram for the experimental sample is shown. The AWS B4.0:2007 and ASTM E606 standards were considered for the specimen configuration and design.

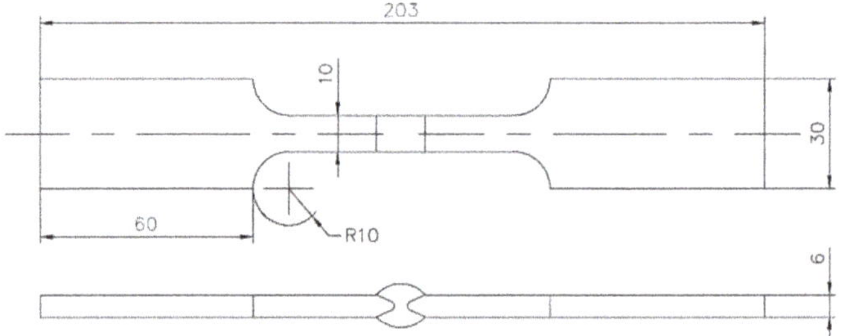

Figure 1. Diagram of tensile and fatigue test specimens (mm).

Subsequently to post-welding procedure, the final geometry of specimens was achieved with a waterjet cutting machine to avoid modifications on post-welding treatment. For uniaxial fatigue experimental test, a non-standard test machine was used as shown in Figure 2. The parameters for the fatigue test used were: frequency of cycle load 4 Hz, peak load 12 kN and a load cycle ratio 0.05, respectively. A load cycle ratio R > 0 is fixed, to guarantee non-compressive contact between the crack surfaces and then avoid loss of information regarding the propagation rate.

Figure 2. Specimen mounted on the uniaxial fatigue machine.

2.5. Uniaxial Tension, Microindentation Hardness, and Charpy Impact Energy Tests

The uniaxial tension tests were performed in a universal test machine Shimadzu UH-500 kNI (Shimadzu Scientific Instruments) following the standard ASTM E8. These tensile tests were carried out with the purpose of obtaining the tensile mechanical properties of the samples studied and finding possible differences between the studied treatments. The micro-indentation hardness test was performed in relation with sample thickness (6 mm) on three equidistant lines and a third of the sample thickness, employing a durometer LECO M-400-G2 Hardness Tester (LECO Corporation, St. Joseph, MI, USA), with a load of 500 g during a time of 10 s (HV 0.5). Specimen fabrication for the impact energy characterization test, its geometry and percentage of fracture due to shearing were carried out by following the standard ASTM A370. The tests were performed at environmental temperature 19 ± 0.1 °C. A macro attack was performed with Nital at 3% over the welding profile following standard ISO 15653. Care was taken to locate the notch in the center of zones HAZ and FZ as shown in Figure 3.

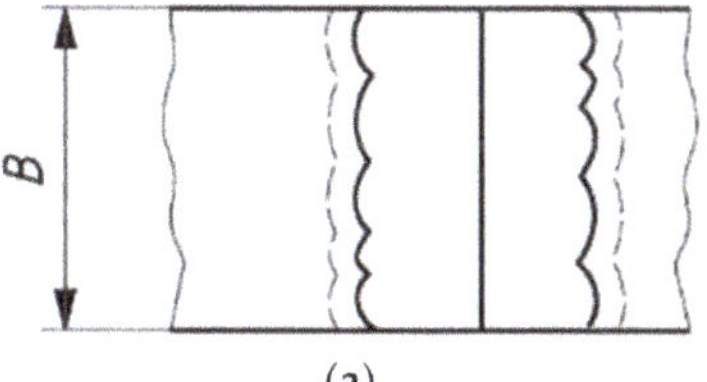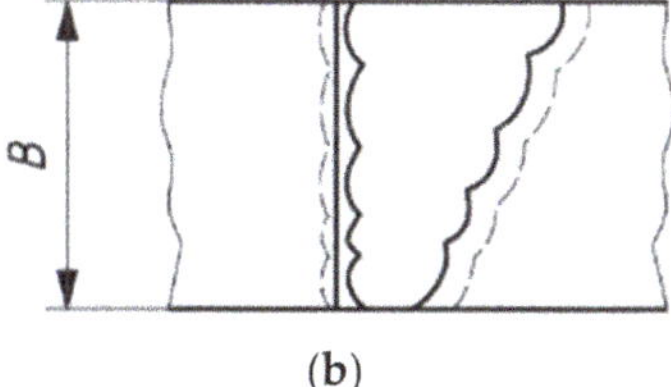

(a) **(b)**

Figure 3. (**a**) Notch location in the fusion zone (FZ) and (**b**) the heat affected zone (HAZ), according to the Standard ISO 15653.

2.6. Microstructure and Fractography

The microstructural characterization of the welding zones was performed by SEM (Quanta 200-r, FEI, Hillsboro, OR, USA) and optical microscope (E600, Nikon, Tokyo, Japan). Likewise, the analysis of failures over crack surface was performed on the failed specimens, to identify the cracks origins, planes of crack stable propagation and final fracture. Through observation with stereoscope, the stable propagation crack zone to readily use the SEM. This last technique was used to counting striations procedure, close to initiation and near the end of stable propagation crack zone, respectively. Details on how to perform this procedure can be found on the study carried out by DeVries et al. [25].

2.7. Crack Growth Modelling

The procedure used in this work, looking forward to obtaining an appropriate numerical model, consisted on the following phases: creation of the model in CAD software (Autocad 2018, Autodesk, CA, USA); residual stresses analysis; field analysis of the stress-strain before the crack growth, incorporation and crack growth and finally, the construction of Paris' curve from the stress intensity factors calculated by the software (FRANC3D) [26].

The model is generated in the 2016 Solidworks CAD software and then imported into the software ANSYS APDL (V17, Ansys Inc, Canonsburg, PA, USA). A preliminary modeling of finite elements of the welding joint model was carried out, taking as a goal a fatigue finite life, thus finding a peak stress of 192 MPa. The final model is loaded with a peak stress tension of 192 MPa on the top, and at the opposite end a built-in restriction type is placed. Furthermore, a linear elastic and isotropic material with a 200 GPa elastic modulus and Poisson coefficient of 0.26. In Figure 4, there is a detail of the meshed weld bead.

Simulations to get the residual stress profile, due to the thermal cycle welding process, were carried out on finite element software Ansys 17.2 (Ansys Inc.). These add-ons were Steady-State Thermal, Transient Thermal and Transient Structural. The utilized add-on in this case in non elasto-plastic.

It required a set of mechanical and thermal properties for the base material and the welded join, the convective cooling behavior, percentages on each phase transformed, as a function of temperature and thermos-dynamic coupling. The aim of this study is to get the residual stress field in a transversal direction to the weld beat. This stress field is over placed to fluctuating fatigue stresses, so that there can be a modeling of the stress intensity factor (SIF).

In the model, the location of the initial micro-crack corresponds to one of the selected specimens tested experimentally. The location corresponds to a 20% eccentricity with regards to the average plane on the test piece, at the weld toe. A semielliptical surface crack was used. Once the location and crack geometry are defined, a type-CDB ANSYS file is generated. The latter should be compatible with FRANC3D software. After the import of the model into this software, the crack is inserted (see details of the crack in Figure 5), and the propagation simulation starts, according to well-defined grow aspects of stable growth (growth model, size, number of iterations, etc.).

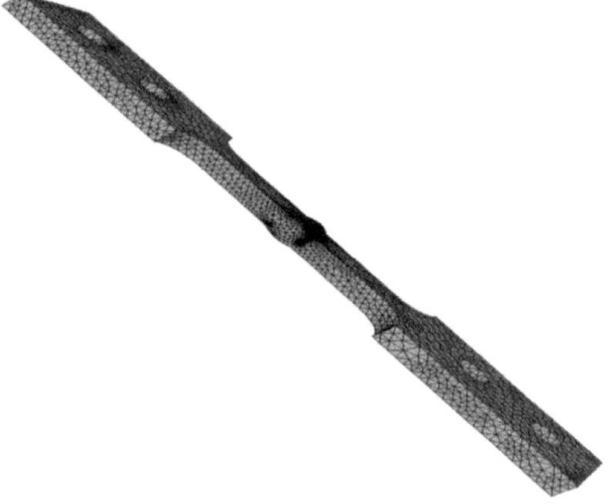

Figure 4. Detail of the mesh generation for the weld joint model.

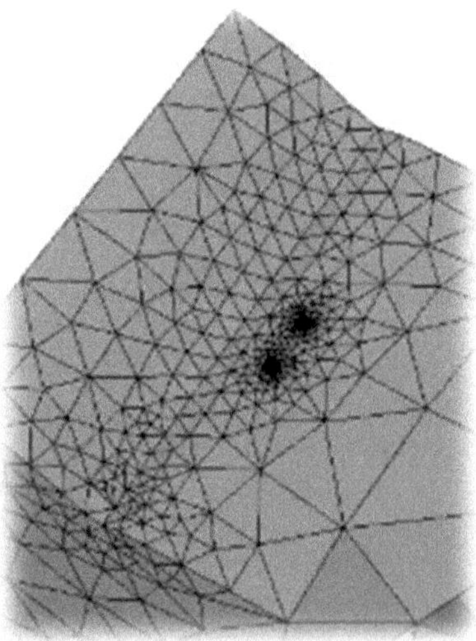

Figure 5. Inserted semielliptical surface crack at the weld toe (FRANC3D).

In Table 6, different semielliptical surface crack sizes used on every interaction are shown. The aspect ratio of the crack remains constant between one iteration and the next.

Table 6. Crack sizes of semi elliptical section using each iteration.

Semi-Major Axis (c) (mm)	Semi-Minor Axis (a) (mm)
0.15	0.06
0.23	0.09
0.30	0.12
0.45	0.18
1.25	0.50
2.50	1.00
5.00	2.00
7.50	3.00

Finally, once all the iterations are run, the next step is extract the information from the SIF's (stress intensity factors), and then build the respective Paris's curves. The procedure described above was performed for test specimens whose thickness on the weld bead corresponded to 1 mm and 3 mm, respectively.

3. Results and Discussion

3.1. Overall Experimental Fatigue Behavior

Using classes FAT 80 and 100 as a reference, the comparison between results for fatigue of butt welded joints for the analyzed treatments are showed in Figure 6. It is appreciated that, treatment 3S shows 50% of its data under limit of class FAT 100. Further, it is seen that only one sample of treatments 3S and 3F are under limit of class FAT 80, while mostly there are data for samples 1F and 1S. The latter represent a major deviation above limit of class FAT 100. Considering these partial results, it is necessary to perform a statistic evaluation for a better interpretation of the results. Based on [3] a statistical procedure to evaluate surviving probability as function of fatigue capacity (log C), which is obtained through (2):

$$logN = logC - m \cdot log \Delta\sigma \tag{2}$$

where for steels m = 3.

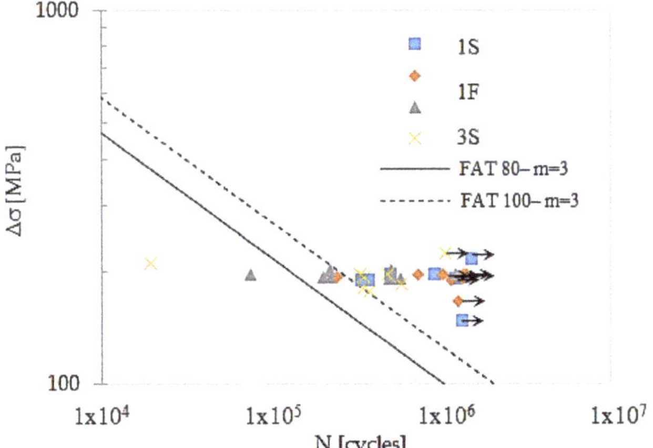

Figure 6. Experimental data and S-N curves according with classes FAT 80 and 100.

Subsequently, the surviving probability plot for the 32 samples is obtained. The latter is presented in the Figure 7, where it is possible to identify the behavior of welded samples in comparison with Figure 6. In Figure 7, it is observable that, in treatment 1S approximately 63% of the samples are above the 70% of surviving probability. The remaining are between 30% and 50% of surviving. In the treatment 1F, the 87.5% of the samples are between the 60% and 90% of surviving and remaining in 19%. In treatment 3S, the 50% of the samples are between 40% and 80% of surviving and the other 50% of the samples are between 2% and 20% of surviving. Finally, in the samples of treatment 3F, the 87.5% of the samples are between 30% and 70% of surviving and remaining in 0.025%. Thus, it is possible to state that treatment 1F is the one which has the highest surviving probability in comparison with the other treatments. On the contrary, the treatment 3S shows the lowest probability of surviving.

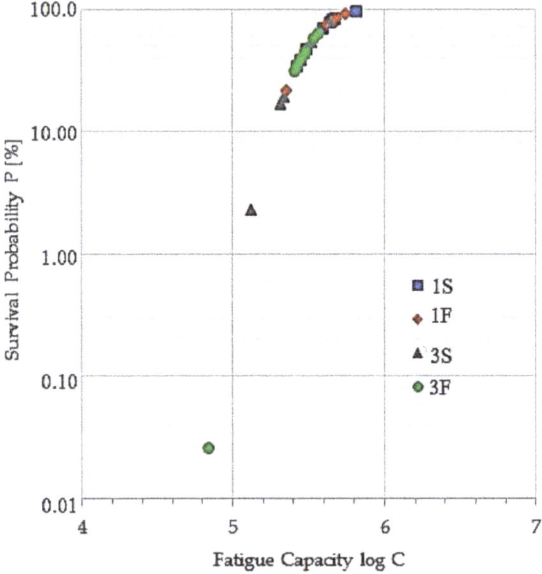

Figure 7. Survival probability for the butt welded joint specimens.

3.2. Crack Growth Model

3.2.1. Convergence Analysis

A computational cost and result variations assessment is carried out for the finite element models, as it should do with the processing time and the relative error. Taking the stress intensity factor (K_I) as a reference, as calculated according to Anderson [27], the input parameters for the computational analysis are larger radii than 1.25 mm, the smaller radii than 0.50 mm, with thickness of 1 mm. The results obtained for each numerical model are shown in Table 7.

Table 7. Relative error and computational cost on determining K_I.

Models	Nodes	Elements	Time CPU (s)	K_I (MPa$\sqrt{m}$)	% Error
Anderson's analytical models	-	-	-	4.41	-
Numerical model 1	68 426	41 758	45	4.34	1.62
Numerical model 2	63 124	38 294	35	4.09	7.04
Numerical model 3	49 892	30 260	25	4.13	6.40

Bearing in mind Figure 8, it can be seen the break-even point between time and central processing unit (CPU) and the error percentage was about the same as the Numerical model 2, which made the selected mesh-type for simulations.

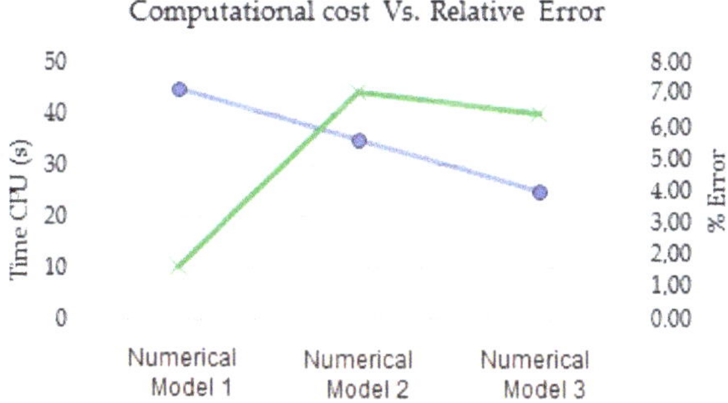

Figure 8. Computational cost vs. relative error.

3.2.2. Results to the Propagation Crack Model

Figure 9 shows the convention used in this paper for describing the residual stress directions longitudinal (x), transverse (y), and through-thickness (z), respectively, relative to a real profile of a butt-welded joint. The residual stresses found in the three directions are different. Residual stresses in the longitudinal and through-thickness directions were smaller than transverse residual stresses. This is caused by the different restraints acting in the three directions. Moreover, transverse residual stresses are the important ones, due to the longitudinal orientation of the superficial semielliptical crack implemented in this study. In Figure 10, the behavior of transverse residual stresses can be seen for the 1 mm and 3 mm weld reinforcements (W.R.), and quiet air and water as cooling media (C.M.). For the four treatments, the resulting residual stresses are compressive for the weld toe position and diminish rapidly after one position about 22 mm far from the weld centerline. It can be concluded that the effect of the welding reinforcement on the residual stresses is weak when quiet air is used as cooling medium. The difference in the residual stresses is small between the quiet air cooling medium treatments; an average of 2 MPa greater for 3 mm of weld reinforcement is observed when compared with 1 mm of weld reinforcement. However, a much more marked influence of the cooling medium on the residual stresses is observed. The treatment with levels of 3 mm for the weld over thickness and water as a cooling medium reaches compressive residual stresses of -125.5 MPa for the position of the weld toe.

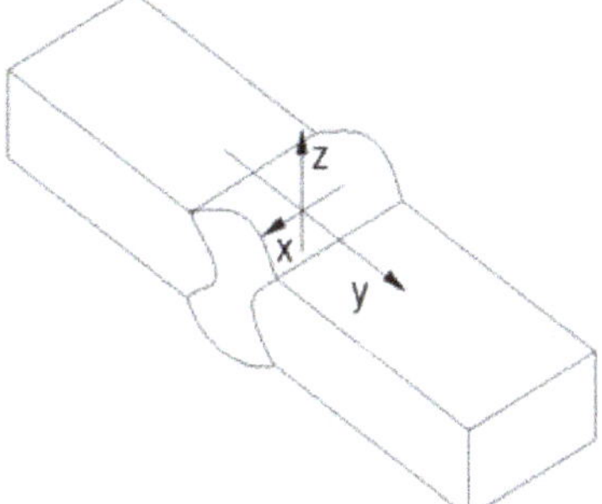

Figure 9. Nomenclature for directions relative to weld: longitudinal (x), transverse (y), and through-thickness (z) respectively.

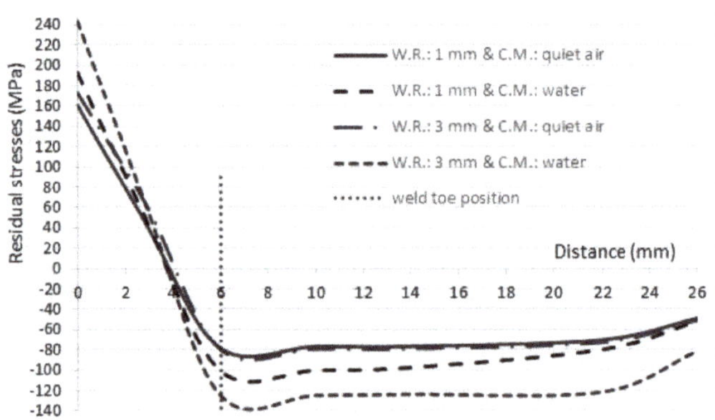

Figure 10. Residual stress profile on transverse direction to weld beat.

In this work, an experimental verification of residual stresses is not performed; although it is possible to make certain comparisons with the results obtained by other authors. For example, researcher Knoedel and colleagues [28] develop a phenomenological numerical model, useful in principle for any type of welded joint of steels after adjusting the process parameters and the properties of the materials dependent on temperature. Their results are also validated with experimental data from other previous works, finding a reasonable similarity. Using this phenomenological model, they find a profile of residual stress similar to that found in the current work for a welded joint of HT-36 steel plates. As a main difference, it appears that in the current work the residual stresses are compressive for the position of the weld toe, while in [28] the residual stresses for this location are tension, decreasing rapidly and moving to compresses in a position a little more away. The differences found are not surprising, given that it is widely known that both experimental measurements and the results obtained from numerical models are often subject to great variability and dispersion [28,29]. The above product, the large number of factors of the welding process, and properties of the materials involved cannot be completely controlled. Additionally, it is difficult to find experimental arrangements in framework that coincide with respect to the materials and procedures used.

Other authors [30,31] have reported that residual stresses are progressively relaxed due to the cyclic plastic deformation characteristic of the fatigue phenomenon. However, in this work, the field of residual stresses was imported into the model for the determination of SIF and remained constant throughout the process of crack growth. This consideration should be re-evaluated for future work.

Figure 11 shows the crack's view when it reaches its critical size for the model, along a 3 mm of weld over thickness.

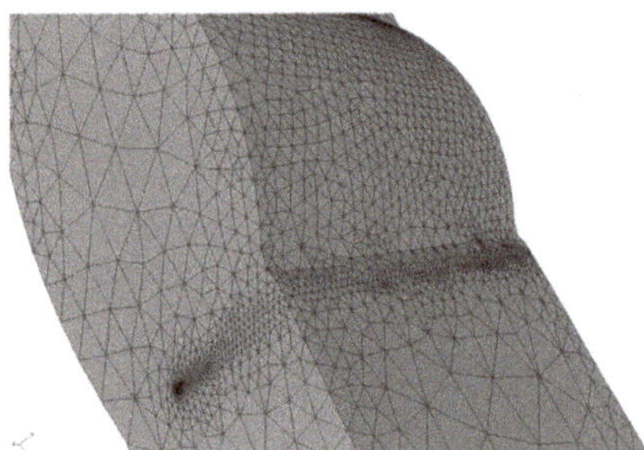

Figure 11. Appearance of the crack for the model with 3 mm of weld reinforcement.

Figures 12 and 13 show dimensionless stress factors values reached, as a function of crack size for the two models used.

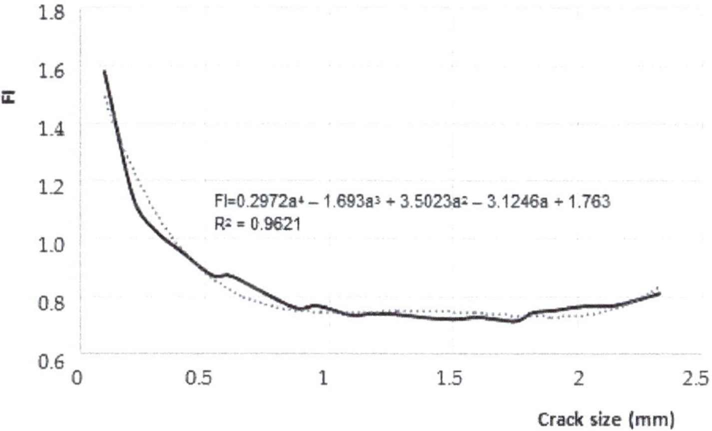

Figure 12. Behavior of F_I as a function of the crack depth (for an over thickness of 1 mm).

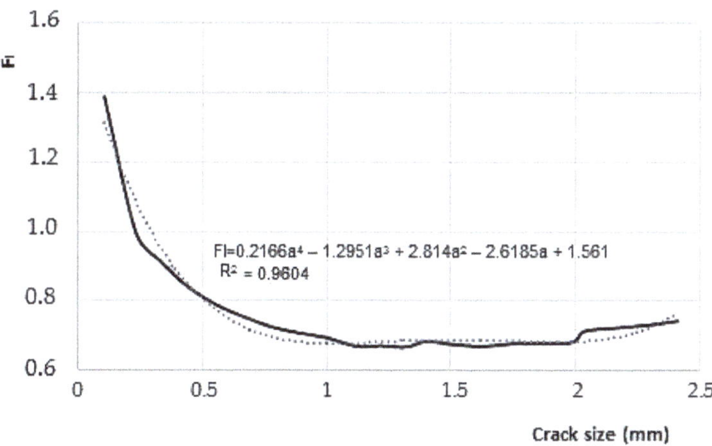

Figure 13. Behavior of F_I as a function of the crack depth (for an over thickness of 3 mm).

On the other hand, behavior of K_I indicates there is a stabilization in its length along the crack; as it grows a little, the formation of two peaks for K_I begins. It is located next to the ends of the crack. However, whenever there was a change on the propagation surface, as in the case with the weld bead on a thickness of 3 mm, the magnitude of the stress intensity factor at the crack tip grows in comparison to the test piece of 1mm, whose crack spread on the surface only.

Using Paris's model, it was possible to state the stable growth rate for the two models. Figure 14 shows the behavior of the crack spread rate depending on the variation range of the stress intensity factor for both models. There have been used typical values of a ferritic–steel for C = 6.89 × 10^{-12} and n = 3, when the da/dN are expressed in m/cycle and ΔK in MPa m$^{1/2}$.

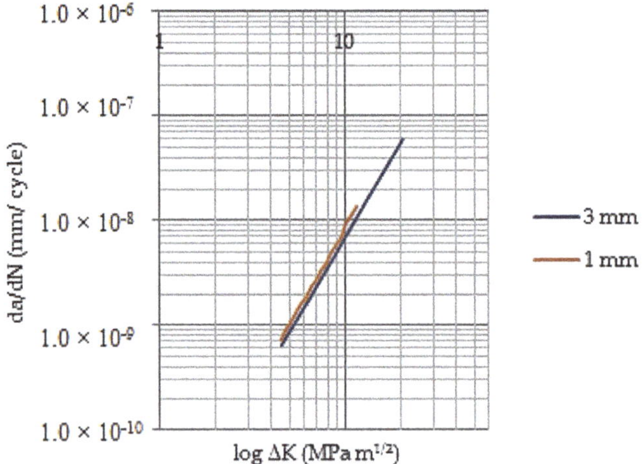

Figure 14. Propagation crack rate versus variation range for SIF.

3.2.3. Experimental Results for Parameter on Paris' Expression

When crack propagation rates are measured at beginning of stable zone and close to its end then the parameters of the expression of Paris can be obtained. The crack propagation rates for both positions are found using SEM by means of separation measurement between striations [16]. In Figure 15a,b, the striations counting performed on sample 3F-H is shown. In addition, the SIF range is calculated using the adjusted finite element model. Then, C and m are obtained by:

$$C = \frac{\left(\frac{da}{dN}\right)_1}{\left(\Delta K_{I,1}\right)^m} \tag{3}$$

$$m = \frac{\log\left(\frac{da}{dN}\right)_1 - \log\left(\frac{da}{dN}\right)_2}{\log \Delta K_{I,1} - \log \Delta K_{I,2}}. \tag{4}$$

In Table 8, the most representative results for C and m are shown. In some of the fatigue surfaces it is not possible to establish the distance among striations, because these are not visible or are confusing among other superficial details. It is observed that maximum value for the propagation rate of the cracks is for sample 1S-G with 1.340×10^{-6} m/cycle, while the sample 1F-H presents a maximum value of 7.970×10^{-7} m/cycle, much lower than the previous one. The latter is consistent with experimental results where fatigue life for the sample 1S-G and 1F-H are 898,000 cycles and 245,000 cycles, respectively.

Table 8. Summary of the parameters calculation of Paris expression for several samples.

Specimen	a_1(m)	a_2(m)	$\left(\frac{da}{dN}\right)_1$ (m/cycle)	$\left(\frac{da}{dN}\right)_2$ (m/cycle)	$\Delta K_{I,1}$ (MPa $\sqrt{m}$)	$\Delta K_{I,2}$ (MPa $\sqrt{m}$)	C	m
1S-G	9.00×10^{-4}	2.45×10^{-3}	6.280×10^{-7}	1.341×10^{-6}	10.2	16.9	6.14×10^{-8}	1.51
1F-H	2.40×10^{-3}	5.29×10^{-3}	2.341×10^{-7}	7.974×10^{-7}	16.7	24.8	1.40×10^{-8}	3.10
3S-D	1.09×10^{-3}	5.41×10^{-3}	5.411×10^{-7}	1.280×10^{-6}	11.2	25.1	4.81×10^{-8}	1.08
3F-H	6.50×10^{-4}	4.42×10^{-3}	3.482×10^{-7}	1.141×10^{-6}	8.7	22.7	4.01×10^{-8}	1.24

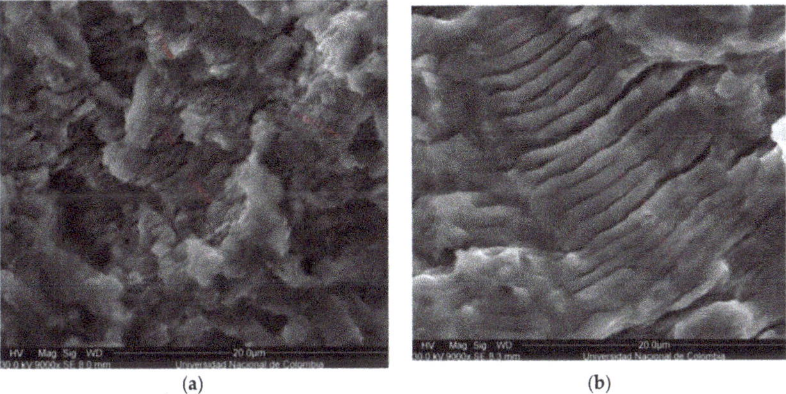

Figure 15. Striation counting close to the crack origin and close to the end of stable crack propagation for specimen 3F-H. (**a**) Position 0–0.001 m, 3.48×10^{-7} m/cycle (9000×); (**b**) Position 0.004–0.0054 m, 1.17×10^{-6} m/cycle (9000×).

3.3. Other Mechanical Property Results

The average results of tension test for each treatment are shown in Table 9. The differences among treatments are not rather significant in means of tension strength and elastic limit. It is important to note the area under the curve for treatment 1F in contrast with treatment 3S present a well-defined difference. This indication allows one to state that, regardless of the post welding cooling medium, specimens with 1 mm of welding reinforcement will have a higher resistance to crack propagation than specimens with 3 mm of welding reinforcement, as it is observed in results from S-N curve and the surviving probability from Figures 6 and 7, respectively.

The mean results for the Charpy impact energy test is shown in Table 10. The fracture toughness of the material was calculated indirectly with the Rolfe-Barsom correlation [32]. In Figure 16, it is presented a comparison between the percent shear fracture for the highest and lowest percentages obtained by following the standard ASTM A370-14 [33], which suggests that percent shear is proportional to material ductility.

Table 9. Results for the mechanical properties obtained by uniaxial tensile tests (average values).

Specimen	Ultimate Tensile Strength (MPa)	Yield Stress 0.20% (MPa)	Percent Reduction of Area (%)	Elongation (%)
1S	462	332	31.4	19.4
1F	461	332	32.1	23.3
3S	455	322	32.1	18.3
3F	465	322	28.9	19.6

Table 10. Results of Charpy impact energy tests (average values).

Specimen	CVN (J)	K_{Ic} (MPa $\sqrt{m}$)	Shear Fracture (%)	Specimen	CVN (J)	K_{Ic} (MPa $\sqrt{m}$)	Shear Fracture (%)
1S-FZ	100.29	163.3	33	1S-HAZ	157.04	157.4	56
1F-FZ	81.17	146.2	33	1F-HAZ	138.67	147.8	40
3S-FZ	75.97	141.2	21	3S-HAZ	125.98	140.7	45
3F-FZ	66.91	131.9	21	3F-HAZ	187.01	172.0	31
BM	204.63	180.0	27	-	-	-	-

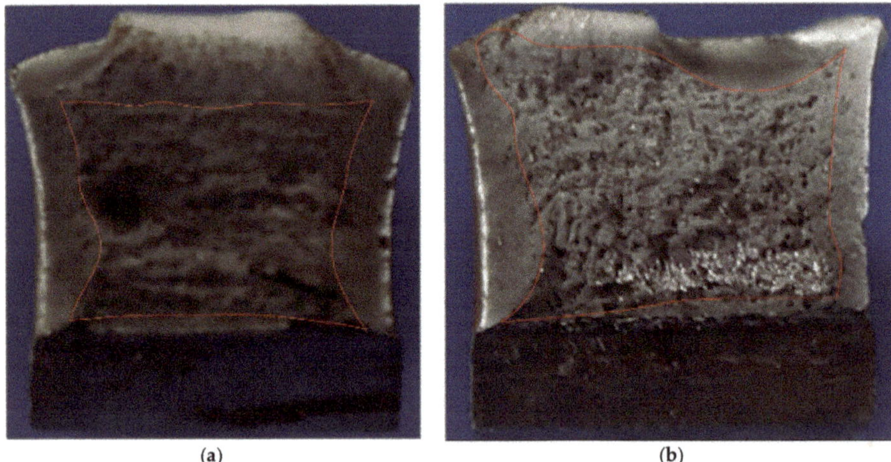

(a)   (b)

Figure 16. Comparison of shear fracture appearance between two specimens, (**a**) 1S-CI-HAZ (56%) and (**b**) 3F-CI-FZ (21%).

The results obtained from microindentation hardness testing are shown in Figure 17, for the four studied treatments. The presented profiles are the real ones. These were obtained by means of digitalization of macro structure image and later processing on CAD software. A strong tendency to present softness for zone HAZ is presented in these results, due to the coarse grains, even though in samples 1F and 3F this behavior is present in a more noteworthy way. Although in Figure 17a–d, the hardness increase is observed on rows 1 and 3 for the samples 1F and 3F, respectively, the hardness value increases for sample 1F on the interfaces FZ-HAZ and HAZ-BM, while in sample 3F that happens on the interfaces HAZ-BM. The possible reasons for the previous behavior could be attributed to fast cooling rate and number of welding applied passes, thus generating microstructural variations in size and shape of the grains.

In Figure 18, a comparison between samples 1S-G and 3S-E is carried out in terms of appearance of the failure zone due to fatigue. It is observed that sample 1S-G presents a single visible crack origin due to the radial marks point toward a single direction, in opposition to sample 3S-E contains three dots which indicate the crack origin of superficial semielliptical kind, based on Figure 4. In sample 3S-E (Figure 18b), it is clearly observed how the three crack fronts, as crack propagation move along, tend to convert into a single crack front, while in sample 1S-G (Figure 18a) the only crack front present propagates nearly in a concentric way around its origin. The bigger size of stable crack growth for a same load cycle means a higher tolerance to damage of the welded joint.

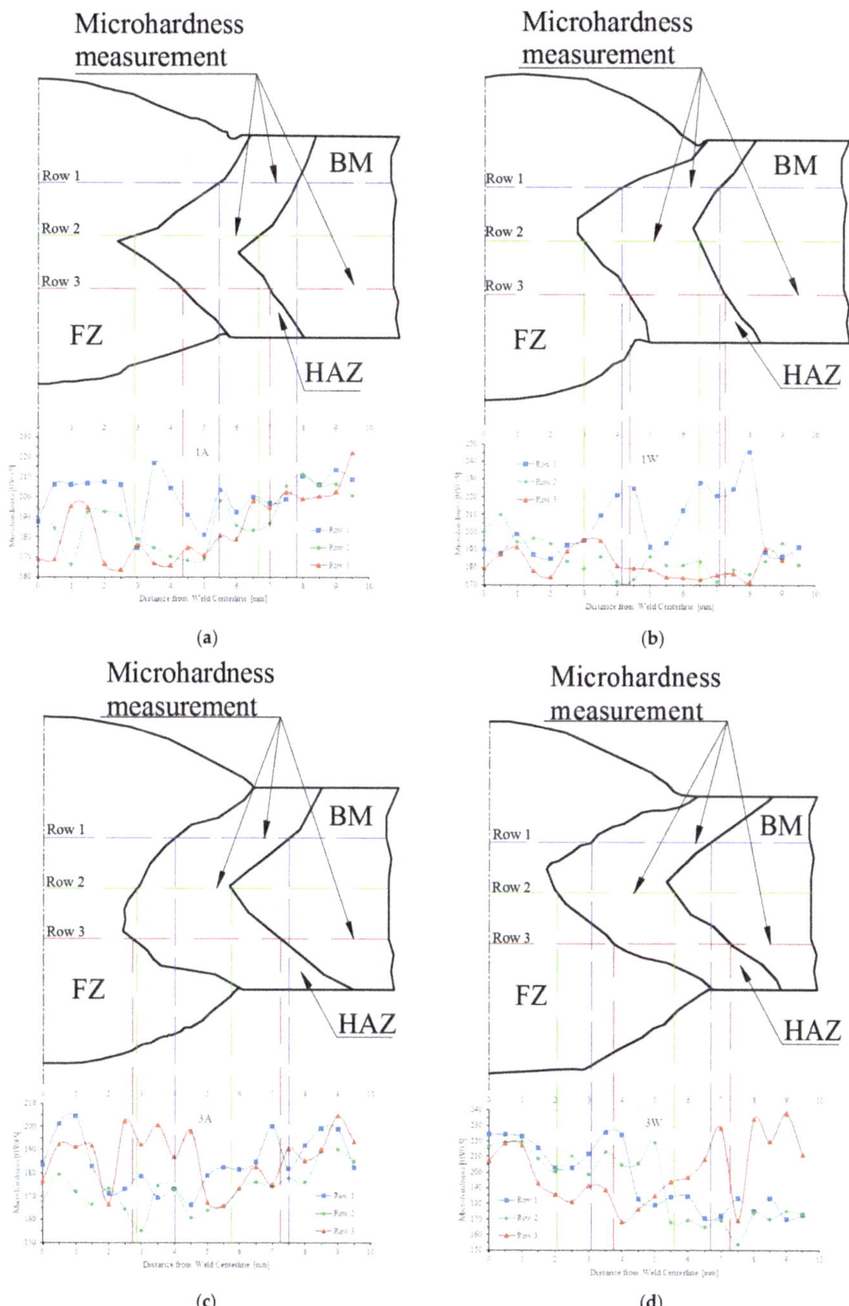

Figure 17. Microindentation hardness profiles for specimens of each treatment: (**a**) specimen 1S; (**b**) specimen 1F; (**c**) specimen 3S; and (**d**) specimen 3F.

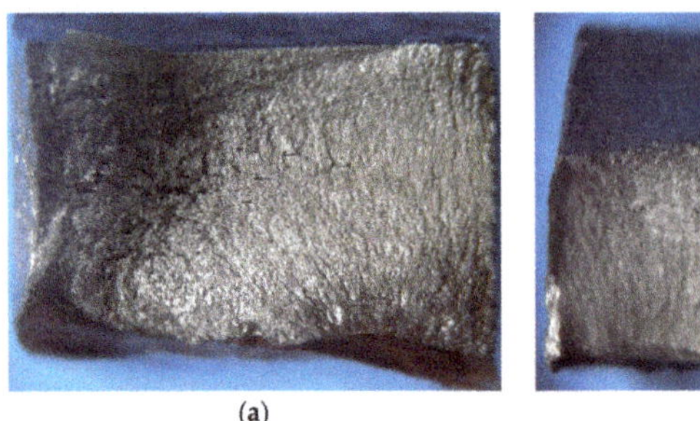
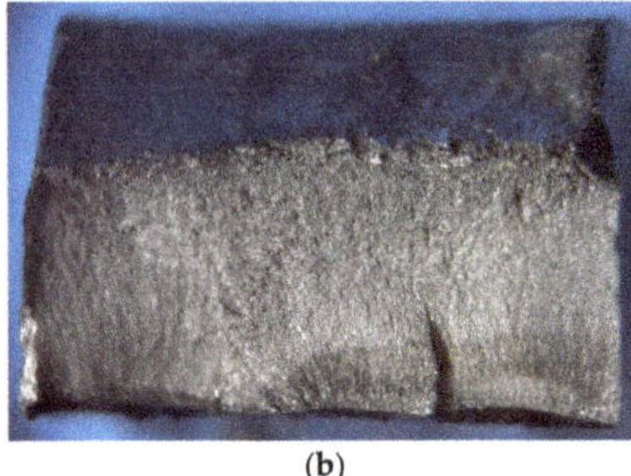

(a)   (b)

Figure 18. Stereo microscopy of fatigue surfaces. Sample 1S-G, with 898,000 cycles to failure (**a**) and sample 3S-E, with 75,000 cycles to failure (**b**).

Finally, the test specimens dimensionally stayed with the tolerances required using abrasive waterjet (AWJ) this manufacturing process did not affect the mechanical properties on the lateral surfaces since all the specimens failed due to cracks that started from weld toe. Therefore, the effect of the machining process on experimental fatigue results is ruled out. In the research [34], as in the present research, no dimensional deviations were detected using this manufacturing procedure.

4. Conclusions

According with the results experimentally obtained in this research, welded joints with a welding reinforcement of 1 mm and cooled in water reached the highest probability of survival to fatigue. While the welded joints with a welding reinforcement of 3 mm and cooled in calm air, reached the worst probability of survival to fatigue. In general, fatigue strength in butt welded joints is strongly influenced by the welding reinforcement. In this research, 93.8% of the specimens presented a fatigue strength superior to the one of class FAT 90, while about 78.1% of the specimens reached a fatigue strength higher than the recommended one by class FAT 100.

The use of the finite element method, along with the fracture mechanics, allowed one to have an adequate simulation of the stable propagation of cracks. The integration of the results obtained for the variation range for stress intensity factor and the separation between striations, obtained through SEM observation, allowed the estimation of the parameters C and m from Paris' expression. For some measurements, C values and m atypical were obtained. The latter could be due to the high gradient change for the local mechanical properties next to the welding bead and the inherent distance difficulties of the measuring method between striations. However, the numerical propagation rates in the order of those found through experimental measurements were obtained.

The crack front on the failure surface presented an approximated semielliptical shape. This allows one to establish the finite element modelling for butt welded joints to establish correlations between experimental results and theoretical calculations. On the other hand, the importance of performing additional studies where the crack nucleation stage is included is highlighted. Where a fatigue model for this first stage should cover the influence of the microstructural phases and the grains shape and size. Moreover, further improvement of this assessment is expected by implementing other specific models for predicting changes in the microstructure and mechanical properties during the thermal cycle of the welding process.

In consistency with welding and post-welding processes applied to butt welded joints, it was observed that the post welding cooling cycle and the quantity of weld passes influence the material ductility and, thus the fracture toughness of the material, as well. In general, higher post-welding cooling speed caused significant lower measured values for the Charpy impact energy. Lower values of the Charpy impact energy are reported for the FZ and the HAZ compared to the base metal. On the other hand, no significant differences were found between experimental treatments for the mechanical properties obtained by the tension tests.

Acknowledgments: The authors would like to thank the research office at the Universidad de Ibagué and the Universidad Nacional de Colombia for their support received for the development of this project number 15-364-INT.

Author Contributions: In this work Oscar Araque, Nelson Arzola, and Edgar Hernandez conceived and designed the experiments and analyzed the results obtained.

Conflicts of Interest: The authors declare no conflict of interest.

References

1. Araque de los Ríos, O.J.; Arzola de la Peña, N. Estado del arte sobre la integridad estructural de uniones soldadas y modelos de propagación de grietas para la gestión de vida en estructuras. Ingeniare. *Rev. Chil. Ing.* **2013**, *21*, 279–292. [CrossRef]
2. Grajales, J.A.; Vanegas, L.V. Métodos para determinar el factor de intensidad de esfuerzos. *Entre Cienc. Ing.* **2016**, *19*, 70–80.
3. Hobbacher, A.F. *Safety Considerations. Recommendations for Fatigue Design of Welded Joints and Components;* Springer: Cham, Switzerland, 2016; pp. 107–111.
4. Blandford, G.E.; Ingraffea, A.R.; Liggett, J.A. Two-dimensional stress intensity factor computations using the boundary element method. *Int. J. Numer. Methods Eng.* **1981**, *17*, 387–404. [CrossRef]
5. Lim, I.L.; Johnston, I.W.; Choi, S.K. Stress intensity factors for semi-circular specimens under three-point bending. *Eng. Fract. Mech.* **1993**, *44*, 363–382. [CrossRef]
6. Paris, P.; Fazil, E. A critical analysis of crack propagation laws. *J. Basic Eng.* **1963**, *85*, 528–533. [CrossRef]
7. Ávila, J.A.; Franco, F.; Jaramillo, H.E. Evaluación de la resistencia a la tensión y la fatiga de uniones soldadas por fricción agitación de la aleación de magnesio az31b. *Rev. Latinoam. Metal. Mater.* **2012**, *32*, 71–78.
8. Amu, B.M.; Franco, A.F. Microestructura y propiedades mecánicas en la zona afectada por el calor de la unión soldada de la aleación 6261-T5. *Supl. Rev. Latinoam. Metal. Mater.* **2009**, *1*, 767–772.
9. Ahmed, S.R.; Agarwal, L.A.; & Daniel, B.S.S. Effect of Different Post Weld Heat Treatments on the Mechanical properties of Cr-Mo Boiler Steel Welded with SMAW Process. *Mater. Today Proc.* **2015**, *2*, 1059–1066. [CrossRef]
10. Bhadeshia, H.K.D.H.; Svensson, L.E.; Gretoft, B. A model for the development of microstructure in low-alloy steel (Fe-Mn-Si-C) weld deposits. *Acta Metal.* **1985**, *33*, 1271–1283. [CrossRef]
11. Yamada, M. Low Cycle Fatigue Fracture Limit as the Evaluation Base of Ductility. In Proceedings of the Stability and Ductility of Steel Structures (SDSS'97), Nagoya, Japan, 29–31 July 1997; pp. 391–399.
12. Madia, M.; Schork, B.; Bernhard, J.; Kaffenberger, M. Multiple crack initiation and propagation in weldments under fatigue loading. *Procedia Struct. Integr.* **2017**, *7*, 423–430. [CrossRef]
13. Bullón, W.; Acosta, J.; Franco, R.; Valverde, Q. Simulación de un proceso de soldadura mediante un modelo termo-mecánico considerando el efecto de esfuerzos residuales utilizando el método de los elementos finitos. In Proceedings of the Memorias del 8° Congreso Iberoamericano de Ingeniería Mecánica, Cuzco, Peru, 23–25 October 2007; p. 13.
14. Almaguer-Zaldivar, P.M.; Estrada-Cingualbres, R.A. Evaluación del comportamiento a fatiga de una unión soldada a tope de acero AISI 1015. *Ing. Mech.* **2015**, *18*, 31–41.
15. Nikitin, I.; Besel, M. Correlation between residual stress and plastic strain amplitude during low cycle fatigue of mechanically surface treated austenitic stainless steel AISI 304 and ferritic–pearlitic steel SAE 1045. *Mater. Sci. Eng.* **2008**, *491*, 297–303. [CrossRef]
16. Berrios, J.A.; Staia, M.H.; Puchi-Cabrera, E.S. Comportamiento a la fatiga de un acero al carbono AISI 1045 recubierto con depósitos autocatalíticos de Ni-P. *Revista de la Facultad de Ingeniería Universidad Central de Venezuela* **2005**, *20*, 5–16.
17. Araque, O.; Arzola, N. Weld Magnification Factor Approach in Cruciform Joints Considering Post Welding Cooling Medium and Weld Size. *Materials* **2018**, *11*, 81. [CrossRef] [PubMed]
18. Valiente, A.; Toribio, J.; Elices, M. Vida en fatiga de un recipiente a presión bajo cargas cíclicas de origen térmico. *Rev. Int. Métodos Numér. Cálc. Diseño Ing.* **1991**, *7*, 437–454.
19. Lewandowski, J.; Rozumek, D. Cracks growth in S355 steel under cyclic bending with fillet welded joint. *Theor. Appl. Fract. Mech.* **2016**, *86*, 342–350. [CrossRef]

20. Yang, Y.; Vormwald, M. Fatigue crack growth simulation under cyclic non-proportional mixed mode loading. *Int. J. Fatigue* **2017**, *102*, 37–47. [CrossRef]

21. Xiao, X.; Yan, X.A. Numerical analysis for cracks emanating from a surface semi-spherical cavity in an infinite elastic body by FRANC3D. *Eng. Fail. Anal.* **2008**, *15*, 188–192. [CrossRef]

22. Chin, P.L. Stress Analysis, Crack Propagation and Stress Intensity Factor Computation of a Ti-6Al-4V Aerospace Bracket Using ANSYS and FRANC3D. Master's Thesis, Rensselaer Polytechnic Institute, Hartford, CT, USA, 2011.

23. Welding, A. Qualify procedures and personnel according to AWS D1. 1/D1. 1M. In *Structural Welding Code-Steel*; American Welding Society: Miami, FL, USA, 2010; p. 60.

24. American Society of Mechanical Engineers (ASME). *Section, IX: Qualify processes and Operators According to ASME*; American Society of Mechanical Engineers (ASME): New York, NY, USA, 2009; Volume 1.

25. DeVries, P.H.; Ruth, K.T.; Dennies, D.P. Counting on fatigue: Striations and their measure. *J. Fail. Anal. Prev.* **2010**, *10*, 120–137. [CrossRef]

26. Varon, O. Análisis Computacional de Propagación de Grietas en Uniones Soldadas a Tope de un Acero ASTM A36 Utilizando un Electrodo E-6013. Ph.D. Thesis, University of Ibagué, Tolima, CA, USA, 2017.

27. Anderson, T.L. *Fracture Mechanics: Fundamentals and Applications*; (CRC) Taylor and Francis: London, UK, 2005.

28. Knoedel, P.; Gkatzogiannis, S.; Ummenhofer, T. Practical aspects of welding residual stress simulation. *J. Constr. Steel Res.* **2017**, *132*, 83–96. [CrossRef]

29. Leggatt, R.H. Residual stresses in welded structures. *Int. J. Press. Vessels Pip.* **2008**, *85*, 144–151. [CrossRef]

30. Shiraiwa, T.; Briffod, F.; Enoki, M. Development of integrated framework for fatigue life prediction in welded structures. *Eng. Fract. Mech.* **2017**. [CrossRef]

31. Lee, C.H.; Chang, K.H.; Van Do, V.N. Finite element modeling of residual stress relaxation in steel butt welds under cyclic loading. *Eng. Struct.* **2015**, *103*, 63–71. [CrossRef]

32. Barsom, J.M.; Rolfe, S.T. Correlations between K IC and Charpy V-notch test results in the transition-temperature range. In *Impact Testing of Metals*; ASTM International: West Conshohocken, PA, USA, 1970.

33. ASTM International. *Standard test methods and Definitions for Mechanical Testing of Sheet Product*; ASTM A370; ASTM International: West Conshohocken, PA, USA, 2011.

34. Krahmer, D.M.; Polvorosa, R.; de Lacalle, L.L.; Alonso-Pinillos, U.; Abate, G.; Riu, F. Alternatives for specimen manufacturing in tensile testing of steel plates. *Exp. Tech.* **2016**, *40*, 1555–1565. [CrossRef]

 *materials* 

Article

ANN Surface Roughness Optimization of AZ61 Magnesium Alloy Finish Turning: Minimum Machining Times at Prime Machining Costs

Adel Taha Abbas [1],*, Danil Yurievich Pimenov [2], Ivan Nikolaevich Erdakov [3], Mohamed Adel Taha [4], Mahmoud Sayed Soliman [1] and Magdy Mostafa El Rayes [1]

[1] Department of Mechanical Engineering, College of Engineering, King Saud University, P.O. Box 800, Riyadh 11421, Saudi Arabia; solimanm@ksu.edu.sa (M.S.S.); melrayes@ksu.edu.sa (M.M.E.R.)

[2] Department of Automated Mechanical Engineering, South Ural State University, Lenin Prosp. 76, Chelyabinsk 454080, Russia; danil_u@rambler.ru

[3] Foundry Department, South Ural State University, Lenin Prosp. 76, Chelyabinsk 454080, Russia; wissenschaftler@bk.ru

[4] Department of Mechanical Design and Production, Faculty of Engineering, Zagazig University, Ash Sharqiyah 44519, Egypt; eng_mohamed_2017@yahoo.com

* Correspondence: aabbas@ksu.edu.sa

Received: 5 April 2018; Accepted: 14 May 2018; Published: 16 May 2018

Abstract: Magnesium alloys are widely used in aerospace vehicles and modern cars, due to their rapid machinability at high cutting speeds. A novel Edgeworth–Pareto optimization of an artificial neural network (ANN) is presented in this paper for surface roughness (Ra) prediction of one component in computer numerical control (CNC) turning over minimal machining time (T_m) and at prime machining costs (C). An ANN is built in the Matlab programming environment, based on a 4-12-3 multi-layer perceptron (MLP), to predict Ra, T_m, and C, in relation to cutting speed, v_c, depth of cut, a_p, and feed per revolution, f_r. For the first time, a profile of an AZ61 alloy workpiece after finish turning is constructed using an ANN for the range of experimental values v_c, a_p, and f_r. The global minimum length of a three-dimensional estimation vector was defined with the following coordinates: Ra = 0.087 μm, T_m = 0.358 min/cm^3, C = \$8.2973. Likewise, the corresponding finish-turning parameters were also estimated: cutting speed v_c = 250 m/min, cutting depth a_p = 1.0 mm, and feed per revolution f_r = 0.08 mm/rev. The ANN model achieved a reliable prediction accuracy of ±1.35% for surface roughness.

Keywords: artificial neutral network; cutting parameters; magnesium alloys; optimization; prime machining costs; surface roughness

---

1. Introduction

Today, many industries such as mechanical engineering, automobile manufacturing, machine-tool building, and aerospace industries, among others, all employ turning. One of the main quality parameters in finish turning [1–7], milling [8–11], and grinding [12,13] is surface roughness. AZ 61 magnesium alloys are widely used in industry, due to their lightweight structure [14,15]. Their basic properties mainly depend on their hexagonal mesh structure [16]. These alloys are used for many cast components in the automotive industry [17,18], such as cast magnesium engine bodies, because plastic deformation in hexagonal metal materials is of greater complexity than than in cubic metals [19]. AZ 61 magnesium alloys are also widely used in many aerospace vehicles and modern cars, in part due to the high cutting speeds of these alloys [20,21]. As well as surface quality, minimization of the use of resources is also an important objective when machining expensive materials such as AZ 61 magnesium alloys. It is essential to assure minimum machining times of unit volume and minimum surface roughness, *Ra*, simultaneously.

Plenty of research has covered the prediction of surface roughness in turning. Risbood et al. [22] established that neural networks can be used to predict surface roughness with reasonable accuracy, by using tool-holder radial vibration under acceleration as feedback. Özel and Karpat [23] used neural network modeling to predict surface roughness and tool flank wear during machining times under various cutting conditions for the finish turning of hardened AISI 52100 steel. Bajić et al. [24] examined the influence of cutting speed, feed rate, and depth of cut on surface roughness and cutting force components in longitudinal turning. Regression analysis and neural networks were applied to the surface roughness prediction model. Muthukrishnan and Davim [25] obtained a model for predicting the roughness of a machined surface using analysis of variance (ANOVA) and artificial neural network (ANN) techniques in the turning of Al/SiC-MMC workpieces. Natarajan et al. [26] described a surface roughness prediction model using Matlab-based ANN processing data on C26000 brass in turning operations under dry cutting conditions. Svalina et al. [27] analyzed the influence of cutting depth, feed rate, and the number of revolutions for ANN surface roughness prediction. Pontes et al. [28] presented a study on the applicability of radial base function (RBF) neural networks for the prediction of roughness average (*Ra*) in the turning process of SAE 52100 hardened steel, applying Taguchi orthogonal arrays as a tool to design network parameters. Hessainia et al. [29] developed surface roughness models for hard (finishing) turning of 42CrMo4 steel with an Al_2O_3/TiC ceramic cutting tool using the response surface methodology (RSM). Krolczyk et al. [30] identified surface integrity of the turned workpieces using fused deposition modeling (FDM). Nieslony et al. [31] presented the problem of precise turning of 55NiCrMoV6 hardened steel mould parts and demonstrated a topographic inspection of the machined surface quality. Acayaba and Escalona [32] developed a model for predicting surface roughness in the low-speed turning of AISI316 stainless steel using multiple linear regression and ANN methodologies. D'Addona and Raykar [33] studied the influence of hard turning parameters—speed, feed rate, depth of cut, and nose radius (for wipers and regular inserts)—on surface roughness. Mia and Dhar [34] obtained an ANN model for predicting average surface roughness when turning EN 24T hardened steel. Jurkovic et al. [35] compared three machine-learning methods in predicting the observed parameters of high-speed turning (surface roughness (*Ra*), cutting force (*Fc*), and tool life (*T*)). Tootooni et al. [36] used a non-contact, vision-based online measurement method for measuring surface roughness while turning the external diameter of the workpiece. Mia et al. [37] focused on developing predictive models of average surface roughness, chip-tool interface temperature, chip reduction coefficient, and average tool flank wear when turning a Ti-6Al-4V alloy. Mia et al. [38] investigated the plain turning of hardened AISI 1060 steel and examined the effect of three sustainable techniques and the traditional flood cooling system on the following machining indices: cutting temperature, surface roughness, chip characteristics, and tool wear.

Even though some studies [22–38] presented surface roughness prediction models, they were unable to solve the problem of establishing the cutting parameters that would yield optimal surface roughness.

Looking at the papers that describe optimal surface roughness parameters in turning, Al-Ahmari [39] developed empirical models for tool life, surface roughness, and cutting force in turning operations. Jafarian et al. [40] proposed a method for determining optimal machining parameters on the basis of three separate neural networks, both to minimize surface roughness and resultant cutting forces and to maximize tool life in the turning process. Mokhtari Homami et al. [41] used neural networks to determine optimum flank wear and surface roughness parameters when turning an Inconel 718 superalloy. Sangwan et al. [42] used ANN and the genetic algorithm (GA) to establish the optimal machining parameters as a function of minimum surface roughness, turning a Ti-6Al-4V titanium alloy. Gupta et al. [43] focused on optimization of certain process parameters of the turning operation: surface roughness, tool flank wear, and power consumption. Venkata Rao and Murthy [44] developed statistical models to investigate the effect of cutting parameters on surface roughness and root mean square of work piece vibration in the boring of AISI 316 stainless steel with physical vapor deposition (PVD)-coated carbide tools. Cutting parameters were optimized for minimum surface roughness and root mean square of work piece vibration using a multi-response optimization technique. Zerti et al. [45] solved an optimization problem of minimizing surface roughness, peripheral force, specific cutting force, and cutting power in the dry turning of AISI D3 steel. Mia et al. [46] presented optimization of cutting forces, average surface roughness, cutting temperature, and chip minimizing coefficient when turning a Ti-6Al-4V alloy under dry conditions and with high pressure coolant (HPC) simultaneously applied to the rake and the flank surfaces. Mia and Dhar [47] evaluated the effects of material hardness and high-pressure coolant jet over dry machining with respect to surface roughness and cutting temperature using a Taguchi L 36 orthogonal array.

However, studies with the objective of establishing the optimal cutting modes are limited [39–47] in that they only take into account surface roughness and not its interconnection with processing performance and unit-volume machining time, which is unacceptable when processing such expensive materials such as AZ 61 magnesium alloys.

Following the above, the papers that describe optimal turning parameters using multi-objective optimization may be considered [48–52]. Basak et al. [48] described two types of Pareto optimization: minimization of production time and minimization of the cost of machining. Surface roughness was considered to be a limitation. Karpat and Özel [49] used neural networks and multi-objective Pareto optimization to establish machining parameters in longitudinal turning of hardened AISI H13 steel. The optimization criteria were defined as follows: minimize surface roughness values and maximize productivity, maximize tool life, and material removal rate, and minimize machining induced stresses on the surface and, likewise, surface roughness. Raykar et al. [50] used grey relational analysis (GRA) to investigate the high-speed turning of Al 7075 high-strength aluminium alloy. As a result of GRA-based multipurpose optimization, the optimum conditions were established for the given surface roughness, energy consumption, material removal rate, and cutting time while turning. Yue et al. [51] used multi-objective Pareto optimization to establish the relation between surface roughness, thickness of plastic deformation, and cutting conditions in the hard turning of Cr12MoV die steel. Abbas et al. [52] obtained a Pareto frontier for Ra and T_m of the finished workpiece from high-strength steel using the ANN model that was later used to determine the optimum finishing cutting conditions. There is, therefore, little research dedicated to multi-objective optimization in turning. The most efficient approach to solving such problems is Pareto optimization. However, studies [48–52] are not concerned with multiobjective optimization of machining AZ61 magnesium alloys. Taking into account the high cost of this material, it is necessary to ensure the design roughness value of the machined surface and the minimum processing time of the material volume at minimal processing costs.

The objective of this study is, therefore, to establish the turning conditions of AZ61 magnesium alloys that provide the minimum unit-volume machining time, T_m, the minimum surface roughness, Ra, and the minimum cost of machining one part, C.

2. Experiment

Magnesium alloy AZ 61 contains aluminum (nominally 6%), zinc (nominally 1%), and other trace elements. Table 1 summarizes the chemical composition of AZ 61.

Table 1. Chemical composition of AZ 61 magnesium alloy.

Element	Aluminum	Zinc	Copper	Silicon	Iron	Nickel	Magnesium
Mass %	6	0.90	0.02	0.008	0.007	0.003	Balance

The CNC lathe used to conduct the experiments was an EMCO Concept Turn 45 (Emco, Salzburg, Austria) equipped with Sinumeric 840-D, a SVJCL2020K16 tool holder, and a VCGT 160404 FN-ALU insert. The cutting edge angle (k_r), back angle (α), and nose radius (r_o) were set at 35°, 5°, and 0.4 mm, respectively. Workpiece length (L), workpiece diameter (D) and tolerance (l_1) were set at 20, 40, and 2 mm, respectively. Experiments were carried out under wet cutting conditions. A TESA Rugosurf 90-G surface roughness tester (TESA, Bugnon, Switzerland) was used. Figure 1 illustrates the test rig used to measure surface roughness.

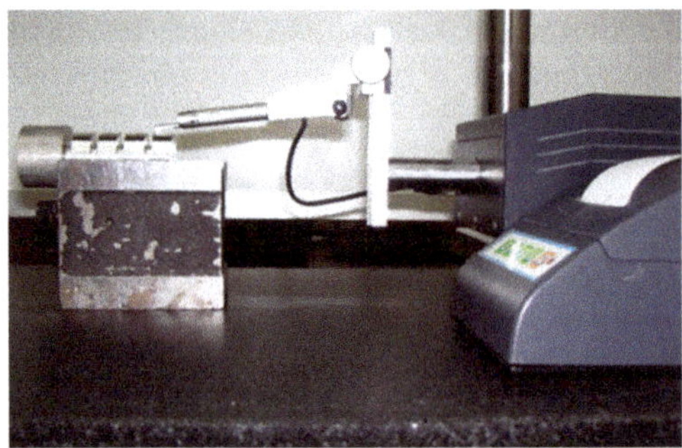

Figure 1. Test rig for measuring surface roughness.

The test plan was implemented through 64 turning runs divided into 16 groups. For each set of four groups, one common cutting speed, v_c, was used (100, 150, 200, and 250 m/min). Each set of experiments was machined using four different depths of cut a_p (0.25, 0.50, 0.750, and 1.00 mm). Each depth of cut was processed using four levels of feed rate f_r (0.04, 0.08, 0.12, and 0.16 mm/rev). The surface roughness values for the different cutting conditions are presented in Table 2.

Table 2. Surface roughness values under different cutting conditions.

Cutting Speed: v_c, (m/min)	Feed: f_r, (mm/rev)	Surface Roughness: Ra (µm)			
		Depth of Cut: a_p, (mm)			
		0.25	0.5	0.75	1.0
100	0.0400	0.1730	0.1660	0.1500	0.1290
100	0.0800	0.3880	0.3610	0.3530	0.4400
100	0.1200	0.8720	0.9520	1.0470	1.0200
100	0.1600	1.6780	2.1040	2.1790	2.6290
150	0.0400	0.1460	0.1320	0.1160	0.1890
150	0.0800	0.3440	0.3480	0.3150	0.4130
150	0.1200	0.9310	1.0540	0.9840	0.9990
150	0.1600	1.6370	1.7640	1.7020	1.8840
200	0.0400	0.1820	0.1800	0.2040	0.1500
200	0.0800	0.3670	0.3860	0.3970	0.3550
200	0.1200	0.8450	1.0240	1.0340	1.2140
200	0.1600	1.9760	1.9220	1.9350	2.0140
250	0.0400	0.1230	0.1830	0.1370	0.2240
250	0.0800	0.3590	0.3890	0.3580	0.3250
250	0.1200	0.9370	0.9680	0.9500	1.0000
250	0.1600	2.0880	1.9540	2.0170	1.8930

Table 3, below, summarizes the basic economic parameters for optimizing the turning of an AZ61 aluminum alloy workpiece.

Table 3. Summary of basic economic parameters.

Mater.	Cost of Machining/Hour (SR 400), *CMh*: $	Cost of Tool Holder, *CToolh*: $	Tool Holder Life: *LTToolh* min	Cost of Insert, *CIn*: $	Setup Insert: k	Unit Cost of Work-Piece: *Cw*: $	Tool Life: *T* Min	Cost of Tool Minute: *CToolmin*, $ *CToolmin* = (*CIn*/(*T* × *k*)) + (*CToolhLTToolh*)
AZ61	106	85	5 Year × 365 Day × 24 h × 60 min = 2,628,000	10	2	8	60	0.083

3. System Adaptation Procedure

The procedure for system adaptation is described as follows:

Step one: Posing a multiobjective optimization problem, i.e., establishing the criteria, limitations, and boundary conditions.

Step two: Establishing a relationship between the parameters of the cutting tool–workpiece system, i.e., build and train an ANN.

Step three: Graphically interpreting the surfaces of normalized three-dimensional space, determine the states of the system in which the values of each particular indicator cannot be improved without aggravating others, i.e., the Pareto frontier.

Step four. Establishing the optimum turning conditions for the workpiece, i.e., to adapt the cutting tool–workpiece system to the given conditions.

Before we begin the procedure, the nomenclature that we will use is introduced. DM—decision maker; m—a number of criteria; I = {1, 2, ... , m}—a set of criteria numbers; X—a set of possible decisions; $f = (f_1, f_2, \ldots , f_m)$—vector-valued criterion; $Y = f(X)$—a set of possible vectors (estimates); R^m—Euclidean space of m-dimensional vectors with real components; $>_X$—preference relation of DM specified in the set X; $>_Y$—preference relation of DM, induced on the set with $>_X$ and specified in the set Y; $>$—relation $>_Y$ continued in the entire space R^m; Sel X—a set of selected decisions; Sel Y—a set of selected vectors (estimates); Ndom X—a set of non-dominated decisions; Ndom Y—a set of non-dominated vectors (estimates); P_f (X)—a set of Pareto optimal decisions; P(Y)—a set of Pareto optimal vectors (Pareto optimal estimates).

4. Formulation of an Optimization Problem

This investigation of machining operations has the objective of resolving the following optimization problem criteria: f_1—surface roughness (*Ra*, μm) and f_2—unit-volume machining time volume in one cutting tool pass (T_m, min/cm^3); and, f_3—the cost price of processing one component part (*C*, $), i.e., $m = 3$. Relatively, a set of possible Y estimates in the two-dimensional space, R^3, is formed with vectors $f = (f_1, f_2, f_3)$. The search is performed for a set of estimates with the minimum length of vector f, which is a vector from the coordinate origin to a point on the estimate surface. The criteria are presented in a normalized dimensionless form with index 1 assigned to the maximum actual numbers.

As the adaptation of the system takes place, the parameters are varied in accordance with the following experimental table (see Table 2): $x_1 = [100 \div 250]$; cutting speed, v_c, m/min; $x_2 = [0.25 \div 1.0]$; depth of cut, a_p, mm; $x_3 = [0.04 \div 0.16]$; and, feed rate, f_r, mm/rev.

We will evaluate the state of the system based on four criteria (Tables 4–7). The first criterion is surface roughness *Ra* (μm), and dimensionless surface roughness *Ra**(f_1). The second criterion is unit-volume machining time, T_m (min/cm^3), and unit-volume dimensionless machining time T_m*(f_2). The third criterion is the cost price of processing one component part, *C* ($), or the dimensionless cost price of processing one component part, *C** (f_3). The fourth criterion is the dimensionless vector of estimates in a three-dimensional normalized space, f.

Table 4. Optimization criteria for the variable machining parameters at a fixed depth of cut—a_p = 0.25 mm.

Variable Parameters						Optimization Criteria					
x_1 Cutting Speed: v_c, (m/min)	x_2 Depth of Cut: a_p, (mm)	x_3 Feed: f_r, (mm/rev)	Surface Roughness: Ra (μm)	Dimensionless Surface Roughness: f_1 (Ra^*), u	Unit Volume Machining Time: T_m (min/cm³)	Dimensionless Volume Machining Time: f_2 ($T_m{}^*$), u	Unit cost Price of Processing One Part: C, ($)	Dimension-less Cost Price of Processing One Part: f_3 (C^*), u	Length of Estimates Vector: f, u	Length of Estimates Vector: f^*, u
---	---	---	---	---	---	---	---	---	---	---
0.4	0.25	0.25	0.1730	0.0660	1.0000	1.0000	9.2729	1.0000	1.4160	1.0000
0.4	0.25	0.5	0.3880	0.1480	0.5000	0.5000	8.6374	0.9310	1.1780	0.8319
0.4	0.25	0.75	0.8720	0.3320	0.3333	0.3330	8.4237	0.9080	1.1260	0.7952
0.4	0.25	1.0	1.6780	0.6380	0.2500	0.2500	8.3187	0.8970	1.2090	0.8538
0.6	0.25	0.25	0.1460	0.0560	0.6667	0.6670	8.8492	0.9540	1.2570	0.8877
0.6	0.25	0.5	0.3440	0.1310	0.3333	0.3330	8.4237	0.9080	1.0840	0.7655
0.6	0.25	0.75	0.9310	0.3540	0.2222	0.2220	8.2837	0.8930	1.0700	0.7556
0.6	0.25	1.0	1.6370	0.6230	0.1667	0.1670	8.2118	0.8860	1.1580	0.8178
0.8	0.25	0.25	0.1820	0.0690	0.5000	0.5000	8.6374	0.9310	1.1710	0.8270
0.8	0.25	0.5	0.3670	0.1400	0.2500	0.2500	8.3187	0.8970	1.0360	0.7316
0.8	0.25	0.75	0.8450	0.3210	0.1667	0.1670	8.2118	0.8860	1.0270	0.7253
0.8	0.25	1.0	1.9760	0.7520	0.1250	0.1250	8.1584	0.8800	1.2100	0.8545
1.0	0.25	0.25	0.1230	0.0470	0.4000	0.4000	8.5084	0.9180	1.1160	0.7881
1.0	0.25	0.5	0.3590	0.1370	0.2000	0.2000	8.2542	0.8900	1.0050	0.7097
1.0	0.25	0.75	0.9370	0.3560	0.1333	0.1330	8.1695	0.8810	1.0180	0.7189
1.0	0.25	1.0	2.0880	0.7940	0.1000	0.1000	8.1271	0.8760	1.2240	0.8644

Table 5. Optimization criteria values for the variable parameters of machining at a fixed depth of cut—a_p = 0.5 mm.

Variable Parameters						Optimization Criteria					
x_1 Cutting Speed: v_c, (m/min)	x_2 Depth of Cut: a_p, (mm)	x_3 Feed: f_r, (mm/rev)	Surface Roughness: Ra (μm)	Dimensionless Surface Roughness: f_1 (Ra^*), u	Unit Volume Machining Time: T_m (min/cm³)	Dimensionless Volume Machining Time: f_2 ($T_m{}^*$), u	Unit Cost Price of Processing One Part: C, ($)	Dimensionless Cost Price of Processing One Part: f_3 (C^*), u	Length of Estimates Vector: f, u	Length of Estimates Vector: f^*, u
---	---	---	---	---	---	---	---	---	---	---
0.4	0.5	0.25	0.1660	0.0630	0.5000	0.5000	9.2729	1.0000	1.2260	0.8658
0.4	0.5	0.5	0.3610	0.1370	0.2500	0.2500	8.6374	0.9310	1.0660	0.7528
0.4	0.5	0.75	0.9520	0.3620	0.1667	0.1670	8.4237	0.9080	1.0590	0.7479
0.4	0.5	1.0	2.1040	0.8000	0.1250	0.1250	8.3187	0.8970	1.2530	0.8849
0.6	0.5	0.25	0.1320	0.0500	0.3333	0.3330	8.8492	0.9540	1.1160	0.7881
0.6	0.5	0.5	0.3480	0.1320	0.1667	0.1670	8.4237	0.9080	1.0040	0.7090
0.6	0.5	0.75	1.0540	0.4010	0.1111	0.1110	8.2837	0.8930	1.0340	0.7302
0.6	0.5	1.0	1.7640	0.6710	0.0833	0.0830	8.2118	0.8860	1.1480	0.8107
0.8	0.5	0.25	0.1800	0.0680	0.2500	0.2500	8.6374	0.9310	1.0590	0.7479
0.8	0.5	0.5	0.3860	0.1470	0.1250	0.1250	8.3187	0.8970	0.9750	0.6886
0.8	0.5	0.75	1.0240	0.3900	0.0833	0.0830	8.2118	0.8860	1.0100	0.7133
0.8	0.5	1.0	1.9220	0.7310	0.0625	0.0630	8.1584	0.8800	1.1710	0.8270

Table 5. *Cont.*

Variable Parameters			Optimization Criteria							
x_1 Cutting Speed: v_c, (m/min)	x_2 Depth of Cut: a_p, (mm)	x_3 Feed: f_r, (mm/rev)	Surface Roughness: Ra (µm)	Dimensionless Surface Roughness: f_1 (Ra^*), u	Unit Volume Machining Time: T_m (min/cm^3)	Dimensionless Volume Machining Time: f_2 (T_m^*), u	Unit Cost Price of Processing One Part: C, ($)	Dimensionless Cost Price of Processing One Part: f_3 (C^*), u	Length of Estimates Vector: f, u	Length of Estimates Vector: f^*, u
1.0	0.5	0.25	0.1830	0.0700	0.2000	0.2000	8.5084	0.9180	1.0240	0.7232
1.0	0.5	0.5	0.3890	0.1480	0.1000	0.1000	8.2542	0.8900	0.9560	0.6751
1.0	0.5	0.75	0.9680	0.3680	0.0667	0.0670	8.1695	0.8810	0.9890	0.6984
1.0	0.5	1.0	1.9540	0.7430	0.0500	0.0500	8.1271	0.8760	1.1700	0.8263

Table 6. The values of optimization criteria for the variable parameters of machining at fixed depth of cut—a_p = 0.75 mm.

Variable Parameters			Optimization Criteria							
x_1 Cutting Speed: v_c, (m/min)	x_2 Depth of Cut: a_p, (mm)	x_3 Feed: f_r, (mm/rev)	Surface Roughness: Ra (µm)	Dimensionless Surface Roughness: f_1 (Ra^*), u	Unit Volume Machining Time: T_m (min/cm^3)	Dimensionless Volume Machining Time: f_2 (T_m^*), u	Unit Cost Price of Processing One Part: C, ($)	Dimensionless Cost Price of Processing One Part: f_3 (C^*), u	Length of Estimates Vector: f, u	Length of Estimates Vector: f^*, u
0.4	0.75	0.25	0.1500	0.0570	0.3333	0.3330	9.2729	1.0000	1.1560	0.8164
0.4	0.75	0.5	0.3530	0.1340	0.1667	0.1670	8.6374	0.9310	1.0260	0.7246
0.4	0.75	0.75	1.0470	0.3980	0.1111	0.1110	8.4237	0.9080	1.0460	0.7387
0.4	0.75	1.0	2.1790	0.8290	0.0833	0.0830	8.3187	0.8970	1.2550	0.8863
0.6	0.75	0.25	0.1160	0.0440	0.2222	0.2220	8.8492	0.9540	1.0650	0.7521
0.6	0.75	0.5	0.3150	0.1200	0.1111	0.1110	8.4237	0.9080	0.9750	0.6886
0.6	0.75	0.75	0.9840	0.3740	0.0741	0.0740	8.2837	0.8930	1.0060	0.7105
0.6	0.75	1.0	1.7020	0.6470	0.0556	0.0560	8.2118	0.8860	1.1220	0.7924
0.8	0.75	0.25	0.2040	0.0780	0.1667	0.1670	8.6374	0.9310	1.0200	0.7203
0.8	0.75	0.5	0.3970	0.1510	0.0833	0.0830	8.3187	0.8970	0.9540	0.6737
0.8	0.75	0.75	1.0340	0.3930	0.0556	0.0560	8.2118	0.8860	0.9980	0.7048
0.8	0.75	1.0	1.9350	0.7360	0.0417	0.0420	8.1584	0.8800	1.1650	0.8227
1.0	0.75	0.25	0.1370	0.0520	0.1333	0.1330	8.5105	0.9180	0.9890	0.6984
1.0	0.75	0.5	0.3580	0.1360	0.0667	0.0670	8.2553	0.8900	0.9370	0.6617
1.0	0.75	0.75	0.9500	0.3610	0.0444	0.0440	8.1702	0.8810	0.9750	0.6886
1.0	0.75	1.0	2.0170	0.7670	0.0333	0.0330	8.1276	0.8760	1.1780	0.8319

Table 7. The values of optimization criteria for the variable parameters of machining at fixed depth of cut—a_p = 1.0 mm.

Variable Parameters						Optimization Criteria				
x_1 Cutting Speed: v_c, (m/min)	x_2 Depth of Cut: a_p, (mm)	x_3 Feed: f_r, (mm/rev)	Surface Roughness: Ra (μm)	Dimensionless Surface Roughness: f_1 (Ra^*), u	Unit Volume Machining Time: T_m (min/cm^3)	Dimensionless Volume Machining Time: f_2 (T_m^*), u	Unit Cost Price of Processing One Part: C, ($)	Dimensionless Cost Price of Processing One Part: f_3 (C^*), u	Length of Estimates Vector: f, u	Length of Estimates Vector: f^*, u
0.4	1.0	0.25	0.1290	0.0490	0.2500	0.2500	9.2729	1.0000	1.1190	0.7903
0.4	1.0	0.5	0.4400	0.1670	0.1250	0.1250	8.6374	0.9310	1.0100	0.7133
0.4	1.0	0.75	1.0200	0.3880	0.0833	0.0830	8.4237	0.9080	1.0290	0.7267
0.4	1.0	1.0	2.6290	1.0000	0.0625	0.0630	8.3187	0.8970	1.3670	0.9654
0.6	1.0	0.25	0.1890	0.0720	0.1667	0.1670	8.8492	0.9540	1.0400	0.7345
0.6	1.0	0.5	0.4130	0.1570	0.0833	0.0830	8.4237	0.9080	0.9650	0.6815
0.6	1.0	0.75	0.9990	0.3800	0.0556	0.0560	8.2837	0.8930	0.9990	0.7055
0.6	1.0	1.0	1.8840	0.7170	0.0417	0.0420	8.2118	0.8860	1.1580	0.8178
0.8	1.0	0.25	0.1500	0.0570	0.1250	0.1250	8.6374	0.9310	0.9980	0.7048
0.8	1.0	0.5	0.3550	0.1350	0.0625	0.0630	8.3187	0.8970	0.9410	0.6645
0.8	1.0	0.75	1.2140	0.4620	0.0417	0.0420	8.2118	0.8860	1.0200	0.7203
0.8	1.0	1.0	2.0140	0.7660	0.0313	0.0310	8.1584	0.8800	1.1800	0.8333
1.0	1.0	0.25	0.2240	0.0850	0.1000	0.1000	8.5084	0.9180	0.9750	0.6886
1.0	1.0	0.5	0.3250	0.1240	0.0500	0.0500	8.2542	0.8900	0.9260	0.6540
1.0	1.0	0.75	1.0000	0.3800	0.0333	0.0330	8.1695	0.8810	0.9770	0.6900
1.0	1.0	1.0	1.8930	0.7200	0.0250	0.0250	8.1271	0.8760	1.1450	0.8086

The values of the first criterion are taken from the experimental table and the rest are calculated on the basis of Formulas (1)–(6):

$$T_m = 1/(1000 \times v_c \times a_p \times f_r);$$ (1)

$C_i = (C_{Mh} \times T') + (C_{Toolmin} \times T') + C_w$, where is Machining Time in Turning $T' = (L + l_1)/(n \times f_r)$, where

$$n = (1000 \cdot v_c)/(3.141 \cdot D);$$ (2)

$$Ra^* = Ra_i/Ra_{max};$$ (3)

$$T_m{}^* = T_{m\,i}/T_{m\,max};$$ (4)

$$C^* = C_i/C_{i\,max};$$ (5)

$$f = \sqrt{f_1^2 + f_2^2 + f_3^2} + \sqrt{R_a^{*2} + T_m^{*2} + C^{*2}},$$ (6)

where, Ra_i is surface roughness for the current combination of X... and f_r; Ra_{max} is the maximum surface roughness value of all the v_c, a_p, and f_r combinations; $T_{m\,i}$ is the unit-volume machining time for the current values of v_c, a_p, and f_r; $T_{m\,max}$ is the maximum unit-volume machining time of all the v_c, a_p, and f_r combinations; C_i is the cost price of processing one part for a given combination of v_c, a_p, and f_r; $C_{i\,max}$—the maximum value.

The optimum search procedure involves a non-negative set of vector estimates, and eliminates the variation of parameter values below zero. The boundary condition is, therefore, that all the variables in this model are non-negative.

Now that the optimization problem is formulated, we shall build and train the neural network that should become the functional operator of the three variables $f(f_1, f_2, f_3)$ and $f(f_1, f_2, f_3)$, as well as the functional Q to the plane $f(f_1, f_2, f_3)$. The ANN complex was constructed using the Skif AURORA-SUSU supercomputer cluster (South Ural State University, Chelyabinsk, Russia) [53].

5. Building a Neural Network Model

Matlab today outperforms other well-known software packages—Maple, Mathematica, and Mathcad—in terms of fundamental quality and versatile numerical calculations. Neural networks can be designed, modeled, and trained easily with the Matlab neural network toolbox. A clear advantage of Matlab is its programming language that can be used to write algorithms and programmes. Many tasks can be achieved with its versatile language, including: data collection, analysis, and structuring, adding to algorithms, system modeling, debugging, object-oriented programming, and graphical user interface development. Matlab applications may also be converted to either C or C++ code.

The programming environment Matlab R2010b, a parallel version of Matlab, was selected in this study. A multi-layer perceptron (MLP) using the Levenberg–Marquardt algorithm was used to train the controlled feedforward neural network. Sigmoid neurons in a hidden layer and output neurons in a linear layer form the network structure; the best structure for multidimensional mapping problems.

The network was trained with only the maximums of the normalized values. Training efficiency was improved with these values within the [0, 1] range.

Improvements to the generalization performance of the network corrected overfitting through the use of a pair of data sets: a training set that, if undesired events took place, updated weights and offsets and a validation set that could stop the training.

The final network configuration (total neurons in the hidden layer) was defined by the lowest mean squared error of the validation set.

A hidden layer with 11, 12, and 13 neurons, with 10% of the tabular data assigned to the validation set, was first used to train the multilayer perceptrons. The following configurations had the lowest error values: MLP 3-11-4, the MLP 3-12-4, and MLP 3-14-4. These are shown in Figures 2–4, respectively.

Analysis of the graphical functions presented in Figures 2–4 showed that the MLP 3-12-4 configuration had the lowest error, 0.002%, in the validation set. The coefficient of determination of the model with respect to criterion f was 0.986, which reflects its high accuracy in predicting surface roughness ($\pm$1.35%). The same model appeared to be the best at generalization performance in the cases of assigning 5% or 15% in the validation set of tabular data, Figure 5. In the case of assigning 5% of the training set, the error was 0.004% (see Figure 5b), and in the case of assigning 15%, it was 0.027% (see Figure 5c).

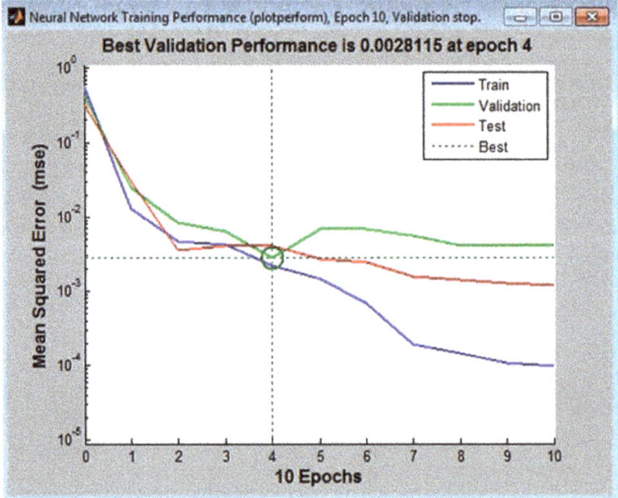

Figure 2. The lowest mean squared error for the validationset in the multi-layer perceptron (MLP) 3-11-4 configuration (calculated in Matlab).

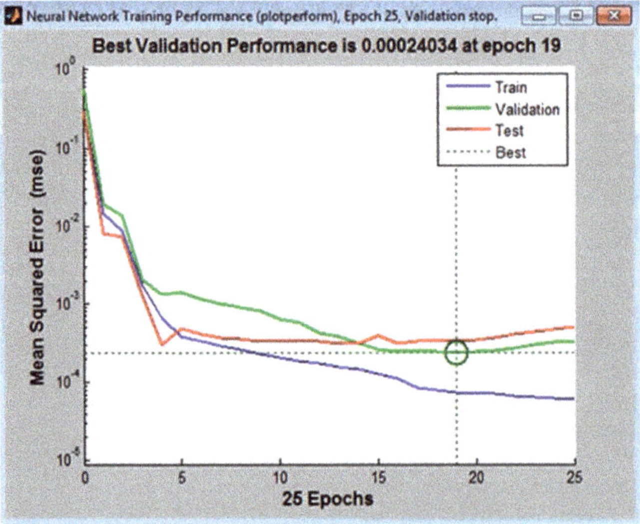

Figure 3. The lowest mean squared error for the validationset in the MLP 3-12-4 configuration (calculated in Matlab).

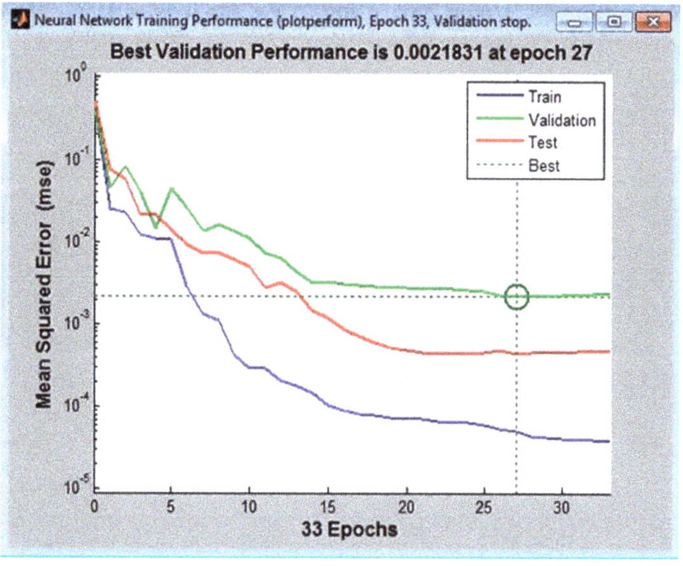

Figure 4. The lowest mean squared error for the validation set in the MLP 3-13-4 configuration (calculated in Matlab).

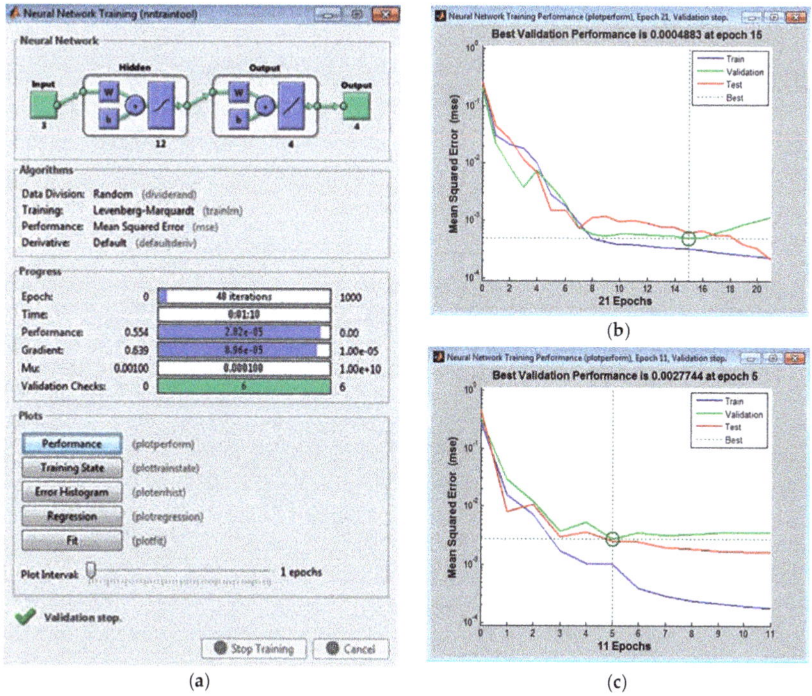

Figure 5. The lowest mean squared error in generalizing experimental data in MLP 3-12-4 (**a**) with various validation sets: (**b**) 5%; (**c**) 15% (calculated in Matlab).

6. Graphical Representation of the Surface of Vector Estimates (D)

The first step was to build the surface of vector estimates, D. The MLP 3-12-4 model and the experimental values of x_1, x_2, and x_3 were used to calculate f_1, f_2 and f_3.

A projection (a wafer map of Ra^* values) of non-linear surface D was plotted on the plane $f_2 f_3$ ($Tm^* C^*$) and is shown Figure 6.

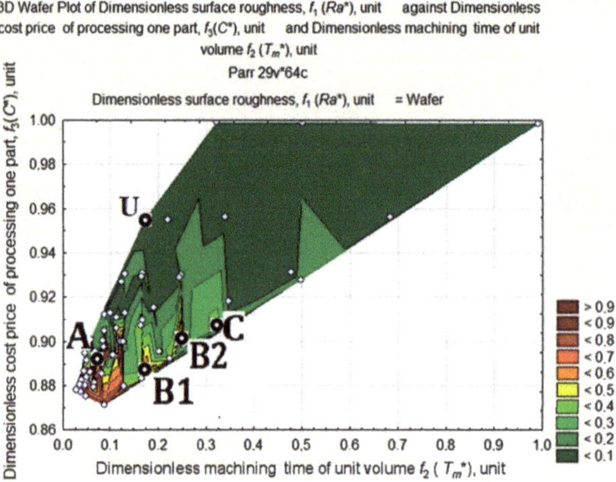

Figure 6. Wafer map of workpiece Ra^* after machining with respect to changes in values of T_m^* and C^*.

An analysis of the wafer map of Ra^* shows elements, forms, and complexes that can all be located. We can clearly see the apex A in the area of minimal values of C^* and T_m^* and the quasi-horizontal plane with a dent U in the opposite area of C^* and T_m^* values. We can also clearly see apexes B1, B2, and C with slopes towards the dent.

A closer look at the apexes on the map can be seen in detail in Figure 7.

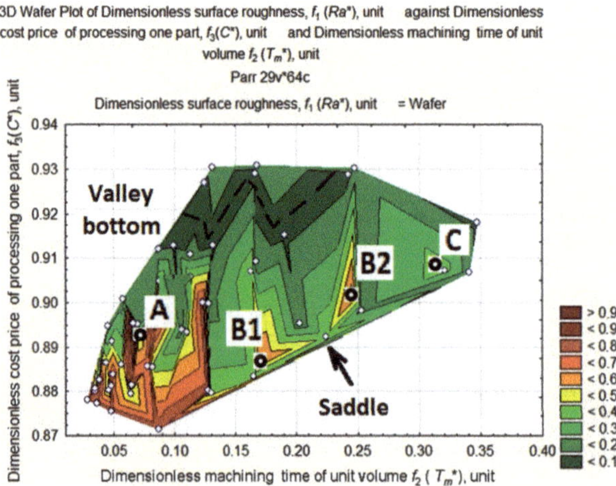

Figure 7. Apexes of the wafer map of Ra^* values of the machined workpiece depending on the change in values of T_m^* and C^*.

In Figure 7, we can see that the highest apex A (0.0667; 0.8953; 0.9868) appears as a ridge, B1 (0.1712; 0.8873; 0.5861) and B2 (0.2481; 0.9014; 0.6630) as mountains, and the lowest apex C (0.3154; 0.9082; 0.3721) appears to be a hill. B1 and B2 are similar in shape and form something like a mountain system with a saddle. A thalweg is plotted in the valley hollow.

7. Establishment of a Pareto Frontier

The target function is represented by the length of the vector in a normalized space and connects the origin of the coordinates with the point of the three-dimensional surface of estimates. Our aim is to identify the smallest of its lengths at the foot of the Ra^* ridge in the area with the smallest values of C^* and T_m^* (see Figure 8). For this purpose, we shall consider surface projection estimates at fixed depths of cut—$a_p = 1$ mm, $a_p = 0.75$ mm, $a_p = 0.5$ mm, and $a_p = 0.25$ mm (Figures 8–11).

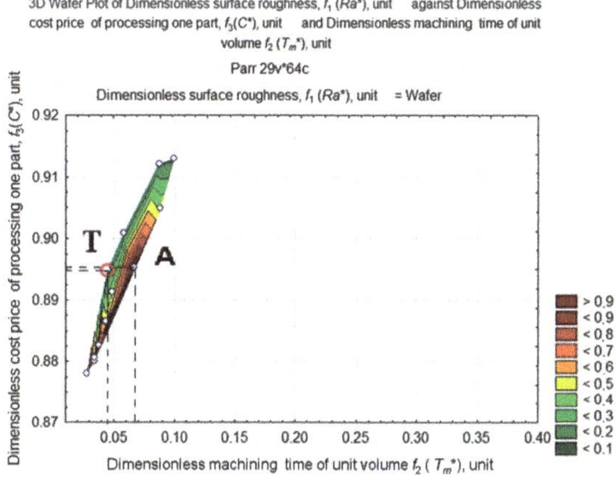

Figure 8. Surface projection of Ra^* values depending on the change in the values of T_m^* and C^* at fixed depth of cut, $a_p = 1$ mm.

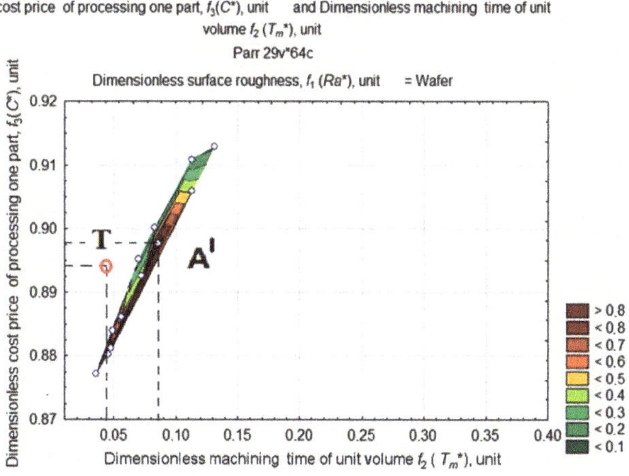

Figure 9. Surface projection of Ra^* values dependent on the change in the values of T_m^* and C^* at a fixed depth of cut, $a_p = 0.75$ mm.

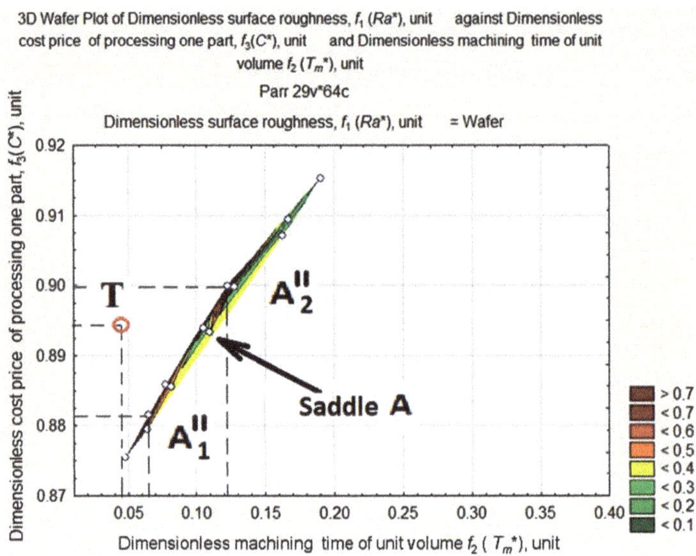

Figure 10. Surface projection of *Ra** values depending on the change in the values of $T_m{}^*$ and C^* at a fixed depth of cut, a_p = 0.5 mm.

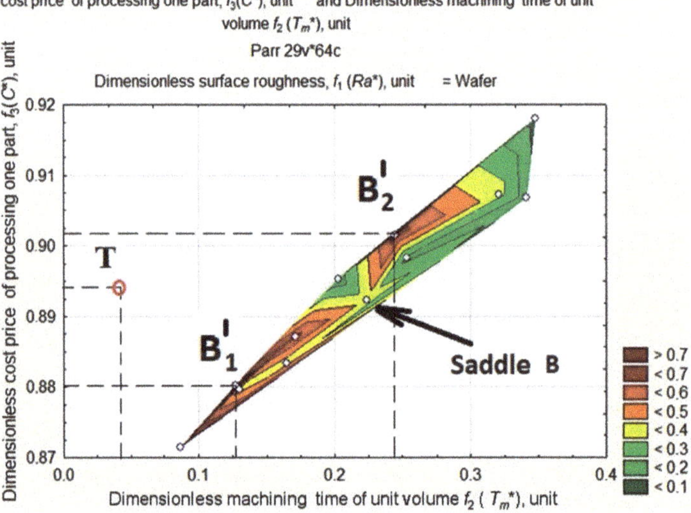

Figure 11. Surface projection of *Ra** values depending on the change in the values of $T_m{}^*$ and C at a fixed depth of cut, a_p = 0.25 mm.

The above figures (see Figures 8–11) illustrate how the area of maximum values of dimensionless roughness is transformed as the depth of cut decreases from a_p = 1.0 to a_p = 0.25 mm. It changes from a merged apex A and A′ to a maximum with two peaks in A1″, A″ with saddle A and B1′, and B2′ with saddle B. The largest displacement of the peaks occurs at the $T_m{}^*$ coordinate with the distance between them increasing. The ridge (see Figure 10) is formed by peaks A, A′, A1″ and A2″, the mountain system, by peaks B1′and B2′, and the hill, with the long slope of peak B2′ and the dent, U.

As a result of analyzing an MLP 3-12-4 neural network model with the coefficient of determination, R^2 = 0.986 (accuracy ±1.35%), a pattern was revealed for the AZ61 alloy. When the other two parameters were fixed and the cutting speed, v_c^*, was increased by 0.1 units, the values of Ra^*, T_m^*, C^*, and f^* decreased by 0.001 units, 0.007 units, 0.003 units, and 0.005 units, respectively. When the other two parameters were fixed and the depth of cut, a_p^*, was increased by 0.1 units, the values of Ra^*, T_m^*, C^*, and f^* decreased by 0.007 units, 0.010 units, 0.0001 units, and 0.005 units, respectively. When the other two parameters were fixed and the feed rate, f_r^*, was increased by 0.1 units, T_m^*, and C^* decreased by 0.004 and 0.006 units, respectively, while Ra^* and f^* increased by 0.103 and 0.015 units, respectively. Compared to v_c, surface roughness (Ra) was 7.3 times more affected by a_p and 102.9 times more affected by f_r; machining time (T_m) was 1.5 times less affected by a_p and 1.5 times more affected by fr; cost of production (C) was 26 times less affected by a_p and 1.9 times more affected by f_r; the integrated optimization criterion (f) was 1.04 times more affected by a_p and 2.9 times more affected by f_r. Hence, we should look for the optimal cutting conditions at the maximum cutting speed and depth of cut and the minimum feed rates.

The optimum must be located at the foot of the ridge in the area of high-speed turning conditions that are most likely to lead to maximum tool wear. According to Figures 8–11, we limit ourselves to the cutting conditions at the depth of cut a_p = 1 mm, as in this case all values of Ra^* are located near the minimum vector estimation F (0.0449; 0.8948; 0.1253), which is marked by point T.

At this depth of cut, a_p = 1 mm, four graphic dependencies, $Ra^* = f(C^*, T_m^*)$, were constructed, corresponding to the fixed v_c = 250 m/min, v_c = 200 m/min, v_c = 150 m/min, v_c = 100 m/min, and variable, f_r, value. After matching the obtained curves with the projection (see Figure 8), we obtained the seven reference points of the Pareto frontier. They are shown in Figure 12: p_1 (0.0281; 0.8782; 0.8135); p_2 (0.0343; 0.8824; 0.8273); p_3 (0.0449; 0.8948; 0.1253); p_4 (0.0561; 0.9012; 0.1072); p_5 (0.0860; 0.9123; 0.0982); p_6 (0.1710; 0.9547; 0.0514); p_7 (0.2500; 1.0000; 0.7903).

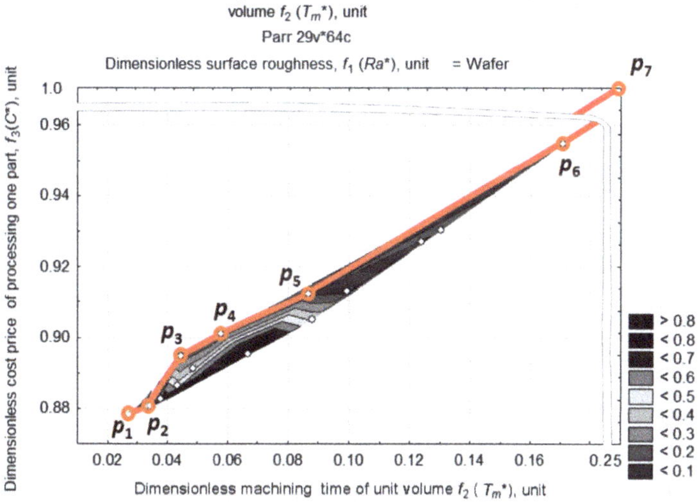

Figure 12. Pareto frontier and seven reference points: p_1 (0.0281; 0.8782; 0.8135); p_2 (0.0343; 0.8824; 0.8273); p_3 (0.0449; 0.8948; 0.1253); p_4 (0.0561; 0.9012; 0.1072); p_5 (0.0860; 0.9123; 0.0982); p_6 (0.1710; 0.9547; 0.0514); p_7 (0.2500; 1.0000; 0.7903).

Six sections are visible on the Pareto frontier.

Section 1 between point p_1 and point p_2, corresponds to v_c = 250–200 m/min, a_p = 1.00 mm, f_r = 0.16 mm/rev. Section 2 between point p_2 and point p_3 corresponds to v_c = 250 m/min, a_p = 1.00 mm, f_r = 0.16–0.08 mm/rev. Section 3 between point p_3 and point p_4 corresponds to v_c = 250–200 m/min, a_p = 1.00 mm, f_r = 0.08 mm/rev. Section 4 between points p_4 and p_5 corresponds to v_c = 200–150 m/min, a_p = 1.00 mm, f_r = 0.08 mm/rev. Section 5 between point p_5 and point p_6 corresponds to v_c = 250 m/min, a_p = 1.00 mm, f_r = 0.12–0.16 mm/rev. Section 6 between point p_6 and point p_7 corresponds to v_c = 150–100 m/min, a_p = 1.00 mm, f_r = 0.16 mm/rev. p_3 and p_6 are special points on the Pareto curve. These points correspond to absolute minimums; p_3 is the absolute minimum of the length of vector f; and p_6 is the absolute minimum of surface roughness.

8. The Optimum Settings

Finally, the procedure to adapt the system requires us to establish the optimum settings. The Pareto optimal decisions have to be narrowed down to a set of Pareto non-dominated decisions. Expert assessment defined the lesser importance of the dimensionless criterion of surface roughness, compared to the unit-volume machining time, Tm^*, and the processing cost price, C_i. In consequence, all vectors located above the blue vector that has the lowest f vector, plotted on the $f_3 f_2$ plane, at an angle of 7.89°, represent the Pareto non-dominated estimates (Figure 13). Point 3 on the Pareto curve coincides with the end point of this vector that was the global minimum in the case of unconditional optimization with the ranking $f_1{:}f_2{:}f_3$ = 1.0:19.9:2.7. Using the real coordinates, the global minimum corresponds to T_m = 0.358 min/cm^3, C = \$8.2973, R_a = 0.087 µm, v_c = 250 rpm, a_p = 1.0 mm, and f_r = 0.08 m/min.

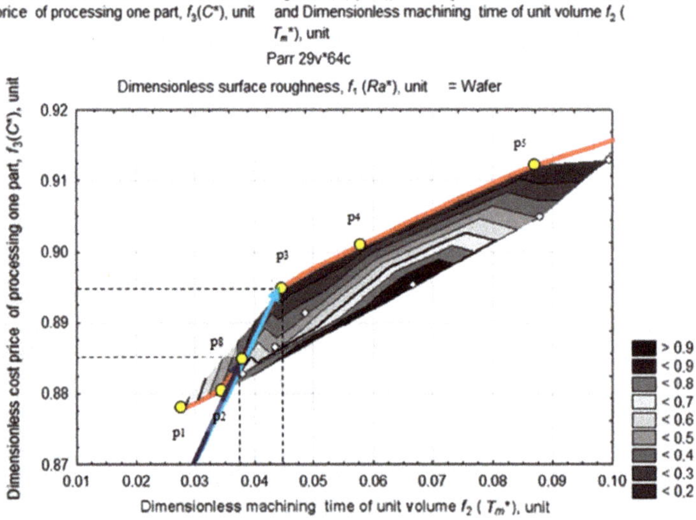

Figure 13. Global and local minimums on the Pareto curve.

Further restrictions were imposed in the form of design documentation requirements and the minimum acceptable surface roughness value was defined at 0.800 µm, which is point p_8 on the Pareto frontier curve, having the coordinates (0.0372, 0.8851, 0.300) in Figure 13. In this case (see Figure 13, blue estimates vector), the optimization criteria has a valid relation of importance that is–$Ra^*/T_m^*/C^*$

= 1.0/27.6/9.3, and the valid preference for points eight and three becomes $y_8 >_Y y_3$ with an induced preference of $x_8 >_X x_3$.

Therefore, the selected estimates as a set, Sel Y, was restricted to the blue vector, the actual end coordinates of which were 0.0372, 0.8851, 0.300, and the set of selected decisions, Sel X, was restricted to the three-dimensional vector of the optimum cutting parameters (v_c = 248 rpm, a_p = 1.0 mm, f_r = 0.10 mm/min).

The optical microscopy results and the profile of surface roughness graphs for the global optimum (cutting speed v_c = 250 m/min, depth of cut a_p = 1.0 mm, and feed rate f_r = 0.08 mm/rev) are presented in Figures 14 and 15.

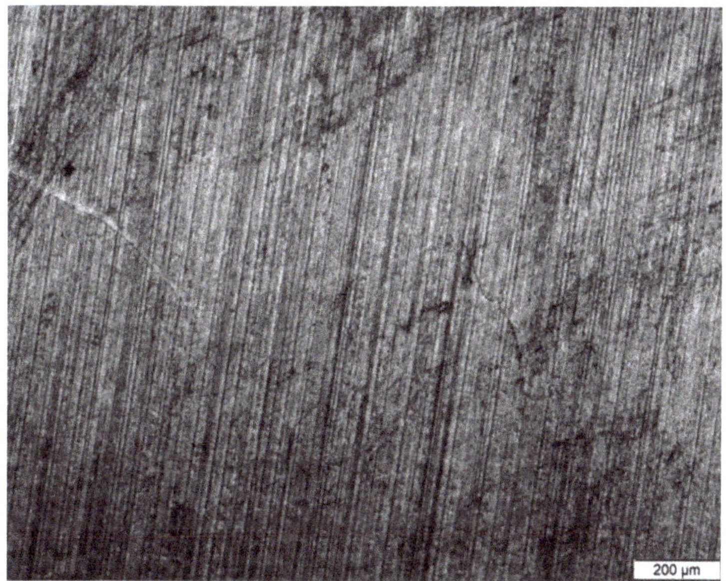

Figure 14. Optical microscopy results for the optimal machining parameters.

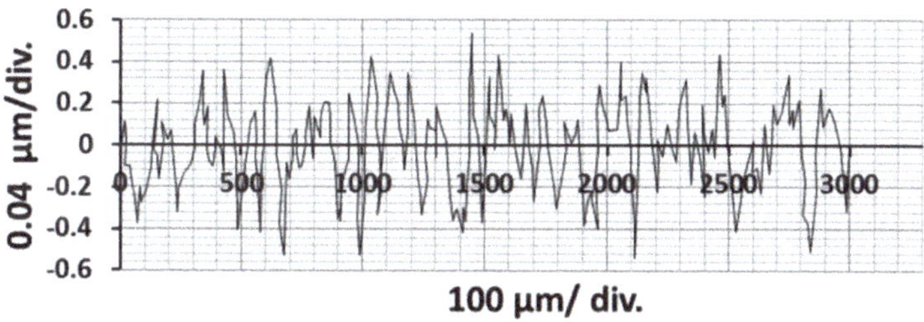

Figure 15. Profile of surface roughness graph from the surface roughness tester for the optimal machining parameters.

All in all, note that the Pareto curve, the correct values of its points, and the vector coordinates were all automatically calculated with a customized function in Matlab based on a neural network model. Implementation of the strategy in Matlab permits rapid calculation of the cutting tool–workpiece system and its local optimums in computer-aided production systems over the entire range of speeds and cutting depths.

9. Conclusions

(1) For the first time in the turning of magnesium alloy, the Edgeworth–Pareto methodology has been used for adapting the cutting tool–workpiece system to the state of the minimal value of the three-dimensional estimates of vector f in a normalized space: $f_1 f_2 f_3$ using an artificial intelligence-based model.

(2) An artificial neural network has been created in the Matlab programming environment based on an MLP 4-12-3 multi-layer perceptron that predicts the values of f_1, f_2, f_3, f in the finishing turning of the AZ61 magnesium alloy workpiece with a width of X mm, a length of X mm, and a height of X mm, at a cutting speed of 100–250 m/min, a depth of cut from 0.25 to 1.0 mm, and a feed rate of 50–150 mm/rev with an accuracy of $\pm 1.35\%$.

(3) According to the neural network model for the AZ61 alloy in finish turning, the value of the integrated optimization criterion, f, has mainly been influenced by feed rate, f_r. Vector f is 2.9 times more influenced by feed rate than by cutting speed and depth of cut. Increasing the feed rate led to an increase in f, and increasing v_c and a_p led to a decrease in f.

(4) For the first time, an AZ61 magnesium alloy workpiece wafer plot of surface roughness after finishing turning has been generated at cutting speeds of 100–250 m/min, at a depth of cut from 0.25–1.0 mm, and at a feed rate of 50–150 mm/rev.

(5) The global optimum in the finish turning of the alloy workpiece has been set as follows: the minimum length of 3D vector estimates with the coordinates $Ra = 0.087$ μm, $T_m = 0.358$ min/cm^3, and $C = \$8.2973$ corresponded to the following optimum conditions of finishing turning: cutting speed $v_c = 250$ m/min, depth of cut $a_p = 1.0$ mm, and feed rate $f_r = 0.08$ mm/rev.

(6) Automated calculation with the Industry 4.0 Framework has been performed in the Matlab environment, to define the optimal turning conditions for magnesium alloy workpieces as products of intelligent computer-aided manufacturing systems.

Author Contributions: For research articles with several authors, a short paragraph specifying their individual contributions must be provided. The following statements should be used "Conceptualization, A.T.A., D.Y.P. and I.N.E.; Methodology, A.T.A., D.Y.P. and I.N.E.; Software, I.N.E.; Validation, A.T.A., D.Y.P. and I.N.E.; Formal Analysis, A.T.A., D.Y.P. and I.N.E.; Investigation, A.T.A., M.A.T., M.S.S. and M.M.E.R.; Resources, A.T.A., M.S.S. and M.M.E.R.; Data Curation, A.T.A., D.Y.P. and I.N.E.; Writing-Original Draft Preparation, A.T.A., D.Y.P. and I.N.E.; Writing-Review & Editing, A.T.A., D.Y.P. and I.N.E.; Visualization, A.T.A., D.Y.P. and I.N.E.; Supervision, A.T.A., D.Y.P. and I.N.E.; Project Administration, A.T.A.; Funding Acquisition, A.T.A.", please turn to the CRediT taxonomy for the term explanation. Authorship must be limited to those who have contributed substantially to the work reported.

Acknowledgments: The authors extend their appreciation to the Deanship of Scientific Research at King Saud University for funding this work through research group No (RG-1439-020). The research was also supported through Act 211 Government of the Russian Federation, contract Nr 02.A03.21.0011.

References

1. López De Lacalle, L.N.; Pérez-Bilbatua, J.; Sánchez, J.A.; Llorente, J.I.; Gutiérrez, A.; Albóniga, J. Using high pressure coolant in the drilling and turning of low machinability alloys. *Int. J. Adv. Manuf. Technol.* **2000**, *16*, 85–91. [CrossRef]

2. Jurkovic, Z.; Cukor, G.; Andrejcak, I. Improving the surface roughness at longitudinal turning using the different optimization methods. *Teh. Vjesn.* **2010**, *17*, 397–402.

3. Fernández-Abia, A.I.; García, J.B.; López de Lacalle, L.N. High-performance machining of austenitic stainless steels (Book Chapter). *Mach. Mach. Tools Res. Dev.* **2013**, 29–90. [CrossRef]

4. Józwik, J.; Mika, D. Diagnostics of workpiece surface condition based on cutting tool vibrations during machining. *Adv. Sci. Technol. Res. J.* **2015**, *26*, 57–65. [CrossRef]

5. Pereira, O.; Rodríguez, A.; Fernández-Valdivielso, A.; Barreiro, J.; Fernández-Abia, A.I.; López-De-Lacalle, L.N. Cryogenic Hard Turning of ASP23 Steel Using Carbon Dioxide. *Proced. Eng.* **2015**, *132*, 486–491. [CrossRef]

6. Feldshtein, E.; Józwik, J.; Legutko, S. The influence of the conditions of emulsion mist formation on the surface roughness of AISI 1045 steel after finish turning. *Adv. Sci. Technol. Res. J.* **2016**, *30*, 144–149. [CrossRef]

7. Nath, C.; Kapoor, S.G.; Srivastava, A.K. Finish turning of Ti-6Al-4V with the atomization-based cutting fluid (ACF) spray system. *J. Manuf. Process.* **2017**, *28*, 464–471. [CrossRef]

8. Karkalos, N.E.; Galanis, N.I.; Markopoulos, A.P. Surface roughness prediction for the milling of Ti-6Al-4V ELI alloy with the use of statistical and soft computing techniques. *Measurement* **2016**, *90*, 25–35. [CrossRef]

9. Arnaiz-González, Á.; Fernández-Valdivielso, A.; Bustillo, A.; López de Lacalle, L.N. Using artificial neural networks for the prediction of dimensional error on inclined surfaces manufactured by ball-end milling. *Int. J. Adv. Manuf. Technol.* **2016**, *83*, 847–859. [CrossRef]

10. Selaimia, A.-A.; Yallese, M.A.; Bensouilah, H.; Meddour, I.; Khattabi, R.; Mabrouki, T. Modeling and optimization in dry face milling of X2CrNi18-9 austenitic stainless steel using RMS and desirability approach. *Measurement* **2017**, *107*, 53–67. [CrossRef]

11. Pimenov, D.Y.; Bustillo, A.; Mikolajczyk, T. Artificial intelligence for automatic prediction of required surface roughness by monitoring wear on face mill teeth. *J. Intell. Manuf.* **2018**, *29*, 1045–1061. [CrossRef]

12. Sutowski, P.; Nadolny, K.; Kaplonek, W. Monitoring of cylindrical grinding processes by use of a non-contact AE system. *Int. J. Precis. Eng. Manuf.* **2012**, *13*, 1737–1743. [CrossRef]

13. Novák, M.; Náprstková, N.; Józwik, J. Analysis of the surface profile and its material share during the grinding Inconel 718 alloy. *Adv. Sci. Technol. Res. J.* **2015**, *26*, 41–48. [CrossRef]

14. Tahreen, N.; Chen, D.L.; Nouri, M.; Li, D.Y. Influence of aluminum content on twinning and texture development of cast Mg-Al-Zn alloy during compression. *J. Alloys Compd.* **2015**, *623*, 15–23. [CrossRef]

15. Olguín-González, M.L.; Hernández-Silva, D.; García-Bernal, M.A.; Sauce-Rangel, V.M. Hot deformation behavior of hot-rolled AZ31 and AZ61 magnesium alloys. *Mater. Sci. Eng. A* **2014**, *597*, 82–88. [CrossRef]

16. Tsao, L.C.; Huang, Y.-T.; Fan, K.-H. Flow stress behavior of AZ61 magnesium alloy during hot compression deformation. *Mater. Des.* **2014**, *53*, 865–869. [CrossRef]

17. Watari, H.; Haga, T.; Koga, N.; Davey, K. Feasibility study of twin roll casting process for magnesium alloys. *J. Mater. Process. Technol.* **2007**, *192–193*, 300–305. [CrossRef]

18. Zhang, Z.; Wang, Z.; Yin, S.; Bao, L.; Chen, B.; Le, Q.; Cui, J. Effect of compound physical field on microstructures of semi-continuous cast AZ61 magnesium alloy billets. *Xiyou Jinshu Cailiao Yu Gongcheng/Rare Met. Mater. Eng.* **2016**, *45*, 2385–2390.

19. Chen, T.; Xie, Z.-W.; Luo, Z.-Z.; Yang, Q.; Tan, S.; Wang, Y.-J.; Luo, Y.-M. Microstructure evolution and tensile mechanical properties of thixoformed AZ61 magnesium alloy prepared by squeeze casting. *Trans. Nonferrous Met. Soc. China* **2014**, *24*, 3421–3428. [CrossRef]

20. Mustea, G.; Brabie, G. Influence of cutting parameters on the surface quality of round parts made from AZ61 magnesium alloy and machined by turning. *Adv. Mater. Res.* **2014**, *837*, 128–134. [CrossRef]

21. Santos, M.C.; Machado, A.R.; Sales, W.F.; Barrozo, M.A.S.; Ezugwu, E.O. Machining of aluminum alloys: A review. *Int. J. Adv. Manuf. Technol.* **2016**, *86*, 3067–3080. [CrossRef]

22. Risbood, K.A.; Dixit, U.S.; Sahasrabudhe, A.D. Prediction of surface roughness and dimensional deviation by measuring cutting forces and vibrations in turning process. *J. Mater. Process. Technol.* **2003**, *132*, 203–214. [CrossRef]

23. Özel, T.; Karpat, Y. Predictive modeling of surface roughness and tool wear in hard turning using regression and neural networks. *Int. J. Mach. Tools Manuf.* **2005**, *45*, 467–479. [CrossRef]

24. Bajić, D.; Lela, B.; Cukor, G. Examination and modelling of the influence of cutting parameters on the cutting force and the surface roughness in longitudinal turning. *Stroj. Vestn. J. Mech. Eng.* **2008**, *54*, 322–333.

25. Muthukrishnan, N.; Davim, J.P. Optimization of machining parameters of Al/SiC-MMC with ANOVA and ANN analysis. *J. Mater. Process. Technol.* **2009**, *209*, 225–232. [CrossRef]

26. Natarajan, C.; Muthu, S.; Karuppuswamy, P. Prediction and analysis of surface roughness characteristics of a non-ferrous material using ANN in CNC turning. *Int. J. Adv. Manuf. Technol.* **2011**, *57*, 1043–1051. [CrossRef]

27. Svalina, I.; Sabo, K.; Šimunović, G. Machined surface quality prediction models based on moving least squares and moving least absolute deviations methods. *Int. J. Adv. Manuf. Technol.* **2011**, *57*, 1099–1106. [CrossRef]

28. Pontes, F.J.; Paiva, A.P.D.; Balestrassi, P.P.; Ferreira, J.R.; Silva, M.B.D. Optimization of Radial Basis Function neural network employed for prediction of surface roughness in hard turning process using Taguchi's orthogonal arrays. *Expert Syst. Appl.* **2012**, *39*, 7776–7787. [CrossRef]

29. Hessainia, Z.; Belbah, A.; Yallese, M.A.; Mabrouki, T.; Rigal, J.-F. On the prediction of surface roughness in the hard turning based on cutting parameters and tool vibrations. *Measurement* **2013**, *46*, 1671–1681. [CrossRef]
30. Krolczyk, G.; Raos, P.; Legutko, S. Experimental analysis of surface roughness and surface texture of machined and fused deposition modelled parts | [Eksperimentalna analiza površinske hrapavosti i teksture tokarenih i taložno očvrsnutih proizvoda]. *Teh. Vjesn.* **2014**, *21*, 217–221.
31. Nieslony, P.; Krolczyk, G.M.; Wojciechowski, S.; Chudy, R.; Zak, K.; Maruda, R.W. Surface quality and topographic inspection of variable compliance part after precise turning. *Appl. Surf. Sci.* **2018**, *434*, 91–101. [CrossRef]
32. Acayaba, G.M.A.; Escalona, P.M.D. Prediction of surface roughness in low speed turning of AISI316 austenitic stainless steel. *CIRP J. Manuf. Sci. Technol.* **2015**, *11*, 62–67. [CrossRef]
33. D'Addona, D.M.; Raykar, S.J. Analysis of Surface Roughness in Hard Turning Using Wiper Insert Geometry. *Proced. CIRP* **2016**, *41*, 841–846. [CrossRef]
34. Mia, M.; Dhar, N.R. Prediction of surface roughness in hard turning under high pressure coolant using Artificial Neural Network. *Measurement* **2016**, *92*, 464–474. [CrossRef]
35. Jurkovic, Z.; Cukor, G.; Brezocnik, M.; Brajkovic, T. A comparison of machine learning methods for cutting parameters prediction in high speed turning process. *J. Intell. Manuf.* **2016**. [CrossRef]
36. Tootooni, M.S.; Liu, C.; Roberson, D.; Donovan, R.; Rao, P.K.; Kong, Z.J.; Bukkapatnam, S.T.S. Online non-contact surface finish measurement in machining using graph theory-based image analysis. *J. Manuf. Syst.* **2016**, *41*, 266–276. [CrossRef]
37. Mia, M.; Khan, M.A.; Dhar, N.R. Performance prediction of high-pressure coolant assisted turning of Ti-6Al-4V. *Int. J. Adv. Manuf. Technol.* **2017**, *90*, 1433–1445. [CrossRef]
38. Mia, M.; Gupta, M.; Singh, G.; Krolczyk, G.; Pimenov, D.Y. An approach to cleaner production for machining hardened steel using different cooling-lubrication conditions. *J. Clean. Prod.* **2018**, *187*, 1069–1081. [CrossRef]
39. Al-Ahmari, A.M.A. Prediction and optimisation models for turning operations. *Int. J. Prod. Res.* **2008**, *46*, 4061–4081. [CrossRef]
40. Jafarian, F.; Taghipour, M.; Amirabadi, H. Application of artificial neural network and optimization algorithms for optimizing surface roughness, tool life and cutting forces in turning operation. *J. Mech. Sci. Technol.* **2013**, *27*, 1469–1477. [CrossRef]
41. Mokhtari Homami, R.; Fadaei Tehrani, A.; Mirzadeh, H.; Movahedi, B.; Azimifar, F. Optimization of turning process using artificial intelligence technology. *Int. J. Adv. Manuf. Technol.* **2014**, *70*, 1205–1217. [CrossRef]
42. Sangwan, K.S.; Saxena, S.; Kant, G. Optimization of machining parameters to minimize surface roughness using integrated ANN-GA approach. *Procedia CIRP* **2015**, *29*, 305–310. [CrossRef]
43. Gupta, A.K.; Guntuku, S.C.; Desu, R.K.; Balu, A. Optimisation of turning parameters by integrating genetic algorithm with support vector regression and artificial neural networks. *Int. J. Adv. Manuf. Technol.* **2015**, *77*, 331–339. [CrossRef]
44. Venkata Rao, K.; Murthy, P.B.G.S.N. Modeling and optimization of tool vibration and surface roughness in boring of steel using RSM, ANN and SVM. *J. Intell. Manuf.* **2016**. [CrossRef]
45. Zerti, O.; Yallese, M.A.; Khettabi, R.; Chaoui, K.; Mabrouki, T. Design optimization for minimum technological parameters when dry turning of AISI D3 steel using Taguchi method. *Int. J. Adv. Manuf. Technol.* **2017**, *89*, 1915–1934. [CrossRef]
46. Mia, M.; Khan, M.A.; Rahman, S.S.; Dhar, N.R. Mono-objective and multi-objective optimization of performance parameters in high pressure coolant assisted turning of Ti-6Al-4V. *Int. J. Adv. Manuf. Technol.* **2017**, *90*, 109–118. [CrossRef]
47. Mia, M.; Dhar, N.R. Optimization of surface roughness and cutting temperature in high-pressure coolant-assisted hard turning using Taguchi method. *Int. J. Adv. Manuf. Technol.* **2017**, *88*, 739–753. [CrossRef]
48. Basak, S.; Dixit, U.S.; Davim, J.P. Application of radial basis function neural networks in optimization of hard turning of AISI D2 cold-worked tool steel with a ceramic tool. *Proc. Inst. Mech. Eng. B. J. Eng. Manuf.* **2017**, *221*, 987–998. [CrossRef]
49. Karpat, Y.; Özel, T. Multi-objective optimization for turning processes using neural network modeling and dynamic-neighborhood particle swarm optimization. *Int. J. Adv. Manuf. Technol.* **2007**, *35*, 234–247. [CrossRef]

50. Raykar, S.J.; D'Addona, D.M.; Mane, A.M. Multi-objective optimization of high speed turning of Al 7075 using grey relational analysis. *Procedia CIRP* **2015**, *33*, 293–298. [CrossRef]
51. Yue, C.; Wang, L.; Liu, J.; Hao, S. Multi-objective optimization of machined surface integrity for hard turning process. *Int. J. Smart Home* **2016**, *10*, 71–76. [CrossRef]
52. Abbas, A.T.; Pimenov, D.Y.; Erdakov, I.N.; Mikolajczyk, T.; El Danaf, E.A.; Taha, M.A. Minimization of turning time for high strength steel with a given surface roughness using the Edgeworth-Pareto optimization method. *Int. J. Adv. Manuf. Technol.* **2017**, *93*, 2375–2392. [CrossRef]
53. Kostenetskiy, P.S.; Safonov, A.Y. SUSU supercomputer resources. *CEUR Workshop Proc.* **2016**, *1576*, 561–573.

 *materials*

Article

Influence of the Regime of Electropulsing-Assisted Machining on the Plastic Deformation of the Layer Being Cut

Saqib Hameed [1], Hernán A. González Rojas [1], José I. Perat Benavides [2], Amelia Nápoles Alberro [1] and Antonio J. Sánchez Egea [3,4,*]

[1] Department of Mechanical Engineering (EPSEVG), Universitat Politécnica de Catalunya, Av. de Víctor Balaguer Vilanova i la Geltrú, 08800 Barcelona, Spain; hameeds@tcd.ie (S.H.); hernan.gonzalez@upc.edu (H.A.G.R.); amelia.napoles@upc.edu (A.N.A.)

[2] Department of Electrical Engineering (EPSEVG), GAECE group, Universitat Politécnica de Catalunya, Av. de Víctor Balaguer Vilanova i la Geltrú, 08800 Barcelona, Spain; iperat@ee.upc.edu

[3] Department of Mechanical and Metallurgical Engineering, Pontificia Universidad Catolica de Chile, Av. Vicuña Mackenna 4860, Region Metropolitana 7820436, Chile

[4] Department of Mechanical Engineering, Aeronautics Advanced Manufacturing Center (CFAA), Faculty of Engineering of Bilbao, Alameda de Urquijo s/n, 48013 Bilbao, Spain

* Correspondence: antonio.egea@ing.puc.cl

Received: 27 April 2018; Accepted: 23 May 2018; Published: 25 May 2018

Abstract: In this article, the influence of electropulsing on the machinability of steel S235 and aluminium 6060 has been studied during conventional and electropulsing-assisted turning processes. The machinability indices such as chip compression ratio ζ, shear plane angle ϕ and specific cutting energy (SCE) are investigated by using different cutting parameters such as cutting speed, cutting feed and depth of cut during electrically-assisted turning process. The results and analysis of this work indicated that the electrically-assisted turning process improves the machinability of steel S235, whereas the machinability of aluminium 6060 gets worse. Finally, due to electropluses (EPs), the chip compression ratio ζ increases with the increase in cutting speed during turning of aluminium 6060 and the SCE decreases during turning of steel S235.

Keywords: electropulsing; machinability; chip compression ratio; current density; specific cutting energy

1. Introduction

Studying the behaviour of metallic alloys during machining is very important in the manufacturing industry. The aspects of good machinability are usually low energy consumption, short chips, smooth surface finish and long tool life [1]. The high process efficiency can be achieved by minimizing the energy required to remove the material as plastically deformed chip [2]. The work done in the plastic deformation of the chip is influenced by the cutting speed through the chip temperature, dimensions of deformation zone adjacent to the cutting edge and velocity of deformation [3]. An increase in cutting speed tends to decrease of chip thickness and the region of plastic deformation becomes smaller which ultimately reduces the energy consumption [4]. It was generally observed that in metal cutting processes only 30–50% of energy is spent for useful work, while 25–60% of energy consumed during cutting is simply wasted [5]. Therefore, reducing the energy spent in metal cutting as much as possible by selecting properly the workpiece material, machining regime and process parameters is of great importance. Generally, the energy consumption in metal cutting processes is a function of cutting speed, feed rates and workpiece material [6]. The SCE increases as the hardness and mechanical strength of cutting materials increase [7]. The high cutting speed reduces the tool life of the most difficult-to-cut materials [8]. During turning of medium carbon steel, the SCE decreased by increasing the feed rate and depth of cut [9]. Fernandez-Abia et al. [10] analyzed that high cutting speeds reduce the cutting forces, which implies less power consumption and the chip thickness is significantly reduced, which implies low chip compression ratio ζ and high shear plane angle ϕ. It was generally observed that chip compression ratio ζ is the measure of plastic deformation of material which decreases with the increase in cutting speed [3].

During machining of ductile materials such as aluminium, a large tool chip contact and high chip compression ratio ζ increases the cutting forces, machining power and heat generation [1]. Materials with more ductility have higher degree of plastic deformation and high temperature of chip than that of less ductility during cutting [11]. Yousef et al. [12] suggested that an excessive increase in deformation rate may increase the machining forces in aluminium alloys. Becze et al. [13] indicated that the adiabatic temperature rise presented considerable heating of shear band which will decrease the flow stress of material. When the material passes through the primary deformation zone, the plastic deformation starts and is sheared at a rapidly increasing strain rates until reaches its maximum value [14]. Lee et al. [15] studied the procedure of determining the large strain distribution in the primary deformation zone by using particle image velocimetry (PIV) technique. They examined that the shear strain rate varies linearly with the cutting speed and the shear zone is narrower for higher shear strain rates. Wang et al. [16] used similar image processing technology to investigate the thermal softening behaviours in several metals alloys electrically assisted under tensile test. They found the current density threshold to achieve an effective thermal softening for each metallic alloy. Previously, they also published electrically-assisted micro-tension results to describe the thermomechanical behaviour of the AZ31 magnesium alloy under different current densities and, consequently, different joule effects [17]. As a result, the thermal softening was attributed only to the contribution of the joule effect and the change of the material resistivity. Following this topic, Sanchez et al. [18] introduced a thermomechanical model to describe the flow stress in an electropulsing wire drawing process and showed the microstructure changes associated to an ultra-fast annealing treatment that promotes a detwinning mechanism. All these experimental studies used low strain rates where it is easy to induce a high amount of energy to affect the material misconstruction; on the contrary, for high strain rates at high cutting speed in machining, the energy required to complete the process is low [19]. Accordingly, Gui et al. [20] investigated that, due to the increase in cutting speed, the temperature rise in primary shear zone decreases the shear strength of material which ultimately reduces the cutting forces. They also validated that decreasing the uncut chip thickness decreases the temperature in the primary shear zone which increases the shear strength of material and hence increases the SCE. Bakkal et al. [21] demonstrated that high cutting speeds significantly reduce the SCE of materials with low thermal conductivity and high thermal softening effect.

Electropulsing, as an instantaneous high energy input method, is recognized as a novel technique in metal cutting processes. Baranov et al. [22] discovered the effect of current pulses on metal cutting area which reduces the static force of cutting and hence the plastic deformation of metals. Recently, Brandt et al. [23] examined that, due to electric current, the reduction in cutting forces is higher for higher strength materials. Sanchez et al. [24] observed that the SCE and the power consumption are decreased when the turning process is assisted with EPs. In a previous study [25], it was found that, at lower feeds, when the higher current density is induced, the shear angle ϕ decreases and the chip compression ratio ζ increases during drilling process. There was no study or few attempts were made to evaluate the machinability of materials with respect to chip compression ratio ζ, shear plane angle ϕ and SCE while cutting with EPs during turning process.

The main purpose of the present study was to investigate the influence of EPs on the plastic deformation and energy consumption of materials such as Steel S235 and aluminium 6060 during turning process. Since the chip compression ratio ζ can be considered as the true measure of plastic deformation in metal cutting [3], and SCE can be used to compare the machinability of different materials, an experimental investigation was performed to measure the chip compression ratio ζ and

SCE by using different cutting parameters to understand the behaviour of metallic materials when machined with and without pulses.

2. Experimental Setup

The turning process was carried out by using WEISS WMP280V-F round turning machine (ORPI SL, Zaragoza, Spain). A carbide tool DIN4980-ISO6 P20 fitted with diamond ground carbide inserts, rake angle of 6° and nose radius of 0.2 mm was used and held by standard tool holder to machine the metallic bars. Commercial steel alloys (S235) and aluminium alloys (Al 6060) of 20 mm diameter were chosen as workpiece materials for test specimens. The chemical composition of the studied metallic materials is shown in Table 1 and mechanical properties of both alloys are listed in Table 2.

Table 1. Chemical composition of studied materials.

Aluminium 6060	% Al	% Zn	% Mg	% Cu	% Fe	% Si	% Mn	% Cr
	balance	0.15	0.35	0.1	0.1	0.3	0.1	0.05
Steel S235	% C	% P	% Cu	% Fe	% Si	% Mn	% S	
	0.24	0.04	0.20	balance	0.04	0.9	0.05	

Table 2. Mechanical properties of studied materials.

Materials	ρ (Ωm)	C_p (J/Kg °C)	E (GPa)	α (°C^{-1})	D (Kg/m^3)	Hardness (95% Data Interval)
Aluminium 6060	3.2×10^{-8}	900	70	23×10^{-6}	2700	71.4 ± 1.8 (HB)
Steel S235	1.42×10^{-7}	470	190	12×10^{-6}	7800	62.8 ± 1.5 (HRB)

A polymeric material was used to electrically isolate the workpiece and tool holder from the lathe. The power consumed was continuously measured by a self made monophasic energy analyzer linked to the motor of the machine. A self made short duration electric pulse generator was developed to discharge multiple positive pulses. The machining parameters and the parameters of current pulses such as voltage, frequency and pulse duration, which were monitored by an osciloscope, are listed in Table 3.

Table 3. Turning and electropulsing operation parameters.

Material	Spindle Speed (rpm)	Cutting Feed (mm/rev)	Depth of Cut (mm)	Current Intensity (A)	Pulse Duration (µs)	Frequency (Hz)
Aluminium 6060 Steel S235	600/900	0.07/0.14	0.2/0.4	140	250	300

A schematic illustration of electrically assisted turning process is shown in Figure 1. The workpieces performed round turning process to make sure the surface was smooth and symmetric. The carbon clamps were attached with workpiece and connected on one side with cutting tool and on the other side with generator through wire. The turning operation was performed without EPs (conventional process). Subsequently, the same procedure was performed with EPs assisted process and it was confirmed that the workpiece was already in contact with the tool to avoid electro-erosion.

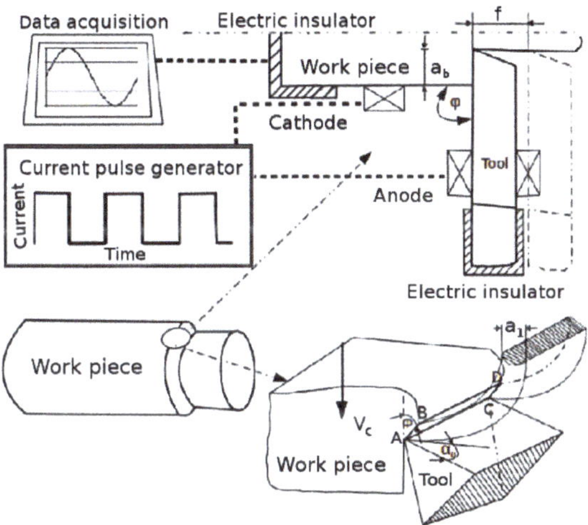

Figure 1. Schematic of electrically assisted turning process.

3. Results and Discussions

3.1. Current Density, Chip Thickness Ratio and Cutting Configurations

In machining, chip formation occurs by imparting shear within a narrow deformation zone called primary shear zone in which the effective strain rates are much larger [26]. The current density which is defined as current intensity through the cross sectional area of the material during cutting, can be considered as an important factor in changing the deformation resistance of material in primary shear zone. The shear cutting area is defined by the segment AB and the length of cutting edge BD (a_b) as shown in Figure 1. The segment AB is described as:

$$AB = \frac{f}{sin(\phi)} \tag{1}$$

where f is the cutting feed (mm per rev) and ϕ is the shear plane angle (°).

The current density (J_e) in the primary deformation zone can be calculated by the following equation:

$$J_e = \frac{I \cdot sin(\phi)}{f \cdot a_b} \tag{2}$$

where a_b is the depth of cut (mm) and I is the current intensity (A). The shear plane angle ϕ proposed in [27] can be expressed geometrically by chip compression ratio ζ as:

$$\phi = \arctan\left(\frac{\cos \alpha_0}{\zeta - \sin \alpha_0}\right) \tag{3}$$

Chip thickness coefficient is defined as a ratio $\zeta = a_1/a$, where a_1 is deformed chip thickness and a is undeformed chip thickness, which in this case is the same value as the cutting feed. α_0 is the rake angle which is 6° for the particular tool used during experiments. Chips were collected randomly at the end of each cutting test to measure the thickness of the chip. The chip thickness was measured by using micrometer with 0.01 mm precision. At least five measurements were taken to get the average chip thickness. The spindle velocity was measured in rpm with a conventional tachometer. Table 4 gives the values of current densities, chip compression ratio ζ and shear plane angle ϕ with and without pulses as a function of cutting feed, cutting speed and depth of cut for steel S235.

Table 4. Values of chip compression ratio, shear plane angle and current densities as a function of cutting feed and cutting speed for steel S235.

Material	Workpiece Diameter	Cutting Feed	Cutting Speed	Depth of Cut	Compression Ratio	Shear Plane Angle Density	Current Density
Steel		(f)	(V_c)	(a_b)	(ζ)	(ϕ)	(J_e)
S235	(mm)	(mm/rev)	(m/min)	(mm)		(°)	(A/mm^2)
Without pulses	20	0.07	38.213	0.2	6.357	9.049	-
		0.14	38.213	0.2	4.321	13.356	-
		0.07	57.461	0.2	3.685	15.663	-
		0.14	57.461	0.2	2.957	19.484	-
		0.07	38.213	0.4	5.257	10.964	-
		0.14	38.213	0.4	3.343	17.262	-
		0.07	57.461	0.4	4.085	14.123	-
		0.14	57.461	0.4	3.057	18.856	-
With pulses	20	0.07	38.213	0.2	3.100	18.599	3190
		0.14	38.213	0.2	2.492	23.007	1954
		0.07	57.461	0.2	3.171	18.186	3121
		0.14	57.461	0.2	2.557	22.449	1909
		0.07	38.213	0.4	4.214	13.697	1184
		0.14	38.213	0.4	4.228	13.651	590
		0.07	57.461	0.4	4.128	13.982	1208
		0.14	57.461	0.4	3.378	17.081	734

It can be seen in Table 4 that chip compression ratio ζ decreases with the increase in cutting feed and cutting speed during turning of steel S235. In addition, the shear plane angle ϕ increases as chip compression ratio ζ decreases. However, as compared to conventional turning process, the values of chip compression ratio ζ is lower during electrically assisted turning process. The low values of chip thickness ratio ζ (high ϕ values) mean low shear strain in the shear plane [28]. Since chip compression ratio ζ is a measure of the plastic deformation of materia, which decreases with the increase in cutting speed [3], an increase in cutting speed leads to a decrease of plastic deformation in chip formation zone. It was generally observed that the chip thickness decreased as the cutting speed increased and the region of plastic deformation becomes smaller which ultimately reduces the energy consumption [4].

The current density decreases with the increase in cutting feed during electrically assisted turning process of steel S235 as shown in Table 4. However, the results are in agreement with the previous study [25], in which current density decreases with the increase in cutting feed during electrically assisted drilling process. As the chip formation zone decreases with increase cutting speed, this may be the reason why current density values are higher at high cutting speeds for steel S235 during electrically assisted turning process. Table 5 gives the values of chip compression ratio ζ, shear plane angle ϕ and current densities with and without pulses as a function of cutting feed, cutting speed and depth of cut for aluminium 6060. Table 5 shows that chip compression ratio ζ decreases with increase in cutting feed and increases with the increase in cutting speed. In addition, shear plane angle ϕ increases with the increase in cutting feed and decreases with the increase in cutting speed. The high values of chip compression ratio ζ mean large amount of strain in shear plane [28]. An increase in the cutting speed leads to an increase in the temperature of the chip so its plastic deformation increases [3]. Hence, during turning of aluminium 6060, an increase in cutting speed tends to increase chip compression ratio ζ which indicates severe plastic deformation in the chip formation zone. It is also seen in Table 5 that, during electrically assisted turning process, the values of chip compression ratio ζ are higher and that of shear plane angle ϕ are lower than conventional turning process in aluminium 6060. In addition, the current density decreases with the increase in cutting speed while increases with the increase in cutting feed.

Table 5. Values of chip compression ratio, shear plane angle and current densities as a function of cutting feed and cutting speed for Aluminium 6060.

Material	Workpiece Diameter	Cutting Feed	Cutting Speed	Depth of Cut	Compression Ratio	Shear plane Angle	Current Density
Aluminium		(f)	(V_c)	(a_b)	(ξ)	(ϕ)	(J_e)
6060	(mm)	(mm/rev)	(m/min)	(mm)		(°)	(A/mm^2)
Without pulses	20	0.07	38.213	0.2	2.371	24.571	-
		0.14	38.213	0.2	1.214	45.312	-
		0.07	57.461	0.2	3.771	15.375	-
		0.14	57.461	0.2	2.214	26.287	-
		0.07	38.213	0.4	4.028	14.372	-
		0.14	38.213	0.4	2.085	27.866	-
		0.07	57.461	0.4	5.285	10.877	-
		0.14	57.461	0.4	2.971	19.600	-
With pulses	20	0.07	38.213	0.2	3.514	16.526	2844
		0.14	38.213	0.2	1.114	48.470	3743
		0.07	57.461	0.2	5.001	11.515	1996
		0.14	57.461	0.2	2.914	19.999	1709
		0.07	38.213	0.4	3.971	14.583	1259
		0.14	38.213	0.4	2.085	27.321	1147
		0.07	57.461	0.4	8.628	6.583	573
		0.14	57.461	0.4	2.914	19.999	854

Figure 2 demonstrates the variation of chip thickness with cutting feed, cutting speed and depth of cut for steel S235 by using factorial analysis. The chip thickness increases with the increase in cutting feed and decreases with the increase in cutting speed. The increase in undeformed chip thickness with increasing feed rates will result in increase of shear plane area [29]. The reduction in chip thickness will result in shorter shear plane and longer shear plane is associated with thicker chip produced during the cutting process [30]. However, depth of cut has very little effect on chip thickness during conventional turning process but chip thickness increases with the increase in depth of cut during EPs assisted turning process. Figure 3 shows that the cutting feeds have very little effects on chip thickness, however chip thickness increases with the increase in cutting speed and depth of cut during turning of aluminium 6060. As compared to conventional process, the chip thickness values are higher during electrically assisted turning process. When the chip thickness increases, the chip compression ratio ξ also increases which means plastic deformation of material increases.

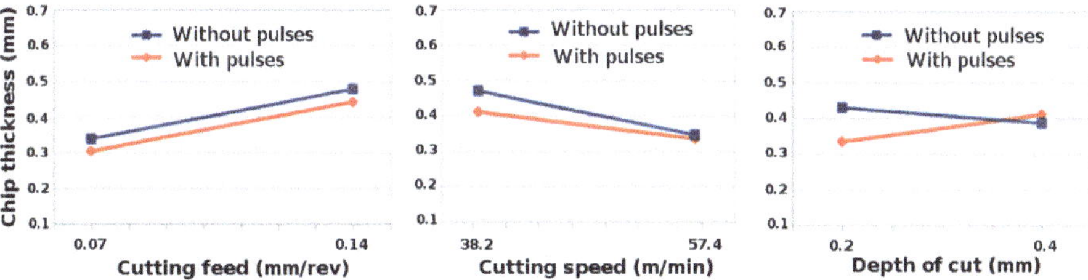

Figure 2. Variation of chip thickness with cutting feed, cutting speed and depth of cut while turning steel S235 with and without pulses.

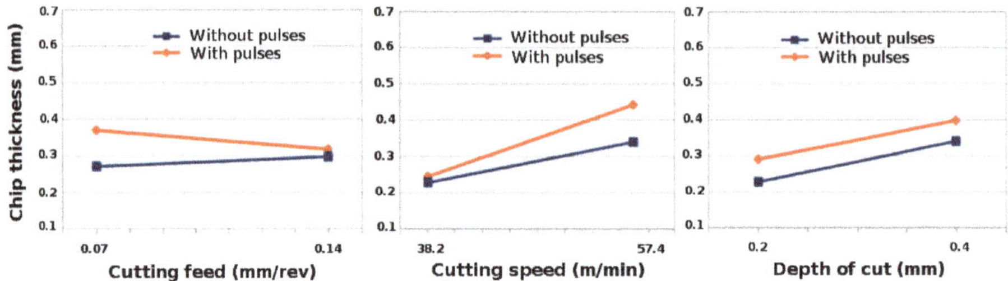

Figure 3. Variation of chip thickness with cutting feed, cutting speed and depth of cut while turning aluminium 6060 with and without pulses.

3.2. Specific Cutting Energy (SCE)

The machinability of materials can be estimated by comparing the SCE during EPs assisted and conventional turning processes. The SCE is the quotient between the cutting power consumption and the material removal rate as described in the previous work [25]. The material removal rate, Q_c, is defined by the following equation:

$$Q_c = V_c \cdot f \cdot a_b \left(1 - \frac{a_b}{D}\right) \tag{4}$$

where V_c is cutting speed (m/min) and D is the workpiece diameter (mm).

The mechanical power consumed was evaluated indirectly during turning by measuring the active electrical power consumed by motor of lathe Pa. The device that measured the active electrical power was constructed with an analog chip, multiplier of 4 quadrants AD633, which gives the product of supply voltage and current consumed. This signal was digitalized with digital analog converter. The mechanical power delivered by the motor multiplied by mechanical efficiency of turning machine allows obtaining cutting power. The mechanical efficiency was not known. The relationship between the active electrical power and mechanical power was also unknown. To obtain the relationship between the cutting power and active electrical power, an experiment was carried out in which a mechanical load was imposed by dynamometer brake and active electrical power was measured for different spindle speeds. With the data of experiment, a linear regression was obtained giving an expression that allowed to relate the active electrical power with mechanical power consumed by the spindle of lathe. Finally, the cutting power obtained as the difference between the mechanical power during cutting and mechanical power without load in conventional turning process was then compared with the cutting power obtained in electrically assisted turning process.

Figure 4 show that the SCE decreased with the increase in cutting feeds and depth of cut for steel S235 by using factorial analysis. It was observed that the effect of cutting speed on SCE is negligible for steel S235. However, the SCE decreased due to the application of EPs during cutting. The reduction in SCE was observed: 7% for high feeds and 14.6% for low feeds; 12.4% for high cutting speed and 10.5% for low cutting speed; and 10.5% for high depth of cut and 12.1% for low depth of cut. Hence, due to thermal contribution of EPs, the deformation resistance decreases [31], which ultimately decreases the SCE and improves the plasticity of material during EP assisted turning process. The results are also in agreement with Sánchez et al. [24] and Hameed [25] where the SCE decreases due to the application of EPs during machining processes.

Figure 5 presents that SCE decreased with the increase in cutting feed and depth of cut for aluminium 6060 by using factorial analysis. However, as compared to steel S235, a different trend was observed when the cutting speed increased the SCE also increased. It is also noted that SCE increased due to the application of EPs during cutting. Since aluminium 6060 has higher thermal conductivity and less thermal softening effect as compared to steel S235, at higher cutting speeds and due to the application of EPs, the strain rate in the shear zone is expected to be high. Thus, more heat energy will be generated and time for heat dissipation decreases, which ultimately increases the temperature [32]. In addition, an excess of electrons can increase the flow stress during superplastic deformation of aluminium alloy [33]. Hence, due to increase of temperature and flow stress, the plastic deformation of material increases which increases the SCE in aluminium 6060. Furthermore, it can be observed that, by increasing the cutting speed for steel S235, the chip thickness decreases and current density increases, while for aluminium 6060, by increasing the cutting speed, chip thickness increases and current density decreases, due to which SCE increases during turning of aluminium 6060 due to the application of EPs. It is assumed that area of deformation zone decreases by increasing the cutting speed for steel S235 and increases by increasing the cutting speed for aluminium 6060. Thus, it is expected that, as compared to steel S235, plastic deformation increases with the increase in cutting speed for aluminium 6060 because of increase of temperature in the chip formation zone.

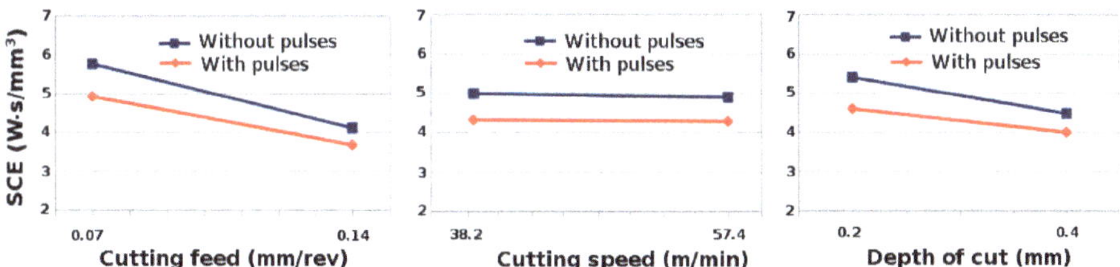

Figure 4. Variation of the SCE respect to the cutting feed, cutting speed and depth of cut while turning steel S235 with and without electropulsing.

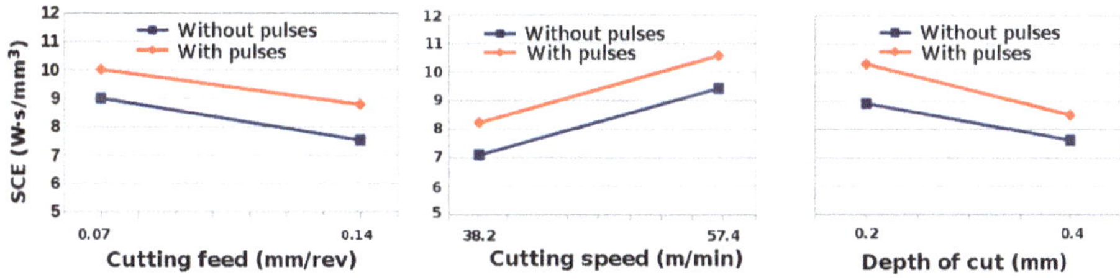

Figure 5. Variation of SCE with cutting feed, cutting speed and depth of cut while turning aluminium 6060 with and without pulses.

4. Conclusions

The effect of electropulsing was observed during turning of steel S235 and aluminium 6060. Correlation among chip compression ratio ζ, shear plane angle ϕ, current density, chip thickness and specific cutting energy was obtained by using different cutting parameters for both conventional and electrically assisted processes. The main findings can be summarized as following:

- By increasing the cutting speed, the chip compression ratio ζ decreases and shear plane angle ϕ increases during conventional turning of steel S235. However, increase of cutting speed seems to have no effect on chip compression ratio ζ and shear plane angle ϕ during electrically assisted turning of steel S235. In contrast, chip compression ratio ζ increases and shear plane angle ϕ decreases with the increase in cutting speed while turning of aluminium 6060 during conventional and electrically assisted processes.

- The current density increases with the increase in cutting speed and decreases with the increase in cutting feed for steel S235. However, the current density has high values at higher feeds and decreases with the increase in cutting speed for aluminium 6060.
- The SCE decreases with the increase in cutting speed and depth of cut for both conventional and electrically assisted turning of steel S235. In addition, as compared to conventional process, the SCE values are lower during electrically assisted turning of steel S235. However, during conventional turning of aluminium 6060, the SCE decreases with the increase in cutting feed and depth of cut, while increases with the increase of cutting speed. In addition, the SCE values are higher during electrically assisted turning of aluminium 6060 as compared to conventional process.
- The electrically assisted turning processes tends to have influence in improving the machinability of steel S235 but for aluminium 6060, the machinability seems to be poor, which is probably due to its high ductility and higher degree of plastic deformation as compared to steel S235. Further investigations are required to further analyze the effect of EPs on machinability of materials by studying the cutting parameters and chip morphology.

Author Contributions: Conceptualization: S.H. and H.A.G.R. Data curation: S.H., A.N.A. and A.J.S.E. Funding acquisition: J.I.P.B., A.J.S.E. and H.A.G.R. Methodology: S.H., A.N.A. and H.A.G. Validation: J.I.P.B. and A.N.A. Writing original draft: S.H. and H.A.G.R. Writing, review and editing: J.I.P.B., A.N.A. and A.J.S.E.

Acknowledgments: This work is supported by the Ministry of Economy and Competitiveness of Spain (reference project: FJCI-2016-29297) and the National Fund for Scientific and Technological Development of Chile (Fondecyt project 3180006).

Conflicts of Interest: All the authors who sign this manuscript do not have any conflict of interest to declare. Furthermore, the corresponding author certifies that this work has not been submitted to or published in any other journal.

References

1. Edward, M.T.; Paul, K.W. *Metal Cutting*, 4th ed.; Butterworth–Heinemann: Woburn, MA, USA, 2000; pp. 24–26, ISBN 0-7506-7069-X.
2. Astakhov, V.P. *Geometry of Single-Point Turning Tools and Drills: Fundamentals and Practical Applications*; Springer: London, UK, 2010; pp. 6–9, ISBN 978-1-84996-052-6.
3. Astakhov, V.P.; Shvets, S. The assessment of plastic deformation in metal cutting. *J. Mater. Process. Technol.* **2004**, *146*, 193–202. [CrossRef]
4. Thamizhmanii, S.; Hasan, S. Machinability study using chip thickness ratio on difficult to cut metals by CBN cutting tool. *Key Eng. Mater.* **2012**, *504–506*, 1317–1322. [CrossRef]
5. Astakhov, V.P. *Tribology of Metal Cutting: Tribology and Interface Engineering Series*; Briscoe, B.J., Ed.; Elsevier Ltd.: London, UK, 2006; Volume 52, pp. 83–87, ISBN 978-0-444-52881-0.
6. Simoneau, A.; Meehan, J. The Impact of machining parameters on peak power and energy consumption in CNC endmilling. *Energy Power* **2013**, *3*, 85–90. [CrossRef]
7. Ng, C.K.; Melkote, S.N.; Rahman, M.; Senthil, K.A. Experimental study of micro and nano-scale cutting of aluminum 7075-T6. *Int. J. Mach. Tools Manuf.* **2006**, *46*, 929–936. [CrossRef]
8. Beranoagirre, A.; Olvera, D.; de Lacalle, L.N.L. Milling of gamma titanium—Aluminum alltoys. *Int. J. Adv. Manuf. Technol.* **2012**, *62*, 83–88. [CrossRef]
9. Sourabh, P.; Bandyopadhyay, P.P.; Paul, S. Minimisation of specific cutting energy and back force in turning of AISI 1060 steel. *Proc. Inst. Mach. Eng. Part B J. Eng. Manuf.* **2017**, 1–11. [CrossRef]
10. Fernàndez-Abia, A.I.; Barreiro, J.; de Lacalle L.N.L.; Martìnez, S. Effect of Very High Cutting Speeds on Shearing, Cutting Forces and Roughness in Dry Turning of Austenitic Stainless Steels. *Int. J. Adv. Manuf. Technol.* **2011**, *57*, 61–71. [CrossRef]
11. Astakhov, V.P.; Xinran, X. A methodology for practical cutting force evaluation based on the energy spent in the cutting system. *Mach. Sci. Technol.* **2008**, *12*, 325–347. [CrossRef]
12. Yousefi, R.; Ichida, Y. A study on ultra–high-speed cutting of aluminium alloy: Formation of welded metal on the secondary cutting edge of the tool and its effects on the quality of finished surface. *Precis. Eng.* **2000**, *24*, 371–376. [CrossRef]

13. Becze, C.E.; Worswick, M.J.; Elbestawi, M.A. High strain rate shear evaluation and characterization of AISI D2 tool steel in its hardened state. *Mach. Sci. Technol.* **2001**, *5*, 131–149. [CrossRef]

14. Boothroyd, G.; Bailey, J.A. Effects of strain rate and temperature in orthogonal metal cutting. *J. Mech. Eng. Sci.* **1966**, *8*, 264–275. [CrossRef]

15. Lee, S.; Hwang, J.R.S.M.; Chandrasekar, S.; Compton, W.D. Large strain deformation field in machining. *Metall. Mater. Trans.* **2006**, *37*, 1633–1643. [CrossRef]

16. Wang X.; Xu, J.; Shan, D.; Guo, B.; Cao, J. Effects of specimen and grain size on electrically-induced softening behavior in uniaxial micro-tension of AZ31 magnesium alloy: Experiment and modeling. *Mater. Des.* **2017**, *127*, 134–143. [CrossRef]

17. Wang, X.; Xu, J.; Shan, D.; Guo, B.; Cao, J. Modeling of thermal and mechanical behavior of amagnesium alloy AZ31 during electrically-assistedmicro-tension. *Int. J. Plast.* **2016**, *85*, 230–257. [CrossRef]

18. Sànchez, E.A.J.; González, R.H.A.; Celentano, D.J.; Jorba, P.J.; Cao, J. Thermomechanical analisys of an electrically-assisted wire drawing process. *J. Manuf. Sci. Eng.* **2017**, *139*, 111017. [CrossRef]

19. Astakhov, V.P.; Shvets, S.V. A novel approach to operating force evolution in high strain rate metal-deforming technological processes. *J. Mater. Process. Technol.* **2001**, *117*, 226–237. [CrossRef]

20. Gui, G.Y.; Shi, F.X.; Xing, H.T.; Lan, H.D. Influence of cutting conditions on the cutting performance of TiAl6V4. *Adv. Mater. Res.* **2011**, *337*, 387–391. [CrossRef]

21. Bakkal, M.; Shih, A.J.; Scattergood, R.O. Chip formation, cutting forces and tool wear in turning of Zr-based bulk metallic glass. *Int. J. Mach. Tools Manuf.* **2004**, *44*, 915–925. [CrossRef]

22. Baranov, S.A.; Staschenko, V.I.; Sukhov, A.V.; Troitskiy, O.A.; Tyapkin, A.V. Electroplastic Metal Cutting. *Russ. Electr. Eng.* **2011**, *82*, 477–479. [CrossRef]

23. Brandt, J.R.; Elizabeth, G.; Farbod, A.N.; Laine, M. Electroplastic Drilling of Low and High Strength Steels. *J. Manuf. Sci. Eng.* **2018**, *140*, 061017. [CrossRef]

24. Sànchez, E.A.J; Gonzàlez, R.H.A.; Montilla, C.A.; Echeverri, V.K. Effect of electroplastic cutting on the manufacturing process and surface properties. *J. Mater. Process. Technol.* **2015**, *222*, 327–334. [CrossRef]

25. Hameed, S.; Gonzàlez, R.H.A.; Sànchez, E.A.J.; Alberro, A.N. Electroplastic cutting influence on power consumption during drilling process. *Int. J. Adv. Manuf. Technol.* **2016**, *87*, 1835–1841. [CrossRef]

26. Deng, W.J.; Lin, P.; Xie, Z.C.; Li, Q. Analysis of large-strain extrusion machining with different chip compression ratios. *J. Nanomater.* **2012**, *2012*, 1–12. [CrossRef]

27. Feng, K.; Jun, N.; Stephenson, D.A. Chip thickening in deep-hole drilling. *Int. J. Mach. Tools Manuf.* **2006**, *46*, 1500–1507. [CrossRef]

28. Silva, L.R.; Abrão, A.M.; Faria, P.; Davim, J.P. Machinability study of steels in precision orthogonal cutting. *Mater. Res.* **2012**, *15*, 589–595. [CrossRef]

29. Adam, U.A.; Noordin, M.Y.; Mohamed, H.S.E.; Hasan, F. Influence of cutting conditions on chip formation when turning ASSAB DF-3 Hardened tool steel. *Int. J. Mater. Mech. Manuf.* **2013**, *1*, 76–79. [CrossRef]

30. Natasha, A.R.; Othman, H.; Ghani, J.A.; Haron, C.H.C.; Syarif, J. Chip formation and coefficient of friction in turning S45C medium carbon steel. *Int. J. Mech. Mechatron. Eng.* **2014**, *14*, 89–92.

31. Magargee, J.; Morestin, F.; Cao, J. Characterization of flow stress for commercially pure titanium subjected to electrically assisted deformation. *J. Eng. Mater. Technol.* **2013**, *135*, 041003. [CrossRef]

32. Thakur, D.G.; Ramamoorthy, B.; Vijayaraghavan, L. A study on the parameters in high-speed turning of superalloy Inconel 718. *J. Mater. Manuf. Process.* **2009**, *24*, 497–503. [CrossRef]

33. Conrad, H.; Sprecher, A.F.; Cao, W.D.; Lu, X.P. Electroplasticity—The effect of electricity on the mechanical properties of metals. *JOM* **1990**, *42*, 28–33. [CrossRef]

Article

Effects of Cutting Edge Microgeometry on Residual Stress in Orthogonal Cutting of Inconel 718 by FEM

Qi Shen [1,2], Zhanqiang Liu [1,2,*], Yang Hua [1,2], Jinfu Zhao [1,2], Woyun Lv [1,2] and Aziz Ul Hassan Mohsan [1,2]

[1] School of Mechanical Engineering, Shandong University, Jinan 250061, China; 201612804@mail.sdu.edu.cn (Q.S.); sduhuayang@gmail.com (Y.H.); sduzhaojinfu@gmail.com (J.Z.); sdulvwoyun@gmail.com (W.L.); hassansdu@yahoo.com (A.U.H.M.)

[2] Key Laboratory of High Efficiency and Clean Mechanical Manufacture of MOE/Key National Demonstration Center for Experimental Mechanical Engineering Education, Jinan 250061, China

* Correspondence: melius@sdu.edu.cn; Tel.: +86-531-88393206; Fax: +86-531-88392045

Received: 15 May 2018; Accepted: 13 June 2018; Published: 14 June 2018

Abstract: Service performance of components such as fatigue life are dramatically influenced by the machined surface and subsurface residual stresses. This paper aims at achieving a better understanding of the influence of cutting edge microgeometry on machined surface residual stresses during orthogonal dry cutting of Inconel 718. Numerical and experimental investigations have been conducted in this research. The cutting edge microgeometry factors of average cutting edge radius $\overline{S}$, form-factor K, and chamfer were investigated. An increasing trend for the magnitudes of both tensile and compressive residual stresses was observed by using larger $\overline{S}$ or introducing a chamfer on the cutting edges. The ploughing depth has been predicted based on the stagnation zone. The increase of ploughing depth means that more material was ironed on the workpiece subsurface, which resulted in an increase in the compressive residual stress. The thermal loads were leading factors that affected the surface tensile residual stress. For the unsymmetrical honed cutting edge with $K = 2$, the friction between tool and workpiece and tensile residual stress tended to be high, while for the unsymmetrical honed cutting edge with $K = 0.5$, the high ploughing depth led to a higher compressive residual stress. This paper provides guidance for regulating machine-induced residual stress by edge preparation.

Keywords: cutting edge microgeometry; residual stress; finite element model; cutting edge preparation; Inconel 718

---

1. Introduction

More research attention has been paid to high performance and high process reliability in production due to the increasing demand for difficult-to-machine materials such as nickel alloys in the aerospace industry [1,2]. Aero-engine components such as turbine disks and shafts are subjected to huge and complex alternating loads during service conditions. Their service performance depends on their surface low cycle fatigue life, which is mainly influenced by the machine-induced surface residual stresses [3]. Specifically, the residual stresses impose an additional stress state on the surface of machined components during operation. The tensile residual stress tends to engender crack initiation and propagation, which then contributes to the diminution of fatigue life, whereas the compressive residual stress is conducive to fatigue life by suppressing crack propagation. Therefore, regulating the final stress condition during the cutting process is of paramount importance.

To improve the state of residual stress, a number of attempts have been made concerning the cutting parameters, cutting conditions, and macroscopic geometric parameters of cutting tools [4–9]. However, when these conditions are determined, there is still a consensus that the state of residual stress can be further improved. Recent research has proved that the cutting edge microgeometry has a direct influence on the deformation zone, cutting temperature distribution, and ploughing force; therefore, the cutting edge microgeometry plays a crucial role in the formation of residual stress [10–12].

The commonly applied edge preparation includes honed cutting edge (symmetrical honed edge and waterfall shape edge) and chamfered cutting edge. Ozel and Ulutan [13] investigated the influence of honed radius combined with the coating condition by three-dimensional FEM (finite element method). Their findings revealed that larger honed radii have the tendency to intensify both the magnitude of tensile residual stress and compressive residual stress. The influence of chamfered edge on residual stress was reported by Varela et al. [14]. Their findings revealed that the chamfered edge facilitates the compressive residual stress during hard turning. The formation of residual stress was also found to be closely related to the thermal properties of workpiece materials. Nasr et al. [15] observed that the thickness of tensile stress layer was not affected by applying a larger honed radius. This may be caused by the low thermal conductivity of the workpiece material, which restricted the effects of temperature rise. Attempts have also been made to figure out the causes behind these results. Ventura et al. [16] explained the larger and deeper compressive residual stress introduced by waterfull honed edge with smaller form-factor K through more intense friction in the deformation zone. The contact length between edge geometries and the material for honed edges with different K were investigated. Results indicated that the magnitude and the depth both increased with the contact length, which confirmed his explanation. The increase of tensile residual stress for larger honed radius has been explained by Nasr et al. [15]. The tensile residual stresses by-depth distribution was compared with workpiece temperature by-depth distribution for variate honed edge radii. High linear correlation was found in both parameters, which demonstrated that the increased thermal loads caused by tool/workpiece friction play a dominant role in increasing the tensile residual stress. Schulze et al. [17] investigated the scale effect of cutting edge radius on residual stress profiles in micro-cutting. Their findings revealed that the larger honed radius caused a deeper penetration of tensile residual stress due to the deeper plastic deformation. However, the explanation was given empirically and no concrete proof was applied through FE analysis.

Although numerous studies have been conducted focusing on the influences of cutting edge microgeometry on machined surface residual stress, there is still a misunderstanding about the influence of unsymmetrical honed cutting edge on machined surface residual stress. Understanding the causes of machined surface residual stress will help us better predict and control the states of residual stress during processing. The formation of residual stress is subjected to the coupled thermo-mechanical phenomenon during the cutting process. FEM can help give a deep insight into the cutting progress through the simulation of residual stress formation and distribution on machined surface [17,18]. However, few studies have attempted to further investigate the influence of microgeometry on formation mechanism of residual stress by FE analysis.

Thus, the following research aims to investigate the influences of cutting edge microgeometry on residual stress in orthogonal turning of Inconel 718 and try to provide physical explanations based on the cutting temperature by-depth distribution and the phenomenon of stagnation zone. Considering the variation of cutting edge microgeometry parameters is severely limited to the complexity of cutting edge preparation, a hybrid method of FE and cutting experiments is adopted. Unsymmetrical honed cutting edges are investigated based on the form–factor characterization method. The edge preparation through micro-blasting and the subsequent microgeometry measurements by Laser Scanning Confocal Microscopy (LSCM) are shown in this paper. To verify the reliability of FE model, the residual stresses of machined surface layer over a range of depth were measured by X-ray diffraction to compare with the simulation results.

2. Cutting Edge Characterization and Edge Preparation

For an unsymmetrical cutting edge, it is difficult to adjust the cutting edge profiles with a suitable circle. Characterization ambiguity and errors are inevitable with a honed edge radius r_β in this case.

Only the parameters of cutting edge are characterized accurately; its impact on cutting process can be reasonably predicted. The form-factor method established by Denkena et al. [19] is more appropriate for describing the honed cutting edge microgeometry. Four basic parameters used to measure the shape of the cutting edge are shown in Figure 1. S_γ, S_α are the distances from the separation point of the honed cutting edge to the tool tip of an ideal sharp cutting edge at the rake face and flank face, respectively. Average cutting edge rounding $\overline{S}$ is used to describe the dimension of honed rounding. Form-factor K (kappa) is introduced to indicate the offset level of the honed rounding to the flank face or the rake face. Profile flattening Δr and apex angle ϕ, which are measured by the shortest distance and the shift between ideal sharp cutting edge tip and the actual shape of rounding, respectively, are used to characterize the tools' bluntness.

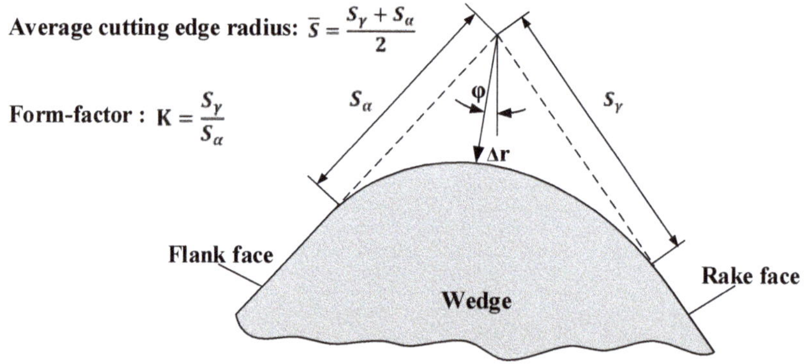

Figure 1. Cutting edge microgeometry characterized by *K*-factor.

The wet micro-blasting experiments were carried out on a customized reciprocating type wet micro-blasting machine called SY-WF4W during edge preparation procedure. SECO tools LCGF160604-0600-GS with an initial average cutting edge radius $\overline{S}$ of about 10.5 μm, a wedge angle of 55° were adopted. Nickel-based superalloy is a typically difficult-to-machine material. The high shear strength, low thermal conductivity, and high hardness of its strengthening phases γ'' cause high cutting force and high cutting temperature during machining. Considering the problems above, non-coated tools with ultra-fine grain carbide matrix of 890 type from SECO Tools, which show excellent toughness and high hardness, were chosen. The abrasive particles were Al_2O_3 with an average size of 220 #. The injection pressure was set to 0.35 MPa. The blast gun sprayed from two different positions at an interval of 30° to process two different K of cutting edge microgeometries, as shown in Figure 2a. Laser scanning confocal microscopy (VK200, KEYENCE, Osaka, Japan) was used to measure the parameters of cutting edge microgeometry. Each cutting edge was measured 10 times repeatedly in a two-dimensional profile and an average value of S_γ and S_α were calculated. Two different values of K 0.524 and 1.096 were achieved, as shown in Figure 2b,c.

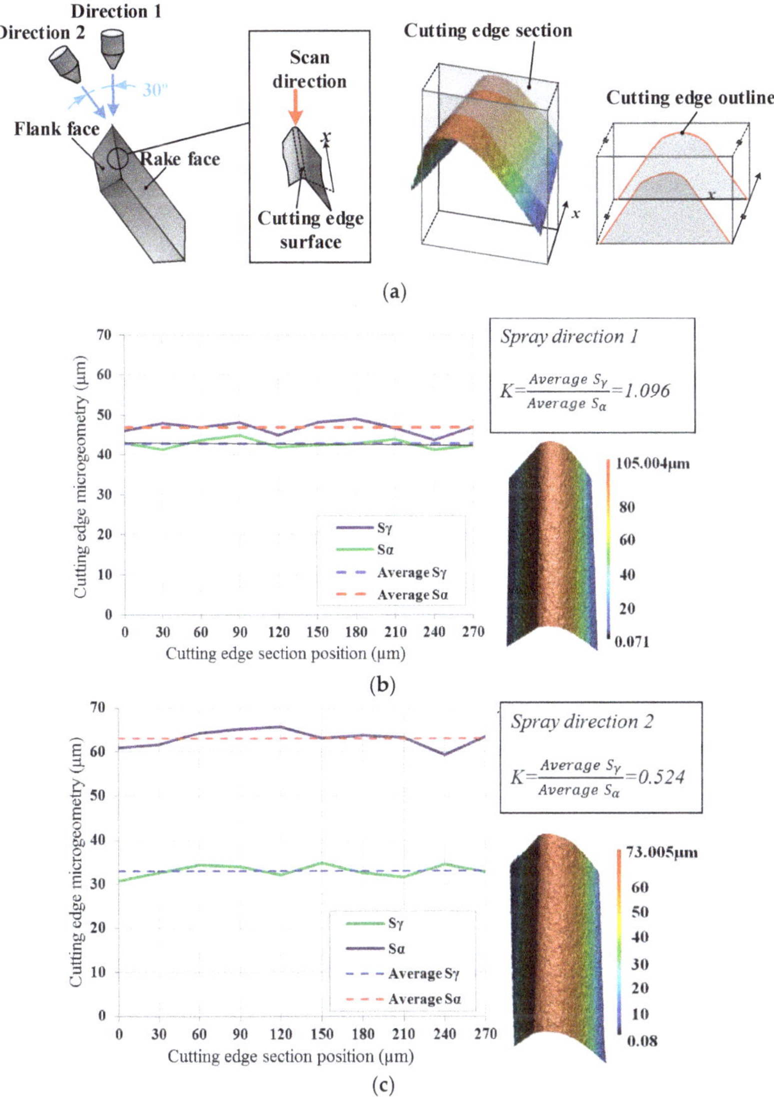

Figure 2. (**a**) Schematic of micro-blasting and the LSCM measurements of cutting edge microgeometry. (**b**) Cutting edge microgeometry measurements of spray direction 1 and (**c**) spray direction 2.

3. Orthogonal Experiments and Measurements of Residual Stress

The orthogonal turning experiments with Inconel 718 were conducted on a CNC lathe center (PUMA 200, DAEWOO, Changwon, Korea). Inconel 718 bars with an outline diameter of 70 mm and an average thickness of 3.5 mm were adopted. The micro-blasted cutting tools were used throughout the experimentation. The tool holder of SECO CFIR2525M06JET (SECO tools, Fagersta, Sweden) was adopted. After installation, the tools had a rake angle of 28°. The cutting speed was set to 29 m/min and the depth of cut was 0.15 mm.

As shown in Figure 3, the machined surface samples were peeled off from the Inconel 718 plates by electrical discharge machining (EDM) after orthogonal cutting. The height of the samples h_s was 15 mm and the width was 20 mm. Electrolytic polishing method was adopted to remove the surface material. After that, each surface of the samples was etched to a certain depth of Δh. By this method, the subsurface emerged for the measurements of residual stress. The parameters of the electrolytic polishing are shown in Table 1. EDM and electrolytic polishing are chemical processing methods that have little influence on the residual stress.

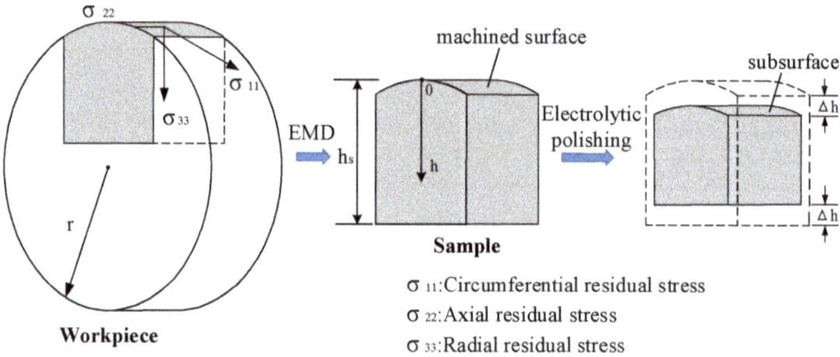

Figure 3. Measurements of residual stress in Inconel 718 sub-surface.

Table 1. Parameters of electrolytic polishing.

Electrolyte Composition	CH_3OH/HNO_3	200 mL/100 mL
Electrolytic Parameters	Voltage	10 V
	Current density	0.5~0.7 A/cm^2
Environment	Room temperature (20 °C)	

A Pulstec μ-X360 X-ray diffraction apparatus (Pulstec, Hamamatsu, Japan) was used to measure the machined surface layer residual stresses. The X-ray diffractometer (XRD) technique is based on the measurement of the crystallographic lattice deformation. The cosα method, also called the single exposure method, was used to calculate the residual stress [20]. As shown in Figure 4, after diffraction in the atomic planes of crystal structure (also called crystallographic plane *hkl*), the diffraction cone of a single incident X-ray beam is formed. The diffraction cone is captured via a 2D detector in the image plane. The Debye ring is a regular circle for unstressed sample, while the Debye ring will be deformed for a stressed sample. The stress is calculated by comparing the amount of deformation between the Debye ring before and after stress. The lattice strain $\bar{\varepsilon}_\alpha^{(hkl)}$ is calculated as follows:

$$\bar{\varepsilon}_\alpha^{(hkl)} = \frac{1}{2}\left[\left(\varepsilon_\alpha^{(hkl)} - \varepsilon_{\pi+\alpha}^{(hkl)}\right) + \left(\varepsilon_{-\alpha}^{(hkl)} - \varepsilon_{\pi-\alpha}^{(hkl)}\right)\right], \tag{1}$$

where $\varepsilon_\alpha^{(hkl)}$, $\varepsilon_{\pi+\alpha}^{(hkl)}$, $\varepsilon_{-\alpha}^{(hkl)}$, $\varepsilon_{\pi-\alpha}^{(hkl)}$ are the strain of four points at an interval of 90° on the Debye ring. When α increases from 0 to 90°, each point on the Debye ring is contained. The values of $\bar{\varepsilon}_\alpha^{(hkl)}$ show a linear relationship with cosα. The stress calculated by cosα method is expressed as in Equation (2):

$$\sigma_\varphi = \frac{E^{(hkl)}}{1 + v^{(hkl)}} \frac{1}{\sin 2\eta \sin 2\psi_0} \frac{\partial \bar{\varepsilon}_\alpha^{(hkl)}}{\partial \cos \alpha} = \frac{1}{1/2S^{(hkl)}} \frac{1}{\sin 2\eta \sin 2\psi_0} \frac{\partial \bar{\varepsilon}_\alpha^{(hkl)}}{\partial \cos \alpha} \tag{2}$$

$$S^{(hkl)} = \frac{2(1 + v^{(hkl)})}{E^{(hkl)}}, \tag{3}$$

where 2η is the Debye ring semi-angle, Ψ_0 is the constant tilt angle of X-ray beam. The elastic constant S in crystallographic plane (hkl) determined by Poisson' ratio $v^{(hkl)}$ and elastic modulus $E^{(hkl)}$ of the specimen material is expressed as Equation (3).

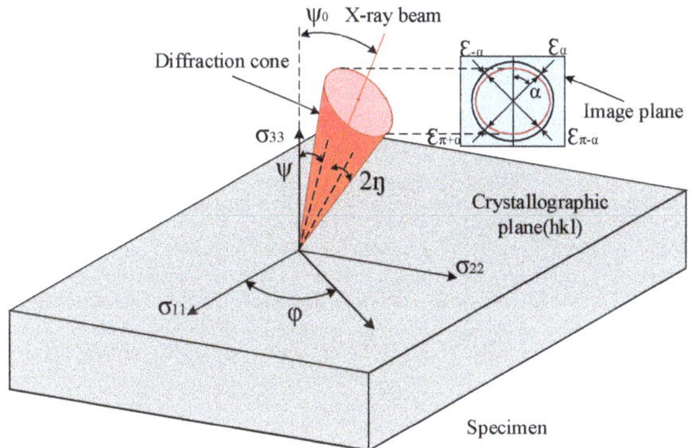

Figure 4. X-ray diffraction (XRD) schematic diagram.

X-ray tube current and voltage are 1 mA and 30 kV, respectively. X-ray incidence angle is $30°$. Diffraction lattice angle 2η is $29.124°$. The value of Poisson's ratio $v^{(hkl)}$ and elastic modulus $E^{(hkl)}$ of Inconel 718 are 0.305 and 214.580 GPa, respectively. The elastic constant $S^{(hkl)}$ for the crystallographic plane is 1.216×10^{-5} MPa^{-1} by calculation.

4. Numerical Modeling

4.1. Geometry Modeling and Mesh Controlling

A two-dimensional plane strain FE model based on ABAQUS/Explicit was built to simulate the orthogonal dry turning of Inconel 718 superalloy with continuous chips. The symmetrical and unsymmetrical cutting edge models based on the form-factor method in the FE models are illustrated in Figure 5. The symmetrical cutting edge with $\overline{S}$ of 15 μm was used to represent the sharp edge. Moreover, a honed plus chamfer cutting edge was built to investigate the effect of chamfer on residual stress. In later discussions, the symmetrical cutting edges with $\overline{S}$ of 15 μm, 45 μm, 75 μm, 105 μm, the unsymmetrical cutting edge with K of 0.5, 2 and the honed plus chamfer cutting edge will be referred to as S15, S45(K1), S75, S105, K0.5, K2, Chamfer, respectively. The arbitrary Lagrangian–Eulerian (ALE) method was applied in the model. The ALE formulation combines the Lagrangian and Eulerian formulation during the re-meshing procedure to accommodate large deformation calculations. Two-dimensional triangle reduction integration element CPE3T was used in tool model. Considering the huge deformation of the workpiece, a relatively stable quadrilateral reduction integration element, CPE4RT, was used. The whole model was set as plane strain thermally coupled element, linear displacement, and temperature. To improve the simulation accuracy and shorten the simulation time, the mesh in the part of tool tip and the machined part of workpiece were locally refined (see Figure 6). In the final mesh models, about 140,000 quad elements were produced for the workpiece model and 900 tri elements were generated in the cutting tool model.

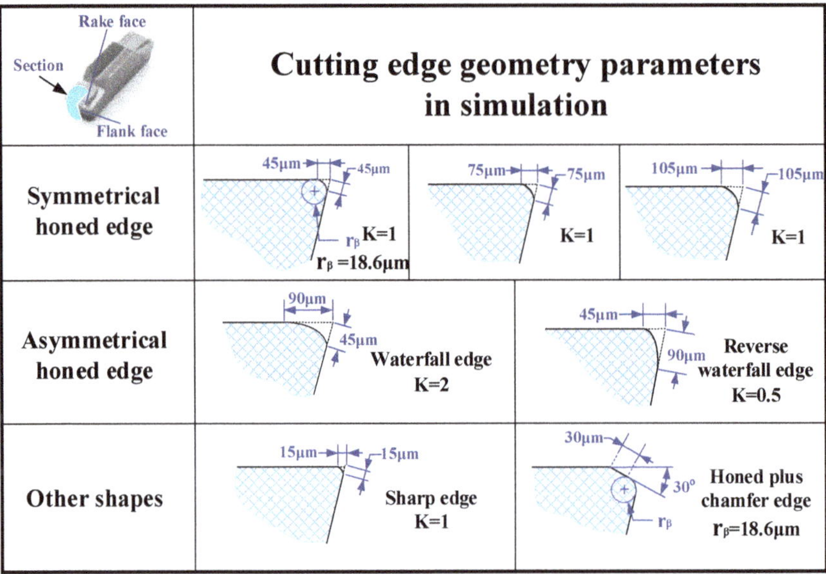

Figure 5. Cutting tool microgeometry parameters in simulations.

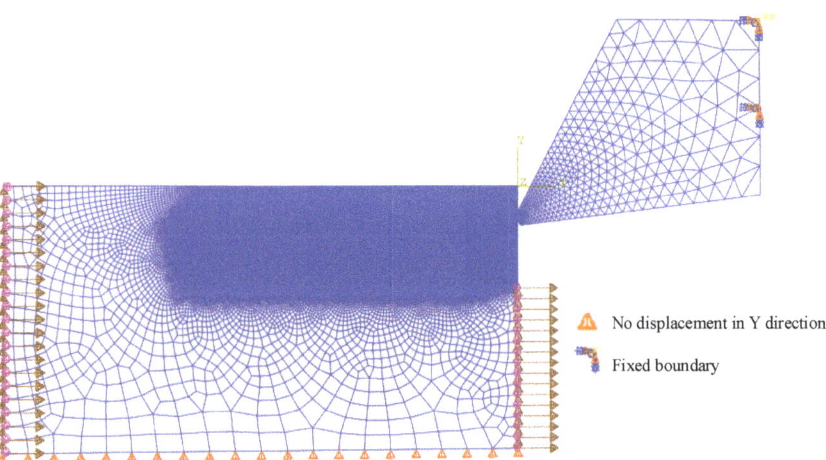

Figure 6. The boundary conditions and element configuration of the FE models in the initial position.

4.2. Initial Conditions

The implementation of boundary conditions is also shown in Figure 6. The cutting speed has to be applied to the workpiece when ALE is applied [21]. The cutting tool was fixed and the workpiece moved in the opposite direction to the cutting direction. The freedom of workpiece in Y direction was restricted. The initial temperature is set to be 20 °C (room temperature).

The interactions can be divided into three zones with different tribological properties: sticking zone, adhesion zone, and sliding zone. The sticking zone is distributed at the honed circle with high pressure, while the sliding zone is located between the sticking zone and the stagnation point in a low-pressure sliding friction state. To model the different friction zones, the Coulomb friction law, which has been widely adopted in FE simulation investigations of metal cutting [15,17,22], was used in the current work. The frictional shear stress τ_f can be calculated by Equation (4):

$$\tau_f = \begin{cases} \tau, & \tau = \mu\sigma < \tau_c \, (\text{Stickingzone}) \\ \tau_c, & \tau = \mu\sigma > \tau_c \, (\text{Slidingzone}) \end{cases} , \tag{4}$$

where μ is the friction coefficient, σ is the normal stress, τ is the shear stress in tool/chip interface, and τ_c is the limited shear stress. The friction coefficient μ was set to 0.3.

For the semi-empirical Johnson–Cook constitutive model, which reacts the thermal viscoplastic deformation behavior of material at high levels of strain, the strain rate was commonly chosen in the FE model of metal cutting [7,15,17,22]. The constitutive law is expressed with Equation (5). A, B, C, m, n are the yield stress, hardening modulus, strain rate dependency coefficient, thermal softening coefficient, and strain hardening coefficient, respectively; T_m is the melting temperature of the material, T_t is room temperature, T is the bulk temperature of workpiece, ε is equivalent strain rate, ε_0 is the reference strain rate, $\bar{\varepsilon}$ is the equivalent strain, and σ is the equivalent stress. The Johnson–Cook constitutive parameters are specified in Table 2. The physical and thermomechanical properties of the tool and the workpiece material are presented in Table 3.

$$\sigma = \left(A + B\bar{\varepsilon}^n\right)\left[1 + C\ln\left(\frac{\varepsilon}{\varepsilon_0}\right)\right]\left[1 - \left(\frac{T - T_r}{T_m - T_r}\right)^m\right] \tag{5}$$

Table 2. Johnson–Cook constitutive model parameters of Inconel 718 [23].

A (MPa)	B (MPa)	C	n	m
450	1700	0.017	0.65	1.3

Table 3. The physical and thermomechanical properties of the tool and the workpiece material.

Properties	Density	Young's Modulus	Poisson's Ratio	Thermal Expansion	Conductivity	Specific Heat
Tool	14,800 Kg/m^3	640 GPa	0.22	4.5 μm/mK	50.24 W/mK	220 J/kgK
Workpiece	8250 Kg/m^3	214.580 GPa	0.305	14.8 μm/mK	17.8 W/mK	526.3 J/kgK

The definition of residual stress is the internal equilibrium stress that remains in a component after eliminating the external force or inhomogeneous temperature field. In order to abide by this principle, the stress relaxation procedure was considered by setting the workpiece as the thermal convection region when the cutting tool is removed. The film coefficient was 10 W/m^2 °C and the sink temperature was 20 °C.

5. Results and Discussions

5.1. Validation of FE Model

As shown in Figure 7, to validate the dependability of the numerical model, the simulated subsurface residual stresses profiles for unsymmetrical tools in cutting direction σ_{11} was compared with the experimental ones in the stable cutting stage. The residual stress profiles from experiments show more tensile and compressive stress than the simulated ones. The simplification of the material model and friction model in simulation, tool wear, and the inhomogeneity of workpiece material in experiments all cause error [15,24]. Although no agreement was achieved between the measured stress profiles and the predicted ones, which was also expected in advance, the FE model accurately predicted the residual stress contour shape and trends. Furthermore, the predicted values of maximum compressive residual stress and its depth were very close to the experimental ones. This indicates that the FE model can provide reliable and valid prediction of residual stress.

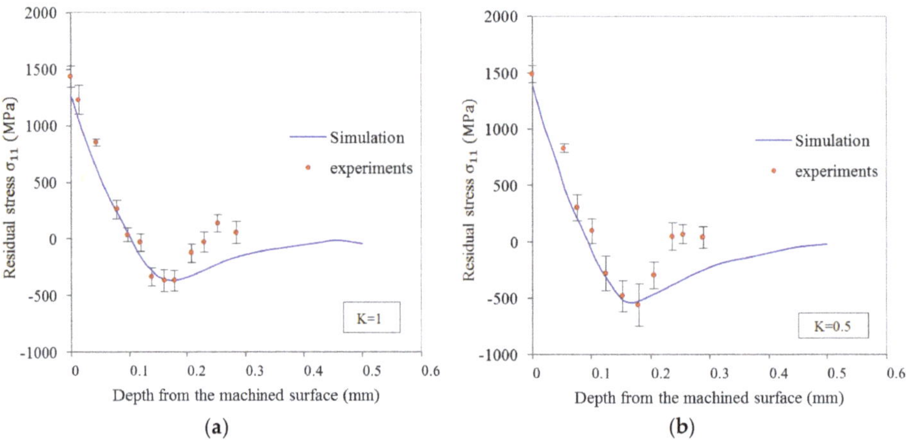

Figure 7. Comparison of residual stress by-depth profiles in cutting direction between simulation and experiments for different form factors K: (**a**) $K = 1$; (**b**) $K = 0.5$.

5.2. Formation Mechanism of Residual Stress

In order to better predict the residual stress states, a thorough understanding of the formation mechanism of machine-induced residual stress is needed. Figure 8 briefly illustrates the formation of machined surface residual stress. When cutting with a non-sharp edge, friction and ironing functions between the tool tip and material will be generated, which causes the ploughing effect. Due to the frictional function of ploughing force component F_{fr} and the press function of ploughing force component F_{pr}, a certain depth of material grain in the workpiece subsurface will be elongated in the cutting direction. Consequently, the tensile plastic strain and tensile elastic strain will be produced in the workpiece subsurface layer. After springing back, the release of the elastic strain tends to introduce compressive stress in the surface layer. It is believed that the magnitude of compressive stress is closely related to the depth of the ironed material, which will be demonstrated in a later section. Different degrees of thermal expansion deformation on the subsurface layer caused by cutting heat introduce a different degree of compressive strain in the surface layer. During cooling down, the retraction of the expansion is limited, which contributes to the formation of tensile stress. Since the formation of residual stresses is always subjected to the identical effect of thermomechanical phenomena, the residual stresses of machined surfaces conventionally have a homogeneous profile as shown in Figure 9. Tensile residual stresses on the surface layer of machined components were found to accelerate the expansion of cracks in some studies [3,25]. Inducing a higher depth or magnitude of compressive residual stresses in the surface or subsurface region will be favorable for fatigue life [26,27]. Furthermore, Guo et al. [27] suggested that the deep region of compressive residual stresses in the subsurface layer might be more beneficial to the fatigue life of bearing than shallower stresses region of greater magnitude. Based on the above research, the surface residual stress σ_s, maximum compressive residual stress σ_c, and its depth h_r will be extracted as analytical indicators because of their influence on fatigue life. Several investigations [7,15,28] have proven that the generation of residual stress in feed direction σ_{33} mostly depends on σ_{11} and σ_{33} always shows the same changes with σ_{11}. So the later analysis will focus on σ_{11} to discuss the effect of cutting tool microgeometry on residual stress in this research.

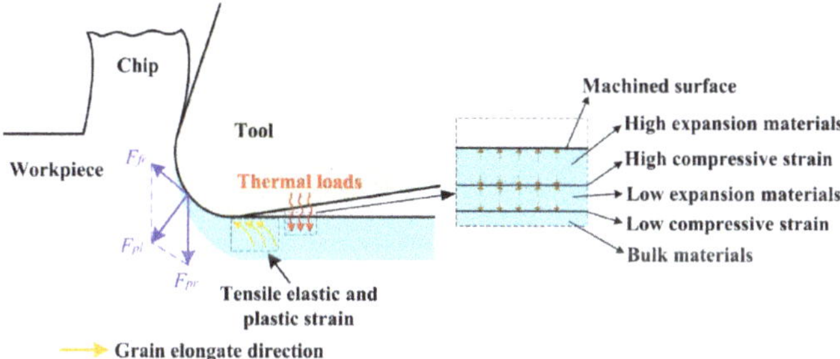

Figure 8. Schematic of the machined surface residual stress formation.

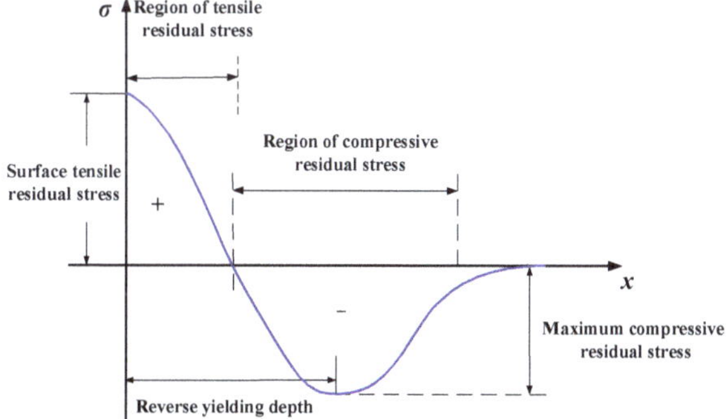

Figure 9. Representative features of the residual stress profile.

5.3. Effects of Cutting Edge Microgeometry on the Variation of Residual Stress

Figure 10 showed the by-depth residual stress profiles in the cutting direction obtained from simulations. Compared to the honed cutting edge, it was clear that both the magnitude of surface tensile residual stress σ_s and the maximum compressive residual stress σ_c experience obvious growth when using a honed plus chamfer edge (see Figure 10a). The magnitude of the tensile and compressive residual stresses was smallest when using cutting tools with a sharp edge. The lowest tensile residual stress was induced when using the sharp cutting edge in simulations. However, using the sharp cutting edge is prone to cause deterioration of the surface quality due to rapid tool wear in the actual cutting process. Figure 10b,c showed that the form-factor K and the average cutting edge radius $\overline{S}$ have a predominant effect on the profiles of residual stress. For the unsymmetrical honed cutting edge of K0.5 and K2, the magnitude of residual stresses in both the tensile and compressive residual stress region is higher than K1. Furthermore, for K0.5 σ_c showed a considerable increase while σ_s showed an obvious increase for K2. It can be seen from Figure 10c that a larger $\overline{S}$ (or honed edge radius r_β) produced a higher σ_s and σ_c. The h_r goes slightly deeper into the workpiece for a cutting edge with a larger $\overline{S}$. The same phenomenon was observed in the experiments of turning AISI 316L [15]. However, in the research of orthogonal turning of AISI 52100, Hua et al. reported that h_r remains almost unchanged for cutting tools with different r_β [29]. This is mainly caused by the different thermal conductivity of the workpiece material. In the case of an asymmetrical cutting edge with $\overline{S} = 45$ μm, changing its K to 0.5 or increasing $\overline{S}$ to 75 μm both intensify the σ_s and σ_c to a similar value (about 1400 MPa and −570 MPa, respectively). However, in the compressive stress depth region, when an asymmetrical cutting edge K0.5 is applied, a deeper region of compressive residual stress (about 315 μm) is shown compared to the 271 μm produced by symmetrical cutting edge S75. This has been proven more conducive to enhancing fatigue life [27].

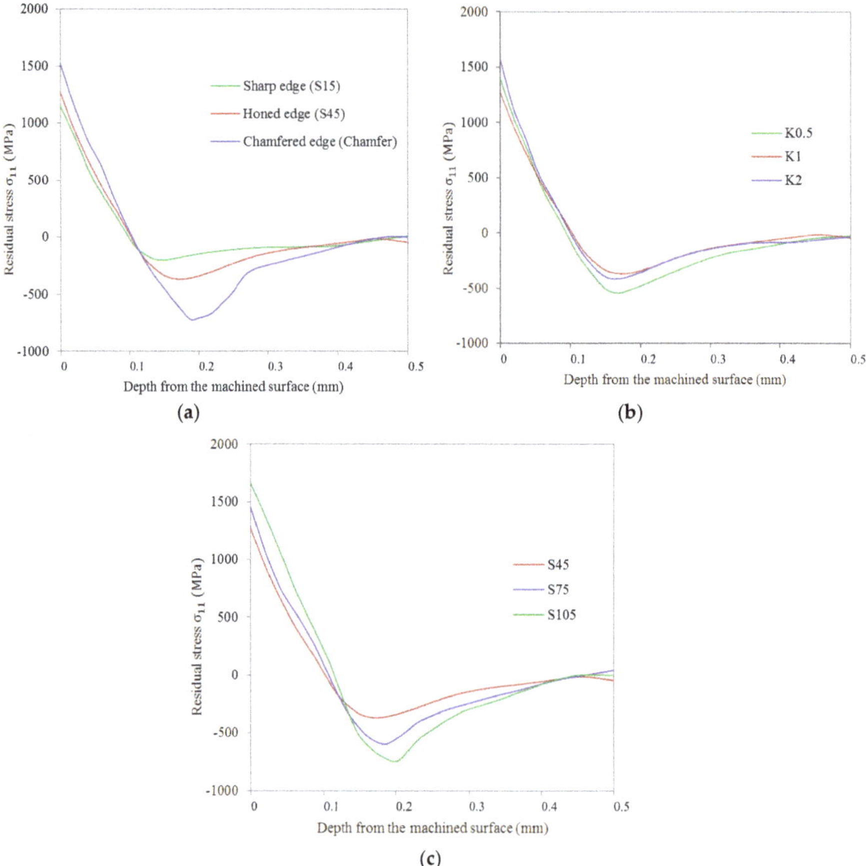

Figure 10. Residual stress profiles in cutting direction σ_{11} for (**a**) different cutting edge type, (**b**) honed cutting edge with different form-factor K, (**c**) symmetrical honed cutting edges with different average cutting edge radius $\bar{S}$.

5.4. Temperature Distribution and Tensile Residual Stress

The values of cutting temperature varied with depth in the stable cutting period, as seen in Figure 11. It is easy to find that the temperature by-depth curves for all cutting edge microgeometries intersected at one point at a depth of 220 μm and approached the ambient temperature. This means the temperature influence depth for all cutting edge microgeometries was almost identical. In this case, a higher surface temperature is equivalent to a higher temperature gradient. It is seen from this figure that tools with larger $\bar{S}$ always produced a higher temperature, while the honed plus chamfer cutting edge provided the second largest cutting temperature. Moreover, for unsymmetrical honed edge K2 the cutting temperature is higher than K0.5 and K1. The contact length between the workpiece and the tool increased due to the larger $\bar{S}$ or chamfer, which led to an increase in the friction. As a result, a larger amount of cutting heat was generated. Meanwhile, the increase in contact length between the tool and workpiece also led to more cutting heat dissipating into the tools, which caused a decreasing trend of workpiece subsurface temperature. Obviously, the increased dissipated cutting heat is not sufficient to affect the increasing tendency of the subsurface temperature. The geometry of waterfall cutting edge K2 has a similar effect, increasing the tool/workpiece friction with honed plus chamfer edge, which explains the significant increase in σ_s for K2 in Figure 10b. The increased temperature partially results from the heat released by the increased plastic deformation due to the increased ploughing depth, which will be introduced in the next section. Because of the significant influence of σ_s on fatigue life, more attention was paid to the reason behind the change of σ_s with different cutting edge microgeometries. The value of σ_s is the result of thermal loads and mechanical loads. Figure 12a illustrates that the changes of surface temperature T_s closely correspond to the changes of σ_s with different cutting edge microgeometries. From Figure 12b, we can observe that σ_s varies almost linearly with T_s and the determination coefficient reaches 0.942. Judging from this phenomenon, it can be concluded that the thermal load plays the dominant role in determining the surface residual stress σ_s.

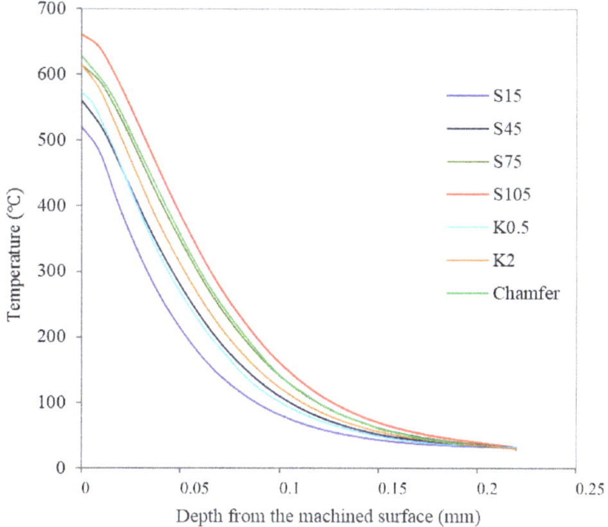

Figure 11. Temperature by-depth distribution with different cutting edge microgeometry.

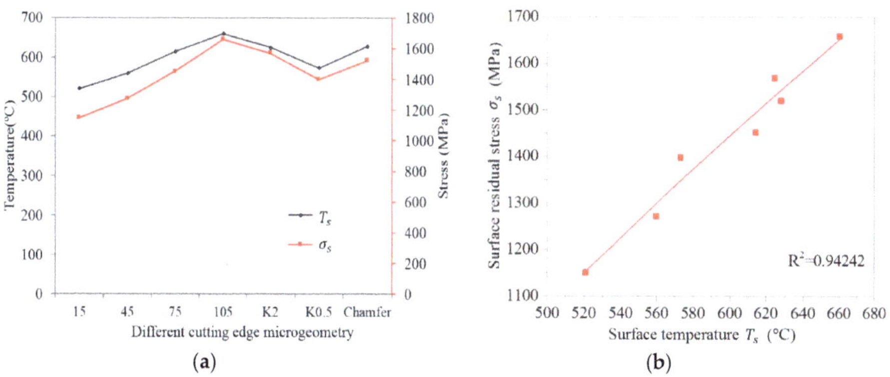

Figure 12. (a) The T_s and σ_s for different cutting edge microgeometry. (b) Changes of σ_s with the increase of T_s.

5.5. Stagnation Zone and Compressive Residual Stress

During cutting with non-sharp edges, there is a special area in front of the tool tip where the material flow rate is stationary relative to the cutting tool. This area is called the "stagnation zone" or "dead metal zone". The formation of a stagnation zone is caused by the accumulation and obstruction of material underneath the tool tip. Figure 13 demonstrates that the characteristics of the stagnation zone depend on the microgeometry of the cutting edge. The distribution of workpiece material velocity near the tool tip with different cutting edge microgeometries during the stable cutting period is shown in Figure 13, where the stagnation zone is distinguished by a dark blue color. As reported in many studies [15,30], the size of the stagnation zone increased when introducing a chamfer or using a cutting edge with a bigger $\overline{S}$ (or r_β). Bassett et al. [11] also demonstrated that when using a honed cutting edge with $K > 1$, a stagnation zone can be observed, while for a cutting edge with $K < 1$ the stagnation zone cannot be recognized. The stagnation zone acts as an extension of the cutting edge and plays a protective role to the cutting edge during machining.

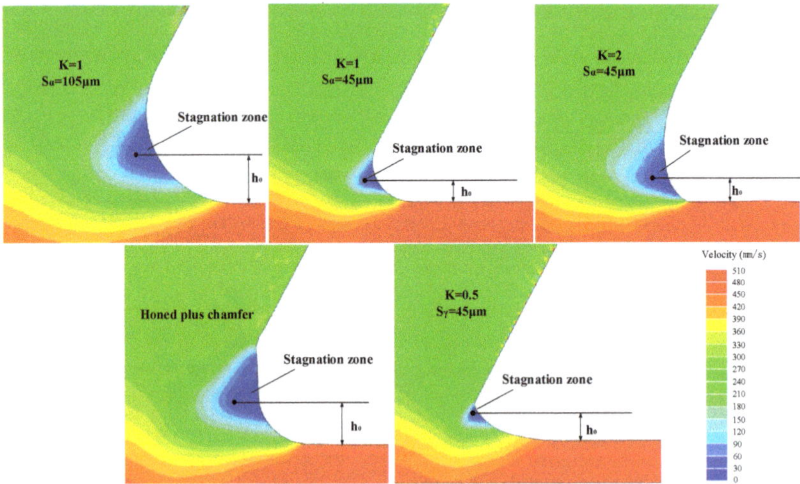

Figure 13. Stagnation zone in front of the cutting edge.

Since the stagnation zone affects the ploughing effect, which has a dramatic influence on the machined surface residual stress, more attention should be paid to understanding of it. Figure 14 briefly illustrates the phenomenon of uncut chip thickness reducing in ploughing effect. The stagnation zone appears as a triangle with one vertex O acting as the separation point of the workpiece material. In the range of theoretical uncut chip thickness, the material below point O will flow downwards and be ironed by the cutting edge. Meanwhile, strong elastic and plastic tensile strain has been generated. This ironed material depth is the ploughing depth h_0. As shown in Figure 15a, when K0.5 and K2 were applied, h_0 increases compared to K1, while for K0.5 h_0 shows a more significant increase. Increasing $\overline{S}$ or introducing a chamfer results in increasing the value of h_0. The increase of h_0 leads to more material below point O being ironed downwards and further results in the generation of stronger plastic deformation on the machined surface layer. Since mechanical plastic deformation tends to induce compressive residual stress, the increasing trend observed in the values of σ_c as h_0 increased, as shown in Figure 15b, can be explained. Furthermore, it is noteworthy that the maximum compressive residual stress σ_c has an almost proportional relationship with h_0, which convincingly demonstrates that the mechanical loads play a dominant role in the formation of residual stress at the h_r. The h_r shows an increasing trend on the whole with the increasing of h_0 in Figure 15b. This is because the higher ploughing depth leads to deeper ranging plastic deformation and therefore higher h_r.

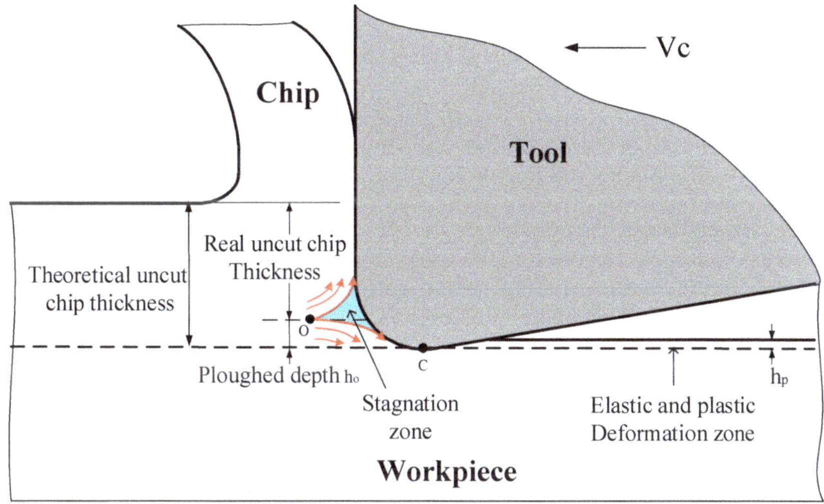

Figure 14. Material flow and phenomenon of uncut chip thickness reducing in ploughing effect.

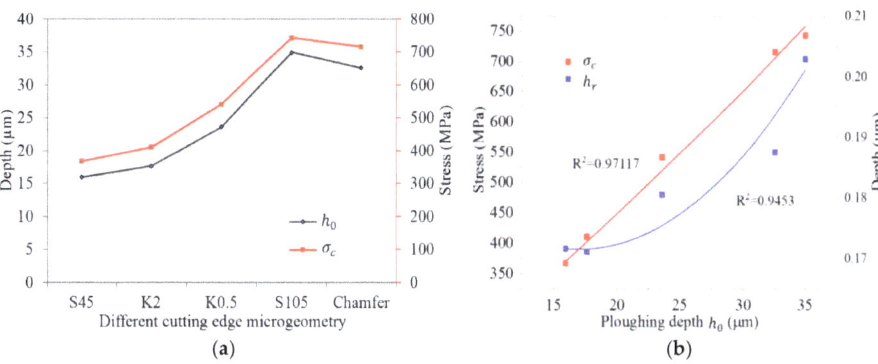

Figure 15. (a) h_o and σ_c for different cutting edge microgeometry. (b) Changes of h_r and σ_c with different h_o.

It should be noted that the profile of residual stress is not a direct consequence of either the thermal load or the mechanical load, but a complex result of both of them. Increasing $\overline{S}$ means the tool tip becomes bigger, which in turn contributes to two conflicting phenomena. On the one hand, the deeper ranging ironed workpiece material (as shown in Figure 13) leads to a higher compressive residual stress. On the other hand, the larger friction between the tool tip and workpiece causes more heat generation and therefore higher tensile residual stress (as shown in Figure 11). No cutting temperature gradient is significantly deeper than 220 μm below the surface. The thermal effect is only remarkable in the near-surface layer and declines quickly with depth. With reference to Figures 12 and 15, we can conclude that the tensile stress caused by thermal loads has a leading effect on residual stress profiles above h_r while the compressive stress caused by mechanical loads plays a major role below the h_r. When h_o increases, the elasto-plastic deformation on the machined surface will also increase and more heat will be released. This is another important cause of the increase in cutting temperature. Therefore, T_s tends to be higher for K0.5 compared to for K1.

6. Conclusions

An FE model for predicting the effect of microgeometry on residual stress profile of machined surface is investigated. According to the numerical results, the conclusions can be listed as follows.

1. With the increase in average cutting edge radius $\overline{S}$, the values of the surface tensile residual stress σ_s and the maximum compressive residual stress σ_c present an increasing trend. Using the honed plus chamfer cutting edge plays a similar role with using a cutting edge with a larger average cutting edge radius $\overline{S}$, increasing the magnitude of residual stress.

2. The increasing surface tensile residual stress σ_s is attributed to the higher temperature gradient in the workpiece subsurface layer for a cutting edge with a bigger average cutting edge radius $\overline{S}$ or honed plus chamfer cutting edge, as more heat is generated due to the increasing frictional contact length between the cutting edge and the workpiece. Another reason for the temperature increasing is the deeper plastic deformation caused by an increase of ploughing depth h_0, which releases more heat.

3. The phenomenon of a stagnation zone is analyzed as a limitation of workpiece material diversion to predict ploughing depth h_0. The increase of h_0 means more subsurface material is ironed on the deformation zone, which provides a specific explanation for the intensification of maximum compressive residual stress σ_c.

4. When a cutting edge with form-factor $K = 2$ is used, the friction between the tool tip and workpiece increased, which consequently contributes to the increase in cutting temperature and σ_s. For a cutting edge with $K = 0.5$, h_0 increased and a higher σ_c can be achieved.

Author Contributions: Q.S. and Z.L. conceived and designed the experiments; Q.S., J.Z., W.L. performed the experiments; Q.S. wrote the paper; Z.L., Y.H. and A.U.H.M. reviewed and edited the manuscript.

Funding: This research was funded by National Natural Science Foundation of China grant numbers 51425503 (51705293), Taishan Scholar Foundation of Shandong Province grant number TS20130922.

Acknowledgments: The authors would like to acknowledge the technical support from Collaborative Innovation Center for Shandong's Main Crop Production Equipment and Mechanization.

Conflicts of Interest: The authors declare no conflict of interest.

References

1. Kim, H.; Cong, W.L.; Zhang, H.-C.; Liu, Z.C. Laser engineered net shaping of nickel-based superalloy Inconel 718 powders onto AISI 4140 alloy steel substrates: Interface bond and fracture failure mechanism. *Materials* **2017**, *10*, 341. [CrossRef] [PubMed]
2. Wang, B.; Liu, Z.; Hou, X. Influences of Cutting Speed and Material Mechanical Properties on Chip Deformation and Fracture during High-Speed Cutting of Inconel 718. *Materials* **2018**, *11*, 461. [CrossRef] [PubMed]
3. Fernández-Valdivielso, A.; López de Lacalle, L.N.; Urbikain, G. Detecting the key geometrical features and grades of carbide inserts for the turning of nickel-based alloys concerning surface integrity. *Proc. Inst. Mech. Eng. Part C J. Mech. Eng. Sci.* **2016**, *230*, 3725–3742. [CrossRef]
4. Huang, X.; Zhang, X.; Ding, H. An enhanced analytical model of residual stress for peripheral milling. *Procedia CIRP* **2017**, *58*, 387–392. [CrossRef]
5. Devillez, A.; Coz, G.L.; Dominiak, S.; Dudzinski, D. Dry machining of inconel 718, workpiece surface integrity. *J. Mater. Process. Technol.* **2011**, *211*, 1590–1598. [CrossRef]
6. Kundrák, J.; Szabó, G.; Markopoulos, A.P. Numerical Investigation of the Influence of Tool Rake Angle on Residual Stresses in Precision Hard Turning. *Key Eng. Mater.* **2016**, *686*, 68–73. [CrossRef]
7. Mohsan, A.U.H.; Liu, Z.; Padhy, G.K. A review on the progress towards improvement in surface integrity of Inconel 718 under high pressure and flood cooling conditions. *Int. J. Adv. Manuf. Technol.* **2016**, *91*, 107–125. [CrossRef]
8. Ji, X.; Li, B.; Zhang, X.; Liang, S.Y. The effects of minimum quantity lubrication (MQL) on machining force, temperature, and residual stress. *Int. J. Prescis. Eng. Manuf.* **2014**, *18*, 547–564. [CrossRef]
9. Qin, M.Y.; Ye, B.Y.; Wu, B. Investigation into influence of cutting fluid and liquid nitrogen on machined surface residual stress. *Adv. Mater. Res.* **2012**, *566*, 7–10. [CrossRef]

10. Denkena, B.; Köhler, J.; Mengesha, M.S. Influence of the cutting edge rounding on the chip formation process: Part 1. Investigation of material flow, process forces, and cutting temperature. *Prod. Eng. Res. Dev.* **2012**, *6*, 329–338. [CrossRef]

11. Bassett, E.; Köhler, J.; Denkena, B. On the honed cutting edge and its side effects during orthogonal turning operations of AISI1045 with coated WC-Co inserts. *CIRP J. Manuf. Sci. Technol.* **2012**, *5*, 108–126. [CrossRef]

12. Denkena, B.; Biermann, D. Cutting edge geometries. *CIRP Ann. Manuf. Technol.* **2014**, *63*, 631–653. [CrossRef]

13. Özel, T.; Ulutan, D. Prediction of machining induced residual stresses in turning of titanium and nickel based alloys with experiments and finite element simulations. *CIRP Ann. Manuf. Technol.* **2012**, *61*, 547–550. [CrossRef]

14. Varela, P.I.; Rakurty, C.S.; Balaji, A.K. Surface Integrity in Hard Machining of 300 M Steel: Effect of Cutting-edge Geometry on Machining Induced Residual Stresses. *Procedia CIRP* **2014**, *13*, 288–293. [CrossRef]

15. Nasr, M.N.A.; Ng, E.G.; Elbestawi, M.A. Modelling the effects of tool-edge radius on residual stresses when orthogonal cutting AISI 316L. *Int. J. Mach. Tools Manuf.* **2007**, *47*, 401–411. [CrossRef]

16. Ventura, C.E.; Breidenstein, B.; Denkena, B. Influence of customized cutting edge geometries on the workpiece residual stress in hard turning. *Proc. Inst. Mech. Eng. Part B J. Eng. Manuf.* **2017**. [CrossRef]

17. Schulze, V.; Autenrieth, H.; Deuchert, M.; Weule, H. Investigation of surface near residual stress states after micro-cutting by finite element simulation. *CIRP Ann. Manuf. Technol.* **2010**, *59*, 117–120. [CrossRef]

18. Fan, Y.H.; Wang, T.; Hao, Z.P. Surface residual stress in high speed cutting of superalloy Inconel718 based on multiscale simulation. *J. Manuf. Process.* **2018**, *31*, 480–493. [CrossRef]

19. Denkena, B.; Reichstein, M.; Brodehl, J.; Leon, G.L. Surface Preparation, Coating and Wear Performance of Geometrically Defined Cutting Edges. In Proceedings of the 8th CIRP Intternational Workshop on Modeling of Machining Operations, Chemnitz, Germany, 10–11 May 2005.

20. Noyan, I.C.; Cohen, J.B. *Residual Stress: Measurement by Diffraction and Interpretation*; Springer: New York, NY, USA, 1987.

21. Grissa, R.; Zemzemi, F.; Fathallah, R. Three approaches for modeling residual stresses induced by orthogonal cutting of AISI316L. *Int. J. Mech. Sci.* **2018**, *135*, 253–260. [CrossRef]

22. Agmell, M.; Ahadi, A.; Stahl, J.E. A fully coupled thermomechanical two-dimensional simulation model for orthogonal cutting: Formulation and simulation. *Proc. Inst. Mech. Eng. Part B J. Eng. Manuf.* **2010**, *225*, 1735–1745. [CrossRef]

23. Uhlmann, E.; von der Schulenburg, M.G.; Zettier, R. Finite element modeling and cutting simulation of Inconel 718. *CIRP Ann. Manuf. Technol.* **2007**, *56*, 61–64. [CrossRef]

24. Fernández-Abia, A.I.; Barreiro, J.; de Lacalle, L.N.L. Behavior of austenitic stainless steels at high speed turning using specific force coefficients. *Int. J. Adv. Manuf. Technol.* **2012**, *62*, 505–515. [CrossRef]

25. Avilés, R.; Albizuri, J.; Rodríguez, A. Influence of low-plasticity ball burnishing on the high-cycle fatigue strength of medium carbon AISI 1045 steel. *Int. J. Fatigue* **2013**, *55*, 230–244. [CrossRef]

26. Matsumoto, Y.; Hashimoto, F.; Lahoti, G. Surface integrity generated by precision hard turning. *CIRP Ann. Manuf. Technol.* **1999**, *48*, 59–62. [CrossRef]

27. Guo, Y.B.; Warren, A.W.; Hashimoto, F. The basic relationships between residual stress, white layer, and fatigue life of hard turned and ground surfaces in rolling contact. *CIRP J. Manuf. Sci. Technol.* **2010**, *2*, 129–134. [CrossRef]

28. M'Saoubi, R.; Outeiro, J.C.; Changeux, B.; Lebrun, J.L.; Morão, D.A. Residual stress analysis in orthogonal machining of standard and resulfurized AISI 316l steels. *J. Mater. Process. Technol.* **1999**, *96*, 225–233. [CrossRef]

29. Hua, J.; Shivpuri, R.; Cheng, X. Effect of feed rate, workpiece hardness and cutting edge on subsurface residual stress in the hard turning of bearing steel using chamfer + hone cutting edge geometry. *Mater. Sci. Eng. A* **2005**, *394*, 238–248. [CrossRef]

30. Agmell, M.; Ahadi, A.; Gutnichenko, O.; Ståhl, J.E. The influence of tool micro-geometry on stress distribution in turning operations of AISI 4140 by FE analysis. *Int. J. Adv. Manuf. Technol.* **2017**, *89*, 3109–3122. [CrossRef]

 *materials*

Article

A Consistent Procedure Using Response Surface Methodology to Identify Stiffness Properties of Connections in Machine Tools

Jesus-Maria Hernandez-Vazquez [1,*], Iker Garitaonandia [1], María Helena Fernandes [1], Jokin Muñoa [2] and Luis Norberto López de Lacalle [1]

[1] Department of Mechanical Engineering, Faculty of Engineering, University of the Basque Country UPV/EHU, Plaza Ingeniero Torres Quevedo 1, E-48013 Bilbao, Spain; iker.garitaonandia@ehu.eus (I.G.); mariahelena.fernandes@ehu.eus (M.H.F); norberto.lzlacalle@ehu.eus (L.N.d.L.)
[2] IK4-IDEKO, Arriaga Kalea 2, E-20870 Elgoibar, Spain; jmunoa@ideko.es
* Correspondence: jesusmaria.hernandez@ehu.eus; Tel.: +34-94-601-7781

Received: 30 May 2018; Accepted: 11 July 2018; Published: 16 July 2018

Abstract: Accurate finite element models of mechanical systems are fundamental resources to perform structural analyses at the design stage. However, uncertainties in material properties, boundary conditions, or connections give rise to discrepancies between the real and predicted dynamic characteristics. Therefore, it is necessary to improve these models in order to achieve a better fit. This paper presents a systematic three-step procedure to update the finite element (FE) models of machine tools with numerous uncertainties in connections, which integrates statistical, numerical, and experimental techniques. The first step is the gradual application of fractional factorial designs, followed by an analysis of the variance to determine the significant variables that affect each dynamic response. Then, quadratic response surface meta-models, including only significant terms, which relate the design parameters to the modal responses are obtained. Finally, the values of the updated design variables are identified using the previous regression equations and experimental modal data. This work demonstrates that the integrated procedure gives rise to FE models whose dynamic responses closely agree with the experimental measurements, despite the large number of uncertainties, and at an acceptable computational cost.

Keywords: stiffness properties; parameter identification; connections; machine tool; response surface methodology; design of experiments; modal testing

---

1. Introduction

Machine tools are stationary, power-driven industrial devices used to manufacture workpieces under user and technological requirements. The most demanded requirements are accuracy and precision, which mainly depend on the static deformation and dynamic behavior of the machine tool under variable cutting forces. Assembly errors, tool trajectory errors, and the effect of thermal sources are also important issues [1]. Therefore, machine tool manufacturers devote strong efforts to perform the appropriate static, modal, and dynamic analyses of the machines, in order to determine the stresses and displacements, natural frequencies, and mode shapes. The final aim is to identify and analyze the vibration sources under different operating conditions, in order to minimize their effects on the surface finishing of the workpieces, stop the appearance of regenerative vibrations or chatter, and slow down the swift wear of the tools [2,3].

Today, the design process of modern machine tools is developed under virtual environments, where the finite element method (FEM) is widely used and particularly advised. The FEM provides a discretized model of the machine tool, whose purpose is to reproduce the real behavior of the structure.

Unfortunately, this approximate model shows physical uncertainties in the material properties and loads, and numerical uncertainties in the modeling and meshing processes, limiting the quality and reliability of the results achieved by this method. In addition, the dynamic modeling of the machine tool connections is quite complicated, because of their non-linear characteristics, which are functions of the interface pressure, contact area, and surface finishes. Therefore, it is essential to devote efforts so as to improve these models, so that their dynamic characteristics resemble the real ones in the frequency range of interest.

Updating techniques [4,5] are appropriate for achieving this objective, as they allow for improving the finite element (FE) models of mechanical systems by using experimental modal data. Bais et al. [6], Houming et al. [7], and Garitaonandia et al. [8,9] have successfully applied these techniques to machine tools. Nevertheless, when the number and range of uncertainties in the FE model are large, which leads to a poor correlation with experimental data, ill-conditioning problems and non-uniqueness solutions may arise, and definitely lead to a failure in the model updating procedure. Moreover, the updating techniques are associated with high computational costs, especially those based on sensitivity calculations.

In order to solve these problems, first, it is convenient to find out which design variables have the greatest influence on the dynamic characteristics of the mechanical system. In this respect, a review of the state of the art machine tools is presented by Brecher et al. [10]. Also, in the literature [11–13], the most significant design variables for different types of machine tools are introduced.

An adequate technique to perform this task is the design of experiments (DoE) methodology [14] and subsequent analysis of variance (ANOVA). The DoE statistically analyzes the effect of several factors and their combinations on a process or system, and allows for determining the significant ones. Also, it is a powerful tool to bring out the interactions between the variables.

On the other hand, an alternative option to address the time-consuming and numerical problems inherent to any iterative updating process involving FE computations, is to replace that model by an approximate model, a so-called surrogate or meta-model, which provides a more simple mathematical relationship between design variables and model responses. For instance, the coefficients of these mathematical formulations are a matter of concern for Lamikiz et al. [15], and are focused on complex new approaches for alternative processes on ruled surfaces [16].

The models developed through the response surface methodology (RSM) are widely used as meta-models [17]. This methodology is very useful in the design and optimization of new processes and products [18–20], especially if is affected by several variables, due to its low computational effort. Also, it can be used in an inverse sense, for the system identification applications, to find out the true values of the design variables that are inaccurately defined in a finite element model, with the help of experimental responses.

Some research related to the previous application of RSM can be found in the literature. Guo and Zhang [21] introduced the general procedure and applied it so as to update the stiffness values of three elements of the FE model of an H-shaped structure. In comparison with the traditional sensitivity-based model updating methods, the RSM-based method was found to be much more cost-efficient, providing, at the same time, accurate results. Later, Rutherford et al. [22] used RSM to perform stiffness and damping identification in two simple five degree of freedom systems, one linear and the other nonlinear. The final purpose was to demonstrate the suitability of this methodology for damage identification in civil structures. In conclusion, the procedure worked efficiently to determine the stiffness and damping coefficients in the simple linear system, while limited success was achieved in the nonlinear case.

Ren and Chen [23] updated the elastic modulus of the FE model of a full-size precast continuous box girder bridge and the cross-sectional area of two connections elements using response surface methodology. The results showed that the frequencies of the updated model were closer to the experimental ones, but there were still differences (up to 12%). Afterwards, Fang and Perera [24] used RSM to identify the structural damages in civil engineering structures. The procedure was tested on two real civil structures, a reinforced concrete frame and the I-40 bridge, and, in conclusion, it was found that the damage predictions in both structures agreed well with the experimental observations.

Recently, Sun and Cheng [25] updated the shear moduli of a honeycomb sandwich plate. That work focused on the analysis of the optimum number and position of the DoE samples, which, in conjunction with an adequate approximation algorithm, led to building the most accurate response surface model. In the end, the updated moduli were included in the FE model and the results of the dynamic analysis of this model corresponded with the experimental ones.

A common feature of these approaches is that few design variables are identified. In this regard, Ren and Chen [23] state that RSM is still not well tested in complex structures, such as machine tools, where there are a large number of uncertain parameters and the relationships between these parameters and responses are more intricate. Another drawback is that the number of responses is small, and when the modal frequencies are selected, it is always assumed that the identified values of the design variables leads to a preservation or, even, an improvement of the correlation features between the numerical and experimental mode shapes. This is true in the modeling processes were beam elements, spring elements, and lumped masses are primarily used, leading to simple mode shapes. However, according to Fang and Perera [24], in complex structures, the correlation between the mode shapes must be taken into consideration, because in these systems, multiple (coupled) modes or different sequences between the experimental and FE modes shapes could appear. Furthermore, Gallina et al. [26] state that when changes in the values of the design variables are introduced in a mechanical system, its modal responses may be affected by degenerative phenomena, such as, mode crossing, mode veering, and mode coalescence. Therefore, it is necessary to keep this problem in mind, because otherwise the quality of the RS model could be greatly affected and, as a result, could lead to important difficulties when comparing numerical and experimental mode shapes.

Finally, in these approaches, it is assumed that all of the selected design variables affect all of the responses. This is not necessarily true, and could cause an erroneous estimation of the response surfaces, due to the presence of redundant terms in the polynomial functions.

Therefore, the aim of this paper is to present a consistent methodology so as to identify the values of the design parameters that better reproduce the dynamic responses in the mechanical systems, with a large number of uncertainties, at a reasonable computational cost, and maintaining the correlation characteristics. Firstly, parameter screening using two-level fractional factorial designs is conducted in order to determine the design variables that specifically affect each modal response, because when there are many variables, not all of them influence all of the responses. Then, second order regression equations relating the design variables and responses are attained by means of central composite design-based RSM methodology and least squares techniques. In order to look simultaneously for the best adequacy and predictive capability of these functions, the non-significant terms are removed. For that purpose, a procedure based on statistic indicators, coefficients of determination R^2, and the t-statistic, is performed and, as a result, the number of terms in these equations is optimized. Next, using these functions, which temporarily replace the FE model, the updated values of design variables are identified by minimizing the residuals between numerical and experimental responses. In this study, a particular application of the so-called desirability function has been used to accomplish this task. Finally, the identified values are placed in the finite element model and the new dynamic responses are determined.

The proposed methodology is applied on a machining center and the comparison of the obtained results to the experimental ones demonstrates its efficiency and efficacy to update the FE models of complex mechanical systems with numerous uncertainties.

2. Dynamic Characteristics of the Machining Center

2.1. Finite Element Model

In this section, the dynamic characteristics of the DANOBATGROUP DS630 (DANOBATGROUP, Elgoibar, Spain) high speed horizontal machining center are presented. This machine tool has three linear axes and is made up of four main modules, namely, a bed frame, column, framework, and ram, which slide over roller type linear guideways. Two servo motors, directly coupled to ball-screws supported by bearings at both ends, provide the displacement along Y- and Z-axes, while the movement in the X-axis is performed by a linear motor. The machine is joined to a concrete basement by anchor bolts and leveling elements adjust and align the bed frame.

Firstly, a FE model of the machine tool has been defined (Figure 1), which consists of 12,804 nodes and 14,983 elements, mainly shell and solid brick elements. The connections between the different components and the connection to the foundation have been modeled by linear spring elements. In this way, the contact elements and friction coefficients in the FE formulation are avoided, reducing the complexity and keeping the model linear. These linear springs characterize the previously mentioned linear guideways, ball-screws, and bolts, and are incorporated into the FE model in their locations. The anchor bolts connecting the bed frame to the basement behave rigidly, so high stiffness spring elements have been used in the modeling process. Linear guideways have been modeled assigning average stiffness values in two directions, perpendicular and transverse to the direction of movement, based on the stiffness curves provided by the guideway supplier [27], and very low stiffness values along the directions where the movement is developed. A similar modeling has been followed for the ball-screws, although, in this case, low stiffness values have been set in perpendicular and transverse directions to the main movement [2,28,29]. Moreover, the tool holder has been simulated as a beam, and spindle, servo motors and the face milling cutter as lumped masses. Finally, solid brick elements have been used to model the primary and secondary sections of the linear motor with a spring element between them. Table 1 describes the main parameter values of the FE model.

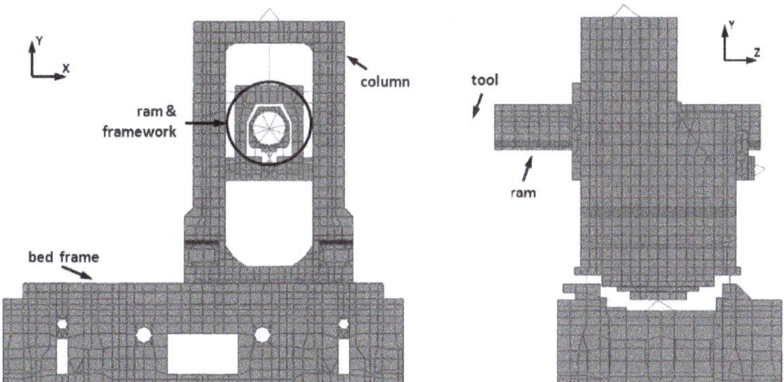

Figure 1. Finite element (FE) model of the machine tool.

Table 1. Main parameter values of the finite element (FE) model.

Parameter	Value(s)	Description
Stiffness X,Y,Z	750,750,750 N/μm	Connections foundation-bed frame.
Stiffness X,Y,Z	1,720,750 N/μm	Connections bed frame-column (guideway).
Stiffness X,Y,Z	720,1,750 N/μm	Connections column-framework (guideway).
Stiffness X,Y,Z	560,750,1 N/μm	Connections framework-ram (guideway).
Stiffness X	110 N/μm	Connection between primary and secondary sections of the linear motor.
Lumped mass	120 kg	Spindle.
Lumped mass	1.5 kg	Face milling cutter.
Stiffness Y	176.7 N/μm	Y ball-screw.

Table 1. *Cont.*

Parameter	Value(s)	Description
Lumped mass	100 kg	Servo motor Y.
Stiffness Z	172.7 N/μm	Z ball-screw.
Lumped mass	100 kg	Servo motor Z.
E, ρ	125 GPa, 7100 kg/m^3	Young's modulus (E) and mass density (ρ) of the bed frame and column (cast iron).
E, ρ	175 GPa, 7100 kg/m^3	Young's modulus (E) and mass density (ρ) of the framework and ram (cast iron GGG70).
E, ρ	210 GPa, 7850 kg/m^3	Young's modulus (E) and mass density (ρ) of specific parts of the machine tool.

Then, using the Lanczos solver, the free motion of the structure has been analyzed by calculating the natural frequencies and mode shapes from the assembled mass and stiffness matrices of the numerical model. As the connection between the bed frame and the basement is considered in the FE model, these modal parameters correspond to the in situ configuration of the machine. According to several tests developed under chatter conditions [30], the frequency range of interest has been defined as 10 Hz to 120 Hz. The natural frequencies are shown in Table 2.

Table 2. Natural frequencies of the initial FE model. FEA—finite element analysis.

f_{FEA1}	f_{FEA2}	f_{FEA3}	f_{FEA4}	f_{FEA5}	f_{FEA6}
33.7	60.4	69.7	73.9	87.5	112.3

2.2. Experimental Modal Analysis

In order to experimentally determine the dynamic characteristics of the machining center, a multiple reference impact test was performed, using, as references, point 5 along X and Y directions and by exciting the system with an instrumented hammer. The translational acceleration responses in the X-, Y-, and Z-axes were measured in 75 points using triaxial accelerometers, so the accelerance frequency response functions (FRFs) corresponding to 225 degrees of freedom were obtained. The total number of measured FRFs was 450. Figure 2 shows a wire frame model representation of the test structure. The references are identified with arrows.

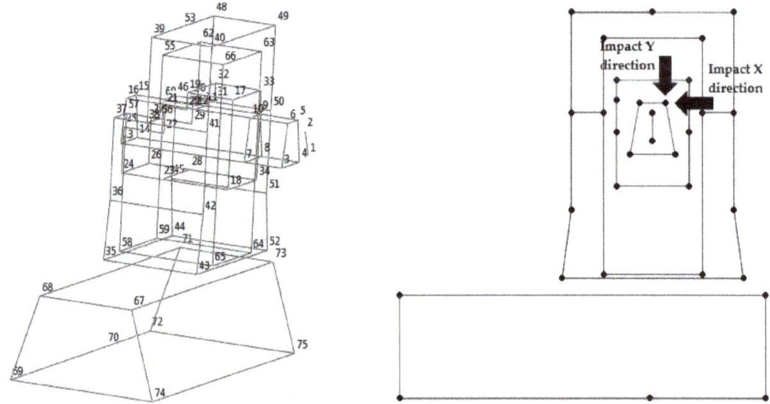

Figure 2. Experimental model of the machining center.

From the measured FRFs, a polyreference Least Squares Complex Frequency (pLSCF) estimator was used to extract the system modal parameters. Table 3 shows the natural frequencies and a brief description of the different mode shapes.

Table 3. Natural frequencies and mode shapes obtained by experimental modal analysis.

Mode Order	f_{exp} (Hz)	Damping Ratio (%)	Description of the Mode Shape
1	33.7	4.8	Rotation of the whole structure around the X-axis.
2	60.5	3.3	Translation along Y of the framework and ram.
3	65.9	3.5	Rotation of the upper part of the machine around the Y-axis.
4	77.2	5.4	Rotation of the upper part of the machine around the Y-axis, but now ram is in counter-phase.
5	84.0	5.1	Rotation of framework and ram around the X-axis.
6	106.5	3.3	Rotation of the whole structure around the Y-axis. Ram is in counter-phase.

2.3. Comparison between FE and Experimental Modal Data

At this point, there are two sets of different results, related to the numerical and experimental models. The next step is to evaluate the correspondence between them, because it is necessary that both models show a considerable degree of correlation, in order to improve the FE model successfully.

Firstly, the geometrical correlation has been developed to match the different coordinate and unit systems used in the models, and, then, the mode shape correlation has been performed to establish a reliable pairing between the numerical and experimental modes. A very useful indicator to compare and contrast the modal vectors from the different sources is the modal assurance criterion (MAC) [31]. The MAC shows the degree of linearity between two modal vectors, φ_{FEA} (FEA—finite element analysis) and φ_{exp}, as follows:

$$MAC\left(\varnothing_{FEA}, \varnothing_{exp}\right) = \frac{\left(\varnothing_{FEA}^{T} \cdot \varnothing_{exp}\right)^{2}}{\left(\varnothing_{FEA}^{T} \cdot \varnothing_{FEA}\right) \cdot \left(\varnothing_{exp}^{T} \cdot \varnothing_{exp}\right)} \tag{1}$$

The MAC can take on values from 0, showing a lack of correspondence between the modal vectors, to 1, which means that the modal vectors are the same.

Table 4 shows the frequency differences and MAC values between the FE and experimental responses, wherein the MAC values corresponding to the paired mode shapes have been bolded.

Table 4. Frequency differences and modal assurance criterion (MAC) values.

FEA Order	f_{FEA} (Hz)	f_{exp1} 33.7	f_{exp2} 60.5	f_{exp3} 65.9	f_{exp4} 77.2	f_{exp5} 84.0	f_{exp6} 106.5	Diff. (Hz)	Diff. (%)	Pair Number
1	33.7	**96.6**	0.6	0.3	0.0	1.9	0.1	0.0	0.0	1
2	60.4	1.7	**98.8**	1.2	0.0	1.3	0.1	−0.1	−0.2	2
3	69.7	0.0	0.0	**76.3**	4.3	0.0	1.5	3.8	5.8	3
4	73.9	0.1	0.0	30.3	**89.0**	1.3	3.7	−3.3	−4.3	4
5	87.5	1.9	0.1	1.2	0.9	**91.0**	0.1	3.5	4.2	5
6	112.3	0.0	1.0	0.1	0.0	0.1	**70.3**	5.8	5.4	6

These correlation results can be considered sufficient for a large number of practical applications [32,33], as the mean frequency difference is 3.3%, and the mean MAC value is 87.0%. Nevertheless, there are still moderate differences between several natural frequencies, which confirm that it is necessary to improve the FE model. Therefore, the main goal of the following procedure will be to match the numerical frequencies to the experimental ones, while maintaining or improving the MAC pairing values.

3. Methods: Design of Experiments, Response Surface Methodology, and Desirability Functions

3.1. Two-Level Designs for Parameter Screening

Among the different types of experimental designs [14], the factorial designs are widely used to identify, at the initial stages, the main variables that affect any process or system (i.e., as screening experiments). The basic design is a two-level or 2^k design, where k is the number of variables and each of them takes an upper and a lower level. A complete trial of such a design needs 2^k runs and allows for estimating the linear effects of the k variables and their interactions.

Nevertheless, as the number of variables, k, increases, the number of runs in the trial also increases, but dramatically, and interactions between three, four, and more variables appear. Assuming that the highest interactions are negligible, it would be possible to obtain information concerning the effects of the variables and low-order interactions by running a part or fraction of the complete factorial design, 2^{k-p}, where p indicates the fraction chosen ($1/2^p$). The so-called resolution V design is especially interesting, which provides information about the contribution of variables and two-factor interactions, mixed with higher-order interactions. As these are negligible, the fractional designs are better than the complete factorial designs, because the number of runs diminishes considerably.

Once the trial has been finished, the next step is to identify the significant factors and interactions by performing an analysis of variance on the results. According to ANOVA, the variability of the results in an experiment that is dependent on several variables, is the sum of variability due to each factor, plus that contributed by the interaction between the factors, and that added by the internal error. Also, using ANOVA, the sum of squares (SS) can be used as a measure of the overall variability, so that the greater the SS due to a factor, the larger its importance on the process or system. Thus, it will be possible to find out which variables and interactions are the most significant.

3.2. Response Surface Methodology to Develop an Optimal Mathematical Model

The purpose of response surface methodology is to build an explicit function to approximate the actual relationship between the variables, x_i, and a response, y, involved in an engineering problem. That function, preferably a low-order polynomial, is in fact a regression model, less expensive to evaluate, which can be used to predict the response developed in the system under a specific combination of variables.

In general, the behavior of the industrial processes and mechanical systems cannot be explained by linear functions [19,23,34], so, in the following section, the second-order models (Equation (2)) and the experimental designs that are preferable to adequately estimate these models will be examined.

$$y = \beta_0 + \sum_{i=1}^{k} \beta_i \cdot x_i + \sum_{i=1}^{k} \beta_{ii} \cdot x_i^2 + \sum_{i=1}^{k-1} \sum_{j=i+1}^{k} \beta_{ij} \cdot x_i \cdot x_j + \varepsilon \tag{2}$$

In Equation (2), β_0 is the average value of response; y, β_i, β_{ii}, and β_{ij} are the partial regression coefficients; ε is the error term; and k is the number of variables.

One of the most popular designs for fitting second-order models is the central composite design (CCD). It is built in a sequential way, based on a two-level factorial (2^k) design, plus $2k$ axial and n_C center points. The points added to factorial design allow an efficient estimation of the possible curvature of the model.

Firstly, a set of responses **y** is obtained on the completion the experiments of the central composite design. Then, the values of these responses and design variables are substituted in Equation (2), and rewritten in matrix form as follows:

$$\mathbf{y} = \mathbf{X} \cdot \mathbf{\beta} + \mathbf{\varepsilon} \tag{3}$$

Equation (3) is solved using the least squares method, by minimizing the sum of the squares of the errors ε_i. That leads to a least squares estimator of β, as follows:

$$\mathbf{b} = (\mathbf{X}^T \cdot \mathbf{X})^{-1} \cdot \mathbf{X}^T \cdot \mathbf{y} \tag{4}$$

At this point, an initial second-order model is completely defined using all of the design variables and interactions. Then, it is necessary to perform an analysis of variance to check the significance of each parameter and the adequacy of the regression model. For this last purpose, various statistical parameters can be used, such as, the coefficient of determination R^2, the adjusted R^2, and the predicted R^2 [17]. These coefficients are all expected to be close to 1.0, which would mean that the regression model, y_{RSM}, explains the response, y, properly and that it also predicts adequately new responses.

Nevertheless, if there are substantial differences between them, the least significant parameter is removed using the t-test, and a new regression model, Equation (2), is built and the analysis is repeated until the remaining parameters are all significant. On the completion of the iteration process, the optimum response surface model can be considered as adequate to carry on the next stage of the improvement procedure.

3.3. Identification of Updated Values of the Design Variables using the Optimum RS Model

Once the mathematical relationships between the design variables and responses have been established, the final step is to identify those values of the design variables that lead to the responses that better fit the experimental ones. This is actually an inverse multi-objective constrained optimization problem, and nonlinear programming techniques can be used to solve it.

Another alternative approach is based on the so-called desirability function [35], which is explained in the following. Firstly, each estimated response, y_{RSMi}, is turned into a desirability function, d_i, as follows:

$$d_i = \left(\frac{y_{RSMi} - y_{LOWi}}{y_{OBJi} - y_{LOWi}} \right)^S, \; y_{LOWi} < y_{RSMi} < y_{OBJi} \tag{5}$$

$$d_i = \left(\frac{y_{RSMi} - y_{UPi}}{y_{OBJi} - y_{UPi}} \right)^T, \; y_{OBJi} < y_{RSMi} < y_{UPi} \tag{6}$$

$$d_i = 0, \; y_{RSMi} < y_{LOWi} \text{ and } y_{UPi} < y_{RSMi} \tag{7}$$

where y_{OBJi} is the target experimental response, and y_{LOWi} and y_{UPi} are the lower and upper limits for each response (Figure 3).

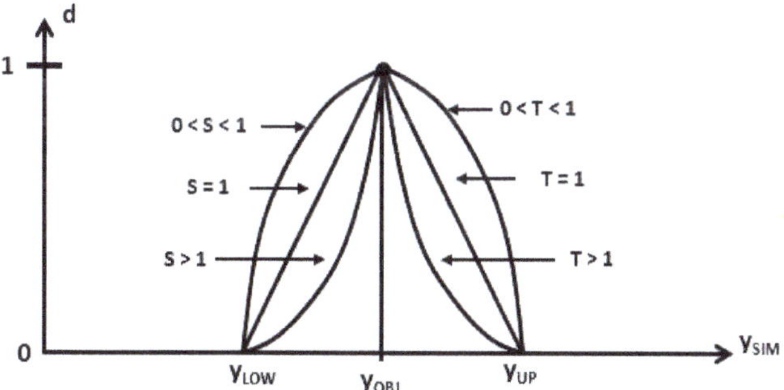

Figure 3. Individual desirability function.

Then, a global desirability function, D, is built as the geometric mean of individual desirabilities, d_i, as follows:

$$D = (d_1 \cdot d_2 \cdot \ldots \cdot d_u)^{1/u} \tag{8}$$

where u is the total number of the experimental responses.

Finally, the results are ranked in decreasing desirability order and the values of the design variables that maximize the global desirability D are selected.

4. Case Study

4.1. Initial Selection of Candidate Design Variables

In order to improve the FE model, first, it is necessary to select the design variables to work with. There are a large number of design parameters to be considered in this machining center, but, in fact, the main uncertainties in the FE model are concentrated on connections, as follows:

1. Stiffness values of the connection elements between main components of the machine tool (Figure 4);
2. Stiffness values assigned to the elements that attach the machine tool to the foundation (Figure 5); and
3. Stiffness value along X direction of the connection element, between the primary and secondary sections of the linear motor.

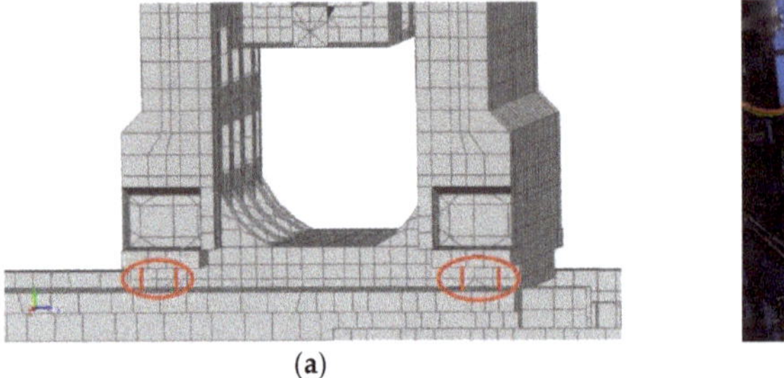

(a)

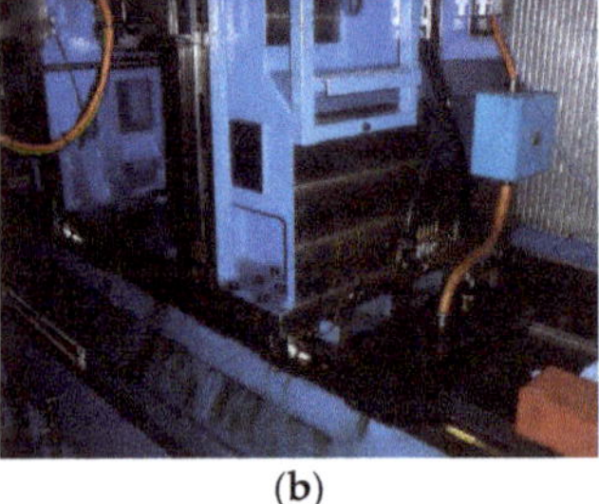

(b)

Figure 4. Connections between bed frame and column in a (**a**) FE model and (**b**) photograph.

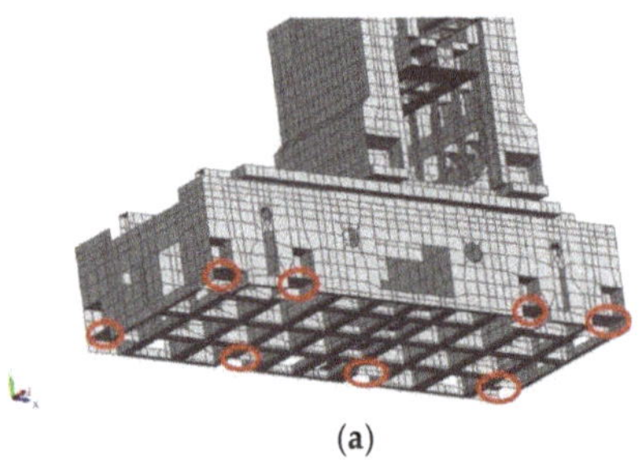

(a)

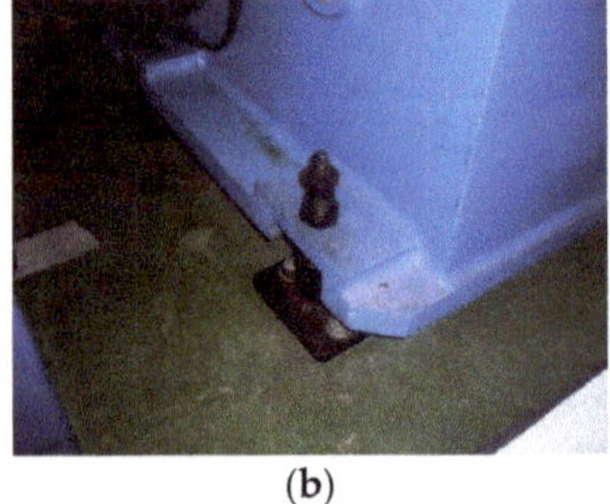

(b)

Figure 5. Supporting mounts of the machining center in a (**a**) FE model and (**b**) photograph.

Nevertheless, the number of variables is still large, three stiffness values for the joints to the foundation, the stiffness value for the inner connection of the linear motor, and six stiffness values for the connections between the modules of the machining center (Table 5). Therefore, first, it is necessary to determine which variables affect, to a large extent, each model response and, hence, remove those whose influence is negligible. For this purpose, a resolution V design would be the most convenient (i.e., in our case, a 2^{10-3} fractional design with 128 runs) [14]. However, it is still too laborious to manage such a number of runs. Furthermore, some parameter combinations could lead to inappropriate model responses, due to the presence of the constraints among them. Thus, in order to facilitate the analysis and gain a progressive comprehension of the significance of each design variable and interaction, instead of a single 2^{10-3} fractional design, an alternative trial with seven 2^{5-1} designs (16 runs each) has been performed (Table 5). Each variable has been paired up with the rest at least once, so that after the completion of the whole set of 2^{5-1} experiments, it has been possible to look into the effects of all of the design variables and two-factor interactions by means of ANOVA.

Table 5. List of variables used in fractional factorial designs 2^{5-1}.

Connection	Design Variable	Code	Lower Bound	Nominal Value	Upper Bound	2^{5-1} Design
Foundation—bed frame	Stiffness X	k_{X21}	600	750	1050	1,3,5,6
Foundation—bed frame	Stiffness Y	k_{Y22}	600	750	1500	1,3,5,6
Foundation—bed frame	Stiffness Z	k_{Z63}	600	750	1050	1,3,5,6
Linear motor (inner)	Stiffness X	k_{X210}	80	110	160	2,4,6
Bed frame—column	Stiffness Y	k_{Y3}	450	720	1125	2,4,5,7
Bed frame—column	Stiffness Z	k_{Z4}	400	750	900	2,4,5,6
Column—framework	Stiffness X	k_{X11}	450	720	1125	2,3,7
Column—framework	Stiffness Z	k_{Z13}	400	750	900	2,3,7
Framework—ram	Stiffness X	k_{X8}	210	560	900	1,4,7
Framework—ram	Stiffness Y	k_{Y9}	450	750	1000	1,4,7

Prior to conducting the fractional designs, the range of each variable was decided (Table 5) according to load–deformation curves [27] and previous works [30].

On completion of the trial, the total corrected sum of squares, SS_T (Figure 6), and the sum of squares of each factor and the two-factor interactions, mixed with higher interactions (SS_i and SS_{ij}) (Figures 7 and 8), have been obtained as a measure of the variability, for all of the frequencies and MAC values. Firstly, for each response, the SS_T has been examined, as some designs add much more variability than others, because of the variables involved. Thus, Figure 6 shows that MAC_1 and MAC_2 are not affected by the changes in the design variables, and that the variability of f_{FEA2} is negligible. As this frequency matches its experimental pair (Table 4), it has been omitted in later analyses.

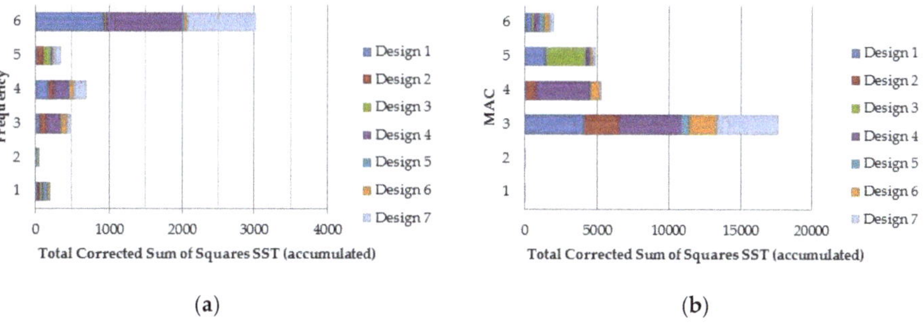

Figure 6. Total corrected sum of squares (SS_T) for natural frequencies (**a**) and modal assurance criterion (MAC) values (**b**).

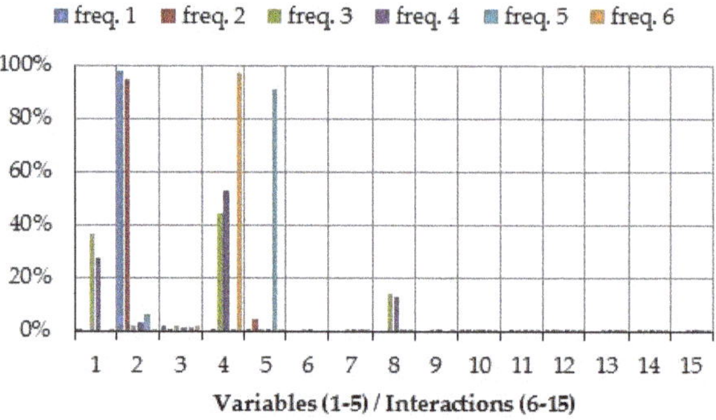

Figure 7. Percentage contribution for natural frequencies in design number 4: 1–5, design variables k_{X210}, k_{Y3}, k_{Z4}, k_{X8}, and k_{Y9}, respectively; 6–15, two factor interactions k_{X21}–k_{Y3}, k_{X210}–k_{Z4}, k_{X210}–k_{X8}, k_{X210}–k_{Y9}, k_{Y3}–k_{Z4}, k_{Y3}–k_{X8}, k_{Y3}–k_{Y9}, k_{Z4}–k_{X8}, k_{Z4}–k_{Y9}, and k_{X8}–k_{Y9}, respectively.

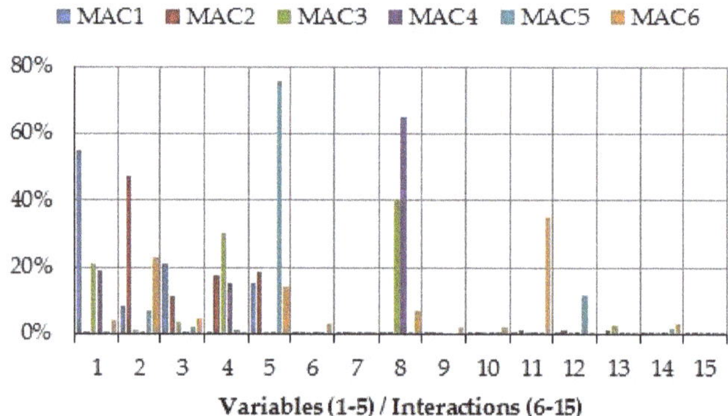

Figure 8. Percentage contribution for MAC values in design number 4: 1–5, design variables k_{X210}, k_{Y3}, k_{Z4}, k_{X8}, and k_{Y9}, respectively; 6–15, two factor interactions k_{X210}–k_{Y3}, k_{X210}–k_{Z4}, k_{X210}–k_{X8}, k_{X210}–k_{Y9}, k_{Y3}–k_{Z4}, k_{Y3}–k_{X8}, k_{Y3}–k_{Y9}, k_{Z4}–k_{X8}, k_{Z4}–k_{Y9}, and k_{X8}–k_{Y9}, respectively.

In order to determine which variables and two-factor interactions provide the largest influence on the variability of natural frequencies and mode shapes, SS_i and SS_{ij} have been gradually analyzed in each 2^{5-1} design. As an example, Figures 7 and 8 illustrate the percentage contribution of each parameter in design number 4.

From Figure 7, it is found that the design variables (1–5) have a larger influence on the natural frequencies than the two-factor interactions (6–15). For instance, design variable 2 (k_{Y3}) dominates the 1st and 2nd natural frequencies, design variable 4 (k_{X8}) has a huge influence on the 6th natural frequency and an important weight on the 3rd and 4th natural frequencies, and design variable 5 (k_{Y9}) governs the 5th natural frequency. However, only parameter 8 (interaction k_{X210}–k_{X8}) slightly affects the 3rd and 4th natural frequencies. Also, design variable 3 (k_{Z4}) does not seem to have a notable influence on any natural frequency.

On the other hand, from Figure 8, some of the interactions play significant roles in MAC values (3, 4, and 6, mainly). In fact, MAC_4 is heavily affected by parameter 8 (interaction k_{X210}–k_{X8}), which also provides an important contribution to the variance of MAC_3. Also, parameter 11 (interaction k_{Y3}–k_{X8}) causes approximately 35% of the total variability to MAC_6. Nevertheless, in general, the individual design variables have a greater influence than the interactions.

This analysis has been repeated for each 2^{5-1} design, so that it has been possible to gradually gain a better insight into the influence of the design variables and two-factor interactions on the responses. Finally, those factors providing more than 99% of the total variability for the frequencies, and 97.5% for the MAC values have been selected (Tables 6 and 7) to continue with the improvement procedure.

Table 6. Summary of variables affecting responses.

Connection	Design Variable	Code	f_{FEA1}	f_{FEA3}	f_{FEA4}	f_{FEA5}	f_{FEA6}	MAC_3	MAC_4	MAC_5	MAC_6
Foundation—bed frame	Stiffness X	k_{X21}									
Foundation—bed frame	Stiffness Y	k_{Y22}	X			X				X	X
Foundation—bed frame	Stiffness Z	k_{Z63}				X				X	X
Linear motor (inner)	Stiffness X	k_{X210}		X	X			X	X		X
Bed frame—column	Stiffness Y	k_{Y3}	X	X	X		X	X	X	X	X
Bed frame—column	Stiffness Z	k_{Z4}	X	X	X		X	X	X		X
Column—framework	Stiffness X	k_{X11}									
Column—framework	Stiffness Z	k_{Z13}				X				X	X
Framework—ram	Stiffness X	k_{X8}		X	X		X	X	X		X
Framework—ram	Stiffness Y	k_{Y9}				X				X	X

Table 7. Summary of two-factor interactions affecting responses.

Responses	Interactions
f_{FEA1}	k_{Y22}–k_{Y3}
f_{FEA3}	k_{X210}–k_{Y3}, k_{X210}–k_{Z4}, k_{X210}–k_{X8}, k_{Y3}–k_{X8}
f_{FEA4}	k_{X210}–k_{Z4}, k_{X210}–k_{X8}, k_{Z4}–k_{X8}
f_{FEA5}	k_{Y22}–k_{Z63}, k_{Y22}–k_{Y9}, k_{Z63}–k_{Z13}, k_{Z63}–k_{Y9}
f_{FEA6}	k_{Y3}–k_{Z4}, k_{Y3}–k_{X8}, k_{Z4}–k_{X8}
MAC_3	k_{X210}–k_{Y3}, k_{X210}–k_{Z4}, k_{X210}–k_{X8}, k_{Y3}–k_{Z4}, k_{Y3}–k_{X8}, k_{Z4}–k_{X8}
MAC_4	k_{X210}–k_{Y3}, k_{X210}–k_{Z4}, k_{X210}–k_{X8}, k_{Y3}–k_{X8}
MAC_5	k_{X21}–k_{X11}, k_{X21}–k_{X8}, k_{Y22}–k_{Z63}, k_{Y22}–k_{Z13}, k_{Y22}–k_{Y9}, k_{Z63}–k_{Z13}, k_{Z63}–k_{Y9}, k_{Y3}–k_{Y9}, k_{Z13}–k_{Y9}
MAC_6	k_{Y22}–k_{Z63}, k_{Y22}–k_{Z4}, k_{X210}–k_{Y3}, k_{X210}–k_{X8}, k_{Y3}–k_{Z13}, k_{Y3}–k_{X8}, k_{Z4}–k_{Y9}, k_{X11}–k_{Y9}, k_{Z13}–k_{X8}

Several conclusions can be inferred from Tables 6 and 7, as follows:

- The fractional factorial experiments have allowed for finding out the design variables and interactions that affect the responses. Therefore, the screening experiment has satisfactorily achieved the initial goal.
- Two design variables do not influence the natural frequencies, namely the stiffness k_{X11} and horizontal stiffness k_{X21} of the connections to the foundation.
- Three design variables are only significant for one natural frequency (f_{FEA5}), namely the transverse stiffness k_{Z63} between the bed frame and foundations, and two stiffnesses between the modules of the machine tool, k_{Z13} and k_{Y9}.
- MAC_5 and MAC_6 are affected by the largest number of interactions. Furthermore, some of them include design variables that do not influence them individually, for example, interaction k_{X11}-k_{X21}. This situation only appears in these two responses. In addition, the total number of design variables, considered both individually and in interactions, which affect each of these responses is nine (i.e., almost all). Nevertheless, along the complete set of fractional designs, the MAC_5 values were always larger than 80% and the MAC_6 values ranged from 68% to 73%. Thus, it has been decided not to carry on with the study of these responses, because the number of involved variables would lead to a costly analysis in the next step, while the benefits would be quite poor.
- The natural frequencies f_{FEA3} and f_{FEA4} and the corresponding MAC values depend on the same group of design variables, k_{X210}, k_{Y3}, k_{Z4}, and k_{X8}. In addition, the natural frequency f_{FEA6} is dependent on three of these variables, k_{Y3}, k_{Z4}, and k_{X8}. Therefore, in the next step of the improvement process, these three natural frequencies will be analyzed together, so as to reduce

the number of experiments necessary to define their meta-models. In addition, it is interesting to note that the mode shapes associated to these frequencies take place in plane XZ.

- The natural frequency f_{FEA5} is affected by four variables that do not have any influence on the frequencies f_{FEA3}, f_{FEA4}, and f_{FEA6}, and, vice versa, the variables that affect these three frequencies do not provide any variability to the natural frequency f_{FEA5}. Moreover, some of the design variables representing stiffness in the X direction, k_{X8} and k_{X210}, do not affect the 1st and 5th mode shapes, whose principal movement is in plane YZ. Thus, it is concluded that the design variables are working collectively.

4.2. Development of Explicit Relationships between Design Variables and Responses

The next step of the improvement procedure is the definition of the mathematical functions that relate the variables and responses of Tables 6 and 7, using response surface methodology.

Taking into consideration the conclusions drawn in the previous section, referring to the collective influence of the design variables on te responses, three different central composite designs have been developed, as follows:

- Central composite (CC) design 1: including f_{FEA1} and design variables k_{Y22}, k_{Y3}, and k_{Z4}. Although it would seem unnecessary to search for this relationship, because f_{FEA1} is already matched, as it is influenced by the design variables that also influence other frequencies, any change on them would affect this frequency too. So, it is indispensable to know this relationship.
- CC design 2: with the following responses f_{FEA3}, f_{FEA4}, f_{FEA6}, MAC_3, and MAC_4, and design variables k_{X210}, k_{Y3}, k_{Z4}, and k_{X8}.
- CC design 3: including f_{FEA5} and design variables k_{Y22}, k_{Z63}, k_{Z13}, and k_{Y9}.

Each central composite design has been developed through 2^k points from the factorial design with k factors; $2k$ axial points face centered, where one variable takes the upper and lower limits and the others have mean values; and finally one central point. Thus, a total number of 65 experiments (15, 25, and 25, respectively) have been completed. Also, prior conducting the experiments, the design variables must be normalized to values (-1), (0), and $(+1)$, which stand for the lower bound, mean value, and upper bound of each variable, respectively (Equation (9)).

$$X_i = 2 \cdot \left(\frac{k_i - k_{LOWi}}{k_{UPi} - k_{LOWi}} \right) - 1 \tag{9}$$

where k_{UPi} and k_{LOWi} are the upper and lower limits defined in Table 5.

Firstly, the study has focused on the relationship between f_{FEA1} and the variables that affect it. Using the results obtained from the central composite design, an initial second-order model with all of the design variables and interactions have been developed, namely Equation (10), as follows:

$$\begin{aligned} f_{RSM1} = {} & 34.9265 + 1.1893 \cdot X_{Y3} + 1.3810 \cdot X_{Y22} + 0.1806 \cdot X_{Z4} + 0.1531 \cdot X_{Y3} \cdot X_{Y22} + \\ & + 0.0065 \cdot X_{Y3} \cdot X_{Z4} + 0.0177 \cdot X_{Y22} \cdot X_{Z4} - 0.4593 \cdot (X_{Y3})^2 - 0.5171 \cdot (X_{Y22})^2 - 0.0559 \cdot (X_{Z4})^2 \end{aligned} \tag{10}$$

where f_{RSM1} is the estimated response corresponding to the first natural frequency, f_{FEA1}.

Then, the significance and the predictive capability of the regression model as well as the significance of the individual regression coefficients have been examined by means of the coefficients of determination, R^2 and t-tests (Tables 8 and 9).

Table 8. Significance and predictive capability of the regression models.

Coef.	Initial Model	Model 2	Model 3
R^2	0.9997	0.9997	0.9996
$_{adj}R^2$	0.9984	0.9986	0.9985
$_{pred}R^2$	0.9974	0.9982	0.9981

Table 9. Significance of individual regression coefficients: *t*-statistic.

Term	Coef.	Initial Model	Model 2	Model 3
Constant	b_0	1410.7606	1521.1830	1482.9372
k_{Y3}	b_1	81.6470	88.0377	85.8242
k_{Y22}	b_2	94.8073	102.2280	99.6577
k_{Z4}	b_3	12.3961	13.3664	13.0303
k_{Y3}–k_{Y22}	b_{12}	9.4043	10.1404	9.8854
k_{Y3}–k_{Z4}	b_{13}	0.4007	-	-
k_{Y22}–k_{Z4}	b_{23}	1.0838	1.1686	-
$(k_{Y3})^2$	b_{11}	−5.9900	−17.2415	−16.8080
$(k_{Y22})^2$	b_{22}	−18.0039	−19.4131	−18.9250
$(k_{Z4})^2$	b_{33}	−1.9461	−2.0984	−2.0457

In Table 8, the R^2 coefficients for the initial model show that the regression function explains the observed responses in the central composite design experiment quite well. Also, $_{pred}R^2$ suggests that the model will fit new responses remarkably.

In order to test the significance of the different terms of Equation (10), the *t*-statistics [17] for coefficients b_j of the initial model have been calculated (Table 9). Using a 95% confidence level ($\alpha = 0.05$), these terms must be larger than the value of the *t*-distribution $t_{0.025,5} = 2.571$, and it is shown that the corresponding *t*-statistics for b_{13}, b_{23}, and b_{33} are smaller. Thus, these three terms are non-significant in the regression model and can be removed. As it is convenient to eliminate one term in each step, b_{13} has been picked out first, as its *t*-statistic was the smallest one.

Then, the same procedure, explained in previous paragraphs, has been repeated for this new model, without the term b_{13}. The results in Table 8 (model 2) show that both $_{adj}R^2$ and $_{pred}R^2$ have increased slightly. Therefore, as expected, removing the non-significant terms in the regression model has led to a more adequate model. Nevertheless, in the regression equation still there are non-significant terms (Table 9), as some *t*-statistics are smaller than $t_{0.025,6} = 2.447$. So, coefficient b_{23} has been removed, and a new model (model 3) has been made. In this case, $_{adj}R^2$ and $_{pred}R^2$ have reduced slightly. Although the differences are totally negligible, considering that this regression model would lead to poorer results than the previous one, the iteration process has been stopped and the preceding regression model has been selected. In Table 10, the results of analysis of variance (ANOVA) of the final model are summarized.

Table 10. Analysis of variance (ANOVA) for the response surface model.

Source	Sum of Squares	Degree of Freedom	Mean Squares	F Value	p Value
Regression	36.114	8	4.514	2473.8	0.000
Residual	0.011	6	0.002	-	-
Total	36.125	14	2.580	-	-

From Table 10, it is shown that the model is highly significant ($p < 0.001$), and confirms that it can be used to simulate the response adequately.

So, the final regression equation for the first natural frequency is as follows:

$$f_{\text{RSM1}} = 34.9265 + 1.1893 \cdot X_{Y3} + 1.3810 \cdot X_{Y22} + 0.1806 \cdot X_{Z4} + 0.1531 \cdot X_{Y3} \cdot X_{Y22} + \\ + 0.0177 \cdot X_{Y22} \cdot X_{Z4} - 0.4593 \cdot (X_{Y3})^2 - 0.5171 \cdot (X_{Y22})^2 - 0.0559 \cdot (X_{Z4})^2 \tag{11}$$

In this case, only one b_j element has been removed and the optimum model is very similar to the initial one. As it will be shown later, in some regression equations, more b_j elements will be eliminated, mainly those referred to in second-order terms (see, for example, Equation (17)).

A similar procedure has been followed for f_{FEA5}. In this case, the regression equation is as follows:

$$f_{\text{RSM5}} = 87.4450 + 0.7551 \cdot X_{Y9} + 0.5414 \cdot X_{Y22} + 2.3563 \cdot X_{Z13} + 0.7345 \cdot X_{Z63} + \\ + 0.0618 \cdot X_{Y9} \cdot X_{Y22} + 0.1096 \cdot X_{Y9} \cdot X_{Z63} + 0.0785 \cdot X_{Y22} \cdot X_{Z63} + 0.0725 \cdot X_{Z13} \cdot X_{Z63} - \\ - 0.2318 \cdot (X_{Y9})^2 - 0.1674 \cdot (X_{Y22})^2 - 0.8645 \cdot (X_{Z13})^2 - 0.2321 \cdot (X_{Z63})^2 \tag{12}$$

The coefficients of determination, $R^2 = 0.9989$, $_{\text{adj}}R^2 = 0.9977$, and $_{\text{pred}}R^2 = 0.9984$ (Table 11), have led again to a reliable model.

Table 11. Summary of coefficients of determination. RSM—response surface methodology.

Coefficients of Determination	f_{RSM1}	f_{RSM3}	f_{RSM4}	f_{RSM5}	f_{RSM6}	MAC_{RSM3}	MAC_{RSM4}
R^2	0.9997	0.9910	0.9870	0.9989	0.9995	0.9304	0.9271
$_{\text{adj}}R^2$	0.9986	0.9845	0.9805	0.9977	0.9993	0.8956	0.9080
$_{\text{pred}}R^2$	0.9982	0.9746	0.9750	0.9948	0.9986	0.8692	0.8816

Finally, the rest of the responses, f_{FEA3}, f_{FEA4}, and f_{FEA6}, along with MAC_3 and MAC_4, have been analyzed altogether, because the variables that affect them were the same. The final regression equations are shown in Equations (13)–(17), and the coefficients of determination in Table 11.

$$f_{\text{RSM3}} = 71.1723 + 2.3373 \cdot X_{X8} + 2.2295 \cdot X_{X210} + 0.4979 \cdot X_{Y3} + 0.4722 \cdot X_{Z4} + \\ + 1.3206 \cdot X_{X8} \cdot X_{X210} + 0.2019 \cdot X_{X8} \cdot X_{Y3} - 0.1702 \cdot X_{X210} \cdot X_{Y3} + 0.1562 \cdot X_{X210} \cdot X_{Z4} - \\ - 2.1869 \cdot (X_{X8})^2 - 0.6804 \cdot (X_{X210})^2 \tag{13}$$

$$f_{\text{RSM4}} = 76.7443 + 2.6736 \cdot X_{X8} + 1.7965 \cdot X_{X210} + 0.6450 \cdot X_{Y3} + 0.5125 \cdot X_{Z4} - \\ - 1.3133 \cdot X_{X8} \cdot X_{X210} + 0.3977 \cdot X_{X8} \cdot X_{Z4} - 0.1550 \cdot X_{X210} \cdot X_{Z4} - 1.2361 \cdot (X_{X8})^2 \tag{14}$$

$$f_{\text{RSM6}} = 118.5952 + 8.1498 \cdot X_{X8} + 0.3207 \cdot X_{Y3} + 1.0787 \cdot X_{Z4} - 0.1247 \cdot X_{X8} \cdot X_{Y3} - \\ - 0.2232 \cdot X_{X8} \cdot X_{Z4} + 0.1697 \cdot X_{Y3} \cdot X_{Z4} - 3.2813 \cdot (X_{X8})^2 - 0.2193 \cdot (X_{Y3})^2 - 0.3628 \cdot (X_{Z4})^2 \tag{15}$$

$$MAC_{\text{RSM3}} = 63.4984 - 9.4610 \cdot X_{X8} + 8.9519 \cdot X_{X210} + 2.1027 \cdot X_{Y3} - 3.8195 \cdot X_{Z4} + \\ + 10.4715 \cdot X_{X8} \cdot X_{X210} + 1.1365 \cdot X_{X8} \cdot X_{Y3} - 2.6332 \cdot X_{X8} \cdot X_{Z4} + 3.0198 \cdot (X_{Z4})^2 \tag{16}$$

$$MAC_{\text{RSM4}} = 84.2937 + 5.9049 \cdot X_{X8} - 5.2622 \cdot X_{X210} - 1.0676 \cdot X_{Z4} + 12.2980 \cdot X_{X8} \cdot X_{X210} - \\ - 9.2492 \cdot (X_{X8})^2 \tag{17}$$

In Table 11, the coefficients of determination for f_{RSM1}, f_{RSM5}, and f_{RSM6} are very close to 1.0, while the coefficients for f_{RSM3} and f_{RSM4} are slightly lower, although greater than 0.974, and all of them are similar or better than those attained by the authors of [22–24]. On the other hand, the coefficients of determination for MAC_{RSM3} and MAC_{RSM4} are lower, in some cases under 0.9. In this case, it is not possible to compare them to others, because, to the best of our knowledge, in the literature, there are no results using RSM to simulate MAC responses. Nevertheless, those values are also superior to the coefficients obtained by the authors of [22–24] for other responses. So, in conclusion, the approximate functions in Equations (11)–(17) were judged as good enough to accurately relate the design variables and responses, and are adequate to use in the subsequent phase of the improvement procedure.

4.3. Determination of Updated Values of Design Variables

Once the explicit relationships between the design variables and model responses have been determined, the next step is to identify the most adequate stiffness values for the connection elements of the FE model, so that the new model simulates accurately the experimental dynamic behavior.

However, prior to performing this step, it is interesting to have a look at Table 12, which shows the responses obtained and the combinations of variables in the CC design 2, and wherein the frequencies inside the range $(f_{expi} - 1\ \text{Hz}) < f_{FEAi} < (f_{expi} + 1\ \text{Hz})$, and where MAC values higher than the initial ones have been bolded.

Table 12. Variables and responses in central composite design 2.

Run	f_{FEA3}	MAC_3	f_{FEA4}	MAC_4	f_{FEA6}	k_{X8}	k_{X210}	k_{Y3}	k_{Z4}
1	75.2	71.8	80.0	86.8	124.3	900	160	1125	900
2	73.8	**84.0**	78.8	88.1	121.9	900	160	1125	400
3	74.2	64.4	78.8	83.6	123.3	900	160	450	900
4	73.0	**81.3**	**77.4**	87.6	122.1	900	160	450	400
5	68.3	37.5	79.1	75.0	124.2	900	80	1125	900
6	67.9	47.6	**76.9**	78.7	121.8	900	80	1125	400
7	**66.4**	32.0	**78.2**	71.9	123.2	900	80	450	900
8	**66.1**	40.0	75.9	74.9	122.0	900	80	450	400
9	67.5	73.0	**76.8**	50.2	108.5	210	160	1125	900
10	**66.3**	72.8	**76.8**	51.3	**105.7**	210	160	1125	400
11	67.2	72.8	75.3	49.2	**107.5**	210	160	450	900
12	**66.0**	72.5	75.3	50.2	104.9	210	160	450	400
13	**66.1**	**79.1**	70.5	88.3	108.3	210	80	1125	900
14	**65.0**	**80.3**	70.3	86.4	**105.5**	210	80	1125	400
15	**65.2**	75.2	69.0	88.1	**107.3**	210	80	450	900
16	64.3	**79.8**	68.6	**89.3**	104.7	210	80	450	400
17	**66.7**	**76.4**	73.0	67.7	**107.2**	210	120	787.5	650
18	67.5	43.9	**76.3**	77.7	118.5	555	80	787.5	650
19	70.4	56.8	76.0	81.8	118.0	555	120	450	650
20	70.7	74.8	75.8	88.3	117.5	555	120	787.5	400
21	71.6	52.8	78.4	80.4	123.5	900	120	787.5	650
22	73.8	**84.1**	**78.0**	88.7	118.6	555	160	787.5	650
23	71.7	66.8	**77.1**	86.0	118.8	555	120	1125	650
24	71.5	58.3	**77.3**	82.8	119.0	555	120	787.5	900
25	71.3	63.8	**76.8**	84.7	118.5	555	120	787.5	650
Initial	69.7	76.3	73.9	89.0	112.3				
Obj	65.9	100	77.2	100	106.5				

From Table 12, several conclusions can be drawn, as follows:

- The natural frequency f_{FEA3} approximately matches its experimental pair and, at the same time, the corresponding MAC value is higher than the initial one, only when the design variable k_{X8} is at its lower boundary. If k_{X8} takes the central or upper values, it is not possible to adequately accomplish the pairing.
- Also, the natural frequency, f_{FEA6}, needs lower k_{X8} values to match its experimental pair.
- However, on the other side, at lower k_{X8} values, it is not viable to adjust the natural frequency f_{FEA4} while maintaining accurate values of MAC. Intermediate or upper values of k_{X8} are necessary to improve f_{FEA4}, although they give rise to MAC values slightly poorer than initially.

These facts suggest that it is not possible to develop a FE model that fits those three frequencies and that provides acceptable MAC values with a unique value of design variable k_{X8}. In fact, Wu et al. [36]

have also addressed a similar behavior in other machine tool with roller type linear guideways. Therefore, it will be necessary to identify one k_{X8} value to match, in combination with k_{X210}, k_{Y3}, and k_{Z4}; natural frequencies f_{FEA3}, f_{FEA6}; and necessarily MAC_3, as well as other k_{X8} value to match f_{FEA4} and MAC_4, taking into account that design variable k_{X8} does not affect the rest of the responses (Table 6).

For that purpose, the desirability function has been used, as explained in Section 3.3. Two types of desirability functions have been defined (Figure 9), one for natural frequencies and the other one for MAC values.

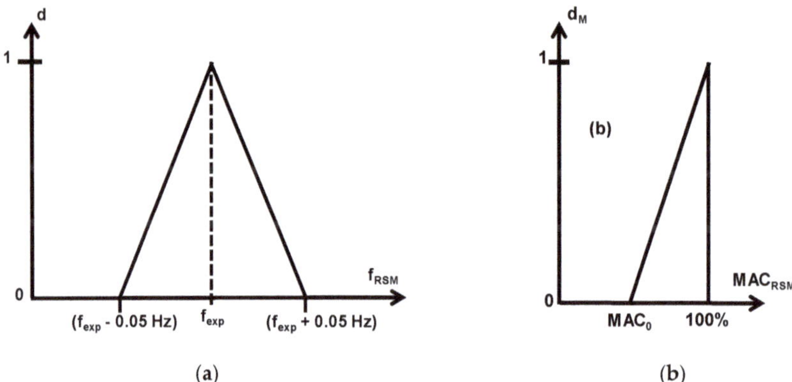

Figure 9. Desirability functions for natural frequencies (**a**) and MAC values (**b**).

These functions can be expressed in mathematical form as follows:

$$d_i = 1 - 20 \cdot \text{ABS}(f_{RSMi} - f_{expi}),\ (f_{expi} - 0.05\ \text{Hz}) < f_{RSMi} < (f_{expi} + 0.05\ \text{Hz}) \tag{18}$$

$$d_i = 0,\ f_{RSMi} < (f_{expi} - 0.05\ \text{Hz}),\ f_{RSMi} > (f_{expi} + 0.05\ \text{Hz}) \tag{19}$$

$$d_{Mi} = 1 - \frac{100 - MAC_{RSMi}}{100 - MAC_0},\ MAC_0 < MAC_{RSMi} < 0 \tag{20}$$

$$d_{Mi} = 0,\ MAC_{RSMi} < MAC_0 \tag{21}$$

where MAC_0 has been selected taking into consideration Tables 4 and 12.

Finally, the global desirability function D, Equation (22), is composed by combining one global desirability function for the frequencies, D_f, and another function for the MAC values, D_M, and applying weighting coefficients w_f and w_M to each of them.

$$D = w_f \cdot D_f + w_M \cdot D_M = w_f \cdot (d_1 \cdot d_3 \cdot d_4 \cdot d_5 \cdot d_6)^{1/5} + w_M \cdot (d_{M3} \cdot d_{M4})^{1/2} \tag{22}$$

Table 13 shows the updated values of the normalized design variables X_i, and the corresponding natural values k_i. As mentioned before, two stiffness values k_{X8} have been estimated, namely: (1) is valid for the frequency ranges from 0 Hz to 72 Hz, and from 100 Hz to the upper limit of the range of interest, and (2) is adequate for the remaining range, which includes the 4th natural frequency.

Table 13. Updated values of the design variables. Units of k_i are N/μm.

X_{Y22}	X_{Z63}	X_{X210}	X_{Y3}	X_{Z4}	X_{Z13}	X_{X8} (1)	X_{Y9}	X_{X8} (2)
−0.920	−0.250	−0.695	1.000	−0.540	−0.800	−1.000	−0.300	0.485
k_{Y22}	k_{Z63}	k_{X210}	k_{Y3}	k_{Z4}	k_{Z13}	k_{X8} (1)	k_{Y9}	k_{X8} (2)
636	769	92	1125	515	450	210	642	722

These values have been driven into the FE model and the posterior FE analysis has led to the frequencies and MAC values, indicated in Table 14. For the sake of comparison, the simulated responses, $f_{\mathrm{RSM}i}$ and $MAC_{\mathrm{RSM}i}$, obtained in Equations (11)–(17), when the updated values of the design variables are substituted, are also shown.

Table 14. Final frequency and MAC values.

FEA Order	f_{RSM}	f_{FEA}	f_{exp}	Diff. (Hz)	Diff. (%)	MAC	MAC_{RSM}	k_{X8}
1	33.7	33.7	33.7	0.0	0.0	96.7	-	(1)
2	-	60.5	60.5	0.0	0.0	98.7	-	(1)
3	65.9	65.8	65.9	−0.1	−0.2	78.9	76.5	(1)
4	77.2	77.0	77.2	−0.2	−0.3	80.6	83.9	(2)
5	84.0	84.1	84.0	0.1	0.1	80.9	-	(2)
6	106.5	106.1	106.5	−0.4	−0.4	70.0	-	(1)

From Table 14, it can be seen that the quadratic regression equations have provided values of the simulated frequencies that almost coincide with the values obtained after the completion of the FE analysis. In fact, the maximum distance is 0.4 Hz in the 6th frequency, which is really insignificant. The difference in the simulated MAC values is greater, which is in accordance with Table 11, where it was suggested that the predictive capability of the MAC regression equations was inferior. Therefore, both the regression meta-models and also the statistic indicators, R^2 and t-statistic, have performed adequately.

Finally, once the identified values of design variables have been incorporated into the FE model, the resultant dynamic responses have shown a closer match to the experimental results, proving the adequacy of the conducted procedure. Thus, in Figure 10, two synthesized FRFs obtained from the updated FE model are compared to the corresponding experimental FRFs in reference point 5 (Figure 2), and the agreement is quite reasonable.

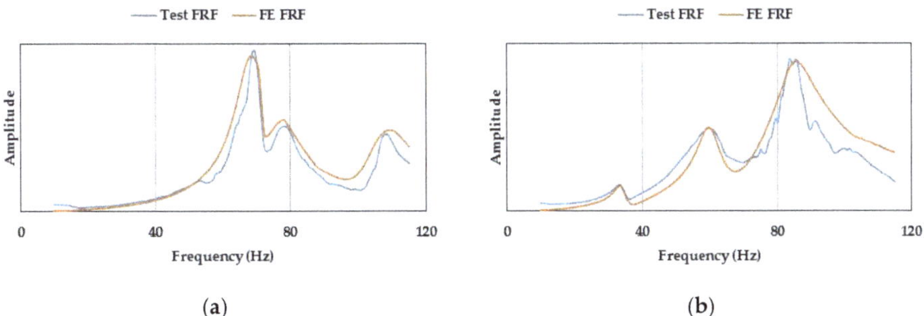

Figure 10. Test frequency response functions (FRFs) and synthetized FE FRF in point 5. (**a**) Response X, impact X; (**b**) response Y, impact Y.

5. Conclusions

This paper presents a consistent methodology to improve the FE models of complex mechanical systems using, in an integrated way, different numerical and experimental techniques. The procedure is applied in a machining center with numerous uncertainties in the internal connections and supporting conditions. In this methodology, the complete design space is encompassed, so that the selection of the initial values of the design variables, which is one of the major drawbacks of the sensitivity-based methods, due to the variable sensitivity values along the design space, is avoided.

Firstly, it is demonstrated that the two-level fractional factorial design is an effective tool to perform parameter screening, as the most significant parameters and two-factor interactions are detected. For this purpose, instead of using one cumbersome resolution V design, more simple fractional designs with fewer parameters are gradually completed and examined. This procedure allows for removing high-order interactions and to circumvent the presence of constraints between the variables, and leads to a better comprehension of the influence of the design variables on the behavior of the mechanical system.

Also, in this step, it is shown that the design variables perform a kind of collective work, as groups of them affect groups of responses. In addition, some of them can be satisfactorily removed, in contrast to other findings reported in literature, where it is assumed that the complete set of the selected design variables is significant. This is a key feature of the proposed methodology, as it allows for diminishing the complexity of the subsequent regression equations, due to a substantial drop in the number of terms.

In the second step, it is demonstrated that the relationships between the stiffness parameters and the modal responses of the machine tool can be accurately expressed by second-order functions. A combined procedure using coefficients of determination and the t-statistic is applied to remove the non-significant terms, and thus the accuracy of regression equations is increased. At this point, the assessment of the predictive capability of the regression meta-models plays an important role. The regression equations for the MAC values are also used to ensure the correspondence between the numerical and experimental responses, because if only the frequency values are matched, it could lead to unacceptable MAC values. So far, this issue has been overlooked in the literature.

Also, the use of central composite designs allows for developing quadratic regression equations at an acceptable cost. In addition, these designs have led to detecting that the stiffness of one connection is dependent on the relative movement between the modules of the machining center.

It is proved that the quadratic regression equations are adequate to accurately identify improved values of the design variables, because when these values are included in the FE model, minimal differences between the FE and experimental responses are found. Also, because of the substitution of the full FE model by polynomial functions, the identification step, usually a costly iterative procedure, is accelerated. The use of desirability functions and weighting factors facilitates the progress of this step.

Finally, the presented methodology can be generalized to any machine tool and any design variable (damping in connections, Young's modulus, mass density, etc.) and will allow for obtaining an updated finite element model, which would serve as a starting point to optimize the machine design and eliminate stability problems under operating conditions.

Potential future research directions include the analysis and implementation of other designs (orthogonal, Latin Hypercube, and D-optimal) for fitting second-order models with constraints in the design space. Another possible direction is the use of techniques of model reduction or transformation to modal space to diminish the time-consuming DoE runs.

Author Contributions: Conceptualization: J.-M.H.-V., I.G., and M.H.F.; methodology: J.-M.H.-V., I.G., and M.H.F.; software: J.-M.H.-V., I.G., and M.H.F.; validation: J.M. and L.N.L.d.L.; formal analysis: J.-M.H.-V., I.G., and M.H.F.; investigation: J.-M.H.-V., I.G., and M.H.F.; resources: M.H.F. and J.M.; writing (original draft preparation): J.-M.H.-V.; writing (review and editing): J.-M.H.-V., I.G., and M.H.F.; visualization: J.-M.H.-V.,

I.G., and M.H.F.; supervision: J.M. and L.N.L.d.L.; project administration: L.N.L.d.L.; funding acquisition: I.G., M.H.F., and L.N.L.d.L.

Funding: This work has been supported by the University of the Basque Country UPV/EHU under programs PPGA17/04 and US17/16, and the Spanish Ministry of Economy, Industry, and Competitiveness under program DPI2016-74845-R.

Conflicts of Interest: The authors declare no conflict of interest.

References

1. Lopez de Lacalle, L.N.; Lamikiz, A. Machine Tools for Removal Processes: A General View. In *Machine Tools for High Performance Machining*; Lopez de Lacalle, L.N., Lamikiz, A., Eds.; Springer: London, UK, 2009; pp. 1–45. ISBN 978-1848003798.

2. Altintas, Y.; Brecher, C.; Weck, M.; Witt, S. Virtual machine tool. *CIRP. Ann. Manuf. Technol.* **2005**, *54*, 115–138. [CrossRef]

3. Quintana, G.; Ciurana, J. Chatter in machining process: A review. *Int. J. Mach. Tools Manuf.* **2011**, *51*, 363–376. [CrossRef]

4. Friswell, M.I.; Mottershead, J.E. *Finite Element Model Updating in Structural Dynamics*; Kluwer Academic Publishers: Dordrecht, The Netherlands, 1995; ISBN 978-0792334316.

5. Janter, T. Construction Oriented Updating of Dynamic Finite Element Models Using Experimental Modal Data. Ph.D. Thesis, Katholieke Universiteit Leuven, Leuven, Belgium, 1989.

6. Bais, R.S.; Gupta, A.K.; Nakra, B.C.; Kundra, T.K. Studies in dynamic design of drilling machine using updated finite element models. *Mech. Mach. Theory* **2004**, *39*, 1307–1320. [CrossRef]

7. Houming, Z.; Chengyang, W.; Zhenyu, Z. Dynamic characteristics of conjunction of lengthened shrink-fit holder and cutting tool in high-speed milling. *J. Mater. Process. Technol.* **2008**, *207*, 154–162. [CrossRef]

8. Garitaonandia, I.; Fernandes, M.H.; Albizuri, J. Dynamic model of a centerless grinding machine based on an updated FE model. *Int. J. Mach. Tools Manuf.* **2008**, *48*, 832–840. [CrossRef]

9. Garitaonandia, I.; Fernandes, M.H.; Hernandez-Vazquez, J.M.; Ealo, J.A. Prediction of dynamic behavior for different configurations in a drilling-milling machine based on substructuring analysis. *J. Sound. Vib.* **2016**, *365*, 70–88. [CrossRef]

10. Brecher, C.; Esser, M.; Witt, S. Interaction of manufacturing process and machine tool. *CIRP Ann. Manuf. Technol.* **2009**, *58*, 588–607. [CrossRef]

11. Zhou, Y.D.; Chu, L.; Bi, D.S. Structural optimization for hydraulic press frame. *China Metal. Equip. Manuf. Technol.* **2008**, *2*, 90–92.

12. Li, Y.B.; Wu, D.H.; Huang, M.H.; Lu, X.J. Design of Parallel Bearing Structure for 800 MN Forging Press with Consideration of Manufacturing Errors. *Appl. Mech. Mater.* **2011**, *52–54*, 2157–2163. [CrossRef]

13. Markowski, T.; Mucha, J.; Witkowski, W. FEM analysis of clinching joint machine's C-frame rigidity. *Eksploat. Niezawodn. Maint. Reliab.* **2013**, *15*, 51–57.

14. Montgomery, D.C. *Design and Analysis of Experiments*, 6th ed.; John Wiley & Sons: New York, NY, USA, 2005; ISBN 978-0471661597.

15. Lamikiz, A.; Lopez de Lacalle, L.N.; Sanchez, J.A.; Bravo, U. Calculation of specific cutting coefficients and geometrical aspects in sculptured surface machining. *Mach. Sci. Technol.* **2005**, *9*, 411–436. [CrossRef]

16. Calleja, A.; Bo, P.; Gonzalez, H.; Barton, M.; Lopez de Lacalle, L.N. Highly accurate 5-axis flank CNC machining with conical tools. *Int. J. Adv. Manuf. Technol.* **2018**, *97*, 1605–1615. [CrossRef]

17. Myers, R.H.; Montgomery, D.C.; Anderson-Cook, C.M. *Response Surface Methodology: Process and Product Optimization Using Designed Experiments*, 4th ed.; John Wiley & Sons: New York, NY, USA, 2016; ISBN 978-1118916018.

18. Bonte, M.H.A.; van den Boogaard, A.I.; Huétink, J. An optimization strategy for industrial metal forming processes. *Struct. Multidiscip. Optim.* **2008**, *35*, 571–586. [CrossRef]

19. Lü, H.; Yu, D. Brake squeal reduction of vehicle disc brake system with interval parameters by uncertain optimization. *J. Sound. Vib.* **2014**, *333*, 7313–7325. [CrossRef]

20. Wang, R.; Lim, P.; Heng, L.; Mun, S.D. Magnetic Abrasive Machining of Difficult-to-Cut Materials for Ultra-High-Speed Machining of AISI 304 Bars. *Materials* **2017**, *10*, 1029:1–1029:11. [CrossRef]

21. Guo, Q.; Zhang, L. Finite element model updating based on Response Surface Methodology. In Proceedings of the 22nd International Modal Analysis Conference, Dearborn, MI, USA, 26–29 January 2004; Society for Experimental Mechanics: Bethel, CT, USA, 2004. Paper No. 93.

22. Rutherford, A.C.; Inman, D.J.; Park, G.; Hemez, F.M. Use of response surface metamodels for identification of stiffness and damping coefficients in a simple dynamic system. *Shock. Vib.* **2005**, *12*, 317–331. [CrossRef]

23. Ren, W.-X..; Chen, H.-B. Finite element model updating in structural dynamics by using the response surface method. *Eng. Struct.* **2010**, *32*, 2455–2465. [CrossRef]

24. Fang, S.-E.; Perera, R. Damage identification by response surface based model updating using D-optimal design. *Mech. Syst. Signal Process.* **2011**, *25*, 717–733. [CrossRef]

25. Sun, W.-Q.; Cheng, W. Finite element model updating of honeycomb sandwich plates using a response surface model and global optimization technique. *Struct. Multidiscip. Optim.* **2017**, *55*, 121–139. [CrossRef]

26. Gallina, A.; Pichler, L.; Uhl, T. Enhanced meta-modelling technique for analysis of mode crossing, mode veering and mode coalescence in structural dynamics. *Mech. Syst. Signal Process.* **2011**, *25*, 2297–2312. [CrossRef]

27. Schaeffler Group. Available online: http://medias.schaeffler.de/medias/en!hp.ec.br.pr/RUE..-E-HL*RUE55-E-HL (accessed on 15 June 2009).

28. Van Brussel, H.; Sas, P.; Németh, I.; de Fonseca, P.; van den Braembussche, P. Towards a mechatronic compiler. *IEEE/ASME. Trans. Mechatron.* **2001**, *6*, 90–105. [CrossRef]

29. Law, M.; Altintas, Y.; Phani, A.S. Rapid evaluation and optimization of machine tools with position-dependent stability. *Int. J. Mach. Tools Manuf.* **2013**, *68*, 81–90. [CrossRef]

30. Muñoa, J. Desarrollo de un modelo general para la predicción de la estabilidad del proceso de fresado. Aplicación al fresado periférico, al planeado convencional y a la caracterización de la estabilidad dinámica de fresadoras universales. Ph.D. Thesis, Mondragon University, Arrasate/Mondragón, Spain, 2007. (In Spanish)

31. Allemang, R.J. Investigation of Some Multiple Input/Output Frequency Response Experimental Modal Analysis Techniques. Ph.D. Thesis, University of Cincinnati, Cincinnati, OH, USA, 1980.

32. Link, M.; Hanke, G. Model Quality Assesment and Model Updating. In *Modal Analysis and Testing*; Maia, N.M.M., Silva, J.M.M., Eds.; Kluwer Academic Publishers: Dordrecht, The Netherlands, 1999; pp. 305–324. ISBN 978-0-7923-5893-7.

33. Mottershead, J.E.; Link, M.; Friswell, M.I. The sensitivity method in finite element model updating: A tutorial. *Mech. Syst. Signal Process.* **2011**, *25*, 2275–2296. [CrossRef]

34. Ealo, J.A.; Garitaonandia, I.; Fernandes, M.H.; Hernandez-Vazquez, J.M.; Muñoa, J. A practical study of joints in three-dimensional Inverse Receptance Coupling Substructure Analysis method in a horizontal milling machine. *Int. J. Mach. Tools Manuf.* **2018**, *128*, 41–51. [CrossRef]

35. Derringer, G.; Suich, R. Simultaneous optimization of several response variables. *J. Qual. Technol.* **1980**, *12*, 214–219. [CrossRef]

36. Wu, J.; Wang, J.; Wang, L.; Li, T.; You, Z. Study on the stiffness of a 5-dof hybrid machine tool with actuation redundancy. *Mech. Mach. Theory* **2009**, *44*, 289–305. [CrossRef]

Article

Numerical Simulation and Experimental Investigation of Cold-Rolled Steel Cutting

Jarosław Kaczmarczyk [1] and Adam Grajcar [2,*]

[1] Institute of Theoretical and Applied Mechanics, Silesian University of Technology, 18A Konarskiego Street, 44-100 Gliwice, Poland; jaroslaw.kaczmarczyk@polsl.pl

[2] Institute of Engineering Materials and Biomaterials, Silesian University of Technology, 18A Konarskiego Street, 44-100 Gliwice, Poland

[*] Correspondence: adam.grajcar@polsl.pl; Tel.: +48-32-237-2933

Received: 28 May 2018; Accepted: 20 July 2018; Published: 23 July 2018

Abstract: The paper presents results of the investigations on numerical computations and experimental verification concerning the influence of selected parameters of the cutting process on the stress state in bundles of cold-rolled steel sheets being cut using a guillotine. The physical model and, corresponding to it, the mathematical model of the analysed steel sheet being cut were elaborated. In this work, the relationship between the cutting depth and the values of reduced Huber–Mises stresses as well as the mechanism of sheet separation were presented. The numerical simulations were conducted by means of the finite element method and the computer system LS-DYNA. The results of numerical computations are juxtaposed as graphs, tables, and contour maps of sheet deformation as well as reduced Huber–Mises strains and stresses for selected time instants. The microscopic tests revealed two distinct zones in the fracture areas. The ductile and brittle zones are separated at the depth of ca. 1/3 thickness of the cut steel sheet.

Keywords: plastic zone; fracture mechanism; steel sheet; cutting process; Huber–Mises stress; finite element method; microscopic analysis

---

1. Introduction

Guillotines are conveniently used for cutting of metal sheets because one can cut not only single sheets, but also bundles. Cutting of bundles is more efficient compared to cutting individual sheets, because it allows many sheets to be cut at a single cutting tool passage. The bundle cutting is used to separate many sheets in order to reach high-quality cut surfaces. The most frequently occurring defects are bends of cut sheet edges, burrs, and vertical scratches. To improve the cutting process, the numerical investigations have been carried out concerning mainly the application of finite element method for numerical simulation of cut sheet separation and experimental research aimed at verifying the obtained results. There are many interesting items in the literature regarding mainly machining [1–7], but much less papers cover the mechanical cutting on guillotines [8]. Different approaches are used for simulating failure, metal separation, etc. The most important is to select an appropriate FEM model [9–11]. Machine deformation can also affect significantly the cutting process [12,13].

It can be concluded that cutting on guillotines is a niche subject appearing in the literature rather sporadically when the problems related to machining processes are discussed. Research concerning cutting of single sheets using a guillotine shear can be found in work [14]. The authors formulated physical models and, corresponding to them, mathematical ones using the computer program LS-DYNA [15]. The obtained results indicate that it is possible to model the process of cutting a single sheet on a guillotine shear.

This paper presents physical models and, corresponding to them, mathematical models in the case of cutting a single sheet on a guillotine intended for bundle cutting, the mechanism of which is different from the guillotine shear cutting. The important difference means that during cutting on a guillotine shear, the blade of a cutting tool simultaneously performs a progressive and angular movement similar to one which is observed during cutting a sheet of paper with scissors. Additionally, the blade of a cutting tool passes through the working surface of the table in the cutting process. However, during cutting of sheet bundles on a guillotine, the cutting tool blade performs only vertical movement and the cutting edge of the cutting tool is parallel to the working table surface [8,16]. It should be mentioned that the surface of the table is horizontal and the blade of the cutting tool, after cutting the last sheet in the bundle, stops on the working surface of the table and then returns to its original position.

For experimental investigations, C75S cold-rolled steel was chosen because it is broadly used in the industry for thin sheets. Cold-rolled strips are gaining popularity in various industries. They are increasingly being implemented through a fairly large number of small-component manufacturers as parts and machine components. In many cases, C75S cold-rolled steel requires mechanical separation, which can be achieved by means of guillotines. The cutting on guillotines has great advantages, therefore it is possible to make the separation without a considerable increase of temperature, which ensures saving of unchanged microstructure unlike in, for example, laser cutting.

The objective of the current paper was to improve the understanding of the mechanism of the cutting process and the influence of selected parameters on the state of deformation as well as on the state of strain and, corresponding to it, stress in bundles of cold-rolled steel sheets being cut. The numerical results were obtained using the finite element method and compared with experimental investigations by means of scanning electron microscopy (SEM).

2. Physical Model of the Cutting Process

Industrial practice shows that the most common problems of the cutting pieces are: quality of cut surfaces of sheets, arising burrs, and occurrence of numerous defects in the form of vertical scratches on cut sheet metal surfaces and therefore it was decided to develop a model consisting of a deformable sheet and ideally rigid cutting tool, as well as pressure beam and working surface of the table. The physical model of the sheet metal cutting process is shown in Figure 1. On a motionless, perfectly rigid table, a sheet being cut is placed and then it is pressed from the top using a perfectly rigid pressure beam made in the form of a wedge with convergence 1:30 to produce a specific stress concentration on the cutting line. The influence of this concentration of stress in the sheets on the cutting line is extremely important because after exceeding the value of permissible stress, the process of separation of cutting pieces begins. Extensive research on this phenomenon is discussed in detail in [16]. From these scientific investigations, one can conclude that it is easier to cut a single sheet with prestress compared to cutting the same sheet without prestress. It is caused by the fact that on the blade of the cutting tool in the first case, the force acts with a lower value than in the second case. Accordingly, the state of stress in the first case (with prestress) is higher than in the second case (without prestress). The gradual increase in the value of the force on the blade causes an increase of the stress state in the sheet metal, until it reaches such a state which enables the separation of a sheet being cut. There is a small gap between the blade of a cutting tool and the pressure beam. It turns out that the size of this gap has a large impact on the value of the prestress on the cutting line. The numerical simulations indicate that the smaller the gap is, the higher the stresses produced are [16].

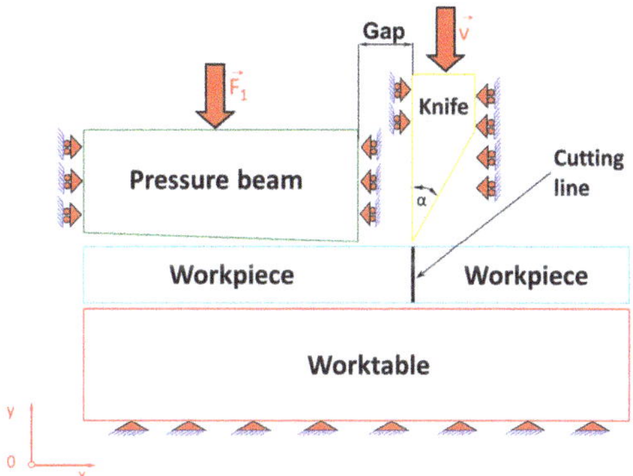

Figure 1. Physical model of a cutting process.

Summarising, both the loading force of the pressure beam and the size of the gap affect the value of the prestress state and, consequently, the cutting process. Increasing the load of the pressure beam and reducing the size of the gap increase the prestress state in the sheets, simultaneously decreasing the value of the force acting at the cutting tool blade and required to separate the sheets, which is extremely important on account of extending the life of a cutting tool blade [8].

In this work, it is assumed that the cutting tool blade has an apex angle $\alpha = 30°$ and the gap between the cutting tool and the pressure beam is 0.1 mm. The elaborated physical model presented in Figure 1 is divided into finite elements corresponding to a plane state of strain, which is illustrated in Figure 2. The Coulomb's friction law was used to describe tool–workpiece, pressure beam–workpiece, and workpiece–worktable interfaces friction with the assumed coefficient of friction as for steel. The unilateral constraints in the form of the so-called contact have been imposed on the surfaces of the sheet being cut, worktable, pressure beam, and cutting tool. The contact is based on the condition of impenetrability, namely the condition that two bodies cannot interpenetrate. Impenetrability cannot be expressed as a single equation, so several simplified approaches have been developed by means of the finite element method. In order to enforce a constrained condition on the governing equations, many well-known algorithms have been applied such as: the Lagrange multiplier method, the penalty method, the augmented Lagrangian method, and the perturbed Lagrangian method [17–22]. In this paper, the contact penalty method for modelling of the contact between early mentioned surfaces in the cutting process has been exploited.

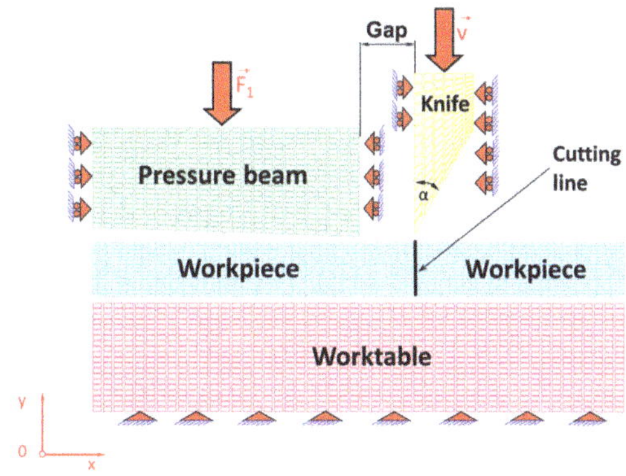

Figure 2. Discretisation into finite elements.

The cutting process is accompanied by separation of material along a cutting line (Figures 1 and 2). In literature, there are many well-known methods concerning fracture mechanics, which is the field of mechanics that studies the propagation of cracks in materials [6]. It uses approaches of analytical solid mechanics to calculate the driving force on a crack and those of experimental solid mechanics to characterize the material's resistance to fracture [23–25]. It should be mentioned that in modern materials science, fracture mechanics is an important tool applied to improve the performance of mechanical components. It uses the physics of stress and strain behaviour of materials, in particular, the theories of elasticity and plasticity [26] to the microscopic defects found in real materials, in order to predict the macroscopic mechanical behaviour of those bodies [27,28].

In this work, for the criterion of material separation, the extreme strain value at which the sample breaks during its uniaxial stretching was found experimentally. In the model on both sides of the cutting line, corresponding mutual pairs of nodes are created, in which the Huber–Mises reduced strain values are checked for each iteration in the numerical simulation of the cutting process. If the Huber–Mises reduced strain is greater than the limit value strain determined from experimental investigation by the uniaxial tensile test $\varepsilon_{(f} = 0.15)$, then the corresponding pairs of nodes will be separated on the cutting line. In the opposite case, when Huber's reduced strains are smaller than $\varepsilon_f = 0.15$, the nodes will not be separated.

Such elaborated two-dimensional model of the cutting process is correct because it is loaded by forces applied parallel to the plane of the sheet and distributed uniformly over the thickness of the sheet and therefore satisfies the conditions of the plane state of strain [26]. The component of the strain along the normal to the plane strain equals zero. There are infinitely many such planes, and the results of numerical calculations in these planes are repetitive and, as a consequence, comparable with those presented in this work regarding the plane state of strain. Furthermore, it should be noticed that the plane state of strain is simpler in the physical interpretation of the obtained numerical results concerning the modelling of issues related to the mechanical separation of cold-rolled steel sheets and then subsequently cut on the guillotines.

Table 1 presents the data concerning the properties of the components of a physical model, such as the numbers of nodes and finite elements, into which the individual components of the physical model are divided (Figure 2). The numerical calculations were carried out using the computer cluster Ziemowit (http://www.ziemowit.hpc.polsl.pl).

Table 1. Juxtaposition of data concerning the components of a physical model.

No	Name of the Part	Kind of Part	Number of Nodes	Number of Elements
1.	Worktable	Rigid	1071	1000
2.	Workpiece	Deformable	1232	1125
3.	Pressure beam	Rigid	546	500
4.	Cutting tool (knife)	Rigid	112	90
	Total Number		**2961**	**2715**

3. Material Properties

The material being cut is a C75S cold-rolled sheet steel containing a eutectoid carbon content. Following cold rolling, the steel sheets are submitted to spheroidizing annealing to reduce the hardness to ca. 200 HV. The sheets are homogeneous across their cross sections (Figure 3a). The microstructure is very fine-grained. It consists of ferrite matrix and globular cementite precipitates (Figure 3b). The carbide particles are uniformly arranged within the microstructure. They are separated from each other (Figure 3c). The ferrite grain size is from 3 to 5 μm whereas particle diameters are within a range from 1 to 2 μm (Figure 3d).

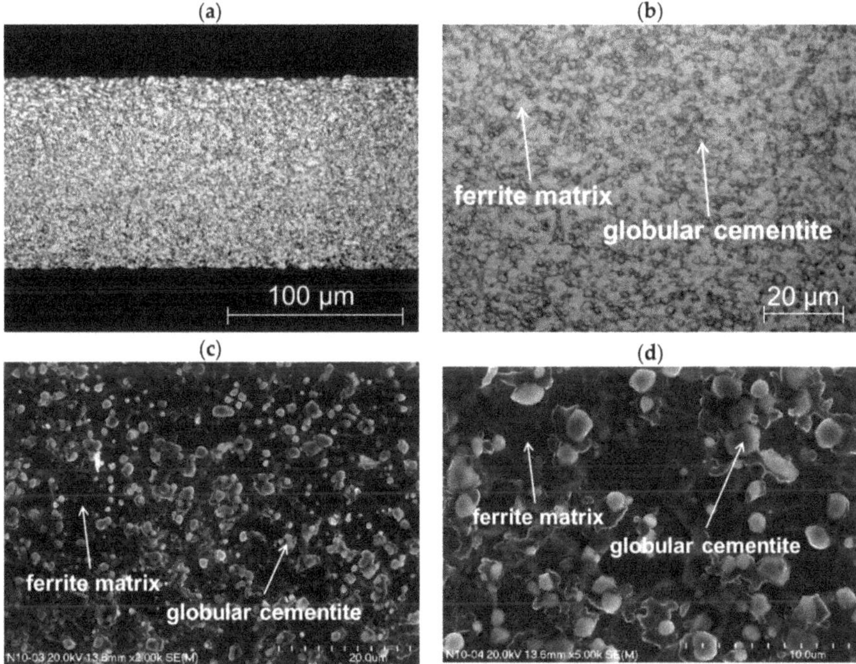

Figure 3. Microstructure of the cut steel shown with the use of light microscope (**a**,**b**) and scanning electron microscope (**c**,**d**).

In order to model the cutting process, it is necessary to adopt the appropriate physical model and, corresponding to it, the mathematical one, which will allow the obtaining of satisfactory results. In practice, some simplified material models that give good results are often used. In this work, a bilinear elastic–plastic material model with plastic strengthening is adopted. Detailed material properties are juxtaposed in Table 2 and graphically presented in Figure 4.

Table 2. Material properties.

No	Name of the Material Properties	Symbol	Value
1.	Young's modulus	E	2.05×10^5 MPa
2.	Poison's ratio	ν	0.28
3.	Kirchhoff's modulus	G	8×10^4 MPa
4.	Tangent modulus	E_T	867 MPa
5.	Failure strain	ε_f	0.15
6.	Yield stress	R_e	510 MPa
7.	Ultimate tensile strength	R_m	640 MPa

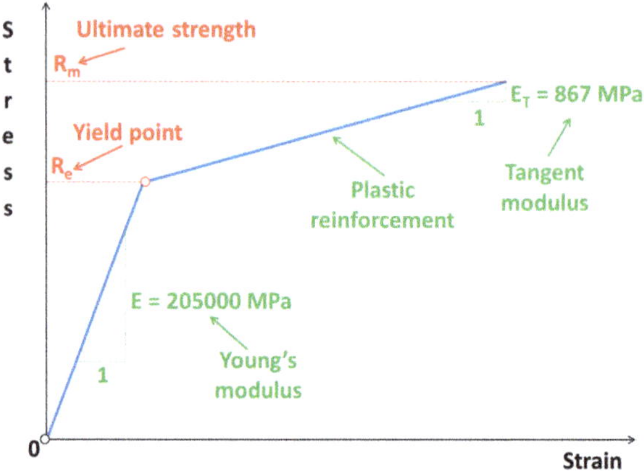

Figure 4. Bilinear elastic–plastic physical material model.

4. Results and Discussion

The cutting process has been designed in such a way that it reflects the real conditions of thin steel sheet cutting on the production line in the best possible manner. In the first stage of the process, the sheets are placed on the working surface of the table and then the preloading of the sheet being cut is carried out using a pressure beam. At the initial moment of the cutting process, the pressure beam is released and the movement in vertical direction is realized in order to press the sheet from the top. This movement takes place in three consecutive stages. In the first stage, the pressure beam is released and gently begins to move at a velocity increasing from zero until reaching the maximum constant value of v = 0.012 mm/s, as shown in Figure 5. In the next stage, the pressure beam moves with a constant velocity, and in the final third stage, the pressure beam gently slows down to a halt, simultaneously pressing the sheets with the desired force. For the purposes of this paper, a cutting process with zero pressing force is considered. This does not mean, however, that the pressure beam is unnecessary, but only that it is loaded with a small force, which can be treated as negligible. During the cutting process, it prevents the sheet from sliding or buckling or being put into any possible self-excited vibrations.

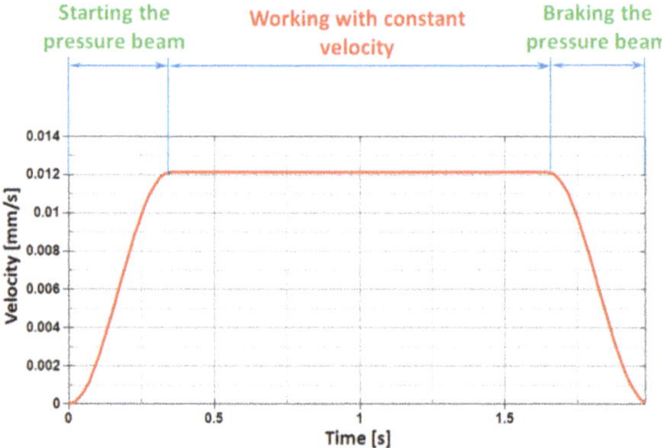

Figure 5. Individual stages of pressure beam motion.

In the next stage, the cutting tool is released and starts moving in the vertical direction from zero to a certain limit with constant velocity (v = 0.01 mm/s). After the sheet metal has been cut, the blade of the cutting tool stops at the horizontal working surface of the table and returns to its original position. During the cutting process, the generator of the tip blade is parallel to the horizontal working surface of the table.

Individual stages concerning the cutting process related respectively to the Huber–Mises reduced stresses of the sheet being cut (Figure 6) and, corresponding to them, states of strains (Figure 7) for selected time instants of the cutting process are juxtaposed and then discussed in detail.

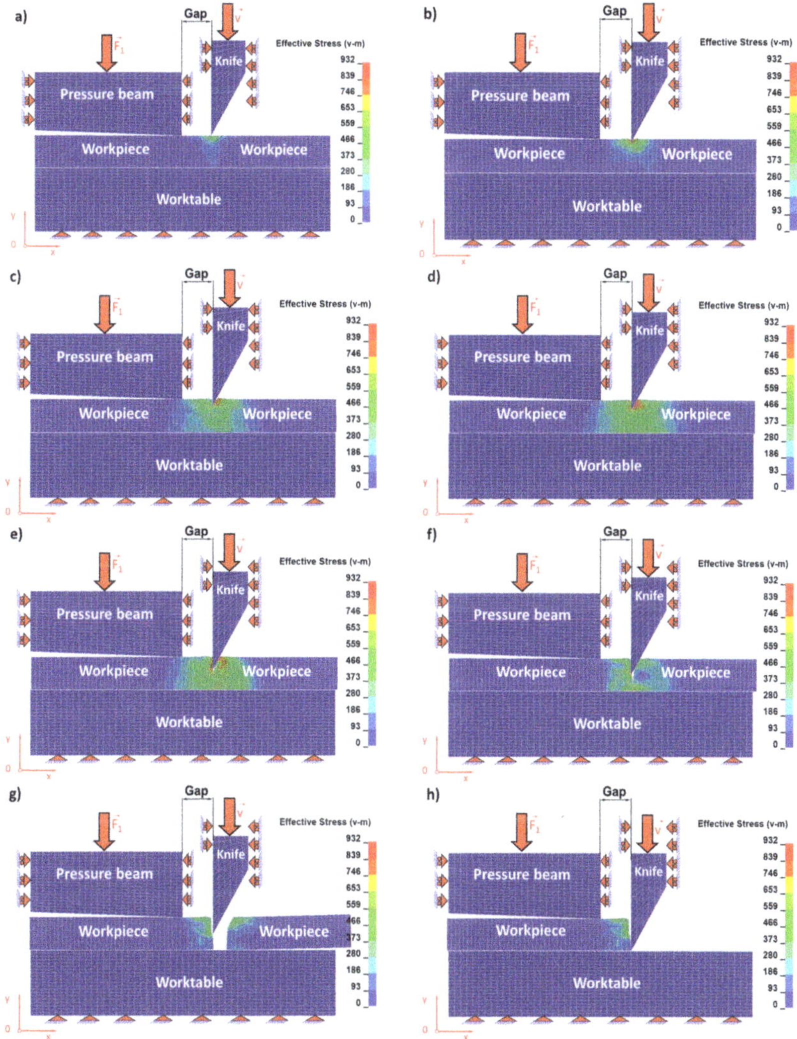

Figure 6. Individual stages of reduced Huber–Mises stresses during numerical simulation of a cutting process for selected time instants: (**a**) t_a = 2.07 s; (**b**) t_b = 3.0 s; (**c**) t_c = 4.0 s; (**d**) t_d = 5.0 s; (**e**) t_e = 6.0 s; (**f**) t_f = 7.1 s; (**g**) t_g = 7.2 s; (**h**) t_h = 12.0 s.

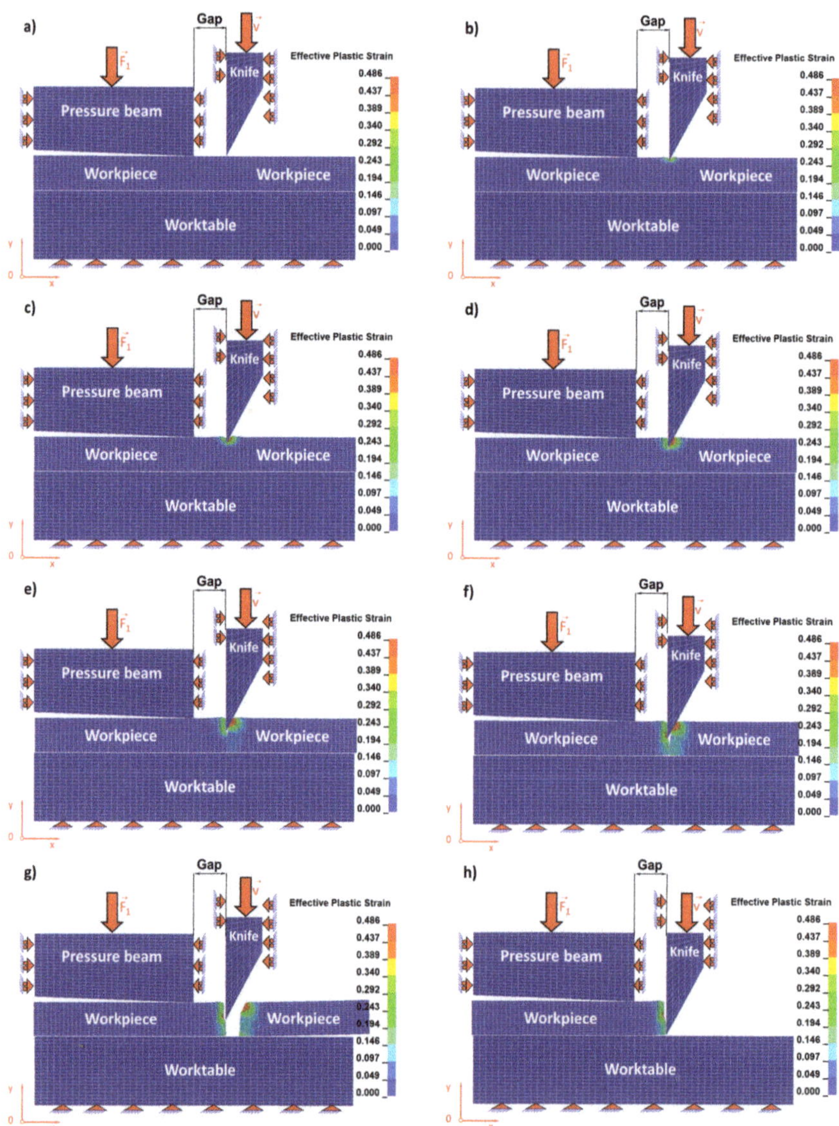

Figure 7. Individual stages of reduced Huber–Mises strains during numerical simulation of a cutting process for selected time instants: (**a**) t_a = 2.07 s; (**b**) t_b = 3.0 s; (**c**) t_c = 4.0 s; (**d**) t_d = 5.0 s; (**e**) t_e = 6.0 s; (**f**) t_f = 7.1 s; (**g**) t_g = 7.2 s; (**h**) t_h = 12.0 s.

The blade of the cutting tool exerts pressure on the upper surface of a steel sheet and three characteristic zones of the cutting process are formed (Figures 6 and 7). The first is associated with accumulating of stresses and the formation of the elastic zone until the elastic limit is reached (Figures 6a and 7a). The second zone begins when the elastic limit is exceeded and is associated with the formation of a plastic zone. In the second stage, the material from which the sheet is made (Figures 6b,c and 7b,c) is submitted to shearing. In the third zone, the material being cut is separated. This is the most characteristic stage, which can be divided into two consecutive phases. The first initial phase consists in separating of the material being cut as a result of the dominant shearing process with a small participation of stretching (Figures 6d,e and 7d,e) and the second phase consists in separating the cut material as a result of the dominant stretching process, similar to what we observe in the case of uniaxial tensile test but with a vanishing participation of material shearing (Figures 6f,g and 7f,g).

In Figure 8, the final stage of the phase related to the ripping of the sheet being cut consisting in separating the material and creating free surfaces as a result of the progressive cracking process is presented. As a result of observation under the microscope (Figure 9a), it can be seen that the area of the cross section of the steel sheet being cut corresponding to ca. 1/3 of its height measured from the top is subjected to plastic shearing and the remaining cross section area corresponding to ca. 2/3 of the sheet height measured from the bottom is subjected to brittle cracking. The characteristic horizontal crack forming the transition between the plastic and brittle zones is marked with a curved line and shown in Figure 9a. The surfaces of the plastic shearing region are deformed permanently and show large unevenness at the left upper edge (Figure 9b), while those of the brittle fracture region present flat and smooth features without bends or burrs.

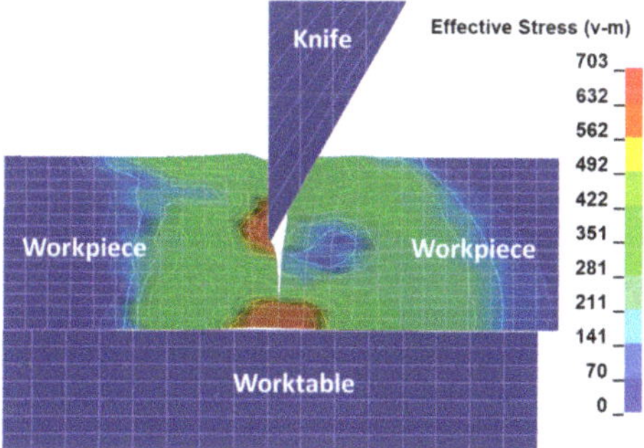

Figure 8. Final phase of a cutting process (cracking of sheet being cut).

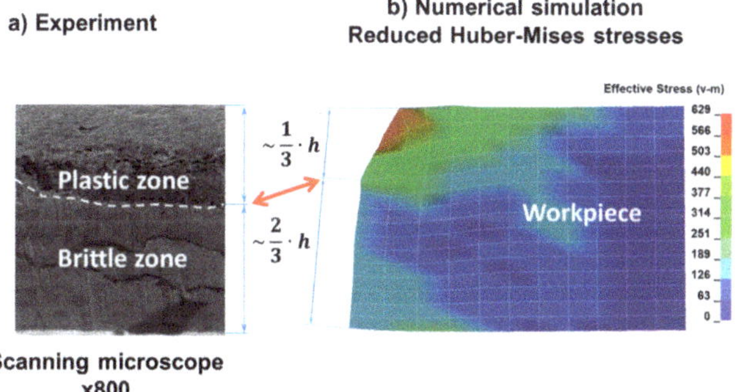

Figure 9. Comparison of the experimental and numerical results: (**a**) view from the forefront of the steel sheet being cut; (**b**) reduced Huber–Mises stresses (the permanent plastic deformation at the left upper edge is visible).

Further microscopic details are visible using the magnifications typical for the scanning electron microscope. Figure 10a indicates that there are two distinct zones (i.e., the upper one has features of plastic deformation and the lower one shows a brittle character). At the higher magnification, it is visible that the upper area shows typical dimples of various size characteristic of ductile fracture (Figure 10b). This part of the cross section is separated from the other one by a curved line. Hence, the separation is not linear. The lower part of the sample shows the typical flat fracture without any dimples. Only brittle walls can be visible. It confirms that the lower part of the steel sheet experiences the brittle cracking.

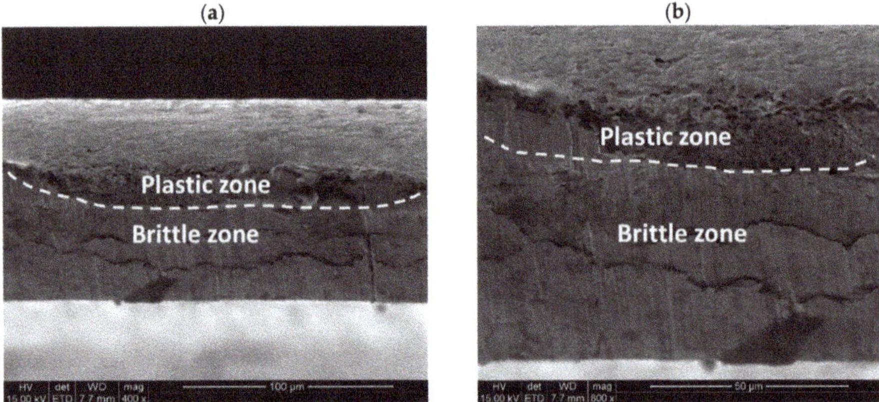

Figure 10. Cross section of the steel sheet being cut showing ductile and brittle zones (**a**) and the magnified part of the cross section showing microscopic details (**b**).

Summarising the cutting process, one may state that it involves the bending of edges of steel sheets being cut in the plastic shearing zone and the occurrence of burrs and large permanent deformations, while brittle surfaces are characterized by a high degree of smoothness (low roughness), flatness of the cross section, lack of burrs, and thus desirable features of metal sheet surfaces being cut. Then the question arises whether the technological parameters of the cutting process can be selected in the meaning of machine settings to obtain a brittle cross section over the entire height of the sheet instead of 2/3 of the brittle cross-sectional height which is presented in this paper. It seems that the height of the brittle fracture can be increased, while at the same time the height of the plastic zone undergoes reduction. It can be achieved by using classical, evolutionary, or hybrid optimization methods that are a combination of classical optimization methods with genetic or evolutionary ones [29–32].

In Tables 3 and 4, the maximum values of component stress and Huber's reduced stress for selected time instants are juxtaposed. It can be noticed that with the passage of time in the initial stage of the cutting process, these stresses increase, and in the final stage of the process, they decrease, which is the result of the progressive separation of the material being cut.

Table 3. Juxtaposition of maximum reduced Huber–Mises stress and strain values in a sheet being cut for selected time instants during numerical cutting process.

Figures 6 and 7	Time Instant (s)	Reduced Huber–Mises Stress (MPa)	Reduced Huber–Mises Plastic Strain (-)
(a)	$t_a = 2.07$	512	0.003
(b)	$t_b = 3.0$	717	0.376
(c)	$t_c = 4.0$	817	0.425
(d)	$t_d = 5.0$	913	0.465
(e)	$t_e = 6.0$	874	0.486
(f)	$t_f = 7.1$	720	0.486
(g)	$t_g = 7.2$	622	0.486
(h)	$t_h = 12.0$	622	0.486

Table 4. Juxtaposition of maximum constituent stress values in a sheet being cut for selected time instants during numerical cutting process.

Time Instant (s)	Stress along x Axis (MPa)	Stress along y Axis (MPa)	Shear Stress in xy Plane (MPa)
$t_a = 2.07$	60	379	271
$t_b = 3.0$	309	268	358
$t_c = 4.0$	471	404	436
$t_d = 5.0$	625	784	520
$t_e = 6.0$	690	472	497
$t_f = 7.1$	745	644	226
$t_g = 7.2$	472	568	309
$t_h = 12.0$	467	572	251

In Figure 11, the selected finite elements along with their numbering belonging to the sheet being cut in front of the knife (on the left side of the cutting line) and in the back of the knife (on the right side of the cutting line), depending on the height on which they occur, are presented. In contrast, in Figures 12–14, the mean reduced Huber–Mises stresses for several selected heights (h_1–h_3) as a function of time for the entire cutting process are juxtaposed. It turns out that at a small height measured from the upper surface of the sheet (h_1), stresses on the right side of the cutting line are significantly higher compared to the stresses on the left side of the cutting line (Figure 12). The high values of the mean reduced Huber–Mises stress obtained as a result of the numerical simulation exceeding the yield point ($R_e = 510$ MPa) indicate that the sheet edge on the right side of the cutting line is subjected to the large plastic deformations compared to the edge of the sheet on the left side of the cutting line, where the maximum stresses slightly exceed the yield point and therefore the corresponding plastic deformations are insignificant in this case. A similar situation is maintained at a slightly higher depth corresponding to the height h_2, with the only difference, in comparison to the variant concerning the height h_1 (Figure 12), is that the difference in stresses on the left and right side of the cutting line is relatively small (Figure 13). The influence of different parameters on the cutting process was the subject of research of Nouari and Makich [2], who presented the analysis of mechanisms involved during the machining process of titanium alloys. They investigated the effect of cutting parameters on the tool wear behaviour and stability of the cutting process. Their results showed that during machining and the action of the cutting tool edge, the workpiece material underwent a strong compression and deformed plastically similarly to the cutting process of cold-rolled steel sheet (Figure 10).

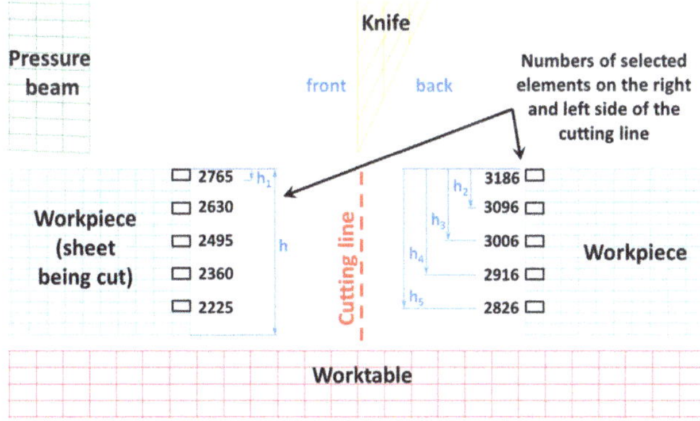

Figure 11. Sheet with selected marked numbers of finite elements accordingly in front of the cutting tool (on the left of the cutting line) and in the back of the cutting tool (on the right of the cutting line).

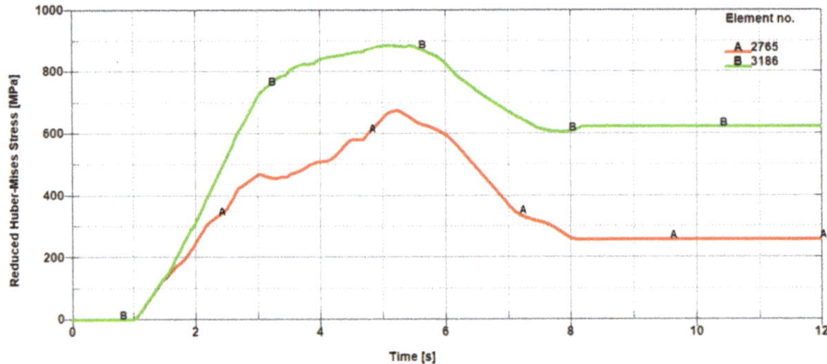

Figure 12. Mean reduced Huber–Mises stresses from the front (A 2765) and from the back (B 3186) of a cutting tool, which correspond to finite element numbers 2765 and 3186, respectively (Figure 11).

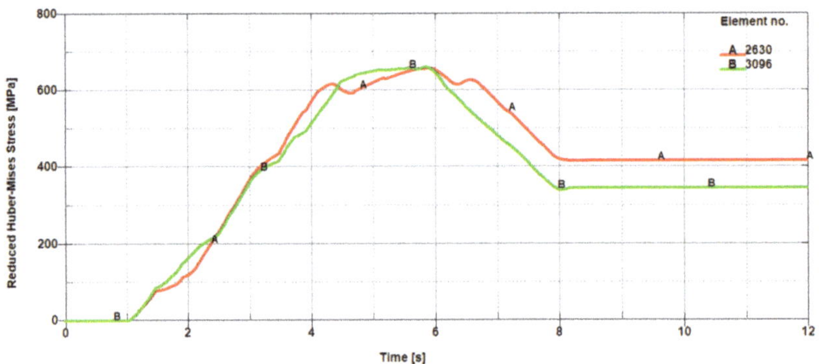

Figure 13. Mean reduced Huber–Mises stresses from the front (A 2630) and from the back (B 3096) of a cutting tool, which correspond to finite element numbers 2630 and 3096, respectively (Figure 11).

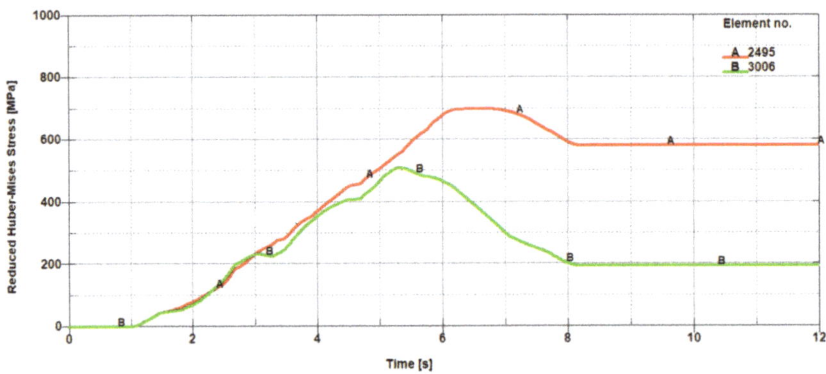

Figure 14. Mean reduced Huber–Mises stresses from the front (A 2495) and from the back (B 3006) of a cutting tool, which correspond to finite element numbers 2495 and 3006, respectively (Figure 11).

With increasing height of delving of the cutting tool blade from h_2 to h_3, there is still a tendency for the occurrence of high stress values in the direct cutting zone exceeding the yield point (R_e). At the depth corresponding to the height h_3, there is a characteristic change in the cutting process involving the change of the location of the extreme in the mean reduced Huber–Mises stresses. For the considered height of delving of the cutting tool blade into the material being cut, the mean reduced Huber–Mises stresses change the location of the extreme occurrence. Extreme stress values start to dominate on the left side of the cutting line (in the front of the knife), which was presented in Figure 14. This change can be explained by a variation in the nature of the cutting process.

In the initial phase, the cutting process consisting in the pure shear, which is inevitably accompanied by the occurrence of large plastic deformations, goes into a more complex process involving the simultaneous occurrence of shear with dominant stretching such as can be observed in the uniaxial tensile test. The mechanism responsible for stretching in the cutting process is the blade of the cutting tool in the shape of the wedge with an apex angle of around 30°. The wedge moving with translational motion from the top towards the bottom is the cause of the occurrence of dominant shear and slight stretching in the first phase and of fading shear with the dominant share of tearing in the next phase. The latter state consequently leads to a brittle cracking resulting in the separation of the sheets being cut.

The observed change in the location of extremes first on the left and then on the right side of the cutting line constitutes a certain characteristic state, which is responsible for creating the transition of a so-called bridge between the plastic and brittle state induced by the dominant tearing state, which is illustrated in Figure 8. The change in the location of extremes of the mean reduced Huber–Mises stresses coincides with the curve that forms the boundary between the plastic and brittle zones. At successive depths corresponding to the heights h_4 and h_5, one can see that the stresses on the left side of the cutting line are slightly higher than the stresses on the right side of the cutting line. The extreme stress values are only slightly higher than the yield stress, which means that the final phase associated with the separation process of sheets takes place as a result of negligible low shear with a dominant share of stretching accompanied by cracking in the final stage of the cutting process.

5. Conclusions

The numerical simulations concerning the mechanism of the cutting process on guillotines by means of the computer program LS-DYNA using an explicit nonlinear dynamic analysis have been conducted. The experimental investigations of C75S cold-rolled steel cutting have been performed on the microstructure and compared with the numerical results obtained by the finite element method. The detailed analysis of the numerical results and metallographic observations allowed the authors to formulate the following conclusions:

- During sheet cutting, the reduced Huber–Mises stresses increase above the limit of steel strength in the direct cutting zone, which causes the material to be destroyed by its separation as a result of the progressive cutting process.
- The stresses in steel sheet grow from zero to the limit of elasticity and then, after exceeding the elastic limit, the plastic zone is formed and the process of material destruction begins along with its separation to the cutting depth of about 1/3 of the sheet height.
- After delving of the cutting tool blade into the metal sheet to a depth greater than 1/3 of its height, the mechanism responsible for separating the material changes. The shear ceases to dominate and stretching domination begins, which causes further progressive destruction of the material by ripping it, resulting in brittle fracture. This stage is advantageous in terms of obtaining sheets with a high smoothness without burrs.
- The stage of plastic shearing causes bending of the edges of the sheets, mainly as a result of large plastic deformations occurring in the first phase of the cutting process. The occurrence of a characteristic line pointing to the border between two areas, plastic and brittle, in microscopic images, as well as the values of the reduced Huber–Mises stresses resulting from numerical calculations, implying the permanent plastic deformation, indicate the correctness of the assumed physical model of the cutting process.

Author Contributions: J.K. conceived, designed, and performed the experiments and simulations; A.G. performed microscopic tests, analysed the corresponding data, and reviewed the paper; both authors discussed the paper.

Funding: The financial support from the statutory funds of the Faculty of Mechanical Engineering of the Silesian University of Technology in 2018 is gratefully acknowledged.

Acknowledgments: The authors would like to express their gratitude to MSc Eng. Anna Łatkiewicz, an expert from the Division of Field Emission Scanning Electron Microscopy and Microanalysis at the Institute of Geological Sciences at the Jagiellonian University in Cracow, for technical assistance during the experimental investigation, her kindness, and valuable advice. Numerical calculations were carried out using the computer cluster Ziemowit (http://www.ziemowit.hpc.polsl.pl) funded by the Silesian BIO-FARMA project No. POIG.02.01.00-00-166/08 in the Computational Biology and Bioinformatics Laboratory of the Biotechnology Centre in the Silesian University of Technology.

Conflicts of Interest: The authors declare no conflict of interest.

References

1. Shaw, M.C. *Metal Cutting Principles*; Oxford University Press: New York, NY, USA, 2005.
2. Nouari, M.; Makich, H. On the Physics of Machining Titanium Alloys: Interactions between Cutting Parameters, Microstructure and Tool Wear. *Metals* **2014**, *4*, 335–338. [CrossRef]
3. Razak, N.H.; Chen, Z.W.; Pasang, T. Effects of Increasing Feed Rate on Tool Deterioration and Cutting Force during and End Milling of 718Plus Superalloy Using Cemented Tungsten Carbide Tool. *Metals* **2017**, *7*, 441. [CrossRef]
4. Caminero, M.A. Experimental Study of the Evolution of Plastic Anisotropy in 5754 Al-Mg Cold Rolled Sheets. *Exp. Tech.* **2015**, *39*, 35–42. [CrossRef]
5. Koklu, U.; Basmaci, G. Evaluation of Tool Path Strategy and Cooling Condition Effects on the Cutting Force and Surface Quality in Micromilling Operations. *Metals* **2017**, *7*, 426. [CrossRef]
6. Paggi, M. Crack Propagation in Honeycomb Cellular Materials: A Computational Approach. *Metals* **2012**, *2*, 65–78. [CrossRef]
7. Haddag, B.; Atlati, S.; Nouari, M.; Moufki, A. Dry Machining Aeronautical Aluminum Alloy AA2024-T351: Analysis of Cutting Forces, Chip Segmentation and Built-Up Edge Formation. *Metals* **2016**, *6*, 197. [CrossRef]
8. Kaczmarczyk, J.; Gąsiorek, D.; Mężyk, A.; Skibniewski, A. Connection Between the Defect Shape and Stresses which Cause it in the Bundle of Sheets Being Cut on Guillotines. *Model. Optim. Phys. Syst.* **2007**, *6*, 81–84.
9. Martínez Krahmer, D.; Polvorosa, R.; López de Lacalle, L.N.; Alonso-Pinillos, U.; Abate, G.; Riu, F. Alternatives for Specimen Manufacturing in Tensile Testing of Steel Plates. *Exp. Tech.* **2016**, *40*, 1555–1565. [CrossRef]
10. Eder, M.A.; Vollum, R.L.; Elghazouli, A.Y.; Abdel-Fattah, T. Modelling and Experimental Assessment of Punching Shear in Flat Slabs with Shearheads. *Eng. Struct.* **2010**, *32*, 3911–3924. [CrossRef]
11. Uriarte, L.; Azcárate, S.; Herrero, A.; Lopez de Lacalle, L.N.; Lamikiz, A. Mechanistic Modelling of the Micro end Milling Operation. *Proc. Inst. Mech. Eng. B-J. Eng.* **2008**, *222*, 23–33. [CrossRef]
12. Del Pozo, D.; López de Lacalle, L.N.; López, J.M.; Hernández, A. Prediction of Press/Die Deformation for an Accurate Manufacturing of Drawing Dies. *Int. J. Adv. Manuf. Technol.* **2008**, *37*, 649–656. [CrossRef]
13. Lamikiz, A.; Lopez de Lacalle, L.N.; Sanchez, J.A.; Bravo, U. Calculation of the Specific Cutting Coefficients and Geometrical Aspects in Sculptured Surface Machining. *Mach. Sci. Technol.* **2005**, *9*, 411–436. [CrossRef]
14. Bohdal, Ł. Application of a SPH Coupled FEM Method for Simulation of Trimming of Aluminum Autobody Sheet. *Acta Mech. Autom.* **2016**, *10/1*, 56–61. [CrossRef]
15. Livermore Software Technology Corporation. *LS-DYNA Keyword User's Manual*; Livermore Software Technology Corporation: Livermore, CA, USA, 2017.
16. Kaczmarczyk, J. Numerical Simulations of Preliminary State of Stress in Bundles of Metal Sheets on the Guillotine. *Arch. Mater. Sci. Eng.* **2017**, *85*, 14–23. [CrossRef]
17. Bathe, K.J.; Chaudhary, A. A Solution Method for Planar and Axisymmetric Contact Problems. *Int. J. Numer. Method Eng.* **1985**, *21*, 65–88. [CrossRef]
18. Belytschko, T.; Liu, W.K.; Moran, B.; Elkhodary, K.I. *Nonlinear Finite Elements for Continua and Structure*; John Wiley & Sons, Ltd.: Southern Gate, CA, USA, 2014.
19. Fish, J.; Belytschko, T.A. *First Course in Finite Elements*; John Wiley & Sons, Ltd.: Southern Gate, CA, USA, 2007.

20. Mohammadi, S. *Discontinuum Mechanics Using Finite and Discrete Elements*; WIT Press Southampton: Boston, MA, USA, 2003.

21. Hughes, T.J.R. *The Finite Element Method. Linear Static and Dynamic Finite Element Analysis*; Manufactured in the United States by RR Donnelley: Chicago, IL, USA, 2016.

22. Zienkiewicz, O.C.; Taylor, R.L. *The Finite Element Method, Solid Mechanics*; Butterworth-Heinemann: Oxford, UK, 2000.

23. Fedeliński, P.; Górski, R.; Czyż, T.; Dziatkiewicz, G.; Ptaszny, J. Analysis of Effective Properties of Materials by Using the Boundary Element Method. *Arch. Mech.* **2017**, *66/1*, 19–35.

24. Saouma, V.E. *Lecture Notes in: Fracture Mechanics*; Department of Civil Environmental and Architectural Engineering, University of Colorado: Boulder, CO, USA, 2000.

25. Perez, N. *Fracture Mechanics*; Kluwer Academic Publishers: Boston, MA, USA, 2004.

26. Timoshenko, S.P.; Goodier, J.N. *Theory of Elasticity*; McGraw-Hill Education: New York, NY, USA, 2017.

27. Grajcar, A.; Radwanski, K. Microstructural Comparison of the Thermomechanically Treated and Cold-Deformed Nb-Microalloyed TRIP Steel. *Mater. Technol.* **2004**, *48*, 679–683.

28. Lisiecki, A.; Piwnik, J. Tribological Characteristic of Titanium Alloy Surface Layers Produced by Diode Laser Gas Nitriding. *Arch. Metall. Mater.* **2016**, *61*, 543–552. [CrossRef]

29. Rothwell, A. *Optimisation Methods in Structural Design*; Springer International Publishing AG: Delft, The Netherlands, 2017.

30. Bhatti, M.A. *Practical Optimization Methods*; Springer-Verlag Inc.: New York, NY, USA, 2000.

31. Goldberg, D.E. *Genetic Algorithms in Search, Optimization, and Machine Learning*; Addison-Wesley Publishing Company, Inc.: New York, NY, USA, 1989.

32. Michalewicz, Z. *Genetic Algorithms + Data Structures = Evolution Programs*; Springer: Berlin/Heidelberg, Germany, 1996.

 *materials*

Article

Investigation of Cutting Temperature during Turning Inconel 718 with (Ti,Al)N PVD Coated Cemented Carbide Tools

Jinfu Zhao [1,2], Zhanqiang Liu [1,2,*], Qi Shen [1,2], Bing Wang [1,2] and Qingqing Wang [1,2]

1   School of Mechanical Engineering, Shandong University, Key Laboratory of High Efficiency and Clean Mechanical Manufacture of MOE, Jinan 250061, China; sduzhaojinfu@gmail.com (J.Z.); sdushenqi@gmail.com (Q.S.); sduwangbing@gmail.com (B.W.); sduwangqingqing@gmail.com (Q.W.)
2   Key National Demonstration Center for Experimental Mechanical Engineering Education, Jinan 250061, China
*   Correspondence: melius@sdu.edu.cn; Tel.: +86-531-8839-3206

Received: 15 May 2018; Accepted: 20 July 2018; Published: 25 July 2018

Abstract: Physical Vapor Deposition (PVD) $Ti_{1-x}Al_xN$ coated cemented carbide tools are commonly used to cut difficult-to-machine super alloy of Inconel 718. The Al concentration x of $Ti_{1-x}Al_xN$ coating can affect the coating microstructure, mechanical and thermo-physical properties of $Ti_{1-x}Al_xN$ coating, which affects the cutting temperature in the machining process. Cutting temperature has great influence on the tool life and the machined surface quality. In this study, the influences of PVD (Ti,Al)N coated cemented carbide tools on the cutting temperature were analyzed. Firstly, the microstructures of PVD $Ti_{0.41}Al_{0.59}N$ and $Ti_{0.55}Al_{0.45}N$ coatings were inspected. The increase of Al concentration x enhanced the crystallinity of PVD $Ti_{1-x}Al_xN$ coatings without epitaxy growth of TiAlN crystals. Secondly, the mechanical and thermo-physical properties of PVD $Ti_{0.41}Al_{0.59}N$ and $Ti_{0.55}Al_{0.45}N$ coated tools were analyzed. The pinning effects of coating increased with the increasing of Al concentration x, which can decrease the friction coefficient between the PVD $Ti_{1-x}Al_xN$ coated cemented carbide tools and the Inconel 718 material. The coating hardness and thermal conductivity of $Ti_{1-x}Al_xN$ coatings increased with the increase of Al concentration x. Thirdly, the influences of PVD $Ti_{1-x}Al_xN$ coated tools on the cutting temperature in turning Inconel 718 were analyzed by mathematical analysis modelling and Lagrange simulation methods. Compared with the uncoated tools, PVD $Ti_{0.41}Al_{0.59}N$ coated tools decreased the heat generation as well as the tool temperature to reduce the thermal stress generated within the tools. Lastly, the influences of $Ti_{1-x}Al_xN$ coatings on surface morphologies of the tool rake faces were analyzed. The conclusions can reveal the influences of PVD $Ti_{1-x}Al_xN$ coatings on cutting temperature, which can provide guidance in the proper choice of Al concentration x for PVD $Ti_{1-x}Al_xN$ coated tools in turning Inconel 718.

Keywords: PVD $Ti_{0.41}Al_{0.59}N$/$Ti_{0.55}Al_{0.45}N$ coating; cutting temperature; Inconel 718

---

1. Introduction

Inconel 718 has been widely used in aeronautical applications due to its high temperature strength and high corrosion resistance. However, great heat is generated in the cutting process of Inconel 718 by cutting tools due to its low thermal conductivity and high hardness. The heat can decrease the tool life and impair the machined surface quality of a workpiece [1–5]. Coating technology was commonly used to protect the tool substrate, which was an effective way to increase the tool life [6,7]. At present, coated cemented carbide tools account for 80% of total tool production [8]. Physical Vapor Deposition (PVD) $Ti_{1-x}Al_xN$ coatings were widely used in machining Inconel 718 alloys due to the high hardness [9], excellent wear resistance [10,11] and super heat stability [12] of the coatings. The thermo-physical properties of PVD $Ti_{1-x}Al_xN$ coating were closely associated with the Al concentration x [13–17]. It is necessary to study the influences of Al concentration on the microstructure, mechanical properties and thermo-physical properties of PVD $Ti_{1-x}Al_xN$ coatings. This can provide some guidance when choosing proper Al concentration x of PVD $Ti_{1-x}Al_xN$ coated tools in turning Inconel 718.

Researchers commonly carried out experimental observation to analyze the microstructures, mechanical properties and thermo-physical properties of $Ti_{1-x}Al_xN$ coatings. Hultman [15] found that PVD $Ti_{1-x}Al_xN$ coating was a combination of the mostly metallic character of cubic TiN with the semi-conducting behavior of hexagonal AlN. The Al was insoluble in TiN, and Ti was insoluble in AlN. The PVD deposition conditions were far from the thermodynamic equilibrium that was required to synthesize the supersaturated metastable $Ti_{1-x}Al_xN$ coatings. Wahlström et al. [16] analyzed the microstructures of polycrystalline $Ti_{1-x}Al_xN$ alloy coatings, which were made by ultra-high-vacuum dual-target magnetron sputtering technologies. They found that coatings with an AlN concentration $x \leq 0.40$ were single-phase with a face-centered cubic structure of $Ti_{1-x}Al_xN$ coating. The interplanar distance and intergranular void density decreased with the increase of AlN concentration x. As detected by selected-area electron diffraction (SAED), the coatings with an AlN concentration $0.4 < x \leq 0.9$ consisted of wurtzite-structure AlN-rich grains and face-centered cubic structure Al depleted $Ti_{1-x}Al_xN$ grains. Plan-view transmission electron microscopy (TEM) also revealed a dramatic decrease in the average grain size from 65 nm to 30 nm and an increase in the intergranular void density to accompany the phase separation. Yoon et al. [17] researched the influences of Al concentration x on the microstructure and mechanical properties of WC-$Ti_{1-x}Al_xN$ super hard composite coatings. With the increase of Al concentration, the crystal grain interfaces of WC-$Ti_{0.47}Al_{0.53}N$ coatings showed a complete nano-crystalline structure with a grain size of 10 nm. The highest hardness value of WC-$Ti_{1-x}Al_xN$ coatings was obtained from the WC-$Ti_{0.43}Al_{0.57}N$ nano-composite coating. Paldey et al. [13] found that the hardness, oxidation resistance and thermal conductivity of $Ti_{1-x}Al_xN$ coating rose with the increase of the Al atom replacing Ti atom within the TiAlN cubic cell. Ding et al. [14] adopted time domain reflectometry to measure the variation curve of the thermal conductivity of PVD $Ti_{1-x}Al_xN$ coating with the increase of Al concentration x at room temperature. They found that the thermal conductivity of PVD $Ti_{1-x}Al_xN$ coating decreased with the increase of Al concentration x ($x < 0.42$), but the thermal conductivity of PVD $Ti_{1-x}Al_xN$ coating increased with the increase of Al concentration x ($x > 0.42$). The thermal conductivity 4.63 W/(m·K) of PVD $Ti_{0.58}Al_{0.42}N$ coating was the lowest value among the PVD $Ti_{1-x}Al_xN$ coatings. The research [14] can offer evidence to determine the thermal conductivities of PVD $Ti_{0.41}Al_{0.59}N$ and $Ti_{0.55}Al_{0.45}N$ coatings in the current study. The research showed that the coating microstructures were closely associated with the thermal conductivities of PVD $Ti_{1-x}Al_xN$ coatings. Barsoum et al. [18] measured the thermal conductivity of $Ti_4AlN_{2.9}$ coating made by the reactive hot isostatic pressing method. The thermal conductivity of $Ti_4AlN_{2.9}$ was 12 W/(m·K), which increased with the temperature to 20 W/(m·K) (1300 K). Rachbauer et al. [19] found that the thermal conductivity of $Ti_{1-x}Al_xN$ coating initially increased, then decreased with increasing temperature. The temperature dependent thermo-physical properties of $Ti_{1-x}Al_xN$ coatings were closely associated with the crystallinity states of coating and the scattering effects of the grain boundaries.

Researchers commonly utilized the mathematical analysis method [20], experimental test method [13,21] and numerical simulation method [22,23] to analyze the influences of tool coating on the cutting temperature. Du et al. [20] used the single domain and multi domain boundary element methods to calculate the temperature distribution within the coated tool. The internal temperature distribution within the tools can be influenced by the coating material and coating thickness. The calculated temperature values of the tool substrate for the TiN coated tool were slightly lower than those of the corresponding point within the uncoated tools. Paldey et al. [13] indicated that the thermal barrier effect of $Ti_{1-x}Al_xN$ coating was due to its lower thermal conductivity than that of the substrate during a high speed dry machining process. Müller et al. [21] applied a two-color pyrometer to measure the cutting temperature in machining 42CrMo4V by uncoated tools, TiN and TiAlN coated tools. For the cutting speed 150 m/min, feed 0.12 mm/rev and cutting depth 2.5 mm, the maximum temperature of chip formed by uncoated tool was about 480 °C. The maximum temperatures in chips formed by TiN and TiAlN coated tools were about 520 °C and 500 °C, respectively. Davoudinejad et al. [22] proposed 3D finite element modeling of micro end-milling Al6082-T6 to analyze the temperature distribution under different cutting conditions. The simulation results were verified by infrared thermal camera. Grzesik [23] found that the microstructure of coating can be optimized to reduce the friction coefficient between the coated tools and workpiece materials. The microstructure of coatings can be optimized to decrease the heat generated in the cutting zone. For cutting easy-to-machine materials with coated tools, the generated heat can be easily dissipated into the chip by the coatings with lower thermal conductivity. For cutting the difficult-to-machine materials with coated tools, the generated heat can be quickly dissipated into the tool substrate by the coatings with higher thermal conductivity.

However, the influences of PVD $Ti_{1-x}Al_xN$ coated tools with different Al concentration x on the cutting temperature in machining Inconel 718 were not clear. In this study, $Ti_{0.55}Al_{0.45}N$ and $Ti_{0.41}Al_{0.59}N$ coatings with the coating thickness 2 μm were deposited on WC-Co cemented carbide substrate. The deposition technique was arc ion plating deposition technology. The microstructure and grain orientation of the two coatings were observed by high resolution transmission electron microscopy/focused ion beam (HRTEM/FIB) techniques. The mechanical properties of the two coatings were researched by measuring the coating hardness, critical loads between the coating and substrate, and the friction coefficient between the coating and Inconel 718. The influences of Al concentration x on the thermal conductivity of PVD $Ti_{1-x}Al_xN$ coatings were analyzed with the finite element method. The cutting temperatures of PVD $Ti_{1-x}Al_xN$ coated tools were measured with the buried thermocouples. The influences of PVD $Ti_{1-x}Al_xN$ coating on heat generation in the cutting zone were analyzed with mathematical analysis models.

2. Mathematical Analysis Model of Cutting Temperature in Turning Inconel 718 by PVD $Ti_{1-x}Al_xN$ Coated Tools

The orthogonal mathematical analysis models proposed by Komanduri-Hou [24,25] were used to investigate the cutting temperature of the primary cutting zone in turning Inconel 718 with PVD $Ti_{1-x}Al_xN$ coated cemented carbide tools. Figure 1 shows the mathematical analysis model of the primary and imaginary shear heat sources in orthogonal cutting of Inconel 718 by PVD $Ti_{1-x}Al_xN$ coated tools.

The separating point of the tool tip and workpiece is assumed as the coordinate origin O as shown in Figure 1. The direction parallel to the cutting speed is assumed as the direction of the x axis, and the direction perpendicular to the cutting speed is assumed as the direction of the z axis. The two-dimensional planar coordinate xOz is established. φ is the shear angle. V is the cutting speed. L is the length of shear heat source. R_1 and R_2 are the polar coordinates of point M(x, z). The imaginary heat source was introduced into the model to compensate for the heat loss due to the assumption of the semi-infinite medium for the workpiece [24,25].

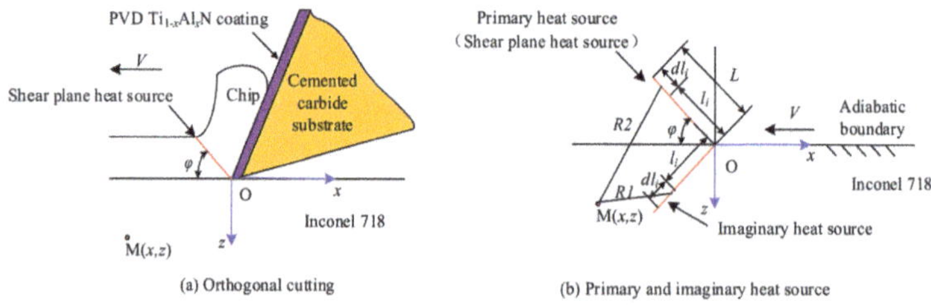

Figure 1. Mathematical analysis model of the primary and imaginary shear heat sources.

The temperature rise at any point M(x, z) was due to the combined effects of the primary and imaginary shear heat sources. Each of these heat sources can be considered as a combination of numerous infinitesimal segments dl_i, with each again as an infinitely long moving line heat source. l_i is the location of the differential small segment of the shear band heat source dl_i relative to the upper end of it and along its width. K_0 is the modified Bessel function of second kind of order zero. The temperature of point M(x, z) within the workpiece induced by the primary and imaginary shear heat sources can be calculated by Equation (1):

$$
\begin{aligned}
T^{shear}_{M(x,z)} &= T_{primary} + T_{imaginary} \\
&= B \frac{q_{shear}}{2\pi\lambda} \int_{l_i=0}^{L} e^{-(x - l_i \cdot \cos(\varphi)) \cdot V / 2\alpha} \cdot \\
&\quad \left\{ K_0 \left[\frac{V}{2\alpha} \sqrt{(x - l_i \cos(\varphi))^2 + (z - l_i \sin(\varphi))^2} \right] + \right. \\
&\quad \left. K_0 \left[\frac{V}{2\alpha} \sqrt{(x - l_i \cos(\varphi))^2 + (z + l_i \sin(\varphi))^2} \right] \right\} dl_i
\end{aligned}
\tag{1}
$$

where $T_{primary}$ and $T_{imaginary}$ are temperatures generated by the primary and imaginary shear heat sources, respectively. The temperatures can be calculated with Equations (2) and (3). B is the coefficient under specific cutting parameters in machining Inconel 718, which can be calculated with Equation (4). q_{shear} is the heat liberation intensity of a moving shear plane heat source, which can be calculated with Equation (5). λ is the thermal conductivity of Inconel 718. α is the thermal diffusivity of Inconel 718. K_0 can be calculated with Equation (6).

$$
T_{primary} = B \frac{q_{shear}}{2\pi\lambda} \int_{l_i=0}^{L} e^{-(x - l_i \cdot \cos(\varphi)) \cdot V / 2\alpha} \times K_0 \frac{V}{2\alpha} \sqrt{(x - l_i \cos(\varphi))^2 + (z - l_i \sin(\varphi))^2} \, dl_i
\tag{2}
$$

$$
T_{imaginary} = B \frac{q_{shear}}{2\pi\lambda} \int_{l_i=0}^{L} e^{-(x - l_i \cdot \cos(\varphi)) \cdot V / 2\alpha} \times K_0 \frac{V}{2\alpha} \sqrt{(x - l_i \cos(\varphi))^2 + (z + l_i \sin(\varphi))^2} \, dl_i
\tag{3}
$$

$$
B = 0.60361 \times N_{th}^{-0.37101} = 0.60361 \times \left(\frac{t_c \times V}{\alpha} \right)^{-0.37101}
\tag{4}
$$

$$
q_{shear} = \frac{(F_c \cos(\varphi) - F_t \sin(\varphi)) \cdot V \cdot \cos(\gamma_0)}{1000 t_c \cdot w \cdot \csc(\varphi) \cdot \cos(\varphi - \gamma_0)}
\tag{5}
$$

where F_c is the tangential force, F_t is the radial force, γ_0 is the rake angle, t_c is the undeformed chip thickness, and w is the width of shear plane heat source.

$$
K_0 = \frac{1}{2} \int_0^{\infty} \frac{d\omega}{\omega} e^{(-\omega - \frac{u^2}{4\omega})}
\tag{6}
$$

where ω is the variable of cutting time, which can be calculated by Equation (7), and u is the integral variable of shear band length, which can be calculated by Equation (8).

$$
\omega = \frac{V^2 t}{4\alpha}
\tag{7}
$$

$$
u = \frac{V}{2\alpha \cdot l_i \cdot \cos(\varphi)}
\tag{8}
$$

3. Experimental Procedure

3.1. Characterization of Surface and Microstructure for PVD $Ti_{1-x}Al_xN$ Coatings

The surface roughness of PVD $Ti_{1-x}Al_xN$ coating was obtained with a laser confocal microscope VK-H1XMC (Keyence, Osaka, Japan). The tool rake face was set perpendicular to the measuring lens. The amplification of the lens used was $10\times$, which could cover the valid area of the tool rake face.

PVD $Ti_{1-x}Al_xN$ coatings were characterized structurally by X-ray diffraction (XRD, Shimadzu, Kanagawa, Japan) and HRTEM/FIB techniques. For the XRD analyses, a wide-range goniometer with a proportional-counter detector was used, with a 2θ accuracy of $0.0001°$. Non-monochromatic Cu radiation was used and K_α peaks were numerically stripped from the spectra using an EVA and TOPAS 4.2 software package. The pole figures were obtained using a X ray diffractometer with the type of D8 advance. The FIB preparation was performed with a FEI Helios 600 Dual Beam, consisting of a liquid gallium ion source operating at 30 kV for sample milling and a field emission electron source operating at 5 kV for secondary electron imaging. The thin lamellae were imaged with a FEI S-TWIN TECNAI G^2 F20-TEM (FEI, Hillsboro, OR, America) operating at 200 kV, to carry out experiments with HRTEM, selected area electron diffraction (SAED), and dark-field imaging.

The cross sections of PVD $Ti_{1-x}Al_xN$ coatings were observed with the field emission scanning electron microscope JSM-7610F (JEOL, Tokyo, Japan). The accelerating voltage was 10 kV. The detailed cross sections of PVD $Ti_{1-x}Al_xN$ coatings were observed at the magnification $20,000\times$.

3.2. Characterization of Mechanical Properties for PVD $Ti_{1-x}Al_xN$ Coated Tools

As referred to in [14,15], the hardness of PVD $Ti_{1-x}Al_xN$ coating was measured with the Vickers hardness tester FM-800. The applied method was the indentation method. It was noted that the indentation depth of the indenter cannot exceed 10–15% of the coating thickness to assure the validity of the measured results. The applied load of the indenter was 50 g.

The Anton Par Revetest was used to apply the scratching tests to obtain the critical loads between the PVD $Ti_{1-x}Al_xN$ coating and carbide substrate. The tool rake faces were set perpendicular to the Rockwell diamond indenter C, then the tools were fixed. The cone angle of the Rockwell diamond indenter C was $120°$, the curvature radius of which was 200 μm. The applied load was in the range of 0–60 N for the Rockwell diamond indenter C. The scratching tests met the requirements of ASTMC1624-05 standard [26]. The scratching speed was 10 mm/min, and the loading rate was 100 N/min. The rupturing sound of the coating and the three-dimensional topography of the scratch were used to find the critical loads between the PVD $Ti_{1-x}Al_xN$ coating and carbide substrate. The detailed measuring process can be referred to the research [27].

The friction coefficient between the PVD $Ti_{1-x}Al_xN$ coated tool and workpiece material Inconel 718 was measured with pin-on-disc testing experiments. The experimental setup was with a type of UMT-TriboLab (Bruker, Billerica, MA, USA). The Inconel 718 was made as the disc with the diameter of 80 mm. The loads applied on the tool were a constant of 5 N at the room temperature.

3.3. Cutting Experiment Procedure

The experimental setup of orthogonal cutting Inconel 718 by PVD $Ti_{1-x}Al_xN$ coated tools without cutting lubricant is shown in Figure 2. The used machine tool was CNC PUMA 200 M. The PVD $Ti_{1-x}Al_xN$ coated cemented carbide tools with the type NG3125R KC5025 (PVD $Ti_{0.55}Al_{0.45}N$ coated tool) and KC5010 (PVD $Ti_{0.41}Al_{0.59}N$ coated tool) were obtained from the KENNAMETAL company. As referred to the methodologies of research [15,17], the thickness of the PVD $Ti_{1-x}Al_xN$ coatings were measured as 2 μm from the cross-sectional views of the coated tools using scanning electron microscopy (SEM) techniques. PVD $Ti_{1-x}Al_xN$ coated tools were compared with the uncoated cemented carbide tools with the NG3125R K313. The three type cutting tools had the same rake angle γ_0 $0°$, relief angle β $11°$ and geometric dimensions. PVD $Ti_{0.55}Al_{0.45}N$ and $Ti_{0.41}Al_{0.59}N$ coated tools had the same substrate material as the uncoated tool. The cutting tool arbor with the type of NSR 3232P3 was shown in the illustration in the Figure 2. The workpiece of Inconel 718 was a cylindrical bar with the diameter of φ24 mm. The ring grooves had been machined on the cylindrical bar. The depth of the ring grooves was 4 mm. The distance was 2 mm between two approached ring grooves.

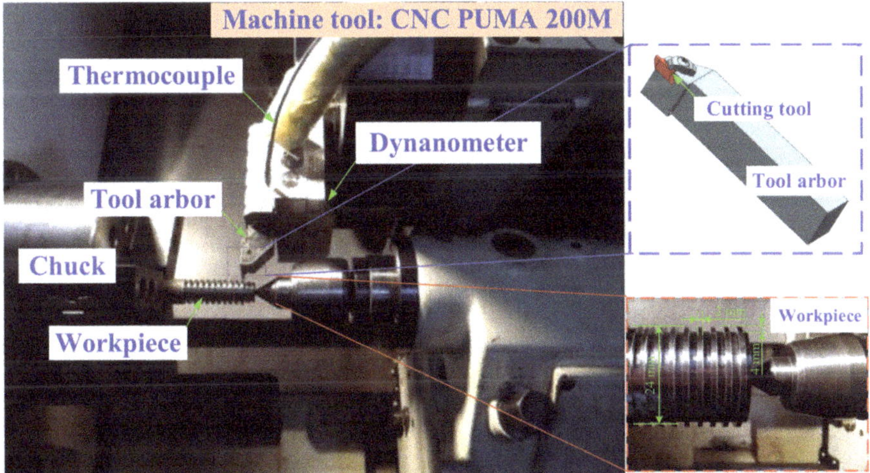

Figure 2. Experimental setup of orthogonal cutting Inconel 718 by PVD $Ti_{1-x}Al_xN$ coated tools without cutting lubricant, the illustration shows the coordinate axis of the machine tool.

The applied cutting parameters of PVD $Ti_{1-x}Al_xN$ coated tool and uncoated tool were the same. The cutting speed used was 20 m/min. The feed used was 0.025, 0.05 and 0.075 mm/rev, respectively. As shown in Figure 2, the cutting forces were measured with the Type 9129A of 3-Component Measuring System. The cutting temperature of tools was measured with the buried K type thermocouple, which was combined with a multiple channel USB data acquisition module OM-DAQ-2401. The schematic of temperature measurement was plotted in Figure 3. As shown in Figure 3, the diameter of the K-type thermocouple was φ0.5 mm. The diameter of the un-displayed drilled hole was φ0.75 mm, which was made by electric discharge machining. The response time of the multiple channel USB data acquisition module OM-DAQ-2401 was 2 ms. The collected electric signals were transferred into the personal computer. Thus, the temperature profiles of tools were obtained accurately during the Inconel 718 turning process.

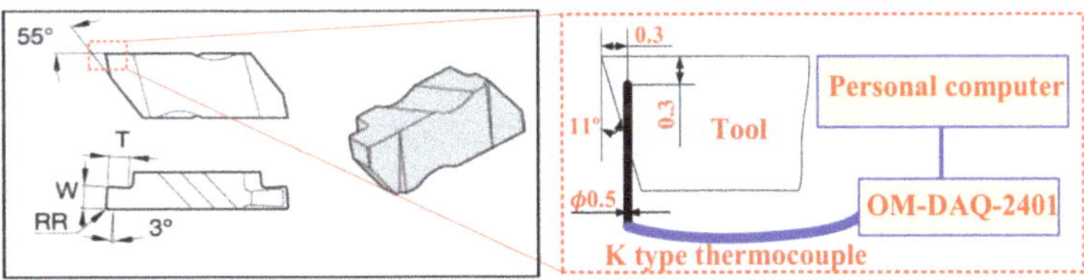

Figure 3. Schematic of temperature measurement with buried K-type thermocouple.

4. Finite Element Simulation of Cutting Temperature in Turning Inconel 718 with PVD $Ti_{1-x}Al_xN$ Coated Tools

The schematic diagram of the experimental setup of the orthogonal cutting of Inconel 718 by PVD $Ti_{1-x}Al_xN$ coated tools is plotted in Figure 4a and the schematic diagram of the related simulation model is plotted in Figure 4b. As in [28–31], the finite element simulation models were established by AdvantEdge software V5.0 without lubricants. As shown in Figure 4b, the cutting tool tip was set at a distance away from the top surface of workpiece. The distance was equal to the feed value. The workpiece was fixed and the tool moved in the cutting direction. The moving speed was equal to the cutting speed value. The feed and cutting speed values were set as shown in Table 1. The influences of PVD $Ti_{0.41}Al_{0.59}N$ and $Ti_{0.55}Al_{0.45}N$ coated tools on the cutting temperature in turning Inconel 718 were analyzed by applying the coatings on the cutting tool. The specific thermo-physical properties of PVD $Ti_{0.41}Al_{0.59}N$ and $Ti_{0.55}Al_{0.45}N$ coatings and the carbide substrate materials were assumed to be the values shown in Table 2. The initial temperature was set to be 20 °C, the same as room temperature. The friction coefficient between PVD $Ti_{1-x}Al_xN$ coated tools and Inconel 718 was assumed to be the values shown in Table 1.

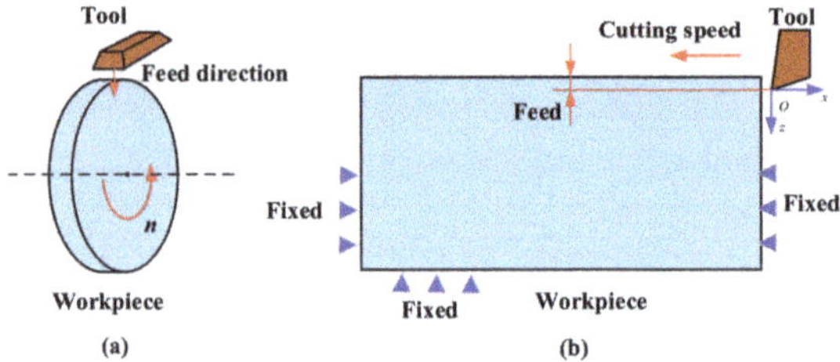

Figure 4. Schematic diagram of experimental set-up and the related simulation models. (**a**) Experimental model; (**b**) simulation model.

Table 1. The mechanical properties of PVD $Ti_{1-x}Al_xN$ coatings at the room temperature.

x	$HV_{0.025}$ (GPa)	Critical Loads (N)	Roughness Ra (µm)	Friction Coefficient (Dry)
0.45	13.18	31.67 ± 2.37	0.561 ± 0.003	0.35
0.59	15.05	38.27 ± 0.67	0.516 ± 0.023	0.33

As shown in Figure 5, the two-dimensional finite element simulation model of cutting temperature field in turning Inconel 718 by a PVD $Ti_{0.41}Al_{0.59}N$ coated tool at feed 0.05 mm/rev and cutting speed 20 m/min were given. $T_{\text{max-workpiece}}$ is the maximum temperature of workpiece. $T_{\text{max-tool}}$ is the maximum temperature of tool. $T_{\text{max-substrate}}$ is the maximum temperature of tool substrate. $T_{\text{tool-corresponding measured point}}$ is the temperature at the measured point of the buried K type thermocouple.

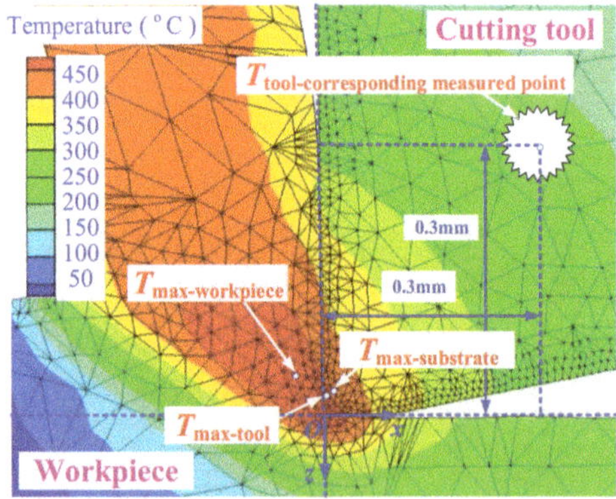

Figure 5. Two-dimensional finite element simulation model of cutting temperature field in turning Inconel 718 with PVD $Ti_{0.41}Al_{0.59}N$ coated tool at feed 0.05 mm/rev and cutting speed 20 m/min.

5. Results and Discussion

5.1. Microstructure of PVD Ti$_{1-x}$Al$_x$N Coatings

Figure 6 illustrates the XRD patterns from PVD Ti$_{0.41}$Al$_{0.59}$N (KC5010) and Ti$_{0.55}$Al$_{0.45}$N (KC5025) coatings. PVD Ti$_{0.41}$Al$_{0.59}$N (KC5010) and Ti$_{0.55}$Al$_{0.45}$N (KC5025) coatings were all face-centered cubic structure Ti$_{(1-x)}$Al$_x$N with cubic lattice. The space groups of two type PVD Ti$_{1-x}$Al$_x$N coatings were the same as Pm-3m (211). For the thin coating thickness and high penetrating capacity of Cu-Ka radiation used in the XRD experiment, the diffraction peaks for the WC-Co carbide substrate were identified clearly. In this research, the crystals of PVD Ti$_{1-x}$Al$_x$N coatings grew in preferred orientations (111) and (200). The grain preferred orientations (111) and (200) for the PVD Ti$_{1-x}$Al$_x$N coating became more evident with the increase of Al concentration.

As shown in Figure 7, the interface areas between the PVD Ti$_{1-x}$Al$_x$N coating and carbide substrate were observed with SEM and HRTEM. Differences between cross-sectional SEM topographies were not evident for the Ti$_{0.41}$Al$_{0.59}$N (KC5010) coated tool and Ti$_{0.55}$Al$_{0.45}$N (KC5025) coated tool. Plan-view HRTEM images showed that the disordering areas and lattice distortions existed around the interface areas between the PVD Ti$_{1-x}$Al$_x$N coating and carbide substrate. The lattice fringe orientations of two PVD Ti$_{1-x}$Al$_x$N coatings were not consistent with the lattice fringe orientations of the WC phase and Co phase in the WC-Co carbide substrate. It was illustrated that the epitaxy growth of TiAlN crystal did not exist at the interface between the PVD Ti$_{1-x}$Al$_x$N coating and carbide substrate. The epitaxy growth of TiAlN crystal did not exist in the physical vapor deposition process. The deposition temperature of physical vapor deposition process was lower than that of the chemical vapor deposition process. Thus, the re-nucleation and growth of TiAlN crystals occurred independently in the physical vapor deposition process of PVD Ti$_{1-x}$Al$_x$N coatings without the influences of the WC phase and Co phase.

Attention was paid to the internal microstructures of PVD Ti$_{1-x}$Al$_x$N coatings. As shown in Figure 8, the internal microstructures of PVD Ti$_{0.41}$Al$_{0.59}$N (KC5010) and Ti$_{0.55}$Al$_{0.45}$N (KC5025) coatings were observed by HRTEM and SAED. The crystallizations of Ti$_{0.41}$Al$_{0.59}$N (KC5010) and Ti$_{0.55}$Al$_{0.45}$N (KC5025) coatings were all nano-crystalline. The corresponding SAED patterns showed that the continuity of diffraction facula for Ti$_{0.41}$Al$_{0.59}$N (KC5010) coating was not as good as that of Ti$_{0.55}$Al$_{0.45}$N (KC5025) coating. As referred to in [16,17], this can be explained by the nano-crystalline of Ti$_{0.41}$Al$_{0.59}$N (KC5010) coating being larger than that of Ti$_{0.55}$Al$_{0.45}$N (KC5025) coating. The crystallization states of Ti$_{0.41}$Al$_{0.59}$N (KC5010) coating were better than that of Ti$_{0.55}$Al$_{0.45}$N (KC5025) coating.

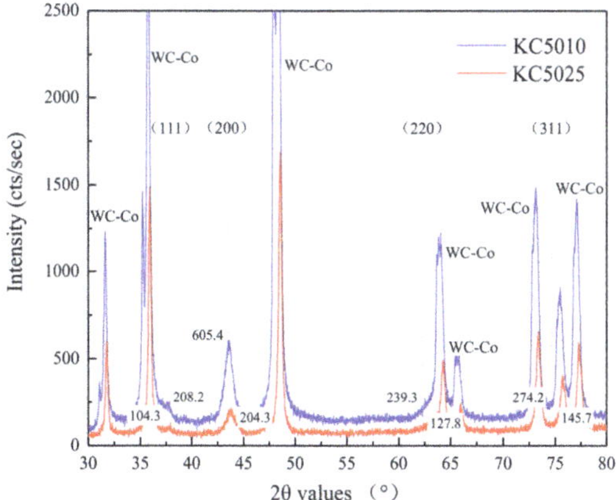

Figure 6. X-ray diffraction (XRD) patterns from PVD Ti$_{0.41}$Al$_{0.59}$N (KC5010) and Ti$_{0.55}$Al$_{0.45}$N (KC5025) coatings.

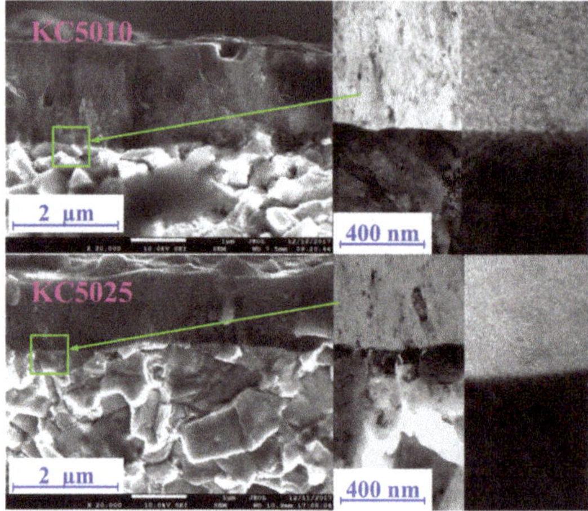

Figure 7. Cross-sectional scanning electron microscopy (SEM) topographies and plan-view high resolution transmission electron microscopy (HRTEM) images from the interface areas between the PVD $Ti_{1-x}Al_xN$ coating and carbide substrate for $Ti_{0.41}Al_{0.59}N$ (KC5010) coated tool and $Ti_{0.55}Al_{0.45}N$ (KC5025) coated tool.

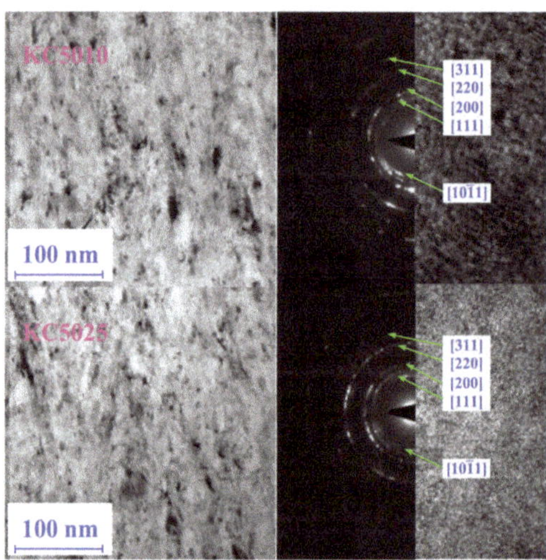

Figure 8. Plan-view HRTEM images and corresponding selected-area electron diffraction (SAED) patterns from PVD $Ti_{0.41}Al_{0.59}N$ (KC5010) coating and PVD $Ti_{0.55}Al_{0.45}N$ (KC5025) coating.

5.2. Mechanical and Thermo-Physical Properties of PVD $Ti_{1-x}Al_xN$ Coatings

The mechanical properties of PVD $Ti_{1-x}Al_xN$ coatings were listed in Table 1. Each experiment was conducted more than three times to obtain average values. As referred to in [32,33], the replacement of Ti atoms by Al atoms induced the lattice distortion of $Ti_{1-x}Al_xN$ coating. The lattice distortion induced internal stress within the $Ti_{1-x}Al_xN$ coatings. The increase of Al concentration increased the pinning effect in the PVD $Ti_{1-x}Al_xN$ coating. The pinning effects prevented the dislocation movement of $Ti_{1-x}Al_xN$ crystals. The friction coefficients between the $Ti_{1-x}Al_xN$ coating and Inconel 718 decreased slightly [34]. Thus, the friction forces between the PVD $Ti_{0.41}Al_{0.59}N$ (KC5010) coated tools and Inconel 718 decreased slightly.

The temperature-dependent mechanical and thermo-physical properties of tungsten-based cemented carbide are referenced to Akbar et al. [29]. Barsoum et al. [18] and Finkel et al. [34] found that the thermal conductivity of Ti_4AlN_3 increased linearly with temperature. Ti_4AlN_3 is another expression of (Ti,Al)N material, defined by the Ti/Al atomic ratio. Akbar et al. [29] summarized the research results of Barsoum et al. [18] and Finkel et al. [34] and obtained the temperature-dependent mechanical and thermo-physical properties of $Ti_{1-x}Al_xN$ coating. To investigate the influences of $Ti_{1-x}Al_xN$ coating on the cutting temperature in turning Inconel 718, the temperature-dependent mechanical and thermo-physical properties of $Ti_{0.41}Al_{0.59}N$ coating were assumed to be the same as the research results of Akbar et al. [29]. According to the research of Ding et al. [14], the thermal conductivity of PVD $Ti_{0.55}Al_{0.45}N$ coating is referred to as 4.6 W/(m·K) at room temperature. The thermal conductivity of PVD $Ti_{0.41}Al_{0.59}N$ coating was inferred as 6.6 W/(m·K) at room temperature. Thus, the thermal conductivity of PVD $Ti_{0.55}Al_{0.45}N$ coating was assumed to be less than that of PVD $Ti_{0.41}Al_{0.59}N$ coating about 2 W/(m·K) with the variation of temperature. The temperature-dependent mechanical and thermo-physical properties of $Ti_{0.41}Al_{0.59}N$ coating, $Ti_{0.55}Al_{0.45}N$ coating and tungsten-based cemented carbide are given as Table 2.

Table 2. Temperature dependent mechanical and thermo-physical properties of $Ti_{0.41}Al_{0.59}N$ coating, $Ti_{0.55}Al_{0.45}N$ coating and tungsten-based cemented carbide.

	Values at the Following Various Temperatures				
	100 °C	300 °C	500 °C	700 °C	900 °C
Properties of $Ti_{0.41}Al_{0.59}N$ coating [14,27,33]					
Young's modulus, GPa	370 (assumed as unchanged with temperature)				
Poisson's ratio	0.22 (assumed as unchanged with temperature)				
Density, kg/m^3	1892 (assumed as unchanged with temperature)				
Thermal conductivity, W/(m·K)	12.61	14.01	15.41	16.81	18.21
Specific heat, J/(kg·K)	639.89	727.28	769.46	794.29	810.67
Properties of $Ti_{0.55}Al_{0.45}N$ coating [14,27,33]					
Young's modulus, GPa	370 (assumed as unchanged with temperature)				
Poisson's ratio	0.22 (assumed as unchanged with temperature)				
Density, kg/m^3	1892 (assumed as unchanged with temperature)				
Thermal conductivity, W/(m·K)	10.61	12.01	13.41	14.81	16.21
Specific heat, J/(kg·K)	639.89	727.28	769.46	794.29	810.67
Properties of tungsten-based cemented carbide [27]					
Young's modulus, GPa	534 (assumed as unchanged with temperature)				
Poisson's ratio	0.22 (assumed as unchanged with temperature)				
Density, kg/m^3	11900 (assumed as unchanged with temperature)				
Thermal conductivity, W/(m·K)	40.15	48.55	56.95	65.35	73.75
Specific heat, J/(kg·K)	346.01	370.01	394.01	418.01	442.01

5.3. Influences of PVD $Ti_{1-x}Al_xN$ Coated Tools on Cutting Temperature

As shown in Figure 9, the cutting temperature profiles with variations of cutting time of $Ti_{0.41}Al_{0.59}N$ coated tool (KC5010), $Ti_{0.55}Al_{0.45}N$ coated tool (KC5025), uncoated tool (K313) were measured by the buried K-type thermocouple at feed 0.05 mm/rev and cutting speed 20 m/min. Compared with the uncoated tools, the existence of PVD $Ti_{1-x}Al_xN$ coating increased the measured cutting temperature. As referred to in the research of Grezsik [23], the heat generated in the cutting zone can be prevented from dissipating from the tool body into the environment quickly by PVD $Ti_{1-x}Al_xN$ coated tools. Thus, the heat that accumulated within the tool body increased the measured temperature. The measured temperatures of PVD $Ti_{0.55}Al_{0.45}N$ coated tool were higher than those of the PVD $Ti_{0.41}Al_{0.59}N$ coated tool. This was associated with the thermal conductivity of $Ti_{(1-x)}Al_xN$ coating being increased with the increase of Al concentration [18,19]. The Al concentration of $Ti_{0.41}Al_{0.59}N$ coating was higher than that of the $Ti_{0.55}Al_{0.45}N$ coating. Thus, the thermal conductivity of $Ti_{0.41}Al_{0.59}N$ coating was higher than $Ti_{0.55}Al_{0.45}N$ coating as referred to in [14,18,19]. Compared with the PVD $Ti_{0.55}Al_{0.45}N$ coated tools, the heat generated can be dissipated quickly from the tool body into the environment and thus decrease the measured temperature for PVD $Ti_{0.41}Al_{0.59}N$ coated tools.

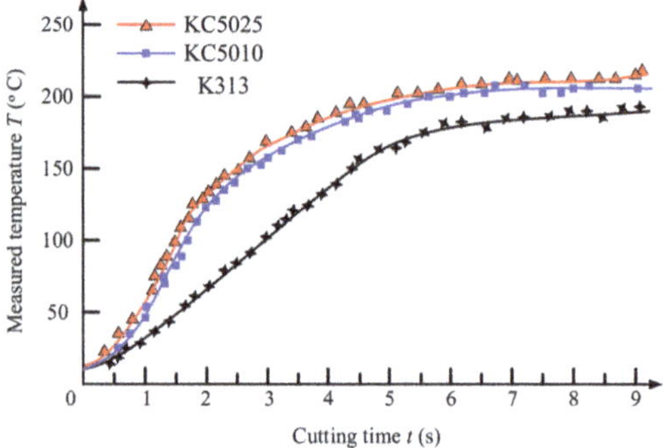

Figure 9. Cutting temperature profiles with variations of cutting time of PVD $Ti_{0.41}Al_{0.59}N$ coated tool (KC5010), $Ti_{0.45}Al_{0.55}N$ coated tool (KC5025) and uncoated tool (K313) measured by the buried K-type thermocouple at feed 0.05 mm/rev and cutting speed 20 m/min.

Comparisons between the actual temperature measured by the buried K-type thermocouple and the corresponding finite element simulation temperature at the same point within the substrate of tools with variations of feed are illustrated in Figure 10. The variations of finite element simulation temperature were consistent with the measured temperatures at the same point within the tool substrates. The finite element analysis model can show the temperature field in the cutting zone directly. Compared with the uncoated tools, the existence of PVD $Ti_{(1-x)}Al_xN$ coating increased the steady temperature in turning Inconel 718. The measured temperature of PVD $Ti_{0.41}Al_{0.59}N$ coated tools was lower than that of PVD $Ti_{0.55}Al_{0.45}N$ coated tools. The $Ti_{0.41}Al_{0.59}N$ coating with higher thermal conductivity accelerated the heat generated in the cutting zone dissipating from the tool body into the environment.

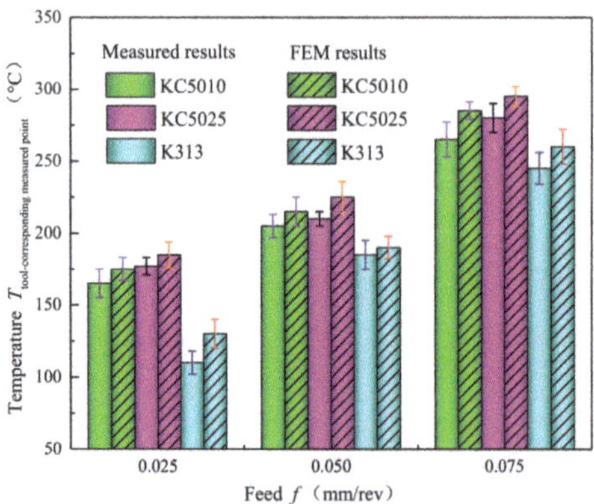

Figure 10. Comparisons between the actual temperature measured by the buried K-type thermocouple and the corresponding finite element simulation temperature at the same point within the substrate of $Ti_{0.41}Al_{0.59}N$ coated tool (KC5010), $Ti_{0.45}Al_{0.55}N$ coated tool (KC5025) and uncoated tool (K313) with variations of feed at cutting speed 20 mm/min.

To analyze the phenomenon, the tangential forces (*Ft*) and radial forces (*Fc*) of PVD $Ti_{0.41}Al_{0.59}N$ coated tool (KC5010), PVD $Ti_{0.55}Al_{0.45}N$ coated tool (KC5025), uncoated tool (K313) with variations of feed at a cutting speed 20 mm/min were obtained, as shown in Figure 11. The tangential forces (*Ft*) and radial forces (*Fc*) increased with the increase of feed. The incremental rate of radial force (*Fc*) was bigger than that of the tangential force (*Ft*). The radial forces decreased with the increase of Al concentration in PVD $Ti_{(1-x)}Al_xN$ coating in turning Inconel 718. The tangential forces of PVD $Ti_{0.41}Al_{0.59}N$ coated tools were lower than that of the PVD $Ti_{0.55}Al_{0.45}N$ coated tools at low feed 0.025 mm/rev. The difference between the tangential forces of the PVD $Ti_{0.41}Al_{0.59}N$ coated tool and the tangential forces of the PVD $Ti_{0.55}Al_{0.45}N$ coated tool were not evident at higher feeds.

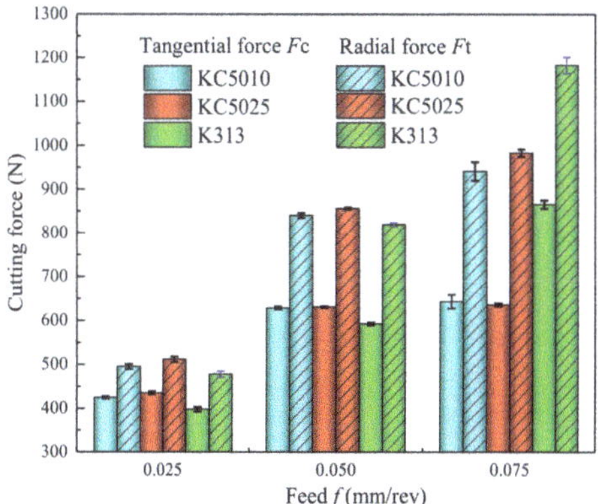

Figure 11. Tangential forces (*Ft*) and radial forces (*Fc*) of PVD $Ti_{0.41}Al_{0.59}N$ coated tool (KC5010), $Ti_{0.45}Al_{0.55}N$ coated tool (KC5025), uncoated tool (K313) with variations of feed at cutting speed of 20 mm/min.

To analyze the influence of PVD $Ti_{(1-x)}Al_xN$ coated tools on the heat generation in turning Inconel 718, the special parameters of adiabatic shear fracture in the cutting zone were calculated with the measured cutting forces, according to Equations (4)–(8). The special parameters of adiabatic shear fracture in the cutting zone for PVD $Ti_{0.41}Al_{0.59}N$ coated tool (KC5010), $Ti_{0.55}Al_{0.45}N$ coated tool (KC5025) and uncoated tool (K313) tools are shown in Table 3. The confidence intervals of the obtained values were 95%. According to the variation of heat liberation intensity q_{shear} of a moving shear plane heat source with feeds, the existence of PVD $Ti_{(1-x)}Al_xN$ coating increased the heat generation in the cutting zone to dissipate more heat into the tool body. This was consistent with the phenomenon that the actual measured temperatures of PVD $Ti_{(1-x)}Al_xN$ coated tool were higher than the uncoated tools. Compared with the PVD $Ti_{0.55}Al_{0.45}N$ coated tools, PVD $Ti_{0.41}Al_{0.59}N$ coated tools decreased the heat liberation intensity q_{shear} of a moving shear plane heat source and decreased the heat dissipating into the tool under the same cutting parameters. This was consistent with the phenomenon of the actual measured temperatures of the PVD $Ti_{0.41}Al_{0.59}N$ coated tool being lower than that of PVD $Ti_{0.55}Al_{0.45}N$ coated tools. The thermal conductivity of PVD $Ti_{0.41}Al_{0.59}N$ coating was higher than that of the PVD $Ti_{0.55}Al_{0.45}N$ coating [35]. PVD $Ti_{0.55}Al_{0.45}N$ coated tools generated more heat in the cutting zone and dissipated more heat from the tool body into the environment. Thus, the cutting temperatures of the PVD $Ti_{0.41}Al_{0.59}N$ coated tool measured by the buried K type thermocouple were lower than that of the PVD $Ti_{0.55}Al_{0.45}N$ coated tool.

As shown in Figure 12, the influences of PVD $Ti_{(1-x)}Al_xN$ coatings on the maximum temperature of the workpiece ($T_{max-workpiece}$), the maximum temperature of the tool ($T_{max-tool}$) and the maximum temperature of the tool substrate ($T_{max-substrate}$) in the finite element simulation models were plotted with variations of feed at a cutting speed of 20 m/min. Compared with the uncoated tool, the existence of PVD $Ti_{(1-x)}Al_xN$ coatings increased the maximum temperature of the workpiece and increased the maximum temperature of the tool. This phenomenon was different from the former research results of Grezsik et al. [36]. Grezsik et al. [36] found that coated tools can decrease the cutting temperature during cutting 45 steel with calculation and simulation methods. Compared with 45 steel, Inconel 718 was used as the hard-to-machine material for its low thermal conductivity and the difficult deformation characteristics. The deformation of Inconel 718 can generate massive heat in the cutting zone. The generated heat cannot be quickly dissipated by chips like the 45 steels. The generated heat in the cutting zone accumulated to increase the maximum temperature of the workpiece in machining Inconel 718 by PVD $Ti_{(1-x)}Al_xN$ coated tools.

Table 3. The special parameters of adiabatic shear fracture in the cutting zone for $Ti_{0.41}Al_{0.59}N$ coated tool (KC5010), $Ti_{0.45}Al_{0.55}N$ coated tool (KC5025) and uncoated tool (K313) tools with variations of feed at a cutting speed of 20 m/min [32].

Type	f (mm/rev)	φ (°)	L (mm)	B	q_{shear} ($J/(mm^2 \cdot s)$)
	0.025	28.11 ± 0.22	0.0843 ± 0.0105	1.129	67.2738 ± 0.0351
KC5010	0.050	31.04 ± 0.15	0.1522 ± 0.0078	0.873	63.3973 ± 0.0530
	0.075	32.23 ± 0.11	0.2203 ± 0.0156	0.751	50.6745 ± 0.0236
	0.025	26.21 ± 0.19	0.0838 ± 0.0245	1.129	70.0889 ± 0.2371
KC5025	0.050	29.13 ± 0.15	0.1538 ± 0.0218	0.873	65.4005 ± 0.1052
	0.075	30.52 ± 0.24	0.1891 ± 0.0027	0.751	54.8375 ± 0.0032
	0.025	27.84 ± 0.11	0.0831 ± 0.0178	1.129	66.6349 ± 0.0261
K313	0.050	29.53 ± 0.21	0.1564 ± 0.0205	0.873	63.4259 ± 0.0331
	0.075	29.21 ± 0.15	0.2143 ± 0.0137	0.751	60.5648 ± 0.0082

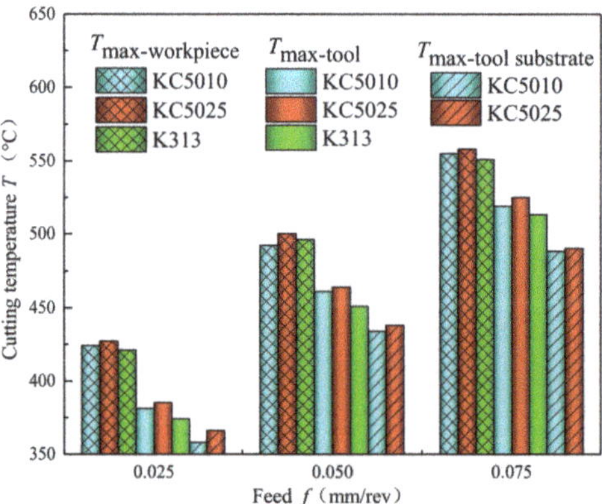

Figure 12. Comparisons of the maximum temperature of the workpiece ($T_{max\text{-}workpiece}$), the maximum temperature of the tool ($T_{max\text{-}tool}$) and the maximum temperature of the tool substrate ($T_{max\text{-}substrate}$) in the finite element simulation models with variations of feed at a cutting speed of 20 m/min for $Ti_{0.41}Al_{0.59}N$ coated tool (KC5010), $Ti_{0.45}Al_{0.55}N$ coated tool (KC5025) and uncoated tool (K313).

Compared with the uncoated tools, the existence of PVD $Ti_{(1-x)}Al_xN$ coating also increased the maximum temperature of the coated tool. Compared with the uncoated tools, PVD $Ti_{(1-x)}Al_xN$ coated tools decreased the maximum temperature of the tool substrate to reduce the thermal stresses within the tools. This was due to the thermal barrier effects of PVD $Ti_{(1-x)}Al_xN$ coating [23]. Compared with PVD $Ti_{0.55}Al_{0.45}N$ coated tools, PVD $Ti_{0.41}Al_{0.59}N$ coated tools decreased the maximum temperature of the workpiece, the maximum temperature of the tool, and the maximum temperature of the tool substrate to reduce the heat generated in the cutting zone. The heat dissipated from the cutting zone into the tool body also decreased in turning Inconel 718 by the PVD $Ti_{0.41}Al_{0.59}N$ coated tool. The decreased maximum temperature of tool substrate for machining Inconel 718 by the PVD $Ti_{0.41}Al_{0.59}N$ coated tool also decreased the generated thermal stresses in the tools under the same cutting parameters.

5.4. Influences of PVD $Ti_{1-x}Al_xN$ Coating on the Surface Topographies of the Tool Rake Faces

The surface topographies of the tool rake face for the $Ti_{0.41}Al_{0.59}N$ coated tool (KC5010), $Ti_{0.45}Al_{0.55}N$ coated tool (KC5025) and uncoated tool (K313) were characterized with a laser confocal microscope of the type VK-H1XMC. Surface topographies of the tool rake faces for $Ti_{0.41}Al_{0.59}N$ coated tool (KC5010), $Ti_{0.45}Al_{0.55}N$ coated tool (KC5025) and uncoated tool (K313) with variations of feed at a cutting speed of 20 m/min were obtained as shown in Figure 13a–i. It was seen that the shiny components were the Inconel 718 workpiece material adhered on the tool rake face.

The adhesion of workpiece material on the tool rake face increased with the increase of feed. The existence of PVD $Ti_{(1-x)}Al_xN$ coating can increase the wear resistance of tools. The wear of uncoated tools (K313) were evident at the feed 0.075 mm/rev and a cutting speed of 20 m/min. The wear of uncoated tools increased the cutting forces and increased the heat liberation intensity q_{shear} of a moving shear plane heat source. As shown in Table 2, the heat liberation intensity q_{shear} of a moving shear plane heat source of an uncoated tool was higher than that of the coated tools at a feed 0.075 mm/rev and cutting speed 20 m/min. As shown in Figure 11, the tangential forces and radial forces of uncoated tool (K313) increased severely from the feed 0.05 mm/rev to 0.075 mm/rev compared with that of the coated tools. But the cutting temperature of the uncoated tool (K313) did not increase severely from the feed 0.05 mm/rev to 0.075 mm/rev compared with the coated tools in the simulation models. This phenomenon can be explained by that the critical wear of the uncoated tools (K313) at feed 0.075 mm/rev was not considered in the finite element simulation. The results were consistent with the research of Devillez et al. [37].

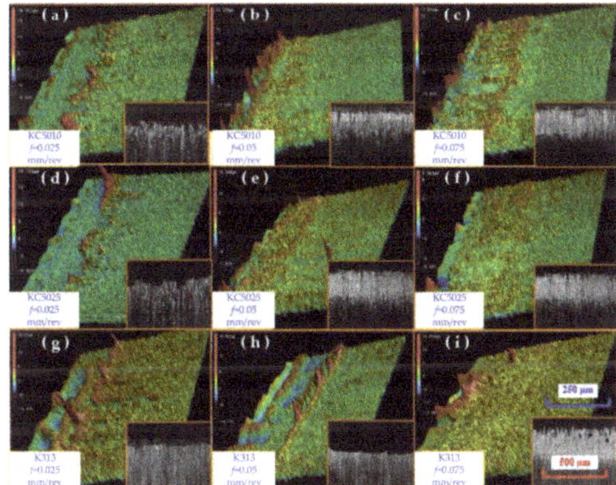

Figure 13. Surface topographies of the tool rake faces for PVD $Ti_{0.41}Al_{0.59}N$ coated tool (KC5010), $Ti_{0.45}Al_{0.55}N$ coated tool (KC5025) and uncoated tool (K313) with variations of feed at a cutting speed of 20 m/min. (**a**) KC5010, f = 0.025 mm/rev, (**b**) KC5010, f = 0.05 mm/rev, (**c**) KC5010, f = 0.075 mm/rev, (**d**) KC5025, f = 0.025 mm/rev, (**e**) KC5025, f = 0.05 mm/rev, (**f**) KC5025, f = 0.075 mm/rev, (**g**) K313, f = 0.025 mm/rev, (**h**) K313, f = 0.05 mm/rev, (**i**) K313, f = 0.075 mm/rev.

6. Conclusions

The influences of PVD $Ti_{(1-x)}Al_xN$ coated tools on the cutting temperature in turning Inconel 718 were analyzed with mathematical analysis model and a finite element simulation model. The results were verified with an orthogonal cutting experiment. The cutting forces and the heat generation in the cutting zone were obtained to analyze the influences of PVD $Ti_{0.55}Al_{0.45}N$ coating and PVD $Ti_{0.41}Al_{0.59}N$ coating on the cutting temperature. The main conclusions can be drawn as follows:

(1) The grain preferred orientations (111) and (200) of PVD $Ti_{0.41}Al_{0.59}N$ coating were more evident compared with that of PVD $Ti_{0.55}Al_{0.45}N$ coating. The epitaxy growth of TiAlN crystals did not exist in PVD $Ti_{1-x}Al_xN$ coated cemented carbide tool for low deposition temperature. PVD $Ti_{0.41}Al_{0.59}N$ coating had better crystallinity than PVD $Ti_{0.55}Al_{0.45}N$ coating.

(2) The pinning effect of coating increased with the increase of Al concentration, which can help to decrease the friction coefficient between cutting tool and Inconel 718 materials. Compared with PVD $Ti_{0.55}Al_{0.45}N$ coated tools, PVD $Ti_{0.41}Al_{0.59}N$ coated tools increased the coating hardness, critical loads and thermal conductivity.

(3) Compared with PVD $Ti_{0.55}Al_{0.45}N$ coated tools, PVD $Ti_{0.41}Al_{0.59}N$ coated tools increased the maximum temperature of the workpiece and the maximum temperature of the coated tool compared with the uncoated tool. PVD $Ti_{0.41}Al_{0.59}N$ coated tools decreased the heat generation and the temperature of the tool body to reduce the thermal stresses generated in the tools.

(4) In this experiment, the PVD $Ti_{0.41}Al_{0.59}N$ and $Ti_{0.55}Al_{0.45}N$ coated tools used can improve the wear resistance of tools.

Author Contributions: J.Z. and Z.L. conceived and designed the experiments; J.Z., Q.S., B.W. and Q.W. performed the experiments; J.Z. and Z.L. analyzed the data and wrote the paper.

Funding: This research was funded by National Natural Science Foundation of China grant numbers 51425503, Taishan Scholar Foundation of Shandong Province grant number TS20130922, and the Major Science and Technology Program of High-end CNC Machine Tools and Basic Manufacturing Equipment grant number 2015ZX04005008.

Acknowledgments: The authors would like to acknowledge the technical support of the Collaborative Innovation Center for Shandong's Main Crop Production Equipment and Mechanization.

Conflicts of Interest: The authors declare no conflict of interest.

Nomenclature

B	Fraction of the shear plane heat conducted into the workpiece material
q_{shear}	Heat liberation intensity of a moving shear plane heat source, $J/(mm^2 \cdot s)$
α	Thermal diffusivity, m^2/s
x, y, z	Spatial coordinate axis, mm
X, Y, Z	Spatial coordinate axis of machine tool, mm
l_i	Location of the differential small segment of the shear band heat source dli relative to the upper end of it and along its width, mm
t	Time, s
V	Cutting speed, m/min
f	Feed, mm/rev
a_p	Depth of cut, mm
t_c	undeformed chip thickness, mm
w	Width of shear plane heat source, mm
L	Length of shear plane heat source, mm
F_t	Radial force or feed force, N
F_c	Tangential force, N
K_0	Modified Bessel function of second kind of order zero
T	Temperature, °C
$T_{max\text{-}workpiece}$	Maximum temperature of workpiece during cutting process, °C
u	Integral variable of shear band length
$T_{primary}$	Temperature generated by the primary shear heat source, °C
$T_{imaginary}$	Temperature generated by the imaginary shear heat source, °C
$T_{max\text{-}tool}$	Maximum temperature of tool during cutting process, °C
$T_{max\text{-}substrate}$	Maximum temperature of substrate during cutting process, °C

Greek symbols

φ	Shear angle, °
γ_0	Rake angle, °
λ	Thermal conductivity, W/(m·K)
ρ	Density, kg/m^3
ω	Function of the time variable t
β	Relief angle, °

References

1. Grzesik, W.; Niesłony, P.; Habrat, W.; Sieniawski, J.; Laskowski, P. Investigation of tool wear in the turning of Inconel 718 superalloy in terms of process performance and productivity enhancement. *Tribol. Int.* **2018**, *118*, 337–346. [CrossRef]
2. Polvorosa, R.; Suárez, A.; López de Lacalle, L.N.; Cerrillo, I.; Wretland, A.; Veiga, F. Tool wear on nickel alloys with different coolant pressures: Comparison of Alloy 718 and Waspaloy. *J. Manuf. Process.* **2017**, *26*, 44–56. [CrossRef]
3. Yan, P.; Chen, K.; Wang, Y.B.; Zhou, H.; Peng, Z.Y.; Li, J.; Wang, X.B. Design and Performance of property gradient ternary nitride coating based on process control. *Materials* **2018**, *11*, 758. [CrossRef] [PubMed]
4. Fernández-Valdivielso, A.; López de Lacalle, L.N.; Urbikain, G.; Rodriguez, A. Detecting the key geometrical features and grades of carbide inserts for the turning of nickel-based alloys concerning surface integrity. *Proc. Inst. Mech. Eng. Part C J. Mech. Eng. Sci.* **2015**, *230*, 1989–1996. [CrossRef]
5. Beranoagirre, A.; Olvera, D.; López de Lacalle, L.N. Milling of gamma titanium–aluminum alloys. *Int. J. Adv. Manuf. Technol.* **2012**, *62*, 83–88. [CrossRef]
6. Palanisamy, S.; Rahman Rashid, R.A.; Brandt, M.; Dargusch, M.S. Tool life study of coated/uncoated carbide inserts during turning of Ti6Al4V. *Adv. Mater. Res.* **2014**, *974*, 136–140. [CrossRef]
7. Palanisamy, S.; Rahman Rashid, R.A.; Brandt, M.; Sun, S.; Dargusch, M.S. Comparison of endmill tool coating performance during machining of Ti6Al4V Alloy. *Adv. Mater. Res.* **2014**, *974*, 126–131. [CrossRef]
8. Kupczyk, M.J. Cutting edges with high hardness made of nanocrystalline cemented carbides. *Int. J. Refract. Met. Hard Mater.* **2014**, *974*, 126–131. [CrossRef]
9. Wojciechowski, S.; Twardowski, P.; Wieczorowski, M. Surface texture analysis after ball end milling with various surface inclination of hardened steel. *Metrol. Meas. Syst.* **2014**, *21*, 145–156. [CrossRef]
10. Zaleski, K. A study on the properties of surface–active fluids used in burnishing and shot peening processes. *Adv. Sci. Technol. Res. J.* **2016**, *10*, 235–239. [CrossRef]
11. Rodríguez-Barrero, S.; Fernández-Larrinoa, J.; Azkona, I.; López de Lacalle, L.N.; Polvorosa, R. Enhanced performance of nanostructured coatings for drilling by droplet elimination. *Mater. Manuf. Process.* **2016**, *31*, 593–602. [CrossRef]
12. Feldshtein, E.; Józwik, J.; Legutko, S. The influence of the conditions of emulsion mist formation on the surface roughness of AISI 1045 steel after finish turning. *Adv. Sci. Technol. Res. J.* **2016**, *10*, 144–149. [CrossRef]
13. Paldey, S.; Deevi, S.C. Single layer and multilayer wear resistant coatings of (Ti, Al) N: A review. *Mater. Sci. Eng. A* **2003**, *342*, 58–79. [CrossRef]
14. Ding, X.Z.; Samani, M.K.; Chen, G. Thermal conductivity of PVD TiAlN films using pulsed photothermal reflectance technique. *Appl. Phys. A* **2010**, *101*, 573–577. [CrossRef]
15. Hultman, L. Thermal stability of nitride thin films. *Vacuum* **2000**, *57*, 1–30. [CrossRef]
16. Wahlström, U.; Hultman, L.; Sundgren, J.E.; Adibi, F.; Petrov, I.; Greene, J.E. Crystal growth and microstructure of polycrystalline Ti$_{1-x}$Al$_x$N alloy films deposited by ultra-high-vacuum dual-target magnetron sputtering. *Thin Solid Films* **1993**, *235*, 62–70. [CrossRef]
17. Yoon, J.S.; Leea, H.Y.; Hana, J.G.; Yangb, S.H.; Musilc, J. The effect of Al composition on the microstructure and mechanical properties of WC–TiAlN superhard composite coating. *Surf. Coat. Technol.* **2001**, *142*, 596–602. [CrossRef]
18. Barsoum, M.W.; Rawn, C.J.; El-Raghy, T.; Procopio, A.T.; Porter, W.D.; Wang, H.; Hubbard, C.R. Thermal properties of Ti$_4$AlN$_3$. *J. Appl. Phys.* **2000**, *87*, 8407–8414. [CrossRef]

19. Rachbauer, R.; Gengler, J.J.; Voevodin, A.A.; Resch, K.; Mayrhofer, P.H. Temperature driven evolution of thermal, electrical, and optical properties of Ti–Al–N coatings. *Acta Mater.* **2012**, *60*, 2091–2096. [CrossRef] [PubMed]

20. Du, F.; Lovell, M.R.; Wu, T.W. Boundary element method analysis of temperature fields in coated cutting tools. *Int. J. Solids. Struct.* **2001**, *38*, 4557–4570. [CrossRef]

21. Müller, B.; Renz, U. Time resolved temperature measurements in manufacturing. *Measurement* **2003**, *34*, 363–370. [CrossRef]

22. Davoudinejad, A.; Tosello, G.; Parenti, P.; Annoni, M. 3D Finite element simulation of micro end-milling by considering the effect of tool run-out. *Micromachines* **2017**, *8*, 187. [CrossRef]

23. Grzesik, W. The role of coatings in controlling the cutting process when turning with coated indexable inserts. *J. Mater. Process. Technol.* **1998**, *79*, 133–143. [CrossRef]

24. Komanduri, R.; Hou, Z.B. Thermal modeling of the metal cutting process: Part I—Temperature rise distribution due to shear plane heat source. *Int. J. Mech. Sci.* **2000**, *42*, 1715–1752. [CrossRef]

25. Huang, K.; Yang, W.Y. Analytical model of temperature field in workpiece machined surfacelayer in orthogonal cutting. *J. Mater. Process. Technol.* **2016**, *229*, 375–389. [CrossRef]

26. *Standard Test Method for Adhesion Strength and Mechanical Failure Modes of Ceramic Coatings by Quantitative Single Point Scratch Testing*; ASTM C1624-05; ASTM International: West Conshohocken, PA, USA, 2015.

27. Yang, Y.; Zhao, S.; Gong, J. Effect of heat treatment on the microstructure and residual stresses in (Ti,Al)N films. *J. Mater. Sci. Technol.* **2011**, *27*, 385–392. [CrossRef]

28. Saif, M.T.A.; Hui, C.Y.; Zehnder, A.T. Interface shear stresses induced by non-uniform heating of a film on a substrate. *Thin Solid Films* **1993**, *224*, 159–167. [CrossRef]

29. Akbar, F.; Mativenga, P.T.; Sheikh, M.A. On the heat partition properties of (Ti, Al) N compared with TiN coating in high-speed machining. *Proc. Inst. Mech. Eng. Part B J. Eng.* **2009**, *223*, 363–375. [CrossRef]

30. Lamikiz, A.; López De Lacalle, L.N.; Sánchez, J.A.; Salgado, M.A. Cutting force integration at the CAM stage in the high-speed milling of complex surfaces. *Int. J. Comput. Integr. Manuf.* **2007**, *18*, 586–600. [CrossRef]

31. Ucun, İ.; Aslantas, K.; Bedir, F. Finite element modeling of micro-milling: Numerical simulation and experimental validation. *Mach. Sci. Technol.* **2016**, *20*, 148–172. [CrossRef]

32. Savkova, J.; Blahova, O. Scratch resistance of TiAlSiN coatings. *Chem. Listy* **2011**, *105*, 214–217.

33. Hsieh, J.H.; Liang, C.; Yu, C.H.; Wu, W. Deposition and characterization of TiAlN and multi-layered TiN/TiAlN coatings using unbalanced magnetron sputtering. *Surf. Coat. Technol.* **1998**, *108*, 132–137. [CrossRef]

34. Finkel, P.; Barsoum, M.W.; El-Raghy, T. Low temperature dependencies of the elastic properties of Ti_4AlN_3, $Ti_3Al_{1.1}C_{1.8}$, and Ti_3SiC_2. *J. Appl. Phys.* **2000**, *87*, 1701–1703. [CrossRef]

35. Gu, L.Y.; Kang, G.Z.; Chen, H.; Wang, M.J. On adiabatic shear fracture in high-speed machining of martensitic precipitation-hardening stainless steel. *J. Mater. Process. Technol.* **2016**, *234*, 208–216. [CrossRef]

36. Grezsik, W.P. Experimental investigation of the cutting temperature when turning with coated indexable inserts. *Int. J. Mach. Tools Manuf.* **1999**, *39*, 355–369. [CrossRef]

37. Devillez, A.; Schneider, F.; Dominiak, S.; Dudzinski, D.; Larrouquere, D. Cutting forces and wear in dry machining of Inconel 718 with coated carbide tools. *Wear* **2007**, *262*, 931–942. [CrossRef]

 *materials* 

Article

Experimental Parametric Model for Adhesion Wear Measurements in the Dry Turning of an AA2024 Alloy

Moises Batista Ponce *, Irene Del Sol Illana, Severo Raul Fernandez-Vidal and Jorge Salguero Gomez

Department of Mechanical Engineering & Industrial Design, Faculty of Engineering, University of Cadiz, Av. Universidad de Cadiz 10, E-11519 Puerto Real-Cadiz, Spain; irene.delsol@uca.es (I.D.S.I); raul.fernandez@uca.es (S.R.F.V.); jorge.salguero@uca.es (J.S.G.)
* Correspondence: moises.batista@uca.es; Tel.: +34-956-483-406

Received: 2 August 2018; Accepted: 30 August 2018; Published: 3 September 2018

Abstract: Adhesion wear is the main wear mechanism in the dry turning of aluminium alloys. This type of wear produces an adhesion of the machining material on the cutting tool, decreasing the final surface quality of the machining parts and making it more difficult to maintain industrial tolerances. This work studies the influence of the cutting parameters on the volume of material adhered to the cutting tool surface for dry machining of AA2024 (Al-Cu). For that purpose, a specific methodology based on the automatic image processing method that can obtain the area and the thickness of the adhered material has been designed. This methodology has been verified with the results obtained through 3D analysis techniques and compared with the adhered volume. The results provided experimental parametric models for this wear mechanism. These models are analytic approximations of experimental data. The feed rate mainly results in low cutting speed, while low depths of cut presents a different behaviour due to the low contact pressure. The unstable behaviour of aluminium adhesion on the cutting tool produces a high variability of results. This continuous change introduces variation in the process caused by the continuous change of the cutting tool geometry.

Keywords: cutting tool wear; secondary adhesion wear; turning; machining; aluminium

1. Introduction

The increase of cutting tool life is a highly attractive objective in industry for improving the profitability of industrial machining processes. Tool wear and control are a critical challenge in the machining process nowadays, although they have been addressed in classic studies [1–4].

Different approaches have been used for aluminium alloys in order to minimise this issue. On the one hand, coatings are proposed to improve the cutting tool life, with the most recommended one being diamond coatings [5,6], because they have a higher hardness like intermetallic particles [6]. However, they have refractory properties that increase the temperature of the cutting zone [7], making the machining of Al parts more difficult, especially when high cutting speeds (up to 600 m/min) cannot be achieved. Moreover, the use of diamond coatings increases the cost of the cutting tool and, in several cases, makes it unprofitable to the industry.

On the other hand, different lubricant solutions have been proposed in recent years, mainly based on the use of Minimum Quantity Lubrication (MQL) [8–10], high-pressure coolant [5,11] and cryogenic systems [12,13]. Nevertheless, due to its low environmental impact and associated costs, dry machining must also be considered for the machining of Al alloys [14,15], as it is the most sustainable solution for the turning process [8,16]. It does not involve pollution of air or water resources but is necessary for a cutting tool with low friction coefficient and adequate heat resistance [8].

However, dry machining of aluminium alloys is usually associated with high adhesion rates on the cutting tool surface. This wear mechanism is related to the increase of the friction coefficient [17]

caused by the contact between the surfaces. This type of contact creates a rough surface and a low sliding contact between tool and work materials [18], increasing the contact pressure and particle transference from one surface to another [19]. This can be classified as primary—from the tool to the chip—or secondary, from the workpiece to the tool. In the latter case, the material can adhere to the cutting tool, giving rise to a Built-Up Edge (BUE), presenting adhesion in the closest areas of the cutting edge, and the Built-Up Layer (BUL), located in the rake face [19–21].

Traditionally, BUE formation is related with low cutting speed, while BUL is associated with high cutting speed due to the conventional behaviour of wear in different steel machining, as was studied by Krolczyk et al. [22] using a methodology for predicting the cutting tool life in stainless steel and its relation to surface roughness [23]. However, this behaviour change is dependent on the machined material.

BUL is also considered in turning of nickel alloys as one of the main causes of poor surface finishing, affecting residual stresses [24]. Cutting forces can also be affected, and some models would help to define the mechanistic of parameters modelling [25]. Cutting tool wear by adhesion mechanisms changes cutting forces and cutting temperature, affecting the quality of the machined workpieces. Nevertheless, only few works in the field of modelling address these aspects [26].

In the particular case of aluminium alloys, different authors [18,19,27–30] define the development of secondary adhesion wear in several steps:

- *Build of Primary BUL.* A layer of near-pure aluminium, with a very low quantity of copper, is adhered due to the thermomechanical conditions. Under high pressure and temperature, the matrix is softened and welded to the cutting tool rake face by pressure.
- *Build of BUE.* It is produced by the mechanical adhesion of the workpiece alloy close to the cutting edge.
- *Build of secondary BUL.* A mechanical extrusion of the workpiece material over the rake face takes place. The BUE and secondary BUL usually grow to a critical size. Once it is reached they are detached—at least partially—from the cutting tool, taking with them cutting tool particles, thereby damaging the tool.

This process is unstable and usually cyclic, being repeated several times during long-term tests. It is a dynamic wear mechanism in which successive chip material layers are welded and hardened on the surface, or are deposited on the cutting tool by thermic or mechanical actions. BUE changes continuously because the workpiece material is deposited and the tool particles are moved away. This phenomenon is usually related to the onset of an over cut, due to the added material that modifies the initial cutting edge geometry and, consequently, the initial cutting conditions [4]. In some cases, it can also be followed by a weakening of the cutting tool, causing in some cases a complete or catastrophic and unexpected fracture.

Trujillo et al. [27] explain that this complexity in the machining of aluminium, especially due to the dynamic behaviour, makes measuring adhesion more difficult. The measurement has to consider the modification of the initial cutting geometry, taking into account volume loss and volume increase, as proposed by Gomez-Parra et al. in [18]. There are several technologies to measure tool wear, such as Focus-Variation Microscopy (FMV) [31,32], SEM-3D or tomography [33], image pattern comparison [34], direct measurements based on optical methods [27,35] or intelligent prediction systems [36]. All of these methods have usually been implemented in wear control, and few of them have been implemented for aluminium machining [27,32,33]. Moreover, none of them are widely implemented in industrial applications due to the equipment cost, complex set-ups and long processing times.

For this reason, different techniques to monitor or evaluate tool wear [37–39] have been developed, using indirect measurements of the cutting tool wear through monitoring systems [40] or simulations [41], and mainly focused on orthogonal cutting and chip formation [42].

There are different image processing methods applied to the control of tool wear [43], but the most recent method proposed came from Trujillo et al. [27], in whose study the area affected for tool wear is selected manually and measured with specific software. This is a simple technique and industrial implementation is possible at low cost.

In this research paper, the evolution of the adhesion wear mechanism in the dry turning of AA2024 (Al-Cu) is reported. The mechanism induces a positive volume deviation of the cutting tool and uses an automated image processing technique that allows for easy, quick and cheap evaluation.

Turning is a well-known machining process with two characteristics that make it ideal for this study. It is a process close to orthogonal cutting conditions and the cutting tool geometry is simple. It is also widely used in industries where light alloys are considered as strategic materials, such as the aerospace industry, where the application of eco-friendly processes is a challenge nowadays.

2. Materials and Methods

Tests were performed in cylindrical samples of aluminium alloy AA2024-T3 (Al-Cu), and carried out in a CNC Lathe EMCOturn 242 (EMCO, Hallein, Austria), with 13 kW power and EMCOtronic TM02 numerical control.

Tests were designed to study the evolution of the wear elements depending on the cutting parameters, using combinations of four different cutting speeds, four different feed rates, and three different depths of cut. The machining time was established in 10 s in all cases and the tests were executed in the absence of cutting fluids (dry machining). Therefore, 48 different tests were carried out. Table 1 collects the different parameters used. The cutting parameters have been selected from those industrially used, according to previous works [27]. The short machining time was considered in order to analyse the first instant of machining [44].

Table 1. Cutting parameters.

Cutting Speed (v_c) (m/min)	50	100	150	200
Feed (f) (mm/rev)	0.05	0.1	0.2	0.3
Depth of cut (a_p) (mm)	0.5	1	2	-
Time of cut (t) (s)		10		

In order to reduce the interference of the cutting tool geometry in the development of the machining tests, a close situation to orthogonal cut is chosen. An uncoated commercial carbide tungsten insert with 10% Co from SECO (ref. ISO DCMT11T308) [45] was chosen. The selection of the uncoated insert is done in order to prevent masked tool wear behaviour by coating. The main cutting insert features are: lip angle 55°, clearance angle 7°, rake angle 17°, and classic chip-breaker geometry for fine cutting application and general purpose.

The cutting tool characterisation was carried out in three steps (Figure 1). The two first steps of the evaluation were done using a Stereoscopic Optical Microscope (SOM) Nikon SMZ-800 (Tokyo, Japan) connected to a 5 Mpx optical camera Optikam B5 (Optika, Ponteranica, Italy). Additionally, an optic-fibre lighting ring with low reflectance was used as the illumination system. After calibration, a precision of 0.94 μm/pixel was calculated.

The first step is a phase analysis. In this case a Perfect Image v7 (Clara Vision, Verrières le Buisson, France) software was used. The image is filtered and colour patterns are recognised automatically. This technique allows us to establish the extension of the adhered material. It is calculated as a percentage of the affected area on the rake face by the adhesion. This extension is evaluated in terms of percentage occupation from a zenith image processing based on projection. All the images are taken using the same magnification and the same area of the tool is examined. Figure 2 shows an example of the difference of colour corresponding to the adhered material (grey). The same image processing procedure was used for every tool, with three iterations for each image.

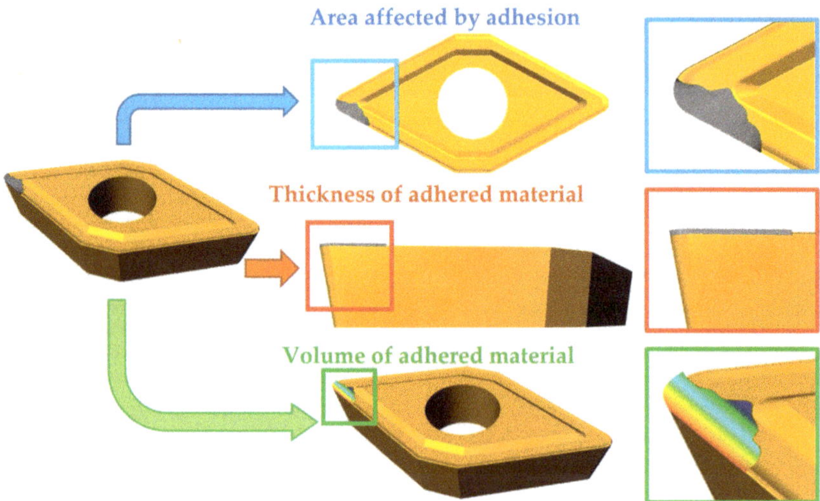

Figure 1. Scheme of the experimental methodology.

The second analysis is of the thickness of the material adhered on the cutting edge (BUE). A minimum of 10 equally spaced images was taken in Zone B, according to the actual standard ISO 3685 [46] before calculating the mean value. This measurement process is shown in Figure 3.

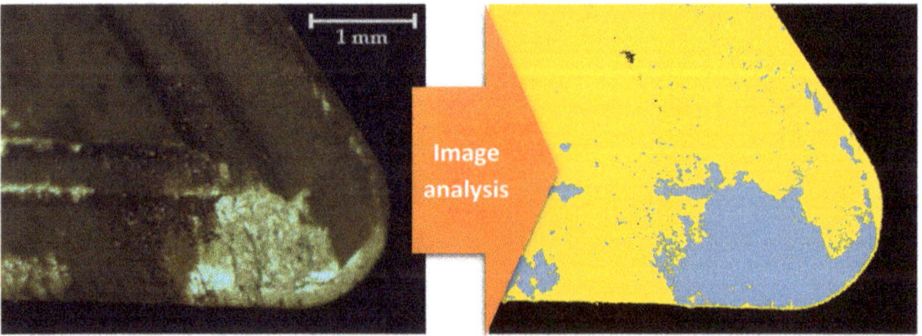

Figure 2. Example of the affected area by adhesion (v_c = 192 m/min, f = 0.3 mm/rev, a_p = 1 mm, t = 10 s).

The area affected by adhesion and its thickness of cutting tool are two-dimensional parameters that define the volume of material adhered of the cutting tool. In order to establish a three-dimensional relationship with the tool wear, the difference in volume of the cutting tool before and after the cutting process was studied. The technique used to study this variation is contact profilometry, using a profilometer TALYSURF CLI 1000 (Taylor Hobson, Leicester, UK). The profiles defined by the topography generate a three-dimensional image of the material adhered to the tool. Adhered volume can be compared with the topography of a new cutting tool using the software Talymap Platinum v5 (Taylor Hobson, Leicester, UK) in order to obtain the volume of adhered material.

The acquired area was 6.15, 6.25 and 8.75 mm² for 0.5, 1 and 2 mm depth cuts, respectively. Five thousand data points are taken in each profile and the space between them is 0.0005 mm.

A section of the tool cut on the perpendicular plane to the cutting edge is studied using a Scanning Electron Microscope (SEM).

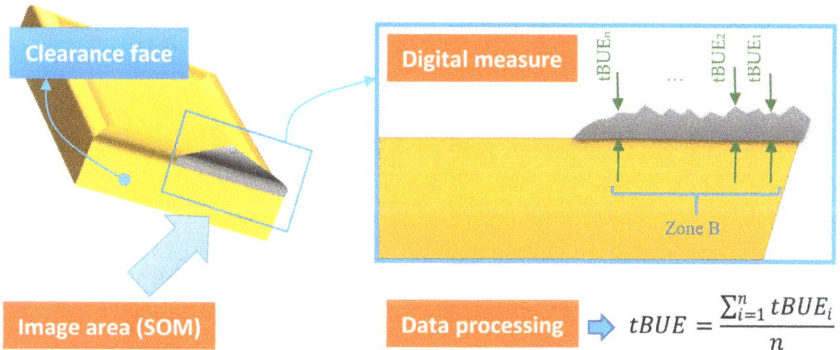

$$tBUE = \frac{\sum_{i=1}^{n} tBUE_i}{n}$$

Figure 3. Built-Up Edge (tBUE) thickness measurement.

3. Results

3.1. Evaluation of the Thickness of the Adhered Material onto the Cutting Edge

Images were taken perpendicularly to the rake face in order to obtain the real thickness of the material adhered to the cutting edge. This adhered material is considered the Built-Up Edge or BUE. Marks of chip fluency are also observed in the clearance face but disappear with a high depth of cut, so may be related to the cross-sectional area of the chip. It can be appreciated that a lower cutting speed is more suitable for presenting a Built-Up Edge (Figure 4), while an increase in the feed rate can reduce this effect (Figure 5).

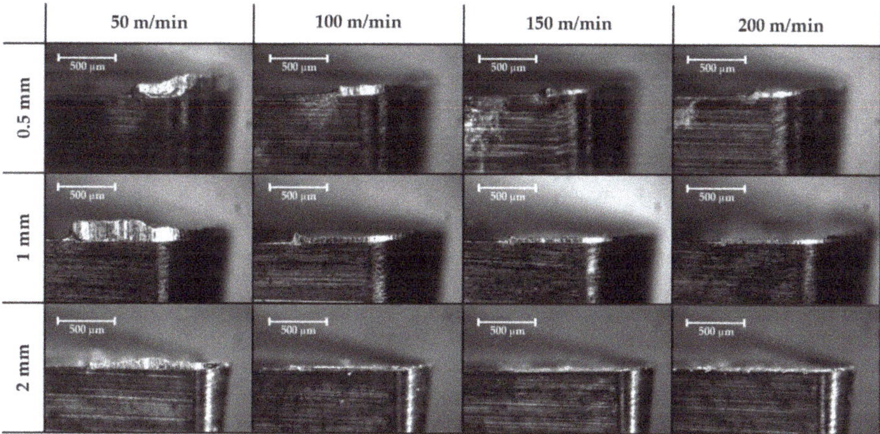

Figure 4. Effect of the feed rate cutting speed and the depth of cut on the secondary adhesion wear in the lateral plane for a 0.2 mm/rev feed rate (63×).

Numerical results show a similar trend to the visual ones. Figure 6 represents the average thickness of the adhered material depending on the cutting speed and the feed rate for a 0.5 mm depth of cut. It has been verified that this behaviour is similar for the other depths of cut.

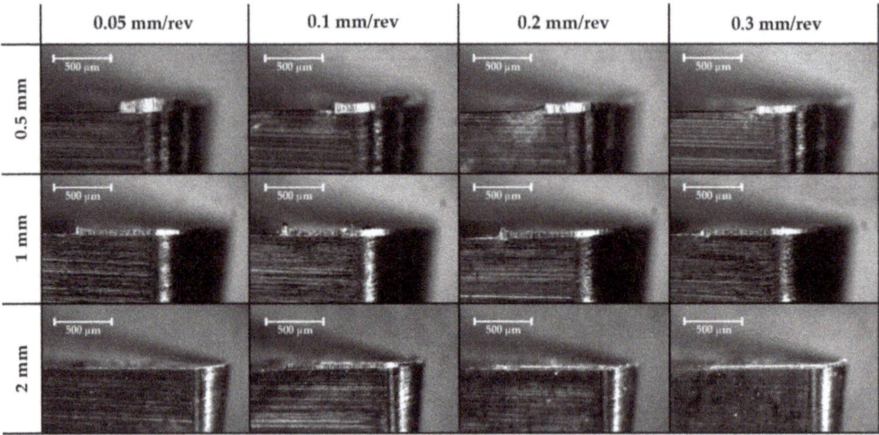

Figure 5. Effect of the feed rate and the depth of cut on the secondary adhesion wear in the cenital plane for cutting speed 100 m/min (63×).

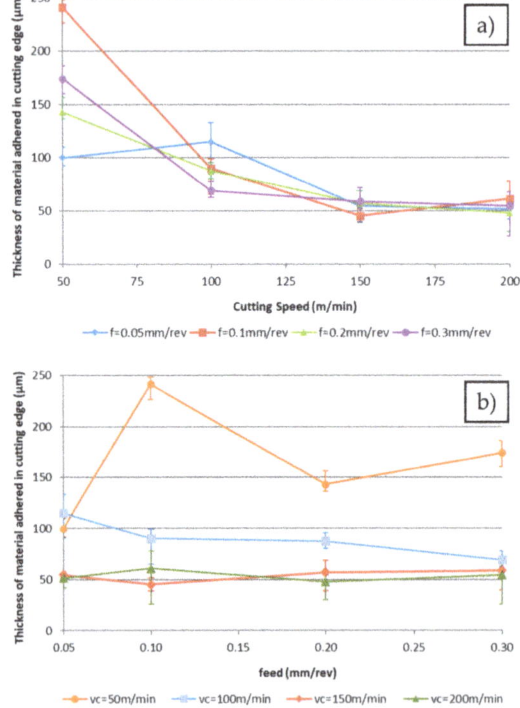

Figure 6. Effect of the thickness of the material adhered on to the tool edge on the secondary adhesion wear for 0.5 mm depth of cut depending on the (**a**) cutting speed; (**b**) feed rate.

The highest values of adhered material are observed for lower cutting speeds, having a similar behaviour for all depths of cut. When the depth decreases, the thickness of the material adhered to the tool is reduced. There is also a decreasing trend related to the cutting speed that plateaus in a horizontal line.

Therefore, the trends are affected by the depth of cut. A lower depth of cut increases the thickness of the adhering material. The slope of this trend decreases when the depth of the cut increases. This effect can be related to the chip forming process. In this case, for the highest feed rates, the cross section of the chip is higher and the associated chip tension is higher too. Due to this, the chip form is long or snarled, in agreement with Rubio et al. [47].

Also, the thickness of the adhered material decreases slightly with the feed rate, while an increase in the cutting speed reduces the sensibility of the feed change. For a high cutting speed, the slope of this trend is higher. This behaviour is related to the temperature of the process, which promotes the mechanical adhesion [48].

Moreover, for extreme values of cutting speed and feed rates, there are higher differences due to the increase of cutting forces. Chip width sense is smaller when decreasing those forces, considering the chip projection over the cutting edge. This behaviour was observed in the cutting force by Carrilero et al. [49].

The experimental results suggest that it is possible to obtain a potential parametric model of $X = f(v_c, f, a_p)$ with a better fit than other models. This was in accordance with the statistical treatment proposed for similar studies [27,50,51].

Potential models have been proposed for each depth of cut:

$$tBUE\ (a_p = 0.5\ mm\ ) = 3985.58 \cdot f^{-0.0002} \cdot v_c^{-0.832} \tag{1}$$

$$tBUE\ (a_p = 1\ mm\ ) = 1002.24 \cdot f^{0.356} \cdot v_c^{-0.461} \tag{2}$$

$$tBUE\ (a_p = 2\ mm\ ) = 384.76 \cdot f^{0.327} \cdot v_c^{-0.432} \tag{3}$$

Those models present the little variations analysed before as well as the high variability of the results. A combined model has been developed from the previous models:

$$tBUE\ = 1244.205 \cdot f^{0.266} \cdot v_c^{-0.567} \cdot a_p^{-0.806} \tag{4}$$

Figure 7 presents three-dimensional models and the combination of them, showing the trends previously mentioned.

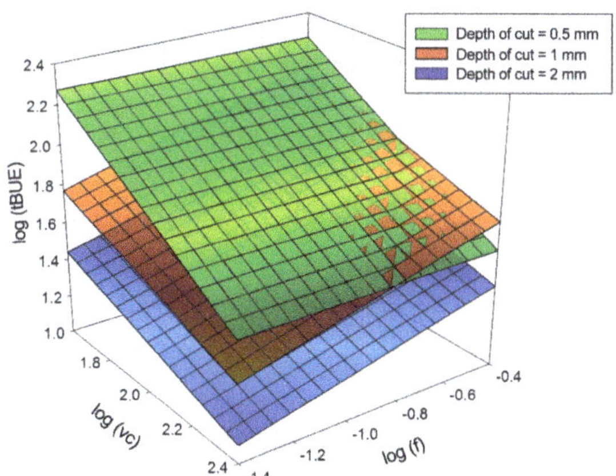

Figure 7. Graphic representations of the models calculated for the thickness of the material adhered on the cutting edge.

It is remarkable that the material adhered on the cutting edge (BUE) changes the geometry of the cutting tool. This change modifies the clearance angle and the rake angle. An example of the new geometry is seen in the transversal section of the tool (Figure 8). In this case, the chip flows out of the rake face and the contact pair in this cutting condition is aluminium–aluminium. This affects the friction coefficient and the temperature of the process, while the contact area decreases [17]. Also, as is shown in Figure 8, the thickness of the adhered material on the cutting edge is higher than that of the material adhered on the rake face.

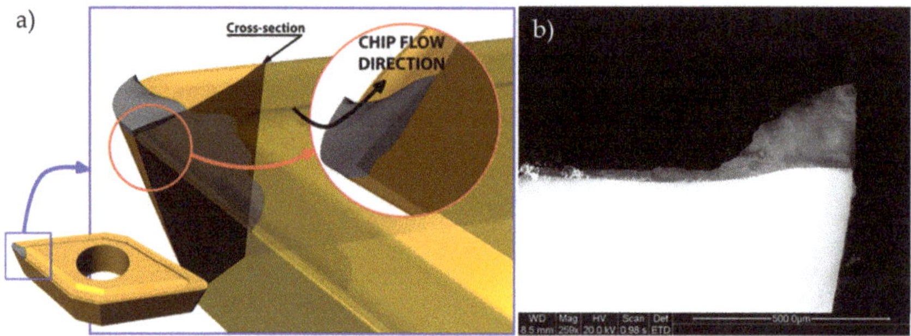

Figure 8. (**a**) Scheme of the cross section of the tool; (**b**) SEM image of this cross section.

3.2. Evaluation of the Area Affected by Adhesion

The images taken in the position perpendicular to the rake face show a layer adhered on this face, considered the Built-Up Layer or BUL. Also, it is possible to see a decrease in the affected area by adhesion when the cutting speed increases (Figure 9). Additionally, there is an increase in the affected area related to the increase of the feed rate (Figure 10). This behaviour is independent of the depth of cut.

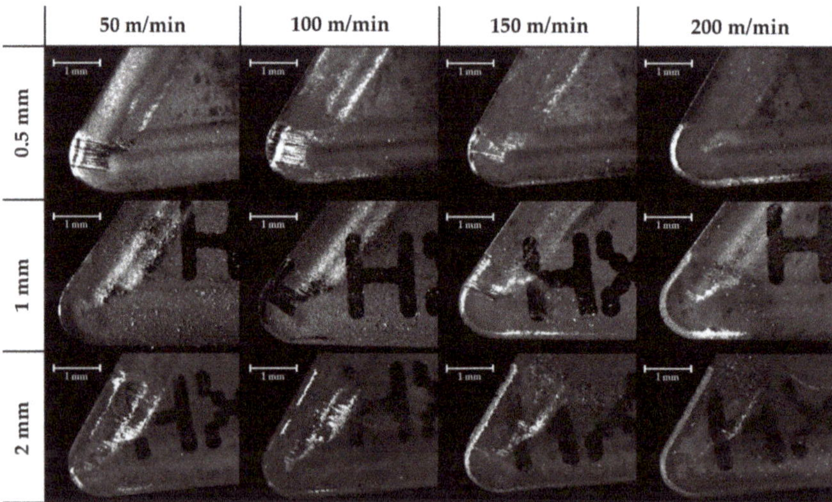

Figure 9. Effect of the cutting speed and the depth of cut on the secondary adhesion wear in the zenith plane for 0.05 mm/rev feed rate (30×).

Figure 11 shows the evolution of the % area affected by adhesion BUE wear, according to the cutting speed and the feed rate using 0.5 mm depth of cut.

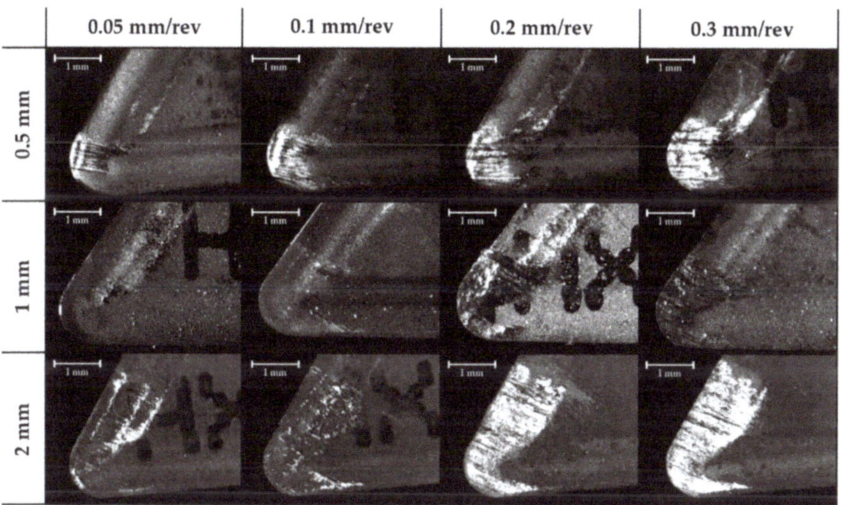

Figure 10. Effect of the feed rate and the depth of cut on the secondary adhesion wear area in the zenith plane for cutting speed 50 m/min (30×).

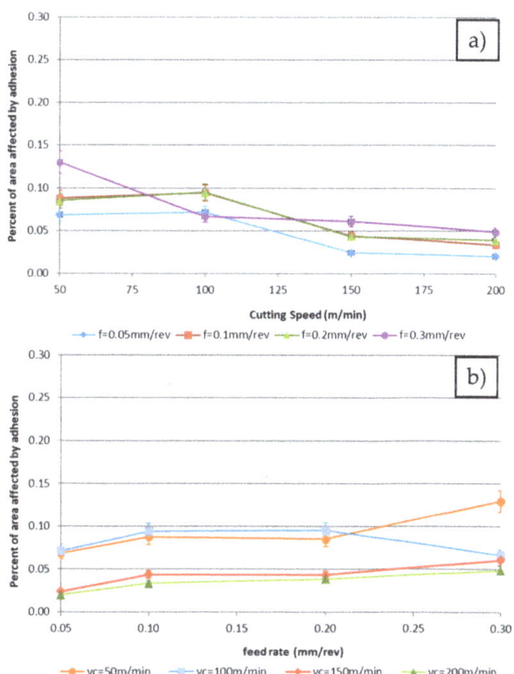

Figure 11. Effect on the area affected by secondary adhesion (%A) for 0.5 mm depth of cut depending on the: (**a**) cutting speed; (**b**) feed rate. Data measured by image processing techniques.

Analytical data show the same trends observed in the images. There is a decreasing trend of adhesion area when the cutting speed increases for all the studied feed rates. These trends seem to be constant and unaffected by the feed rate.

This fact is related to the cutting temperature increment that appears with the rise in the speed, as was analysed by [44,52]. Moreover, a horizontal trend is noted for the highest cutting speed values, which can be related to a stabilisation of this parameter. This result matches the results obtained in the test performed by [29] in a high-speed machining test.

In this case, the reduction of the affected area can be related to the tool geometrical change explained before. An increase in the cutting edge adhesion induces a rise in the contact length between the chip and the tool, reducing the adhesion effect. This phenomenon has been verified in orthogonal cut by Atlati et al. [53]. This effect is related to the resilience of the material, obtaining opposite tendencies to those observed in Trujillo et al.'s [27] experiments. Those differences are due to the use of aluminium with the highest resilience in Trujillo's tests. Also, it has to be considered that the method used by Trujillo et al. is manual and an automatic one can be more robust.

The surface of the chip varies due to the lack of contact between tool and material. For the highest cutting speed, the number of scratches, marks and defects on the chip surface are reduced (Figure 12). Those defects are caused by the type of flow over the rake face. It can be noted that on the contact area the grains are elongated in the longitudinal sense (Figure 13).

We expected a higher affected area because of the increase in the depth of cut caused by the increase of the contact area between the part and the tool. For this reason, for a smaller depth of cut, the area affected is smaller. However, the parametric trends seem to be constant for every depth of cut studied.

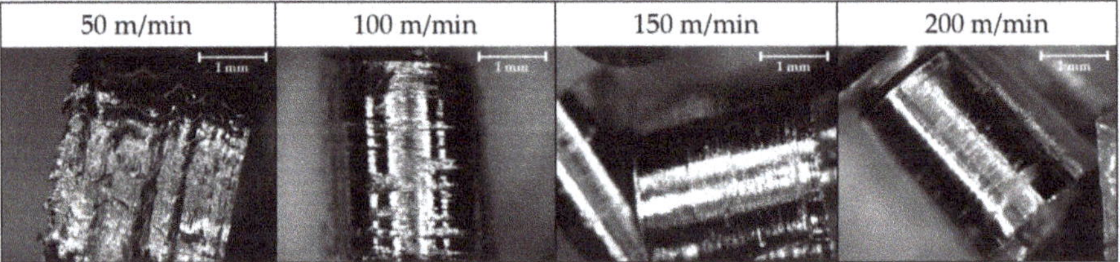

Figure 12. Chip contact face for different cutting speed at 0.2 mm/rev feed rate and 2 mm depth of cut (30×).

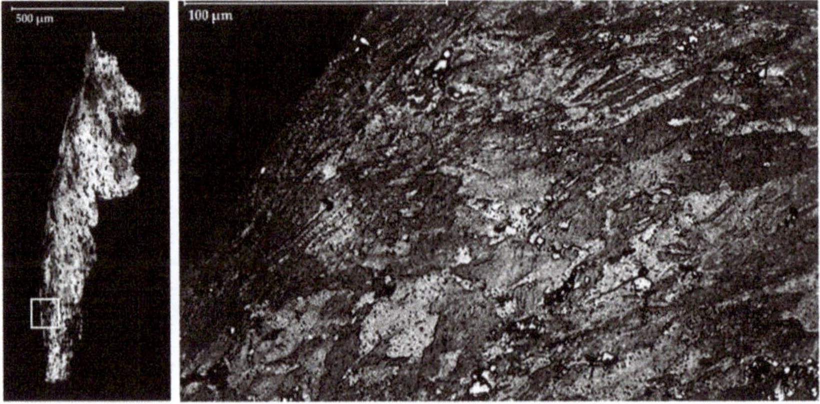

Figure 13. Metallographic optical microscopy of the chip cross section.

Despite the fact that adhesion is a cyclical process and the adhered material is removed and reformed in the evaluated period [18], the deviation of the data is not remarkable. This phenomenon could hide other effects of adhesion on the tool rake face, like a possible crater. For this reason, this method does not draw any distinction between primary and secondary BUL, which limits the evaluation to the material extension.

From the obtained data, several marginal models of the depth of cut have been calculated. The objective is to make the graphic representation of the joining model easier:

$$\%A \left(a_p = 0.5 \text{ mm}\right) = 359.317 \cdot f^{0.301} \cdot v_c^{-0.752} \tag{5}$$

$$\%A \left(a_p = 1 \text{ mm}\right) = 120.462 \cdot f^{0.529} \cdot v_c^{-0.317} \tag{6}$$

$$\%A \left(a_p = 2 \text{ mm}\right) = 57.968 \cdot f^{0.850} \cdot v_c^{-0.006} \tag{7}$$

A combined model based on the overlap of these models is established:

$$\%A = 146.657 \cdot f^{0.551} \cdot v_c^{-0.371} \cdot a_p^{0.421} \tag{8}$$

It is observed that the exponent of the cutting speed points, a horizontal trend for the highest values of the depth of cut and the increase of depth of cut is shown as a high exponent of the feed rate. Figure 14 presents the superposition model. This figure reflects the trends of behaviour explained previously. The planes are cut because the trends are inverted when the depth of cut increases.

As presented in the model, the sensitivity of the geometry change depends on the depth of cut. With a higher value of this parameter, the tool is more sensitive to changes caused by an increase in the feed rate and less so for those caused by an increase in the cutting speed, due to an increase of the chip section and an increase in the specific cutting tension. This is related with a highest stress to perfume the cutting process, which is also linked to an increase in the temperature in the cutting zone.

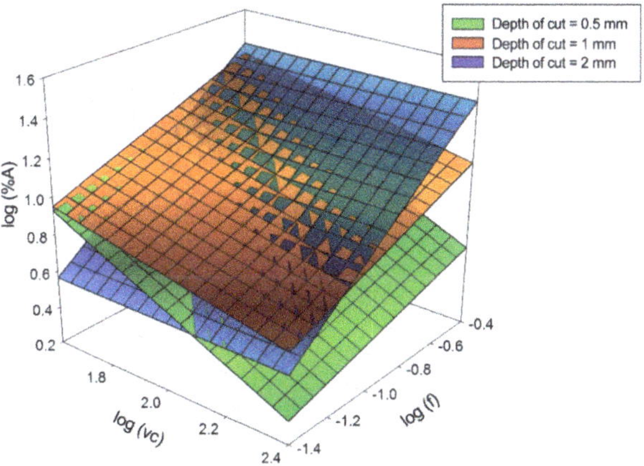

Figure 14. Graphic representations of the models calculated for the secondary adhesion-affected area.

3.3. Volume of the Material Adhered of the Cutting Tool

In this case, similar to in a previous analysis, we studied the tools obtained for each depth of cut. Figure 15 shows the topographies obtained for a 1 mm depth of cut. However, the same trends were found for the other depths of cut.

This is in accordance with what was observed in previous cases and can be explained by the tension of the chip and the lack of contact with the tool.

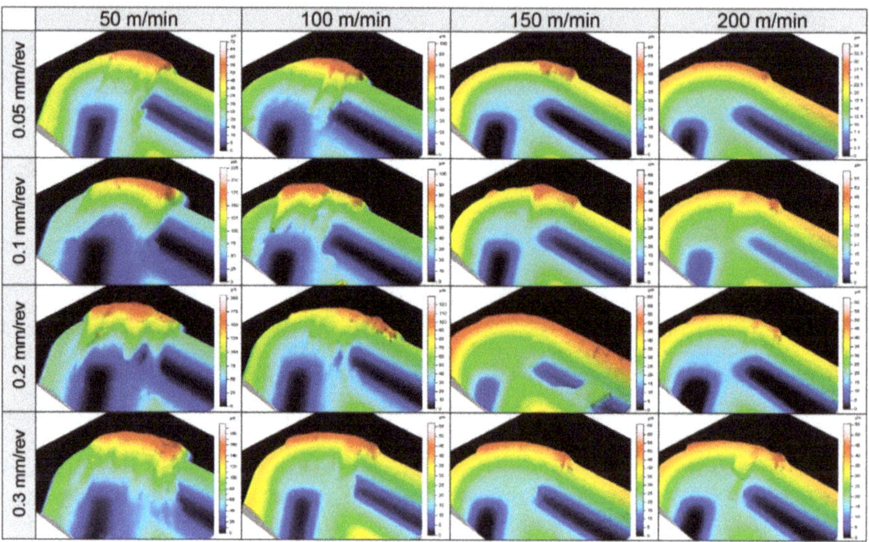

Figure 15. Comparative topography of the tools used in 0.5 mm depth of cut machining tests.

Figure 16 shows the effect of the cutting parameters on the volume of the material adhered to the cutting tool for 0.5 mm depth of cut. Others depths of cut shows the same trends.

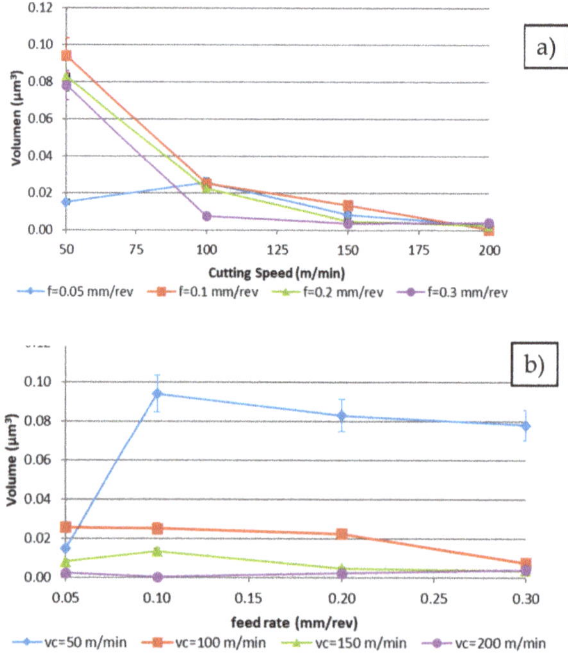

Figure 16. Effect of the (**a**) cutting speed; (**b**) feed rate on the volume of the material adhered to the cutting tool (V) for 0.5 mm depth of cut.

It is difficult to establish homogeneity given the dynamic behaviour of the adhesion mechanism. As shown in Figure 17, the adhering material can be easily removed (the whole adhesion layer or a part of it). This continuously alters the behaviour of the tool, changing the performance of the process.

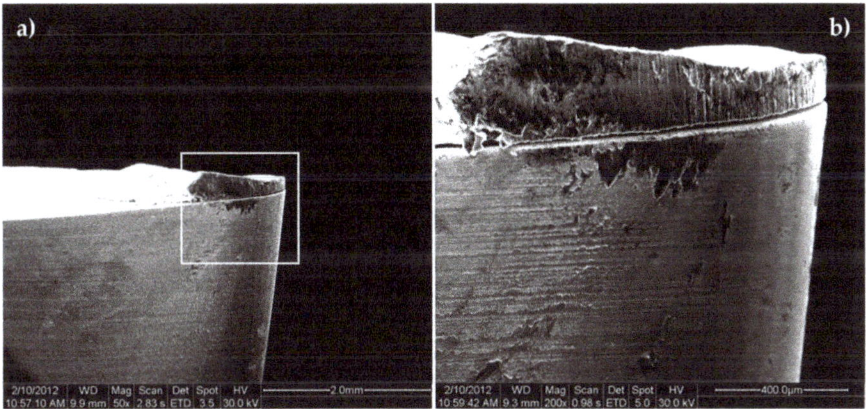

Figure 17. SEM of a tool in the previous instants to the detachment of the adhering layer. (**a**) 50× magnification; (**b**) 200× magnification.

According to the data obtained, a potential model has been developed as shown in Equation (9):

$$V = 2.121 \cdot f^{1.463} \cdot v_c^{0.501} \cdot a_p^{0.111} \tag{9}$$

Figure 18 shows the regression planes of the marginal models and the joint model. The model represents the union of the two effects already studied. It is also noted that the most material is adhered to the cutting edge, while the material adhered on the rake face has a lower thickness and volume.

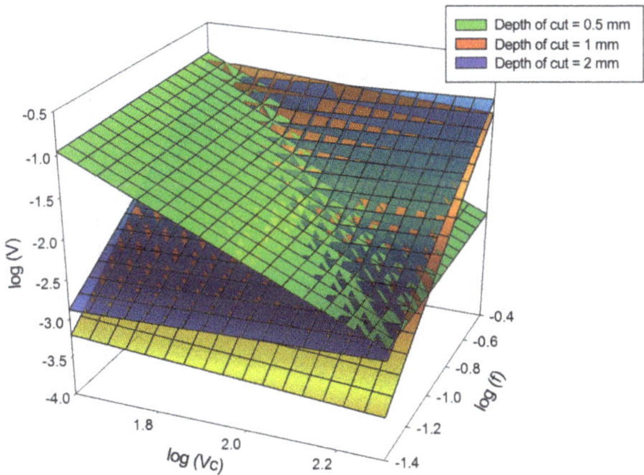

Figure 18. Graphic representations of the models calculated for the volume of adhering material.

4. Conclusions

This paper has evaluated the secondary adhesion wear in dry turning of an AA2024 aluminium alloy using a specific methodology. The three effects of the secondary adhesion wear were studied:

the affected area in the rake face (BUL), the thickness of material adhered over the cutting edge (BUE), and the total volume adhered on the tool (BUL/BUE). Those parameters are studied under two-dimensional optical techniques and with a 3D profilometer. Parametric models for the three effects are proposed by combining different cutting parameters.

The thickness of material adhered on the cutting edge shows more dependence on the cutting speed than the feed rate. A layer of higher thickness appears with a lower cutting speed. Similar trends are noted with depth of cut. The geometry of this adhering material also changes the geometry of the tool; this change does not have linear behaviour.

The affected area in the rake face is influenced by this geometrical change. For this reason, when the thickness of the material adhered on the cutting edge is increased, the contact length of the chip–tool pair is lower and the affected area of the rake face is decreased. Thus, the influence of the cutting speed is lower and the feed rate is higher. The surface of the chip shows this behaviour with a clear surface.

The adhesion wear is a dynamic mechanism. For this reason, there is instability in the wear mechanism. This instability is observed in some tools when the material adhered can be removed easily. This may affect the instantaneous results, but the general trend is not affected.

However, the study of the volume of the material adhered to the tool shows a higher dependence on the material adhered in the cutting edge because it presents a lower thickness to a rake face. In this way, the volume adhered on the tool shows a similar trend to the thickness of the material adhered on the cutting edge.

This study uses two different techniques, a basic optical technique and a complex metrological technique. The optical technique allows for the measurement of the tool wear with high precision, making it a suitable technique for industrial implementation due its simplicity, low cost and precision. This study also sets up a foundation to establish an automatic calibration system for a tool based on optical techniques. This can minimise the negative effects of secondary adhesion wear.

Author Contributions: M.B.P. and J.S.G. conceived and designed the experiments; M.B.P. and J.S.G. performed the experiments; M.B.P. and S.R.F.-V. analysed the data; M.B.P. and I.D.S.I. wrote the paper.

Funding: This work has received financial support from the Spanish Government through the Ministry of Economy, Industry and Competitiveness, the European Union (FEDER/FSE) and the Andalusian Government (PAIDI).

Acknowledgments: A special acknowledge to Mariano Marcos Barcena, a great scientific and engineer, and personally our father in research works. In memoriam.

Conflicts of Interest: The authors declare no conflict of interest.

References

1. Merchant, M.E.; Krabacher, E.J. Radioactive tracers for rapid measurement of cutting tool life. *J. Appl. Phys.* **1951**, *22*, 1507–1508. [CrossRef]
2. Ernst, H.; Martellotti, M. The formation of the Build-Up Edge. *ASME Mech. Eng.* **1938**, *57*, 487–498.
3. Bailey, J.A. Friction in metal machining-mechanical aspects. *Wear* **1975**, *31*, 243–275. [CrossRef]
4. Komanduri, R. Machining and grinding: A historical review of the classical papers. *Appl. Mech. Rev.* **1993**, *46*. [CrossRef]
5. Da Silva, R.B.; Machado, Á.R.; Ezugwu, E.O.; Bonney, J.; Sales, W.F. Tool life and wear mechanisms in high speed machining of Ti-6Al-4V alloy with PCD tools under various coolant pressures. *J. Mater. Process. Technol.* **2013**, *213*, 1459–1464. [CrossRef]
6. Nouari, M.; List, G.; Girot, F.; Coupard, D. Experimental analysis and optimisation of tool wear in dry machining of aluminium alloys. *Wear* **2003**, *255*, 1359–1368. [CrossRef]
7. Santos, M.; Machado, A.; Sales, W.; Barrozo, M.; Ezugwu, E. Machining of aluminum alloys: A review. *Int. J. Adv. Manuf. Technol.* **2013**, *86*, 3067–3080. [CrossRef]
8. Debnath, S.; Reddy, M.M.; Yi, Q.S. Environmental friendly cutting fluids and cooling techniques in machining: A review. *J. Clean. Prod.* **2014**, *83*, 33–47. [CrossRef]

9. Mia, M.; Gupta, M.K.; Singh, G.; Królczyk, G.; Pimenov, D.Y. An approach to cleaner production for machining hardened steel using different cooling-lubrication conditions. *J. Clean. Prod.* **2018**, *187*, 1069–1081. [CrossRef]

10. Krolczyk, G.M.; Maruda, R.W.; Krolczyk, J.B.; Nieslony, P.; Wojciechowski, S.; Legutko, S. Parametric and nonparametric description of the surface topography in the dry and MQCL cutting conditions. *Measurement* **2018**, *121*, 225–239. [CrossRef]

11. Polvorosa, R.; Suárez, A.; López de Lacalle, L.; Cerrillo, I.; Wretland, A.; Veiga, F. Tool wear on nickel alloys with different coolant pressures: Comparison of alloy 718 and waspaloy. *J. Manuf. Process.* **2017**, *26*, 44–56. [CrossRef]

12. Sartori, S.; Moro, L.; Ghiotti, L.; Bruschi, S. On the tool wear mechanisms in dry and cryogenic turning Additive Manufactured titanium alloys. *Tribol. Int.* **2017**, *105*, 264–273. [CrossRef]

13. Mia, M. Multi-response optimization of end milling parameters under through-tool cryogenic cooling condition. *Measurement* **2017**, *111*, 134–145. [CrossRef]

14. Shokrani, A.; Dhokia, V.; Newman, S.T. Environmentally conscious machining of difficult-to-machine materials with regard to cutting fluids. *Int. J. Mach. Tools Manuf.* **2012**, *57*, 83–101. [CrossRef]

15. Goindi, G.; Sarkar, P. Dry machining: A step towards sustainable machining-challenges and future directions. *J. Clean. Prod.* **2017**, *165*, 1557–1571. [CrossRef]

16. Krolczyk, G.; Nieslony, P.; Maruda, R.; Wojciechowski, S. Dry cutting effect in turning of a duplex stainless steel as a key factor in clean production. *J. Clean. Prod.* **2017**, *142*, 3343–3354. [CrossRef]

17. Salguero, J.; Vazquez-Martinez, J.M.; Del Sol, I.; Batista, M. Application of Pin-On-Disc techniques for the study of tribological interferences in the dry machining of A92024-T3 (Al-Cu) alloys. *Materials* **2018**, *11*, 1236. [CrossRef] [PubMed]

18. Gómez-Parra, A.; Alvarez-Alcon, M.; Salguero, J.; Batista, M.; Marcos, M. Analysis of the evolution of the Built-Up Edge and Built-Up Layer formation mechanisms in the dry turning of aeronautical aluminium alloys. *Wear* **2013**, *302*, 1209–1218. [CrossRef]

19. Carrilero, M.S.; Bienvenido, R.; Sánchez, J.M.; Álvarez, M.; González, A.; Marcos, M. A SEM and EDS insight into the BUL and BUE differences in the turning processes of AA2024 Al–Cu alloy. *Int. J. Mach. Tools Manuf.* **2002**, *42*, 215–220. [CrossRef]

20. Trent, E.M.; Wright, P.K. *Metal Cutting*; Butterworth-Heinemann: Oxford, UK, 2000; ISBN 978-0-7506-7069-2.

21. Klocke, F.; Eisenblätter, G.; Krieg, T. Machining: Wear of Tools. In *Encyclopedia of Materials: Science and Technology*; Elsevier Ltd.: New York, NY, USA, 2001; pp. 4708–4711. ISBN 978-0-08-043152-9.

22. Krolczyk, G.; Gajek, M.; Legutko, S. Predicting the tool life in the dry machining of duplex stainless steel. *Maint. Reliab.* **2013**, *15*, 62–65.

23. Krolczyk, G.; Legutko, S.; Gajek, Y.S. Predicting the surface roughness in the dry machining of duplex stainless steel (DSS). *Metalurgija* **2013**, *52*, 259–262.

24. Fernández-Valdivielso, A.; Lopez de Lacalle, L.; Urbikain, G.; Rodriguez, A. Detecting the key geometrical features and grades of carbide inserts for the turning of nickel-based alloys concerning surface integrity. *J. Mech. Eng. Sci.* **2016**, *230*, 3725–3742. [CrossRef]

25. Lamikiz, A.; Lopez de Lacalle, L.; Sanchez, J.; Bravo, U. Calculation of the specific cutting coefficients and geometrical aspects in sculptured surface machining. *Mach. Sci. Technol.* **2005**, *9*, 411–436. [CrossRef]

26. Calleja, A.; Bo, P.; Gonzalez, H.; Barton, H.; Lopez de Lacalle, L. Highly accurate 5-axis flank CNC machining with conical tools. *Int. J. Adv. Manuf. Technol.* **2018**, *97*, 1605–1615. [CrossRef]

27. Trujillo, F.; Sevilla, L.; Marcos, M. Experimental parametric model for indirect adhesion wear measurement in the dry turning of UNS A97075 (Al-Zn) alloy. *Materials* **2017**, *10*, 152. [CrossRef] [PubMed]

28. Batista, M.; Salguero, J.; Gómez, A.; Carrilero, M.S.; Álvarez, M.; Marcos Bárcena, M. Identification, analysis and evolution of the mechanisms of wear for secondary adhesion for dry turning processes of Al-Cu alloys. *Adv. Mater. Res.* **2010**, *107*, 141–146. [CrossRef]

29. Salguero, J.; Batista Ponce, M.; Sánchez-Carrilero, M.; Álvarez, M.; Marcos-Bárcena, M. Sustainable manufacturing in aerospace industry—Analysis of the viability of intermediate stages elimination in sheet processing. *Adv. Mater. Res.* **2010**, *107*, 9–14. [CrossRef]

30. Salguero, J.; Carrilero, M.S.; Batista, M.; Álvarez, M.; Marcos, M. Analysis of the influence of thermal treatment on the dry turning of Al-Cu alloys. In Proceedings of the Third Manufacturing Engineering Society International Conference: Mesic-09, Alcoy, Spain, 17–19 June 2009.

31. Bleither, F.; Reiter, M. Wear reduction on cutting inserts by additional internal cooling of the cutting edge. *Procedia Eng.* **2018**, *21*, 518–524. [CrossRef]

32. Maine, J.; Batista, M.; García-Jurado, D.; Shaw, L.; Marcos, M. FVM based methodology for evaluating adhesion wear of cutting tools. *Procedia CIRP* **2013**, *8*, 552–557. [CrossRef]

33. Gontard, L.; Batista, M.; Salguero, J.; Calvino, J. Three-dimensional chemical mapping using non-destructive SEM and photogrammetry. *Sci. Rep.* **2018**, *8*, 11000. [CrossRef] [PubMed]

34. Castejon, M.; Alegre, E.; Barreiro, J.; Hernandez, L. On-line tool wear monitoring using geometric descriptors from digital images. *Int. J. Mach. Tools Manuf.* **2007**, *47*, 1847–1853. [CrossRef]

35. Shahabi, H.; Ratnam, M. In-cycle detection of Built-Up Edge (BUE) from 2-D images of cutting tools using machine vision. *Int. J. Adv. Manuf. Technol.* **2010**, *46*, 1179–1189. [CrossRef]

36. Sen, B.; Mandal, U.; Mondal, S. Advancement of an intelligent system based on ANFIS for predicting machining performance parameters of Inconel 690—A perspective of metaheuristic approach. *Measurement* **2017**, *109*, 9–17. [CrossRef]

37. Dan, L.; Mathew, J. Tool wear and failure monitoring techniques for turning—A review. *Int. J. Mach. Tools Manuf.* **1990**, *30*, 579–598. [CrossRef]

38. Cook, N.H. Tool wear sensors. *Wear* **1980**, *62*, 49–57. [CrossRef]

39. Rivero, A.; Lacalle, L.; Penalva, M. Tool wear detection in dry high-speed milling based upon the analysis of machine internal signals. *Mechatronics* **2008**, *18*, 627–633. [CrossRef]

40. Lauro, C.; Brandao, L.; Baldo, D.; Reis, R.; Davim, J. Monitoring and processing signal applied in machining processes—A review. *Measurement* **2014**, *58*, 73–86. [CrossRef]

41. Haddag, B.; Atlati, S.; Nouari, M.; Moufki, A. Dry machining aeronautical aluminum alloy AA2024-T351: Analysis of cutting forces, chip segmentation and Built-Up Edge formation. *Metals* **2016**, *6*, 197. [CrossRef]

42. Melkote, S.N.; Liu, R.; Fernandez-Zelaia, P.; Marusich, T. A physically based constitutive model for simulation of segmented chip formation in orthogonal cutting of commercially pure titanium. *CIRP Ann.-Manuf. Technol.* **2015**, *64*, 65–68. [CrossRef]

43. Davim, J.P. *Modern Mechanical Engineering*; Springer-Verlag: Berlin/Heidelberg, Germany, 2014; ISBN 978-3-642-45176-8.

44. Batista, M.; Salguero, J.; Gómez-Parra, A.; Alvarez, M.; Marcos, M. Image based analysis evaluation of the elements of secondary image based analysis evaluation of the elements of secondary. *Adv. Mater. Res.* **2012**, *498*, 133–138. [CrossRef]

45. International Organization for Standardization (ISO). *1832:2012 Indexable Inserts for Cutting Tools—Designation*; ISO: Genève, Switzerland, 2012.

46. International Organization for Standardization (ISO). *3685:1993 Tool-Life Testing with Single-point Turning Tools*; ISO: Genève, Switzerland, 1993.

47. Rubio, E.; Camacho, A.; Sanchez-Sola, J.; Marcos, M. Chip arrangement in the dry cutting of aluminium alloys. *J. Achievements Mater. Manuf. Eng.* **2006**, *16*, 164–170.

48. Davis, J.R. *ASM Handbook Volume 16: Machining*, 9th ed.; ASM International: Metals Park, OH, USA, 1989; ISBN 978-0-87170-022-3.

49. Carrilero, M.; Marcos, M.; Sanchez, V. Feed, cutting speed and cutting forces as machinability parameters of Al-Cu alloy. *J. Mech. Behav. Mater.* **1996**, *7*, 167–178. [CrossRef]

50. Carrilero, M.; Marcos, M. On the machinability of aluminium and aluminium alloys. *J. Mech. Behav. Mater.* **1996**, *7*, 179–193. [CrossRef]

51. Sebastian, M.; Sanchez-Sola, J.; Carrilero, M.; Gonzalez, J.; Alvarez, M.; Marcos, M. Parametric model for predicting surface finish of machined UNS A92024 alloy bars. *J. Manuf. Sci. Prod.* **2011**, *4*, 181–188. [CrossRef]

52. Agustina, B.; Rubio, E.; Villeta, M.; Sebastián, A.M. Analysis of the surface roughness obtained during the dry turning of UNS A97050-T7 aluminium alloys. In Proceedings of the Third Manufacturing Engineering Society International Conference: Mesic-09, Alcoy, Spain, 17–19 June 2009.

53. Atlati, S.; Haddag, B.; Nouari, M.; Moufki, A. Effect of the local friction and contact nature on the Built-Up Edge formation processin machining ductile metals. *Tribol. Int.* **2015**, *90*, 217–227. [CrossRef]

Article

Comparison of Flank Super Abrasive Machining vs. Flank Milling on Inconel® 718 Surfaces

Haizea González [1,*], Octavio Pereira [2], Asier Fernández-Valdivielso [2], L. Norberto López de Lacalle [2] and Amaia Calleja [3]

[1] Department of Mechanical Engineering, University of the Basque Country (UPV/EHU), Plaza Ingeniero Torres Quevedo 1, 48013 Bilbao, Spain

[2] CFAA—University of the Basque Country (UPV/EHU), Parque Tecnológico de Zamudio 202, 48170 Bilbao, Spain; octaviomanuel.pereira@ehu.eus (O.P.); asier.fernandezv@ehu.eus (A.F.-V.); norberto.lzlacalle@ehu.eus (L.N.L.d.L.)

[3] Department of Mechanical Engineering, University of the Basque Country (UPV/EHU), Nieves Cano 12, 01006 Vitoria, Spain; amaia.calleja@ehu.eus

* Correspondence: haizea.gonzalez@ehu.eus; Tel.: +34-94-601-3932

Received: 19 July 2018; Accepted: 5 September 2018; Published: 6 September 2018

Abstract: Thermoresistant superalloys present many challenges in terms of machinability, which leads to finding new alternatives to conventional manufacturing processes. In order to face this issue, super abrasive machining (SAM) is presented as a solution due to the fact that it combines the advantages of the use of grinding tools with milling feed rates. This technique is commonly used for finishing operations. Nevertheless, this work analyses the feasibility of this technique for roughing operations. In order to verify the adequacy of this new technique as an alternative to conventional process for roughing operations, five slots were performed in Inconel® 718 using flank SAM and flank milling. The results showed that flank SAM implies a suitable and controllable process to improve the manufacture of high added value components made by nickel-based superalloys in terms of roughness, microhardness, white layer, and residual stresses.

Keywords: flank super abrasive machining (*SAM*); flank milling; Inconel® 718; roughness; residual stress

---

1. Introduction

Thermoresistant superalloys, such as titanium- and nickel-based alloys, are an actual challenge for manufacturing technologies. These alloys are widely used for many applications that require stability of material properties working under extreme conditions and temperatures up to 400 °C and 600 °C, respectively [1]. One of the main characteristic of these materials, among others, is the optimal combination of hardness and good ductility with low thermal conductivity [2,3]. Superalloys, like Inconel® 718, has multiple applications as a consequence of its mechanical and physical properties, it is spreading to industries such as petrochemical plants, marine equipment, food processing equipment, and nuclear reactors [4]. Nevertheless, these alloys are known as difficult-to-cut materials, implying premature tool wear and high cutting forces [5,6]. Moreover, the challenge lies in the low machinability combined with difficult geometries and finishing requirements that leads to optimizing traditional manufacturing processes, improving cutting strategies, new tools design [7], and cooling techniques on milling Inconel® 718 [8,9].

Nonetheless, traditional methods such as milling, grinding or broaching presents some difficulties due to the low thermal conductivity and extreme high strength of these superalloys [10,11]. An extra critical adverse effect of conventional techniques consists of tool wear; due to the fact that it is conditioned by cutting parameters and cutting fluids [12,13]. For this reason, non-conventional manufacturing technologies were considered as an alternative or complementary, such as electrochemical machining (ECM), linear friction welding (LFW), electro-discharge machining (EDM) [14]. On the other hand, among abrasive machining, it could be found non-conventional technologies, such as abrasive flow machining (AFM), magnetic abrasive finishing (MAF), or magneto-rheological abrasive flow finishing (MRAFF). These processes are known for high surface quality with low-medium removal rates [15,16]. The main advantage of these non-conventional technologies is that they are able to achieve tough dimensional accuracy and excellent finishing surfaces working on complex geometrical cavities where milling has no accessibility [17,18]. However, these technologies offer low material removal rates, translated to higher manufacturing time and costs [19].

In this line, it is important to find more efficient processes that optimize manufacturing time and machining quality. Among these technologies, super abrasive machining (SAM) was presented in [20,21] as a solution to increase machining efficiency during the production of blades and turbine disks [22]. This method consists of applying grinding processes with machining rates and conditions. Besides, under similar cutting conditions of single point machining it offers finishing precisions closest to grinding technology, what makes this process more versatile than grinding or milling techniques. Among grinding techniques with super abrasive grinding tools, it is found the creep feed grinding (CFG) defined as grinding process with larger cutting depths and higher feed rates resulting on higher removal rates decreasing machining time [23]. Petrilli et al. [24] defined creep fatigue grinding as the closest rival for SAM achieving higher speeds and specific material removal rates up to 1000 mm^3/mm s [20]. Major benefits from SAM reside on higher material removal rates at higher speed along with the near-shape surface obtaining more accurate dimensional tolerances [24,25]. The main limitation for the optimal use of this technology is found in spindle speed requirements, up to 90,000 rpm [26].

Previous work was carried out following this trend, comparing conventional milling with the SAM technique using a conventional machining centre limited to 18,000 rpm of spindle rotary speed. Haizea et al. [27] analysed these techniques, manufacturing a complex geometry made of Inconel® 718, what added an extra challenge related to complex geometry and non-developable surfaces. In terms of surface roughness, dimensional deviation, and force measurement, SAM presented competitive results compared to conventional milling. Figure 1 shows obtained results for roughness and dimensional deviation comparing SAM and milling.

Nevertheless, critical parameters need to be analysed to make this technique a feasible alternative to milling technique or a complementary technique using the same equipment, avoiding clamping and unclamping additional errors, among others. For instance, the appearance of residual stress is considered crucial to be controlled. Some residual stresses lead to plastic deformations or structural modifications [28]. Depending on the industrial applications of some components, compressive residual stresses present improvements on material behaviour; owing to this type of stress prevents from brittle fracture and fatigue failure [29].

Therefore, the novelty of this work stems from the idea of studying the material behaviour about using SAM technology instead of conventional milling technology for roughing using conventional machining centre. This technique is becoming a good alternative for finishing, however this work analyses SAM feasibility for full slot roughing. Based on the concept of multitasking machines, it offers the possibility of applying this technique using conventional machining centres. In order to extend the knowledge about SAM behaviour at roughing operations, five full slots were machined with both techniques (flank SAM and flank milling); selected material was Inconel® 718. In order to analyse the adequacy of flank SAM the following parameters were measured: roughness, surface irregularities, white layer, residual stresses, and microhardness.

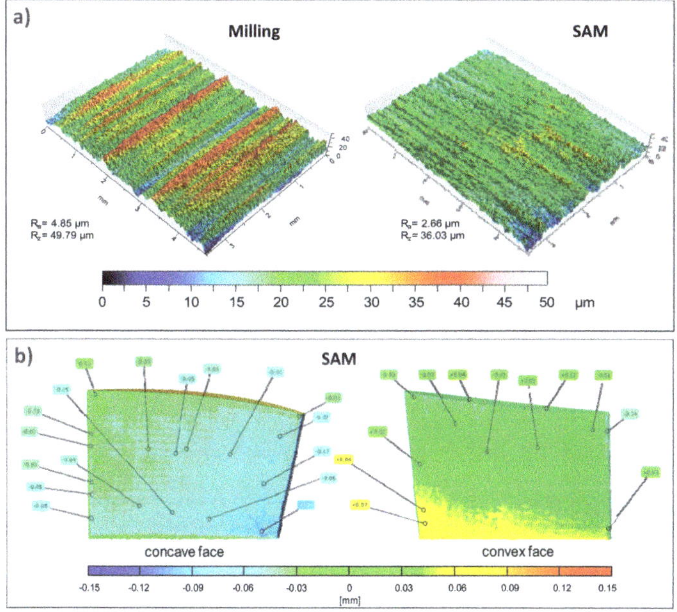

Figure 1. SAM vs. conventional milling obtained results: roughness (**a**) and dimensional deviation (**b**) [27].

2. Experimental Setup and Procedure

With the aim of analysing the adequacy of flank SAM compared to flank milling in terms of surface integrity applied to thermo-resistant super alloys, a serial of experiments were designed. The experiments were performed as five full slots with the total effective tool length for flank SAM and for flank milling.

According to previous work, Inconel® 718 was selected as a challenging material for these processes. This material consists of a nickel-based hardened alloy through the precipitation of its metallic matrix secondary phases [30]; Figure 2 shows the microstructure of selected material obtained with an optical microscopy at 50×. It was observed a fine grained structure (ASTM6) with a discrete carbide phase scattered inside the grains, usually presented in Inconel® 718. This superalloy was selected due to the fact that it constitutes one of the most utilized materials in the aeronautic industry, more concretely for engines and turbines [31].

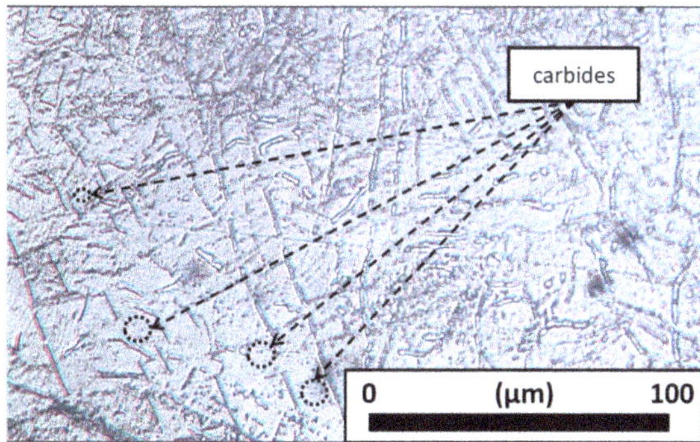

Figure 2. Microstructure of Inconel® 718.

This thermo-resistant super alloy is characterized by good resistance to fatigue and creep combined with high corrosion resistance under extreme working conditions at high temperatures. Nevertheless, it is considered a difficult-to-cut material due to the magnitude of cutting forces, low material removal rates, built-up edges, and extreme tool wear during machining [32,33]. Thus, Inconel® 718 was selected as the material for experimental trials in order to provide data for both manufacturing techniques (flank SAM and flank milling); showing their behaviour working with low-machinable materials and under aggressive machining conditions.

Table 1 shows chemical composition, mechanical and physical properties for the selected material.

Table 1. Inconel® 718 chemical composition, mechanical and physical properties [34].

Chemical Composition (%)												
Ni	Cr	Co	Fe	Nb	Mo	Ti	Al	B	C	Mn	Si	Others
52.5	19	1	17	5	3	1	0.6	0.01	0.08	0.35	0.35	1.79

Mechanical and Physical Properties						
Hardness	Young's Modulus	Tensile Strength	Density	Specific Heat	Melting Temp.	Thermal Conduct
42 HRc	206 GPa	1.73 GPa	8470 kg/m^3	461 J/(kg·K)	1550 K	15 W/(m·K)

Figure 3 shows the experimental set-up for performing defined trials. A five-axis machining centre was used, three linear axes (X, Y, Z) and two rotary axes (A, C). The main limitation of this machine is the spindle speed capacity, a spindle speed up to 18,000 rpm and 18 KW. The machining centre model was an Ibarmia ZV-25/U600 (IBARMIA INNOVATEK, S.L.U., Guipuzkoa, Spain).

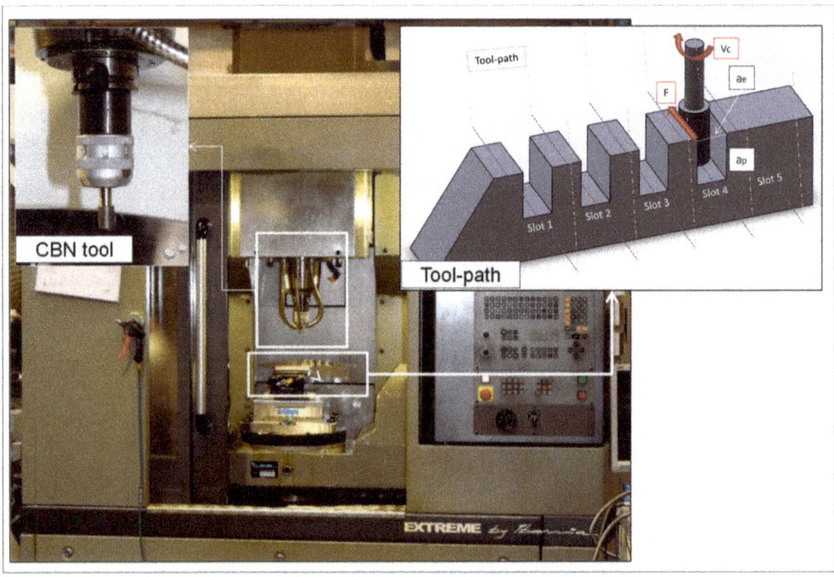

Figure 3. Experimental setup at the University of the Basque Country (UPV/EHU).

Related to manufacturing strategies, flank milling and flank SAM were selected in order to remove the highest amount of material according to tool limitations, using the total cutting effective length. In the case of flank milling, a 16 mm diameter, 20 mm cutting length four-tooth carbide tool coated with AlTiN was selected. This coating was selected because the presence of aluminium implies higher surface hardness and resistance to oxidation [35]. On the other hand, for flank SAM operation a 16 mm diameter, 20 mm cutting length PCBN grinding tool was selected.

Table 2 contains used cutting conditions during trials; cutting parameters and tool type were selected to manufacture Inconel® 718 according to industrial conditions and based on previous experiments [27]. In spite of not using the optimal spindle speed for the SAM technique; in this case, cutting conditions were limited to machine capacities. This leads to the allowance of performing this technique with conventional machining centres.

Table 2. Defined cutting conditions for the experiments.

Cutting Conditions	Flank Milling	Flank SAM
Feed Rate	0.01 mm/tooth	45 mm/min
Cutting Speed	20 m/min	900 m/min
a_p	20 mm	20 mm
a_e	16 mm	16 mm
Cutting Fluid	Synthetic oil emulsion Houghton® 20%	Synthetic oil emulsion Houghton® 20%

Machine spindle speed led to the use of limited cutting conditions for SAM technology. Though, with the aim of comparing different grinding processes and equate performed processes, there are two parameters commonly used for this purpose inside grinding technology: the equivalent chip thickness (h_{eq}) and the specific material removal rate (Q') [36]. Figure 4 shows the grinding parameters involved in this calculus, followed by the equations.

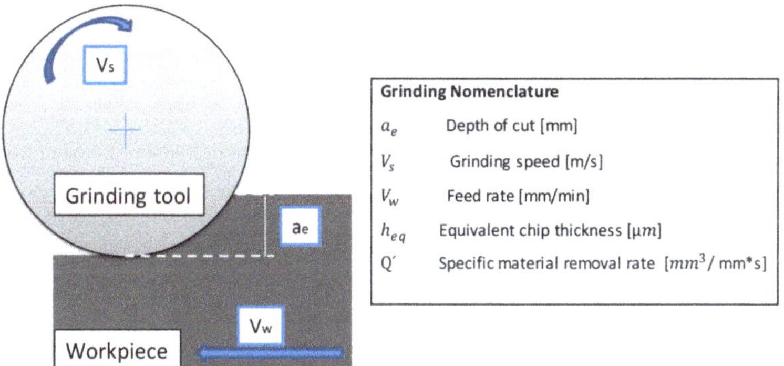

Figure 4. Grinding process parameters definition.

$$Q' = a_e\, v_w, \tag{1}$$

$$h_{eq} = a_e\, \frac{V_w}{v_s}, \tag{2}$$

Through Equations (1) and (2) equivalent chip thickness and specific material removal rate were calculated. These two concepts depend on the following parameters: cutting depth (a_e), feed (v_w), and cutting speed (v_s). The main difference between the proposed roughing strategies compared with conventional grinding conditions is found in the depth of cut since, in this case, the whole tool diameter was used. Consequently, the equivalent chip thickness obtained was 0.8 µm and the specific material removal rate was 12 mm³/mm s. According to Marinescu et al. [37], removal rates in creep feed grinding obtained for Inconel® 989 and Inconel® 718 were around 13 mm³/mm s. This process consists of a grinding process characterized by high stock-removal rates with deep depths of cut as the presented case.

3. Results and Discussions

With the aim of analysing the feasibility of this manufacturing technique compared to the conventional one, roughness, microstructure and white layer, residual stresses, and microhardness were measured after manufacturing.

3.1. Roughness

A Surtronic Duo portable roughness tester from Taylor Hobson® (Ultra Precision Technologies Division of AMETEK Inc., Berwyn, PA, USA) was used to measure roughness during machining trials. It needs to be mentioned that the direction of measurements was perpendicular to the cut. This was carried out in order to obtain the peaks and valleys of the tool teeth, which is the most unfavourable case.

Additionally, after performed tests, a confocal microscope was used to obtain the profile for each sample produced by different techniques. Roughness measuring setting in this case was a 0.8 mm cut-off length and an evaluation length of 4 mm, according to the standard of ISO 4288 [38]. Figure 5 shows obtained data for both the techniques used. Obtained values for flank milling and flank SAM are admissible values for roughing high added-value components in aerospace and aeronautical industries made by this thermo-resistant material [14]. Thus, this implies stable and controllable cutting processes in both cases related to roughness.

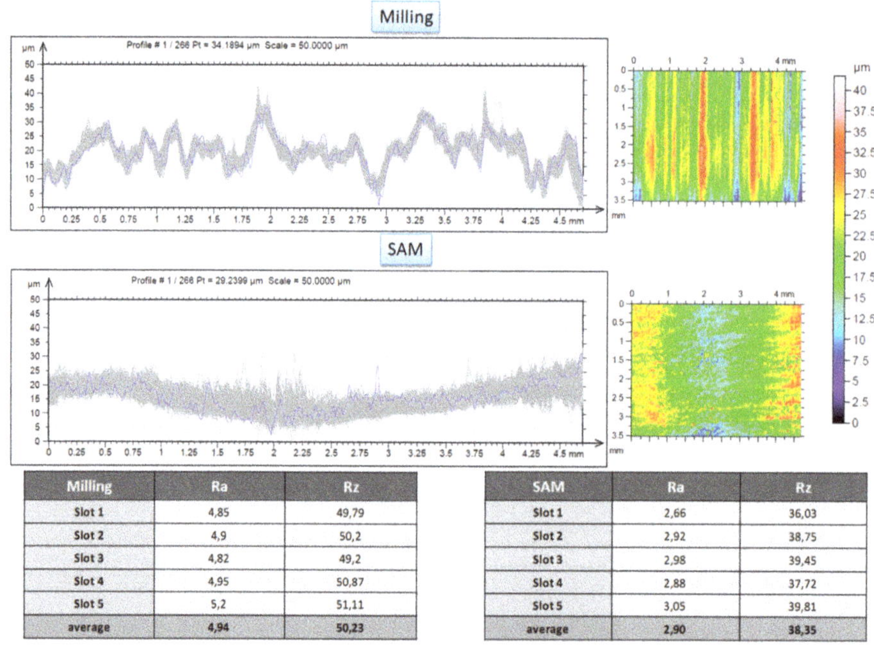

Milling	Ra	Rz
Slot 1	4,85	49,79
Slot 2	4,9	50,2
Slot 3	4,82	49,2
Slot 4	4,95	50,87
Slot 5	5,2	51,11
average	4,94	50,23

SAM	Ra	Rz
Slot 1	2,66	36,03
Slot 2	2,92	38,75
Slot 3	2,98	39,45
Slot 4	2,88	37,72
Slot 5	3,05	39,81
average	2,90	38,35

Figure 5. Roughness values (R_a and R_z) for flank milling and flank SAM.

Notwithstanding, it should be pointed that under similar cutting conditions, the flank SAM method produces a surface which is characterized by lower value roughness parameters in relation to flank milling, improvements up to 172.28 % and 132.33 % of R_a and R_z, respectively. Moreover, the patterns followed by both processes shown in Figure 3 are typical of milling and grinding processes, respectively. In particular, in the first case the pattern is directional due to the milling tool edge. In the second one, the pattern is non-directional because the grinding tool is composed by small grains which are distributed along the tool surface randomly.

3.2. Cross-Section and White Layer

Figure 6 shows the cross-section perpendicular to the workpiece cutting direction for flank milling and flank SAM. Flank milling contained some irregularities on the machined surface with a macroscopic deviation. It is possible that this is as a result of aggressive cutting conditions with thermo-resistant superalloys. Furthermore, as an extra difficulty in the designed tests consisted on opening a full slot, this implied inadequate space for chip removal and refrigeration. Hence, the cutting tool is more prone to suffer from tooth breakage instead of regular wear [39].

On the contrary, the flank SAM top layer presented a regular finished surface with a minimal deviation. This is the consequence of utilizing grinding tools instead of milling tools; these tools have a wear type more regular even in the case of grain detachment [40]. These results explained extensively roughness obtained values. Regarding surface finishing, flank SAM offered a more stable behaviour comparing with conventional technique. This implies the possibility of reducing machining steps in real pieces with the aim of obtaining final surfaces.

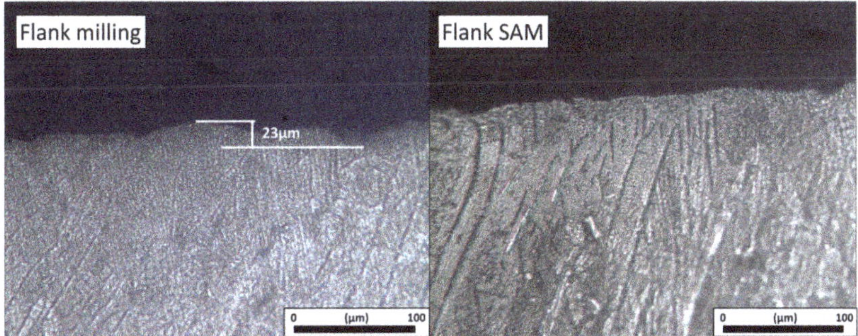

Figure 6. Cross-section of the Inconel® 718 surface after flank milling and flank SAM.

The formation of white layer consists of a hard surface layer formed by ferrous materials whilst using high cutting temperatures that can modify the surface integrity. It is produced by a rapid heating during machining over austenitizing temperature and followed by a quick cooling of this surface [41]. Some experts related the existence of this white layer for nickel-based superalloys not only with the high heat during the process but with the low thermal conductivity property of these materials [42]. In order to detect this layer, it is shown in the microscope as a thin white layer. It is important to mention that the existence of the white layer implies a direct discard of technique or conditions if the size overpass is 2 µm [43]. The white layer depends on the cutting parameters and environment contributing to fatigue failure [29].

Chen-Wei Dai et al. [44] found white layer on the surface whilst grinding Inconel® 718 under extreme conditions due to the quick jump from rapid heating to cooling. In this line, considering roughing strategies for flank SAM and flank milling as hard cutting conditions and suffering considerable thermal changes, it is important to point out that no white layer was found for any of these techniques.

3.3. Residual Stress

With the aim of measuring residual stress in the final fabricated surfaces and analysing differences between both techniques, hole-drilling strain gage method was selected [45]. According to the selected material, properties for determining residual stresses were established as: Young's modulus (206 GPa), Poisson's ratio (0.294), or yield stress (550MPa). The rosette type selected was 062 UL with a mean diameter of 5.13 mm, the hole diameter was 1.94 mm and a limit depth of 0.75 mm. The relation between hole diameter and maximum depth is 0.75/1.94 = 0.386. In line with ASTM standard E837 [45], when applying blind-hole drilling residual stress analysis the relation between hole diameter (D) and maximum hole depth (Z) is specified as Z/D = 0.4. Figure 7 shows the obtained results of principal stresses and each axis stress for flank SAM and flank milling, respectively.

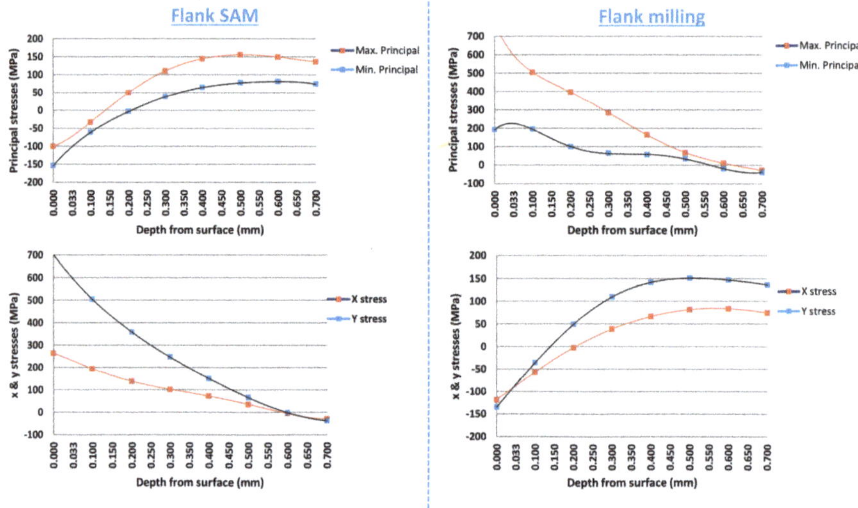

Figure 7. Residual stresses obtained for flank SAM and flank milling on Inconel® 718.

According to residual stresses obtained results, for flank milling technique it was appreciated a tensile residual stress near to machined surface common for this technique [46]. On the contrary, flank SAM measurements showed a compressive pattern residual stress near to the manufactured surface. Additionally to the differences between obtained values magnitude, the most remarkable aspect is that flank SAM values are compressive values, usual for grinding techniques. This implies better behaviour to fatigue failure and prevents brittle fracture.

3.4. Microhardness

Related to material properties, microhardness was measured for material base, flank milling, and flank SAM. Following the ASTM standard E384 [47] that covers microindentation hardness testing, the Vickers test was selected. Figure 8 shows the Vickers indenter used and the obtained results. According to the ASTM standard E140 [48], these values were converted internally into Rockwell hardness, the more commonly used unit for these materials.

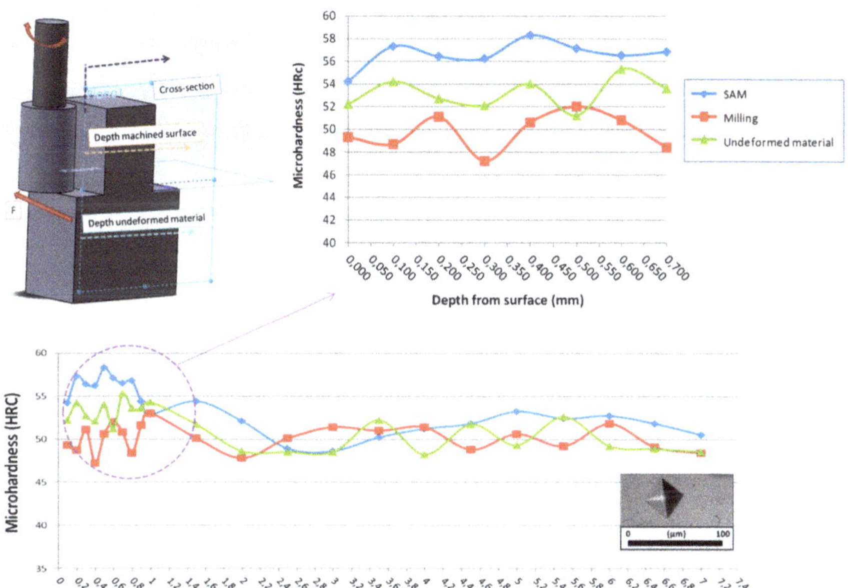

Figure 8. Residual stresses obtained for flank SAM and flank milling on Inconel® 718.

The measuring length along the first 1 mm was 0.1 mm and over this value was 0.5 mm. This was carried out in order to obtain more information close to machined surface. The results showed that undeformed material microhardness presents an average of 52 HRC, with a minimum of 49 HRC and maximum of 55 HRC. These values were set as a reference to study material property variation for both manufacturing techniques. Regarding the machined areas, differences were observed in the first 0.8 mm; approximately the same depth residual stresses were stabilized. In the case of flank milling, microhardness was situated within 46.5 HRC and 53 HRC. Conversely, for flank SAM an increase was observed in microhardness values up to 58 HRC. These values are a direct consequence of the appearance of compressive residual stresses on the machined surface; compressive stresses implied a rise in hardness values. This behavior is in concordance with the study carried out by Hua et al. [49] in which the increase of hardness was related with the maximum hardness value.

4. Conclusions

In this work, full slots were manufactured in Inconel® 718 with the main objective of comparing flank milling and flank SAM techniques in conventional machining centers. The new concept of using SAM as a roughing technique apart from finishing strategies leads to considerer this technique as a robust alternative for traditional milling. Additionally, the possibility of adapting a conventional machine to the use of both technologies fits with the trend of multitasking machines.

For this study, both techniques were compared from a technical point of view. In particular, surface roughness, microstructure, white layer, residual stresses, and microhardness were analyzed. The main conclusions obtained are listed below:

- Regarding surface roughness, flank SAM presented lower values for surface roughness comparing with *flank milling*. These results lead to consider flank SAM technique as an alternative for roughing strategies applied to these difficult-to-cut materials. Furthermore, obtained results conclude to the capability of using this technique for replacing intermediate manufacturing stages as semi-finishing strategies.

- Concerning cross-section, it was observed that conventional technique generated irregularities on the machined surface. This is directly related to the differences presented on roughness values. The reason why surface irregularities was avoided with flank SAM derived from the tool type. In this case, the use of grinding tools maintained a more constant tool wear and consequently a regular surface finishing. Additionally, it needs to be pointed that no white layer was found in any case.

- Related to residual stresses, considering the importance of the appearance of residual stress (both tensile and compressive), it should be highlighted that compressive residual stresses and tensile residual stresses arose on the machined surface by flank SAM and flank milling respectively, obtaining values five times higher for the conventional technique. Furthermore, compressive residual stresses in some processes add value to the machined material properties, such as better behavior to fatigue failure.

- Finally, microhardness showed higher values on the flank SAM surface. This improvement of the material property is a direct product of compressive residual stresses.

Therefore, from a technical point of view related with surface integrity in terms of surface roughness, residual stresses, microstructure and microhardness, the results obtained in this experiment showed that flank SAM technology does not present a limitation for being used with conventional machines; as long as cutting conditions were adequately adapted to spindle rotary capacity. This novel technology implies better results than conventional milling obtaining a suitable, controllable and predictable process to manufacture high added value components made of heat-resistant super alloys, such as Inconel® 718.

Author Contributions: H.G. and O.P. designed and performed the experiments. On the other hand, H.G. and O.P. wrote the paper. Additionally, A.C. analyzed the data related with residual stress and microstructure and A.F.-V. analyzed the data related to microhardness and white layer, and supervised the experiments. Finally, L.N.L.d.L. contributed with the resources (machine, tools, material, etc.) and supervised all the work carried out in this research.

Acknowledgments: The authors wish to acknowledge the financial support received from the Spanish Ministry of Economy and Competitiveness with the project TURBO (DPI2013-46164-C2-1-R), grant number [BES-2014-068874], to HAZITEK program from the Department of Economic Development and Infrastructures of the Basque Government and from FEDER funds, related to the HEMATEX project and Vice-chancellor of Innovation, Social Compromise and Cultural Action from UPV/EHU (the Bizialab program from the Basque Government). Finally, thanks are also addressed to Spanish Project MINECO DPI2016-74845-R and RTC-2014-1861-4.

References

1. Campbell, F. *Manufacturing Technology for Aerospace Structural Material*, 1st ed.; Elsevier Science: Amsterdam, The Netherlands, 2006; pp. 221–272. ISBN 13: 9781856174954.

2. Moussaoui, K.; Mouseigne, M.; Senatore, J.; Chieragatti, R.; Lamesle, P. Influence of milling on the fatigue lifetime of a Ti6Al4V titanium alloy. *Metals* **2015**, *5*, 1148–1162. [CrossRef]

3. Klocke, F.; Zeis, M.; Klink, A.; Veselovac, D. Technological and economical comparison of roughing strategies via milling, sinking-EDM, wire-EDM and ECM for titanium- and nickel-based blisks. *CIRP J. Manuf. Sci. Technol.* **2013**, *6*, 198–203. [CrossRef]

4. Thellaputta, G.R.; Bose, P.S.C.; Rao, C.S.P. Machinability of Nickel Based Superalloys: A Review. *Mater. Today Proc.* **2017**, *4*, 3712–3721. [CrossRef]

5. Klocke, F.; Krämer, K.; Sangermann, H.; Lung, D. Thermo-mechanical tool load during high performance cutting of hard-to-cut materials. *Procedia CIRP* **2012**, *1*, 295–300. [CrossRef]

6. Chen, Z.; Zhou, J.M.; Peng, R.L.; M´Saoubi, R.; Gustafsson, D.; Palmert, F.; Moverare, J. Plastic deformation and residual stress in high speed turning of AD730^{TM} Nickel-based Superalloy with PCBN and WC tools. *Procedia CIRP* **2018**, *71*, 440–445. [CrossRef]

7. Wu, C.Y. Arbitrary surface flank milling & flank SAM in the design and manufacturing of jet engine fan and compressor airfoils. In Proceedings of the ASME Turbo Expo 2012: Turbine Technical Conference and Exposition, Copenhagen, Denmark, 11–15 June 2012; ASME Digital Collection: Copenhagen, Denmark, 2012.

8. Pereira, O.; Rodríguez, A.; Barreiro, J.; Rodríguez, A.; Fernández-Abia, A.I.; López de Lacalle, L.N. Nozzle design of combined use of MQL and cryogenic gas in machining. *Int. J. Precis. Eng. Manuf.-Green Technol.* **2017**, *4*, 87–95. [CrossRef]

9. Pereira, O.; Martín-Alfonso, J.E.; Rodríguez, A.; Calleja, A.; Fernández-Valdivielso, A.; López de Lacalle, L.N. Sustainability analysis of lubricant oils for minimum quantity lubrication based on their tribo-rheological performance. *J. Clean. Prod.* **2017**, *164*, 1419–1429. [CrossRef]

10. Li, L.; Guo, Y.B.; Wei, X.T.; Li, W. Surface integrity characteristics in wireEDM of Inconel 718 at different discharge energy. *Procedia CIRP* **2013**, *6*, 220–225. [CrossRef]

11. Godino, L.; Pombo, I.; Sanchez, J.A.; Alvarez, J. On the development and evolution of wear flats in microcrystalline sintered alumina grinding wheels. *J. Manuf. Process.* **2018**, *32*, 494–505. [CrossRef]

12. Jahanbakhsh, M.; Akhavan Farid, A.; Lotfi, M. Optimal flank wear in turning of Inconel 625 super-alloy using ceramic tool. *Proc. Inst. Mech. Eng. Part B J. Eng. Manuf.* **2016**, *232*, 208–216. [CrossRef]

13. Ghoreishi, R.H.; Roohi, A.; Dehghan Ghadikolaei, A. Analysis of the influence of cutting parameters on surface roughness and cutting forces in high speed face milling of Al/SiC MMC. Mater. *Res. Express* **2018**, *5*, 086521. [CrossRef]

14. Klocke, F.; Schmitt, R.; Zeis, M.; Heidemanns, L.; Kerkhoff, J.; Heinen, D.; Klink, A. Technological and economical assessment of alternative process chains for blisk manufacture. *Procedia CIRP* **2015**, *35*, 67–72. [CrossRef]

15. Dehghan Ghadikolaei, A.; Vahdati, M. Experimental study on the effect of finishing parameters on surface roughness in magneto-rheological abrasive flow finishing process. *Proc. Inst. Mech. Eng. Part B J. Eng. Manuf.* **2014**, *229*, 1517–1524. [CrossRef]

16. Chandra Verma, G.; Kala, P.; Mohan Pandey, P. Experimental investigations into internal magnetic abrasive finishing of pipes. *Int. J. Adv. Manuf. Technol.* **2017**, *88*, 1657–1668. [CrossRef]

17. Sánchez, J.A.; López de Lacalle, L.N.; Lamikiz, A.; Bravo, U. Dimensional accuracy optimisation of multi-stage planetary EDM. *Int. J. Mach. Tools Manuf.* **2002**, *42*, 1643–1648. [CrossRef]

18. Wang, J.; Guo, Y.B.; Fu, C.; Jia, Z. Surface integrity of alumina machines by electrochemical discharge assisted diamond wire sawing. *J. Manuf. Process.* **2018**, *31*, 96–102. [CrossRef]

19. Flaño, O.; Ayesta, I.; Izquierso, B.; Sánchez, J.A.; Zhao, Y.; Kunieda, M. Improvement of EDM performance in high-aspect ratio slot machining using multi-holed electrodes. *Precis. Eng.* **2018**, *51*, 223–231. [CrossRef]

20. Włodzimierz, W.; Jacek, T. Modern technology of the turbine blades removal machining. In Proceedings of the 8 International Conference Advanced Manufacturing Operations, Kranevo, Bulgaria, 18–20 June 2008.

21. Erickson, R.E.; Faughnan, P.R., Jr. Method of Machining Integral Bladed Rotors for a Gas Turbine Engine. U.S. Patent 7967659 B2, 28 June 2011.

22. Schwartz, B.J.; Vaillette, B.D.; Wu, C.Y.; Colacino, G.J.; Packman, A.B.; United Technologies Corp. Flank Superabrasive Machining. U.S. Patent 7,101,263, 5 September 2006.

23. Marinescu, I.; Hitchiner, M.; Uhlmann, E. *Handbook of Machining with Grinding Wheels*; Taylor & Francis Group: Boca Raton, FL, USA, 2006; ISBN 9781574446715.

24. Petrilli, R. Super abrasive machining for PM. *Met. Powder Rep.* **2012**, *67*, 38–41. [CrossRef]

25. Guo, C.; Ranganath, S.; McIntosh, D.; Elfizy, A. Virtual high performance grinding with CBN wheels. *CIRP Ann. Manuf. Technol.* **2008**, *57*, 325–328. [CrossRef]

26. Aspinwall, D.K.; Soo, S.L.; Curtis, D.T.; Mantle, A.L. Profiled superabrasive grinding wheels for the machining of a nickel based superalloy. *CIRP Ann. Manuf. Technol.* **2007**, *56*, 335–338. [CrossRef]

27. González, H.; Calleja, A.; Pereira, O.; Ortega, N.; López de Lacalle, L.N.; Barton, M. Super Abrasive Machining of Integral Rotary Components Using Grinding Flank Tools. *Metals* **2018**, *8*, 24. [CrossRef]

28. Sharman, A.R.C.; Hughes, J.I.; Ridgway, K. An analysis of the residual stresses generated in Inconel 718[TM] when turning. *J. Mater. Process. Technol.* **2006**, *173*, 359–367. [CrossRef]

29. Thakur, A.; Gangopadhyay, S. State-of-the-art in surface integrity in machining of nickel-based super alloys. *Int. J. Mach. Tools Manuf.* **2016**, *100*, 25–54. [CrossRef]

30. Pottlacher, G.; Hosaeus, H.; Kaschnitz, E.; Seifter, A. Thermophysical properties of solid and liquid Inconel 718 Alloy. *Scand. J. Metall.* **2002**, *31*, 161–168. [CrossRef]

31. Artetxe, E.; González, H.; Calleja, A.; Fernández Valdivielso, A.; Polvorosa, R.; Lamikiz, A.; López de Lacalle, L.N. Optimised methodology for aircraft engine IBRs five-axis machining process. *Int. J. Mechatron. Manuf. Syst.* **2016**, *9*, 385–401. [CrossRef]

32. Thakur, D.G.; Ramamoorthy, B.; Vijayaraghavan, L. Study on the machinability characteristics of superalloy Inconel 718 during high speed turning. *Mater. Des.* **2009**, *30*, 1718–1725. [CrossRef]

33. Costes, J.P.; Guillet, Y.; Poulachon, G.; Dessoly, M. Tool-life and wear mechanisms of CBN tools in machining of Inconel 718. *Int. J. Mach. Tools Manuf.* **2007**, *47*, 1081–1087. [CrossRef]

34. Kitagawa, T.; Kubo, A.; Maekawa, K. Temperature and wear of cutting in high-speed machining of Inconel 718 and Ti-6Al-6V-2Sn. *Wear* **1997**, *202*, 142–148. [CrossRef]

35. Fernández de Larrinoa, J. Optimización de procesos de recubrimiento para herramientas de corte. In *Tecnologías de Recubrimiento, Métodos de Caracterización y Optimización de las Propiedades*; University of the Basque Country: Bilbao, Spain, 2015.

36. Vidal, G.; Ortega, N.; Bravo, H.; Dubar, M.; González, H. An Analysis of Electroplated cBN Grinding Wheel Wear and Conditioning during Creep Feed Grinding of Aeronautical Alloys. *Metals* **2018**, *8*, 350. [CrossRef]

37. Marinescu, I.D.; Rowe, W.B.; Dimitrov, B.; Inasaki, I. *Tribology of Abrasive Machining Processe*; William Andrew: Norwich, NY, USA, 2004; pp. 341–412. ISBN 0-8155-1490-5.

38. *Geometrical Product Specifications (GPS), Surface Texture: Profile Method, Rules and Procedures for the Assessment of Surface Texture*; ISO Standard 4288; ISO: Geneva, Switzerland, 1996.

39. López de Lacalle Luis, N.; Lamikiz, A.; Sánchez, J.A.; Arana, J.L. Improving the surface finish in high speed milling of stamping dies. *J. Mater. Process. Technol.* **2002**, *123*, 292–302. [CrossRef]

40. Wenfeng, D.; Linke, B.; Yejun, Z.; Zheng, L.; Yucan, F.; Honghua, S.; Jiuhua, X. Review on monolayer CBN superabrasive wheels for grinding metallic materials. *Chin. J. Aeronaut.* **2017**, *30*, 109–134.

41. Hosseini, S.; Beno, T.; Klement, U.; Kaminski, J.; Ryttberg, K. Cutting temperatures during hard turning Measurements and effects on white layer formation in AISI 52100. *J. Mater. Process. Technol.* **2014**, *214*, 1293–1300. [CrossRef]

42. Herbert, C.R.J.; Kwong, J.; Kong, M.C.; Axinte, D.A.; Hardy, M.C.; Withers, P.J. An evaluation of the evolution of workpiece surface integrity in hole making operations for a nickel-based super alloy. *J. Mater. Process. Technol.* **2012**, *212*, 1723–1730. [CrossRef]

43. Smith, S.; Melkote, S.N.; Lara-Curzio, E.; Watkins, T.R.; Allard, L.; Riester, L. Effect of surface integrity of hard turned AISI 52100 steel on fatigue performance. *Mater. Sci. Eng.* **2007**, *459*, 337–346. [CrossRef]

44. Dai, C.W.; Ding, W.F.; Zhu, Y.J.; Xu, J.H.; Yu, H.W. Grinding temperature and power consumption in high speed grinding of Inconel 718 nickel-based superalloy with a vitrified CBN wheel. *Precis. Eng.* **2018**, *52*, 192–200. [CrossRef]

45. *Determining Residual Stresses by the Hole-Drilling Strain-Gage Method*; ASTM Standard E 837; ASTM International: West Conshohocken, PA, USA, 2013.

46. Denkena, B.; De Leon, L. Milling Induced Residual Stresses in Structural Parts out of Forged Aluminum Alloys. *Int. J. Mach. Machinab. Mater. (IJMMM)* **2008**, *4*, 335–344.

47. *Standard Test Method for Microindentation Hardness of Materials*; ASTM Standard E 384; ASTM International: West Conshohocken, PA, USA, 2017.

48. *Standard Hardness Conversion Tables for Metals Relationship Among Brinell Hardness, Vickers Hardness, Rockwell Hardness, Superficial Hardness, Knoop Hardness, and Scleroscope Hardness*; ASTM Standard E140; ASTM International: West Conshohocken, PA, USA, 2012.

49. Hua, J.; Shivpuri, R.; Cheng, X.; Bedekar, V.; Matsumoto, Y.; Hashimoto, F.; Watkins, T.R. Effect of feed rate, workpiece hardness and cutting edge on subsurface residual stress in the hard turning of bearing steel using chamfer+hone cutting edge geometry. *Mater. Sci. Eng. A* **2005**, *394*, 238–248. [CrossRef]

Article

Using the Machine Vision Method to Develop an On-machine Insert Condition Monitoring System for Computer Numerical Control Turning Machine Tools

Wei-Heng Sun [1] and Syh-Shiuh Yeh [2,*]

[1] Institution of Mechatronic Engineering, National Taipei University of Technology, Taipei 10608, Taiwan; sun245689@gmail.com

[2] Department of Mechanical Engineering, National Taipei University of Technology, Taipei 10608, Taiwan

* Correspondence: ssyeh@ntut.edu.tw; Tel.: +886-2-2771-2171

Received: 21 August 2018; Accepted: 11 October 2018; Published: 14 October 2018

Abstract: This study uses the machine vision method to develop an on-machine turning tool insert condition monitoring system for tool condition monitoring in the cutting processes of computer numerical control (CNC) machines. The system can identify four external turning tool insert conditions, namely fracture, built-up edge (BUE), chipping, and flank wear. This study also designs a visual inspection system for the tip of an insert using the surrounding light source and fill-light, which can be mounted on the turning machine tool, to overcome the environmental effect on the captured insert image for subsequent image processing. During image capture, the intensity of the light source changes to ensure that the test insert has appropriate surface and tip features. This study implements outer profile construction, insert status region capture, insert wear region judgment, and calculation to monitor and classify insert conditions. The insert image is then trimmed according to the vertical flank, horizontal blade, and vertical blade lines. The image of the insert-wear region is captured to monitor flank or chipping wear using grayscale value histogram. The amount of wear is calculated using the wear region image as the evaluation index to judge normal wear or over-wear conditions. On-machine insert condition monitoring is tested to confirm that the proposed system can judge insert fracture, BUE, chipping, and wear. The results demonstrate that the standard deviation of the chipping and amount of wear accounts for 0.67% and 0.62%, of the average value, respectively, thus confirming the stability of system operation.

Keywords: machine vision; on-machine monitoring; tool insert condition; computer numerical control; turning machine tools

---

1. Introduction

The quality of mechanical parts is dependent on the accuracy of the machining tools and the abrasion conditions of cutting tools. For instance, Fernández-Valdivielso et al. [1] analyzed the effects of geometrical features of inserts on workpiece surface integrity and developed an indirect method for determining the geometrical features of inserts that achieve the best performance in machining difficult-to-cut alloys. Pereira et al. [2] considered the abrasion conditions on the interface between an insert and a workpiece, and proposed a coolant structure that combines cryogenic cooling and the minimum quantity of lubrication to improve tool life and workpiece surface integrity. Thus, to improve the quality of products, mechanical part manufacturers must be aware of the service behaviors of cutting tools in the actual machining process, as determined from the on-machine cutting tool condition monitoring system, to be able to analyze tool life and decide whether the cutting tool needs to be changed [3,4]. The insert wear formation mechanism in the turning process comprises abrasion, diffusion, oxidation, fatigue, and adhesion wear. As shown in Figure 1, flank wear, chipping, fracture, and built-up edge (BUE) occur most frequently in general cutting processes and are mostly concentrated at the tool tip and tool flank [5–8]. Therefore, these four conditions are classified in this study, and the insert condition is reviewed by visual inspection. Flank wear gradually occurred at the cutting insert owing to the erosion between the portions of the insert in contact with the workpiece. Excessive cutting force can usually lead to brittle fracture of a cutting insert. However, due to the high temperature at the contact area between the workpiece and the insert during the machining processes, the BUE (the phenomenon that the machined material builds up on the insert edge) occurs and it could break away from the insert edge and could carry a portion of material from the insert, thereby causing fracturing and chipping.

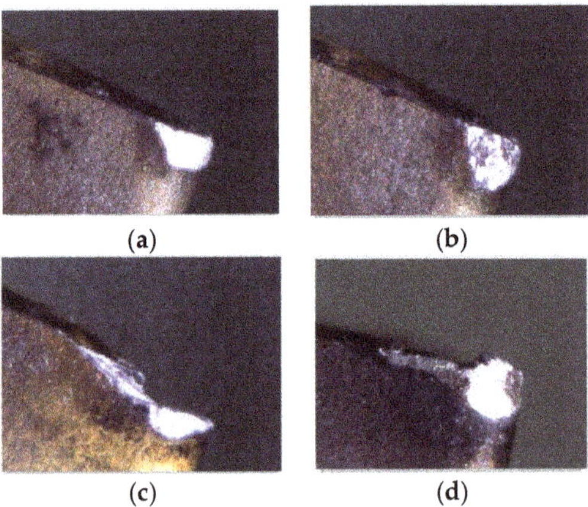

Figure 1. Four insert status forms. (**a**) Flank wear; (**b**) Chipping; (**c**) Fracture; (**d**) BUE.

There are two types of insert condition inspections in turning processes: one is indirect inspection, where the external sensors feedback the analytical machine data [9,10], and the other is direct inspection, where the cutting tool status is measured [11,12]. Indirect inspection analyzes data to estimate the cutting tool status; some machine states are analyzed according to a reference, which means that the cutting status is systematically evaluated, thus replacing the judgment of experienced operators to reduce human errors and enhancing the ability of production automation [13]. For example, the cutting tool wear condition is analyzed based on the difference in the cutting noise or vibration [14,15], the cutting tool is monitored by measuring the changes in cutting temperature and cutting forces [16,17], and the cutting status is analyzed using the machine power or current variation signal [18]. All these methods use sensing signals for inspection analysis. Lately, indirect inspection by a charge-coupled device (CCD) camera has become popular. It analyzes the cutting tools by capturing the workpiece surface texture in images to determine whether the cutting tool is worn, judged according to the changes in the workpiece surface texture and surface roughness [19–23]. Some studies have focused on the fusion of multiple sensors and visual information of images for further tool status monitoring [24,25] or used different algorithm models for analysis to implement more accurate monitoring and evaluation [26–29]. According to the aforementioned references, the status of cutting tools can be obtained by analyzing machine information variations; however, such indirect inspection sometimes reduces the accuracy of the system under the effect of the external sensing environment [30]. Therefore, a direct inspection method is required to analyze the changes in the status of cutting tools.

Direct inspection analyzes the problems in machining by directly observing the practical situation of the cutting tool. Some documents use sound, light, or a probe to build a cutting tool model to observe the status of the cutting tool [25,31,32]. However, such measurement equipment is relatively complicated and unsuitable for onsite inspection. Another method uses a charge-coupled device (CCD) camera to capture tool images and analyze the status of cutting tools. There are two types of analysis regarding the status of cutting tools, one is to analyze the wear condition by outer contour and profile inspection [33], which is generally used to monitor the outer profile wear status to judge whether the cutting tool is still workable. In comparison with the indirect inspection methods that are used to analyze and inspect the surface texture of a machined workpiece to determine the tool status, the direct inspection method is to judge the status of the cutting tools by surface texture or surface roughness analysis of the tool edge after machining [34], which is applied for a more detailed inspection of the cutting tool and the machine states, as it provides detailed machining information. The general visual inspection of a CCD camera analyzes the different locations of a cutting tool, for example, some studies have implemented analysis according to crater wear [35,36], whereas others have implemented it according to the flank wear condition [37]. A majority of the status information regarding a cutting tool can be gathered by visual inspection; in other words, changes in the outer profile can be obtained from the images. Giusti et al. [38] proposed a visual inspection method for cutting tool wear, Rangwala and Dornfeld [39] proposed using a neural network to analyze wear status, and many scholars successively proposed other related inspection methods [26]. Regarding the methods for optimization of wear features, Kurada and Bradley [40] proposed using gradient operators to calculate texture features, where the wear region boundary feature search was calculated using an octagonal-shaped matrix, and the slope was established by the brightness difference and radial distance from the matrix center to determine the location of optimized wear features. The original image was smoothed during preprocessing to reduce the interference of irregularities. For feature calculation, the pixel value was converted by image thresholding to obtain the actual wear intensity and determine the change in wear amount. Yuan et al. [41] proposed a new filtering method to obtain average images and proposed a new edge detection method based on wavelet transform. When the wavelet function is selected, a new wavelet function is generated that describes the gray change of the image. In other words, noise interference can be avoided to obtain better edge features and the abrasion region, width, length, and center location of abrasion region can be measured. Wang et al. [42] proposed an image processing procedure, which is different from the traditional method based on constant thresholding. In this method, a rough-to-fine strategy is considered. First, the thresholding images are obtained for the search candidate's wear bottom edge points. Then, the threshold-independent edge detection method, based on moment invariance, is used to determine the wear-edge. To shorten the computing time, a critical area is defined first, and only this area is taken as the region of interest in subsequent processes; thus, evading the threshold-dependent wear features detection method. Li et al. [43] used the pulse-couple neutral networks (PCNN) of bionics in cutting tool wear monitoring and used the spatial neighbor and similar gray clusters of pixels to segment the binary image of tool wear according to the condition that the gray intensity is higher than the body of the tool and background in the field of tool wear. Shahabi and Ratnam [44] used the external profile of the original image to test the alignment of the tool image and then used median filtering, morphological operations, and thresholding algorithms to reduce the system errors resulting from cutting tool misalignment, the presence of micro-dust particles, vibrations, and the intensity variations of ambient light. The aim was to determine the tool holder position and positioning error to ensure that cutting tool wear could be inspected without precision tool alignment. Pfeifer and Wiegers [45] used light source changes to determine wear-edge features under different light sources. While light changes can influence the shadow changes of the cutting tool wear-edge, the actual edge location does not vary with the light source. Thus, cutting tool wear image information under different light sources can be obtained using high-pass filter and thresholding images and the recurrent edge location can be obtained by overlapping to determine the location of a strong edge to filter out the misrecognition due to contaminants and reduce the effects of contaminants and shadow changes on the inspection system. Barreiro et al. [46] used different moments as descriptors to illustrate the tool wear images and then used a finite mixture MCLUST model to classify tool wear conditions into low-, medium-, and high-wear classes. Furthermore, the monitoring results were validated through the use of linear and quadratic discriminant analyses. Based on the image processing results of the cutting edge, Alegre et al. [47] developed a procedure to determine the time for tool replacement through the use of k-nearest neighbors and a multilayer neural network. D'Addona and Teti [48] used an image standardization process to obtain images with standard size and pixel density during cutting tests; then, the back-propagation neural network was optimized and used to estimate tool wear conditions with standardized cutting tool images.

Differing from existing research findings, this study analyzes insert statuses and uses fusion contour and texture inspection methods to build a more accurate evaluation and judgment system, which is applicable to on-machine automatic inspections and eliminates the environmental problems during inspection. A visual inspection system that can be used in CNC turning machine tools is constructed, which consists of a CCD camera and a lens for capturing insert images, a protection box to protect the photographic equipment, and a peripheral circuit and components, to avoid scrap splashes and cutting fluids during the cutting processes. The visual inspection system has a cleaning air tube, which jets air toward the insert to clean the surface of the inspected insert, thus, reducing the problems of subsequent image processing and increasing the accuracy of the insert condition judgment. The visual inspection system designed in this study has a surrounding light source and a fill-light for the tip of the insert to ensure that the insert condition can be analyzed under changing lighting conditions. When the light source is adjusted to determine the location of the blade and the tip of the insert, it is applicable to insert condition monitoring if onsite tool alignment is not accurate, thus enhancing the feasibility of image recognition in the machine. The light source intensity is adjusted and the insert image is captured under varying intensities for inspection analysis. The effect of any external environment changes on the insert condition monitoring result can be reduced and the system designed in this study can obtain accurate results in different environments. Image underexposure or solarization that generally result from changes in the insert condition are also improved. This study analyzes captured insert images with different features and the common insert conditions of an external turning tool, including fracture, BUE, chipping, and wear, can also be inspected. The analysis of the results can be quantized according to the texture feature distribution. This study conducts on-machine insert condition monitoring experiments with inserts in different states and the results show that the insert condition monitoring system designed in this study is applicable to computer numerical control (CNC) turning machine tools for correct and stable identification of insert fracture, BUE, chipping, and wear conditions. Contributions of this study therefore include

- development of an on-machine insert condition monitoring system that can be used to one-time identify the four insert conditions—fracture, BUE, chipping, and flank wear.
- development of a mountable visual system with different light sources to on-machine capture good-quality insert images that can be exactly analyzed under different lighting conditions.
- development of a contour and texture fusion inspection method to reduce environmental problems and to accurately identify insert conditions during inspection.

The structure of this paper is as follows. Section 2 describes the experimental system and related equipment used in this study, along with the hardware architecture design of the machine vision inspection system. Section 3 describes the insert image capture process designed in this study and the usage of the surrounding light source and fill-light for the insert tip. Section 4 describes the insert condition monitoring classification process designed in this study, including the insert outer profile construction, insert status region capture, and wear region judgment and calculation. Section 5 describes the experimental process and results of insert condition monitoring. The experiment on the on-machine insert condition monitoring by a CNC turning machine tool validates the feasibility and stability of this system. Section 6 summarizes this paper.

2. Introduction to the Experimental System and Equipment

The CNC turning machine tool used in this study, shown in Figure 2, is tested using an external turning tool. The test external turning tool is mounted on the turning machine tool turret and the turning machine tool turret is moved by a computer numerical controller to the visual inspection system placed above the turning machine tool spindle for insert condition monitoring. Here, each insert position is adjusted by moving the turret such that the region of interest is focused in order to reduce the blurring of captured images. Moreover, during the period of experiments, the security door that is usually used to protect operators was closed so that the turning zone environment can reduce the influence from external environments. A GigE DFK 23GP031 color industrial camera, with an image resolution of 2592 × 1944 (15 fps), is used in this study. Figure 3a shows the camera hardware combination; the lens is a Myutron HS3514J CCTV lens, combined with a double lens to capture the feature image and the 90-degree reflecting mirror can adjust the angle of the camera. Due to the space constraints of the internal structure of the machine and considering the potential contamination resulting from the actual machining environment, this study designs a visual inspection system that can be mounted in turning machine tools, as shown in Figure 3b. The protection box for the camera hardware, as shown in Figure 3a, prevents the cutting scrap in the machine from splashing, thus, reducing the contamination of cutting fluid on the lens. To capture sharp insert images, the cleaning air tube jets air toward the insert for cleaning. The protection box is equipped with a surrounding LED light source with adjustable brightness and is covered with epoxy resin for protection. The protection box extends the fill-light for the tip of the insert to be inspected (tip light source). Two magnetic bases are set up at the protection box base to fix the protection box in the machine tool for on-machine insert condition monitoring.

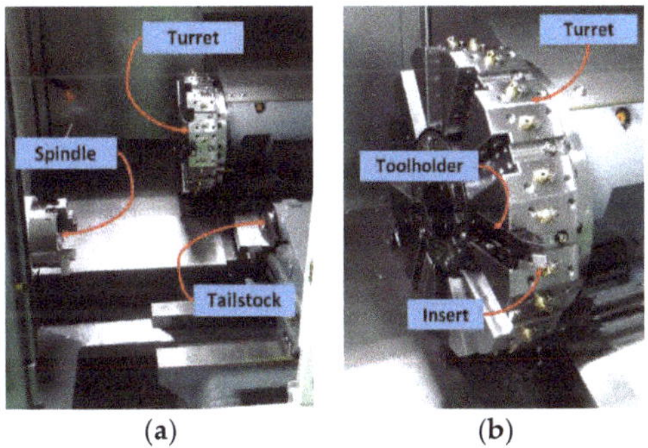

Figure 2. CNC turning machine tool for experiment. (**a**) Turning zone; (**b**) Turret structure.

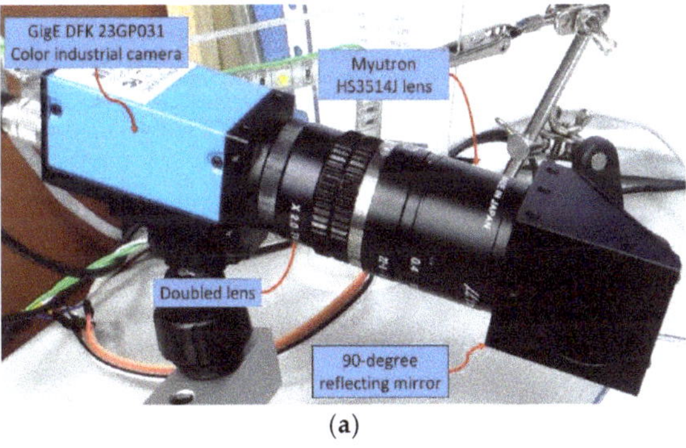

Figure 3. *Cont.*

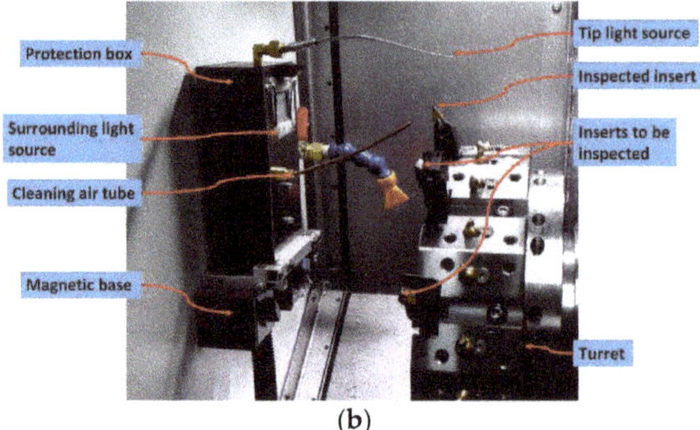

(b)

Figure 3. Visual inspection system mountable inside turning machine tools designed in this study. (**a**) Industrial camera and lens related components; (**b**) Visual inspection system protection box.

3. Insert Image Capture Process

During image capture, the fill-light is used for the inspected insert and the light source intensity is changed to ensure that different inserts have appropriate feature strength. This study uses two light sources in different positions as shown in Figure 3b. In terms of the surrounding light source, a strong light irradiates the test insert to obtain its surface shape and area features. In terms of the fill-light for the tip of an insert, the tip status feature is enhanced to facilitate later processing and analysis of the captured image. In the insert image capture process, as designed in this study, the insert is shot under a high-strength surrounding light source to capture the tool flank exposure image, as shown in Figure 4a. Then the insert image is captured using a high-strength surrounding light source and fill-light for the insert tip, as shown in Figure 4b. Here, the exposure images can be sequentially used to confirm the insert position, enhance geometry features, and strengthen wear features. The featured images are captured after the exposure image capture. First, the fill-light for the tip of an insert is closed and the surrounding light source intensity is adjusted to obtain appropriate flank feature images, as shown in Figure 5a, and then the intensity of the fill-light for the tip of an insert is adjusted to obtain appropriate tip feature images, as shown in Figure 5b. Here, the adjustment of light source intensity is automatically performed depending on the average thresholding value of the captured images. Referring to the exposure images, as shown in Figure 4, the feature images are used to analyze different insert conditions and can be utilized in the classification process of the insert conditions, including insert profile construction, status region capture, and wear judgment and calculation.

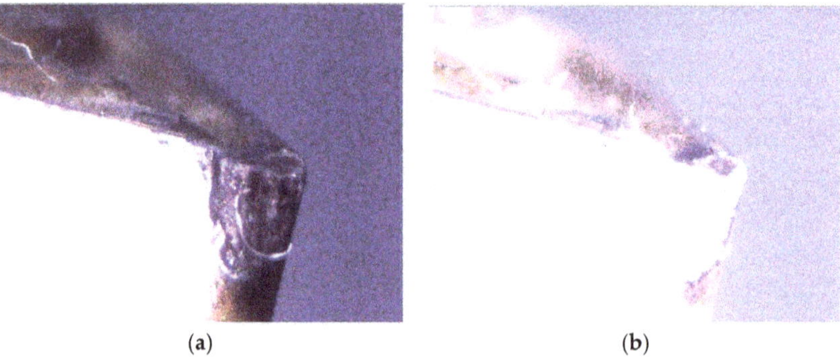

(a)   **(b)**

Figure 4. Captured flank and insert exposure images. (**a**) Flank exposure image; (**b**) Insert exposure image.

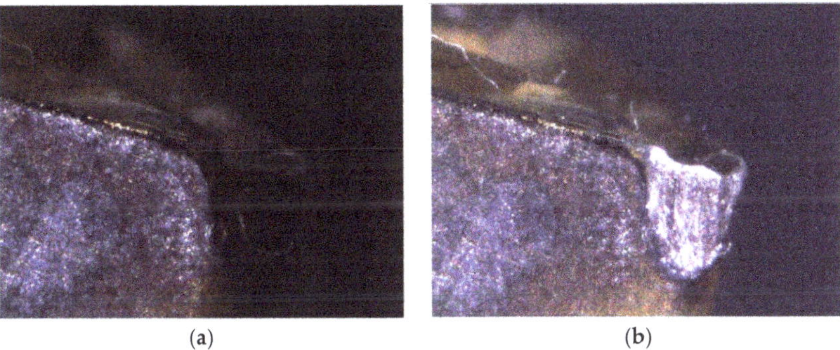

(a)　　　　　　　　　　　　　　　(b)

Figure 5. Captured flank and tip feature images. (**a**) Flank feature image; (**b**) Tip feature image.

4. Insert Condition Monitoring Classification Process

4.1. Insert Outer Profile Construction

First, the flank profile feature is determined using the flank exposure image in Figure 4a. This study uses grayscale image thresholding to determine the flank profile feature, as shown in Figure 6a. Similarly, the insert profile feature in Figure 4b and grayscale image thresholding are used to determine the insert profile feature, as shown in Figure 6b. Here, the thresholding value is 250. The lines in the thresholding images in Figure 6 are determined using straight-line Hough transform, as shown in Figure 7. The flank profile exposure thresholding images determine the vertical flank line and horizontal blade line, while the insert profile exposure thresholding images determine the vertical blade line. The thresholding image can be trimmed and rotated along the horizontal blade line (Figure 7a) and vertical blade line (Figure 7b) in Figure 7 to construct a complete insert outer profile thresholding image, as shown in Figure 8a. According to the vertical flank line in Figure 7a, the complete insert outer profile thresholding image is divided into two blocks, as shown in Figure 8b: tip front-end underside (block B) and insert backend underside (block A) for subsequent insert condition feature recognition. Figure 9 shows the results of the trimmed insert images by referring to the completed insert outer profile thresholding image.

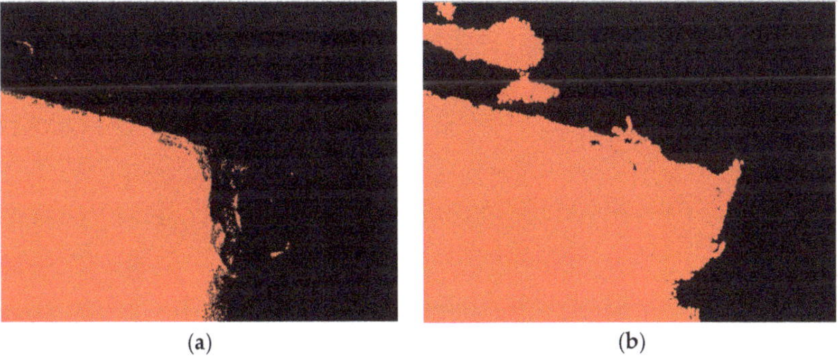

(a)　　　　　　　　　　　　　　　(b)

Figure 6. Thresholding operation result of captured exposure insert grayscale image. (**a**) Flank exposure thresholding image; (**b**) Insert exposure thresholding image.

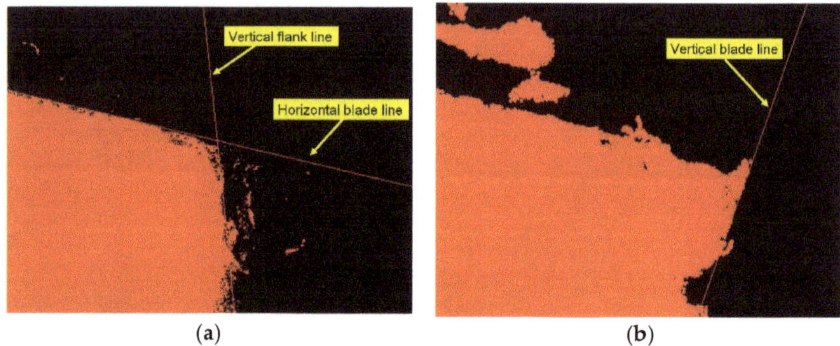

(a)   (b)

Figure 7. Lines of thresholding images. (**a**) Vertical flank line and horizontal blade line of flank profile feature; (**b**) Vertical blade line in insert exposure image.

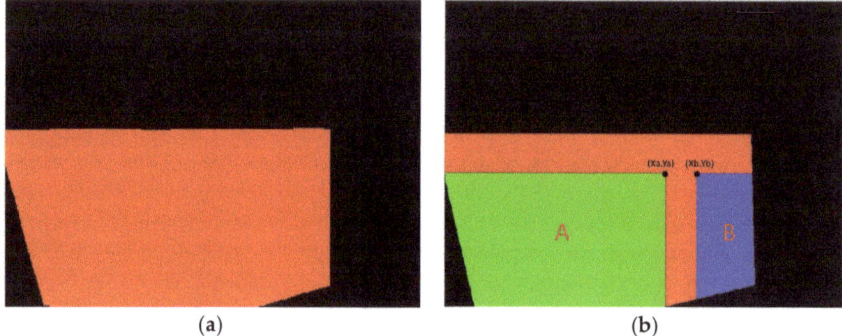

(a)   (b)

Figure 8. Completed insert outer profile thresholding images and block division. (**a**) Completed insert outer profile thresholding image; (**b**) Insert outer profile block division.

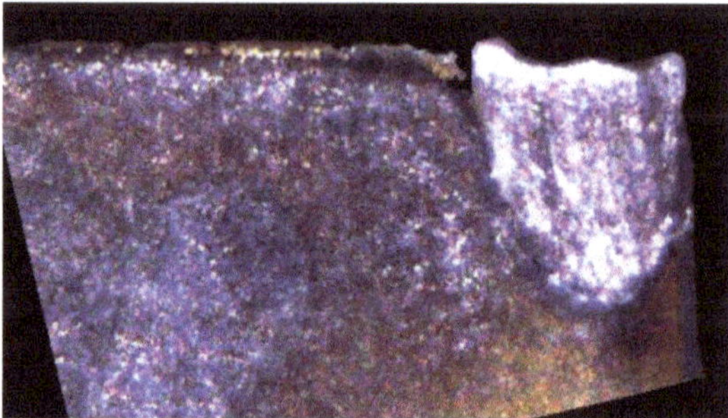

Figure 9. Trimmed insert outer profile image.

4.2. Insert Status Region Capture

The horizontal blade line in Figure 7a can be used to judge whether the insert has a fracture or BUE. The insert thresholding image in Figure 10a is obtained after the grayscale image thresholding process of Figure 9. Here, the erosion and dilation operations with the 11×11 diamond-shaped structuring element are used to clear the geometry features. The insert thresholding image is segmented along the horizontal blade line to obtain the insert fracture zone in Figure 10b and the insert BUE zone in Figure 10c, where the pixel areas of the fracture zone and BUE zone are calculated to judge the insert fracture or BUE status.

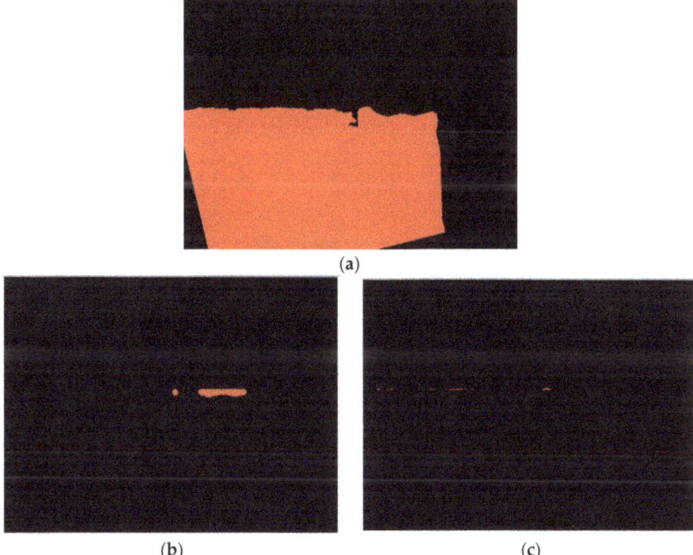

(a)

(b)   (c)

Figure 10. Judgment of insert fracture and BUE statuses. (**a**) Insert thresholding image; (**b**) Insert fracture zone; (**c**) Insert BUE zone.

If the insert condition, as identified by the insert condition monitoring system designed in this study, is not classified as fracture or BUE, the flank wear judgment process begins. First, the grayscale transformation is implemented for the trimmed insert outer profile image in Figure 9. This study uses the average image RGB values for grayscale processing. After the insert outer profile image is converted into a grayscale image, the Sobel operator is used for insert edge detection to obtain a good insert edge feature. The insert outer profile blocks are then segmented, as shown in Figure 8b, and the lower region at the backend of the insert (block A) is removed to segment the location of the flank wear feature, as shown in Figure 11a. To facilitate the trimming of the flank wear part for subsequent judgment and calculation, the computation for noise removal, contrast stretching process, erosion, and dilation operations are implemented, as shown in Figure 9, and the flank wear zone image is obtained, as shown in Figure 11b. Here, the 3×3 box-pattern low-pass filter and the 21×21 box-pattern median filter are used for noise suppression. Finally, the trimmed operation is implemented for the insert outer profile image in Figure 9 according to Figure 11b and the flank wear zone image in Figure 11c is obtained.

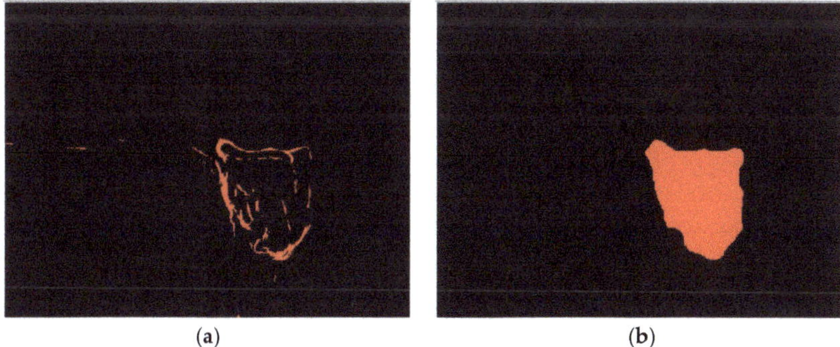

(a)   (b)

Figure 11. *Cont.*

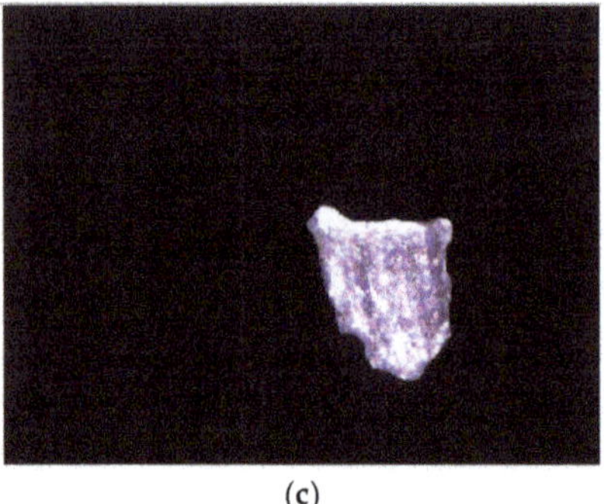

(c)

Figure 11. Actual image of trimmed insert wear region. (**a**) Flank wear feature zone; (**b**) Range of flank wear zone; (**c**) Flank wear zone image.

4.3. Wear Region Judgment and Calculation

The flank wear or chipping wear status can be classified according to the trimmed flank wear zone image, as shown in Figure 11c. Figure 12 shows that there is a significant difference between the flank wear and chipping wear. The flank wear is the tear resulting from the rub between the cutting blade and workpiece in the machining process, thus, the flank wear surface features are mostly continuous and even. However, as chipping wear is the tip breakage resulting from abnormal machining processes, the chipping surface is relatively rough. This study analyzes the continuity of surface features for the actual image of a wear region to identify the insert wear region as flank or chipping wear, as shown in Figure 11c. The grayscale value histogram of all pixels can be obtained after Figure 11c is converted into a grayscale image, as shown in Figure 13a. The number of pixels is obviously larger than the pixel grayscale value histogram distribution of the flank wear image, as shown in Figure 13b. Therefore, the number of pixels larger than the preset threshold value is divided by the calculated value percentage of the number of pixels of the overall wear region to identify the wear region as flank or chipping wear. In other words, the percentage (chipping rate) of the number of pixels larger than the preset threshold value to the number of pixels of the overall wear region is taken as the basis of judgment. Moreover, this study uses the length of the pixels of the upper and lower boundaries of the wear region image to calculate the wear amount. The pixel unit is converted using the wear region image, as shown in Figure 14, where the conversion length of the wear region image pixels is 0.007 mm and the length in the pixels of the upper and lower boundaries of wear region image is 184 pixels, thus, the converted wear amount is 1.288 mm.

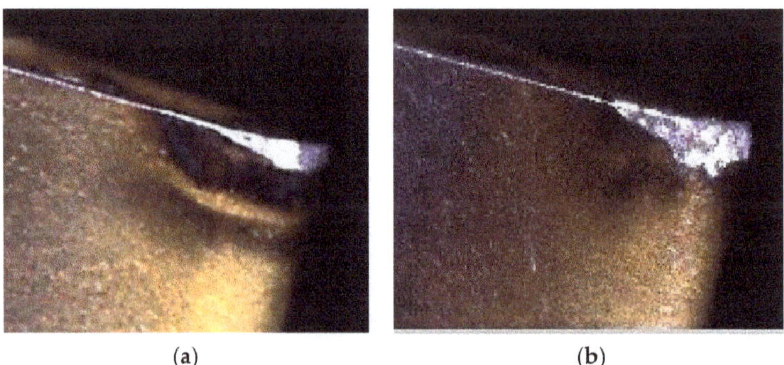

(a)   (b)

Figure 12. Comparison between flank wear and chipping wear. (**a**) Flank wear; (**b**) Chipping wear.

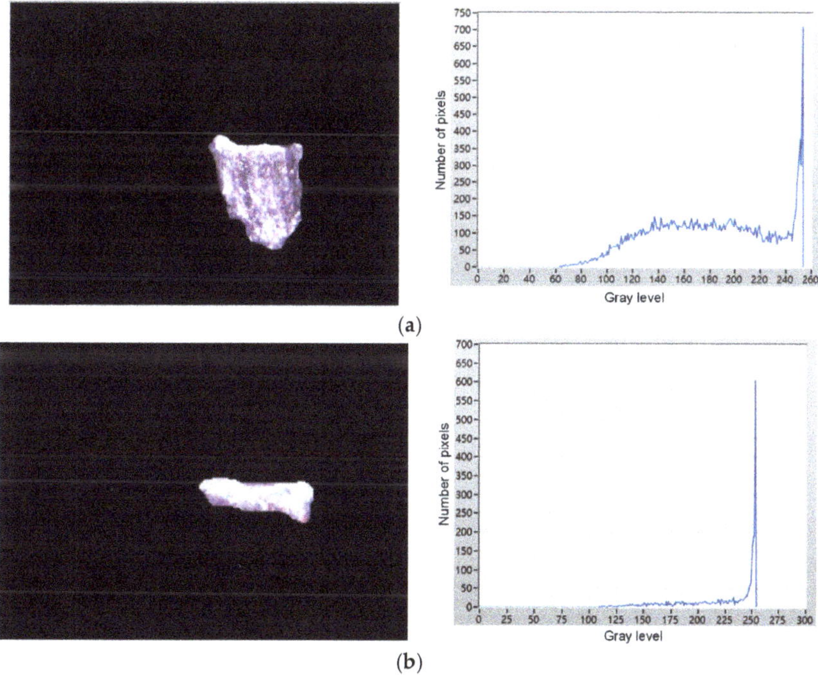

Figure 13. Histogram of grayscale image. (a) Grayscale image of Figure 11c; (b) Grayscale image of flank wear.

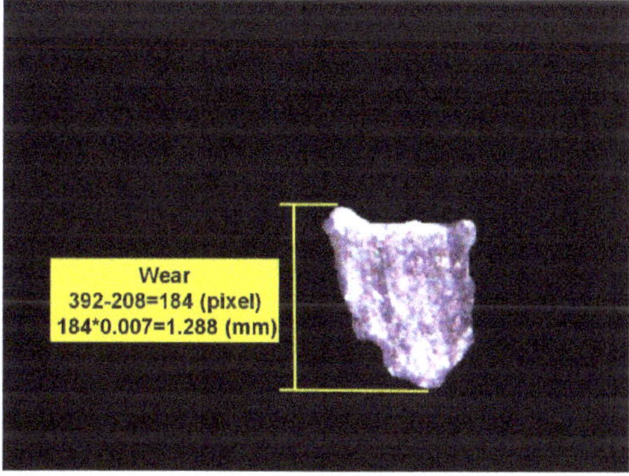

Figure 14. Wear amount result of wear region image.

5. Experiment Monitoring Insert Condition

To validate the feasibility of the on-machine insert condition monitoring system proposed in this study, the visual inspection system mounted on the turning machine tool for insert condition monitoring experiments is shown in Figure 15. This study uses twenty used inserts in various states for experimentation and the results are presented in Table 1 and Figure 16. Here, the used inserts were collected after turning with cutting speed (130–150 m/min), cutting feed rate (0.2–0.3 mm/rev), and depth-of-cut (2–3 mm). The workpiece material is medium carbon steel and the insert material is tungsten carbide. The laptop computer with an Intel Core i7-4720HQ, 2.6-GHz CPU, and 64-bit Microsoft Windows 10 operating system was utilized to implement the whole system so that the time required for each monitoring task is approximately 13 s in which 2.75 s, on average, are required for the identification of insert conditions. To further reduce the time required for each monitoring task, a computer with a faster CPU could be used to implement the system. Table 1 shows the judgment results of the inserts in different states. The chipping rate is set at 50% for monitoring and the wear amount is set as 0.3 mm for identifying over-wear. Based on the results, the system developed in this study can correctly identify the various insert conditions of the test inserts. Table 1 presents three types of BUE inserts, where two of them have slight BUE. Thus, it can be said that this study identifies the BUE status accurately according to the preset threshold of BUE.

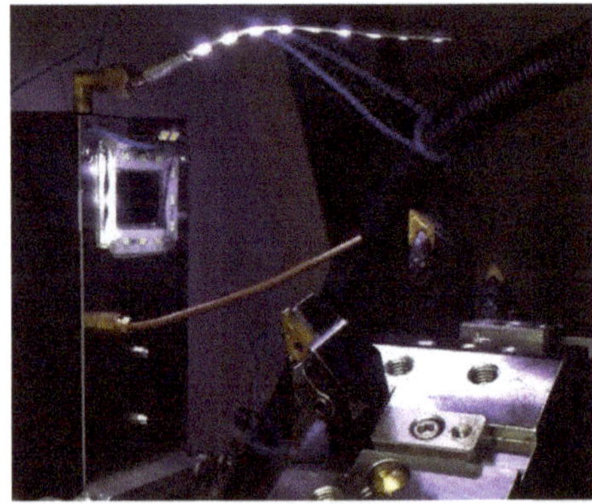

Figure 15. Experimental system setup for insert condition monitoring.

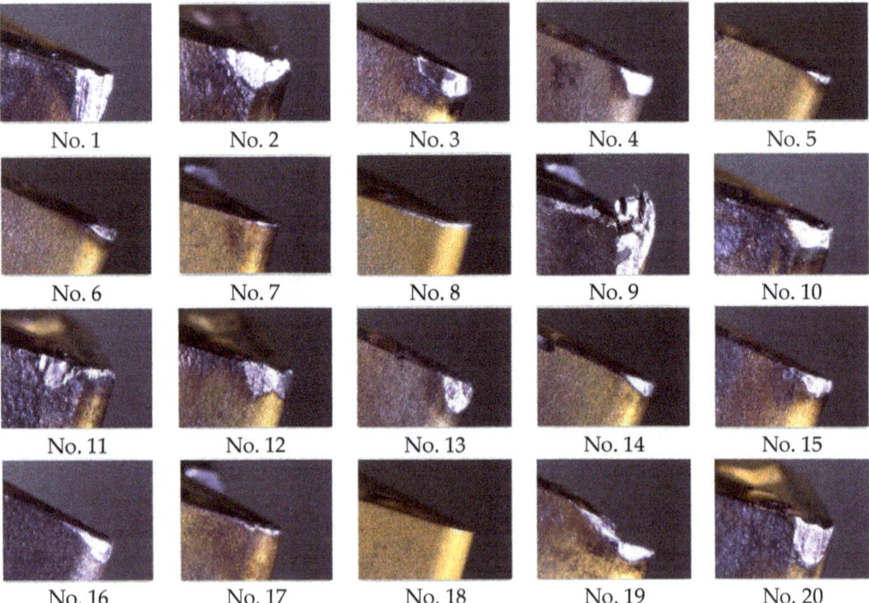

Figure 16. Captured images of different insert conditions (corresponding to insert numbers in Table 1).

Table 1. Experimental results of condition monitoring of different inserts.

No.	Chipping Rate (%)	Wear Amount (mm)	Status Determination	No.	Chipping Rate (%)	Wear Amount (mm)	Status Determination
1	45.76	1.530	over-wear	11	62.47	0.651	chipping
2	0.00	0.000	BUE	12	0.00	0.000	BUE
3	35.14	0.735	over-wear	13	52.42	0.875	chipping
4	20.65	0.658	over-wear	14	19.26	0.427	over-wear
5	26.04	0.238	normal wear	15	37.09	0.532	over-wear
6	33.75	0.287	normal wear	16	31.35	0.903	over-wear
7	12.20	0.105	normal wear	17	29.48	0.161	normal wear
8	26.34	0.252	normal wear	18	0.00	0.000	normal wear
9	0.00	0.000	BUE	19	0.00	0.000	fracture
10	29.26	0.686	over-wear	20	63.06	1.250	chipping

The insert condition monitoring system can identify different insert conditions and its operational stability is a key point of evaluation. Due to the changing external environment and light source intensity, there will be different results for insert condition tests and calculations. This study repeatedly tests the same insert to validate the stability of the insert condition monitoring system and the experimental results are shown in Table 2, where the wear amount of the insert wear region is calculated for comparison analysis. The experiment is repeated 10 times, the chipping rate and wear amount of each experiment are recorded, and system stability is checked using the calculated mean value and standard deviation. The experimental results show that the chipping rate analysis has large standard deviation, which signifies that there is a large variation in the results. The chipping rate is calculated according to the pixel grayscale value histogram distribution of the flank wear zone image, even though the algorithm and light source system operating procedure are identical, each moment of image capture is affected by the light source change and the grayscale value histogram distribution of the wear images changes. Despite all this, the standard deviation of the chipping rate is only 0.67% of the average value and the stability of the chipping rate calculation of this insert condition monitoring system can be calculated. In terms of wear amount results, the standard deviation of wear amount is only 0.62% of the average value, in other words, the standard deviation is lower than two pixels. Hence, the stability of this insert condition monitoring system in calculating wear amount can be calculated. Therefore, the aforementioned experimental results can be used to validate the feasibility and stability of the insert condition monitoring system and calculation method designed in this study.

Table 2. Experimental results of computational stability of insert wear.

	No.	Chipping Rate (%)	Wear Amount (mm)
	1	52.886	1.302
	2	52.993	1.288
	3	52.922	1.288
	4	52.831	1.302
	5	53.335	1.288
	6	53.003	1.288
	7	52.820	1.302
	8	52.746	1.295
	9	51.878	1.281
	10	52.760	1.302
	Average value	52.817	1.294
	Standard deviation	0.352	0.008

This study developed an on-machine insert condition monitoring system to identify four external turning tool insert conditions; fracture, BUE, chipping, and flank wear. The experimental results demonstrate that the developed monitoring system can successfully identify the four insert conditions. Moreover, as shown in Figure 16, the developed system can be used for identifying the insert conditions when it is difficult to measure the wear amount precisely using standard wear measurement methods. However, because the view angle of the developed visual inspection system that is mounted inside the turning machine tools is different from the view angle of standard wear measurement devices, the calculated wear amount, which is used to indicate the degree of wear conditions, cannot be compared with the measurement results obtained using standard wear measurement devices.

6. Conclusions

The status of cutting tools used in the cutting processes of machine tools will obviously influence the manufacturing quality of machine parts. Therefore, this study develops an on-machine insert condition monitoring system for the turning tool insert of CNC turning machine tools and uses the machine vision method to inspect the common flank wear, chipping, fracture, and BUE statuses of turning tool inserts. This study differs from the existing research methods and outcomes as it fuses the machine vision method with contour and texture inspections to analyze the insert status. This eliminates the environmental problems in the insert inspection process to build a more accurate on-machine turning tool insert condition monitoring system.

To fix the CCD camera and lens in the CNC turning machine tool to carry out the on-machine insert condition visual inspection process, a visual inspection system with a protection box, cleaning air tube, and two light sources is designed. The protection box can avoid the scrap splash and contamination of cutting fluid on the lens during the cutting processes, while the cleaning air tube jets air toward the insert to clean off surface contaminants. A surrounding light source and a fill-light for the tip of an insert with variable light intensities are employed to analyze the effect of change in lighting conditions on the visual inspection of the insert status. In the insert image capture process, the intensity of the surrounding light source and fill-light is changed to ensure that the test insert has appropriate feature strength. The surrounding light source uses strong light to irradiate the insert surface to obtain the surface shape and area features, while the fill-light enhances the tip status feature to facilitate subsequent captured image processing and analysis. An insert condition monitoring classification process designed in this study includes insert outer profile construction, insert status region capture, and wear region judgment and calculation. The insert outer profile construction uses the exposure image to determine the outer profile feature, and then the vertical flank line, horizontal blade line, and vertical blade line are established according to this outer profile feature. The insert image can be trimmed for subsequent insert condition feature recognition. In terms of insert status region capture, the insert fracture zone and BUE zone are identified according to the outer profile feature lines and the insert outer profile image is trimmed to obtain the actual image of the insert wear region. For insert wear region judgment and calculation, the flank wear or chipping wear is identified based on the grayscale value histogram of all pixels of the trimmed flank wear zone image. The wear amount is calculated using the pixel length of the upper and lower boundaries of the wear region image, which are used as the reference index to identify the normal wear or over-wear status of the insert. Finally, inserts in different states are used for on-machine insert condition monitoring experimentation to confirm that the system designed in this study can identify insert fracture, BUE, chipping, and wear statuses. In addition, as the changes in external environment and light source sometimes influence the image processing result, the operational stability of the on-machine insert condition monitoring system is tested in this study. The experiment is conducted repeatedly and the average value and standard deviation of the chipping rate and wear amount in the experimental results are recorded as the basis for evaluating the operational stability of the system. The experimental results show that the light source variation does influence the calculation of chipping rate and wear amount. The standard deviation of the chipping rate is only 0.67% of the average value, while the standard deviation of wear amount is 0.62% of the average value (standard deviation lower than 2 pixels), thus validating the stability of system operation.

Author Contributions: Investigation, W.-H.S., S.-S.Y.; Supervision, S.-S.Y.

Funding: This research was funded in part by the Ministry of Science and Technology, Taiwan, R.O.C., under Contract MOST 104-2221-E-027-132 and MOST 103-2218-E-009-027-MY2.

Acknowledgments: The authors would like to thank representatives from the SRAM Taiwan Company for their useful discussions with the research team. The authors especially thank to Meng-Hui Lin (SRAM Taiwan Company) for his beneficial discussions.

Conflicts of Interest: The authors declare no conflict of interest.

References

1. Fernández-Valdivielso, A.; López De Lacalle, L.N.; Urbikain, G.; Rodriguez, A. Detecting the key geometrical features and grades of carbide inserts for the turning of nickel-based alloys concerning surface integrity. *Proc. Inst. Mech. Eng. Part C J. Mech. Eng. Sci.* **2016**, *230*, 3725–3742. [CrossRef]

2. Pereira, O.; Rodríguez, A.; Fernández-Abia, A.I.; Barreiro, J.; López de Lacalle, L.N. Cryogenic and minimum quantity lubrication for an eco-efficiency turning of AISI 304. *J. Clean. Prod.* **2016**, *139*, 440–449. [CrossRef]

3. Yu, J. Machine tool condition monitoring based on an adaptive Gaussian mixture model. *J. Manuf. Sci. Eng. Trans. ASME* **2012**, *134*, 031004. [CrossRef]

4. Jones, B.E. Sensors in industrial metrology. *J. Phys. E Sci. Instrum.* **1987**, *20*, 1113–1116. [CrossRef]

5. Avinash, C.; Raguraman, S.; Ramaswamy, S.; Muthukrishnan, N. An Investigation on Effect of Workpiece Reinforcement Percentage on Tool Wear in Cutting Al-SiC Metal Matrix Composites. In Proceedings of the ASME International Mechanical Engineering Congress and Exposition, Seattle, WA, USA, 11–15 November 2008; pp. 561–566.

6. Ee, K.C.; Balaji, A.K.; Jawahir, I.S. Progressive tool-wear mechanisms and their effects on chip-curl/chip-form in machining with grooved tools: An extended application of the equivalent toolface (et) model. *Wear* **2003**, *255*, 1404–1413. [CrossRef]

7. Nordgren, A.; Melander, A. Tool wear and inclusion behaviour during turning of a calcium-treated quenched and tempered steel using coated cemented carbide tools. *Wear* **1990**, *139*, 209–223. [CrossRef]

8. Akbar, F.; Mativenga, P.T.; Sheikh, M.A. An evaluation of heat partition in the high-speed turning of AISI/SAE 4140 steel with uncoated and TiN-coated tools. *Proc. Inst. Mech. Eng. Part B J. Eng. Manuf.* **2008**, *222*, 759–771. [CrossRef]

9. Ralston, P.A.S.; Ward, T.L.; Stottman, D.J.C. Computer observer for in-process measurement of lathe tool wear. *Comput. Ind. Eng.* **1988**, *15*, 217–222. [CrossRef]

10. Massol, O.; Li, X.; Gouriveau, R.; Zhou, J.H.; Gan, O.P. An exTS based neuro-fuzzy algorithm for prognostics and tool condition monitoring. In Proceedings of the 11th International Conference on Control, Automation, Robotics and Vision, ICARCV 2010, Singapore, 7–10 December 2010; pp. 1329–1334.

11. Rutelli, G.; Cuppini, D. Development of wear sensor for tool management system. *J. Eng. Mater. Technol. Trans. ASME* **1988**, *110*, 59–62. [CrossRef]

12. Novak, A.; Wiklund, H. Reliability improvement of tool-wear monitoring. *CIRP Ann. Manuf. Technol.* **1993**, *42*, 63–66. [CrossRef]

13. Szélig, K.; Alpek, F.; Berkes, O.; Nagy, Z. Automatic inspection in a CIM system. *Comput. Ind.* **1991**, *17*, 159–167. [CrossRef]

14. Downey, J.; Bombiński, S.; Nejman, M.; Jemielniak, K. *Automatic Multiple Sensor Data Acquisition System in a Real-Time Production Environment*; Procedia CIRP: Capri, Italy, 2015; pp. 215–220.

15. Scheffer, C.; Heyns, P.S. Monitoring of turning tool wear using vibration measurements and neural network classification. In Proceedings of the 25th International Conference on Noise and Vibration Engineering, ISMA, Leuven, 13–15 September 2000; pp. 921–928.

16. Prasad, B.S.; Prabha, K.A.; Kumar, P.V.S.G. Condition monitoring of turning process using infrared thermography technique—An experimental approach. *Infrared Phys. Technol.* **2017**, *81*, 137–147. [CrossRef]

17. Fu, P.; Li, W.; Guo, L. *Fuzzy Clustering and Visualization Analysis of Tool Wear Status Recognition*; Procedia Engineering: Shenzhen, China, 2011; pp. 479–486.

18. Hamade, R.F.; Ammouri, A.H. Current Rise Index (CRI) maps of machine tool motors for tool-wear prognostic. In Proceedings of the ASME 2011 International Mechanical Engineering Congress and Exposition, IMECE 2011, Denver, CO, USA, 11–17 November 2011; pp. 867–872.

19. Dutta, S.; Datta, A.; Chakladar, N.D.; Pal, S.K.; Mukhopadhyay, S.; Sen, R. Detection of tool condition from the turned surface images using an accurate grey level co-occurrence technique. *Precis. Eng.* **2012**, *36*, 458–466. [CrossRef]

20. Dutta, S.; Pal, S.K.; Sen, R. On-machine tool prediction of flank wear from machined surface images using texture analyses and support vector regression. *Precis. Eng.* **2016**, *43*, 34–42. [CrossRef]

21. Datta, A.; Dutta, S.; Pal, S.K.; Sen, R. Progressive cutting tool wear detection from machined surface images using Voronoi tessellation method. *J. Mater. Process. Technol.* **2013**, *213*, 2339–2349. [CrossRef]

22. Kwon, Y.; Ertekin, Y.; Tseng, T.L. Characterization of tool wear measurement with relation to the surface roughness in turning. *Mach. Sci. Technol.* **2004**, *8*, 39–51. [CrossRef]

23. Kassim, A.A.; Mannan, M.A.; Jing, M. Machine tool condition monitoring using workpiece surface texture analysis. *Mach. Vis. Appl.* **2000**, *11*, 257–263. [CrossRef]

24. Prasad, B.S.; Sarcar, M.M.M. Experimental investigation to predict the condition of cutting tool by surface texture analysis of images of machined surfaces based on amplitude parameters. *Int. J. Mach. Machinabil. Mater.* **2008**, *4*, 217–236. [CrossRef]

25. Prasad, B.S.; Sarcar, M.M.M.; Ben, B.S. Surface textural analysis using acousto optic emission- and vision-based 3D surface topography-a base for online tool condition monitoring in face turning. *Int. J. Adv. Manuf. Technol.* **2011**, *55*, 1025–1035. [CrossRef]

26. Dutta, S.; Pal, S.K.; Mukhopadhyay, S.; Sen, R. Application of digital image processing in tool condition monitoring: A review. *CIRP J. Manuf. Sci. Technol.* **2013**, *6*, 212–232. [CrossRef]

27. Dutta, S.; Pal, S.K.; Sen, R. Progressive tool flank wear monitoring by applying discrete wavelet transform on turned surface images. *Meas. J. Int. Meas. Confed.* **2016**, *77*, 388–401. [CrossRef]

28. Bhat, N.N.; Dutta, S.; Pal, S.K.; Pal, S. Tool condition classification in turning process using hidden Markov model based on texture analysis of machined surface images. *Meas. J. Int. Meas. Confed.* **2016**, *90*, 500–509. [CrossRef]

29. Mannan, M.A.; Mian, Z.; Kassim, A.A. Tool wear monitoring using a fast Hough transform of images of machined surfaces. *Mach. Vis. Appl.* **2004**, *15*, 156–163. [CrossRef]

30. Bartow, M.J.; Calvert, S.G.; Bayly, P.V. Fiber bragg grating sensors for dynamic machining applications. In Proceedings of the SPIE—The International Society for Optical Engineering, Troutdale, OR, USA, 20 November 2003; pp. 21–31.

31. Wang, W.H.; Wong, Y.S.; Hong, G.S. 3D measurement of crater wear by phase shifting method. *Wear* **2006**, *261*, 164–171. [CrossRef]

32. Dawson, T.G.; Kurfess, T.R. Quantification of tool wear using white light interferometry and three-dimensional computational metrology. *Int. J. Mach. Tools Manuf.* **2005**, *45*, 591–596. [CrossRef]

33. Klancnik, S.; Ficko, M.; Balic, J.; Pahole, I. Computer vision-based approach to end mill tool monitoring. *Int. J. Simul. Model.* **2015**, *14*, 571–583. [CrossRef]

34. Kerr, D.; Pengilley, J.; Garwood, R. Assessment and visualisation of machine tool wear using computer vision. *Int. J. Adv. Manuf. Technol.* **2006**, *28*, 781–791. [CrossRef]

35. Lanzetta, M. A new flexible high-resolution vision sensor for tool condition monitoring. *J. Mater. Process. Technol.* **2001**, *119*, 73–82. [CrossRef]

36. Giusti, F.; Santochi, M.; Tantussi, G. On-line sensing of flank and crater wear of cutting tools. *CIRP Ann. Manuf. Technol.* **1987**, *36*, 41–44. [CrossRef]

37. Bahr, B.; Motavalli, S.; Arfi, T. Sensor fusion for monitoring machine tool conditions. *Int. J. Comput. Integr. Manuf.* **1997**, *10*, 314–323. [CrossRef]

38. Giusti, F.; Santochi, M.; Tantussi, G. A flexible tool wear sensor for NC lathes. *CIRP Ann. Manuf. Technol.* **1984**, *33*, 229–232. [CrossRef]

39. Rangwala, S.; Dornfeld, D. *Integration of Sensors via Neural Networks for Detection of Tool Wear States*; American Society of Mechanical Engineers, Production Engineering Division (Publication) PED: Boston, MA, USA, 1987; pp. 109–120.

40. Kurada, S.; Bradley, C. A machine vision system for tool wear assessment. *Tribol. Int.* **1997**, *30*, 295–304. [CrossRef]

41. Yuan, Q.; Ji, S.M.; Zhang, L. Study of monitoring the abrasion of metal cutting tools based on digital image technology. In Proceedings of the SPIE—The International Society for Optical Engineering, Beijing, China, 11–14 May 2004; pp. 397–402.

42. Wang, W.H.; Hong, G.S.; Wong, Y.S. Flank wear measurement by a threshold independent method with sub-pixel accuracy. *Int. J. Mach. Tools Manuf.* **2006**, *46*, 199–207. [CrossRef]

43. Li, P.; Li, Y.; Yang, M.; Zheng, J.; Yuan, Q. Monitoring technology research of tool wear condition based on machine vision. In Proceedings of the World Congress on Intelligent Control and Automation (WCICA), Chongqing, China, 25–27 June 2008; pp. 2783–2787.
44. Shahabi, H.H.; Ratnam, M.M. On-line monitoring of tool wear in turning operation in the presence of tool misalignment. *Int. J. Adv. Manuf. Technol.* **2008**, *38*, 718–727. [CrossRef]
45. Pfeifer, T.; Wiegers, L. Reliable tool wear monitoring by optimized image and illumination control in machine vision. *Meas. J. Int. Meas. Confed.* **2000**, *28*, 209–218. [CrossRef]
46. Barreiro, J.; Castejón, M.; Alegre, E.; Hernández, L.K. Use of descriptors based on moments from digital images for tool wear monitoring. *Int. J. Mach. Tools Manuf.* **2008**, *48*, 1005–1013. [CrossRef]
47. Alegre, E.; Alaiz-Rodríguez, R.; Barreiro, J.; Ruiz, J. Use of contour signatures and classification methods to optimize the tool life in metal machining. *EST J. Eng.* **2009**, *15*, 3–12. [CrossRef]
48. D'Addona, D.M.; Teti, R. Image data processing via neural networks for tool wear prediction. In Proceedings of the 8th CIRP International Conference on Intelligent Computation in Manufacturing Engineering, ICME, Ischia, Italy, 18–20 July 2012; Elsevier B.V.: Ischia, Italy, 2013; pp. 252–257.

Article

Fast Analytic Simulation for Multi-Laser Heating of Sheet Metal in GPU

Daniel Mejia-Parra [1,2], Diego Montoya-Zapata [1,2], Ander Arbelaiz [2], Aitor Moreno [2,*], Jorge Posada [2] and Oscar Ruiz-Salguero [1]

[1]   Laboratory of CAD CAM CAE, Universidad EAFIT, Cra 49 no 7-sur-50, 050022 Medellín, Colombia; dmejiap@eafit.edu.co (D.M.-P.); dmonto39@eafit.edu.co (D.M.-Z.); oruiz@eafit.edu.co (O.R.-S.)
[2]   Vicomtech, Paseo Mikeletegi 57, Parque Científico y Tecnológico de Gipuzkoa, 20009 Donostia/San Sebastián, Spain; aarbelaiz@vicomtech.org (A.A.); jposada@vicomtech.org (J.P.)
*   Correspondence: amoreno@vicomtech.org; Tel.: +34-943-309-230

Received: 3 October 2018; Accepted: 21 October 2018; Published: 24 October 2018

Abstract: Interactive multi-beam laser machining simulation is crucial in the context of tool path planning and optimization of laser machining parameters. Current simulation approaches for heat transfer analysis (1) rely on numerical Finite Element methods (or any of its variants), non-suitable for interactive applications; and (2) require the multiple laser beams to be completely synchronized in trajectories, parameters and time frames. To overcome this limitation, this manuscript presents an algorithm for interactive simulation of the transient temperature field on the sheet metal. Contrary to standard numerical methods, our algorithm is based on an analytic solution in the frequency domain, allowing arbitrary time/space discretizations without loss of precision and non-monotonic retrieval of the temperature history. In addition, the method allows complete asynchronous laser beams with independent trajectories, parameters and time frames. Our implementation in a GPU device allows simulations at interactive rates even for a large amount of simultaneous laser beams. The presented method is already integrated into an interactive simulation environment for sheet cutting. Ongoing work addresses thermal stress coupling and laser ablation.

Keywords: multi-beam laser; heat transfer analysis; fast simulation; GPU; analytic solution

---

1. Introduction

Multi-beam laser heating of sheet metal is a relevant metalworking technique which has arisen interest of researchers in the last years. In contrast to single-beam heating, multi-beam heating provides two main advantages to the former: (1) the ability to process different locations of the sheet simultaneously [1], and (2) control of thermal stress levels by specific multi-beam configurations [2]. Industrial applications of multi-beam heating of sheet metal include laser forming and bending, laser cutting and additive manufacturing.

Thermal simulation is crucial for temperature and stress analysis of manufactured pieces. An adequate selection of laser parameters and a correct path planning allows for improving the efficiency of the process and minimizes material damage and waste. Current simulation approaches rely on numerical schemes which require fine geometry and time discretizations. Such discretizations imply high computational costs, which limit the application of these approaches in real manufacturing scenarios with large time/space domains and complex laser trajectories [3].

This manuscript presents a simulation approach for the multi-beam laser heating problem based on an analytic solution to the heat equation on rectangular domains. This analytic solution does not require a mesh nor a fine time discretization to solve the problem. As a consequence, our algorithm is able to run complex simulations with large time/space domains and complex multi-laser trajectories at interactive time rates. Furthermore, each laser beam trajectory is allowed to be independent from the others, with different time-dependent parameters, trajectories and time frames (i.e., each laser beam can be turned on/off independently at any point of the simulation).

The remainder of this manuscript is organized as follows: Section 2 discusses the relevant literature. Section 3 presents the mathematical model and describes the implementation of the proposed algorithm. Section 4 discusses the obtained results for different test cases. Finally, Section 5 presents the conclusions and introduces the future work.

2. Literature Review

2.1. Multi-Beam Single Trajectory vs. Multiple-Trajectory Simultaneous Laser Heating

There are currently two main applications for multi-laser heating in laser machining: (1) single-trajectory multi-laser heating and (2) multi-trajectory multi-laser heating.

In single-trajectory multi-beam laser heating, a leading laser is followed by a pattern of secondary finishing ones, all in the same trajectory (or with minimum spatial offset). Experimental evidence has shown that such a configuration reduces the thermal stresses produced by the laser beams in laser cutting (compared to single-beam cutting) [4]. Furthermore, specific configurations of the laser beams have been shown to reduce the required pressure of assisting gas [5]. Each laser beam may be produced by an independent source [6] or by diffraction of a single beam source [7].

On the other hand, in multi-trajectory heating, each laser beam follows an independent trajectory [1]. Multi-trajectory heating is relevant as it improves the machining times by processing different zones of the sheet at the same time. This manuscript focuses on the simulation of multi-trajectory laser heating.

2.2. Thermal Simulation for Sheet Metal Laser Heating

The problem of laser beam heating simulation has been widely researched for single-beam applications. Numerical methods are the most common simulation approach. Methods such as the Finite Differences Method (FDM) [8–11] and the Boundary Element Method (BEM) [12,13] have been used in the literature to study the thermal behaviour of sheet laser heating. However, these methods impose several numerical limitations such as dense rectangular grids in the case of FDM and non-sparse linear systems for BEM, requiring a high amount of computational resources even for simple problems.

The Finite Element Method (FEM) is a standard numerical approach for the simulation of physical problems. Nonlinear FEM has been applied to study the thermal/stress behaviour of the single-beam laser heating of rectangular sheets [14–18]. In contrast to FDM and BEM, FEM works by discretizing the domain using different types of meshes which allow fine discretizations (high level of detail) near interest zones (laser spot and trajectory) and coarse resolutions in other zones resulting in less expensive systems of equations. To address laser ablation and material removal, methods such as element birth and death [19], volume fractions [20], temperature thresholds [21–23], and the enthalpy method [15–18,24] are coupled to the different numerical schemes.

The Finite Volume Method (FVM) has been recently introduced in the literature for the study of laser ablation and sheet heating [25,26]. Contrary to previous numerical methods, the FVM allows to accurately model and simulate interactions between the laser beam and the sheet in its physical states: solid, liquid and gas. However, these interactions are highly nonlinear and, as a consequence, computationally expensive.

Numerical methods provide tools to simulate nonlinear interactions, phase changes, laser ablation and material removal. However, they are also computationally expensive, limiting the applications of the algorithms in real manufacturing scenarios with complex laser trajectories and large space/time domains. Analytical methods do not have such limitations, allowing fast simulation of complex problems at the cost of some simplifications of the physical model. Uni-dimensional analytic models have been implemented for the simulation of laser drilling processes for static laser beams [27,28].

A 3D model for laser heating of sheet metal for straight line trajectories is presented in [29], while 2D models for arbitrary laser trajectories have been presented in [30,31].

Despite the large amount of literature concerning single-beam laser heating, simulation for multi-beam laser processing has been rarely studied. In [6,32], the FDM is used to analyze the thermal and structural impact of two independent laser beams melting a sheet metal. On the other hand, in [2,33], a thermal/stress analysis of multi-beam laser heating is performed using FEM. These simulations show that multi-beam heating reduce the thermal stresses afflicted to the material. A semi-analytic model for the steady multi-beam laser heating problem is presented in [34]. This pseudo-analytic model is also implemented inside an optimization algorithm which estimates the best laser parameters for a given manufacturing process. The use of analytic solutions is crucial for optimization due to the optimization process being expensive per se, requiring multiple evaluations of the temperature fields with different laser parameters and/or trajectories.

2.3. Conclusions of the Literature Review

Contrary to single-beam, multi-beam heating is scarce in the literature. Numerical (as opposed to analytical) methods are computationally expensive. As a consequence, these methods are unable to simulate real world sheet sizes and laser trajectories.

This manuscript offers the implementation of an analytic Fourier-based method to simulate multi-trajectory laser heating. The characteristics of the implemented method are: (a) constant material properties, (b) natural convection, (c) simplification of the sheet into a 2D domain, (d) transient (time-history) temperature, (e) multiple laser head configurations, (f) independent laser trajectories (with independent parameters), and (g) arbitrary power on/off time history on each trajectory.

Our algorithm is heavily based on the previous work presented in [31]. However, our method presents several improvements to this previous work as follows: (a) our algorithm reformulates the analytic solution to allow simultaneously multiple laser beams instead of only one. (b) Our method allows completely asynchronous laser beams (independent parameters, trajectories and time frames) which is not a problem for single-beam approaches and has not been addressed in other multi-beam related work. (c) Instead of computing and storing the full temperature history, our implementation exploits the frequency-based solution approach by interactively providing the temperature only when requested by the user (even in non-monotonic order). (d) Computing the solution in the frequency domain allows arbitrary sheet discretizations for visualization and analysis, including triangular meshes, rectangular grids, polylines or sets of sampled points on the sheet. (e) Our GPU implementation of the algorithm using on-chip (local) memory presents an increased $10x$ speed-up on computation time over previous work.

3. Methodology

3.1. Heat Transfer Equation for Multi-Beam Laser Heating

The temperature distribution $u = u(\vec{x}, t)$ of a 2D rectangular sheet satisfies the following partial differential equation:

$$\rho c_p \frac{\partial u}{\partial t} - \nabla \cdot (\kappa \nabla u) = \frac{F - q}{\Delta z},$$
$$q = h(u - u_\infty),$$

(1)

where ρ, c_p and κ are the material density, specific heat and thermal conductivity, respectively. $\nabla \cdot \nabla = \partial^2/\partial^2 x + \partial^2/\partial^2 y$ is the 2D Laplace operator. Δz is the thickness of the sheet. $F = F(\vec{x}, t)$ is the set of surface heat sources affecting the sheet, and $q = q(u)$ is the temperature-dependent convection on the sheet surface with its respective convection coefficient $h \geq 0$ and ambient temperature $u_\infty \in \mathbb{R}$. For the previous partial differential equation, the following boundary and initial conditions are imposed on the sheet:

$$u|_{x=0} = u|_{x=a} = u|_{y=0} = u|_{y=b} = u_\infty,$$
$$u(\vec{x}, 0) = u_\infty. \tag{2}$$

For the multi-beam approach, the set of heat sources F is defined as the sum of the heat inputs of each laser beam f_k:

$$F(\vec{x}, t) = \sum_{k=1}^{num\_lasers} f_k(\vec{x}, t),$$

$$f_k(\vec{x}, t) = \begin{cases} \frac{P_k(1-R)}{\pi r_k^2}, & \|\vec{x} - \vec{x}_0^k(t)\|_\infty < \frac{r_k\sqrt{\pi}}{2}, \\ 0, & \text{otherwise}, \end{cases} \tag{3}$$

where $P_k \geq 0$ is the power of the laser beam, $r_k > 0$ is the radius of the laser spot and $\vec{x}_0^k(t) \in \mathbb{R}^2$ is the center of the laser spot at time t. $0 \leq R < 1$ is the reflectivity of the material. $\|\vec{x}\|_\infty = \max(x, y)$ is the infinity norm in $\mathbb{R}^2$. The above laser model transforms the circular laser spot with radius r_k to its equivalent squared spot with area πr_k^2. We apply such transformation in order to develop the analytic solution in the next section. Figure 1 presents a scheme of the multi-beam laser heating problem.

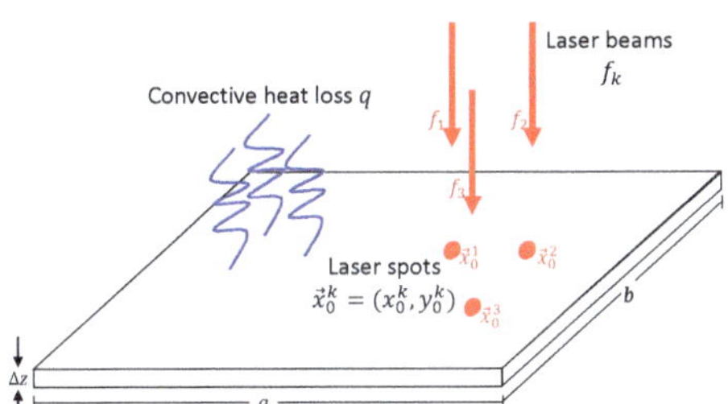

Figure 1. Multi-beam laser heating scheme. The sheet surface is heated by a set of laser beams $f_1, f_2, \ldots$ and cooled down due to natural convection q.

3.2. Analytic Solution

Following the same procedure as in [31], the temperature u on the rectangular sheet can be expressed as a linear combination of Fourier functions:

$$u(\vec{x}, t) = u_\infty + \sum_{i=1}^{\infty} \sum_{j=1}^{\infty} \Theta_{ij}(t) \mathbb{X}_i(x) \mathbb{X}_j(y), \tag{4}$$

with basis:

$$\mathbb{X}_i(x) = \sin\frac{i\pi x}{a},$$
$$\mathbb{Y}_j(y) = \sin\frac{j\pi y}{b}. \tag{5}$$

Applying separation of variables, the Fourier coefficients Θ_{ij} from Equation (4) can be expressed as a sum of the pseudo-coefficients θ_{ij}^k of each independent heat source:

$$\Theta_{ij}(t) = \sum_{k=1}^{num\_lasers} \theta_{ij}^k(t),\tag{6}$$

where each laser beam f_k defines its respective pseudo-coefficient θ_{ij}^k as follows:

$$\theta_{ij}^k(t) = \theta_{ij}^k(t_0)e^{-\omega_{ij}(t-t_0)} + \frac{4}{ab\rho c_p\Delta z}\int_{t_0}^t\int_0^b\int_0^a f_k(\vec{x},\tau)\mathbb{X}_i(x)\mathbb{Y}_j(y)e^{-\omega_{ij}(t-\tau)}dxdyd\tau\tag{7}$$

with corresponding Laplacian eigenvalues:

$$\omega_{ij} = \frac{\kappa}{\rho c_p}\left(\frac{i^2\pi^2}{a^2}+\frac{j^2\pi^2}{b^2}\right) + \frac{h}{\rho c_p\Delta z}.\tag{8}$$

Equation (7) is written recursively in terms of a previous known solution $\theta_{ij}^k(t_0)$. Therefore, each laser trajectory is discretized as a piecewise linear trajectory $\vec{x}_0^k = [\vec{x}_0^k(0), \vec{x}_0^k(t_1), \vec{x}_0^k(t_2), \ldots]$. In contrast to previous work [31], Equations (3), (6) and (7) have been reformulated to account for the multiple asynchronous laser beams. Finally, the closed form for the integral term in Equation (7) has been already presented in [31].

3.3. Algorithm

Figure 2 presents a diagram of the simulation algorithm based on the analytic solution presented in Section 3.2. Each step of the algorithm is discussed in detail below:

1. **Discretize laser trajectories:** As discussed in Section 3.2, the laser beam trajectories $\vec{x}_0^k(t)$ are discretized as sequences of piecewise linear trajectories $[\vec{x}_0^k(0), \vec{x}_0^k(t_1), \ldots, \vec{x}_0^k(T_f)]$, as described in [31]. The only requirement for this discretization is the fact that all laser beam trajectories must share the same time discretization, i.e., $t_0, t_1, \ldots, T_f$ are the same for all trajectories $\vec{x}_0^k$.

2. **Compute the Laplacian eigenvalues:** The Laplacian eigenvalues of the sheet are computed as per Equation (8). Since the eigenvalues ω_{ij} are time-independent, this step is performed before the simulation loop starts.

3. **Initialize time and sheet temperature:** The simulation time is initialized to $t_0 \leftarrow 0$. In order to satisfy the initial temperature condition $u(0) = u_\infty$, the pseudo coefficients are initialized as $\theta_{ij}^k(0) = 0$ (see Equation (4)).

4. **Update current time t:** The current simulation time $t = t_{l+1}$ is updated according to the previous time $t_0 = t_l$, in concordance with the discretization of trajectories from step 1.

5. **For each laser beam k:** This inner loop computes the pseudo-coefficients $\theta_{ij}^k(t)$ for each laser beam ($k = 1\ldots num\_lasers$).

6. **Question: Is laser beam k turned on?:** This step allows for simulating asynchronous laser beams by asking at the current time t if the laser is turned on/off. Therefore, each laser beam might have its own internal time frame $[t_0^k, t_f^k]$, different from the simulation time frame $[0, T_f]$.

7. **Set power P_k/Set null power $P_k \leftarrow 0$:** In the previous step, the program checks the state (on/off) of the current laser beam k. The laser is turned on by the simulation by setting its corresponding power input P_k. In the case of the laser being turned off, the simulation simply sets its power to 0.

8. **Compute the pseudo-coefficients for each laser beam:** The pseudo-coefficients in Equation (7) are solved analytically for each laser beam as described in [31]. The number of coefficients computed is truncated to num_coeffs.

9. **Question: $k < num\_lasers$?:** Check if the pseudo-coefficients have been computed for all the laser beams.

10. **Compute the Fourier coefficients as per Equation (6):** This step computes the real Fourier coefficents $\Theta_{ij}(t)$ from the pseudo-coefficients $\theta_{ij}^k(t)$ as per Equation (6).

11. **If required, compute temperature:** The temperature field is computed on a set of discrete points sampled on the sheet $[(x_0, y_0), (x_1, y_1) \ldots]$ as per Equation (4). The number of coefficients used to compute the solution is truncated to num_coeffs. This step is optional since the result may be stored in the frequency domain ($\Theta_{ij}(t)$). Therefore, the temperature is made available only when requested by the user, allowing for skipping iterations of no interest and allowing non-monotonic access to the transient temperature history, improving the performance in the process.

12. $t < T_f$**?:** Check if the simulation has reached the final step.

13. **END SIMULATION**

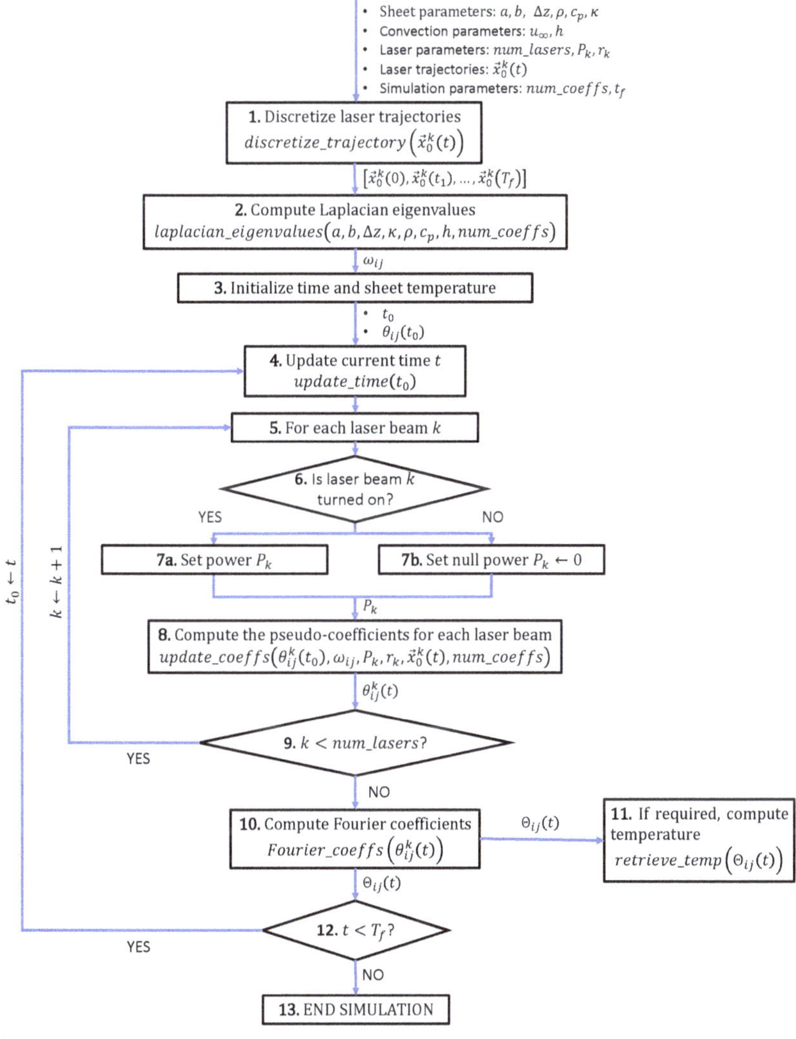

Figure 2. Diagram of the analytic multi-laser simulation algorithm.

There are several concepts in the previous algorithm that help to improve its efficiency, crucial for the interactive nature of the simulation:

1. In step 1 of the previous algorithm, the curved laser beam trajectories $\vec{x}_0^k(t)$ are discretized as piecewise linear ones $[\vec{x}_0^k(0), \vec{x}_0^k(t_1), \ldots, \vec{x}_0^k(T_f)]$, which inherently produces the time discretization $[0, t_1, \ldots, T_f]$. This time discretization does not affect the numerical accuracy of the temperature solution. Therefore, as opposed to FEA (Finite Element Analysis), the time step size $\Delta t = t_{l+1} - t_l$ of our algorithm can be arbitrarily large.

2. Step 6 allows turning on/off any laser beam at any point of the simulation, allowing complete asynchronicity between the multiple laser beams. In addition, the algorithm allows to change any laser parameters at will, resulting in time-dependent parameters $P_k(t)$ (laser power) and $r_k(t)$ (spot radius). For simplicity, this manuscript uses constant parameters.

3. The complete information of the solution is stored in the frequency domain (step 10) and temperature data is computed only when requested. Therefore, the user requests the temperature only at specific times and in specific zones (i.e., at the middle or the end of the simulation). This frequency-based approach has the advantage of providing non-monotonic access to the temperature history, allowing arbitrary simulation time-location requests. Furthermore, since the space discretization does not affect the solution, any sheet sampling can be used to render the temperature (rectangular grid, triangular mesh, a curve or a single point in the sheet).

4. In step 10, each pseudo-coefficient θ_{ij}^k is independent from the rest of the pseudo-coefficients (Equation (6)). Similarly, in step 11, the temperature value u at a given point $\vec{x}$ is independent from the temperature on the rest of the sheet (Equation (4)). Therefore, the computation in both of these steps is parallelized.

4. Results

This section presents the simulation results of our algorithm. For all the simulations, a Fourier discretization of 512×512 coefficients and a spatial (grid) discretization of 512×512 points are used. These resolutions have been chosen as they have shown in our experiments enough accuracy within our desired execution time ranges (see Sections 4.1 and 4.3, respectively). As the ratio $\frac{\max(a,b)}{\min_k(r_k)}$ (sheet size vs. spot size) increases, it could be necessary to increase the number of Fourier coefficients. However, a sensitivity analysis of our algorithm's discretization is out of the scope of this manuscript.

Table 1 presents the geometric and physical parameters of the sheet used in the simulations. All of the experiments presented in this manuscript do not consider the effect of heat reflected by the sheet ($R = 0$). The first study case presents two different laser beams heating the sheet simultaneously. The first laser beam follows a star-shaped trajectory on the sheet while the second laser beam follows a rectangular trajectory. Table 2 presents the parameters of each laser beam. As discussed in Section 3, our algorithm not only enables different parameters for multiple lasers (path, power, speed, spot radius), but it also allows independent time frames for each trajectory. Parameter values for Tables 1 and 2 are taken from state-of-the-art literature [14,31,35]. However, laser beam power has been down-scaled to account for sheet heating instead of cutting or bending.

Figure 3 plots the simulation results for two laser beams heating the sheet surface. Figure 3a shows the star and square laser trajectories planned on the sheet. The laser beam parameters for each trajectory are described in Table 2. At the beginning of the simulation, a unique laser beam (star) heats the surface (Figure 3b). As discussed in Section 3.3, our algorithm allows independent time frames for the multiple laser beams. Therefore, the second laser is introduced in the simulation at $t = 0.065$ s (Figure 3c). Figure 3d plots the temperature when the two lasers are near each other (the trajectories do not intersect). Finally, both laser beam trajectories end at different steps: $t_f^k = 0.13$ s for the square trajectory (Figure 3e) and $t_f^k = 0.16$ s for the star trajectory (Figure 3f).

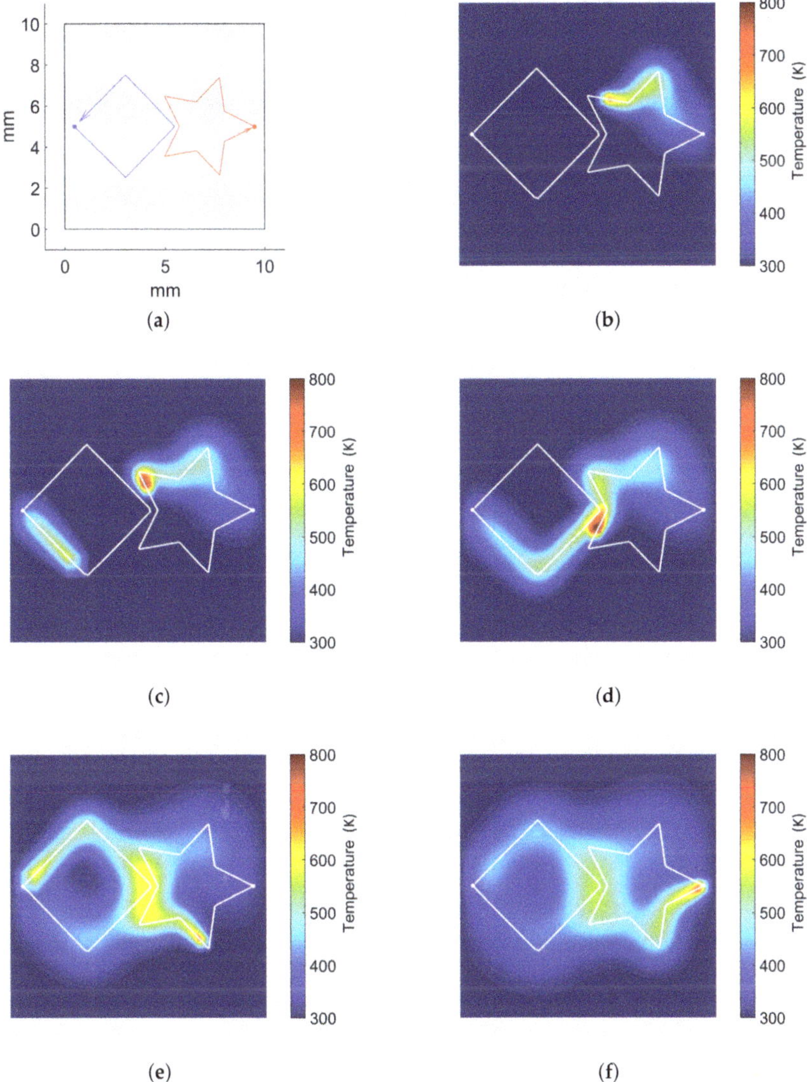

Figure 3. Laser trajectories and sheet temperature history for simultaneous diverse shape laser trajectories. (**a**) laser trajectories for the simulation; (**b**) $t = 0.060$ s. Only Star trajectory occurs; (**c**) $t = 0.075$ s. Square trajectory enters the simulation; (**d**) $t = 0.094$ s. Lasers are simultaneously near each other; (**e**) $t = 0.13$ s. Square trajectory finishes. Star trajectory continues; (**f**) $t = 0.166$ s. Star trajectory finishes.

Table 1. Sheet heat transfer parameters for the simulations (same as in [14,31,35]).

Parameter	Description	Value
Geometry		
a	width	0.01 m
b	height	0.01 m
Δz	thickness	0.001 m
Material	**AISI 304 Steel**	
ρ	density	8030 kg/m^3
c_p	specific heat	574 J/(kg $\cdot$ K)
κ	thermal conductivity	20 W/(m $\cdot$ K)
R	reflectivity	0
Convection Type	**Natural**	
h	convection coefficient	20 W/(m$^2 \cdot$ K)
u_∞	ambient temperature	300 K

Table 2. Parameters for the simultaneous laser heating trajectories of Figure 3a (similar to [14,31,35]).

Parameter	Description	Star Trajectory	Square Trajectory
t_0^k	start time of trajectory	0.00 s	0.06 s
t_f^k	end time of trajectory	0.16 s	0.13 s
P_k	power	100 W	200 W
r_k	spot radius	0.0003 m	0.0005 m
$\|\vec{v}_k\|$	trajectory speed	0.1 m/s	0.2 m/s

4.1. Comparison with FEA

To validate the analytic approach, FEA simulation is performed. The FEA linear system for Equation (1) becomes [31]:

$$\left[\left(\frac{\rho c_p}{\Delta t} + \frac{h}{\Delta z}\right)\mathbf{M} + \kappa\mathbf{L}\right]\mathbf{U}^{(t+\Delta t)} = \mathbf{M}\left(\frac{\rho c_p}{\Delta t}\mathbf{U}^{(t)} + \frac{1}{\Delta z}\sum_{k=1}^{num\_lasers}\int_t^{t+\Delta t}\mathbf{F}_k^{(\tau)}d\tau + \frac{h}{\Delta z}u_\infty\right), \quad (9)$$

where $\mathbf{U}$ and $\mathbf{F}_k$ are the vectors of temperature and heat sources sampled on the sheet, and $\mathbf{L}$ and $\mathbf{M}$ are the stiffness and mass matrices, respectively [31].

The software ANSYS® Academic Research Mechanical, Release 17.2, is used to perform the FEA simulations. ANSYS® element SHELL131 is employed. Element thickness and material properties are set as per Table 1. The elements are configured to have a constant temperature along the thickness and to evaluate convection at the sheet surface, as per Equation (1). To represent the area heated by the laser beams at every time step, ANSYS® surface loads (Heat Fluxes) are applied on the FEA elements that lie inside the heated zone.

Figure 4 plots the FEA results for the study case presented in Table 2. The mesh of the domain is generated so that it is more dense in the neighborhoods of the laser trajectories. Figure 4a shows the initial triangular mesh computed using the FEA software which is then refined several times (Figure 4b) to improve the numerical accuracy of the solution. After five re-meshing iterations, the final mesh contains 126 k triangles and 63 k nodes. Figure 4c plots the FEA temperature at the moment the two laser beams get closer to each other ($t = 0.094$ s). The temperature peak is in the middle of the two trajectories (Figure 4d), due to both paths not intersecting each other.

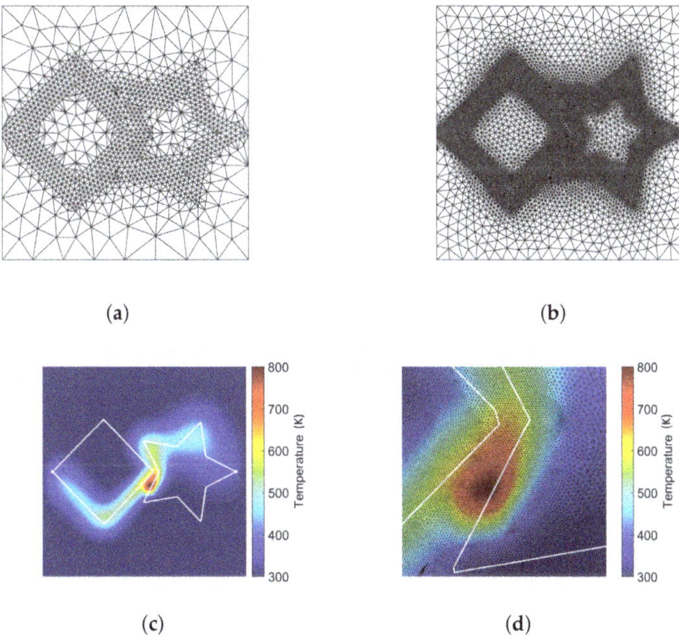

(a)   (b)

(c)   (d)

Figure 4. ANSYS® FEA simulation results for the same test case presented in Figure 3. (**a**) coarse FEA mesh (2.1 k triangles); (**b**) intermediate mesh refinement at laser trajectory (13 k triangles); (**c**) $t = 0.094$ s. FEA temperature; (**d**) $10x$ zoom on Figure 4c.

Figure 5 plots the relative error distribution of our analytic solution (Figure 3d) with respect to FEA (Figure 4c). The maximum relative error is 3.9%, located around the laser spots (Figure 5b). This small deviation between our method and FEA arises because: (a) the number of Fourier coefficients used in implementation of Equation (4) affect our algorithm precision, (b) the precision of the FEA algorithm is dependent on mesh and time resolution, element type, etc., and (c) the FEA approach approximates the laser spots using FEA elements at each time step, not preserving the exact geometry of the laser spots. The small square shape of the error in Figure 5b is due to the squared laser model presented in Equation (3).

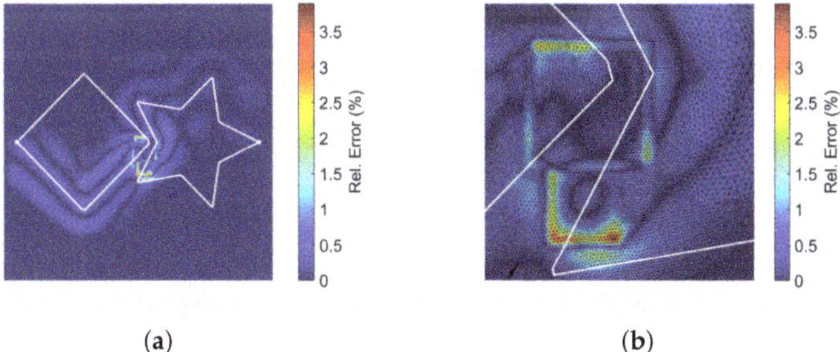

(a)   (b)

Figure 5. Appraisal of our analytic solution (Figure 3d) vs. the FEA solution (Figure 4c). (**a**) relative error. Analytic solution w.r.t. FEA (max. relative error: 3.9%); (**b**) $10x$ zoom on Figure 5.

4.2. Multiple Laser Beams

The algorithm introduced in this manuscript allows multiple laser beams heating the surface at the same time. This section presents additional experiments with more than just two laser beams. Figure 6 presents a study case with four simultaneous laser beams drawing different shapes on the sheet: a square, a star, a spiral and a circle trajectory (Figure 6a). All of the laser beams have the same parameters (power and radius) and start at the same time (Figure 6b). However, they do not finish at the same time (Figure 6c).

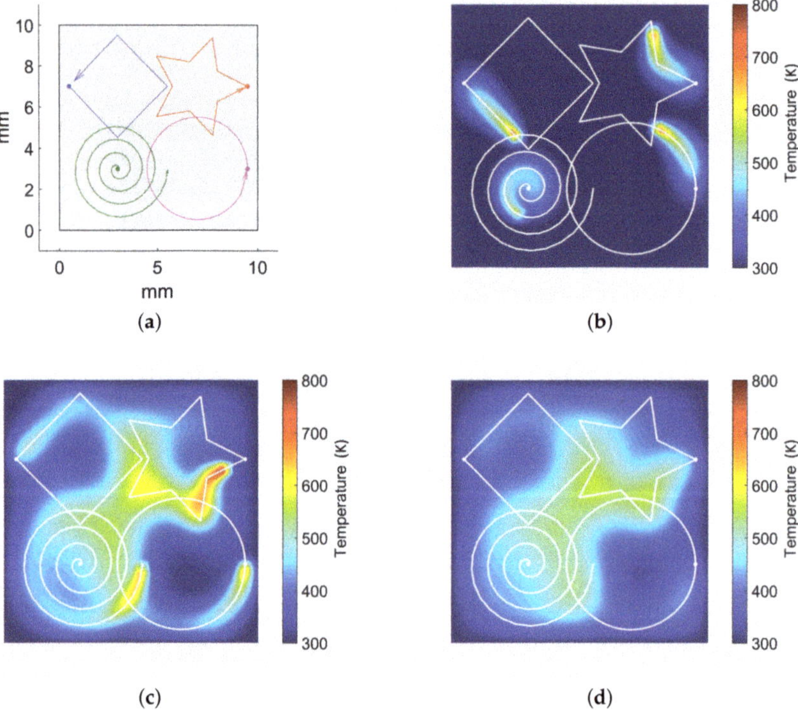

Figure 6. Laser trajectories and temperature history for time and space overlapping trajectories. (**a**) planned laser trajectories: Square, Star, Spiral, Circle; (**b**) $t = 0.03$ s. All trajectories start at $t = 0$ s; (**c**) $t = 0.158$ s. Square and Circle trajectories finished; (**d**) $t = 0.2$ s. All trajectories finished. The sheet cools down.

As discussed in Section 3.3, our algorithm allows for defining different independent time frames for each laser beam by turning on/off specific laser beams. Such approach even allows to turn off all the laser beams and continue the simulation. Figure 6d illustrates this approach by continuing the simulation after all the laser beams have finished their trajectories, where only thermal conduction and thermal convection are taken into account.

In order to visually compare our algorithm with other simulation approaches in the literature, the study case presented in [33] is replicated in this manuscript (Figure 7). In this study case, seven simultaneous lasers are distributed uniformly on the y-axis. Each laser beam follows a straight line trajectory and together they draw an arc in the heating front (Figure 7a,b). Our algorithm is able to produce similar results to [33] despite the simplification of the analytic model (Figure 7c,d).

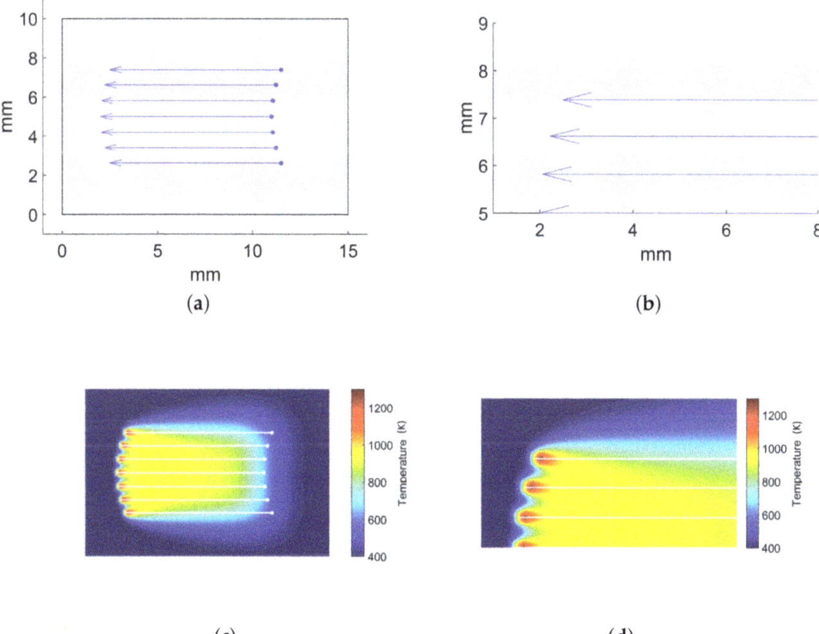

Figure 7. Laser trajectories and temperature results for the simulation case presented in [33]. Similar simulation parameters have been used to replicate the experiment. (**a**) complete sheet. Simultaneous linear trajectories based on [33]; (**b**) laser trajectories (zoom near laser spots); (**c**) temperature map (full sheet); (**d**) temperature map (zoom).

Figure 8 plots the temperature distribution along the arc for different number of laser beams. As the number of laser beams increase (and the arc length remains the same), the oscillations of the temperature are mitigated and the temperature increases. Such a result is consistent with the analysis presented in [33]. As discussed in Section 3.3, our algorithm allows for computing easily the temperature along the arc without even requiring to calculate the temperature on the rest of the sheet, which improves the computation time for this particular case.

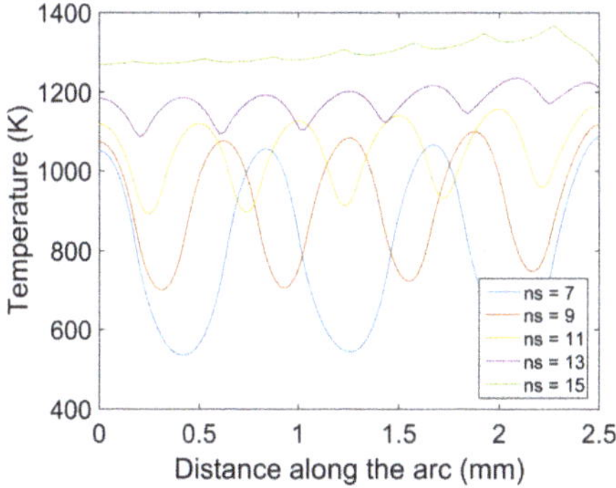

Figure 8. Temperature along the arc for the simulation presented in Figure 7. Our results are similar to those in [33]. *ns* represents the number of lasers.

4.3. Performance Assessment

This section evaluates the performance of our algorithm under hardware-accelerated (GPU) and non-accelerated (CPU) platforms. The presented algorithm has been implemented using the OpenCL framework to create an optimized solution that targets both CPU and GPU. On systems where only a CPU is available, our implementation makes use of multi-core parallelization and vectorization to speed-up computation. On systems where a GPU is available, the memory hierarchy can be explicitly controlled using the OpenCL API. The workload is divided into small groups, in order to exploit reuse of computed data using local memory. In this manner, a high speed-up is achieved due to both efficient use of memory and massive parallelization.

The performance results have been measured with the following test platform: A desktop PC using Windows 10 OS with an Intel® Core™ i5-6500 (CPU), 8 GB RAM and NVIDIA® GeForce® GTX 960 graphics card. Our algorithm is able to simulate any number of laser beams. Figure 9 plots the execution times for the computation of the Fourier coefficients as a function of the number of laser beams. The figure compares the execution times of the CPU against the GPU to compute 512×512 Fourier coefficients. The computational cost increases with a large slope in the CPU approach while being nearly constant in the GPU approach. The more laser beams are added, the more it benefits from GPU parallelization. In addition, the computation of the Fourier coefficients in the GPU is preferable. In this way, there is no need to transfer the coefficients back and forth from host-to-device on each iteration since they always stay in GPU memory. For this analysis, a single time step has been considered instead of the whole simulation.

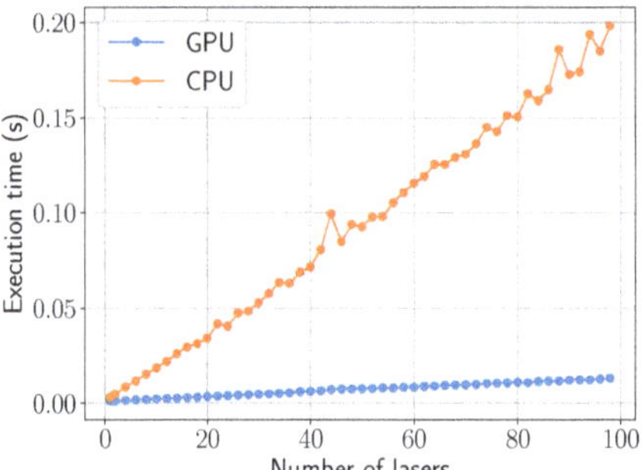

Figure 9. Comparison of CPU and GPU execution times (s) for the computation of 512×512 Fourier coefficients as the number of lasers increase.

Figure 10 shows the computation times of the temperature retrieval in a mesh grid as a function of the grid size. While the curves for Fourier resolutions (number of coeffs) below 512×512 display a computational cost relatively low (<0.5 s) for any spatial resolution, Fourier resolutions above that point (1024×1024) become expensive (>0.5 s) for our simulation purposes. Therefore, on our test platform, we have observed that a good balance between accuracy and computation time can be achieved by setting a resolution of 512×512 for both the frequency (Fourier) and spatial (grid) domains. Compared to the previous work [31], a 10x speed-up is achieved at the temperature evaluation by using local GPU memory.

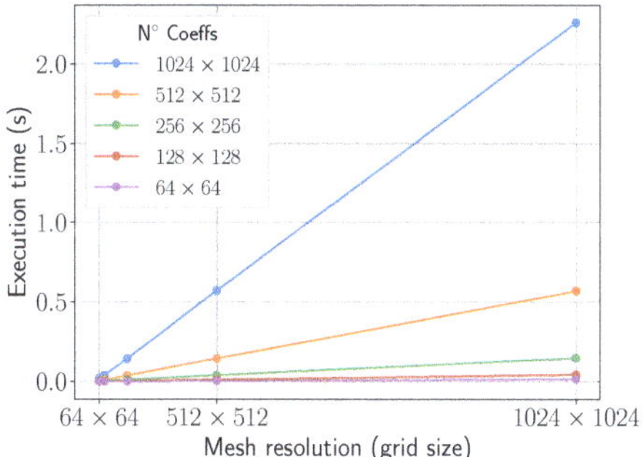

Figure 10. Execution times (s) for the computation of the temperature as the resolution (mesh size) increases. Results for different number of Fourier coefficients are presented.

Additionally, since the computation of the temperature is not compulsory at each iteration of the algorithm (as discussed in Section 3.3), the simulation may ignore irrelevant time steps (defined by the user). Moreover, the independence of the algorithm accuracy from the spatial discretization allows the user to specify specific domains (e.g., a sheet section, a curve or a single point) without requiring the whole sheet temperature. These aspects improve the computation efficiency of the simulation for specific requirements of the user.

4.4. Integration within an Interactive Laser Cutting Simulation Environment

Nowadays, multi-laser machines with multi-trajectory capabilities do not represent a significant share of the market. The current state-of-the-art laser cutting machines can be divided in three main groups: (1) multi-laser machines with independent and disjoint working areas for each laser head, (2) multi-laser machines with parallel heads working simultaneously and (3) multi-laser machines with fully individual and autonomous laser heads (whose trajectories may intersect or get close enough). The first scenario can be reduced to a collection of mono-laser machining scenarios, since the individual working areas for each laser head avoid interference with the neighbouring laser heads and their heat effects. Therefore, the approach introduced in [31] is still valid to simulate the temperature for each laser head of these multi-laser machines.

In the second type of multi-laser machines, all laser heads receive the same trajectories and machining commands, but their separation or offset can be setup and reconfigured during the machining. Additionally, each laser head can be switched off and on individually. The offset between the heads does not guarantee disjoint working zones. Therefore, the simulation approach presented in this work is used to address the potential interference between the multiple heat sources.

The last type of multi-laser machines use mirrors and lenses to move the spot of the available laser heads. Each laser spot can be commanded independently of the others, even allowing two or more spots to converge at the same physical point on the sheet metal. The methodology presented in this work is used to simulate such scenarios.

In the context of the second scenario, this section presents an interactive simulator of a laser cutting machine with three laser heads. All laser heads are arranged side by side in the bridge of the machine (x-axis of the machine). Each individual laser head can be switched on and off individually and their relative positions among them (y-axis of the machine) can be changed by means of specific machining instructions.

The virtual simulator itself uses a contour-based representation to model the geometry of the processed sheet [36,37], presenting a virtual 3D interactive scenario with the multi-laser CNC (Computer Numerical Control) machine that receives the machining instructions. In this virtual scene, the moving and cutting instructions move the corresponding axis of the machines (bridge, laser heads offset along the bridge and laser head height over the sheet metal). The heat sources are then calculated and sent to the heat simulation procedure, updating the temperature of the sheet, which is rendered as a texture over the virtual sheet metal.

Since all the laser heads receive the same machining instructions (although their position along the bridge differ), all movements start and end at the same time. As discussed in Section 3.3, the laser beam trajectories must share the same time discretization, which in this case, is guaranteed by design. If a unrestricted multi-laser machine is used with independent laser heads, i.e., receiving different machining instructions, a global time clock must be used in order to trigger the update of the temperature computation with the correct positions of the moving laser heads.

The introduction of the multi-laser machines in the industrial world has an impact from the design point of view. Even with just one laser, the designer of the NC programs must take into account the expected produced heat and how it spreads over the sheet metal in order to optimize the nesting of the produced parts. With a multi-laser machine, this procedure is even more critical, as the resulting heat of the multiple sources can accumulate in some areas. NC designer is expected to use the presented virtual simulator to visualize how the cutting process produces the parts and to analyze the computed temperature through the sheet. If any situation becomes problematic, the designer would make changes to the NC program in order to address the situation.

The presented multi-laser simulator runs at interactive rates. Therefore, the designer can improve the optimization workflow, as many simulations can be run in a short period of time. In contrast, high quality simulations with FEA software, although being highly precise, are computationally expensive. Thus, the number of test configurations that the designer can test during the design phase is limited. Nevertheless, at the end of the optimization phase, the interactive heat simulations can be complemented with high quality FEA simulations.

From the industrial point of view, this improved design workflow, assisted with the interactive multi-laser simulation, (1) provides better NC programs in terms of the quality of the produced parts, (2) produces economical benefits due to shorter machining times or less wasted resources, and (3) improves the safety of the operators in the factory floor, reducing unnecessary risks due to unexpected behaviour of the NC programs.

Figure 11a shows a machine with three laser heads working simultaneously, i.e., they receive the same machining instructions with fixed spatial offsets). Each laser head machines a star figure. The stars machined by the first and second laser heads overlap, while the third laser head produce an isolated star figure. Figure 11b shows a closer view of the cutting area while Figure 11c shows the same viewpoint, but the cut stars removed from the visualization. The simulation results show that intersecting trajectories present temperature peaks (black zones) where the trajectories overlap (see Figure 11b).

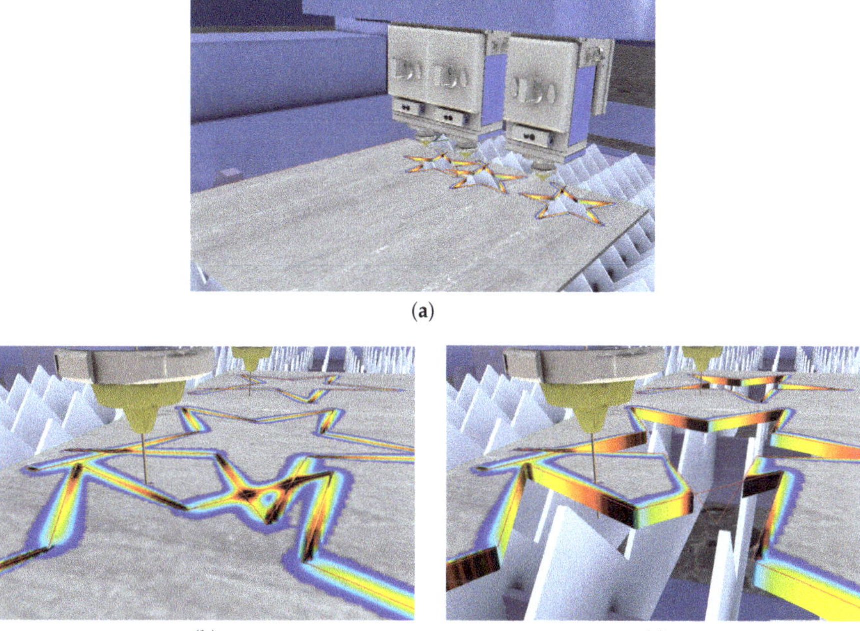

Figure 11. Multi-laser machining in the interactive simulator. A star shape is machined by the three laser heads. (**a**) virtual multi-laser machine; (**b**) temperature shown on the sheet surface; (**c**) geometric cut of the sheet.

5. Conclusions

This manuscript presents a novel methodology for the simulation of heat transfer on rectangular sheet metal under multi-laser beam heating. The algorithm is based on an analytic frequency-based solution to the heat transfer equation which considers some simplifications (2D domain, constant material parameters) in favor of simulation speed. Such simplifications allow the algorithm to solve the transient temperature map on large space/time domains and complex laser trajectories. Furthermore, the algorithm allows many simultaneous laser beams with independent parameters (laser power, speed and spot radius) and time frames (i.e., each laser beam can be turned on/off during the simulation). The numerical accuracy of the algorithm is independent from the space/time discretization, allowing non-monotonic access to the temperature time history and arbitrary discretization of the sheet domain. Our simulation tests show that the algorithm is able to render correctly the temperature maps for several laser beams with different trajectories using mesh grids. A numerical comparison with FEA shows that our algorithm solution deviates from the FEA one a maximum of 3.9% and only around the laser spot. An assessment of the algorithm performance shows that, in an implementation using GPUs, the number of laser beams barely affects the simulation time. In addition, compared to previous work [31], the temperature retrieval time has been reduced by a factor of ten. Finally, the algorithm is implemented into an interactive laser cutting simulation environment for the assessment in real time of laser cutting processes in real manufacturing scenarios.

The presented algorithm simplifies by design the mathematical model of the problem in favor of interactive simulations. As future work, we work on: (1) simulating the nonlinear behaviour of the material properties which arises due to the high temperature gradients, (2) simulating physically the laser ablation and material removal in sheet cutting processes, (3) coupling the model with an analytic stress model in order to evaluate the potential structural damage due to thermal stress, and (4) studying the sensitivity of the Fourier discretization with respect to parameters such as sheet size and radii of the laser spots.

Author Contributions: D.M.-P., D.M.-Z., A.A. and A.M. conceived, designed and performed the simulations and wrote the manuscript. Since August 2015, J.P. and O.R.-S. have supervised the Heat Transfer, Computational Geometry, and Applications aspects and reviewed/edited the manuscript.

Funding: This research was funded by the Basque Government Industry Department under the ELKARTEK program (Project "LANAII", KK-2016/00052).

Acknowledgments: The authors thank Lantek Business Solutions and Lantek Investigación y Desarrollo for providing the laser cutting machining NC programs used in this work.

Conflicts of Interest: The authors declare no conflict of interest.

Abbreviations

The following abbreviations are used in this manuscript:

FEA/FEM	Finite Element Analysis/Finite Element Method.
GPU	Graphics Processing Unit.
$a, b, \Delta z$	Width, height and thickness of the sheet (m).
$\vec{x}, t$	Spatial $\vec{x} = (x, y) \in [0, a] \times [0, b]$ and temporal $0 \leq t \leq T_f$ (s) coordinates for the simulation.
$u = u(\vec{x}, t)$	Temperature distribution along the sheet at any given time (K).
ρ	Sheet density (kg/m^3).
c_p	Sheet specific heat (J/(kg K)).
κ	Sheet thermal conductivity (W/(m K)).
R	Sheet reflectivity ($0 \leq R < 1$).
$q = q(u)$	Temperature-dependent heat convection on the sheet surface (W/m^2).
u_∞	Ambient temperature K.
h	Natural convection coefficient (W/(m^2 K)).
f_k	Heat input from laser beam k (W/m^2). $k = 1 \ldots num\_lasers$.
P_k	Power of laser beam k (W).
r_k	Radius of laser spot k (m).
$\vec{x}_0^k = \vec{x}_0^k(t)$	Laser spot center for laser beam k at time t.
$[t_0^k, t_f^k]$	Simulation time frame in which the laser beam k remains turned on. $0 \leq t_0^k < t_f^k \leq T_f$.
$\vec{v}_k$	Scan speed of laser beam k (m/s).
$F = F(\vec{x}, t)$	Sum of all laser beam heat sources (W/m^2).
$\mathbb{X}_i = \mathbb{X}_i(x)$	Fourier basis function associated to the x coordinate. $i = 1 \ldots \infty$.
$\mathbb{Y}_j = \mathbb{Y}_j(y)$	Fourier basis function associated to the y coordinate. $j = 1 \ldots \infty$.
$\Theta_{ij} = \Theta_{ij}(t)$	Fourier coefficient associated to basis functions $\mathbb{X}_i$ and $\mathbb{Y}_j$.
$\theta_{ij}^k = \theta_{ij}^k(t)$	Pseudo-Fourier-coefficient associated to the k-th laser beam source.
ω_{ij}	ij-th eigenvalue of the heat operator (Laplacian) on the rectangular sheet.

References

1. Wiesner, A.; Schwarze, D. Multi-Laser Selective Laser Melting. In Proceedings of the 8th International Conference on Photonic Technologies LANE 2014, Fürth, Germany, 8–11 September 2014; pp. 1–3.
2. Shuja, S.; Yilbas, B. Laser multi-beam heating of moving steel sheet: Thermal stress analysis. *Opt. Lasers Eng.* **2013**, *51*, 446–452. [CrossRef]
3. Mejia, D.; Moreno, A.; Ruiz-Salguero, O.; Barandiaran, I. Appraisal of open software for finite element simulation of 2D metal sheet laser cut. *Int. J. Interact. Des. Manuf.* **2017**, *11*, 547–558. [CrossRef]
4. Abe, F.; Osakada, K.; Shiomi, M.; Uematsu, K.; Matsumoto, M. The manufacturing of hard tools from metallic powders by selective laser melting. International symposium on advanced forming and die manufacturing technology. *J. Mater. Process. Technol.* **2001**, *111*, 210–213. [CrossRef]
5. Kaakkunen, J.J.J.; Laakso, P.; Kujanpää, V. Adaptive multibeam laser cutting of thin steel sheets with fiber laser using spatial light modulator. *J. Laser Appl.* **2014**, *26*, 032008. [CrossRef]

6. Heeling, T.; Zimmermann, L.; Wegener, K. Multi-Beam Strategies for the Optimization of the Selective Laser Melting Process. In *Solid Freeform Fabrication 2016, Proceedings of the 27th Annual International Solid Freeform Fabrication Symposium, Austin, TX, USA, 8–10 August 2016*; The University of Texas Press: Austin, TX, USA, 2016; pp. 1428–1438. [CrossRef]

7. Olsen, F.O.; Hansen, K.S.; Nielsen, J.S. Multibeam fiber laser cutting. *J. Laser Appl.* **2009**, *21*, 133–138. [CrossRef]

8. Modest, M. Three-dimensional, Transient model for laser machining of ablating/decomposing materials. *Int. J. Heat Mass Transf.* **1996**, *39*, 221–234. [CrossRef]

9. Modest, M. Laser through-cutting and drilling models for ablating/decomposing materials. *J. Laser Appl.* **1997**, *9*, 137–145. [CrossRef]

10. Han, G.; Na, S. A study on torch path planning in laser cutting processes part 1: Calculation of heat flow in contour laser beam cutting. *J. Manuf. Syst.* **1999**, *18*, 54–61. [CrossRef]

11. Xu, W.; Fang, J.; Wang, X.; Wang, T.; Liu, F.; Zhao, Z. A numerical simulation of temperature field in plasma-arc forming of sheet metal. *J. Mater. Process. Tech.* **2005**, *164–165*, 1644–1649. [CrossRef]

12. Kim, M. Transient evaporative laser-cutting with boundary element method. *Appl. Math. Model.* **2000**, *25*, 25–39. [CrossRef]

13. Kim, M. Transient evaporative laser cutting with moving laser by boundary element method. *Appl. Math. Model.* **2004**, *28*, 891–910. [CrossRef]

14. Yilbas, B.S.; Arif, A.F.M.; Abdul Aleem, B.J. Laser cutting of rectangular blanks in thick sheet steel: Effect of cutting speed on thermal stresses. *J. Mater. Eng. Perform.* **2010**. [CrossRef]

15. Akhtar, S. Laser cutting of thick-section circular blanks: Thermal stress prediction and microstructural analysis. *Int. J. Adv. Manuf. Technol.* **2014**, *71*, 1345–1358. [CrossRef]

16. Akhtar, S.; Kardas, O.; Keles, O.; Yilbas, B. Laser cutting of rectangular geometry into aluminum alloy: Effect of cut sizes on thermal stress field. *Opt. Laser Eng.* **2014**, *61*, 57–66. [CrossRef]

17. Yilbas, B.; Akhtar, S.; Karatas, C. Laser cutting of rectangular geometry into alumina tiles. *Opt. Laser Eng.* **2014**, *55*, 35–43. [CrossRef]

18. Yilbas, B.; Akhtar, S.; Keles, O. Laser cutting of triangular blanks from thick aluminum foam plate: Thermal stress analysis and morphology. *Appl. Therm. Eng.* **2014**, *62*, 28–36. [CrossRef]

19. Roberts, I.; Wang, C.; Esterlein, R.; Stanford, M.; Mynors, D. A three-dimensional finite element analysis of the temperature field during laser melting of metal powders in additive layer manufacturing. *Int. J. Mach. Tool. Manuf.* **2009**, *49*, 916–923. [CrossRef]

20. Shi, B.; Attia, H. Integrated process of laser-assisted machining and laser surface heat treatment. *J. Manuf. Sci. Eng.* **2013**, *135*. [CrossRef]

21. Akarapu, R.; Li, B.; Segall, A. A thermal stress and failure model for laser cutting and forming operations. *J. Fail. Anal. Prev.* **2004**, *4*, 51–62. [CrossRef]

22. Nyon, K.; Nyeoh, C.; Mokhtar, M.; Abdul-Rahman, R. Finite element analysis of laser inert gas cutting on Inconel 718. *Int. J. Adv. Manuf. Technol.* **2012**, *60*, 995–1007. [CrossRef]

23. Fu, C.; Sealy, M.; Guo, Y.; Wei, X. Finite element simulation and experimental validation of pulsed laser cutting of nitinol. *J. Manuf. Process.* **2015**, *19*, 81–86. [CrossRef]

24. Yilbas, B.; Akhtar, S.; Keles, O. Laser cutting of aluminum foam: Experimental and model studies. *J. Manuf. Sci. Eng.* **2013**, *135*. [CrossRef]

25. Kheloufi, K.; Hachemi, A.; Benzaoui, A. Numerical simulation of Transient three-dimensional temperature and kerf formation in laser fusion cutting. *J. Heat Transf. ASME* **2015**, *137*. [CrossRef]

26. Yuan, P.; Gu, D. Molten pool behaviour and its physical mechanism during selective laser melting of TiC/AlSi10Mg nanocomposites: Simulation and experiments. *J. Phys. D Appl. Phys.* **2015**, *48*, 035303. [CrossRef]

27. Modest, M.; Abakians, H. Evaporative Cutting of a Semi-infinite Body With a Moving CW laser. *J. Heat Transf. ASME* **1986**, *108*, 602–607. [CrossRef]

28. Zimmer, K. Analytical solution of the laser-induced temperature distribution across internal material interfaces. *Int. J. Heat Mass Transf.* **2009**, *52*, 497–503. [CrossRef]

29. Jiang, H.J.; Dai, H.L. Effect of laser processing on three dimensional thermodynamic analysis for HSLA rectangular steel plates. *Int. J. Heat Mass Transf.* **2015**, *82*, 98–108. [CrossRef]

30. Winczek, J. Analytical solution to Transient temperature field in a half-infinite body caused by moving volumetric heat source. *Int. J. Heat Mass Transf.* **2010**, *53*, 5774–5781. [CrossRef]
31. Mejia, D.; Moreno, A.; Arbelaiz, A.; Posada, J.; Ruiz-Salguero, O.; Chopitea, R. Accelerated thermal simulation for three-dimensional interactive optimization of computer numeric control sheet metal laser cutting. *J. Manuf. Sci. Eng.* **2017**, *140*. [CrossRef]
32. Heeling, T.; Wegener, K. Computational investigation of synchronized multibeam strategies for the selective laser melting process. *Phys. Procedia* **2016**, *83*, 899–908. [CrossRef]
33. Shuja, S.Z.; Yilbas, B.S. Multi-beam laser heating of steel: Temperature and thermal stress analysis. *Trans. Can. Soc. Mech. Eng.* **2012**, *36*, 373–381. [CrossRef]
34. Petzet, V.; Büskens, C.; Pesch, H.J.; Karkhin, V.; Makhutin, M.; Prikhodovsky, A.; Ploshikhin, V. OPTILAS: Numerical Optimization as a Key Tool for the Improvement of Advanced Multi-Beam Laser Welding Techniques. In *High Performance Computing in Science and Engineering, Garching 2004*; Bode, A., Durst, F., Eds.; Springer: Berlin, Germany, 2005; pp. 153–166.
35. Yilbas, B.; Akhtar, S. Laser bending of metal sheet and thermal stress analysis. *Opt. Laser Technol.* **2014**, *61*, 34–44. [CrossRef]
36. Moreno, A.; Segura, Á.; Arregui, H.; Posada, J.; Ruíz de Infante, Á.; Canto, N., Using 2D Contours to Model Metal Sheets in Industrial Machining Processes. In *Future Vision and Trends on Shapes, Geometry and Algebra*; Springer: London, UK, 2014; pp. 135–149. [CrossRef]
37. Velez, G.; Moreno, A.; Infante, A.R.D.; Chopitea, R. Real-time part detection in a virtually machined sheet metal defined as a set of disjoint regions. *Int. J. Comput. Integr. Manuf.* **2016**, *29*, 1089–1104. [CrossRef]

Article

Multi-Objective Optimization for Grinding of AISI D2 Steel with Al$_2$O$_3$ Wheel under MQL

Aqib Mashood Khan [1], Muhammad Jamil [1], Mozammel Mia [2], Danil Yurievich Pimenov [3,*], Vadim Rashitovich Gasiyarov [4], Munish Kumar Gupta [5] and Ning He [1]

[1] College of Mechanical and Electrical Engineering, Nanjing University of Aeronautics and Astronautics, Nanjing 210016, China; dr.aqib@nuaa.edu.cn (A.M.K.); engr.jamil@nuaa.edu.cn (M.J.); drnhe@nuaa.edu.cn (N.H.)

[2] Mechanical and Production Engineering, Ahsanullah University of Science and Technology, Dhaka 1208, Bangladesh; mozammelmiaipe@gmail.com

[3] Department of Automated Mechanical Engineering, South Ural State University, Lenin Prosp. 76, Chelyabinsk 454080, Russia

[4] Department of Mechatronics and Automation, South Ural State University, Lenin Prosp. 76, Chelyabinsk 454080, Russia; gasiyarovvr@gmail.com

[5] Department of Mechanical Engineering, Chandigarh University, Gharuan 140413, Punjab, India; munishguptanit@gmail.com

* Correspondence: danil_u@rambler.ru

Received: 26 October 2018; Accepted: 7 November 2018; Published: 13 November 2018

Abstract: In the present study, the machinability indices of surface grinding of AISI D2 steel under dry, flood cooling, and minimum quantity lubrication (MQL) conditions are compared. The comparison was confined within three responses, namely, the surface quality, surface temperature, and normal force. For deeper insight, the surface topography of MQL-assisted ground surface was analyzed too. Furthermore, the statistical analysis of variance (ANOVA) was employed to extract the major influencing factors on the above-mentioned responses. Apart from this, the multi-objective optimization by Grey–Taguchi method was performed to suggest the best parameter settings for system-wide optimal performance. The central composite experimental design plan was adopted to orient the inputs wherein the inclusion of MQL flow rate as an input adds addition novelty to this study. The mathematical models were formulated using Response Surface Methodology (RSM). It was found that the developed models are statistically significant, with optimum conditions of depth of cut of 15 µm, table speed of 3 m/min, cutting speed 25 m/min, and MQL flow rate 250 mL/h. It was also found that MQL outperformed the dry as well as wet condition in surface grinding due to its effective penetration ability and improved heat dissipation property.

Keywords: minimum quantity lubrication; surface grinding; multi-objective optimization; grey relational analysis; surface topography; sustainable machining

---

1. Introduction

AISI D2 (ENX160CrMoV12) tool steel is regarded as a key material in high performance engineering applications such as in the mold-and-die industry; for use as industrial cutting tools, gauges, and machine parts exposed to wear and injection screws; in aerospace and automotive industries; medical appliances; heavy engineering; and tools in the manufacturing industry [1]. This is due to its superior properties like high wear resistance, compressive strength, temperature resistance, and narrow tolerances as well as its high strength-to-weight ratio. However, the manufactured parts demand for high geometrical and dimensional intricacy—which raises the necessity of using a grinding operation.

Dry grinding is generally applied due to its cost-effectiveness and environmentally-friendly nature with no pollution with cost-saving lubricants. However, it involves major problems i.e., thermal damage, high friction, high residual stress, and high wheel-wear [2]. Moreover, the thermal adversities influence the integrity of the newly generated surface. For instance, as reported by Guo et al. [3], the dry surface grinding has induced a heat affected zone associated with oxidation and cutting marks. Thereby, the alternative(s) to dry cutting is being sought. Conventional flood cooling assisted cutting apparently solved this problem—but raised ecological and economic concerns. This paradoxical state urged for novel and effective solution such as minimum quantity lubrication (MQL).

The flood cutting removes the heat from the cutting zone i.e., wheel-workpiece interface, however, it was not always found to be effective. Moreover, it consumes a vast amount (i.e., 8 L per minutes [3]) of coolant/lubricant, raising a serious question about its economic and environmental sustainability. Howes et al. [4] worked on the environmental effect of grinding fluids and reported that grinding fluids have significant impact on the ecology and internal environment (workshop). Use of petroleum-based lubricants in the manufacturing sector is increasing tremendously with 1% of annual increment, and it is equivalent to 13,726 million tons of oil [5].

In MQL the fluid is delivered in extremely small quantities (10,000 times less than conventional cooling) so that for all manufacturing processes it resembles like dry machining. In this technology, an aerosol (oil–air mixture) is fed into interface of cutting tool and workpiece. This technique is free from problems like consumption of bulk quantity of fluids, their storage, and disposal. It also helps in promoting a green environment [6]. Mia et al. [7–9] studied the beneficial influence of MQL in the milling process, and they claimed that MQL, under different circumstances, showed better results than the dry mode of cutting. Moreover, Dhar et al. [10] have reported the superiority of MQL technology over wet cooling. Similarly, Wang et al. [11] claimed that the vegetable oil-based MQL grinding process revealed a lower specific grinding energy and friction coefficient due to its excellent lubrication property. The works of Abbas et al. [12,13] suggested application of Edgeworth–Pareto optimization of an Artificial Neural Network (ANN) to predict surface roughness (Ra) within minimal machining time (Tm), at the prime cost of machining one part (C).

The cumbersome process of hit and trial to obtain the best experimental run can be avoided by conducting systematic research optimization. Furthermore, this approach can simply identify the effect of process parameters on the ground parts for specific material. From the comprehensive literature review, summarized in Figure 1, it is discernable that researchers have studied alternative approaches of operation on various categories of material using different grinding process. The material grades are plotted on the *x*-axis, and the grinding processes on the *y*-axis. It is noted from Figure 1 that many researchers studied the effect of process parameters of the grinding process on AISI 52100, 100Cr6, Inconel, AISI 4340, AISI 8620, AISI 1045 [14], titanium alloys [15,16], and three-layer metal-composite system [17]. In the past researchers developed a model for cutting force in plunge cylindrical grinding [18], whilstin internal grinding [19], an innovate modeling approach to predict temperature distribution was developed [20]. For measurement of response Naolny et al. has proposed a new SEM-EDS-based analysis of grinding of titanium grade 5 [21]. Sutowski et al. worked on monitoring the cylindrical grinding process using acoustic emission signal (AE) and image analyses [22]. Similarly, the researchers have used the Taguchi method, response surface methodology (RSM), regression analysis as well as the conventional experiments to model responses of grinding process. Prediction of response using optimal combinations of significant process parameter has made RSM the most effective approach.

It can be concluded from the literature review that the process parameters such as workpiece speed, depth of cut, cutting speed and MQL flow rate have a direct influence on the workpiece in grinding process and effect of these process parameter varies from material-to-material. Literature review shows that no work has been reported yet on AISI D2 steel to evaluate, compare, and optimize (surface roughness, temperature, and force) using MQL grinding process. The authors get motivation by the fact that insufficient work has been done on the multi-objective optimization of grinding of AISI

D2 steel. Moreover, the lubrication performance of Accu-Lube 6000 in MQL grinding has not been addressed. Therefore, this experimental investigation shows a systematic approach to evaluate the influence of above-mentioned input parameters on grinding outputs. This research aims to develop an empirical relationship using RSM to predict the normal forces, surface roughness, and temperature and to optimize the process parameters to get the best surface finish, minimum forces, and temperature.

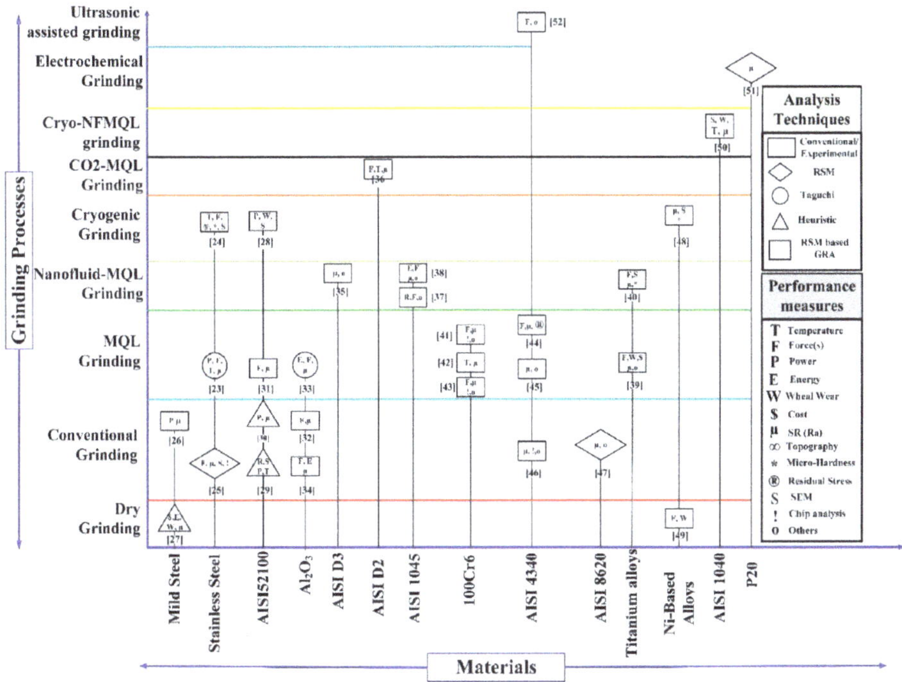

Figure 1. Overview of research on grinding types, the different materials used, and methods used [22–52].

2. Methodology

Experimental Setup

The surface grinding was performed on AISI D2 steel having 1.56% carbon and 12% chromium by weight. The chemical composition of the workpiece material is verified through optical emission spectrometer before experiments. Alumina oxide grinding wheel manufactured by NORTON (Shanghai, China) was used. The grinding process was performed on a surface grinding machine with Model (WAZAA415X-NC) with motor driving having a capacity of 6 kW. Surface grinding was employed on AISI D2 steel having chemical composition as presented in Table 1. Alumina oxide grinding wheel (FE 38A60KV) manufactured by NORTON was used in present work according to the cutting condition, hardness and tool manufactures recommendation. Alumina Oxide vitrified grinding wheel of size (250 mm × 72.5 mm × 31.8 mm) was used as abrasive cutting material.

Table 1. Chemical composition of AISI D2 Steel.

Element	C	Si	Mn	Cr	Mo	V	P	S	Ni	Fe
%	1.56	0.30	0.40	11.90	0.78	0.80	0.023	0.015	0.05	Balance

The experiments were performed under three conditions namely dry, wet, and MQL grinding. Under the dry condition, grinding was performed without coolant. Castrol Syntilo 9954 (Shangai, China) with density 1066 kg/m^3 was used as the coolant for conventional wet grinding. A flow rate of 6 L/min was maintained throughout wet grinding. For experiments involving grinding under MQL condition Accu-Lube 6000 with specific gravity 0.92, flash point 214 °C, and density 8.9 CSt special cutting oil was used at variable flow rates between 50 and 50 mL/h. During the whole process the air pressure was 6 bar, with a nozzle angle of 15°, and nozzle distance of 30 mm, which were kept constant. The schematic diagram for surface grinding process along with responses measurement has been shown in Figure 2.

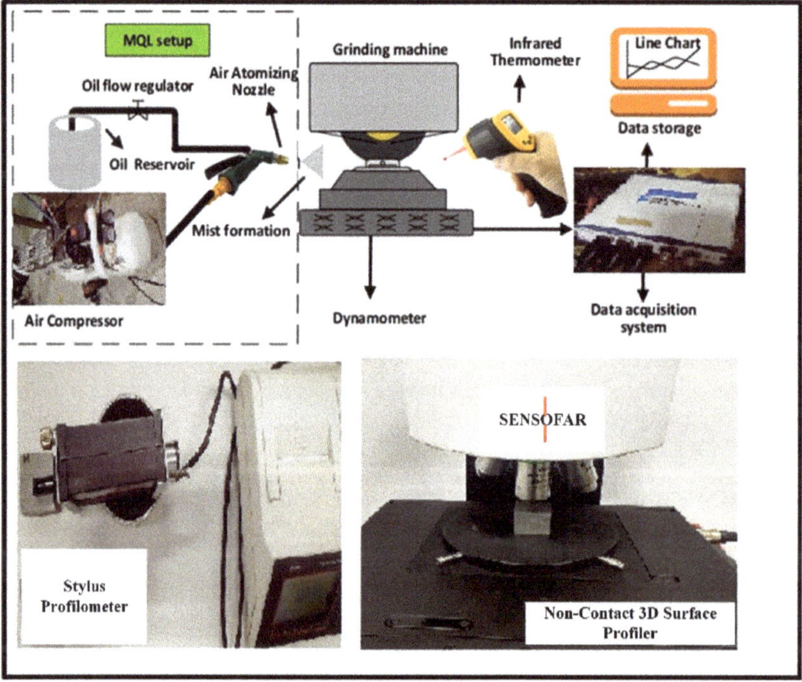

Figure 2. Schematic diagram showing minimum quality lubrication (MQL) working principle and response measurements.

Response surface methodology (RSM) with central composite design (CCD) has been considered for the modeling and analysis of the performance measures of the ground workpiece. In CCD, if some process parameters are represented by k and number of central points by m, the total number of experimentation can be calculated by Equation (1) [53]. Process parameters and their levels for Dry, Wet, and MQL grinding are shown in Table 2.

$$\text{Number of Experiments} = 2k + 2^k + m \tag{1}$$

A total of 30 experiments have been performed for different process parameter combinations (depth of cut, table speed, cutting speed, and lubricant flow rate) in MQL-grinding and 18 experiments have been performed with process parameter (depth of cut, table speed, and cutting speed) for the dry and wet grinding processes. The experimental run, process parameters, and response parameters have been shown in Tables 3 and 4.

Table 2. Process parameters and their levels for Dry, Wet, and MQL grinding.

SL. No.	Parameter (s)	$-\alpha$	Low (−1)	Centre (0)	High (+1)	$+\alpha$
		Dry, Wet, MQL				
1	Depth of cut: a_p, (μm)	15	20	25	30	35
2	Table speed: v_w, (m/min)	3	5	7	9	11
3	Cutting Speed: v_s, (m/s)	15	20	25	30	35
		For MQL only				
4	MQL flow rate: Q, (mL/h)	50	100	150	200	250

Table 3. Design matrix for dry and wet condition grinding.

Exp. No.	Process Parameters			Response Measurements					
				Dry Grinding			Wet Grinding		
	a_p	v_w	v_s	F	T	R_a	F	T	R_a
	(μm)	(m/min)	(m/s)	(N)	(°C)	(μm)	(N)	(°C)	(μm)
1	20	5	20	82	245	0.5	60	69	0.29
2	30	5	20	115	270	0.58	70	83	0.34
3	20	9	20	83	240	1.4	54	80	0.58
4	30	9	20	117	291	1.51	81	84	0.63
5	20	5	30	85	245	1.12	46	58	0.48
6	30	5	30	92	297	0.89	71	94	0.46
7	20	9	30	110	248	1.45	47	71	0.59
8	30	9	30	103	310	1.49	80	92	0.51
9	15	7	25	82	225	0.81	40	62	0.51
10	35	7	25	119	319	1.07	90	109	0.56
11	25	3	25	86	268	0.46	55	74	0.25
12	25	11	25	105	300	1.6	65	82	0.65
13	25	7	15	111	262	1.01	61	70	0.40
14	25	7	35	98	276	1.4	49	80	0.55
15	25	7	25	90	262	1.35	56	69	0.42
16	25	7	25	91	260	1.39	55	70	0.43
17	25	7	25	90	268	1.42	54	68	0.41
18	25	7	25	93	260	1.41	58	71	0.42

Table 4. Design matrix for MQL-assisted grinding.

Exp. No.	Process Parameters				Response Measurements		
	a_p	v_w	v_s	Q	F	T	R_a
	(μm)	(m/min)	(m/s)	(mL/h)	(N)	(°C)	(μm)
1	20	5	20	100	36	104	0.39
2	30	5	20	100	50	121	0.4
3	20	9	20	100	65	114	0.58
4	30	9	20	100	71	131	0.6
5	20	5	30	100	35	109	0.39
6	30	5	30	100	37	121	0.51
7	20	9	30	100	62	117	0.62
8	30	9	30	100	58	133	0.66
9	20	5	20	200	28	82	0.33
10	30	5	20	200	37	95	0.395
11	20	9	20	200	55	85	0.56
12	30	9	20	200	67	97	0.61
13	20	5	30	200	28	85	0.35
14	30	5	30	200	31	99	0.44
15	20	9	30	200	53	94	0.53
16	30	9	30	200	64	101	0.63
17	15	7	25	150	41	97	0.42
18	35	7	25	150	56	118	0.53
19	25	3	25	150	17	85	0.31

Table 4. *Cont.*

Exp. No.	Process Parameters				Response Measurements		
	a_p	v_w	v_s	Q	F	T	R_a
	(μm)	(m/min)	(m/s)	(mL/h)	(N)	(°C)	(μm)
20	25	11	25	150	70	94	0.71
21	25	7	15	150	54	112	0.47
22	25	7	35	150	46	118	0.51
23	25	7	25	50	57	134	0.5
24	25	7	25	250	50	74	0.45
25	25	7	25	150	44	109	0.49
26	25	7	25	150	47	107	0.47
27	25	7	25	150	45	111	0.48
28	25	7	25	150	47	108	0.45
29	25	7	25	150	38	106	0.47
30	25	7	25	150	45	108	0.46

v_w : Table speed; a_p: Depth of cut; v_s: Cutting speed; Q: MQL flow rate; R_a: Average surface roughness.

3. Results and Discussion

In this section, the normal force, grinding temperature, and surface roughness are analyzed by the experimental investigation, statistical analysis, mathematical modeling, and graphical plots.

3.1. Normal Forces

The cutting forces largely influence the performance of grinding. The grinding mechanics are, in turn, altered by the lubrication effects prevailing between the wheel and work surface. The normal forces were measured during the grinding process using dynamometer. To check the influence of dry, wet, and MQL conditions on the normal forces, a comparison has been drawn in Figure 3. Owing to the dominant role of depth of cut on normal forces (discussed later), the comparison has been performed against all combinations of depth of cut, at different values of cutting speed, and table speed. This comparison shows variation in the normal force with the variation in the cutting parameters. Note that a comparison has been performed for a set of seven experiments where cutting conditions were same in each environment, and cutting conditions have been selected from design space provided in Tables 3 and 4. The normal force in dry grinding was kept as a reference point to calculate percent reduction.

It can be seen that a reduction of 37.86 to 79.26% in normal forces was found for MQL grinding; however, a reduction of only 22.33 to 57.27% in forces was achieved for wet grinding. It can be concluded from the results that the MQL environment has generated lower forces than the dry and wet environment.

MQL-assisted grinding confirms effective supply of the mist of oils (droplets) in between the interfaces of workpiece-wheel grain that provides better and efficient lubrication. This efficient lubrication improves the slipping of grain between the workpiece-wheel. Penetration of small oil–mist droplets into the cutting region with high speed explains the effectiveness of the process [54]. It makes a consistent lubricant tribofilm with reduced shearing strength than the base material, and it ensures the lower normal forces in MQL-assisted grinding in comparison with the wet and dry mode.

In case of MQL assisted grinding process, the grinding forces are decreased significantly; this is due to the proper penetration of cutting fluid as in aerosol form. Durable oily films formed during MQL grinding allow proper slipping of grinding wheel grains and offer advanced level lubrication at high pressure. The second reason for MQL having least forces could be excellent and superior lubrication properties of Acculube-6000 than Syntile-9956. Moreover, the abrasive grain sharpness is maintained for a long time due to this excellent lubrication. Larger forces in dry grinding are due to the absence of cutting fluids, which leads to dulling of abrasive grits, premature breakage of grains and side flow of materials [41]. The results are in good agreement with the published literature, such as with the study of Sadeghi et al. [40]. They have also claimed to have better surface quality and lower force while grinding. This was accredited to the lower coefficient of friction.

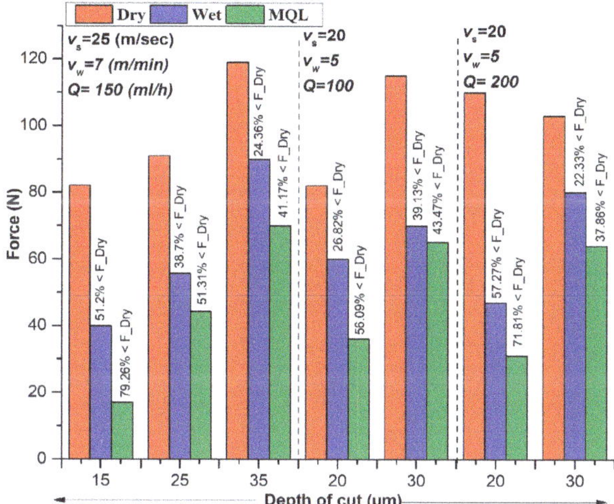

Figure 3. Comparison of force measurements in different cooling environments.

Besides the comparison of the normal force, to make the study complete, here the contribution of the factors on normal force is evaluated by analysis of variance (ANOVA). This statistical analysis was conducted using Design Expert 8.0.7. The fit summary for normal force (dry, wet, and MQL) suggested the quadratic relation as the best fit model. In dry grinding, the main effects of depth of cut, table speed, and cutting speed were identified as the significant model terms for the force. The main parameters that contribute significantly to forces generation in wet grinding includes depth of cut and cutting speed. The ANOVA result suggested that the main effects of depth of cut, table speed, cutting speed, and MQL flow rate were the significant model terms associated with the force for MQL grinding. Model for the prediction of force in dry, wet and MQL environment are provided in Equations (2)–(4) respectively.

The ANOVA includes the significant terms along with adequacy measures R^2, Adj. R^2 and predicted R^2 is close to '1' indicating the adequacy of resulting model for all considered environments. The results of ANOVA indicate that the model is significant with a p-value of less than 0.05.

$$\text{DRY}_{(F)} = +34.5 + (6.4 \times a_p) - (7.8 \times v_w) - (1.7 \times V_s) - (0.16 \times a_p \times v_w) - (0.33 \times a_p \times V_s) + (0.41 \times a_p \times v_w) + (0.096 \times a_p^2) + (0.28 \times v^2) + (0.13 \times v_s^2) \tag{2}$$

$$\text{WET}_{(F)} = +219.3 - (7.6 \times a_p) - (13.09 \times v_w) - (3.8 \times v_s) + (0.31 \times a_p \times v_w) + (0.10 \times a_p \times v_s) + (0.06 \times v_w \times v_s) + (0.1050 \times a_p^2) + (0.34375 \times v_w^2) + (0.005 \times v_s^2) \tag{3}$$

$$\text{MQL}_{(F)} = +95.3 + (2.3 \times a_p) - (1.4 \times v_w) - (7.2 \times v_s) - (0.1 \times Q) - (0.01 \times a_p \times v_w) + (0.03 \times a_p \times v_s) - (3.6 \times 10^{-4} \times a_p \times Q) + (0.1 \times v_w \times v_s) - (0.018 \times v_w \times Q) + (0.004 \times v_s \times Q) - (0.01 \times a_p^2) + (0.25 \times v_w^2) + (0.08 \times v_s^2) + (5.4 \times 10^{-4} \times Q^2) \tag{4}$$

Figure 4a–c demonstrates the contribution of the depth of cut and table speed on the normal force found in dry, wet, and MQL grinding. By comparing the response surface plots for dry, wet and MQL, it is evident that the behavior of force is nearly similar at varying depth of cut and table speed conditions. The force is more sensitive to variation in depth of cut compared to table speed. The similar results are also claimed in a previous paper [55]. The force increases with the increase in table speed in dry grinding; however, in wet MQL grinding table speed has a negligible effect.

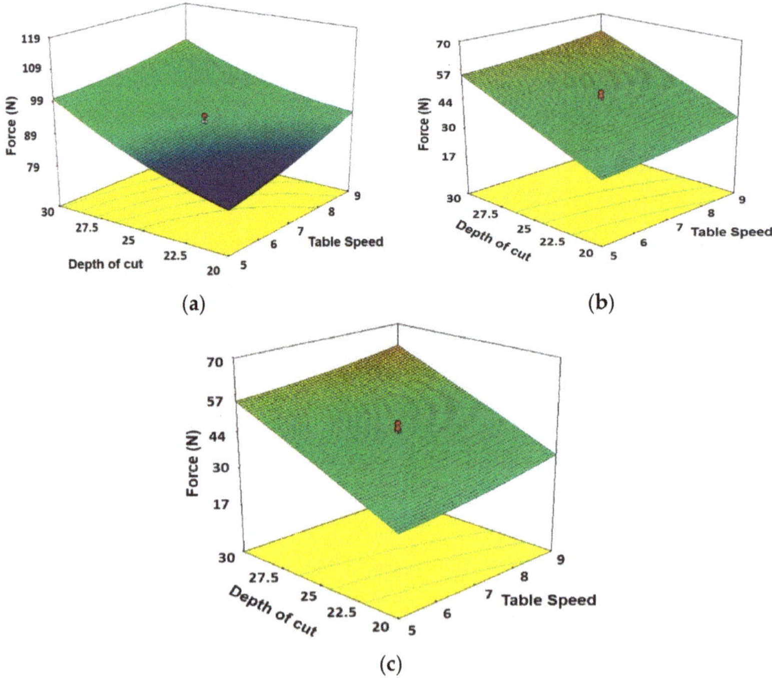

Figure 4. 3D Surface plots showing the effects of depth of cut and table speed on forces for (**a**) Dry, (**b**) Wet, and (**c**) MQL conditions. Units of table speed and depth of cut are 'm/min' and 'μm', respectively.

3.2. Temperature

Temperature is produced due to the interaction of the grinding wheel with the workpiece. Data for temperatures were recorded using an Infrared thermometer (Raytek-Raynger MX4 (Guangdong, China)). Comparison of temperatures allowed us an assessment of the thermal performance of the MQL-assisted grinding compared to dry and wet cases. It gives an insightful analysis of lubrication in wet and MQL techniques. As the depth of cut was found the most significant parameter for temperature (discussed later), therefore, it is plotted on the *x*-axis in the comparison graph in Figure 5. It is depicted in Figure 4a that the temperature reduction in wet grinding is between 65.83% and 72.44% and in MQL grinding is between 55.18% and 67.4%. It is further noticeable that the temperature achieved in MQL environment is much lower than dry and close to the wet temperature; it is even lower than the flash point of oil used which recommends its usage during grinding.

Comparison of measured values of temperatures in dry, wet and MQL mode at workpiece surface is shown in Figure 4a. As expected, the wet mode results in less cutting temperature and dry mode produces the highest temperature. By applying the MQL method, the temperatures values are about 74–134 °C and this is less than those in dry grinding conditions. This is due to effective cooling and lubrication of the oil mist provided by MQL that reduces the cutting temperature. Moreover, a significant amount of heat is removed through convection heat transfer phenomena [42].

The reason for higher temperatures in dry grinding is intrinsically associated with high specific energy requirement. This high specific energy demand is due to grain sliding phenomena and shearing with adverse grit geometry. Such high temperature is usually for thermal damages on the ground surface, such as residual stress, oxidation, and burning.

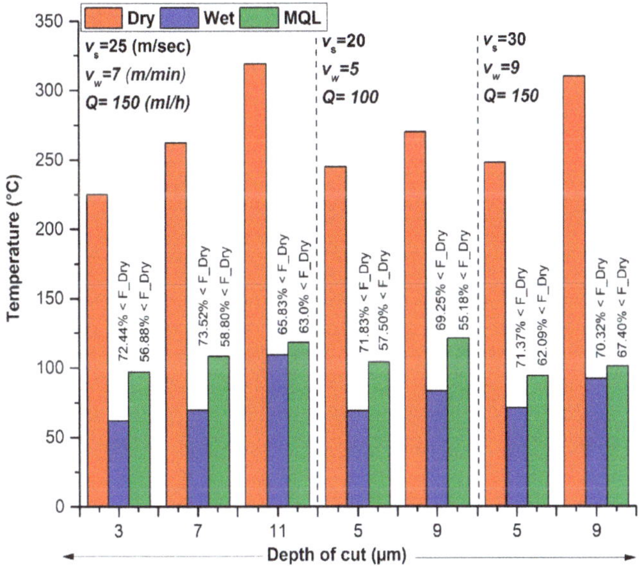

Figure 5. Comparison of temperature measurements in different cooling environments.

From Figure 5, moderately low-temperature values have been seen in all cooling modes. However, the combined effects of lubrication (the lowest normal forces) and cooling are main reasons for the improved cutting efficiency in MQL technology. It is worth noticing that the maximum temperature achieved in MQL is far below the thermal damage temperature. The lowest temperature in the wet cooling mode could be due to the excellent cooling provided by Castrol Syntilo 9954. In addition, the obtained results are in good agreement with the previous literature survey [21].

The wet cooling eliminates the symptoms of the problems but not the cause. The MQL induces a different approach to deal with heat and friction in the machining process. In the MQL-assisted machining process, the interface of the cutting tool and workpiece is coated with the high-quality oil droplets delivered in an atomized spray. This thin layer of lubricant is extremely efficient at reducing friction and therefore cooling the temperature.

The secret of Accu-Lube 6000 success is the polar nature of its molecule. Molecules of the selected lubricant align themselves by their poles making bonds exceptionally strong, both to each other and tometallic surfaces. This strong bond allows Accu-Lube 6000 to create a tough and desirable lubrication barrier. The MQL is not only useful for lubricating the grinding wheel, but also it removes heat from the grinding zone.

For further analysis, the statistical tool ANOVA was used to check the adequacy of the temperature model in each environment. The quadratic relationship was found as the best fit model for temperature. The main and interaction effect that contributes significantly to temperature for dry grinding includes depth of cut, table speed, and cutting speed. The ANOVA for wet grinding revealed that the main effects of depth of cut and table speed were noted as significant model terms associated with temperature. On the other hand, ANOVA result divulged that main process parameter that contributes significantly in temperature for MQL grinding includes depth of cut, table speed, cutting speed, and MQL flow rate. The ANOVA results of temperature for dry, wet, and MQL can be validated from Tables 3 and 4. Finally, the mathematical models for temperature in dry, wet and MQL environment are provided in Equations (5)–(7) respectively.

$$\text{DRY}_{(T)} = +445.08 - (7.4 \times a_p) - (26.2 \times v_w) - (6.4 \times v_s) + (0.4 \times a_p \times v_w) \\ + (0.19 \times a_p \times v_s) + (0.08 \times a_p{}^2) + (1.28 \times v_w{}^2) + (0.05 \times v_s{}^2) \tag{5}$$

$$\text{WET}_{(T)} = +235.9 - (8.82 \times a_p) + (1.46 \times v_w) - (7.5 \times v_s) - (0.31 \times a_p \times v_w) + (0.19 \times a_p \times v_s) - (0.01 \times v_w \times v_s) + (0.16 \times a_p{}^2) + (0.56 \times v_w{}^2) + (0.060 \times v_s{}^2) \tag{6}$$

$$\text{MQL}_{(T)} = +53.8 + (2.4 \times a_p) + (18.9 \times v_w) - (3.5 \times v_s) - (0.03 \times Q) - (0.02 \times a_p \times v_w) - (0.02 \times a_p \times v_s) - (0.004 \times a_p \times Q) + (0.03 \times v_w \times v_s) - (0.015 \times v_w \times Q) + (0.002 \times v_s \times Q) + (0.003 \times a_p{}^2) - (1.1 \times v_w{}^2) + (0.07 \times v_s{}^2) - (3.16 \times 10^{-4} \times Q^2) \tag{7}$$

Furthermore, 3D response surface graphs in Figure 6a–c highlights the effects of depth of cut and table speed on temperature for dry, wet, and MQL environments. It is noted that the interaction effect of depth of cut and table speed is not similar for all given environment. In the case of dry environment (Figure 6a), the temperature increases as the depth of cut increases; also the depth of cut has a direct and significant effect on temperature. However, the table speed has a less significant effect on temperature compared to the effect of depth of cut; therefore, the temperature slightly increases with an increase in table speed. Figure 6b depicts that in wet grinding the temperature increases rapidly at a higher level of table speed. Figure 4c represents the 3D response surface plot for MQL grinding; in that case, the MQL flow rate was the most significant parameter for temperature control. It can be seen that initially the temperature decreases with an increase in MQL flow rate (Q). The minimum temperature was found at the lowest level of MQL flow (Q).

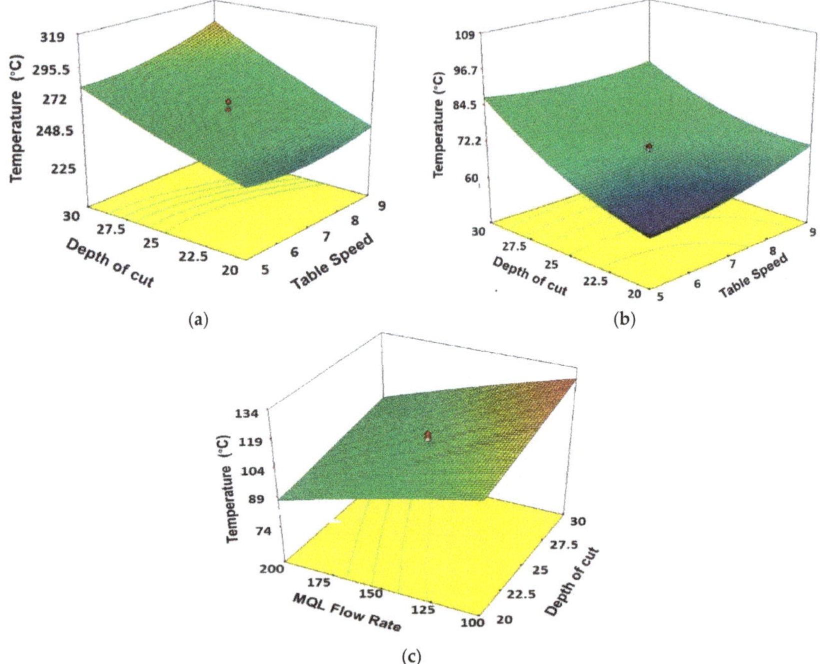

Figure 6. Surface plots showing the effects of depth of cut and table speed on temperature for (**a**) Dry, (**b**) Wet, (**c**) and MQL conditions. Units of table speed, depth of cut, and MQL flow rate are 'm/min', 'μm', and 'mL/h', respectively.

3.3. Surface Roughness

Surface roughness is an important parameter to analyze the quality of the surface of the workpiece. The surface finish largely influences the dimensional accuracy and product life. Surface roughness values were measured by using portable stylus profilometer Surftest SJ-410 (Mitutoyo, Tokyo, Japan). To elaborate the effectiveness of MQL over the wet and dry environment, a comparison of surface roughness in three different environments has been made and presented in Figure 7. Since the table speed was found as the most significant parameter for surface roughness (discussed later), therefore, the comparison has been drawn by varying the table speed and keeping the other process parameters at a constant level. During comparison, the dry grinding process was set as a reference point. It can be noted from Figure 7 that wet grinding resulted in a 42 to 69.7% reduction in surface roughness, whereas MQL grinding was only able to reduce roughness by 32 to 66.1% when compared with the dry grinding process. From these results, it is clear that MQL provides better results than dry and almost similar results with wet conditions.

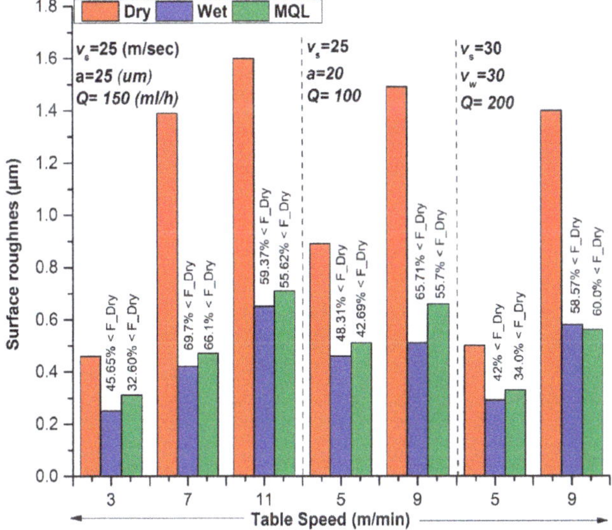

Figure 7. Comparison of surface roughness in different cooling environments.

The better surface roughness generated by MQL is because of excellent control of temperature in grinding zone. An increase in the depth of cut increases grinding temperature which softens the workpiece material. This causes the clogging which produces low-quality surfaces with high surface roughness. In dry grinding, due to the absence of cutting fluid the frictional forces increase and these frictional forces are converted to heat which is further transferred to workpiece material and chips. This phenomenon influences the surface roughness, and it is responsible for surface burns and grinding mark on the workpiece. In wet grinding, these defects are avoided by the intensive lubrication.

In the dry mode, there is no cooling, hence, even at a depth of cut 25 µm, burning traces started to appear. This phenomenon is called grain flattening, and it results in high surface roughness and thermal damages of the ground surface as compared to other cooling modes. It is noted from Figure 7 that under specific conditions MQL competes with, and even performs better than, the wet cooling mode. Moreover, the results achieved are in good agreement with the published literature [21].

From Figure 7, it can be seen that the wet cooling mode provides better surface finish than MQL. The reason for this lies in the fact that when MQL was applied in machining of hardened steel, the frictional coefficient decreased; however, an increase in material deformation leads to poor surface quality. Furthermore, during MQL mode, the lower normal forces have been observed which reduced the self-sharpening of grinding wheel. Hence, a total number of cutting edges and their sharpness are lower than the wet cooling mode that leads to thicker chips and relatively high surface roughness values. These experimental results are in good agreement with the published literature [29].

For further analysis, ANOVA was performed, and it revealed that the main effects of table speed, and cutting speed were found as significant model terms associated with surface roughness for dry grinding. The main influence includes the effects of table speed, cutting speed, depth of cut, cutting speed, and table speed as the significant model terms associated with surface roughness. The empirical models for prediction of surface roughness under dry, wet and MQL grinding are provided in Equations (8)–(10), respectively.

$$
\text{Dry}_{(Ra)} = -8.13 + \left(0.25 \times a_p\right) + \left(0.66 \times v_w\right) + \left(0.24 \times v_s\right) + \left(0.0037 \times a_p \times v_w\right) \\
- \left(0.0019 \times a_p \times v_s\right) - \left(0.011 \times v_w \times v_s\right) - \left(0.004 \times a_p{}^2\right) - \left(0.023 \times v_w{}^2\right) - \left(0.001 \times v_s{}^2\right)
\tag{8}
$$

$$
\text{Wet}_{(Ra)} = -0.55 - \left(0.02 \times a_p\right) + \left(0.16 \times v_w\right) + \left(0.038 \times v_s\right) + \left(7.5 \times 10^{-4} \times a_p \times v_w\right) \\
- \left(0.001 \times a_p \times v_s\right) - \left(0.005 \times v_w \times v_s\right) + \left(0.001 \times a_p{}^2\right) + \left(0.002 \times v^2\right) + \left(5.9 \times 10^{-4} \times v_s{}^2\right)
\tag{9}
$$

$$
\text{MQL}_{(Ra)} = +0.52 - \left(0.014 \times a_p\right) + \left(0.028 \times v_w\right) - \left(0.014 \times v_s\right) - \left(6.78 \times 10^{-4} \times Q\right) - \\
\left(4.6 \times 10^{-4} \times a_p \times v_w\right) + \left(5.1 \times 10^{-4} \times a_p \times v_s\right) + \left(2.8 \times 10^{-5} \times a_p \times Q\right) - \\
\left(5.38 \times 10^{-4} \times V_w \times V_s\right) + \left(2.8 \times 10^{-5} \times v_w \times Q\right) - \left(3.8 \times 10^{-5} \times v_s \times Q\right) + \\
\left(1.3 \times 10^{-4} \times a_p{}^2\right) + \left(0.003 \times v_w{}^2\right) + \left(2.8 \times 10^{-4} \times v_s{}^2\right) + \left(1.3 \times 10^{-6} \times Q^2\right)
\tag{10}
$$

It is significant to know that Figure 8a–c shows the effect of two process parameters (simultaneously) on the output parameters while other process parameters are fixed at their middle levels. The effects of table speed and cutting speed on surface roughness (R_a) for dry, wet, and MQL grinding environments are presented in Figure 8a–c, it is evident from these figures that the surface roughness has a direct relation to table speed and cutting speed, i.e., an increase in speed also increases surface roughness. The impact of changing table speed is more at a low level of cutting speed as compared to the high level of cutting speed. Similarly, the impact of table speed on surface roughness is higher at a low level of table speed as compared to the high level of table speed.

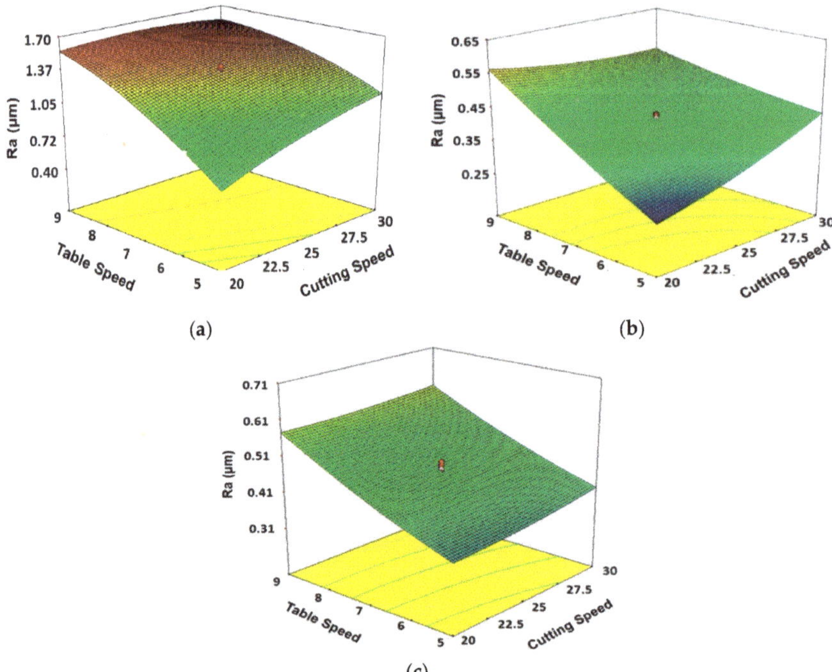

Figure 8. 3D Surface plots showing the effects of cutting speed and table speed on surface roughness for (**a**) Dry, (**b**) Wet, and (**c**) MQL conditions. Units of table speed and cutting speed are 'm/min' and 'm/s', respectively.

3.4. Surface Topography for MQL-Assisted Grinding

Appropriate analysis of surface components, in terms of waviness, form, and roughness and multi-scalar feature topographical features, is very important. Surface topography was studied using noncontact 3D surface profilometer (S NEOX) with magnification $\times 1000$ for AISI D2 steel in the MQL-assisted grinding process (Figure 9a). Figure 9b,c highlights 2D and 3D topography. The surface topography area was considered for 2mm along the grinding width direction and 1.5 mm along the grinding direction. From the experimental data, it was found that the table speed has a significant influence on ground surface roughness and topography. Figure 9 shows the machined surface topography of the ground surface at depth of cut of 15 μm, table speed of 3 m/min, cutting speed of 25 m/s, and MQL flow rate of 250 mL/h; this was obtained using SENSOFAR with magnification $\times 1000$ for AISI D2 steel in the MQL-assisted grinding process. Figure 9 indicates that in the MQL-assisted grinding process the plastic deformation by the mechanical load, distortion of the surface layer, and pull-out of grains are not present on the ground surfaces. MQL technique has fewer defects, especially no plastic deformation, due to higher lubrication conditions. It was noted that in MQL-assisted grinding a minimum of 0.31 μm surface roughness was achieved.

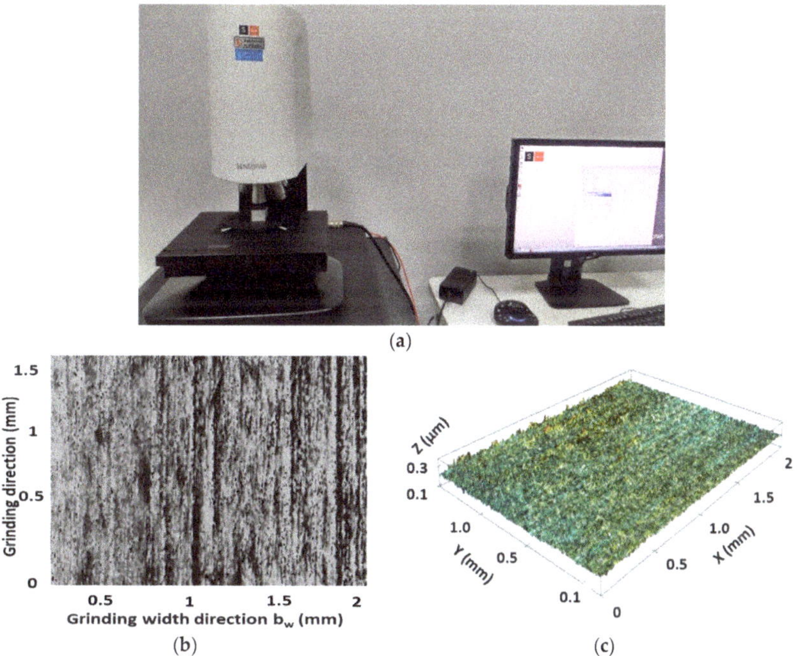

Figure 9. (a) General view of 3D optical profilometer S neox (Sensofar, Barcelona, Spain), (b) micrography of the ground surface, and (c) surface topography of the ground surface.

4. Formulation and Validation of Multi-Objective Optimization

Due to superior nature of MQL-grinding, as delineated earlier, compared to dry as well as conventional wet grinding condition, the optimization is performed only for MQL-grinding. Amulti-objection optimization was performed to concurrently optimize the normal force, surface roughness, and temperature to select optimum cutting parameters in MQL-assisted grinding. It is pertinent to mention that, to avoid over presentation of the data and calculation, only the ANOVA for the Grey Relational Grade (GRG) has been displayed.

4.1. Evaluation of Optimal Parameters for MQL Grinding

Single response optimization is simple to perform and commonly used to address the problems for optimization. However, that method cannot be used to solve multiple-criteria decision-making problems [6]. Therefore, to solve this problem, in this study the response surface methodology (RSM)-based grey relational analysis (GRA) method has been deployed for multiresponse optimization. Grey relational analysis consists of three simple steps: (a) Normalizing measured data for surface roughness, temperature, and force (minimum the better) using Equation (11); (b) calculation of Grey relational coefficient (GRC) using Equation (12); and (c) finally, computing grey relational grade based on equal weights using Equation (13). The detailed method of calculating grey relational coefficient (GRC) and grey relational grade (GRG) can be seen from previous works [41,55].

$$x_i^*(k) = \frac{\max(x_i^0) - x_i^0(k)}{\max(x_i^0(k) - \min(x_i^0(k)))} \tag{11}$$

$$\xi(k) = \frac{\Delta_{min} + \xi\Delta_{max}}{\Delta 0_i(k) + \xi\Delta_{max}} \tag{12}$$

$$\gamma_i = \frac{\sum_{k=1}^{n} w_k \xi_i(k)}{n} \tag{13}$$

Here, $\Delta 0_i(k)$ is known as deviation sequence and it is the absolute value. The minimum and maximum values of $\Delta 0_i(k)$ are denoted by Δ_{min} and Δ_{max} respectively. γ_i represents the grey relational grade of each experiment. The values of ξ lies between 0 and 1 and it is called distinguishing coefficient, normally its value is taken as 0.5. w_k represents the normalized weight of factor k. Table 5 consists of GRC and GRG of MQL grinding.

Table 5. Grey relational coefficient (GRC) and grey relational grade (GRG) for MQL-grinding.

No.	GRC of Surface Roughness	GRC of Temperature	GRC of Normal Force	GRG
1	0.7143	0.5000	0.5870	0.6004
2	0.6897	0.3896	0.4500	0.5098
3	0.4255	0.4286	0.3600	0.4047
4	0.4082	0.3448	0.3333	0.3621
5	0.7143	0.4615	0.6000	0.5919
6	0.5000	0.3896	0.5745	0.4880
7	0.3922	0.4110	0.3750	0.3927
8	0.3636	0.3377	0.3971	0.3659
9	0.9091	0.7895	0.7105	0.8030
10	0.7018	0.5882	0.5745	0.6215
11	0.4444	0.7317	0.4154	0.5305
12	0.4000	0.5660	0.3506	0.4389
13	0.8333	0.7317	0.7105	0.7585
14	0.6061	0.5455	0.6585	0.6034
15	0.4762	0.6000	0.4286	0.5016
16	0.3846	0.5263	0.3649	0.4253
17	0.6452	0.5660	0.5294	0.5802
18	0.4762	0.4054	0.4091	0.4302
19	**1.0000**	**0.7317**	**1.0000**	**0.9106**
20	0.3333	0.6000	0.3375	0.4236
21	0.5556	0.4412	0.4219	0.4729
22	0.5000	0.4054	0.4821	0.4625
23	0.5128	0.3333	0.4030	0.4164
24	0.5882	1.0000	0.4500	0.6794
25	0.5263	0.4615	0.5000	0.4960
26	0.5556	0.4762	0.4737	0.5018
27	0.5405	0.4478	0.4909	0.4931
28	0.5882	0.4688	0.4737	0.5102
29	0.5556	0.4839	0.5625	0.5340
30	0.5714	0.4688	0.4909	0.5104

Bold indicates the optimum run.

4.2. ANOVA for Grey Relational Grade

ANOVA results give useful information about significant grinding parameters at the 5% significance level (95% confidence level). As listed in Table 6, ANOVA has been carried out to analyze the effect of grinding parameters on multiresponses.

Table 6. ANOVA analysis for GRG of MQL grinding.

Source	SS	df	MS	*F*-Value	*p*-Value
Model	0.4686	14	0.0335	61.1807	<0.0001
Depth of Cut (a_p)	0.0476	1	0.0476	86.9630	0.0001
Table Speed (v_w)	0.2664	1	0.2664	487.0265	<0.0001
Cutting Speed (v_s)	0.0011	1	0.0011	2.0556	0.1722
MQL Flow rate (Q)	0.0929	1	0.0929	169.8022	0.0001
$a_p \times v_w$	0.0054	1	0.0054	9.8734	0.0067
$a_p \times v_s$	0.0001	1	0.0001	0.2234	0.6432
$a_p \times Q$	0.0036	1	0.0036	6.6193	0.0212
$v_w \times v_s$	0.0001	1	0.0001	0.2026	0.6591
$v_w \times Q$	0.0032	1	0.0032	5.8058	0.0293
$v_s \times Q$	0.0003	1	0.0003	0.5101	0.4861
$a_p{}^2$	0.0005	1	0.0005	0.9133	0.3544
$v_w{}^2$	0.0359	1	0.0359	65.7029	0.0001
$v_s{}^2$	0.0051	1	0.0051	9.3403	0.008
Q^2	0.0011	1	0.0011	2.0548	0.1722
Residual	0.0082	15	0.0005		
Lack of Fit	0.0071	10	0.0007	3.2625	0.102
Pure Error	0.0011	5	0.0002		
Cor. Total	0.4768	29			
Std. Dev.	0.023		R-Squared		0.9828
Mean	0.53		Adj R-Squared		0.9667
C.V. %	4.44		Pred R-Squared		0.9107
PRESS	0.043		Adeq Precision		30.121

The results show the highest influence was exerted by the table speed on GRG. Moreover, the *p*-value (0.0001), which is less than 0.05, shows the significance of the model. For GRG, the main effects of depth of cut, table speed, and MQL flow rate; interaction effects of depth of cut and cutting speed ($a_p \times v_s$), table speed, and cutting speed ($v_w \times v_s$); depth of cut, MQL flow rate ($a_p \times Q$), table speed, and MQL flow rate $v_w \times Q$; and quadric effects of cutting speed ($v_s{}^2$) and table speed ($v_w{}^2$), were identified as the significant model terms. The ANOVA results comprising of significant model terms along with adequacy measures R^2, adjusted R^2 and predicted R^2 are shown in Table 6. The values of adequacy measures R^2, adjusted R^2 and predicted R^2 close to 1 indicate the adequacy of the resulting models. The empirical models for the prediction GRG has been provided in Equation (14).

$$\begin{aligned} MQL_{(GRG)} = {} & +0.51 - \left(0.045 \times a_p\right) - \left(0.11 \times v_w\right) - \left(6.8 \times 10^{-3} \times v_s\right) + \\ & \left(0.062 \times Q\right) + \left(0.018 \times a_p \times v_w\right) - \left(0.015 \times a_p \times Q\right) - \left(0.014 \times V_w \times Q\right) - \\ & \left(4.2 \times 10^{-3} \times a_p{}^2\right) + \left(0.036 \times v_w{}^2\right) - \left(0.014 \times v_s{}^2\right) + \left(6.4 \times 10^{-3} \times Q^2\right) \end{aligned} \tag{14}$$

4.3. Confirmatory Test

The statistical model has been employed to verify the adequacy of the developed model. In the final step of RSM, to validate the model experimentally, confirmation experiments have been performed using RSM based grey relational analysis GRA technique. For the determination of improvement in responses and to validate the accuracy of optimization, confirmation experiments were employed at initial levels (experiment 19) and optimal levels. These experiments were repeated three times to mitigate the error. It is relevant to mention that multi-objective optimization and confirmation experiments are performed for all environments, however, Table 5 consist of data only for MQL environments. The estimated grey relational grade, which represents optimal cutting conditions, can be computed as follows,

$$\gamma_p = \gamma_m + \sum_{i=1}^{k} (\gamma_i - \gamma_m) \tag{15}$$

where, γ_p is a predicted grey relational grade, γ_m is a total mean of GRG, and γ_i is an optimum average GRG value at specific cutting parameter level, $i = 1, 2, \ldots, k$, where 'k' is the number of parameters significantly affect the GRG [56].

It was noted from Table 7 that percentage improvement in the GRG from initial cutting conditions ($a_p{}^3$-$v_w{}^1$-$v_s{}^3$-Q^3) to the optimum cutting condition ($a_p{}^1$-$v_w{}^1$-$v_s{}^3$-Q^5) was 3.18 and these results are in good agreement with the results from the literature. Pawade and Joshi [55] employed the Taguchi grey relational analysis (TGRA) method in the turning process and they achieved 4.11% achievement in optimum GRG when compared with initial conditions.

Table 7. Confirmation experiments.

Initial Cutting Conditions		Optimal Cutting Conditions	
		Predicted Results	Experimental Results
Levels	$a_p{}^3$-$v_w{}^1$-$v_s{}^3$-Q^3	$a_p{}^1$-$v_w{}^1$-$v_s{}^3$-Q^5	$a_p{}^1$-$v_w{}^1$-$v_s{}^3$-Q^5
Surface roughness (μm)	0.31		0.29
Temperature (°C)	85		81
Normal force (N)	17		16
GRG	0.9106	0.9242	0.9396
Improvement in GRG = 0.029; the % improvement in GRG = 3.18			

5. Conclusions

In this paper, the effects of the cutting parameter and the cooling mode on the grinding zone temperature, normal forces, and ground surface quality of AISI D2 steel were investigated. The influence of process parameter on three different responses has been analyzed based on the developed mathematical model. A mathematical model for each response in each cooling mode has been developed to correlate the process parameter. Finally, multi-objective optimization was performed for MQL cooling mode. Based on the results of the experimental investigation, the following comparative conclusions for MQL grinding can be drawn.

1. In all grinding conditions, the surface roughness is profoundly affected by the table speed, and the surface roughness values achieved in wet and MQL conditions are comparable.
2. Depth of cut has been found as the critical process parameter for temperature in all cutting conditions. Comparison results for temperatures show that overall minimum temperature was achieved in wet grinding followed by MQL-assisted grinding. Moreover, it was noted that MQL grinding results in a temperature reduction of 55.18 to 67.4% as compared to dry grinding.
3. Cutting forces were found increasing as the depth of cut was increased. Comparative results for forces have shown that minimum cutting forces were achieved in MQL environment. When compared with dry grinding, 37.86–79.26% reduction in forces was found in MQL-assisted grinding.
4. During the experiments, it was observed that MQL flow rates values used in the grinding process did not cause in the dispersion of mist, which is environmentally friendly. More interesting to mention, the MQL maintained the cutting zone clean like dry grinding, and it produced a comparable, and sometimes better, result when compared to wet grinding.
5. The modeling and optimization of the MQL-assisted grinding process can be applied in the green manufacturing process. It was noted that RSM prediction is well matched with the experimental results. Also, research findings and a mathematical model developed in this study can be used in

the industry by the machinist to predict the surface roughness, temperature, and normal forces for grinding of AISI D2 steel.

6. The work surface quality in MQL is comparable to wet condition. Therefore, MQL is recommended over flood cooling due it MQL's economic and clean production perspective. However, considerable further research is needed to fully exploit its potential.

Few works have been published to check the effectiveness of vegetable oils in MQL-assisted grinding. Use of a mixture of two or more oils or may be explored with the addition of nanoparticles, which can enhance the lubricating and oxidation stability of these oils. Still, efforts are needed to improve the surface integrity of difficult to cut materials during machining under the MQL-assisted lubri-cooling mode. Moreover, analysis of the surface integrity of tritium and Inconel alloys using hybrid lubri-cooling techniques such as MQL and LN_2 is still not exploded very well. Finally, further research should be carried out on how these newly developed technologies can become sustainable regarding economic and environmental aspects.

Author Contributions: Conceptualization, A.M.K., M.J. and N.H.; Methodology, A.M.K., M.J. and N.H.; Software, A.M.K. and M.J.; Validation, A.M.K., M.J. and N.H.; Formal Analysis, A.M.K., M.J. and N.H.; Investigation, A.M.K., M.J. and N.H.; Resources, A.M.K. and M.J.; Data Curation, A.M.K. and M.J.; Writing-Original Draft Preparation, A.M.K., M.J. and M.M.; Writing-Review & Editing, A.M.K., M.J., M.M., D.Y.P., V.R.G. and M.K.G.; Visualization, A.M.K. and M.J.; Supervision, A.M.K., M.J., M.M., D.Y.P., V.R.G. and M.K.G.; Project Administration, A.M.K. and M.J.; Funding Acquisition, A.M.K., M.J., D.Y.P. and V.R.G.

Funding: The work was supported by the Fundamental Research Funds for the Central Universities, NO. NP2018302.

Acknowledgments: The authors would like to acknowledge the efforts made by Kornel F. Ehmann of Northwestern University (USA) who reviewed the manuscript and gave some valuable suggestions.

References

1. Malkin, S.; Guo, C. *Grinding Technology: Theory and Application of Machining with Abrasives*; Industrial Press Inc.: London, UK, 2008.
2. Rowe, W. *Principles of Modern Grinding Technology*; William Andrew: Liverpool, UK, 2014.
3. Guo, G.; Liu, Z.; An, Q.; Chen, M. Experimental investigation on conventional grinding of Ti-6Al-4V using SiC abrasive. *Int. J. Adv. Manuf. Technol.* **2011**, *57*, 135–142. [CrossRef]
4. Howes, T.D.L.; Tönshoff, H.K.; Heuer, W.; Howes, T. Environmental Aspects of Grinding Fluids. *CIRP Ann. Manuf. Technol.* **1991**, *40*, 623–630. [CrossRef]
5. Amiril, S.A.S.; Rahim, E.A.; Embong, Z.; Syahrullail, S. Tribological investigations on the application of oil-miscible ionic liquids interested in modified Jatropha-based metalworking fluid. *Tribol. Int.* **2018**, *120*, 520–534. [CrossRef]
6. Mashood Khan, A.; Ning, H.; Liang, L.; Jamil, M. Comment to paper entitled "Experimental investigation of machinability characteristics and multiresponse optimization of end milling in aluminium composites using RSM based grey relational analysis". *Measurement* **2018**, *119*, 175–177. [CrossRef]
7. Mia, M.; Gupta, M.K.; Singh, G.; Królczyk, G.; Pimenov, D.Y. An approach to cleaner production for machining hardened steel using different cooling-lubrication conditions. *J. Clean. Prod.* **2018**, *187*, 1069–1081. [CrossRef]
8. Mia, M.; Bashir, M.A.; Khan, M.A.; Dhar, N.R. Optimization of MQL flow rate for minimum cutting force and surface roughness in end milling of hardened steel (HRC 40). *Int. J. Adv. Manuf. Technol.* **2017**, *89*, 675–690. [CrossRef]
9. Mia, M.; Khan, M.A.; Rahman, S.S.; Dhar, N.R. Mono-objective and multi-objective optimization of performance parameters in high pressure coolant assisted turning of Ti-6Al-4V. *Int. J. Adv. Manuf. Technol.* **2017**, *90*, 109–118. [CrossRef]
10. Dhar, N.R.; Islam, M.W.; Islam, S.; Mithu, M.A.H. The influence of minimum quantity of lubrication (MQL) on cutting temperature, chip and dimensional accuracy in turning AISI-1040 steel. *J. Mater. Process. Technol.* **2006**, *171*, 93–99. [CrossRef]

11. Wang, Y.; Li, C.; Zhang, Y.; Yang, M.; Li, B.; Jia, D.; Hou, Y.; Mao, C. Experimental evaluation of the lubrication properties of the wheel/workpiece interface in minimum quantity lubrication (MQL) grinding using different types of vegetable oils. *J. Clean. Prod.* **2016**, *127*, 487–499. [CrossRef]

12. Abbas, A.T.; Pimenov, D.Y.; Erdakov, I.N.; Taha, M.A.; Soliman, M.S.; El Rayes, M.M. ANN surface roughness optimization of AZ61 magnesium alloy finish turning: Minimum machining times at prime machining costs. *Materials* **2018**, *11*, 808. [CrossRef] [PubMed]

13. Abbas, A.T.; Pimenov, D.Y.; Erdakov, I.N.; Taha, M.A.; El Rayes, M.M.; Soliman, M.S. Artificial intelligence monitoring of hardening methods and cutting conditions and their effects on surface roughness, performance, and finish turning costs of solid-state recycled aluminum alloy 6061 chips. *Metals* **2018**, *8*, 394. [CrossRef]

14. De Lacalle, L.N.L.; Rodriguez, A.; Lamikiz, A.; Celaya, A.; Alberdi, R. Five-axis machining and burnishing of complex parts for the improvement of surface roughness. *Mater. Manuf. Process.* **2011**, *26*, 997–1003. [CrossRef]

15. Beranoagirre, A.; López de Lacalle, L.N. Grinding of gamma TiAl intermetallic alloys. *Procedia Eng.* **2013**, *63*, 489–498. [CrossRef]

16. Hao, X.; Cui, W.; Li, L.; Li, H.; Khan, A.M.; He, N. Cutting performance of textured polycrystalline diamond tools with composite lyophilic/lyophobic wettabilities. *J. Mater. Process. Technol.* **2018**, *260*, 1–8. [CrossRef]

17. Pashnyov, V.A.; Pimenov, D.Y. Stress analysis of a three-layer metal composite system of bearing assemblies during grinding. *Mech. Compos. Mater.* **2015**, *51*, 109–128. [CrossRef]

18. Pereverzev, P.P.; Pimenov, D.Y. A grinding force model allowing for dulling of abrasive wheel cutting grains in plunge cylindrical grinding. *J. Frict. Wear* **2016**, *37*, 60–65. [CrossRef]

19. Pereverzev, P.P.; Popova, A.V.; Pimenov, D.Y. Relation between the cutting force in internal grinding and the elastic deformation of the technological system. *Russ. Eng. Res.* **2015**, *35*, 215–217. [CrossRef]

20. Pashnyov, V.A.; Pimenov, D.Y.; Erdakov, I.N.; Koltsova, M.S.; Mikolajczyk, T.; Patra, K. Modeling and analysis of temperature distribution in the multilayer metal composite structures in grinding. *Int. J. Adv. Manuf. Technol.* **2017**, *91*, 4055–4068. [CrossRef]

21. Nadolny, K.; Rokosz, K.; Kapłonek, W.; Wienecke, M.; Heeg, J. SEM-EDS-based analysis of the amorphous carbon-treated grinding wheel active surface after reciprocal internal cylindrical grinding of Titanium Grade 2® alloy. *Int. J. Adv. Manuf. Technol.* **2017**, *90*, 2293–2308. [CrossRef]

22. Sutowski, P.; Nadolny, K.; Kaplonek, W. Monitoring of cylindrical grinding processes by use of a non-contact AE system. *Int. J. Precis. Eng. Manuf.* **2012**, *13*, 1737–1743. [CrossRef]

23. Barczak, L.M.; Batako, A.D.L.; Morgan, M.N. A study of plane surface grinding under minimum quantity lubrication (MQL) conditions. *Int. J. Mach. Tools Manuf.* **2010**, *50*, 977–985. [CrossRef]

24. Ben Fredj, N.; Sidhom, H.; Braham, C. Ground surface improvement of the austenitic stainless steel AISI 304 using cryogenic cooling. *Surf. Coat. Technol.* **2006**, *200*, 4846–4860. [CrossRef]

25. Rabiei, F.; Rahimi, A.R.; Hadad, M.J.; Ashrafijou, M. Performance improvement of minimum quantity lubrication (MQL) technique in surface grinding by modeling and optimization. *J. Clean. Prod.* **2015**, *86*, 447–460. [CrossRef]

26. Ebbrell, S.; Woolley, N.H.; Tridimas, Y.D.; Allanson, D.R.; Rowe, W.B. Effects of cutting fluid application methods on the grinding process. *Int. J. Mach. Tools Manuf.* **2000**, *40*, 209–223. [CrossRef]

27. Saravanan, R.; Asokan, P.; Sachidanandam, M. A multi-objective genetic algorithm (GA) approach for optimization of surface grinding operations. *Int. J. Mach. Tools Manuf.* **2002**, *42*, 1327–1334. [CrossRef]

28. Reddy, P.P.; Ghosh, A. Effect of Cryogenic Cooling on Spindle Power and G-ratio in Grinding of Hardened Bearing Steel. *Procedia Mater. Sci.* **2014**, *5*, 2622–2628. [CrossRef]

29. Hamdi, H.; Zahouani, H.; Bergheau, J.-M. Residual stresses computation in a grinding process. *J. Mater. Process. Technol.* **2004**, *147*, 277–285. [CrossRef]

30. Nandi, A.K.; Pratihar, D.K. Design of a genetic-fuzzy system to predict surface finish and power requirement in grinding. *Fuzzy Sets Syst.* **2004**, *148*, 487–504. [CrossRef]

31. Witold, F.H. Effect of Bond Type and Process Parameters on Grinding Force Components in Grinding of Cemented Carbide. *Proced. Eng.* **2016**, *149*, 122–129.

32. Tawakoli, T.; Hadad, M.J.; Sadeghi, M.H. Influence of oil mist parameters on minimum quantity lubrication—MQL grinding process. *Int. J. Mach. Tools Manuf.* **2010**, *50*, 521–531. [CrossRef]

33. Emami, M.; Sadeghi, M.H.; Sarhan, A.A.D. Investigating the effects of liquid atomization and delivery parameters of minimum quantity lubrication on the grinding process of Al_2O_3 engineering ceramics. *J. Manuf. Process.* **2013**, *15*, 374–388. [CrossRef]
34. Emami, M.; Sadeghi, M.H.; Sarhan, A.A.D.; Hasani, F. Investigating the Minimum Quantity Lubrication in grinding of Al_2O_3 engineering ceramic. *J. Clean. Prod.* **2014**, *66*, 632–643. [CrossRef]
35. Chen, J.; Shen, J.; Huang, H.; Xu, X. Grinding characteristics in high speed grinding of engineering ceramics with brazed diamond wheels. *J. Mater. Process. Technol.* **2010**, *210*, 899–906. [CrossRef]
36. Prabhu, S.; Vinayagam, B.K. AFM investigation in grinding process with nanofluids using Taguchi analysis. *Int. J. Adv. Manuf. Technol.* **2012**, *60*, 149–160. [CrossRef]
37. Sanchez, J.A.; Pombo, I.; Alberdi, R.; Izquierdo, B.; Ortega, N.; Plaza, S.; Martinez-Toledano, J. Machining evaluation of a hybrid MQL-CO_2 grinding technology. *J. Clean. Prod.* **2010**, *18*, 1840–1849. [CrossRef]
38. Zhang, D.; Li, C.; Jia, D.; Zhang, Y.; Zhang, X. Specific grinding energy and surface roughness of nanoparticle jet minimum quantity lubrication in grinding. *Chin. J. Aeronaut.* **2015**, *28*, 570–581. [CrossRef]
39. Zhang, Y.; Li, C.; Jia, D.; Zhang, D.; Zhang, X. Experimental evaluation of MoS_2 nanoparticles in jet MQL grinding with different types of vegetable oil as base oil. *J. Clean. Prod.* **2015**, *87*, 930–940. [CrossRef]
40. Zhao, W.; Gong, L.; Ren, F.; Li, L.; Xu, Q.; Khan, A.M. Experimental study on chip deformation of Ti-6Al-4V titanium alloy in cryogenic cutting. *Int. J. Adv. Manuf. Technol.* **2018**, *96*, 4021–4027. [CrossRef]
41. Sadeghi, M.H.; Haddad, M.J.; Tawakoli, T.; Emami, M. Minimal quantity lubrication-MQL in grinding of Ti-6Al-4V titanium alloy. *Int. J. Adv. Manuf. Technol.* **2009**, *44*, 487–500. [CrossRef]
42. Tawakoli, T.; Hadad, M.J.; Sadeghi, M.H.; Daneshi, A.; Stöckert, S.; Rasifard, A. An experimental investigation of the effects of workpiece and grinding parameters on minimum quantity lubrication-MQL grinding. *Int. J. Mach. Tools Manuf.* **2009**, *49*, 924–932. [CrossRef]
43. Nadolny, K.; Kapłonek, W. The effect of wear phenomena of grinding wheels with sol-gel alumina on chip formation during internal cylindrical plunge grinding of 100Cr6 steel. *Int. J. Adv. Manuf. Technol.* **2016**, *87*, 501–517. [CrossRef]
44. Tawakoli, T.; Hadad, M.J.; Sadeghi, M.H. Investigation on minimum quantity lubricant-MQL grinding of 100Cr6 hardened steel using different abrasive and coolant-lubricant types. *Int. J. Mach. Tools Manuf.* **2010**, *50*, 698–708. [CrossRef]
45. Silva, L.R.; Bianchi, E.C.; Catai, R.E.; Fusse, R.Y.; França, T.V.; Aguiar, P.R. Study on the behavior of the minimum wholesale lubricant—MQL technique under different lubricating and cooling conditions when grinding ABNT 4340 steel. *J. Braz. Soc. Mech. Sci. Eng.* **2005**, *27*, 192–199. [CrossRef]
46. Alves, J.A.C.; De Barros Fernandes, U.; Da Silva, J.C.E.; Bianchi, E.C.; De Aguiar, P.R.; Da Silva, E.J. Application of the minimum quantity lubrication (MQL) technique in the plunge cylindrical grinding operation. *J. Braz. Soc. Mech. Sci. Eng.* **2009**, *31*, 1–4. [CrossRef]
47. Oliveira, D.D.J.; Guermandi, L.G.; Bianchi, E.C.; Diniz, A.E.; De Aguiar, P.R.; Canarim, R.C. Improving minimum quantity lubrication in CBN grinding using compressed air wheel cleaning. *J. Mater. Process. Technol.* **2012**, *212*, 2559–2568. [CrossRef]
48. Neşeli, S.; Asiltürk, I.; Çelik, L. Determining the optimum process parameter for grinding operations using robust process. *J. Mech. Sci. Technol.* **2012**, *26*, 3587–3595. [CrossRef]
49. Pusavec, F.; Hamdi, H.; Kopac, J.; Jawahir, I.S. Surface integrity in cryogenic machining of nickel based alloy—Inconel 718. *J. Mater. Process. Technol.* **2011**, *211*, 773–783. [CrossRef]
50. Rapeti, P.; Pasam, V.K.; Rao Gurram, K.M.; Revuru, R.S. Performance evaluation of vegetable oil based nano cutting fluids in machining using grey relational analysis-a step towards sustainable manufacturing. *J. Clean. Prod.* **2016**, *172*, 2862–2875. [CrossRef]
51. Puri, A.B.; Banerjee, S. Multiple-response optimisation of electrochemical grinding characteristics through response surface methodology. *Int. J. Adv. Manuf. Technol.* **2013**, *64*, 715–725. [CrossRef]
52. Rabiei, F.; Rahimi, A.R.; Hadad, M.J. Performance improvement of eco-friendly MQL technique by using hybrid nanofluid and ultrasonic-assisted grinding. *Int. J. Adv. Manuf. Technol.* **2017**, *93*, 1001–1015. [CrossRef]
53. Montgomery, D.C. *Design and Analysis of Experiments*, 2nd ed.; Journal of the American Statistical Association: Phoenix, AZ, USA, 2000; Volume 16, pp. 241–242.
54. Ben Fathallah, B.; Ben Fredj, N.; Sidhom, H.; Braham, C.; Ichida, Y. Effects of abrasive type cooling mode and peripheral grinding wheel speed on the AISI D2 steel ground surface integrity. *Int. J. Mach. Tools Manuf.* **2009**, *49*, 261–272. [CrossRef]

55. Pawade, R.S.; Joshi, S.S. Multi-objective optimization of surface roughness and cutting forces in high-speed turning of Inconel 718 using Taguchi grey relational analysis (TGRA). *Int. J. Adv. Manuf. Technol.* **2011**, *56*, 47–62. [CrossRef]

56. Sahoo, A.K.; Sahoo, B. Performance studies of multilayer hard surface coatings (TiN/TiCN/Al$_2$O$_3$/TiN) of indexable carbide inserts in hard machining: Part-II (RSM, grey relational and techno economical approach). *Measurement* **2013**, *46*, 2868–2884. [CrossRef]

Article

Drilling Process in γ-TiAl Intermetallic Alloys

Aitor Beranoagirre [1,*]**, Gorka Urbikain** [1,*]**, Amaia Calleja** [2] **and Luis Norberto López de Lacalle** [3]

[1]   Department of Mechanical Engineering, University of the Basque Country (UPV/EHU), Plaza Europa 1, 20018 San Sebastián, Spain

[2]   Department of Mechanical Engineering, University of the Basque Country (UPV/EHU), Nieves Cano 12, 01006 Vitoria, Spain; amaia.calleja@ehu.eus

[3]   CFAA, University of the Basque Country (UPV/EHU), Parque Tecnológico de Zamudio 202, 48170 Bilbao, Spain; norberto.lzlacalle@ehu.es

*   Correspondence: aitor.beranoagirre@ehu.eus (A.B.); gorka.urbikain@ehu.eus (G.U.); Tel.: +34-943-018-636 (A.B.)

Received: 26 October 2018; Accepted: 21 November 2018; Published: 26 November 2018

Abstract: Gamma titanium aluminides (γ-TiAl) present an excellent behavior under high temperature conditions, being a feasible alternative to nickel-based superalloy components in the aeroengine sector. However, considered as a difficult to cut material, process cutting parameters require special study to guarantee component quality. In this work, a developed drilling mechanistic model is a useful tool in order to predict drilling force (F_z) and torque (T_c) for optimal drilling conditions. The model is a helping tool to select operational parameters for the material to cut by providing the programmer predicted drilling forces (F_z) and torque (T_c) values. This will allow the avoidance of operational parameters that will cause excessively high force and torque values that could damage quality. The model is validated for three types of Gamma-TiAl alloys. Integral hard metal end-drilling tools and different cutting parameters (feeds and cutting speeds) are tested for three different sized holes for each alloy.

Keywords: Gamma-TiAl; superalloys; slight materials; drilling; titanium aluminides

---

1. Introduction

Aeronautics is an emerging sector; aircraft manufacturing predictions estimate doubling of the current fleet by the year 2033, in order to satisfy passenger's increment demand that grows at a rate of 4.2% per year [1]. Components manufacturing for the aeronautic sector is a high value added process, and, concretely, motor components require special attention because they represent one of the most expensive (20%) components [2]. The aviation industry will try to satisfy manufacturing demand according to regulation requirements regarding efficiency, noise, fuel consumption and contamination. Therefore, new materials and manufacturing processes are under development by aeronautic sector manufacturers. In this sense, γ-TiAl alloys are a feasible replacement for nickel-based alloys frequently used for compressor blades and stator in gas turbine aeroengines [3–5]. Moreover, titanium alloys are also interesting for bio-medical engineering, automobile sector and chemical industries [6].

Gamma titanium aluminide intermetallics present excellent strength-weight ratio, and corrosion resistance at high temperature [7,8]. However, it is considered a difficult to cut material [9,10] due to its poor tensile and low room temperature ductility (<2%). Besides, high heat values are generated during γ-TiAl machining due to low thermal conductivity o the material. Consequently, tool and workpiece wear is accelerated [11], tool life is reduced [12] and workpiece integrity is affected [13]. Moreover, γ-TiAl reacts chemically with many materials causing material adhesion.

Several studies have considered γ-TiAl a difficult to cut material regarding turning [14,15], grinding [16], high speed milling [17–19], drilling and micro-drilling [20].

In this sense, high machining conditions can cause irreversible consequences such as surface cracking, hardened layers and tensile residual stresses [21]. Optimal machining strategies and parameters are studied for electrochemical machining [22], turning [23,24], and milling [25] related to cutting temperature technique evaluation, and, appropriate machining conditions in order to reduce tool wear and increase tool life. Milling studies have also investigated the effects of operating parameters and conditions on tool life and surface integrity when milling γ-TiAl alloys [26–28].

In relation to lubrication techniques, some studies [29] focus on cryogenic lubrication (liquid nitrogen) obtaining cutting feed value increments while tool life is maintained.

Machining processes modelling is also a hot topic for difficult to cut materials. In this sense, two-dimensional models [30] were developed for cutting parameters such as cutting speed and feed influence determination. Numerical and experimental analyses of residual stresses generated in the machining of Ti6A14V titanium alloy are developed [31]. Empirical models are also used for thrust and torque values prediction in composites. Other techniques such as response surface methodology (RSM) and finite element analysis (FEA) are implemented for milling [32] and drilling [33] process parameter analysis. In addition, finite element models are also developed for performance characteristics such as hole quality prediction [34]. Other techniques are focused on the side of monitoring. For instance, power consumption during drilling is a key magnitude in predictive/preventive techniques [35,36].

Considering γ-TiAl promising application possibilities, drilling machinability studies are a focus of interest due to presented machinability problems, drilling being one of the most frequent processes for component assembly. In this work, the developed drilling mechanistic model is a useful tool in order to predict drilling force (F_z) and torque (T_c) for optimal drilling conditions determination. The model is validated for three types of Gamma-TiAl alloys. Integral hard metal end-drilling tools and different cutting parameters (feeds and cutting speeds) are tested for three different sized holes for each alloy.

2. Experimental Procedure

2.1. Material

Four different titanium alloys are tested in performed drilling tests: Ti-6Al-4V and three different γ-TiAl alloys (TNB, extruded MoCuSi and ingot MoCuSi). Table 1 shows the tested alloys' mechanical properties.

Table 1. Mechanical properties comparison between TiAl alloys.

Property	TNB	Ti–6Al–4V (Annealed)	MoCuSi Extruded	MoCuSi Ingot
Density (g/cm^3)	3.86	4.49	3.74	3.88
Specific modulus (GPa/(mg·m^{-3}))	43	24	43	37
Tensile strength (MPa)	683	1087	607	689
Specific strength (MPa/(g·cm^{-3}))	192	947	198	180
Yield strength (MPa)	589	942	589	570
Ductility (%)	1.9	7.8	1.7	2.4
Fracture toughness (MPa·m$^{1/2}$)	23	52	23	20
Thermal conductivity (W/(m·K))	24	8.6	24	19
Maximum operating temperature (°C)	900	615	900	865

γ-TiAl alloys present higher aluminum percentage in comparison to other titanium alloys such as Ti-6Al-4V, 43–48% in γ-TiAl and 6% in Ti-6Al-4V, improving thermal conductivity in γ-TiAl. On the other hand, ductile transition temperature occurring between 600 and 800 °C, depending on the microstructure and grain size alloys, is increased for alloys with higher titanium percentage [12]. The intermetallic TiAl provides low density [37] as well as high mechanical strength under high temperatures and corrosive environments. The intermetallic γ-TiAl superalloys offer excellent mechanical properties [38], with low density (4 gr/cm^3), high resistance at high temperatures, low electrical and thermal conductivity, oxidation resistance, ultimate strength of 1000 MPa and Young' s modulus of 160 GPa.

MoCuSi alloy [Ti-(43–46) Al-(1–2) Mo-(0.2) Si-Cu] is used at low temperatures with high resistance below 650 °C, and, TNB alloy [Ti-(44–45) Al-(5–10) Nb-(0.2–0.4) C] resists very high temperatures maintaining high resistance and oxidation values. Regarding MoCuSi alloys, the material is presented in both, extruded and ingot structure. Extruded alloys present an oriented structure oriented in the extrusion direction whereas melted alloys present a structure without any preferable orientation, typical of no extruded or laminated materials. The ingot structure is directly obtained from the VAR (Vacuum Arc Remelting) process. For extruded structures, material is extruded at 1200 °C and smaller sized grains are obtained with superior creep strain, yield strength and KIC.

2.2. Equipment

Machining tests were performed in a vertical CNC milling machine, Kondia® model B640 (Kondia, Elgoibar, Spain), with maximum rotational speed of 10,000 rpm and 25 kW. During the tests, process-cutting conditions are measured and recorded. For the axial/thrust (F_z) and radial (F_x, F_y) cutting forces, and, Z axis (T_c) torque measurements, a dynamometric Kistler® equipment 9257B (Kistler, Winterthur, Switzerland) was used (Figure 1).

Figure 1. Kondia® model B640 (**left**) and Kistler® 9257B dynamometer (**right**).

2.3. Machining Conditions

Drilling tests were carried out for the described four titanium alloys (Table 1). As can be seen in Table 2, for each material, tested feed values are 0.05 and 0.1 mm/rev for calibration tests and 0.06 and 0.08 mm/rev for validation tests, and, $D = 3, 4, 5, 6, 7$ mm (pilot holes) and 8.5 mm holes (final holes) were drilled in each case (Figure 2). The drilled depth is 20 mm. $D = 3, 4, 5, 6, 7$ mm (pilot holes) and 8.5 mm holes (final holes) are used for calibration, and, $D = 4$–6 mm (pilot holes) and 8.5 mm holes (final holes) are used for the validation tests.

Table 2. Machining conditions for the 8.5 mm hole.

Material	v_c [m/min]	n [rev/min]	f [mm/rev]	f [mm/rev]
Ti-6Al-4V	50	1874	0.05	0.1
TNB	15	1874	0.05	0.1
MoCuSi Extruded	15	562	0.05	0.1
MoCuSi Ingot	15	562	0.05	0.1

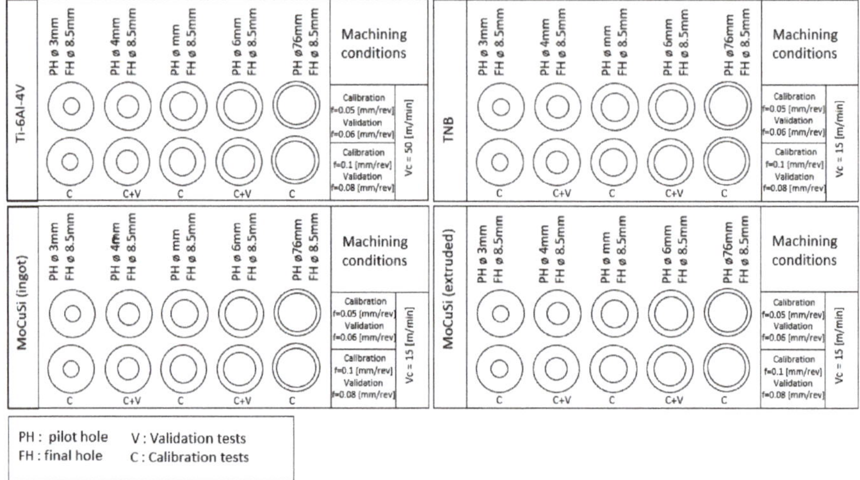

Figure 2. Experimental holes set–up.

The selected tool is a solid carbide drill (Mitsubishi®, Tokyo, Japan, MPS0850S-DIN-C, Figure 3). During the machining operations, cutting forces, torque and power consumption were recorded.

Figure 3. Cutting tool geometry.

One of the critical aspects when drilling γ-TiAl is chip evacuation and heat dissipation. Internal lubrication is directly applied on the cutting edge, due to the poor thermal conductivity of these alloys. Specially designed for low machinability materials, Rhenus® FU W (Table 3, Rhenus, Mönchengladbach, Germany) coolant was used. Internal coolant pressure is set to 8.5 bar.

Table 3. Properties of Rhenus® FU 70 W coolant.

Concentrated		Emulsion	
Viscosity 20 °C (mm^2/s)	Content of mineral oil %	pH Value 5% concentration	Protection against corrosion (DIN 51360/1)
Approx. 150	Approx. 33	Approx. 9,0	Note 0 al 2%

3. Cutting Forces Prediction Mechanistic Model

Predictive cutting forces mechanistic models help programmers with cutting parameter selection. Especially when drilling low machinability titanium alloys, it is interesting to simulate different machining conditions in order to decide whether the estimated cutting forces values lead to an optimum machining process.

This section explains the modelling of cutting forces and torque in drilling. Since the cutting speed varies along the drill's lips, the way the cutting force coefficients are introduced inside the model is a crucial issue. There have been a number of attempts based on orthogonal-oblique transformation [39] and mechanistic models [40].

The cutting edge is divided into discrete elements in the drill axis ($j = 1$ to n elements) and the global cutting force and torque are obtained by summing the elemental contributions along each lip

(i = 1 to Z). As it can be seen in Figure 4, while F_x and F_y are coupled due to tool rotation leading to the radial Fr and tangential Ft cutting force components, F_z is directly obtained from the axial component Fa. Here, we will focus on the thrust force (Equation (1)) in Z direction and torque (Equation (2)), which are critical parameters when characterizing tool life. These magnitudes are calculated as:

$$F_z(t) = \sum_{i=1}^{Z} \sum_{j=1}^{n} K_{c,z}(h,r) \cdot \Delta z \cdot h \tag{1}$$

$$T_c(t) = \sum_{i=1}^{Z} \sum_{j=1}^{n} K_{t,z}(h,r) \cdot \Delta z \cdot h \cdot r \tag{2}$$

where K_z and K_T are the corresponding thrust force and torque coefficients, Δz is the axial width of a differential element, h the chip thickness (here, $h = 0.5f \cdot sin70°$) and r the tool radius. As usual, the thrust cutting force and torque coefficients need to be calibrated. Here, 4 different titanium alloys were tested: (1) Ti-6Al-4V, taken as reference; (2) TNB and MoCuSi types, (3) extruded and (4) ingot types. Under this approach, the cutting force coefficients were assumed a function of the feedrate (or h) and radial distance of the cutting point to drill axis as quadratic functions. First, a_0-a_5 and b_0-b_5 coefficients in Equations (3) and (4) are identified from a linear regression for the four types of materials from a set of experiments.

$$K_z(h,r) = a_0 + a_1 \cdot h + a_2 \cdot r + a_3 \cdot hr + a_4 \cdot r^2 + a_5 \cdot h^2 \tag{3}$$

$$K_T(h,r) = b_0 + b_1 \cdot h + b_2 \cdot r + b_3 \cdot hr + b_4 \cdot r^2 + b_5 \cdot h^2 \tag{4}$$

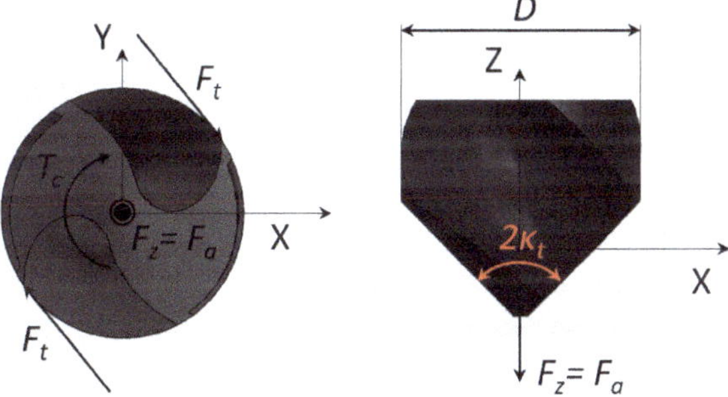

Figure 4. Cutting or tangential force (F_t), thrust force (F_z) and cutting torque (T_c) in the drilling process; Drill bottom view (**left**) and front view (**right**).

4. Results

The mechanistic model is valid for v_c = 50 m/min (Ti-6Al-4V), and v_c = 15 m/min (TNB, and extruded/ingot types), the cutting speed at the tool diameter D. To identify the cutting coefficients, 10 tests were done. After a pilot hole, the hole was finished at D = 8.5 mm. Five different pilot hole diameters (D_0 = 3, 4, 5, 6, 7 mm) and two feed rates (f = 0.05–0.1 mm) were programmed.

The corresponding functions for the cutting coefficients are summarized in Table 8 with the obtained coefficients of polynomial functions K_z and K_T in Equations (3) and (4).

4.1. Results for Ti-6Al-4V Alloy

As can be seen, both experimental and the predicted thrust force and torque values are higher for higher feed rate values (Figure 5). Table 4 shows the experimental data for both variables during the calibration of the cutting coefficients. While the curve F_z reflects almost a linear tendency, the torque is approximated as a curve with an inflexion (third order). At the same time, the higher the dispersion, the lower is the difference between the pilot and final hole diameters.

Table 4. Experimental results for F_z [N] and T_c [Nm] in Ti-6Al-4V alloy.

D_0 [mm]	F_z [N]		T_c [Nm]	
	f [mm/rev]		f [mm/rev]	
	0.05	0.1	0.05	0.1
3	591	723	3.1	3.9
4	460	565	2.4	3.7
5	319	440	2.1	3.4
6	242	332	1.9	3.2
7	169	225	1.8	3.1

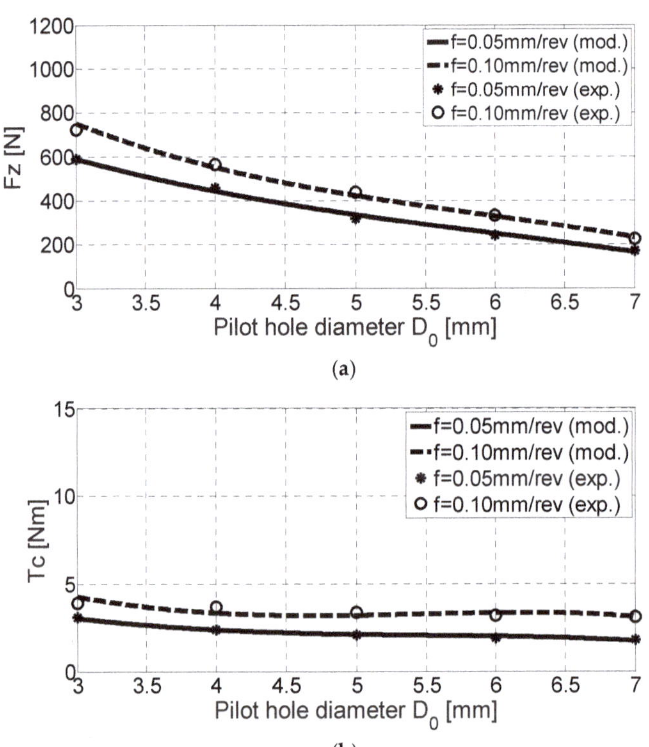

Figure 5. F_z [N] (**a**) and T_c [Nm] (**b**) for Ti-6Al-4V alloy drilling experimental and mechanistic model values.

4.2. Results for TNB Alloy

Thrust force values (Figure 6a) show higher values for higher feed rate values but are not doubled when doubling the feed rate. However, the cutting tool behaved correctly despite that the cutting force increased ×1.33 and cutting torque increased ×1.5–2, with respect to Ti4Al6V. When feed rates values are multiplied ×2, torque values are multiplied ×1.3 approximately. Table 5 shows again the experimental data necessary to obtain the polynomials of the cutting coefficients. A higher dispersion is seen for the cutting torque T_c as the pilot hole diameter approaches the final hole diameter (Figure 6b).

Table 5. Experimental results for F_z [N] and T_c [Nm] in TNB alloy.

D_0 [mm]	F_z [N]		T_c [Nm]	
	f [mm/rev]		f [mm/rev]	
	0.05	0.1	0.05	0.1
3	802	954	5.8	7.0
4	631	765	5.3	6.7
5	476	575	4.3	6.1
6	343	421	4.0	6.0
7	240	259	3.5	5.7

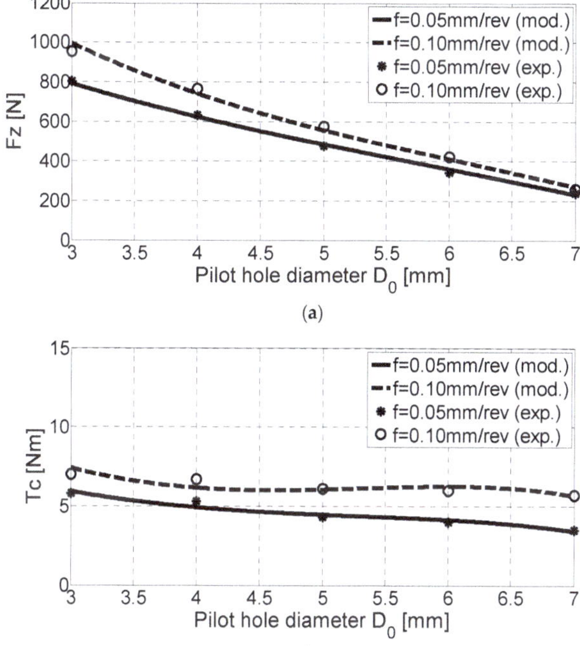

Figure 6. F_z [N] (**a**) and T_c [Nm] (**b**) for TNB alloy drilling experimental and mechanistic model values.

4.3. Results for Ingot MoCuSi Alloy

Ingot MoCuSi alloys are harder Ti-6Al-4V alloys and TNB alloys. Table 6 shows the main values. This statement was verified on sight of Figure 7. The axial force F_z and cutting torque T_c are higher regarding previous alloys. Focusing on cutting torque (Figure 7b), when the feed rate value is multiplied $\times 2$, torque values are multiplied $\times 1.1$ approximately.

Table 6. Experimental results for F_z [N] and T_c [Nm] in ingot MoCuSi alloy.

D_0 [mm]	F_z [N]		T_c [Nm]	
	f [mm/rev]		f [mm/rev]	
	0.05	0.1	0.05	0.1
3	852	906	6.0	8.7
4	702	775	5.5	8.3
5	525	600	4.8	7.5
6	419	466	4.6	7.3
7	275	290	4.0	7.0

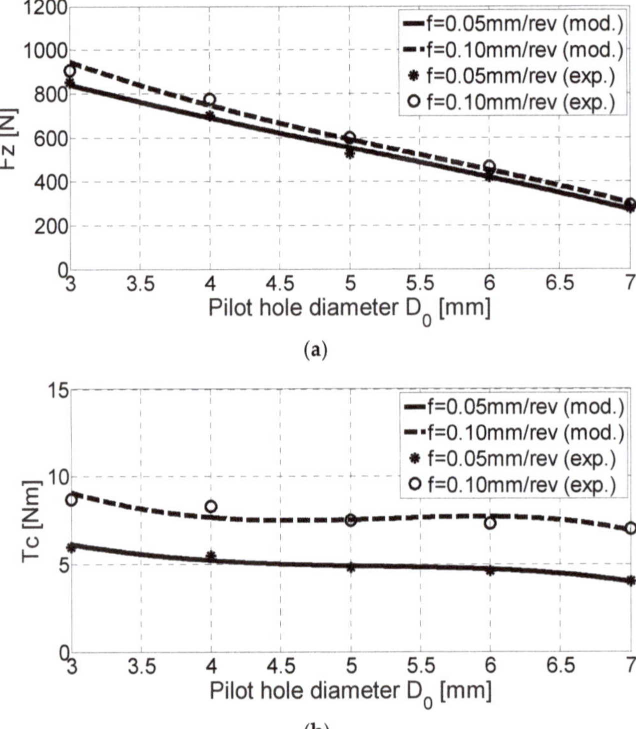

Figure 7. F_z [N] (**a**) and T_c [Nm] (**b**) for ingot MoCuSi alloy drilling experimental and mechanistic model values.

4.4. Results for Extruded MoCuSi Alloy

MoCuSi extruded alloys present the highest hardness values in comparison to the latter ones. While the results for the cutting torque T_c are quite similar to the ingot MoCuSi alloy (both results are almost identical), the thrust force is more dramatic for the extruded MoCuSi alloy. Regarding cutting torque (Figure 8b), when the feed rate value is multiplied by two, torque values are multiplied ×1.2 approximately (see also results in Table 7).

Table 7. Experimental results for F_z [N] and T_c [Nm] in extruded MoCuSi alloy.

D_0 [mm]	F_z [N]		T_c [Nm]	
	f [mm/rev]		f [mm/rev]	
	0.05	**0.1**	**0.05**	**0.1**
3	925	1050	6.5	9.3
4	751	860	6.1	8.6
5	550	621	5.5	8.0
6	427	475	4.7	7.7
7	280	292	4.0	7.5

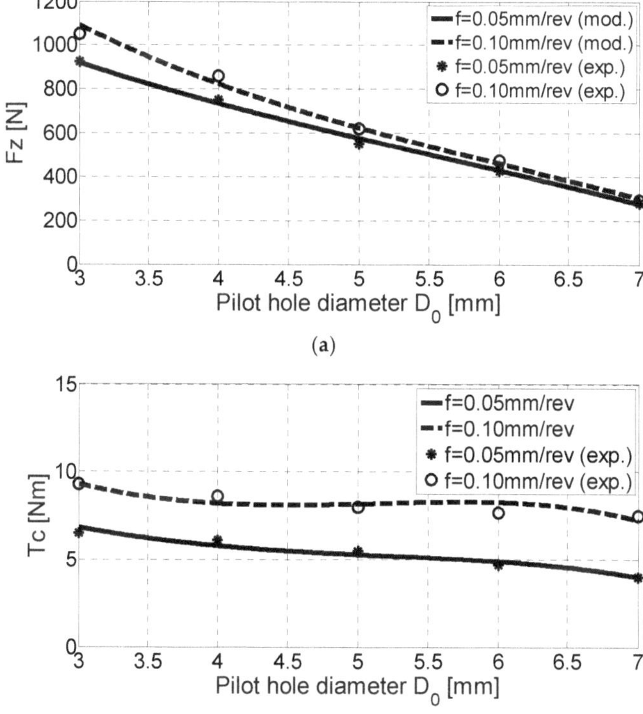

Figure 8. F_z [N] (**a**) and T_c [Nm] (**b**) for extruded MoCuSi alloy drilling experimental and mechanistic model values.

Using the above experimental results, the coefficients in Equations (3) and (4) are obtained and then the corresponding polynomial functions are built (Table 8).

Table 8. Obtained coefficients for the polynomials in Equations (3) and (4).

Materials	$K_z(h,r)$						$K_T(h,r)$					
	a_0	a_1	a_2	a_3	a_4	a_5	b_0	b_1	b_2	b_3	b_4	b_5
Ti-6Al-4V	13,159	−93,861	−3,225	−10,986	977.11	−7,056.0	39.646	−68.280	−21.597	−41.882	5.253	−5.171
TNB	19,860	−217,630	−3,830	17,235	855.42	−16,347.1	77.853	−283.109	−36.204	−61.168	8.334	−21.331
MoCuSi (ingot)	23,445	−270,543	−4,028	20,962	688.07	−20,320	90.152	−274.538	−45.120	−17.808	9.663	−20.703
MoCuSi (extruded)	24,016	−282,289	−4,270	24,114	860.73	−21,202	84.201	−120.895	−45.120	−17.808	9.663	−20.703

For model validation, different combinations of parameters were tested within the window parameters. New pilot hole drills and final drill of $D = 8.5$ mm were used for these tests. Eight different tests were programmed to put the model to work. Table 9 presents the conditions for these validation tests.

Table 9. Cutting conditions for the validation tests.

Materials	V_c [m/min]	D_0 [m/min]	f [mm/rev]	# Test
Ti-6Al-4V	50	4	0.06	1
		6	0.08	2
TNB	15	4	0.06	3
		6	0.08	4
MoCuSi (ingot)	15	4	0.06	5
		6	0.08	6
MoCuSi (extruded)	15	4	0.06	7
		6	0.08	8

Figure 9 shows the results from the simulated values using the model described and the experimentally measured values. The model presents a good correlation with the real thrust forces and cutting torques. For the cutting forces, the maximum relative error between the predicted and measured values was 8% while for the cutting torque the worst case was found to be 13%. The model behaves well in the targeted interval of feed rates, with higher deviations, the higher is the programmed feed. On the other hand, the mechanistic model seems less sensitive to the range of tested pilot hole diameters and final hole diameters. A higher dispersion is seen also in the torque values with respect to thrust forces that can be predicted with very good agreement.

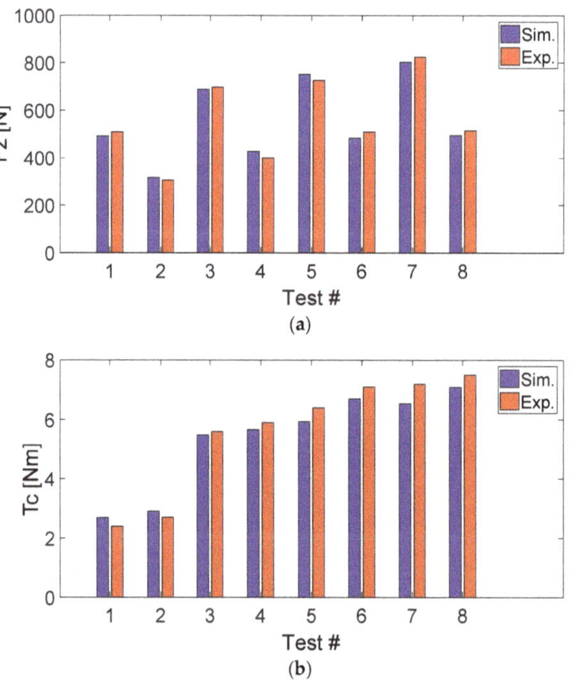

Figure 9. Model validation for F_z (**a**) and T_c (**b**) magnitudes.

5. Conclusions

This work presents the machinability results for four different titanium alloys. Currently there are few data about machining of γ-TiAl. The obtained results reflect that:

- These alloys present a much smaller machinability than conventional titanium alloys.
- The presented mechanistic stationary model is capable of simulating the thrust force and cutting torque in the drilling process. The influence of vibratory effects, lateral vibrations or runout was not considered. After model calibration, validation tests were done, showing a good agreement.
- Experimental drilling tests done to evaluate the machinability of these difficult-to-cut materials (the high cost of the tested materials (approx. 400 €/kg) limited the number of tests and a stronger experimental design) show that the obtained cutting coefficients present a quasi-linear dependence on the pilot hole D_0 for the axial force F_z, while a cubic is more suitable in the case of T_c.
- Machining problems during drilling tests confirm that these are very brittle materials, so operational parameter determination is essential for chipping and cracking avoidance of components during machining processes [41,42].
- Experiments demonstrated large differences between the more conventional Ti6Al4V alloy and the three γ-TiAl intermetallic alloys. As the mechanical properties of extruded/ingot alloys are higher than those of the TNB alloy, this trend is inversely true for the machinability grade. Tests proved that MoCuSi alloys are the toughest materials.
- Despite the good results during the validation, a statistical approach would be desirable to test any differences in material series, machine tools, as well as to build more complete models. Wear studies including process damping effects can be subjects for future research works.

Author Contributions: A.B. designed and performed the experiments; G.U. and A.C. analyzed obtained experimental and mechanistic model predicted results. Finally, L.N.L.d.L. contributed with resources (machine, tools, material, et al.) and general supervision of the work.

Funding: Thanks are addressed to the UFI in Mechanical Engineering of the UPV/EHU for its support to this project, and to Spanish project DPI2016-74845-R, ESTRATEGIAS AVANZADAS DE DEFINICIÓN DE FRESADO EN PIEZAS ROTATIVAS INTEGRALES, CON ASEGURAMIENTO DE REQUISITO DE FIABILIDAD Y PRODUCTIVIDAD and project RTC-2014-1861-4, INNPACTO DESAFIO II.

Acknowledgments: The authors acknowledge the technical assistance from Eng. Garikoitz Goikoetxea at UPV/EHU.

Conflicts of Interest: The authors declare no conflict of interest. The funders had no role in the design of the study; in the collection, analyses, or interpretation of data; in the writing of the manuscript, or in the decision to publish the results.

References

1. Homepage of Boeing. Available online: http://www.boeing.com (accessed on 20 October 2018).
2. Martin, R.; Evans, D. Reducing Costs in Aircraft: The Metals Affordability Initiative Consortium. *JOM* **2000**, *52*, 24–28. [CrossRef]
3. Voice, W.E.; Henderson, M.; Shelton, E.F.J.; Wu, X. Gamma titanium aluminide-TNB. *Intermetallics* **2005**, *13*, 959–964. [CrossRef]
4. Boyer, R.R. An overview on the use of titanium in the aerospace industry. *Mater. Sci. Eng. A* **1996**, *213*, 103–114. [CrossRef]
5. Jha, A.K.; Singh, S.K.; Kiranmayee, M.S.; Sreekumar, K.; Sinha, P.P. Failure analysis of titanium alloy (Ti6Al4V) fastener used in aerospace application. *Eng. Fail. Anal.* **2010**, *17*, 1457–1465. [CrossRef]
6. Murr, L.E.; Quinones, S.A.; Gaytan, S.M.; López, M.I.; Rodela, A.; Martinez, E.Y.; Hernandez, D.H.; Martinez, E.; Medina, F.; Wicker, R.B. Microstructure and mechanical behavior of Ti-6Al-4 V produced by rapid-layer manufacturing for biomedical applications. *J. Mech. Behav. Biomed. Mater.* **2009**, *2*, 20–32. [CrossRef] [PubMed]
7. Yamaguchi, M.; Inui, H.; Ito, K. High temperature structural intermetallics. *Acta Mater.* **2000**, *48*, 307–322. [CrossRef]

8. Jozwik, J. Evaluation of Tribological Properties and Condition of Ti6Al4V Titanium Alloy Surface. *Tehnički vjesn.* **2018**, *25*, 170–175. [CrossRef]

9. Kim, Y.W.; Dimiduk, D.M. Gamma TiAl alloys: Emerging new structural metallic materials. In Proceedings of the ICETS 2000-ISAM, Beijing, China, 1 October 2000.

10. Semiatin, S.L.; Seetharaman, V.; Weiss, I. Hot workability of titanium and titanium aluminide alloys. *Mater. Sci. Eng.* **1998**, *243*, 1–24. [CrossRef]

11. López de Lacalle, L.N.; Perez, J.; Llorente, J.I.; Sanchez, J.A. Advanced cutting conditions for the milling of aeronautical alloys. *J. Mater. Process. Technol.* **2000**, *100*, 1–11. [CrossRef]

12. Kuczmaszewski, J.; Kazimierz, Z.; Matuszak, J.; Pałka, T.; Mądry, J. Studies on the effect of mill microstructure upon tool life during slot milling of Ti6Al4V alloy parts. *Eksploat. I Niezawodn. Maint. Reliab.* **2017**, *19*, 590–596. [CrossRef]

13. Józwik, J.; Ostrowski, D.; Milczarczyk, R.; Krolczyk, G.M. Analysis of relation between the 3D printer laser beam power and the surface morphology properties in Ti-6Al-4V titanium alloy parts. *J. Braz. Soc. Mech. Sci. Eng.* **2018**, *40*, 1678–5878. [CrossRef]

14. Aspinwall, D.K.; Dewes, R.C.; Mantle, A.L. The Machining of γ-TiAl Intermetallic Alloys. *CIRP Ann.* **2005**, *54*, 99–104. [CrossRef]

15. Arrazola, P.J.; Garay, A.; Iriarte, L.M.; Armendia, M.; Marya, S.; Maitre, F.L. Machinability of titanium alloys (Ti6Al4V and Ti555.3). *J. Mater. Process. Technol.* **2009**, *209*, 2223–2230. [CrossRef]

16. Beranoagirre, A.; López de Lacalle, L.N. Grinding of Gamma TiAl Intermetallic Alloys. *Proc. Eng.* **2013**, *63*, 489–498. [CrossRef]

17. Lindemann, J.; Glavatskikh, M.; Leyens, C. Surface effects on the mechanical properties of gamma titanium aluminides. In Proceedings of the International Conference on Processing & Manufacturing of Advanced Materials, Quebec City, QC, Canada, 1–5 August 2011; pp. 1071–1076.

18. Hood, T.; Aspinwall, D.K.; Sage, C.; Voice, W. High speed ball nose and milling of γ-TiAl alloys. *Intermetallics* **2013**, *32*, 284–291. [CrossRef]

19. Thepsonthi, T.; Ozel, T. Experimental and Finite Element Simulation based Investigations on Micro-Milling Ti–6Al–4V Titanium Alloy: Effects of cBN Coating on Tool Wear. *J. Mater. Process. Technol.* **2013**, *213*, 532–542. [CrossRef]

20. Cantero, J.L.; Tardio, M.M.; Canteli, J.A.; Marcos, M.; Miguelez, M.H. Dry drilling of alloy Ti6Al4V. *Int. J. Mach. Tools Manuf.* **2005**, *45*, 1246–1255. [CrossRef]

21. Hood, R.; Aspinwall, D.K.; Soo, S.L.; Mantle, A.L.; Novovic, D. Workpiece surface integrity when slot milling γ-TiAl intermetallic alloy. *CIRP Ann.* **2014**, *63*, 53–56. [CrossRef]

22. Liu, J.; Zhu, D.; Zhao, L.; Xu, Z. Experimental Investigation on Electrochemical Machining of γ-TiAl Intermetallic. *Proc. CIRP* **2015**, *35*, 20–24. [CrossRef]

23. Kosaraju, S.; Anne, V.G. Optimal machining conditions for turning Ti-6Al-4 V using response surface methodology. *Adv. Manuf. Technol.* **2013**, *1*, 329–339. [CrossRef]

24. Mantle, A.L.; Aspinwall, D.K. Temperature Measurement and Tool Wear When turning gamma titanium aluminide intermetallic. In Proceedings of the 13th Conference of the Irish Manufacturing Committee (IMC-13), Limerick, Ireland, 4–6 September 1996; pp. 427–446.

25. Aspinwall, D.K.; Mantle, A.L.; Chan, W.K.; Hood, R.; Soo, S.L. Cutting Temperatures when Ball Nose End Milling γ-TiAl Intermetallic Alloys. *CIRP Ann.* **2013**, *62*, 75–78. [CrossRef]

26. Mantle, A.L.; Aspinwall, D.K. Surface integrity of a high speed milled gamma titanium aluminide. *J. Mater. Process. Technol.* **2001**, *118*, 143–150. [CrossRef]

27. Olvera, D.; Urbicain, G.; Beranoagirre, A.; López de Lacalle, L.N. Hole Making in Gamma TiAl. In *DAAAM International Scientific Book 2010*, 1st ed.; Katalinic, B., Ed.; DAAAM International Publishing: Vienna, Austria, 2010; pp. 337–347. ISBN 978-3-901509-74-2.

28. Priarone, P.C.; Rizzuti, S.; Settineri, L.; Vergnano, G. Effects of Cutting Angle, Edge Preparation and Nano-Structured Coating on Milling Performance of a Gamma Titanium Aluminide. *J. Mater. Process. Technol.* **2012**, *212*, 2619–2628. [CrossRef]

29. Hong, S.Y.; Ding, Y. Cooling approaches and cutting temperatures in cryogenic machining of Ti-6Al-4 V. *Int. J. Mach. Tools Manuf.* **2001**, *41*, 1417–1437. [CrossRef]

30. Marasi, A. Modeling the effects of cutting parameters on the main cutting force of Ti6Al4V alloy by using hybrid approach. *Int. J. Adv. Eng. Appl.* **2013**, *2*, 6–14.

31. Niesłony, P.; Grzesik, W.; Laskowski, P.; Sienawski, J. Numerical and Experimental Analysis of Residual Stresses Generated in the Machining of Ti6Al4V Titanium Alloy. *Procedia CIRP* **2014**, *13*, 78–83. [CrossRef]

32. Kadirgama, K.; Abou-El-Hossein, K.A.; Mohammad, B.; Habeeb, A.; Noor, M.M. Cutting force prediction model by FEA and RSM when machining Hastelloy C-22HS with 90° holder. *J. Sci. Ind. Res.* **2008**, *67*, 421–427.

33. Bagci, E.; Ozcelik, B. Finite element and experimental investigation of temperature changes on a twist drill in sequential dry drilling. *Int. J. Adv. Manuf. Technol.* **2006**, *28*, 680–687. [CrossRef]

34. Chatterjee, S.; Mahapatra, S.S.; Abhishek, K. Simulation and optimization of machining parameters in drilling of titanium alloys. *Simul. Model. Pract. Theory* **2016**, *62*, 31–48. [CrossRef]

35. Franco, A.; Abul, C.; Rashed, A.; Romoli, L. Analysis of energy consumption in micro-drilling processes. *J. Clean. Prod.* **2016**, *137*, 1260–1269. [CrossRef]

36. Hameed, S.; Rojas, H.A.G.; Egea, A.J.S.; Alberro, A.N. Electroplastic cutting influence on power consumption during drilling process. *Int. J. Adv. Manuf. Technol.* **2016**, *87*, 1835–1841. [CrossRef]

37. Azeem, A.; Feng, H.Y.; Wang, L. Simplified and efficient calibration of a mechanistic cutting force model for ball-end milling. *Int. J. Mach. Tools Manuf.* **2004**, *44*, 291–298. [CrossRef]

38. Altintas, Y. *Manufacturing Automation: Metal Cutting Mechanics, Machine Tool Vibrations, and CNC Design*; Cambridge University Press: Cambridge, UK, 2000.

39. Budak, E.; Altintas, Y.; Armarego, E.J.A. Prediction of milling force coefficients from orthogonal cutting data. *ASME J. Manuf. Sci. Eng.* **1996**, *118*, 216–224. [CrossRef]

40. Roukema, J.C.; Altintas, Y. Time domain simulation of torsional–axial vibrations in drilling. *Int. J. Mach. Tools Manuf.* **2006**, *46*, 2073–2085. [CrossRef]

41. Beranoagirre, A.; Olvera, D.; López de Lacalle, L.N.; Urbicain, G. Drilling of intermetallic alloys gamma TiAl. *AIP Conf. Proc.* **2011**, *1315*, 1023–1028.

42. Beranoagirre, A.; Urbicain, G.; Calleja, A.; López de Lacalle, L.N. Hole Making by Electrical Discharge Machining (EDM) of γ-TiAl Intermetallic Alloys. *Metals* **2018**, *8*, 543. [CrossRef]

 *materials*

Article

Experimental Determination of Residual Stresses Generated by Single Point Incremental Forming of AlSi10Mg Sheets Produced Using SLM Additive Manufacturing Process

Cecilio López [1], Alex Elías-Zúñiga [2,*], Isaac Jiménez [2], Oscar Martínez-Romero [2], Héctor R. Siller [3] and José M. Diabb [2]

[1] Honeywell Aerospace, Parque Industrial Ávalos, Vialidad Tabalaopa 8507, Chihuahua 31074, Chih., Mexico; Cecilio.Lopez@honeywell.com

[2] Tecnologico de Monterrey, Escuela de Ingeniería y Ciencias, Ave. Eugenio Garza Sada 2501, Monterrey 64849, NL, Mexico; isaachjc@yahoo.com (I.J.); oscar.martinez@itesm.mx (O.M.-R.); jdiabbzv@itesm.mx (J.M.D.)

[3] Department of Engineering Technology, University of North Texas, 3940 North Elm Street, Denton, TX 76207, USA; hector.siller@unt.edu

* Correspondence: aelias@itesm.mx; Tel.: +52-818-359-1699

Received: 13 November 2018; Accepted: 11 December 2018; Published: 13 December 2018

Abstract: This paper focuses on investigating the residual stress values associated with a part fabricated by Selective Laser Melting technology (SLM) when this is subjected further to forces on single point incremental forming (SPIF) operation of variable wall angle. The residual stresses induced by the SLM manufacturing process on the fabricated AlSi10Mg metallic sheets, as well as those produced during their forming SPIF operation were determined by X-ray diffraction (XRD) measurements. Significant residual stress levels of variation, positive or negative, along the metallic sample were observed because of the bending effects induced by the SPIF processes. It is also shown how the wall thickness varies along the additive manufactured SPIFed part as well as the morphology of the melting pools as a function of the deformation depth.

Keywords: additive manufacturing; single point incremental sheet forming; residual stresses; X-ray diffraction

1. Introduction

Additive Manufacturing (AM) is a process of joining materials to produce components from 3D computer data model that are fabricated layer by layer [1]. The purpose of this technology is to improve or replace manufacturing processes commonly used in rapid prototyping since manufacturers have found that the fabrication of models and prototypes rely on complex processes that hindered the development and design phases of new components. The additive manufacturing technology, used in this research called Selective Laser Melting (SLM), was invented by the Fraunhofer Institute Laser Technology in the mid. 1990s. SLM is one of the fastest growing AM technologies worldwide [2–4]. This technology has allowed the fabrication of complex components that cannot be made by conventional manufacturing processes. Recent advances show that this technology has expanded its application to metallic-based matrix composites such as the titanium-based materials [5] with improved physical properties and significant cost, production time, and weight reduction when compared to traditional production processes [6]. However, there are still some important issues to be addressed such as fatigue properties, porosity, geometrical accuracy, corrosion properties, wear performance, surface finishing, resulting residual stresses, among others, before the component or system produced by SLM can satisfy certification requirements, standards, international regulations, product life cycling, and intellectual property, among others [7]. Therefore, extensive studies are required to evaluate the mechanical performance of the fabricated SLM parts before these are assembled into a final product. One of these focuses on determining residual stress that are created when a part is processed by SLM technology because of the high temperature gradients, thermal expansion, and non-uniform plastic deformation during heating and cooling cycle.

In fact, residual stresses in metals are created from inhomogeneous plastic deformations because of chemical, thermal, and mechanical fabrication methods. The mechanisms for the developing of residual stresses in AM technologies are due to the temperature gradient since the material strength is instantaneously reduced along the laser beam scan layer because of the heat induced, and to the molten top layers cool-down phase [8,9]. Therefore, temperature gradients that induce non-uniform thermal expansions and contractions surrounding the affected zone will result in additional stresses, while the latter mechanism tends to shrink the underlying material because of its thermal contraction. A review of the major methods for measuring residual stresses is summarized by Huang et al. [10], while numerical techniques to predict residual stresses for selective laser melting process are discussed in [11–14].

On the other hand, and in order to reduce the undesirable residual stress effects that influence the part mechanical performance, post-processing operations such as shot-peening, grinding, heat treatment, age hardening, or polishing are used to treat the parts fabricated by SLM [15]. These tend to improve the material ductility, increases the fracture toughness threshold value for crack initiation, and reduce surface roughness that affect fatigue performance [16].

Here, in this article, we investigated how residual stresses of a part fabricated by SLM vary when this is subjected further to a manufacturing post-processing operation called single point incremental forming (SPIF). In general, it is not expected that a part fabricated by AM technologies will be subjected to a forming post-processing to reach its final form, however it is of special interest for industrial applications to understand its structural response when this is subjected to external loads that could affect its residual stress magnitudes induced during SLM additive manufacturing process. In this sense, and in order to quantify residual stress distribution induced by the SLM manufacturing process, an experimental technique based on X-ray diffraction (XRD) measurements is used [17,18]. Then, XRD measurements were carried out on the metallic samples to determine residual stresses on the surfaces of the truncated pyramid shape formed by SPIF. Furthermore, we investigated potential effects such as different residual stress levels of variation, positive or negative, along the sample surfaces that were observed during experimental measurements because of the bending stresses induced by the SPIF processes.

2. Materials and Methods

The fabrication of the aluminum samples was carried out as follows: AM technology based on SLM 280®2.0 with a maximum working volume of $280 \times 280 \times 365$ mm^3 was used to produce a 155 mm × 155 mm aluminum sheet with an average thickness value of 1.2 mm using as based material spherical grain shapes of aluminum metal powder (typically particle size distribution between 20–63 μm). AlSi10Mg powder alloy was used in an argon 5.0 atmosphere to manufacture metallic sheets by considering 200 °C of constant temperature of the building platform. The SLM equipment as well as the powder alloy were supplied by SLM solutions GmbH (Lübeck, Germany). The powder consisted of spherical particles with a size distribution according to the following D-values as measured by laser diffraction (Beckman Coulter LS 13320 PIDS, Lübeck, Germany): $D10 = 30$ μm, $D50 = 43$ μm, and $D90 = 55$ μm.

The SLM process involved a layer thickness of 30–50 μm of base material spread uniformly across a metal platform. Then, a laser beam started to scan across it using high-speed XY galvanometers based on the computer-generated programmed path. The focused laser beam melted the powder and fused it to the layer below, after which the platform descended one layer thickness, and the whole process was repeated until the part was finished. The build platform was located inside a metal chamber and was protected with an argon gas to avoid melt pool oxidation [1]. Once the AlSi10Mg sheets were made by SLM, each of these metal samples were clamped into a fixture system of a Kryle CNC 535 (Monterrey, México) three-axis milling machine to form a truncated pyramid part by considering an effective working sample area of $120 \times 120 \ mm^2$, as shown in Figure 1. During the forming process of the aluminum samples, the following SPIF parameters were considered: incremental step $\Delta z = 0.5$ mm, feed rate of 3000 mm/min, tool free rotation, initial square length path of 100 mm, varying wall angle, α, starting from 45° until samples failure, a forming tool made of steel with a 10 mm hemi-spherical tip, and mineral oil based lubricant to reduce the friction between the tool and the AlSi10Mg sheet during its forming process.

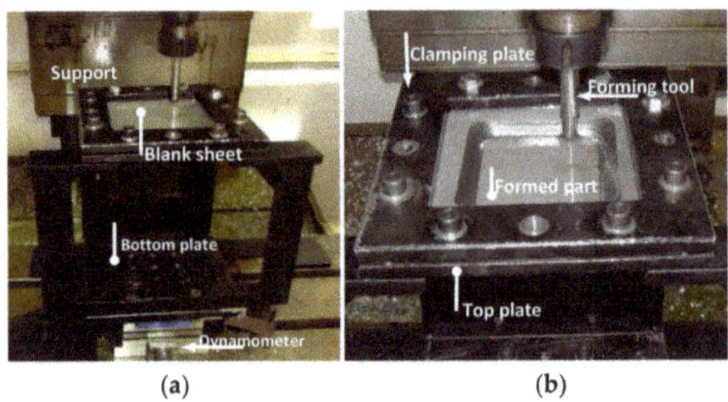

Figure 1. Single point incremental forming experimental setup mounted on a Kryle 9257 CNC milling machine: (**a**) SPIF fixture setup, and (**b**) forming tool with a hemi-spherical tip.

Once the aluminum sheet samples were formed, a portion of the pyramidal frustum surface were divided into 7 sections to measure the inner and outer surface residual stresses by using X-ray diffraction (XRD) system (see Figure 2). To measure the residual stress on the formed part, the sample was cut by using wire-cut electrical discharge Fanuc Robocut α-C600Ia high performance machine (Chihuahua, México), as illustrated in Figure 2c. Then, a portion of the part, shown in Figure 2d, was mounted in a Panalytical XRD system to measure residual stresses. This experimental process will be discussed next.

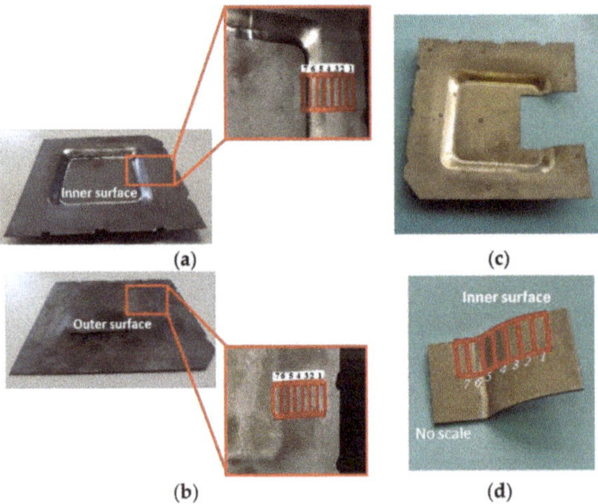

Figure 2. Scanned surface area to determine the residual stress values on (**a**) the inner surface, (**b**) the outer surface, (**c**) cut SPIFed formed part, and (**d**) sample piece used to measure residual stresses.

3. Experimental Measurements

3.1. Determination of Residual Stresses

A Panalytical Empyrean X-ray diffractometer with α-Cu radiation, at 45 kV and 20 mA, and k-β nickel filter was used to measure residual stresses along the inner and outer surfaces of the formed aluminum samples, as illustrated in Figure 3. The methodology discussed in [17] was followed to estimate the residual stress values along the part forming depth, h_i, and wall angle α. The residual stress values were computed on the inner and outer sample sections shown in Figure 4a,b by assuming a Poisson ratio value of ν = 0.33, and Young modulus of E = 69.5 GPa [19].

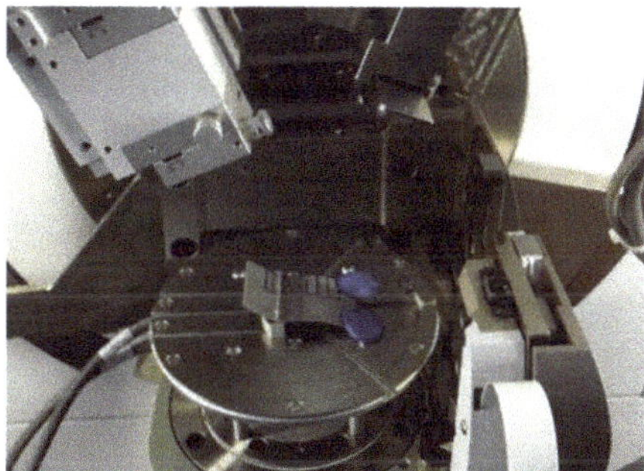

Figure 3. Illustration of the sample regions where residual stress were measured.

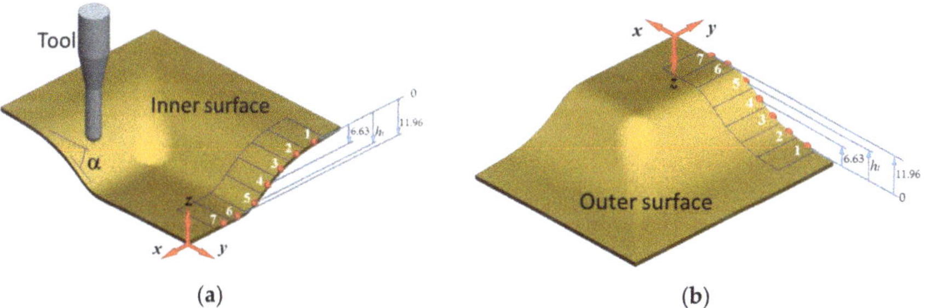

Figure 4. Points on the sample surface at which residual stresses were estimated by XRD technique for (**a**) the inner surface, and (**b**) the outer surface.

As shown in Figure 5, the inner and outer forming surfaces are subjected to residual stresses whose magnitude vary as a function of the forming depth and wall angle values. Their maximum collected residual stress values occur at the sections at which the aluminum sheet does not experience any deformation during SPIF process (Sections 1 and 7). Along the plastically formed regions, the inner surface exhibits residual stresses that vary from compressive to tensile values, which is a clear indication of bending effects [17], as illustrated in Figure 5. Table 1 summarizes the recorded residual stresses along the inner and outer surfaces, and the thickness variation in each of the selected sections. Notice from Table 1 that the maximum compressive residual stresses on the inclined surfaces of the truncated pyramid have the magnitude of −91.7 MPa for the outer surface, and of −83.7 MPa for the inner surface. These stress magnitudes were experimentally recorded from the fabricated SLM part before it was post-processed, at the beginning of the sample forming depth. It is observed from Table 1 that for all forming depths of the AM aluminum samples, the outer surface experiences only compressive residual stresses. This is not the case in the inner surface since it has a maximum tensile residual stress of 46.7 MPa recorded at h_i = 6.63 mm, as illustrated in Figure 5. Figure 5 also illustrates the typical standard deviation error bars attained during the XRD measurements whose magnitude values are listed in Table 1. The common sources of these error values are mostly due to the X-ray elastic constants, the apparatus focusing geometry, diffracted peak location, grain size, microstructure, and surface condition. Additionally, preliminary XRD measurements were carried out on the metallic samples in order to determine the initial residual stresses induced by the SLM manufacturing process on both surfaces of the AlSi10Mg metallic sheets and found only compressive stresses with a maximum average value of −92 MPa. As discussed above, this residual stresses change from tensile to compressive ones along the surfaces of the metallic samples when subjected to the SPIF post-processing because of bending effects.

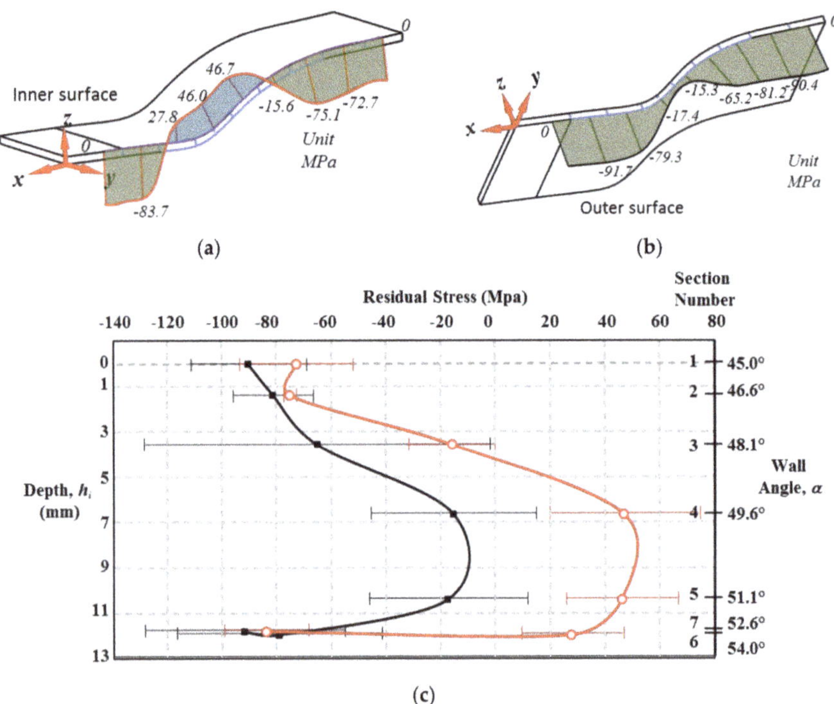

Figure 5. Measured residual stress values on (**a**) the inner surface, (**b**) the outer surface, and (**c**) schematic representation of the residual stress value curves and their standard deviation error bars attained during the XRD measurements as a function of the SPIF process forming depth, h_i, and the wall angle α parameter values.

Table 1. Residual stress values recorded in the inner and outer surfaces of the AM Aluminum sheets as a function of the forming depth.

Section Number	Depth h_i (mm)	Wall Thickness t (mm)	Outer Surface Residual Stress (MPa)	Outer Surface Standard Deviation (MPa) (+/−)	Inner Surface Residual Stress (MPa)	Inner Surface Standard Deviation (MPa) (+/−)
1	0	1.226	−90.4	21.1	−72.7	40.7
2	1.39	1.220	−81.2	14.4	−75.1	4.7
3	3.57	1.060	−65.2	63.6	−15.6	31.5
4	6.63	1.017	−15.3	30.0	46.7	54.4
5	10.41	0.730	−17.4	28.8	46.0	40.5
6	11.96	1.012	−79.3	37.3	27.8	36.4
7	11.81	1.225	−91.7	36.6	−83.7	30.8

3.2. SPIFed Parts Mechanical Modelling

It is well-known that during SPIF process of parts, stretching and bending effects occur. The corresponding equations that describe these phenomena are discussed in [17]. For convenience, we briefly recall some essential relations that describe the mechanical response of parts when subjected to SPIF process.

First, let us consider the elongation strain at which the material samples are subjected during the SPIF process. In this case, thickness strain, better known as the stretching strain at the middle section of the cross sectional area is described by the expression:

$$\varepsilon_{stretching} = \ln \frac{t_0}{t} \tag{1}$$

here t_0 represents the original sheet thickness before deformation, and t is the current sheet thickness that varies as a function of the forming depth and the wall angle. For bending effects on SPIFed parts, Malhotra et al. [20] and Jiménez and co-workers [17] concluded that these effects are not negligible during the mechanical response of forming parts since these are responsible for cracks appearance near the maximum sample forming depth. Therefore, bending effects must be estimated if one wants to avoid crack initiation during SPIF process. In this case, bending strain deformation is described by:

$$\varepsilon_{bending} = \ln \frac{r}{\rho} \tag{2}$$

where $r = \rho + y$, ρ is the theoretical radius of the curvature of the formed part, and y is the distance from the middle surface to the element under consideration.

The bending moment, M, at which the material is subjected during SPIF process, as a function of the wall thickness sample, can be computed from:

$$M = 2K \left(\frac{1}{\rho} \right)^n \int_0^{t/2} y^{1+n} dy = K \left(\frac{1}{\rho} \right)^n \frac{t^{n+2}}{(n+2)2^{n+1}} \tag{3}$$

where K represents the material strength coefficient that for AlSi10Mg has the value of 470 MPa with n equal to 0.14. See References [17,21]. Finally, the true stress value that the metallic sheet experiences during SPIF process is computed from the Hollomon expression [21]:

$$\sigma = K\varepsilon^n \approx K \left(\frac{y}{\rho} \right)^n \tag{4}$$

where y represents the distance from the middle section to the element under consideration, as shown in Figure 6.

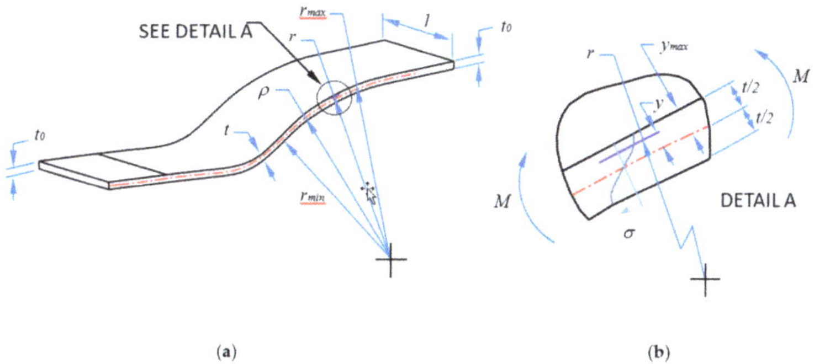

Figure 6. *Cont.*

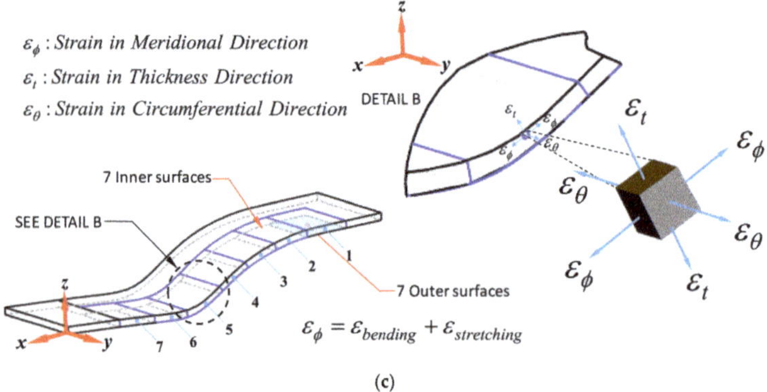

(c)

Figure 6. (**a**) and (**b**) Schematic of bending effects acting on the formed part during SPIF, and (**c**) strain deformations experienced by the metallic sheet.

3.3. Mechanical Behavior of SLM AlSi10Mg SPIFed Samples

The sample wall thickness was measured from the formed part at the points indicated in Figure 7 by using a 3D scan blue light technology equipment ATOS ScanBox 5108. The collected data is summarized in Table 2. Here, a mean value of 1.226 mm of sheet wall thickness was measured before the AM sheet was SPIFed. Figure 8 illustrates the recorded sample wall thickness variation as a function of the sheet forming depth. These collected values were used to quantify stretching and bending effects of the AM sample through Equations (1) to (4).

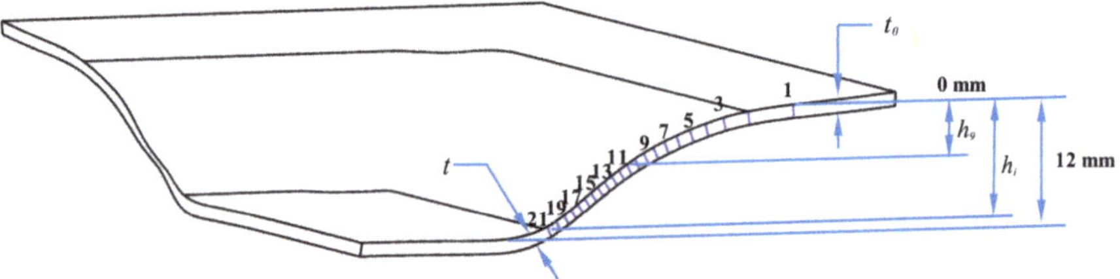

Figure 7. Frontal view of the aluminum sample. Thickness variation along the sample wall was obtained by 3D scan blue light technology equipment.

Table 2. Experimental data that shows wall thickness variation as a function of the sample forming depth.

Point	Depth h_i (mm)	Thickness t (mm)	Point	Depth h_i (mm)	Thickness t (mm)	Point	Depth h_i (mm)	Thickness t (mm)
1	0	1.226	8	3.5	1.181	15	7.0	0.917
2	0.5	1.225	9	4.0	1.159	16	7.5	0.860
3	1.0	1.223	10	4.5	1.134	17	8.0	0.808
4	1.5	1.220	11	5..0	1.104	18	8.5	0.764
5	2.0	1.216	12	5.5	1.064	19	9.0	0.735
6	2.5	1.215	13	6.0	1.022	20	9.5	0.724
7	3.0	1.198	14	6.5	0.973	21	10.0	0.735

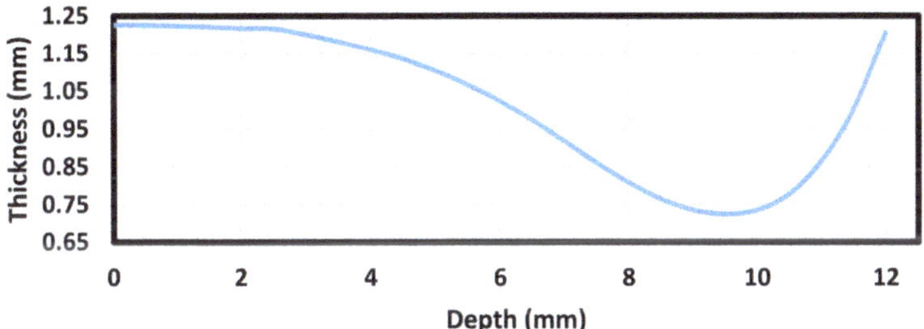

Figure 8. Sample wall thickness variation.

The samples stretching and bending strains were estimated by considering the variation of the sample wall thickness along the formed part. As one can see from Figure 9, the AM sample stretching strain is higher than bending strains. The maximum value of the stretching strain (0.52 mm/mm) is reached at the forming sample depth of 9.5 mm. At this point, the material sample has been stretched close to its forming limit, and then necking formation takes place, as illustrated in Figure 7 (points 17–21). This is consistent with Young and Jeswiet findings in [22] since the accumulation of high plastic deformation in this region can result in sample damage initiation. In contrast, the maximum value of bending strain (0.00611 mm/mm) occurs at the beginning of the SPIF process. Notice that its value gradually decreases until it reaches a minimum value, at the region at which the stretching strain has a maximum value.

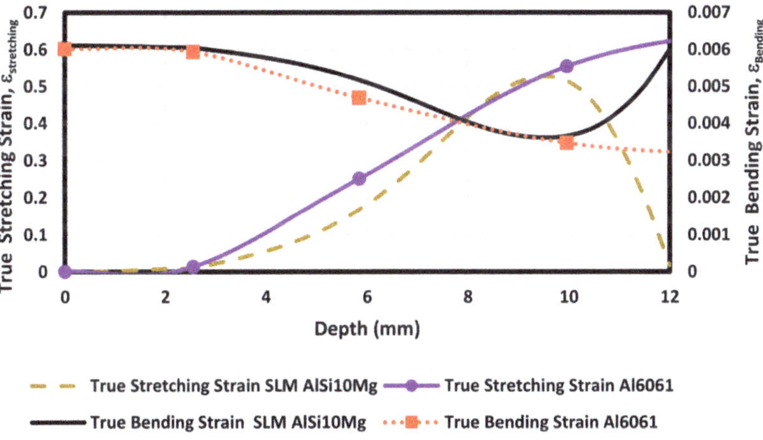

Figure 9. Calculated true stretching and bending strain curves for AlSi10Mg and Al6061.

To have insightful information about the structural behavior of AM aluminum sheets after post processing, bending moment M per unit width, and true stresses were calculated by using Equations (3) and (4), respectively. Figure 10 shows the predicted value curves. Notice that both curves tend to reach their maximum value at the beginning of the forming SPIF process. At this stage, the maximum true stress magnitude value was found to be 230.23 MPa, which is close to the material yield stress value of 275 MPa. It is important to mention that the maximum average forming depth value achieved during the samples SPIF process was 12 mm. For exceeding depth values, the metallic samples tend to fracture since the AM process induces, by itself on the metallic sheets, microstructural defects such as pores and overlapping melting pools that hinders their formability limits. In fact, Figure 2b confirms the occurrence of sample fracture close to this region due to meridional stresses [17,20,23].

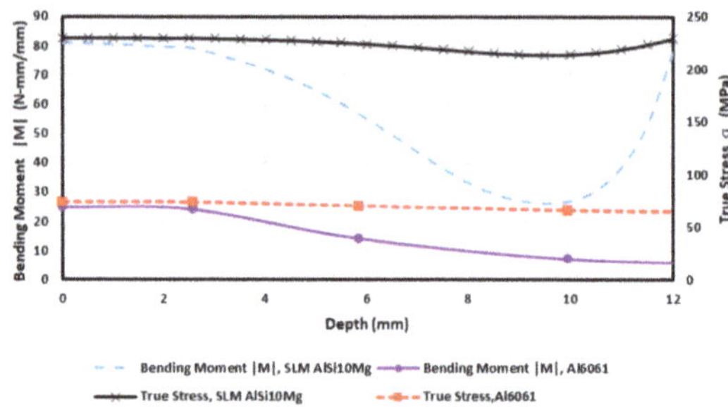

Figure 10. Calculated true stress and bending moment curves.

3.4. Comparison of the Mechanical Behavior of SLM AlSi10Mg and Al6061 SPIFed Samples

This section focuses on addressing the principal findings of the mechanical properties of cast aluminum alloys 6061 and SLM AlSi10Mg samples when subjected to the SPIF process. While SPIFed metallic sheets are from different aluminum batches, their mechanical response under large deformations could be of interest for several engineering applications. Therefore, it is believe that this comparison will provide insightful information about the mechanical behavior of Al6061 with respect to post-processed AM parts.

Measurements of the stretching strains of SLM AlSi10Mg and Al6061 samples exhibit slightly different strain behavior mainly due to the work hardening properties of each material. As a result, Al6061 ($n = 0.2$, $K = 205$ MPa) has better stretch-ability at low forming depths than the SLM AlSi10Mg ($n = 0.14$, $K = 470$ MPa), but soon becomes almost of the same magnitude than the AM aluminum samples for forming depth values close to 10 mm, as observed in Figure 9. Furthermore, both material samples exhibit almost the same bending strain behavior, mainly due to the recorded cross-sectional wall thickness of the AlSi10Mg and Al6061 samples used to compute these curves, as illustrated in Figure 10.

The acting forces on SLM AlSi10Mg and Al6061 samples during the SPIfed process were recorded by using a Kistler piezoelectric dynamometer connected to a data acquisition system. Surprisingly, the measured forming forces on the Al6061 samples are higher than those collected on SLM AlSi10Mg sheets, as illustrated in Figure 11. There are several causes that could explain this situation. One is related to the formation of dislocations during the forming process of the material samples. In an ideal situation in which there is no porosity, it would be expected that the cell boundaries would limit dislocation movement during deformation. This increases material strength since the Al grains contain Si particles surrounded by eutectic region, which is indicative of strong interfacial bonding that hinders dislocation motion. See [24–26]. Porosity in SLM AlSi10Mg samples are the preferential sites for inhomogeneous deformation and consequently for crack initiation that allows deformation at much lower forming forces, as illustrated in Figure 11.

On the other hand, experimental measurements of residual stresses in the inner and outer surfaces of the SLM AlSi10Mg part, recorded after its SPIFed process, proved to have residual stresses with magnitudes 3 times higher than that recorded in Al6061 samples for the forming depth of 12.6 mm [17]. In fact, SML samples fracture by meridional tensile stresses when the forming depth is about 12 mm, while those of Al6061 failed by circumferential fracture near the forming depth value of 52 mm. Both sample fractures took place at the region of critical thickness reduction [23], i.e., at point 20 of Table 2 for SML AlSi10Mg, and between points 13 and 15 of Table 3. Of course, for additive manufactured parts, fatigue cracks could initiate because of some manufactured defects such as pores and overlapped melting pool regions, near the surface or within the sample interior. This AM sample defects will be addressed in the next section.

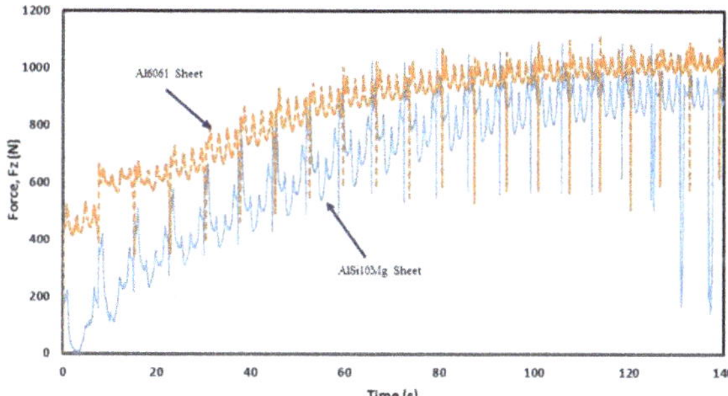

Figure 11. Collected forming forces, F_z, on AM AlSi10Mg and Al6061 SPIFed parts.

Table 3. Experimental data for Al6061 sample thickness variation as a function of the SPIF depth. Adapted by permission from the International Journal of Advanced Manufacturing Technology [17], Copyright 4486470813200, 2018.

Point	Depth h_i (mm)	Thickness t (mm)	Point	Depth h_i (mm)	Thickness t (mm)
1	2.565	1.1981	10	38.213	0.3817
2	5.842	0.9454	11	41.212	0.3594
3	9.957	0.6978	12	45.024	0.3189
4	14.258	0.6204	13	48.553	0.2755
5	18.833	0.5747	14	51.954	0.3303
6	23.714	0.5329	15	54.682	0.6798
7	27.577	0.4918	16	54.818	1.1588
8	31.253	0.4526	17	54.756	1.2102
9	34.562	0.4281	18	54.874	1.2144

4. Microstructural Evolution During SPIF Process

In order to study the evolution of the microstructure of AM metallic sheets and the impact that the SPIF forming process has on the melt-pool morphology, SEM images were obtained along the cross sectional area of the formed aluminum samples. Figure 12 shows the optical micrographs evolution along the formed AlSi10Mg etched sample. Six regions have been selected to experimentally obtain, via SEM images, the sample microstructure along its cross-sectional area to understand how the forming process influences the morphology of the melted pools. Elongated melted pools, as well as some pores, are observed from the micrographs illustrated in Figure 12. It is well-known that pores are critical defects since these could initiate inhomogeneous deformation as well as cracks initiation [19,27]. The insufficient overlap observed in Figure 12 of melt pools is detrimental to fatigue life due to the stress concentration at the sharp edges of the pores. Thus, the existence of pores in additive manufactured parts must be controlled by adjusting SLM parameters [28,29] or by subjecting the fabricated part to a post-processing operation. Region IV shows melted pools that have an interwoven form that is mainly due to the hatch spacing between scans in one layer. It is also observed from Figure 12 that the melted pool dimensions vary along the microstructure.

As the forming sample depth increases, the melted pools of Regions V and VI exhibit different morphology. Note that in Region VI, the melt-pool boundaries are well defined. It is noteworthy that in Regions I, II, and III, bending effects take place, mainly due to the contact forces between the aluminum sheet and the forming tool [17,20,30]. Furthermore, the melt-pools change inclination towards the building scanning direction. In these regions, the outer surfaces experienced tensile residual stresses that cause melt-pool and pores elongations. In Region III, pore alignment is evident along the melt-pool building directions. Elongated pore-form is relevant to the sample mechanical properties because of the potential increase in stress concentration magnitude values that could promote crack initiation and propagation that may lead to part premature failure [19,31,32].

On the other hand, Figures 2b and 12 exhibits the region of the circumferential failure mode of the truncated pyramid part located close to Region III. In this Region III, the sample experience tensile residual stress of about 27.8 MPa, and reaches its maximum formability capacity since the sample cross-section has the lowest thickness value, as illustrated in Figure 10. Therefore, it is concluded that AM parameters such as scanning direction that influences fatigue life because of the alloy microstructure, smaller hatching spacing that can eliminate pores, power, velocity, and layer thickness that are directly link to melt-pool aspect ratio (length/width) must be appropriately adjusted to prevent premature sample failure during its service life [33,34].

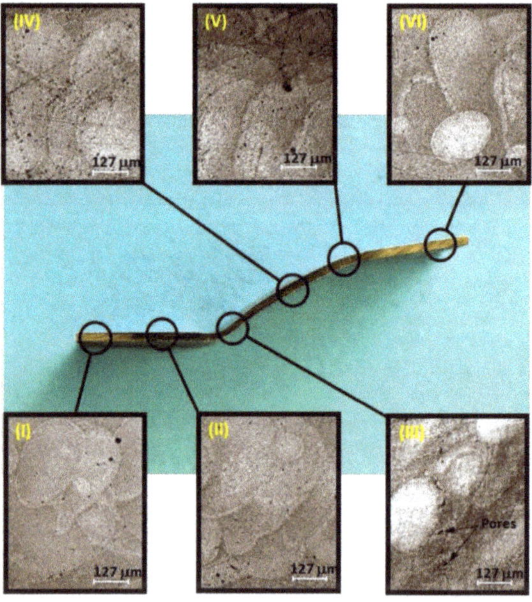

Figure 12. Microstructural evolution of the SPIFed AlSi10Mg part.

5. Conclusions

XDR measurements showed that AlSi10Mg metallic sheets produced by AM technology have only compressive residual stresses on the top and bottom sheet surfaces with a maximum average magnitude value of −92 MPa. However, when these metallic sheets are subjected to SPIF post-process, the magnitude of these stresses change from compressive to tensile values on the inner surface because of bending effects that could initiate part fracture due to meridional stresses. We have also found that the maximum value of the stretching strain occurs for the forming sample depth of 9.5 mm close to the necking formation region for which the material reaches its forming limit. Moreover and during SPIF post-processing of the metallic samples, it was found that after 12 mm of sample forming depth, small cracks appeared on the surface specimens mainly due to AM microstructural defects such as pores and overlapping melting pools that hinders their formability limits. Some defects were observed close to the meridional failure region of the truncated pyramid. At this forming depth, the maximum true stress magnitude value is about 230.23 MPa, which is closed to the material yield stress value of 275 MPa. The experimental samples had sufficient strength to withstand the forming forces that were acting upon them during the SPIF process without failure until they reached a forming depth average value close to 12 mm. However, if sample defects are removed by adjusting the AM parameters and by subjecting the printed metallic parts to the hipping process (hot isostatic pressing for additive manufacturing densification), an increase in the forming depth and fatigue strength will be possible. Therefore, it is concluded that if a part built by metal AM technologies is to be used, a careful examination of its mechanical properties needs to be done before moving it into certification process and full product service. In this sense, this study provides information regarding the variation of residual stresses on the inner and outer surfaces of the metallic samples, and how their magnitude values vary during the SPIF post processing of the fabricated parts in an attempt to emulate their structural response if these are further subjected to external loads during their service life.

Author Contributions: Conceptualization, C.L., A.E.-Z., and I.J.; Methodology, C.L., O.M.-R., I.J., A.E.-Z., and H.R.S.; Validation, C.L., O.M.-R., I.J., A.E.-Z., and H.R.S.; Investigation, C.L., O.M.-R., I.J., A.E.-Z., H.R.S., and J.M.D.; Data Acquisition, I.J., C.L., and J.M.D.; Formal Analysis, C.L., O.M.-R., I.J., and A.E.-Z.; Writing–Original Draft Preparation, C.L., O.M.-R., I.J., A.E.-Z., H.R.S., and J.M.D.; Writing–Review & Editing, C.L., O.M.-R., I.J., and A.E.-Z.; Funding Acquisition, C.L., O.M.-R., I.J., and A.E.-Z.

Funding: This research was funded by Tecnológico de Monterrey through the Research Group of Nanotechnology for Devices Design, and by the Consejo Nacional de Ciencia y Tecnología de México (Conacyt), Project Numbers 242269, 255837, and National Lab in Additive Manufacturing, 3D Digitizing and Computed Tomography (MADiT) LN280867.

Acknowledgments: The authors would like to thank The International Journal of Advanced Manufacturing Technology for allows to use the listed data of Table 3 that was first published in [17], Copyright License Number 4486470813200 issued by the Journal publisher Springer Nature, 2018. We also would like to thank the anonymous reviewers for their careful reading of our manuscript.

Conflicts of Interest: The authors declare no conflict of interest.

References

1. Lee, J.Y.; An, J.; Chua, C.K. Fundamentals and applications of 3D printing for novel materials. *Appl. Mater. Today* **2017**, *7*, 120–133. [CrossRef]

2. Brandt, M. (Ed.) *Laser Additive Manufacturing, Materials, Design, Technologies and Applications*; Woodhead Publishing: Cambridge, UK, 2017; pp. 55–77, ISBN 978-0-08-100433-3.

3. *Wohler Report 2017, 3D Printing and Additive Manufacturing State of the Industry Annual Worldwide Progress Report*; Wohlers Associates, Inc.: Fort Collins, CO, USA, 2017; ISBN 978-0-9913332-3-3.

4. 3D Printer and 3D Printing News. Available online: http://www.3ders.org (accessed on 9 December 2017).

5. Attar, H.; Haghighi, S.E.; Kent, D.; Dargusch, M.S. Recent developments and opportunities in additive manufacturing of titanium-based matrix composites: A review. *Int. J. Mach. Tools Manuf.* **2018**, *133*, 85–102. [CrossRef]

6. Attar, H.; Ehtemam-Haghighi, S.; Kent, D.; Wu, X.; Dargusch, M.S. Comparative study of commercially pure titanium produced by laser engineered net shaping, selective laser melting and casting processes. *Mater. Sci. Eng. A* **2017**, *705*, 385–393. [CrossRef]

7. ASTM International, Technical Committees. 2016. Committee F42 on Additive Manufacturing Technologies. Available online: http://www.astm.org/COMMITTEE/F42.htm (accessed on 12 May 2017).

8. Mercelis, P.; Kruth, J.P. Residual stresses in selective laser sintering and selective laser melting. *Rapid Prototyp. J.* **2006**, *12*, 254–265. [CrossRef]

9. Kruth, J.; Froyen, L.; Van Vaerenbergh, J.; Mercelis, P.; Rombouts, M.; Lauwers, B. Selective laser melting of iron-based powder. *J. Mater. Process. Technol.* **2004**, *149*, 616–622. [CrossRef]

10. Huang, X.; Liu, Z.; Xie, H. Recent progress in residual stress measurement techniques. *Acta Mech. Solida Sin.* **2013**, *26*, 570–583. [CrossRef]

11. Vrancken, B.; Cain, V.; Knutsen, R.; Van Humbeeck, J. Residual stress via the contour method in compact tension specimens produced via selective laser melting. *Scr. Mater.* **2014**, *87*, 29–32. [CrossRef]

12. King, W.; Anderson, A.T.; Ferencz, R.M.; Hodge, N.E.; Kamath, C.; Khairallah, S.A. Overview of modelling and simulation of metal powder bed fusion process at Lawrence Livermore National Laboratory. *Mater. Sci. Technol.* **2015**, *31*, 957–968. [CrossRef]

13. Jabbari, M.; Baran, I.; Mohanty, S.; Comminal, R.; Sonne, M.R.; Nielsen, M.W.; Spangenberg, J.; Hattel, J.H. Multiphysics modelling of manufacturing processes: A review. *Adv. Mech. Eng.* **2018**, *10*, 1–31. [CrossRef]

14. DebRoy, T.; Wei, H.L.; Zuback, J.S.; Mukherjee, T.; Elmer, J.W.; Milewski, J.O.; Beese, A.M.; Wilson-Heid, A.; Ded, A.; Zhang, W. Additive manufacturing of metallic components—Process, structure and properties. *Prog. Mater. Sci.* **2018**, *92*, 112–224. [CrossRef]

15. Yap, C.Y.; Chua, C.K.; Dong, Z.L.; Liu, Z.H.; Zhang, D.Q.; Loh, L.E.; Sing, S.L. Review of selective laser melting: Materials and applications. *Appl. Phys. Rev.* **2015**, *2*, 041101. [CrossRef]

16. Greitemeier, D.; Dalle Donne, C.; Syassen, F.; Eufinger, J.; Melz, T. Effect of surface roughness on fatigue performance of additive manufactured Ti–6Al–4V. *Mater. Sci. Technol.* **2016**, *32*, 629–634. [CrossRef]

17. Jiménez, I.; López, C.; Martínez-Romero, O.; Siller, H.R.; Diabb, J.; Elías-Zúñiga, A. Investigation of residual stress distribution in single point incremental forming of aluminum parts by X-ray diffraction technique. *Int. J. Adv. Manuf. Technol.* **2017**, *91*, 2571–2580. [CrossRef]

18. Wu, J.; Wang, L.; An, X. Numerical analysis of residual stress evolution of AlSi10Mg manufactured by selective laser melting. *Optik* **2017**, *137*, 65–78. [CrossRef]

19. Kempen, K.; Thijs, L.; Van Humbeeck, J.; Kruth, J.P. Mechanical properties of AlSi10Mg produced by selective laser melting. *Phys. Procedia* **2012**, *39*, 439–446. [CrossRef]

20. Malhotra, R.; Xue, L.; Belytschko, T.; Cao, J. Mechanics of fracture in single point incremental forming. *J. Mater. Process. Technol.* **2012**, *212*, 1573–1590. [CrossRef]

21. Kalpakjian, S.; Schmid, S. *Manufacturing Processes for Engineering Materials*, 5th ed.; Pearson: London, UK, 2008; ISBN 978-0132272711.

22. Young, D.; Jeswiet, J. Wall thickness variations in single-point incremental forming. *J. Eng. Manuf. Part B* **2004**, *218*, 1453–1459. [CrossRef]

23. Isik, K.; Silva, M.B.; Tekkaya, A.E.; Martins, P.A.F. Formability limits by fracture in sheet metal forming. *J. Mater. Process. Technol.* **2014**, *214*, 1557–1565. [CrossRef]

24. Wu, J.; Wang, X.Q.; Wang, W.; Attallah, M.M.; Loretto, M.H. Microstructure and strength of selectively laser melted AlSi10Mg. *Acta. Mater.* **2016**, *117*, 311–320. [CrossRef]

25. Li, X.P.; Ji, G.; Chen, Z.; Addad, A.; Wu, Y.; Wang, H.W.; Vleugels, J.; Van Humbeeck, J.; Kruth, J.P. Selective laser melting of nano-TiB$_2$ decorated AlSi10Mg alloy with high fracture strength and ductility. *Acta Mater.* **2017**, *129*, 183–193. [CrossRef]

26. Chen, B.; Moon, S.K.; Yao, X.; Bi, G.; Shen, J.; Umeda, J.; Kondoh, K. Strength and strain hardening of a selective laser melted AlSi10Mg alloy. *Scr. Mater.* **2017**, *141*, 45–49. [CrossRef]

27. Aboulkhair, N.T.; Everitt, N.M.; Ashcroft, I.; Tuck, C. Reducing porosity in AlSi10Mg parts processed by selective laser melting. *Addit. Manuf.* **2014**, *1*, 77–86. [CrossRef]

28. Olakanmi, E.O.; Cochrane, R.F.; Dalgarno, K.W. A review on selective laser sintering/melting (SLS/SLM) of aluminium alloy powders: Processing, microstructure, and properties. *Prog. Mater. Sci.* **2015**, *74*, 401–477. [CrossRef]

29. Ming, T. Inclusions, Porosity, and Fatigue of AlSi10Mg Parts Produced by Selective Laser Melting. Ph.D Thesis, Carnegie Mellon University, Pittsburgh, PA, USA, 18 April 2017. Available online: http://repository.cmu.edu/dissertations/903 (accessed on 20 March 2018).

30. Silva, M.B.; Isik, K.; Tekkaya, A.E.; Martins, P.A.F. Fracture Loci in Sheet Metal Forming: A Review. *Acta Metall. Sin. (Engl. Lett.)* **2015**, *28*, 1415–1425. [CrossRef]

31. Wang, Q.G. Microstructural effects on the tensile and fracture behavior of Aluminum casting alloys A356/357. *Metall. Mater. Trans. A* **2003**, *34*, 2887–2899. [CrossRef]

32. Yadroitsev, I.; Thivillon, L.; Bertrand, P.; Smurov, I. Strategy of manufacturing components with designed internal structure by selective laser melting of metallic powder. *Appl. Surf. Sci.* **2007**, *254*, 980–983. [CrossRef]

33. Franco, B.E.; Ma, J.; Loveall, B.; Tapia, G.A.; Karayagiz, K.; Liu, J.; Elwany, A.; Arroyave, R.; Karaman, I. A sensory material approach for reducing variability in additively manufactured metal parts. *Sci. Rep.* **2017**, *7*, 3604. [CrossRef]

34. Tang, M.; Pistorius, P.C. Oxides, porosity and fatigue performance of AlSi10Mg parts produced by selective laser melting. *Int. J. Fatigue* **2017**, *94*, 192–201. [CrossRef]

Article

Theoretical and Experimental Investigation of Surface Topography Generation in Slow Tool Servo Ultra-Precision Machining of Freeform Surfaces

Duo Li [1,2,*], Zheng Qiao [1], Karl Walton [2], Yutao Liu [1], Jiadai Xue [1], Bo Wang [1,*] and Xiangqian Jiang [2]

[1] Centre for Precision Engineering, Harbin Institute of Technology, Harbin 150006, China; qiaozhengyunlong@126.com (Z.Q.); 16B908095@stu.hit.edu.cn (Y.L.); brucexjd@hit.edu.cn (J.X.)

[2] EPSRC Future Metrology Hub, University of Huddersfield, HD1 3DH Huddersfield, UK; K.Walton@hud.ac.uk (K.W.); x.jiang@hud.ac.uk (X.J.)

* Correspondence: duo.kevin.li@gmail.com (D.L.); bradywang@hit.edu.cn (B.W.); Tel.: +86-451-8641-5244 (D.L.)

Received: 21 November 2018; Accepted: 14 December 2018; Published: 17 December 2018

Abstract: Freeform surfaces are featured with superior optical and physical properties and are widely adopted in advanced optical systems. Slow tool servo (STS) ultra-precision machining is an enabling manufacturing technology for fabrication of non-rotationally symmetric surfaces. This work presents a theoretical and experimental study of surface topography generation in STS machining of freeform surfaces. To achieve the nanometric surface topography, a systematic approach for tool path generation was investigated, including tool path planning, tool geometry selection, and tool radius compensation. The tool radius compensation is performed only in one direction to ensure no high frequency motion is imposed on the non-dynamic axis. The development of the surface generation simulation allows the prediction of the surface topography under various tool and machining variables. Furthermore, it provides an important means for better understanding the surface generation mechanism without the need for costly trial and error tests. Machining and measurement experiments of a sinusoidal grid and microlens array sample validated the proposed tool path generation and demonstrated the effectiveness of the STS machining process to fabricate freeform surfaces with nanometric topography. The measurement results also show a uniform topography distribution over the entire surface and agree well with the simulated results.

Keywords: ultra-precision machining; slow tool servo; surface topography; simulation; microlens array; sinusoidal grid

---

1. Introduction

Owing to the superior optical and physical properties, freeform surfaces are an important catalyst in the development of high-value-added photonics and telecommunication products, such as laser beam printers and scanners, head mounted displays, progressive lens molds, fiber optic connectors, and advanced automotive lighting systems [1–4]. Differing from conventional simple surfaces, freeform surfaces are more geometrically complex and normally have no symmetry in rotation or translation. To ensure the functionality of the freeform components, these surfaces are required to have sub-micrometer form accuracy and nanometer surface topography [5].

With the technical evolution in advanced manufacturing, ultra-precision machining technologies have been developed for the deterministic fabrication of high precision freeform surfaces including tool servo turning, ultra-precision raster milling, ultra-precision grinding and polishing [6–9]. Among them, slow tool servo (STS) machining provides an important means for generating optical freeform surfaces without the need for any subsequent processing. It has the advantages of simpler setup, faster cycle times and better machining accuracy over other techniques. Yi and Li [7] proposed an innovative diamond tool trajectory that allows the entire 5×5 microlens array to be fabricated in a single operation using STS machining. The machined microlens array surface was measured for both curve conformity and surface roughness and 34.5 nm Ra was achieved over 0.7 mm scanning length. Yin et al. [10] investigated the fabrication of off-axis aspheric surfaces using STS techniques. The form error of off-axis paraboloid caused by the tool centering error was analyzed in detail and verified by cutting experiment. RMS form error 0.063 μm of the overall region was achieved after adjusting the tool center. Mukaida and Yan [11] performed an experimental study of STS machining for single-crystal silicon microlens arrays. The surface error, material phase transformation, and cutting force characteristics were investigated experimentally. Spherical microlens arrays with a form error of 300 nm PV (peak to valley) and surface roughness of 6 nm Sa (arithmetical mean height of a surface) were successfully fabricated. Chen et al. [12] developed a model of three dimensional tool shape compensation for generating tool path in STS diamond turning of an asymmetrically toric surface for an astigmatic contact lens. The form accuracy of the freeform surface was evaluated by an ultra-high accuracy 3D profilometer. The form error was less than 0.5 μm both in the X and Y direction after the correction process.

Most studies have focused on the experimental investigation of STS machining of various types of freeform surfaces and the evaluation of form accuracy. However, surface topography should be studied [13,14], which is closely related to its functional performance, such as the diffractive properties of freeform optics [15,16]. Little systematic work has been reported about theoretical and experimental study of surface topography generation in STS machining of freeform surfaces. A successful STS machining depends largely on the selection of machining variables, tool parameters, and machining trajectories. Moreover, surface topography generation simulation offers a cost effective solution to select optimal cutting conditions, to predict the surface quality and to understand the machining phenomenon. A trial-and-error cutting approach is not economic because it is time consuming and costly [17].

This work presents a theoretical and experimental study of surface topography generation in STS machining of freeform surfaces. Several key machining aspects including tool path planning, selection of cutting tool geometries, and the tool radius compensation method, are discussed in detail. Moreover, surface generation simulation is proposed to theoretically investigate topography generation during the machining process. Finally, machining and measurement experiments of typical freeform surfaces are carried out to validate the effectiveness of the proposed methodology.

2. Tool Path Generation

The workflow of STS machining of freeform surfaces is proposed as illustrated in Figure 1. According to the design and specification of freeform surfaces, the first task is to generate machining trajectories. In the initial stage, machining variables are selected to meet the targeted production requirement. Next, tool interference analysis is conducted to check if the diamond cutting tool can fully access the proposed machined features. To eliminate any overcutting phenomenon, tool radius compensation needs to be performed on the ideal tool path. Subsequently, the motion of machine tool axes is analyzed for its reachability of the modified tool path. In the second stage, numerical modelling is carried out which provides an important means to predict theoretical surface generation without the need for costly trial and error tests. Profile topography is generated by the intersection of a tool tip profile along the feeding direction and surface topography is formulated by a combination of all the radial sectional profiles along each angle. The simulated surface error is used as feedback information to guide the tool path generation processes. If it is less than the pre-defined value, the machining operation will be carried out.

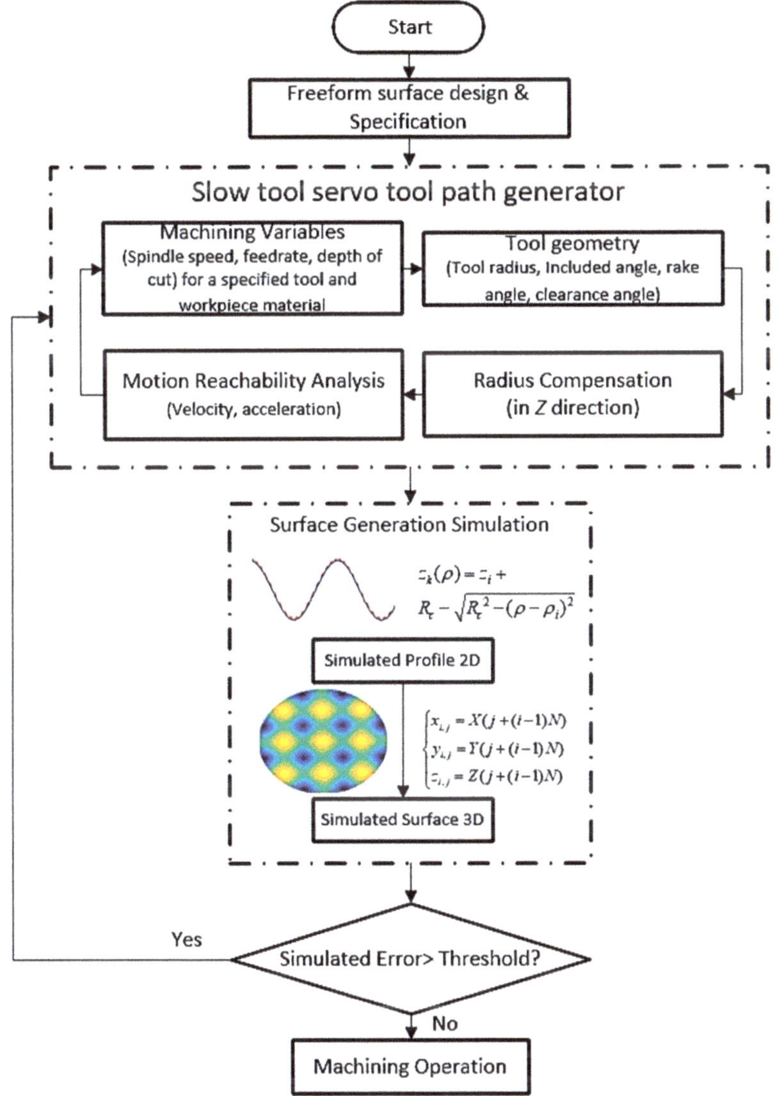

Figure 1. Workflow for slow tool servo (STS) tool path generation.

2.1. STS Machining Principle

The conventional single point diamond turning (SPDT) process utilizes two linear axes for contouring motion with a velocity controlled spindle. Therefore, only rotational symmetric surfaces can be fabricated. As an adaption of conventional SPDT, the STS technique enables the spindle to actuate in a position controlled mode (also called C axis mode). The schematic of the STS machining setup is shown in Figure 2. An arbitrary 3D tool path for non-rotationally symmetric freeform surfaces can be achieved when the X axis, Z axis, and C axis move simultaneously following a given set of numerical motion commands. In most applications, the workpiece is mounted on the C axis while a diamond tool is installed on the Z axis, which needs to oscillate both forward and reverse in servo-synchronization with the angular position of the C axis and translational position of the X axis.

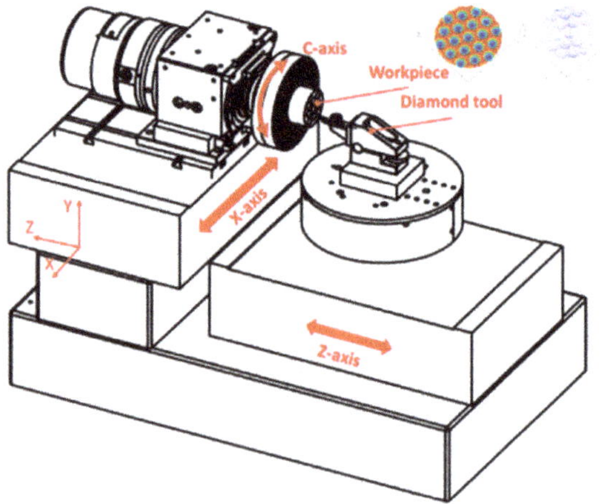

Figure 2. The schematic of the STS machining setup.

In STS machining, the coordinate system is described as a cylindrical coordinate. The tool path projection on the X–Y plane is an equivalent spiral curve no matter how complex the surface is (as shown in Figure 3). The discrete points are equal-angle spaced for simple computation and control.

In the X–Y plane, the spiral curve can be described mathematically as follows:

$$\begin{cases} \rho_i = R_w - (i-1)\frac{f}{S \cdot N_\theta} \\ \theta_i = (i-1)\frac{2\pi}{N_\theta} \\ i = 1, 2, \cdots, \frac{R_w \cdot S \cdot N_\theta}{f} + 1 \end{cases} \tag{1}$$

where ρ_i is the radial distance (cylindrical coordinate) in mm, θ_i is the polar angle (cylindrical coordinate) in radians, R_w is the radius of workpiece in mm, i is the number of control points, f is the feedrate in mm/min, S is the C axis rotational speed in revolutions per minute (rpm), and N_θ is the number of programmed points per revolution. However, the surface model to be fabricated is often expressed in a Cartesian coordinate (x, y, z) system. Under right-hand coordinate convention, the transformation between the two coordinate systems is as follows:

$$\begin{cases} x_i = \rho_i \cos(\theta_i) \\ y_i = \rho_i \sin(\theta_i) \\ z_i = F(x_i, y_i) = F(\rho_i \cos(\theta_i), \rho_i \sin(\theta_i)) \end{cases} \tag{2}$$

where (x_i, y_i, z_i) are the surface model points and $F(\cdot)$ is the surface description. To illustrate the tool path generation principle, an STS ideal tool path (ρ_i, θ_i, z_i) for a typical freeform surface (sinusoidal grid) can be generated. The surface is mathematically expressed as,

$$z = A_x \cos(\frac{2\pi}{\lambda_x} x + \varphi_x) + A_y \cos(\frac{2\pi}{\lambda_y} y + \varphi_y) \tag{3}$$

where A_x and A_y are the amplitudes in the X and Y direction. λ_x and λ_y are the wavelength in the X and Y direction. φ_x and φ_y are the phase in the X and Y direction. Figure 3 shows the generated tool path and its spiral X–Y projection.

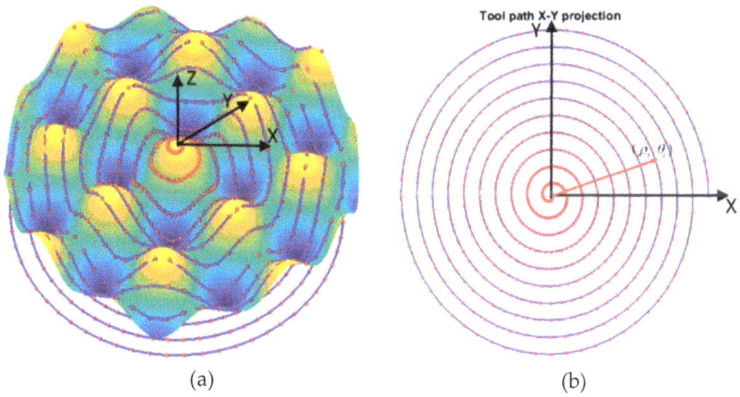

(a)　　　　　　　　　　　　　　　(b)

Figure 3. (**a**) STS ideal tool path and (**b**) X–Y projection.

2.2. Tool Geometry Selection

Tool geometries should be carefully selected to guarantee the accessibility to the features of the proposed freeform surfaces. As shown in Figure 4, geometric parameters of a typical diamond cutting tool include the tool radius R_c, the included angle ψ, the rake angle γ, and the clearance angle α.

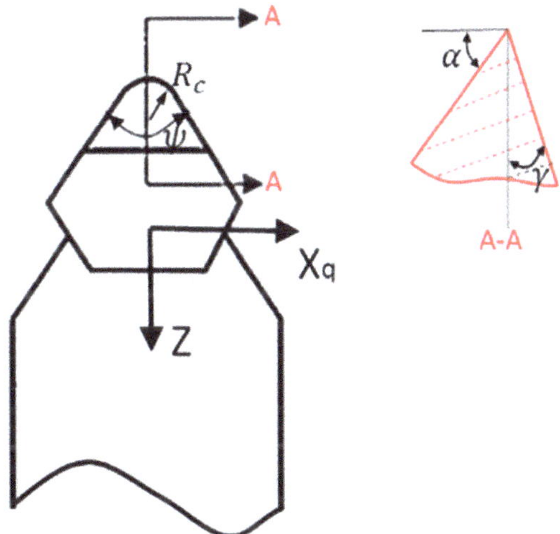

Figure 4. Geometric parameters of a typical diamond cutting tool.

The schematic for tool geometry selection in STS freeform machining is illustrated in Figure 5. For every cutting point (red dot in the plot), a radial cutting plane is determined by the Z axis and the cutting point while a normal plane is perpendicular to the radial plane and crosses the cutting point (shown in Figure 5a). Proper tool parameters should be selected to avoid interference with the machining surfaces in both planes. R_c and ψ are calculated in the radial cutting plane, whereas γ and α are calculated in the normal plane.

As shown in Figure 5b, along each sectional profile $f_\rho(\rho, \theta)$ in the radial plane, the tool tip radius R_c should be small enough so that the tool is accessible to all the profile features while its critical value is determined by the minimum radius of curvature for all the cutting points. The included angle ψ should be large enough to make sure the cutting edge always stays in contact with the machining surface and its critical value is determined by the maximum value of the angle of inclination along the radial intersection profiles. The two conditions can be mathematically expressed as follows:

$$
\begin{cases}
R_c \leq \min\left\{ \dfrac{\left(1+(f'_\rho(\rho,\theta))^2\right)^{\frac{3}{2}}}{f''_\rho(\rho,\theta)} \right\} & 0 < \rho \leq R \\[3mm]
\psi \geq 2\max\left\{ \arctan(f'_\rho(\rho,\theta)) \right\} & 0 < \theta \leq 2\pi
\end{cases}
\tag{4}
$$

where $f'_\rho(\rho,\theta)$ and $f''_\rho(\rho,\theta)$ are respectively the first derivative and second order derivative of the radial intersectional profile $f_\rho(\rho,\theta)$. To calculate the limit of the tool rake angle and the clearance angle, the intersection profile $g_{y_q}(y_q,\rho,\theta)$ in the normal plane is obtained in the normal plane perpendicular to the radial plane (the red curve shown in Figure 5c). The tool rake face and flank should not interfere the machined surface. Thus, the following conditions must be met:

$$
\begin{cases}
\gamma \geq \max\left\{ \arctan(g'_{y_q}(y_q,\rho,\theta)) - \frac{\pi}{2} \right\} \\[3mm]
\alpha \geq \max\left\{ -\arctan(g'_{y_q}(y_q,\rho,\theta)) \right\}
\end{cases}
\tag{5}
$$

where $g_{y_q}{'}(y_q,\rho,\theta)$ is the first derivative of the normal plane intersectional profile. Besides the accessibility issue, the effect of the tool tip on the surface generation needs to be considered, which is discussed in the following section.

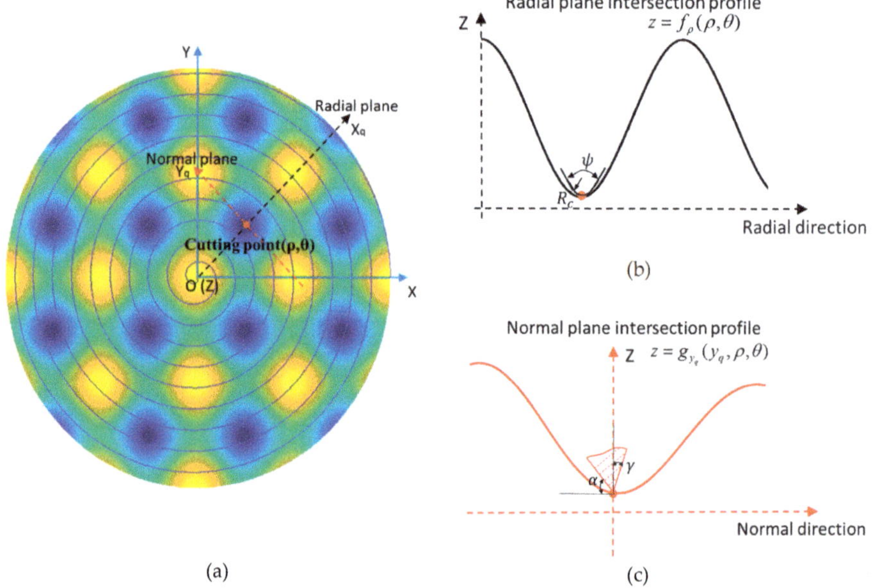

Figure 5. Schematic of tool geometry selection for STS freeform machining. (**a**) Radial and normal cutting plane; (**b**) Radial sectional profile; (**c**) Normal sectional profile.

2.3. Tool Radius Compensation

Due to the circular geometry of the diamond tool tip, the cutting edge will cause overcut on the machined surface if the tool path is programmed based on the ideal infinitely sharp profile. Such an effect is illustrated in Figure 6. The red area shows the overcutting phenomenon, which would deteriorate the surface accuracy.

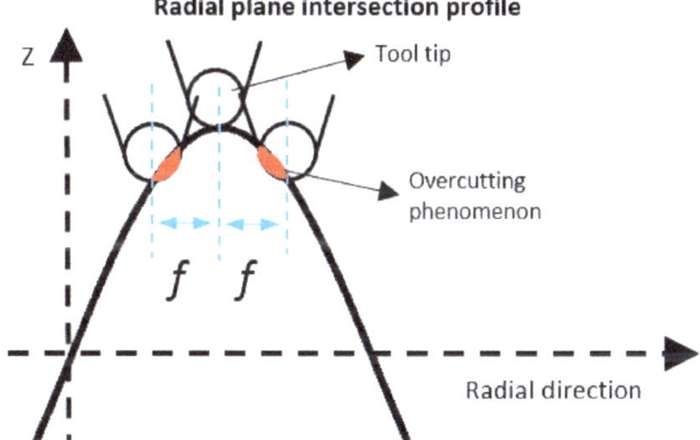

Figure 6. Schematic of overcutting phenomenon caused by circular tool tip.

To avoid overcut on a machined surface, tool radius compensation is performed so that the circular tool edge should always be tangent to the intersection profile in each radial plane. Conventionally, tool radius compensation is performed in the normal direction on the cutting points [18,19]. The normal method is illustrated in Figure 7. The black tool tip shows the original programmed position. To compensate the overcut, the cutting position (red dot) is shifted so that the tool edge profile contacts tangential to the surface profile as indicated by the orange tool tip. The center of the circular tool edge is along the normal direction of the cutting point.

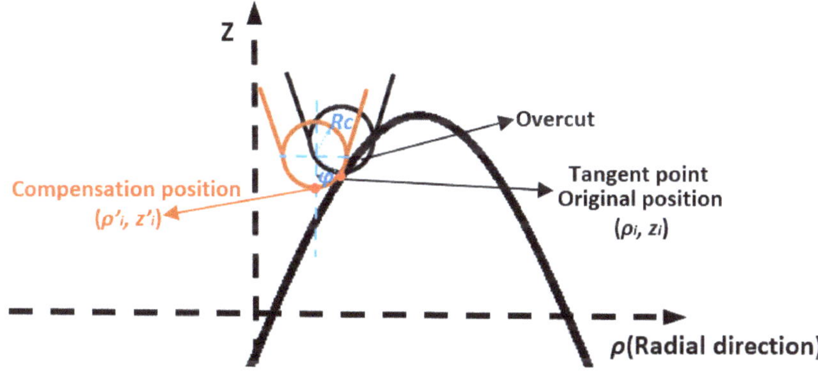

Figure 7. Tool radius compensation using normal direction method.

For a given radial intersection profile $f_\rho(\rho, \theta)$, the relationship between the original cutting point (ρ_i, z_i) and the compensation position (ρ'_i, z'_i) can be expressed as follows:

$$\begin{cases} \rho_i' = \rho_i - R_c \sin(\varphi_i) \\ z_i' = z_i + R_c \cos(\varphi_i) - R_c \\ \tan(\varphi_i) = \frac{df_\rho(\rho)}{d\rho}\big|_{\rho=\rho_i} \end{cases} \tag{6}$$

where φ_i is the slope angle at (ρ_i, z_i) in the intersection profile. Calculation of the slope angle is required at every cutting point as the slope of freeform surfaces varies along the radial direction as well as at different angles.

To illustrate the tool radius compensation process, tool path generation was performed for a sinusoidal grid surface described by Equation (3). The surface design parameters were set to be $A_x = A_y = 1$ μm, $\lambda_x = \lambda_y = 0.5$ mm, $\varphi_x = \varphi_y = 0$. The machining variables were selected to be $f = 5$ mm/min, $S = 100$ rpm, $R_c = 0.5$ mm. The compensated and uncompensated 3D tool path are shown in Figure 8a. Figure 8b indicates the X–Y projection of the compensated tool path and how it deviates from the original spiral trajectories. The motion analysis, illustrated in Figure 8c,d, shows additional motion components appearing on both the X and Z axis after the radius compensation in the normal direction. The disadvantage of the normal compensation method is that the tool tip shift is required to be performed in both the X and Z direction. Moreover, the shift value is not constant as the slope angle varies at different cutting points on freeform surfaces. The X feed will change along the radial direction using the normal compensation method. Therefore, high frequency motion of both the X and Z axis is required for tool radius compensation in the normal direction. For the configuration of the machine tool used in this work (shown in Figure 2), a heavy working spindle is mounted on the X axis and its dynamic response is thus limited.

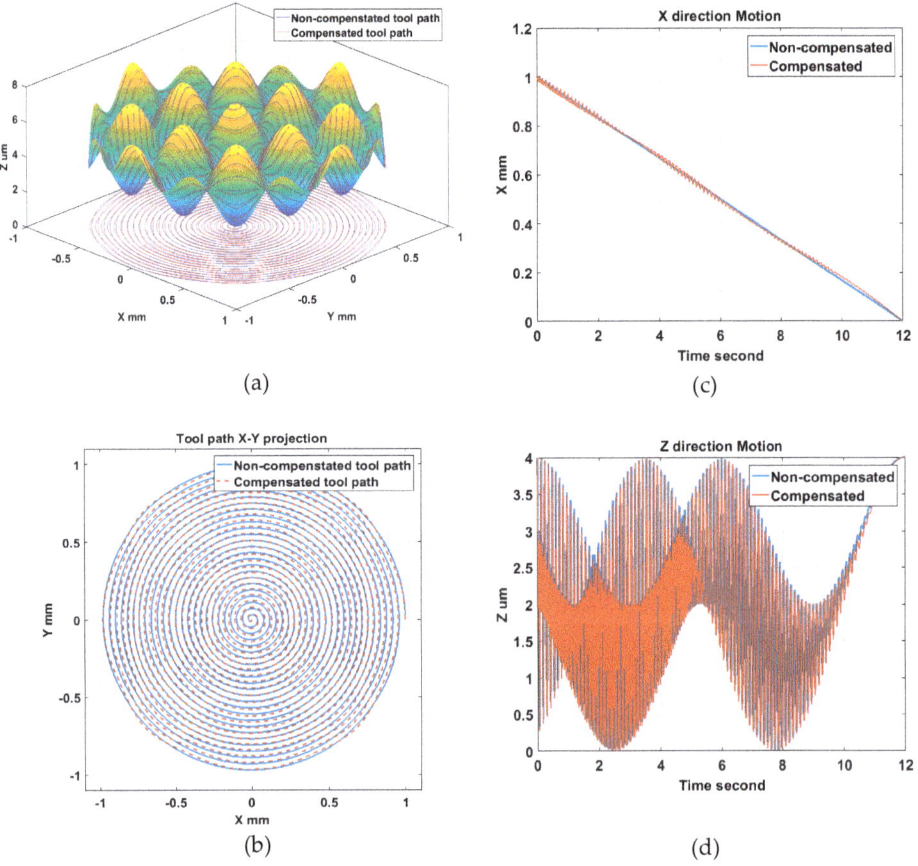

Figure 8. Normal direction compensation method and tool path analysis: (**a**) Tool path; (**b**) Tool path projection; (**c**) X axis motion; (**d**) Z axis motion.

Therefore, a modified tool radius compensation method is proposed as shown in Figure 9. In the Z direction compensation method, the tool tip only needs to shift along the Z direction until the cutting edge is tangential to the surface. The relationship between the original cutting point (ρ_i, z_i) and the compensation position (ρ'_i, z'_i) can be expressed as follows:

$$\begin{cases} \rho_i{}' = \rho_i \\ z_i{}' = z_i + \Delta z \end{cases} \tag{7}$$

$$\begin{cases} \Delta z = \dfrac{Rc}{\cos \varphi'} - Rc \\ \tan(\varphi') = f_\rho'(\rho_i + R_c \sin(\varphi_i)) \end{cases} \tag{8}$$

where ΔZ is the tool shift value in the Z direction and φ' is the slope angle of the new tangential point. φ' is in an implicit equation and solved using Newton's iterative algorithm [20].

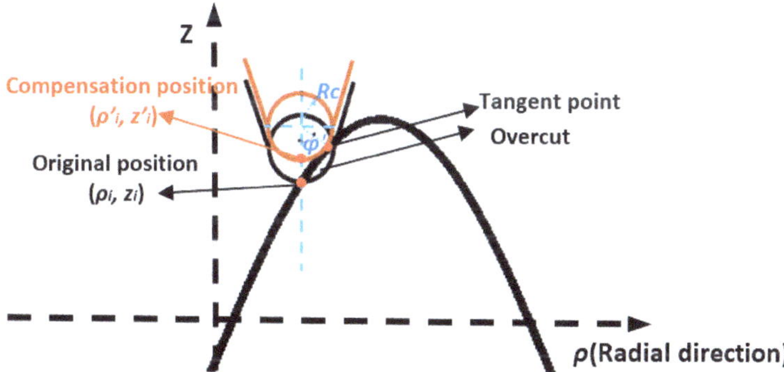

Figure 9. Tool radius compensation using the Z direction method.

The difference between the two compensation methods is simulated along a cosine radial profile and illustrated in Figure 10. The red dots show the tool position using the Z compensation method whereas the black dots represent the tool position using the normal compensation method.

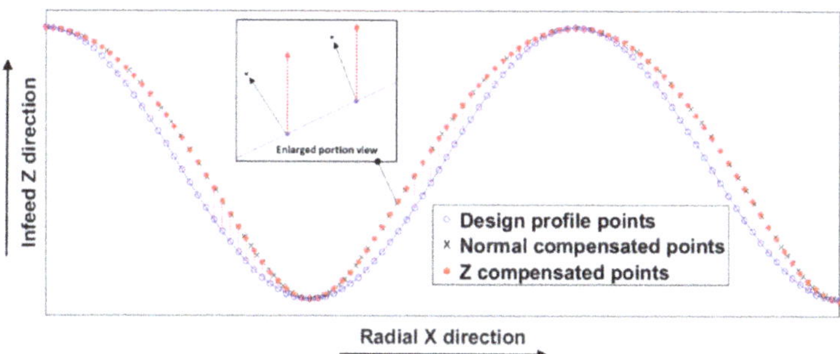

Figure 10. Comparison of two compensation methods along a radial profile.

In addition, tool path simulation using the Z direction compensation method was carried out for the same sinusoidal grid surface presented above. As shown in Figure 11b, the X–Y projection of the compensated tool path coincides with the uncompensated one, which means the tool shift is only performed in the Z direction. The motion analysis in Figure 11c,d also validates that additional high frequency motion is avoided for the low-dynamic X axis. The X feed will not change along the radial direction using the Z compensation method. Therefore, the Z direction compensation method is considered stable and good for dynamic performance in STS machining of freeform surfaces.

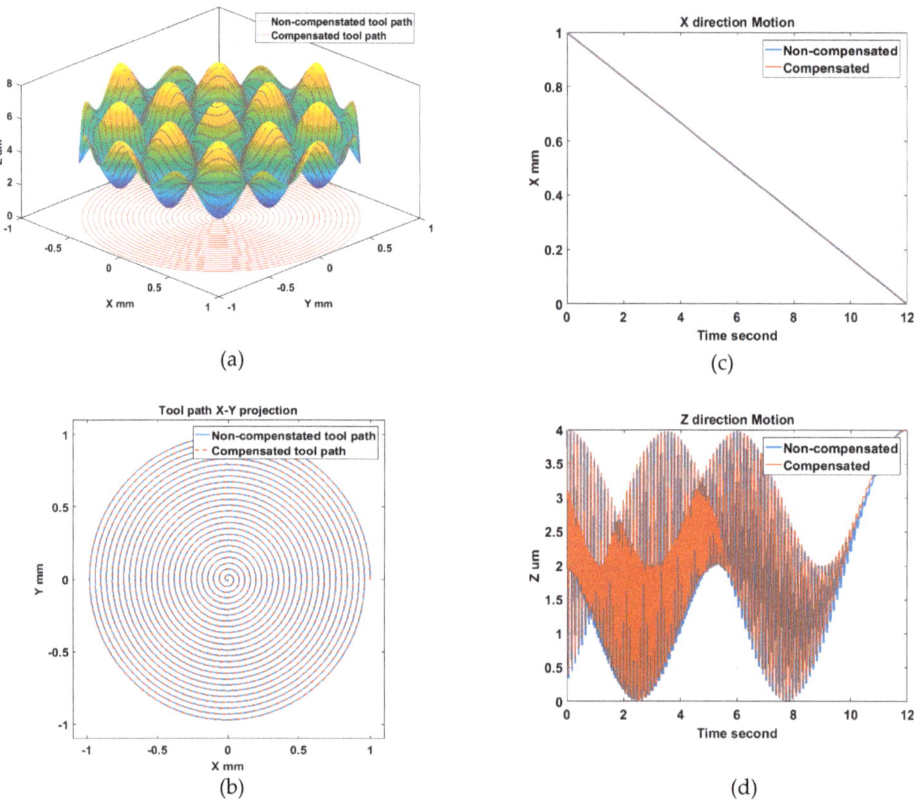

Figure 11. Z direction compensation method and tool path analysis: (**a**) Tool path; (**b**) Tool path projection; (**c**) X axis motion; (**d**) Z axis motion.

3. Surface Generation Simulation and Analysis

3.1. Principles

Surface generation simulation offers a cost effective solution to select optimal cutting conditions, to predict the surface quality and to understand the machining phenomenon without the need for costly trial and error machining tests. As illustrated in Figure 12, the successive tool positions are distributed at the interval of the feedrate along each radial intersection profile curve. Once the tool path is derived (the dashed line), the theoretical surface topography can be formed as the envelope of consecutive tool tip profiles along the feeding trajectory. Between the intersection points, the theoretical surface profile is a section of the circular tool tip profile (shown as black solid line).

Assuming the tool tip radius is R_c and tool tip location is (ρ_i, z_i), the cutting profile can be expressed as:

$$z(\rho) = z_i + R_c - \sqrt{R_c{}^2 - (\rho - \rho_i)^2} \tag{9}$$

Thus, the profile topography height $h_{envelope}$ at radial position ρ can be calculated as the minimum value of all the cutting profiles:

$$h_{envelope}(\rho) = \min\{z_k(\rho)\}, \quad k : i - 1 \, to \, i + 1 \tag{10}$$

Taking a cosine radial profile as an example to validate the topography generation method, the result in Figure 13 clearly shows the successive tool tip profiles along the feeding direction and the resulting topography generation. The areal surface topography can be formulated by combination of all the radial intersection profile topography at each angle. With the above proposed method, generation simulation of the sinusoidal grid surface is performed. The surface parameters are the same as those in Section 2.3. For illustration purposes, the machining variables are selected to be $f = 2$ mm/min, $S = 100$ rpm, $R_c = 0.05$ mm. Figure 14a,b respectively show the simulated areal surface topography and extracted profile topography at 0 degrees. The theoretical turning marks can be clearly seen on the simulated surface.

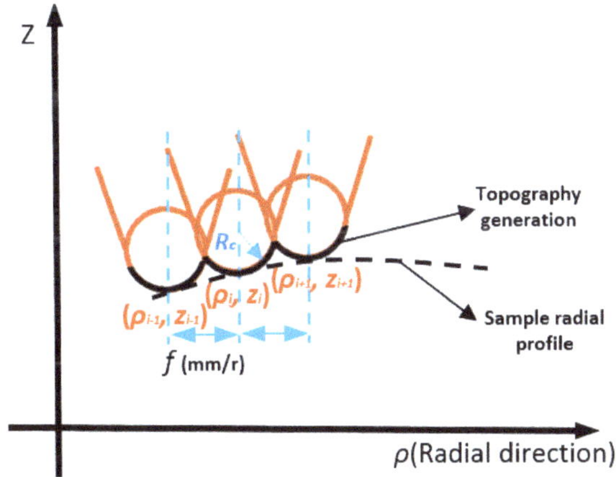

Figure 12. Schematic diagram of profile topography generation.

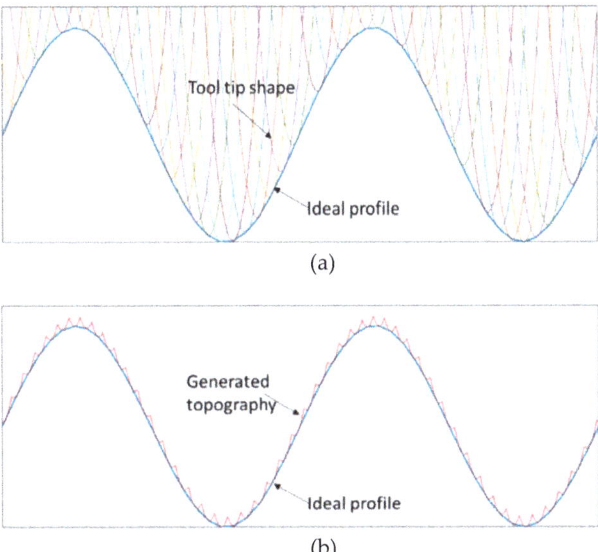

Figure 13. Simulation of topography generation along the radial profile: (**a**) Successive tool tip profiles along the feeding direction; (**b**) Resulting topography generation.

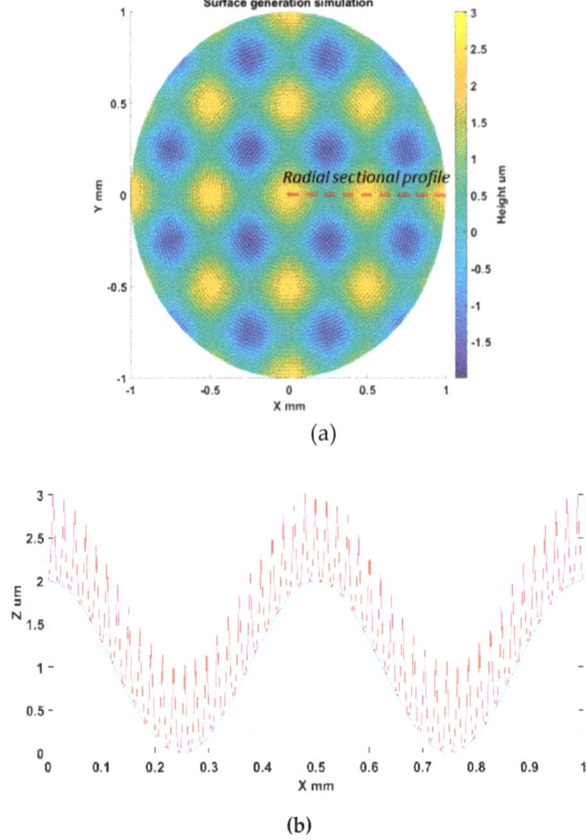

(a)

(b)

Figure 14. Simulation example of areal surface topography generation: (**a**) Surface generation simulation; (**b**) Radial section profile topography (0 degrees).

3.2. Simulation Analysis

With the established surface generation model, simulation analysis using MATLAB software is carried out in this section. The analysis is used to guide the selection of cutting parameters to achieve the targeted surface quality and better understand the machining processes as well.

Without consideration of material effects, there are three processing parameters that influence the theoretical surface generation, which are tool radius R_c (mm), feedrate f (mm/min), and spindle speed S (rpm). The relationship between processing parameters and surface quality is investigated with the aid of surface generation simulation. The investigation range is set as: R_c 0.1–1 mm; f 0.2–1 mm/min; S 50–150 rpm. The root mean square height S_q value (root mean square) is adopted to quantitatively describe the simulated surface quality.

Figure 15a–c illustrates the quantitative relationship between processing parameters and surface quality using the surface generation simulation. From the simulation results, it can be concluded that a better surface finish (lower S_q value) can be obtained under a higher spindle speed, a smaller federate, and a larger tool radius. In practice, it is better to choose a higher spindle speed rather than decreasing the feedrate. A lower feedrate would increase the machining time, decrease the tool life, and make the machining process vulnerable to environmental variations. However, higher spindle speed in STS machining requires a higher motion frequency and servo bandwidth, which is limited by the machine tool configuration and control strategy. The increase of tool radius results in the decrease of the S_q

value, the tool tip accessibility should be taken into consideration, which is discussed in Section 2.2. The relationship graphs are generated with the aid of surface generation simulation without costly trial and error experiments, which are useful to select optimized processing parameters to obtain a targeted surface quality.

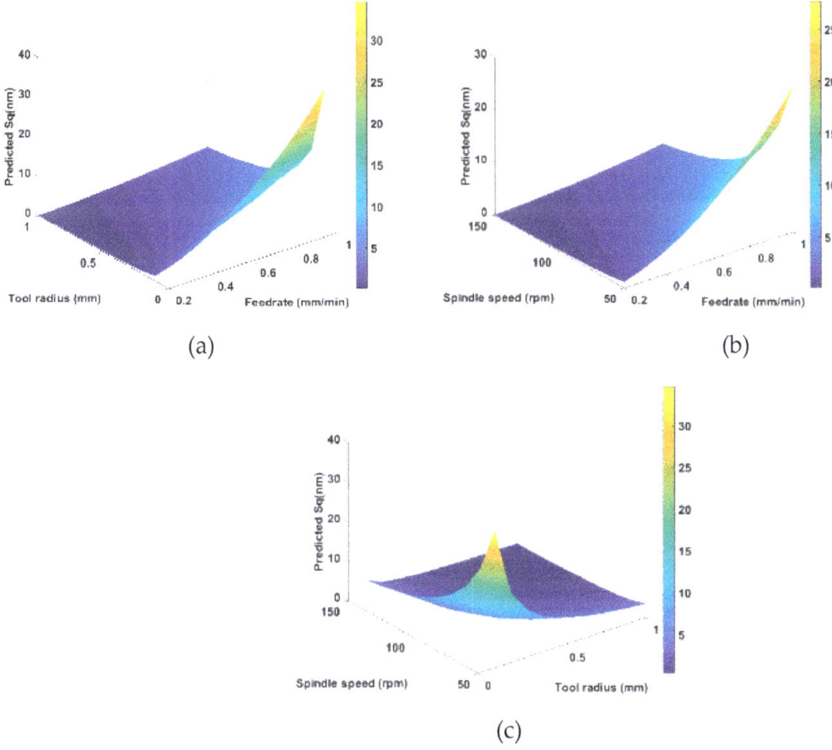

(a)

(b)

(c)

Figure 15. Relationship graphs between processing parameters and surface quality: (**a**) R_c and f vs. S_q (S = 100 rpm); (**b**) S and f vs. S_q (R_c = 0.5 mm); (**c**) S and R_c vs. S_q (f = 0.6 mm/min).

Surface generation simulation also provides an important means for understanding the cutting phenomenon. In the following section, simulation analysis is performed to study the overcutting phenomenon and the effectiveness of tool radius compensation.

The sinusoidal grid surface described by Equation (3) is used in the simulation. The surface design parameters are kept the same as those in Section 2.3. The machining variables are selected to be f = 5 mm/min, S = 50 rpm, R_c = 0.5 mm. The simulation results are shown in Figure 16. Plots on the right and left show the simulation analysis of surface generation with and without tool radius compensation, respectively. As shown in the left plot of Figure 16b,c, the overcutting phenomenon can be clearly observed and waviness error components are induced on the machined surface due to tool tip overcut. In contrast, the overcutting phenomenon is avoided with the proposed tool radius compensation and waviness error components are eliminated, as shown in the right plot of Figure 16b–d which illustrates areal surface topography residual after form removal. The result also indicates that the pattern of induced waviness errors varies with the intersection angles. The study validated the proposed tool radius compensation method and effectiveness of simulation analysis to investigate the cutting phenomenon.

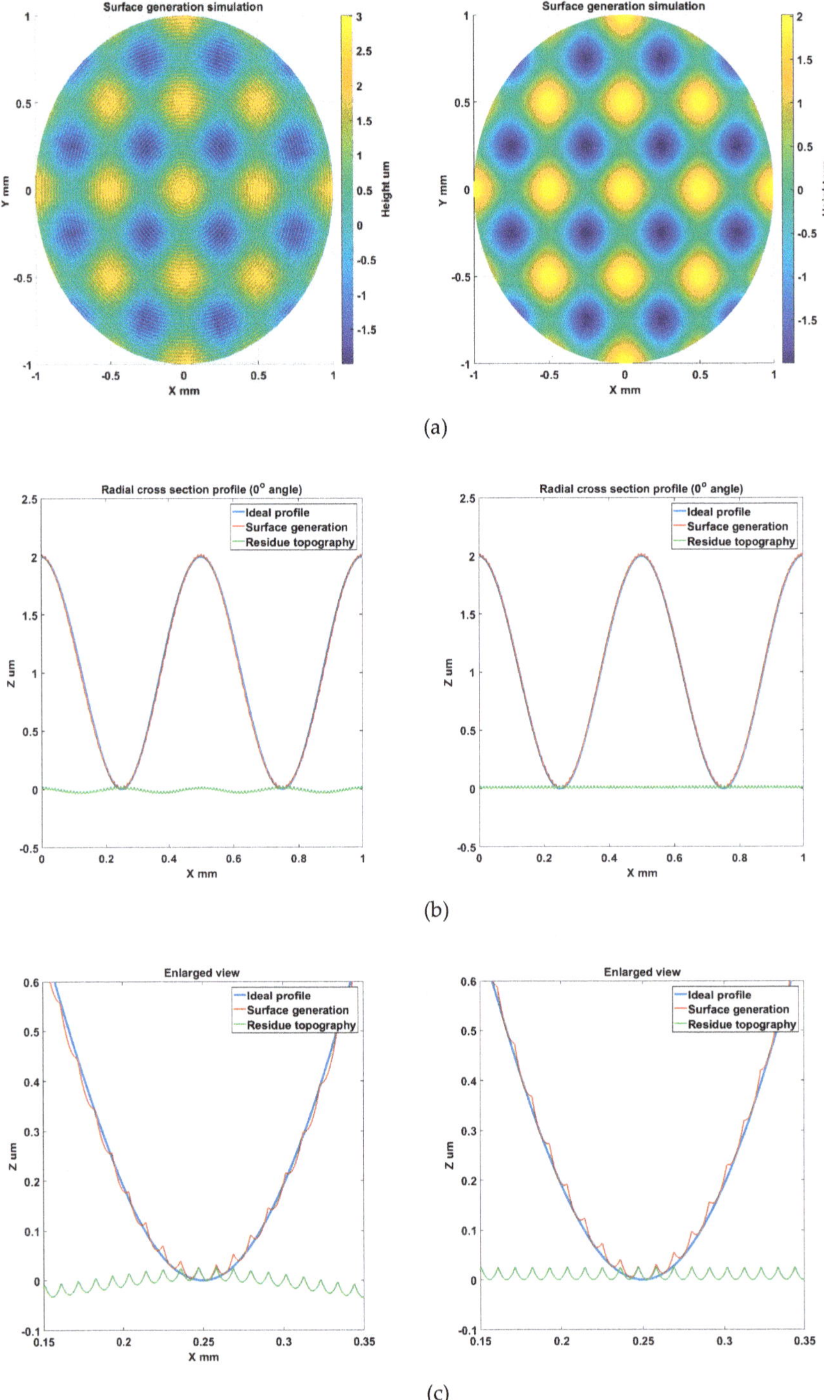

Figure 16. *Cont.*

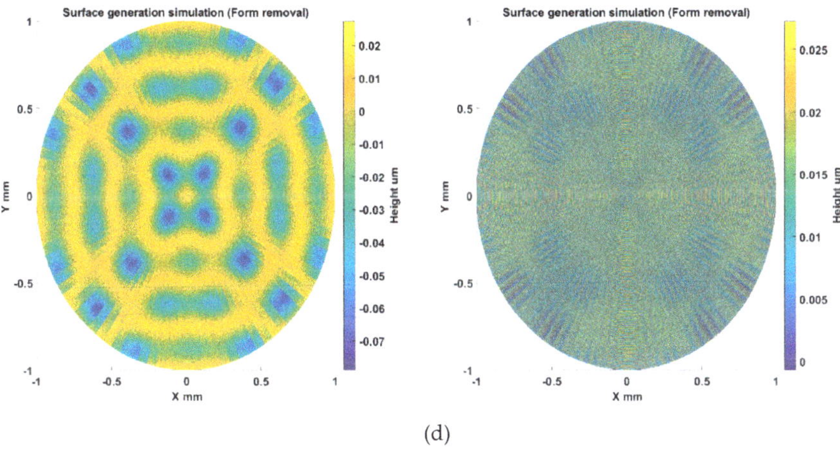

(d)

Figure 16. Simulation analysis of overcutting phenomenon and tool radius compensation (left column: without compensation, right column: with compensation): (**a**) Simulated surface generation; (**b**) Extracted radial profile topography; (**c**) Enlarged view of radial profile topography; (**d**) Simulated surface topography residual (after form removal).

4. Experiments and Discussion

In order to show the effectiveness of the proposed STS machining methodology and surface topography generation, machining and measurement experiments of typical freeform surfaces (a sinusoidal grid and MLA surface) are carried out in the section.

4.1. Experimental Setup

The machine tool used in the machining experiment is a Precitech Nanoform 250 Ultra Grind (Keene, NH, USA), which is shown in Figure 17. It can be used for diamond turning and ultra-precision grinding. The machine tool incorporates a finite element analysis (FEA) optimized dual frame for ultimate environmental isolation. A sealed natural granite base also provides excellent long term stability and vibration damping. Both the X and Z slides are equipped with hydrostatic oil bearing with symmetrical linear motor placement. The position of the X and Z axes is measured with linear laser scale encoders, which are capable of resolving 0.016 nm after signal subdivision. The straightness error for both the X and Z axes over the full travel is less than 0.2 μm. Under position controlled mode, maximum rotational speed of C axis can be 1500 rpm with a feedback resolution of 0.01 arc sec, while maintaining axial and radial error motion of less than 15 nm. The high precision and stability of the machine tool is a prerequisite for the ultra-precision machining process. The sample material used in the experiments is an aluminum alloy (Al6082) with a chemical composition of (0.7% Mn, 0.5% Fe, 0.9% Mg, 1% Si, 0.1% Cu, 0.1% Zn and 0.25% Cr). The material is of good machinability and its mechanical properties are listed in Table 1. The machined samples are shown in Figure 18.

Table 1. Mechanical properties of the sample material (Al6082).

Parameters	Value
Density (g/cm3)	2.70
Modulus of elasticity (GPa)	70
Tensile strength (MPa)	260
Shear strength (MPa)	170
Thermal conductivity (W/m.K)	180

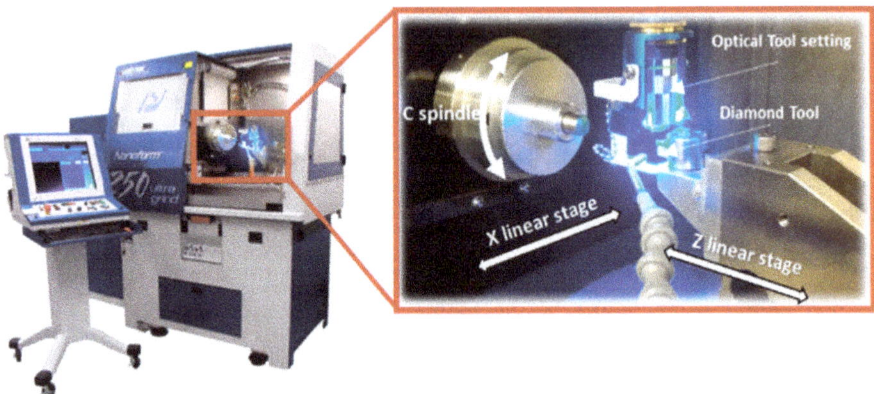

Figure 17. Experimental setup of STS machining.

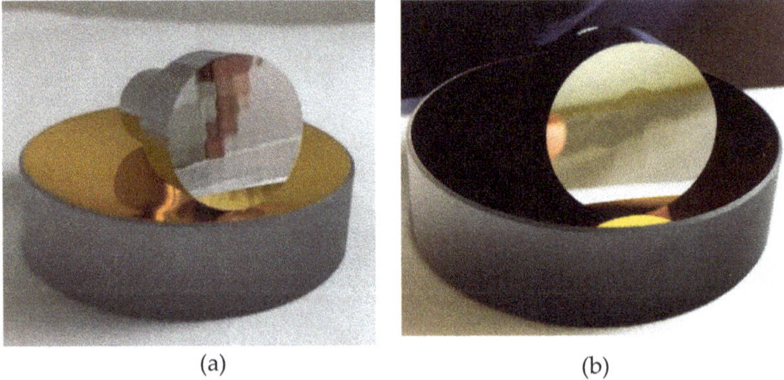

(a)   (b)

Figure 18. (**a**) Machined sinusoidal grid surface sample; (**b**) Machined micro-lens array (MLA) surface sample.

4.2. Sinusoidal Grid Surface Machining

A sinusoidal grid surface can be used for measurement of two-dimensional (2D) planar displacements [21]. The freeform surface is continuous and described mathematically by Equation (3). In the experiment, the design parameters were set to be $A_x = A_y = 2$ µm, $\lambda_x = \lambda_y = 2.5$ mm, $\varphi_x = \varphi_y = 0$. The machining and diamond cutting tool parameters are respectively listed in Tables 2 and 3. With the analysis discussed in Section 2.2, the selected diamond tool can avoid interference with the machined surface. The proposed Z direction tool radius compensation was also performed on the ideal tool path to avoid the overcutting phenomenon. The design and STS tool path of the sinusoidal grid surface are illustrated in Figure 19a,b respectively. The sample was successfully machined, as shown in Figure 18a.

Table 2. Machining variables for sinusoidal grid surface.

Parameters	Value
Machining mode	STS
Spindle speed (rpm)	50
Feedrate (mm/min)	0.5
Cutting depth (µm)	3

Table 3. Diamond tool parameters.

Parameters	Value
Manufacturer	Contour fine tooling
Tool material	Single crystal
Tool tip radius (mm)	0.514
Rake angle (deg)	0
Clearance angle (deg)	10
Included angle (deg)	60

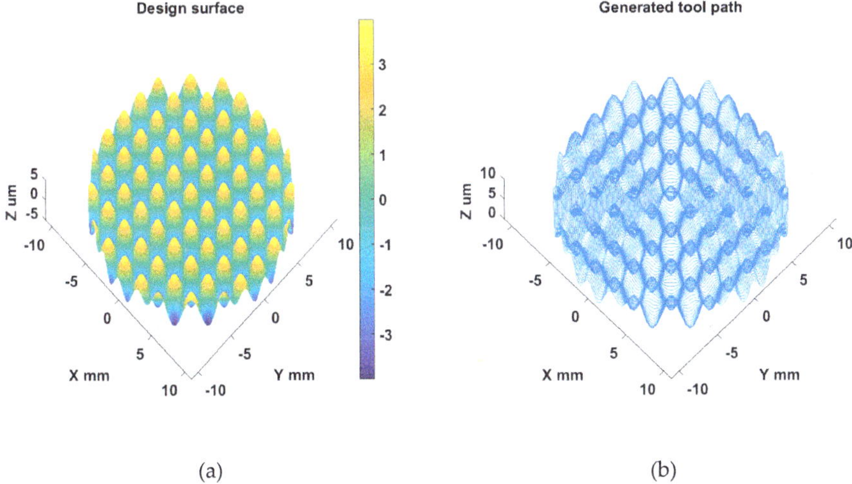

Figure 19. (**a**) Design and (**b**) STS tool path of sinusoidal grid surface.

To inspect the machining quality, the sample was measured using a Talysurf Coherence Correlation Interferometry (CCI) 3000 (Taylor Hobson, Leicester, UK), equipped with a 20X microscope objective. The original and processed measurement results are shown in Figure 20a,b respectively. After filtering out the form component, the turning marks can be clearly observed from the CCI measurement. The surface topography was characterized by S_q.

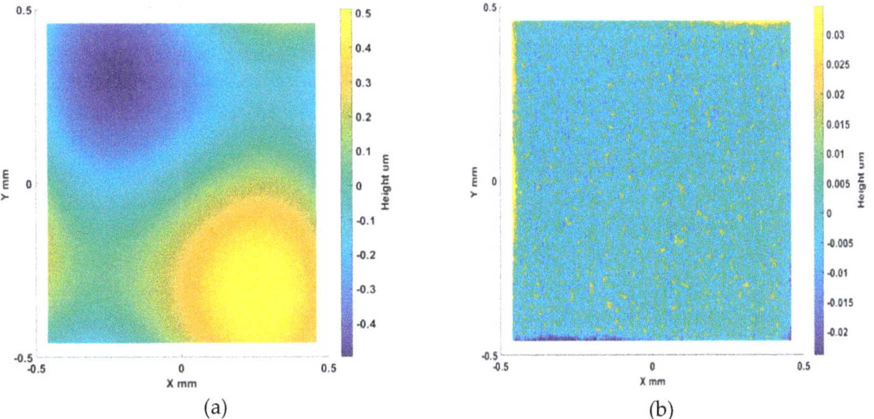

Figure 20. Machined sinusoidal grid surface CCI measurement: original (**a**); after form removal (**b**).

To examine the uniformity of the topography distribution, five areas were measured on the surfaces. The average S_q is calculated as 7.1 nm and standard deviation is 0.30 nm. The measurement results (shown in Figure 21) indicate the machined topography of the continuous freeform surface is uniformly distributed over the surface and less than 10 nm, which meets the requirement for visible light optical application.

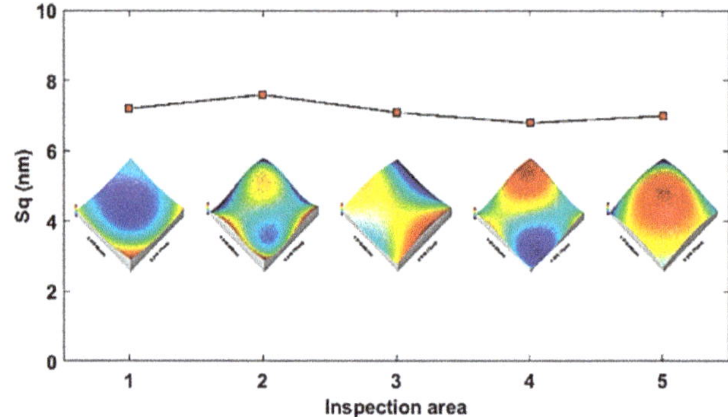

Figure 21. Topography distribution of sinusoid grid surface.

4.3. MLA Surface Machining

Micro-lens arrays (MLAs) play a key role in highly efficient light transmission [22]. The MLA surface is regarded as a type of structured freeform surface. It is composed of multiple elemental lenses, which are distributed in a specific pattern. In this experiment, the design parameters for MLA are listed in Table 4.

Table 4. Micro-lens array (MLA) design parameters.

Parameters	Value
Nominal feature shape	Sphere
Pattern	2×2
Center Spacing (mm)	4.243
Aperture radius (mm)	2
Chord height (μm)	8
Radius of curvature (mm)	250.004

The same machining variables and tool used for the sinusoidal grid sample were used. The design and STS tool path of the MLA surface are respectively illustrated in Figure 22a,b. As shown in Figure 18b, the MLA sample was successfully machined to prove the effectiveness of STS machining of different types of freeform surfaces.

To inspect the machining quality, five areas in a different element lens were measured using Talysurf CCI 3000. The measurement results are shown in Figure 23. The average S_q is calculated as 7.4 nm and the standard deviation is 0.34 nm. The measurement results indicate the machined topography of the structured freeform surface is also less than 10 nm and uniformly distributed.

The measured and simulated results of surface topography are also summarized in Table 5. The S_q value of the actual measurement agrees with the simulated value, which proves the feasibility of the surface generation simulation.

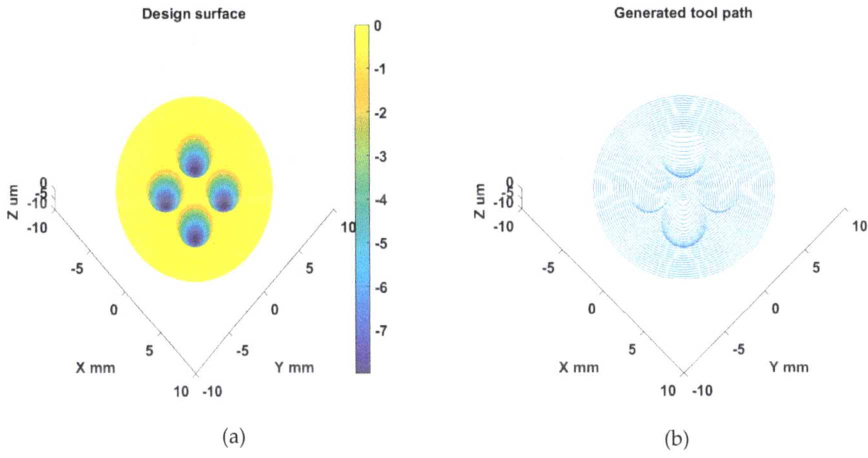

(a)   (b)

Figure 22. Design (**a**) and STS tool path (**b**) of MLA surface.

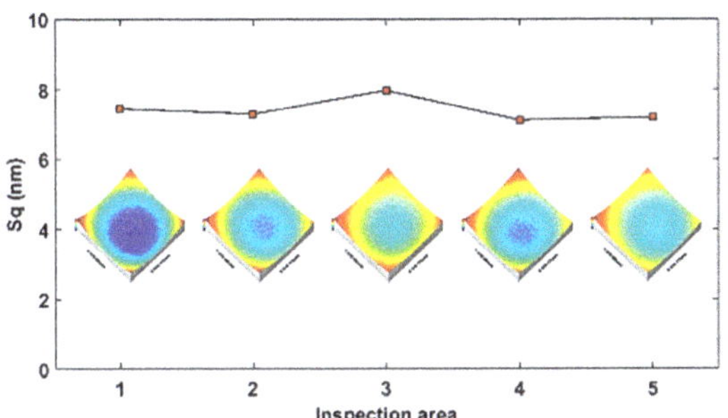

Figure 23. Topography distribution of MLA surface.

Table 5. Surface topography S_q by actual measurement and simulation.

Sample	Measured Average S_q (nm)	Standard Deviation S_q (nm)	Simulated S_q (nm)
Sinusoidal grid	7.1	0.30	6.7
MLA	7.4	0.34	6.7

5. Conclusions

This paper presents the investigation of surface topography generation in STS machining of freeform surfaces. Both theoretical and experimental investigation were conducted to prove the validity of the proposed tool path generation methodology and the effectiveness of surface generation simulation. The main conclusions can be drawn as follows:

(1) To avoid the overcut of a rounded tool tip, tool radius compensation in the Z direction was performed to ensure no high frequency motion is imposed on the dynamic-limited X axis. Tool path motion analysis validated the Z direction compensation method and it was shown to be advantageous over conventional normal direction compensation methods.

(2) The development of surface generation simulation allows prediction of the surface topography under various tool and machining variables. From the simulation results, it can be concluded that a better surface finish (lower S_q value) can be obtained under a higher spindle speed, a smaller feedrate and a larger tool radius. The simulation analysis also reveals the surface generation

mechanism (such as the overcutting phenomenon) without the need for costly trial and error tests. With the proposed tool radius compensation method, waviness error components resulting from the overcut can be totally eliminated.

(3)   Machining experiments of a sinusoidal grid and MLA sample demonstrated the effectiveness of the proposed STS machining to fabricate optical freeform surfaces with a nanometric surface topography (less than 10 nm). The measurement results show uniform topography distribution over the entire surface and agree well with the simulation results.

The presented experimental results demonstrated the validity of the tool radius compensation method and surface generation simulation, which have the potential to be applied to other diamond machining processes (such as fly-cutting, grinding and fast tool servo) for freeform surfaces in order to improve the machining performance. These investigations will be of particular relevance to researchers in the ultra-precision freeform machining field.

Author Contributions: Investigation, D.L. and Z.Q.; Methodology, D.L., Y.L., J.X., and Z.Q.; Validation, D.L.; Writing for original draft, D.L.; Writing for review and editing, D.L., K.W., X.J., and B.W.

Acknowledgments: The authors appreciate Mr. Wei Wang for editing and discussion of the current paper. The authors also would like to sincerely thank the reviewers for their valuable comments on this work.

Funding: This research was funded by the UK's Engineering and Physical Sciences Research Council (EPSRC) funding (Grant Ref: EP/P006930/1), National Science and Technology Major Project of High-end CNC Machine Tools and Basic Manufacturing Equipment of China (grant number 2011ZX04004–021) and China Scholarship Council (CSC).

Conflicts of Interest: The authors declare no conflict of interest.

References

1.   Fuse, K.; Okada, T.; Ebata, K. Diffractive/refractive hybrid F-theta lens for laser drilling of multilayer printed circuit boards. In *SPIE Conference Proceedings*; International Society for Optics and Photonics: Washington, DC, USA, 2003; pp. 95–100.

2.   Cheng, D.; Wang, Y.; Hua, H.; Talha, M. Design of an optical see-through head-mounted display with a low f-number and large field of view using a freeform prism. *Appl. Opt.* **2009**, *48*, 2655–2668. [CrossRef] [PubMed]

3.   Wang, J.; Santosa, F. A numerical method for progressive lens design. *Math. Models Methods Appl. Sci.* **2004**, *14*, 619–640. [CrossRef]

4.   Dross, O.; Mohedano, R.; Hernandez, M.; Cvetkovic, A.; Benitez, P.; Miñano, J.C. Illumination optics: Köhler integration optics improve illumination homogeneity. *Laser Focus World* **2009**, *45*, 135–150.

5.   Jiang, X.J.; Whitehouse, D.J. Technological shifts in surface metrology. *CIRP Ann.-Manuf. Technol.* **2012**, *61*, 815–836. [CrossRef]

6.   Fang, F.Z.; Zhang, X.D.; Weckenmann, A.; Zhang, G.X.; Evans, C. Manufacturing and measurement of freeform optics. *CIRP Ann.-Manuf. Technol.* **2013**, *62*, 823–846. [CrossRef]

7.   Yi, A.; Li, L. Design and fabrication of a microlens array by use of a slow tool servo. *Opt. Lett.* **2005**, *30*, 1707–1709. [CrossRef] [PubMed]

8.   Brinksmeier, E.; Mutlugünes, Y.; Klocke, F.; Aurich, J.; Shore, P.; Ohmori, H. Ultra-precision grinding. *CIRP Ann.-Manuf. Technol.* **2010**, *59*, 652–671. [CrossRef]

9.   Klocke, F.; Brecher, C.; Brinksmeier, E.; Behrens, B.; Dambon, O.; Riemer, O.; Schulte, H.; Tuecks, R.; Waechter, D.; Wenzel, C. Deterministic Polishing of Smooth and Structured Molds. In *Fabrication of Complex Optical Components*; Springer: Berlin/Heidelberg, Germany, 2013; pp. 99–117.

10.   Yin, Z.; Dai, Y.; Li, S.; Guan, C.; Tie, G. Fabrication of off-axis aspheric surfaces using a slow tool servo. *Int. J. Mach. Tools Manuf.* **2011**, *51*, 404–410. [CrossRef]

11.   Mukaida, M.; Yan, J. Ductile machining of single-crystal silicon for microlens arrays by ultraprecision diamond turning using a slow tool servo. *Int. J. Mach. Tools Manuf.* **2016**, *115*, 2–14. [CrossRef]

12.   Chen, C.-C.; Cheng, Y.-C.; Hsu, W.-Y.; Chou, H.-Y.; Wang, P.-J.; Tsai, D.P. *Slow Tool Servo Diamond Turning of Optical Freeform Surface for Astigmatic Contact Lens*; Optical Manufacturing and Testing IX, 2011; International Society for Optics and Photonics: Washington, DC, USA, 2011; p. 812617.

13. Krolczyk, G.M.; Maruda, R.W.; Krolczyk, J.B.; Nieslony, P.; Wojciechowski, S.; Legutko, S. Parametric and nonparametric description of the surface topography in the dry and MQCL cutting conditions. *Measurement* **2018**, *121*, 225–239. [CrossRef]

14. Krolczyk, G.M.; Maruda, R.W.; Nieslony, P.; Wieczorowski, M. Surface morphology analysis of Duplex Stainless Steel (DSS) in Clean Production using the Power Spectral Density. *Measurement* **2016**, *94*, 464–470. [CrossRef]

15. Jiang, X.; Scott, P.J.; Whitehouse, D.J.; Blunt, L. Paradigm shifts in surface metrology. Part II. The current shift. *Proc. R. Soc. A Math. Phys. Eng. Sci.* **2007**, *463*, 2071–2099. [CrossRef]

16. Beaucamp, A.; Freeman, R.; Morton, R.; Ponudurai, K.; Walker, D. *Removal of Diamond-Turning Signatures on x-ray Mandrels and Metal Optics by Fluid-Jet Polishing*; Advanced Optical and Mechanical Technologies in Telescopes and Instrumentation, 2008; International Society for Optics and Photonics: Washington, DC, USA, 2008; p. 701835.

17. Agarwal, S.; Rao, P.V. A probabilistic approach to predict surface roughness in ceramic grinding. *Int. J. Mach. Tools Manuf.* **2005**, *45*, 609–616. [CrossRef]

18. Kong, L.; Cheung, C.; Kwok, T. Theoretical and experimental analysis of the effect of error motions on surface generation in fast tool servo machining. *Precis. Eng.* **2014**, *38*, 428–438. [CrossRef]

19. Zhou, J.; Li, L.; Naples, N.; Sun, T.; Allen, Y.Y. Fabrication of continuous diffractive optical elements using a fast tool servo diamond turning process. *J. Micromech. Microeng.* **2013**, *23*, 075010. [CrossRef]

20. Venkataraman, P. *Applied Optimization with MATLAB Programming*; John Wiley & Sons: New York, NY, USA, 2009.

21. Gao, W.; Dejima, S.; Shimizu, Y.; Kiyono, S.; Yoshikawa, H. Precision measurement of two-axis positions and tilt motions using a surface encoder. *CIRP Ann.-Manuf. Technol.* **2003**, *52*, 435–438. [CrossRef]

22. Chang, S.-I.; Yoon, J.-B.; Kim, H.; Kim, J.-J.; Lee, B.-K.; Shin, D.H. Microlens array diffuser for a light-emitting diode backlight system. *Opt. Lett.* **2006**, *31*, 3016–3018. [CrossRef] [PubMed]

 *materials*

Article

Optimum Selection of Variable Pitch for Chatter Suppression in Face Milling Operations

Alex Iglesias [1,*], **Zoltan Dombovari** [2], **German Gonzalez** [1,3], **Jokin Munoa** [1] and **Gabor Stepan** [2]

[1] Dynamics and Control, IK4-Ideko, 20870 Elgoibar, Basque Country, Spain; jmunoa@ideko.es
[2] Department of Applied Mechanics, Budapest University of Technology and Economics, H-1521 Budapest, Hungary; dombovari@mm.bme.hu (Z.D.); stepan@mm.bme.hu (G.S.)
[3] Karlsruhe Institute of Technology (KIT), wbk Institute of Production Science, Kaiserstr. 12, 76131 Karlsruhe, Germany; ggonzalez@ideko.es
* Correspondence: aiglesias@ideko.es; Tel.: +34-943748000

Received: 19 November 2018; Accepted: 24 December 2018; Published: 31 December 2018

Abstract: Cutting capacity can be seriously limited in heavy duty face milling processes due to self-excited structural vibrations. Special geometry tools and, specifically, variable pitch milling tools have been extensively used in aeronautic applications with the purpose of removing these detrimental chatter vibrations, where high frequency chatter related to slender tools or thin walls limits productivity. However, the application of this technique in heavy duty face milling operations has not been thoroughly explored. In this paper, a method for the definition of the optimum angles between inserts is presented, based on the optimum pitch angle and the stabilizability diagrams. These diagrams are obtained through the brute force (BF) iterative method, which basically consists of an iterative maximization of the stability by using the semidiscretization method. From the observed results, hints for the selection of the optimum pitch pattern and the optimum values of the angles between inserts are presented. A practical application is implemented and the cutting performance when using an optimized variable pitch tool is assessed. It is concluded that with an optimum selection of the pitch, the material removal rate can be improved up to three times. Finally, the existence of two more different stability lobe families related to the saddle-node and flip type stability losses is demonstrated.

Keywords: Milling stability; variable pitch; chatter; self-excitation

1. Introduction

The machining of components for the energy sector implies high material removal rates in heavy duty machining processes. Current trends point towards the development of even higher power components in nuclear plants and wind turbine applications, which results in more demanding machining processes and higher cutting capacity requirements in production means, due to the larger component size and, thus, bigger material stocks to remove.

The material range to machine is very broad ranging from the typical steel or cast iron used for wind turbine components to heat-resistant alloys used in the nuclear industry. The big size of the components of the energy sector to manufacture brings along the need of flexible machines. The floor-type milling machine with an extensible ram is one of the most common architectures for this purpose (Figure 1a).

The higher requirements in terms of productivity pave the way for self-excited vibrations, also commonly known as chatter vibrations, which are one of the main challenges linked to heavy duty milling processes. In this case, the chatter vibrations are usually linked to structural low frequency chatter, which are extremely detrimental. Among the adverse effects of this type of vibrations, bad surface finish, excessive tool wear, and even tool or part damage can be mentioned.

Many works and scientific research have been devoted to chatter avoidance technique development throughout the years [1]. Depending on the source of vibration, which has to be evaluated beforehand, and the stability lobe region the process takes place in, the most appropriate technique can be selected for each specific case [1].

For heavy duty machining operations (Figure 1a), passive [2,3] and active damping [4] of the structure subject to vibrations has been a commonly proposed solution for chatter suppression. Passive dampers usually require big masses and occupied volume due to the high modal masses involved in the critical modes, which usually makes this type of solution unfeasible. Active solutions, on the other hand, are usually rather complex and costly, due to the necessary hardware and controller. Active damping can also be achieved through a machine's own drives [5], although the effectiveness of this type of technique is very limited due to the long actuation distance from the cutting point and the non-collocated control.

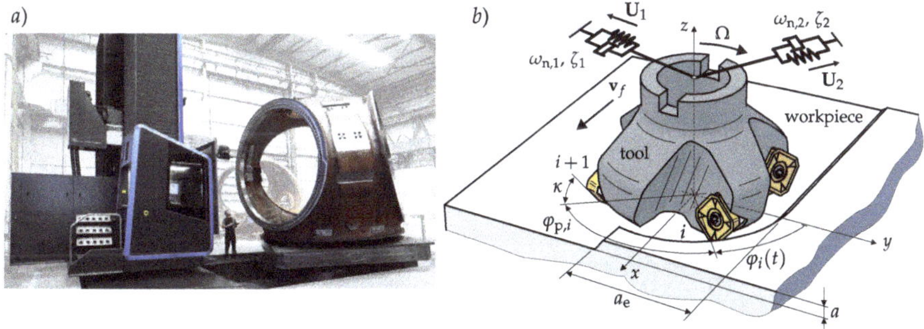

Figure 1. (**a**) Turbine hub machining in a floor-type milling machine with an extensible ram, (**b**) geometry of the considered inserted face milling cutter.

Among the control techniques for chatter avoidance, continuous spindle speed variation [6,7] can be highlighted. However, according to Bediaga et al. [7], this is an effective measure when dealing with chatter occurring at high lobe numbers, which is not the case in heavy duty face milling operations, where low frequency chatter places the process at low lobe numbers.

The spindle speed variation based on a tuning procedure [8] is another useful technique whose key point is to match the milling cutting force intrinsic excitation frequencies with the frequency of the critical mode which is causing chatter. When the dynamics of the cutting system are complex and many modes are involved in stability limitation, the use of the stability lobe diagram (SLD) is useful to identify the stability pockets along the applicable spindle speed range.

However, whereas in materials, such as steel or aluminium, the cutting speed can be broadly varied without a detrimental effect for the tool life, when heat-resistant alloys, such as titanium, Inconel, or stainless steel, are machined, the variation range is very narrow [9]. In some extreme cases, a slight variation of the cutting speed could mean a severe decrease in tool life, therefore, it is very important to adjust the speed to the exact cutting speed recommended by the tool manufacturer. In view of this, the recurring solution of spindle speed tuning lacks applicability for heat-resistant alloy machining. Therefore, instead of acting on the cutting conditions, the selection of the optimum tool is a good alternative solution for chatter removal in heat-resistant alloy face milling processes.

For regular cutters, the tool selection charts developed by Iglesias et al [10] are a useful method for optimum tool diameter to number of inserts (D/Z) ratio determination. On the other hand, Budak and Kops [11] demonstrated the cutting stability leap that can be achieved with the use of variable pitch cutters when cutting aeronautic components. The lack of spindle speed variation flexibility mentioned above makes the variable pitch a reasonable solution for vibration problems for heat-resistant alloy machining, since they offer a good value for money.

The variable pitch, which was first proposed by Hahn [12], alters the regenerative phases between the past and the present surface patterns, which has a disruptive effect in the typical regenerative effect of chatter that increases stability. Milling tool manufacturers offer off-the-shelf variable pitch cutters for general purpose milling. Altintas et al [13] demonstrated that a random selection can be far from the optimum achievable pitch, with the possibility of even decreasing the stability if an unlucky decision is made.

For this reason, in order to take advantage of the full potential of variable pitch cutters, it is very important to have a variable pitch selection tool. Slavicek [14] analysed tools with only two alternating delays, and proposed an analytical formula to suppress chatter at a certain spindle speed and chatter frequency. Vanherck [15] upgraded Slavicek's technique by considering more than two pitch angles on the cutter, and Varterasian [16] included experimental results with randomly distributed pitch angles. After these analytical attempts, Tlusty [17] tried to obtain the best design parameters using time domain milling simulations.

The formulation of the zeroth order approximation (ZOA) approximation in milling permitted the formulation of faster frequency domain procedures for the optimization of the tool geometry [13]. Based on this approximation, Budak [11] introduced an analytical design methodology based on the phase differences between consecutive constant delays. Its effectiveness was confirmed by an industrial application. In the work of Olgac and Sipahi [18], a unique scheme, namely the cluster treatment of characteristic roots paradigm, was used to investigate the effect of two constant delays on the time-averaged dynamics of milling operation with variable pitch cutters. In interrupted milling operations, however, higher harmonics can have significant strength in the regenerative force, causing discrepancies in time averaging methods [19].

The application of time domain based stability models can solve these limitations and they can become an important tool for geometry optimization. Following this approach, a new method based on the semidiscretization stability calculation is proposed [20]. Based on the brute force (BF) iterative method recently developed by Stepan et al [21], a new face milling tool design procedure is deployed and applied to a specific face milling operation. Finally, the advantages of the new method with respect to the traditional optimum pitch estimation method are experimentally demonstrated, as well as the existence of a new double period instability related to the variable pitch pattern repetition.

All these approaches are based in theoretical frequency and time domain stability models, where the regenerative effect is deterministically modelled, and the optimal distribution is obtained analytically or numerically. In the last years, optimization procedures based on experimental studies and multi objective optimization have been proposed for force and vibration reduction and surface roughness improvement [22,23]. In the future, similar procedures could be applied for tool geometry optimization.

In the same way as variable pitch cutters are successfully used in the industry for aeronautic heat-resistant alloy machining, where high frequency chatter related to tools and thin walls limits productivity, this work will evaluate their performance on roughing face milling operations, where they are not typically used nowadays.

2. Stability Lobes for Variable Pitch Cutters

In this section, an overview of the method to predict the performance of variable pitch tools using the general dynamic model of milling is presented. It is important to emphasize that variable pitch cutters are introducing regeneration with multiple constant delays, which must be taken into account in the model development. Depending on the number of pitch angles participating in the optimization, a general optimization scheme can be drawn, although some authors suggest that the optimization of just one angle can already make huge improvement [14].

2.1. Stability Model for Variable Pitch Cutters

The different regenerative delays, τ_i, of a variable pitch tool can be defined with the help of the period of the tool, $T = 2\,\pi/\Omega$, and the pitch angles, $\varphi_{p,l}$, between consecutive edges as Equation (1):

$$\tau_i = \frac{\varphi_{p,i}}{\Omega} \text{ and } \Omega = \frac{2\,\pi\,n}{60\,\frac{s}{min}}, \tag{1}$$

where Ω and n are the spindle speed in rad/s and rpm, respectively.

Unlike regular cutters, the principal period, T_p, of the milling process is not the tooth passing period, $T_Z = T/Z$. Depending on the actual configuration, it is determined by the natural number divisor. The equation (2) shows an elegant mathematical way to count different pitch angles of the variable pitch tool and determine its main (principal) period:

$$N = \frac{Z}{N_\tau} \text{ and } N_\tau = \text{rank}\left[\varphi_{p,(k+l-1)\bmod Z}\right]_{k,l=1}^{Z}, \tag{2}$$

as $T_p = T/N$.

To derive the motion and the load of the cutting edges, the relative motion of the tool is considered $\mathbf{x} = \text{col } x, y, z$ (see Figure 1b). The x direction coincides with the feed direction. An inserted cutter with a regular lead angle, $\kappa: = \kappa_i$, is posed. In case of variable pitch tools, the stationary part of the feeds, $(\mathbf{f}_i)$, is not constant for the different inserts, and it can be approximated for a milling operation by Equation (3) in the x direction ($\mathbf{f}_i = \text{col}(f_i, 0, 0)$) using the complete feed per revolution, f (m/rev), as:

$$f_i = f\frac{\varphi_{p,i}}{2\,\pi} \text{ and } f = v_f\frac{2\,\pi}{\Omega}\ \left(v_f = \left|\mathbf{v}_f\right|\right). \tag{3}$$

This chip thickness is also affected by the actual and previous vibrations ($\Delta\mathbf{x}_i$). Therefore, the local chip thickness of the ith insert can be derived [11,20] considering the chip thickness direction, $\mathbf{n}\ (\varphi_i(t))$ $= \text{col}(\sin\kappa\,\sin\varphi_i(t), \sin\kappa\,\cos\varphi_i(t), -\cos\kappa)$:

$$\begin{aligned}
h_i(t) = \mathbf{n}^T(\varphi_i(t))\,(\mathbf{f}_i + \Delta\mathbf{x}_i(t)) = \\
(f_i + x(t) - x(t - \tau_i))\,\sin\kappa\,\sin\,\varphi_i(t) + \\
(y(t) - y(t - \tau_i))\,\sin\kappa\,\cos\,\varphi_i(t) - (z(t) - z(t - \tau_i))\,\cos\kappa,
\end{aligned} \tag{4}$$

The angle which determines the position of the ith tooth geometrically (see Figure 1b) is $\varphi_I(t) = \Omega\,t$ $+ \sum_{k=1}^{i-1}\varphi_{p,i}$. Assuming (for simplicity) linear cutting force characteristics with edge coefficients, $\mathbf{K}_e$, and cutting coefficients, $K_{c,t}\,\kappa_c = K_{c,t}\,\text{col}(1, \kappa_{c,r}, \kappa_{c,a})$ (see Figure 1b and [24,25]), the regenerative cutting force has the form:

$$\begin{aligned}
\mathbf{F}(t, \mathbf{x}(t), \mathbf{x}(t - \tau_k)) = -\tfrac{a}{\sin\kappa}\sum_{i=1}^{Z} g(\varphi_i(t))\,\mathbf{T}(\varphi_i(t))\,(\mathbf{K}_e + K_{c,t}\kappa_c h_i(t)) = \\
\mathbf{G}(t) + \tfrac{a\,K_{c,t}}{\sin\kappa}\sum_{i=1}^{Z} A_i(t)(\mathbf{x}(t) - \mathbf{x}(t - \tau_i)),
\end{aligned} \tag{5}$$

where the transformation matrix, $\mathbf{T}$, transfers cutting forces of each flute to the general tool coordinate system (xyz), and the screen function, g, considers when the insert is in or out of cutting [20] in the radial engagement, a_e. The regenerative cutting force in Equation (5) depends on the present, $\mathbf{x}(t)$, and the delayed, $\mathbf{x}(t - \tau_k)$ ($k = 1, 2, \dots, N_\tau$), positions of the tool. Furthermore, it can be separated into a periodic state independent part, $\mathbf{G}(t) = \mathbf{G}(t + T_p)$, and linear state dependent part with periodic coefficient matrices, $\mathbf{A}_i\,(t) = \mathbf{A}_i\,(t + T_p)$.

By assuming a general linear nonproportionally damped modal formalism, the cutting force can be projected to the different modes and the system can be analyzed in the space state [26]:

$$\mathbf{q}(t) - [\lambda_k]\,\mathbf{q}(t) = \mathbf{U}^T\mathbf{F}(t, \mathbf{x}(t), \mathbf{x}(t - \tau_k)), \tag{6}$$

where $\mathrm{col}(\mathbf{x}, \mathbf{x}) = \mathrm{col}(\mathbf{U}, \mathbf{U}[\lambda_k])\,\mathbf{q}$. Theoretically, the dynamics in Equation (6) can be described by the following proto transfer function:

$$\mathbf{H}(\lambda) = \mathbf{U}(\lambda\mathbf{I} - [\lambda_k])^{-1}\,\mathbf{U}^{\mathrm{T}}, \tag{7}$$

where the mass normalized modal transformation matrix is $\mathbf{U}$ and $\lambda_k = -\omega_{n,k}\,\zeta_k \pm j\omega_{n,k}\,(1 - \zeta_k^2)^{1/2}$, with damping ratios, ζ_k, and natural frequencies, $\omega_{n,k}$.

To determine the asymptotic (linear) stability of the stationary solution, $\mathbf{x}_p$, the variational equation is derived by using perturbation $\mathbf{y}$ as $\mathbf{x} = \mathbf{x}_p + \mathbf{y}$ ($\mathbf{q} = \mathbf{q}_p + \mathbf{u}$), resulting in the following form from Equation (6):

$$\mathbf{u}(t) - [\lambda_k]\,\mathbf{u}(t) = \mathbf{U}^{\mathrm{T}}\left(\frac{a\,K_{c,t}}{\sin\kappa}\sum_{i=1}^{Z} A_i(t)(\mathbf{y}(t) - \mathbf{y}(t - \tau_i))\right). \tag{8}$$

The asymptotic behaviour of this time domain form can be determined directly by the semidiscretization method [27] or a similar time domain based method [28]. In semidiscretization, the aim is to derive the linear map (transition matrix, $\mathbf{\Phi}$) between the current complete state, $\mathbf{z}_0 = \mathrm{col}_k\,\mathbf{u}(t - (k - 1)\Delta t)$, where ($k = 1, ..., m$), and the next period, $\mathbf{z}_1 = \mathrm{col}_k\,\mathbf{u}(T_p + t - (k - 1)\Delta t)$, such as $\mathbf{z}_1 = \mathbf{\Phi}\,\mathbf{z}_0$. The $\mathbf{\Phi}$ is compiled over the principal period by successively solving up to $T_p = r\,\Delta t$ with $m = \lceil\tau_{\max}/\Delta t\rceil$ ($\tau_{\max} = \max_i \tau_i$). If all eigenvalues ('multipliers', μ_k) of $\mathbf{\Phi}$ have a magnitude less than the unity, the process is stable. The critical multiplier, μ_c, crosses the unit circle in an arbitrary position for a Hopf kind of stability loss, crosses at -1 in the case of flip, while the saddle type of stability loss accompanies $\mu_c = 1$ (see tiny unit circles in Figure 2).

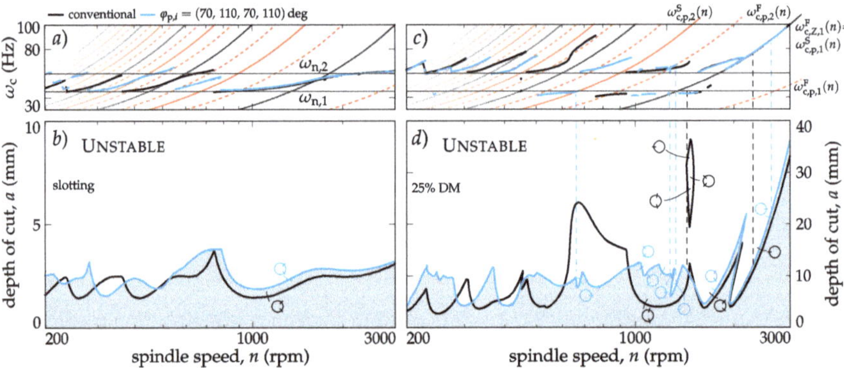

Figure 2. Comparison of milling stability for a four insert ($Z = 4$) regular pitch tool versus a variable pitch tool, $\varphi_{p,i} = (70, 110, 70, 110)$ deg in case of slotting (**a**,**b**) and 25% down milling (DM, **c**,**d**).

Alternatively, to semi discretization, the problem can be analysed in the frequency domain [11,13]. According to the Floquet theory, the following trial solution can be assumed and the Fourier expansion of A_i using $\Omega_p = 2\pi/T_p$ can be performed:

$$\mathbf{y}(t) = \sum_{k=-\infty}^{\infty} Y_k e^{jk\Omega_p t} e^{\lambda t} \quad \text{and} \quad \mathbf{A}_i(t) = \sum_{l=-\infty}^{\infty} A_{l,i} e^{jl\Omega_p t}, \tag{9}$$

Equation (8) can be rewritten in the Fourier domain ($\lambda = j\omega$) by using Equation (7) and Equation (9) in the following form:

$$\left(\mathbf{I} - \frac{a\,K_{c,t}}{\sin\kappa}\mathbf{\Phi}\,(\omega)\sum_{i=1}^{Z}(I - e^{-j\omega\tau_i}E_i)D_i\right)\mathbf{Y} = 0, \tag{10}$$

considering h modulations of the vibration frequency, ω:

$$\boldsymbol{\Phi}(\omega) = \operatorname*{diag}_{l=-h}^{h} \mathbf{H}(\omega + l\,\Omega_{\mathrm{p}}), \quad \mathbf{E}_i = \operatorname*{diag}_{l=-h}^{h} \mathbf{I}\,e^{-\mathrm{j}\,l\,\varphi_{\mathrm{p},i}},$$
$$\mathbf{D}_i = [\mathbf{A}_{l-k,i}]_{l,k=-h}^{h} \text{ and } \mathbf{Y} = \operatorname*{col}_{k=-h}^{h}\mathbf{Y}_k. \tag{11}$$

Non-trivial solution of Equation (10) is granted by:

$$\det\left(\mathbf{I} - \frac{a\,K_{\mathrm{c},t}}{\sin\kappa}\,\boldsymbol{\Phi}\,(\omega)\sum_{i=1}^{Z}\left(I - e^{-\mathrm{j}\,\omega\,\tau_i}\,E_i\right)D_i\right) = 0, \tag{12}$$

whose solution can be determined in various ways [29,30]. Basically, Equation (12) is a truncated Hill's representation, which is known as a multi-frequency solution in combination with the frequency response function (FRF) $\mathbf{H}(\omega):= \mathbf{H}(\mathrm{j}\omega)$ Equation (7). The FRF $\mathbf{H}(\omega)$, e.g., at the tool tip can be measured experimentally [31], and can be used directly without performing any fitting to determine the modal parameters introduced at Equation (7). In case of variable pitch tools, the different regenerations, τ_l, are summed in an exponential term in Equation (12), while their effects are weighted by $\mathbf{E}_i$ in the modulations. Due to the modulation weight, $\mathbf{E}_i$, there is no closed form expression of the eigenvalue problem in Equation (12).

However, a zeroth order solution can be introduced by setting $h = 0$ in Equation (12), resulting in the following form:

$$\det\left(\mathbf{I} - \frac{a\,K_{\mathrm{c},t}}{\sin\kappa}\,\mathbf{H}\,(\omega)\sum_{i=1}^{Z}\left(1 - e^{-\mathrm{j}\,\omega\,\tau_i}\right)A_{i,0}\right) = 0, \tag{13}$$

$$\mathbf{A}_{i,0} = -\frac{1}{T_{\mathrm{p}}}\int_0^{T_{\mathrm{p}}} g(\varphi_i(t))T(\varphi_i(t))\,\kappa_{\mathrm{c}}\mathbf{n}^{\mathrm{T}}(\varphi_i(t))\mathrm{d}t. \tag{14}$$

The problem with the description of Equation (13) is that there is still no way to deliver an analytical formula as an eigenvalue solution if $\mathbf{A}_{i,0}$ is under the sum operation. This can be avoided by lifting out common average directional factors, $\mathbf{A}_0$, presented in [32] and defined as:

$$\mathbf{A}_0 := -\frac{N}{2\,\pi}\int_{\varphi_{en}}^{\varphi_{ex}} T(\varphi)\,\kappa_{\mathrm{c}}n^{\mathrm{T}}(\varphi)\,\mathrm{d}\varphi. \tag{15}$$

Thus, it has the form:

$$\det\left(\mathbf{I} - \frac{a\,K_{\mathrm{c},t}}{\sin\kappa}\sum_{k=1}^{N_\tau}\left(1 - e^{-\mathrm{j}\,\omega\,\tau_k}\right)\mathbf{H}\,(\omega)\,\mathbf{A}_0\right) = 0. \tag{16}$$

This single frequency solution [33] is a commonly used method, which offers a good trade-off between accuracy and speed to solve the eigenvalue problem. The angles, φ_{ex} and φ_{en}. in Equation (15) are the exit and entry angles that define the radial immersion, a_{e} (Figure 1b).

2.2. Stability Property of Variable Pitch Tools

The efficiency of the developed stability model for variable pitch tools has been studied using an example described in the literature [19], where the precision of the different stability models for different cutting engagements was evaluated. It is well known that the effect of the harmonics of the cutting force is not important for large radial immersion (a_{e}, Figure 1b). However, interrupted cutting with small a_{e} boots the periodic nature of the milling process, inducing a period doubling (flip) type of stability loss and mode coupling [19,25]. Since the principal period of the variable pitch tool differs from the tooth passing frequency, a more intricate pattern of period related stability losses are expected. This is demonstrated in Figure 2, where a stability comparison of a regular and a selected variable pitch tool ($\varphi_{p,l}$ = (70, 110, 70, 110) deg) for full immersion (slotting) and for 25% down milling (DM) cases is made.

In the slotting case, since in the full immersion case, the periodicity is averaged out for the conventional tool with $Z = 4$, and stays insignificant for the variable pitch case, only the Hopf type stability loss appears. It is clear that the variable pitch can increase the chatter-free depth of the cut only in certain areas. In fact, the introduction of irregular tools can reduce the stability in other spindle speed areas. Therefore, the need for optimization to avoid trial and error is clear.

In the case of interrupted cutting, the previous conclusions are reinforced. Nevertheless, the variable pitch tool shows intricate flip type stability losses on top of the conventional (inherited) stability limits (Figure 2c,d) originating from the different principal periods.

There are various explanations for the origin of the period doubling instability depending on the different mathematical approaches used for investigating the linear behaviour of the period-one stationary cutting solution ($\mathbf{x}_p$) with the principal period, T_p. From time domain based point of view, the period-one stationary cutting solution of the originally nonlinear system undergoes a period doubling (flip) bifurcation when a period-two orbit emerges with a $2T_p$ period [34]. The local behaviour of this period-two orbit can be investigated on a Poincaré section that is normal to the original period-one orbit. On this abstract section, the onset of the period-two orbit appears as an alternating sequence resembling a sequence that flips between two halves of the Poincaré section [35].

If some of the regular pitch angles of a milling tool are slightly varied, resulting in $T_p = T/N$ (2) principal period, where $N < Z$, the original flip instability of a conventional tool with a tooth passing period, $T_Z = T/Z$, will be altered. Since $Z = N\,N_\tau$ (2), $T_p = N_\tau\,T_Z$ [36], which means the flip critical multiplier, $\mu_c = -1$, is determined over T_p, resulting in a $\mu_c = (-1)^{N_\tau}$ critical multiplier. Depending on whether N_τ is odd or even, both a $\mu_c = -1$ and $\mu_c = 1$ situation can appear. The case, $\mu_c = 1$, corresponds to a saddle type of stability loss of the corresponding system, which was completely unknown to happen in any milling process.

Calculating the corresponding vibration frequencies using [37] as:

$$\omega_{c,k} = \omega_{c,b} + l\,\Omega_p = (\arg \mu_c + 2\pi\,l)/T_p, \tag{17}$$

the modulations, $\omega_{c,l}$, of the chatter frequency, ω_c (dominant $\omega_{c,l}$), can be determined. In this consideration, it is well known that double period (flip) chatter happens when the chatter frequency, ω_c, and one of its modulated frequencies, $\omega_{c,l}$, with the principal period, $T_p = 2\pi/\Omega_p$, are shaking the same critical mode due to $\arg \mu_c = \pm\pi$ ($\mu_c = -1$). However, keeping the same argument, the saddle kind of stability loss of the stationary solution appears when a modulation is located at zero frequency due to $\arg \mu_c = 0$ ($\mu_c = 1$), which was never happening in regenerative milling models. This means the dominant chatter frequencies in flip and saddle cases lies on the following lines extracted from Equation (17):

$$\omega^F_{c,p,l}(n) = (2l - 1)\frac{Z}{N_\tau}\frac{n}{120\frac{s}{\min}}(\text{Hz}), \quad l \in N, \tag{18}$$

$$\omega^S_{c,p,l}(n) = 2l\frac{Z}{N_\tau}\frac{n}{120\frac{s}{\min}}(\text{Hz}), \quad l \in N. \tag{19}$$

In the conventional case, when $N_\tau = 1$, considering that the saddle case instability cannot be present, the flip case can be expressed as:

$$\omega^F_{c,Z,l}(n) = (2l - 1)Z\frac{n}{120\frac{s}{\min}}(\text{Hz}), \quad l \in N. \tag{20}$$

These special lines are presented in Figure 2a,c when $N_\tau = 2$ for the variable pitch case with $\varphi_{p,i}$ = (70, 110, 70, 110) deg.

3. Variable Pitch Tool Design Process

Several methods to estimate the proper tuning in non-constant pitch tools have been proposed by different authors. In this work, the classical analytical tuning criteria proposed by Slavicek [14] and Budak [32] are briefly introduced and compared with the brute force (BF) iterative method [21], which is based on the semidiscretization stability calculation proposed in the previous section [20,27]. In order to compare different analytical methodologies, altered pitch-deviations are considered, that is, $\varphi_{p,i} = (\varphi_p - \Delta\varphi_p, \varphi_p + \Delta\varphi_p, \varphi_p - \Delta\varphi_p, \dots,)$, when Z is even and $N_\tau = 2$ in (2).

3.1. Slavicek´s Methodology

Originally, this method was a graphical way of tuning based on the regenerative phase shifts, $\varepsilon_1 = \omega\,\tau_1 = \omega\,\varphi_{p,1}/\Omega$ and $\varepsilon_2 = \omega\,\tau_2 = \omega\,\varphi_{p,2}/\Omega$, of two subsequent surface undulations based on the scalar single frequency formulation originating from Equation (16) [14]:

$$1 - \frac{a\,K_{c,t}}{\sin\kappa}\sum_{k=1}^{N_\tau}\left(1 - e^{-j\,\omega\,\tau_k}\right)H(\omega)\,A_0 = 0. \tag{21}$$

Although this methodology was originally developed for a simple broaching process, it may be also applied for the milling process case by considering a set of technological parameters, where the averaged multi-degrees of freedom, but one dimensional dynamics can describe the process well.

Considering the regenerative effect of the previous surface waves by regenerative phase angles, ε_i, and the dynamics by $H(\omega) = r\,(\omega)\,e^{j\,\psi(\omega)}$, the following expression is obtained [36]:

$$\left(N_\tau - \left(e^{-j\,\varepsilon_1(\omega)} + e^{-j\,\varepsilon_2(\omega)}\right)\right) = \frac{\sin\kappa}{a\,K_{c,t}\,r(\omega)\,A_0}e^{-j\,\psi(\omega)}, \tag{22}$$

where the average regenerative phase between two consecutive waves is $\varepsilon = (\varepsilon_1 + \varepsilon_2)/2$ and the phase difference is $\Delta := \Delta\varepsilon/2 = (\varepsilon_2 - \varepsilon_1)/2 = \omega\,\Delta\varphi_p/\Omega/2$. Using trigonometric relations, the previous equation can be rearranged as:

$$e^{j\,\delta(\omega)} := \left(N_\tau - 2\,e^{-j\,\varepsilon(\omega)}\cos\Delta(\omega)\right) = \frac{\sin\kappa}{a\,K_{c,t}\,r(\omega)\,A_0}e^{-j\,\psi(\omega)}. \tag{23}$$

The resultant phase shift, δ, in Equation (23) has the least chance (see [14]) to "intersect" with $-\psi$ w.r.t. ω if δ varies in the slightest way, that is $\cos\Delta \approx 0$. In this case, when the system is less prone to lose stability, an analytical formula can be derived for regenerative phase difference:

$$\Delta = m\frac{\pi}{2}, \quad \Delta\varphi_p = \frac{\Omega}{\omega}(\pi + 2k\pi), \tag{24}$$

where ω is the chatter frequency and, e.g., $\Delta\varphi_p = \varphi_{p,2} - \varphi_{p,1}$ is the pitch difference ($m = 1, 3, 5, \dots$ and $k = 1, 2, 3, \dots$).

3.2. Budak´s Method

This method is based on the single frequency solution explained in detail in [32]. It serves a parametric solution originating from the eigenvalue solution of the multi-dimensional averaged model (16). The expression for the limit depth of the cut for variable pitch tools can be presented as:

$$a(\omega) = -\sin\kappa\frac{2\,\pi}{K_{c,t}}\frac{\Lambda_{\mathrm{I}}(\omega)}{S(\omega)}, \quad \Lambda(\omega) = \Lambda_{\mathrm{R}}(\omega) + j\Lambda_{\mathrm{I}}(\omega), \tag{25}$$

where:

$$\Lambda(\omega) = \frac{a}{2\,\pi}\frac{K_{c,t}}{\sin\kappa}\sum_{k=1}^{N_\tau}\left(1 - e^{-j\,\omega\,\tau_k}\right)\text{ and }S(\omega) = \sum_{k=1}^{N_\tau}\sin\omega\,\tau_k. \tag{26}$$

The main concept of the equation (25) is that the depth of the cut will achieve theoretically its maximum when the $S(\omega)$ denominator achieves its minimum, resulting in:

$$
\begin{aligned}
S(\omega) &= \sin \varepsilon_1 + \sin (\varepsilon_1 + \Delta\varepsilon) \\
&= 2 \cos \tfrac{\Delta\varepsilon}{2} \sin \tfrac{\omega}{\Omega}\pi = 0 \qquad \Rightarrow \qquad \cos \tfrac{\Delta\varepsilon}{2} = 0.
\end{aligned}
\tag{27}
$$

This leads to an equivalent definition to Slavicek´s one. By equating (27), the same expression can be determined as in Equation (24), consequently:

$$
\Delta\varphi_{\mathrm{p}} = \frac{\Omega}{\omega}(\pi + 2k\pi).
\tag{28}
$$

Therefore, the criterion for variable pitch tuning is the same as the criterion proposed by Slavicek. Thus, Budak´s and Slavicek´s solutions are considered the same for this case of study, and will be further referred to as the BS tuning method.

3.3. BF Methodology

The brute force (BF) iterative method [21,36] was developed to overcome the limitations of previous methodologies. All analytical optimization methods are based on time-averaged models of the milling process. The BF algorithm utilizes the general modelling of semidiscretization, capable of including general dynamics and geometry in process modelling [36].

By using semidiscretization, the magnitude of the Floquet multiplier, μ [37], represents the abstract 'distance' from the stability boundary (unit circle) in the characteristic complex plane. With this special property of the semidiscretization algorithm, the goal function is to minimize the largest magnitude multiplier, $\mu_{\max}$. To achieve this goal, the BF uses a bisection algorithm on an initial mesh on the design parameters similar to [38]. Possible switches on eigenvalues prevent the use of any gradient algorithm performing efficiently in this kind of parameter optimum search. There are some conditions that can specify the parameter at the optimum regarding the multipliers: (i) The magnitude at the minimum possible value for $\mu_{\max}$ should have a zero derivative with respect to the optimum parameter; (ii) the minimum possible value for $\mu_{\max}$ is set at a parameter point where two multipliers cross. Direct methods are specified separately for (i) and (ii) cases, but they require condition checking, increasing the computational load and the complexity, and losing generality. That is why, to avoid these problems, the robustly convergent bisection algorithm was chosen.

To ease the search for the parameters, the geometry of the regeneration was considered as it was presented first by Comak and Budak [39]. In this case, by using the calculated chatter frequency, ω_{c} (17), the regeneration wavelength is determined. The initial mesh is set along that wavelength, avoiding large portions of the largest possible pitch angle span. This in modulo recurrence can be taken into account in the design phase to ensure good chip evacuation for the cutter.

The method works as follows:

1. Selection of technological parameters in the stability chart;
2. perform pre-calculation for the vibration frequency using [37];
3. bisection algorithm is used to iterate the best possible angle; and
4. perform feasibility analysis to check whether the tool can be manufactured at all considering the in modulo recurrences.

3.4. Stabilizability (SYD) and Optimal Pitch Angle Diagrams

In this section, the tool selection procedure with the introduction of the stabilizability (SYD) and optimal pitch angle diagram (OPD) is described. All achievable topological cases are investigated based on a real industrial case (Figure 3a). The effect of various topologies is compared and the best possible tool geometry that is capable of improving the process using acceptable resources is derived. Different calculation cases based on the real dynamics of a machine tool (see Figure 3a) are presented for the depicted $L = 400$ mm position, while the process and dynamics details are listed later in the experimental part in Tables 1 and 2.

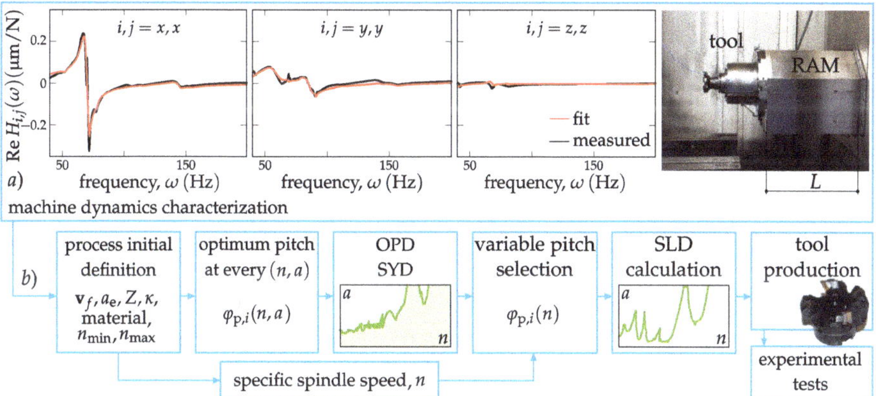

Figure 3. (**a**) Machine dynamics characterization; (**b**) flowchart of the procedure for calculation of the optimal pitch variation.

System dynamics and process characteristics must be known beforehand in order to perform the BF method (see Figure 3). Once the optimum pitch distribution is calculated through the BF method, the SYD is built (see Figure 3b) to have a clear picture for choosing the best pitch angle distribution, $\varphi_{p,l}$, to gain maximum stability in the depth of cut, a. The SYD is the theoretical combined graph, where the optimum pitch distribution is calculated for every spindle speed, n, of the (n, a) plane. Therefore, it shows the hypothetical highest depth of cut a one could achieve if the tool pitch is changed speed by speed in (n, a). Finally, once the final variable pitch is selected, the standard SLD calculation (see Figure 3b) is carried out. This will allow the process planner to perform the final fine tuning of the process, with the possibility of increasing the productivity according to the theoretical simulation and producing the tool physically. Figure 3b explains schematically the steps to follow to determine the proper tuned pitch angle(s).

It must be noted that, in a real industrial case scenario, the optimization is usually carried out for a specific spindle speed, n, due to the fact that the spindle speed is mostly given and considered as a hardly changeable technological parameter in the case of hard-to-cut materials [9].

3.5. The Effect of Different Cutting Tool Topologies

By introducing the BF tuning methodology, a wide range of different tool topologies are available to perform tuning on multiple pitch angle parameters (see Figure 4a). Multiple parameter optimization is possible by applying a successive search for optimum parameters. This theoretically does not tend to the actual global optimum, but it will practically always be near the optimum. In this manner, multiple angles can be iterated to achieve a higher stability for a certain type of milling process. Naturally, the more parameters are included in the optimization, the more time consuming the optimization is going to be. Thus, it is essential to find the balance between the calculation load and the achievable gain in stability by choosing appropriately the number of optimization parameters.

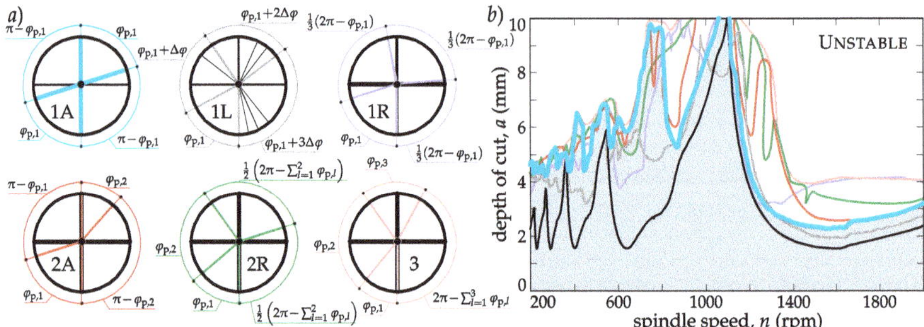

Figure 4. Panel (**a**) presents the different topologies that can be realized easily in a $Z = 4$ milling cutter. Panel (**b**) presents the stabilizability diagram (SYD) for the shown topologies. $\varphi_{p,min} = 70$ deg.

In Figure 4a, different possible topologies for a $Z = 4$ inserted tool are presented. The main aim is to select the best optimization scenario to tune the pitch distribution of the milling tool in order to increase productivity. It is also important to mention that if more different pitch angles are included in the optimization, this can drastically increase the computation time. Thus, the required time to construct a SYD diagram could be about 100–10,0000 times higher than the time needed to construct simple SLD.

According to Figure 4b, large improvements can be achieved excluding the first lobe by adopting any kind of optimization procedure. This general improvement means that the simplest alternated topology (1A) combined with BF (BF-1A) optimization should be the most effective one, requiring the least calculation load and providing the highest gain. However, the choice matters in the first lobe (high spindle speed zone, see Figure 4b), where tiny changes in the regenerative phase induce large angular changes in tool design. In this manner, the one parameter optimization with regular angle distribution (BF-1R) shows the largest improvement.

In summary, in real cases, the optimization of one parameter is enough; it gives the largest gain of stability, which means in most cases, there is no need to perform long multiple parameter optimizations. However, there are areas in the SYD where the inclusion of additional parameters can increase stability (BF-2A, BF-2R in Figure 4b, $n{\sim}1300$ rpm). Three parameter optimization serves as an envelope for all other topologies, presenting a minimal gain meaning that it is not worth performing that BF-3 optimization case in an industrial environment. In every calculation, a minimum production bound ($\varphi_{p,min} = 70$ deg) was kept for the pitch angle to ensure the integrity of the tool and satisfactory chip evacuation grooves.

3.6. Selecting the Optimal Pitch Angle

Choosing the 1A topology according to Section 3.5, the optimization of BF and BS algorithms are compared in this section. The optimizations were performed using the industrial case dynamics (Table 1) and process parameters (Table 2) and the resultant OPD (Figure 5a) and SYD (Figure 5b) are determined both for BF and BS algorithms. In both cases, the relative position of the lobes is crucial for the successful application of the variable pitch tool.

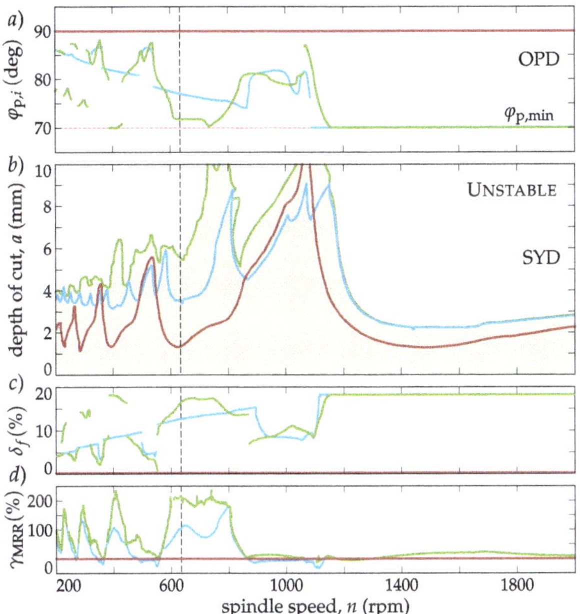

Figure 5. (**a**) Optimal pitch angle chart (OPD), (**b**) stabilizability diagram (SYD), (**c**) relative feed load drop, δ_f, while in (**d**), the relative material removal rate (MRR) gain, γ_{MRR}, are presented (green: BF-1A, red: regular, blue: BS, and $\varphi_{p,min}$ = 70 deg).

On the one hand, it can be observed that in Figure 5a, at very low spindle speeds (high number of lobes), the optimum angle tends to 90 degrees, because of the smaller and smaller regeneration wavelengths. This means that the required optimum angle variation is so small that it is not feasible to manufacture the pitch variation so accurately. On the other hand, in the first lobes (k = 0, 1), the optimum angle, $\varphi_{p,1}$, decreases to really small values (lower than $\varphi_{p,min}$ = 70 deg) and therefore a proper chip evacuation cannot be obtained in some cases. These big variations also unbalance the chip load, and some edges are overloaded (see Equation (3)).

In the OPD and SYD in Figure 5a,b, it can be observed that the optimum pitch angle distribution selected through the BS method (blue) outperforms the regular pitch tool (red) at practically every spindle speed, whereas the BF-method (green) bests the BS method to a considerable extent.

3.7. Productivity of Variable Pitch Tools

Since varying the regular pitch distribution results in larger pitch angle spans, teeth (inserts) will encounter an irregular, but larger feed load per tooth for a given feed speed, v_f (see Figure 1b). However, insert manufacturers define a maximum feed on an insert, $f_{Z,\,max}$, which can only be maintained if f_{max} is dropped with it as:

$$f_{max} = \frac{2\,\pi}{\varphi_{p,\,max}} f_{Z,\,max} \text{ and } v_{f,max} = f_{max}\frac{n}{60\frac{s}{min}},\tag{29}$$

where $\varphi_{p,max}$ = $\max_i \varphi_{p,i}$. This will directly affect the achievable material removal rate (MRR = $v_{f,max}\, a\, a_e$). This effect obviously causes a drop in the complete feed per revolution, f, however, the pitch variation also causes an increase of stability, $\tilde{a}$, too, which eventually improves MRR.

In this manner, the relative drop on the feed (drop on nominal MRR) can be defined as:

$$\delta_f := \frac{f - f_{\max}}{f} = \frac{Z\,\varphi_{\mathrm{p,max}} - 2\pi}{Z\,\varphi_{\mathrm{p,max}}}, \tag{30}$$

using the complete original $f = Z\,f_{Z,\max}$ for the conventional cutter. Figure 5c shows the expected drop in the usable complete feed compared to the conventional tool.

However, if one considers the gain in stability, the relative gain on MRR can be defined as:

$$\gamma_{\mathrm{MRR}} := \frac{\mathrm{MRR}_{\max} - \mathrm{MRR}}{\mathrm{MRR}} = \frac{2\pi\tilde{a} - a\,Z\,\varphi_{\mathrm{p,max}}}{a\,Z\,\varphi_{\mathrm{p,max}}}, \tag{31}$$

which clearly shows (Figure 5d) the productivity improvement of these optimized cutters in MRR. With a properly tuned variable pitch tool, around a 200% improvement can be achieved in MRR at certain speeds using the proper tuning, while keeping the chip load predefined by the insert manufacturer and preserving the minimal pitch angle, $\varphi_{\mathrm{p,min}}$, to ensure chip evacuation. Thus, productivity is increased and the inserts are not excessively overloaded.

4. Experimental Validation

In this section, the experimental validation of the previous theoretical development is provided. Most of the theoretical calculations were performed in the previous sections using actual data originating from the experimental validation, including Figures 3–5. Below, several experimental tests have been carried out, with the following objectives:

1. Show the outperformance of the BF tuned pitch with respect to the constant pitch cutter and the tool selected through BS criterion.

2. Demonstrate that the optimum pitch angle for a particular spindle speed could be the worst if a different spindle speed is used for cutting.

3. Find experimental evidence of the variable pitch pattern related double period instability.

4.1. Machine Dynamics Characterization and Process Definition

An industrial case was characterized on a 3-axis Danobat DS630 ram-type machining centre (see Figure 3a), in which a 4-insert (R245-12 T3 K-MM 1040) and 50 mm-diameter tool (Sandvik R245) were used. The machine position was defined at a $L = 400$ mm ram overhang. The dynamic characterization was performed by fitting the experimental FRFs obtained on the tool tip (see Figure 3a) according to [40] and a non-proportionally damped dynamic model was built, which was used for calculation [21]. The used modes are listed in Table 1 by presenting the proportional approximation of modal stiffness, k, and mode shapes, $\mathbf{U}$.

Table 1. Modal parameters at $L = 400$ mm ram overhang.

	$\omega_{\mathrm{n},k}$ (Hz)	ζ_k (%)	Q_k (μm/N/s)	$\sim k_k$ (N/μm)	$\mathbf{U}_k / |\mathbf{U}_k|$ (1)
1	37.8	4.86	0.09–1.17j	101.36	$[0.07 \quad 0.75 \quad 0.66]^{\mathrm{T}}$
2	58.0	13.27	2.07–3.08j	59.05	$[0.22 \quad 0.97 \quad 0.08]^{\mathrm{T}}$
3	69.6	3.10	1.34–5.30j	41.28	$[0.98 \quad 0.13 \quad 0.11]^{\mathrm{T}}$
4	86.6	5.85	−0.90–2.68j	101.32	$[0.09 \quad 0.97 \quad 0.22]^{\mathrm{T}}$
5	144.8	3.02	−0.48–1.08j	422.24	$[0.83 \quad 0.55 \quad 0.10]^{\mathrm{T}}$

4.2. Objective 1: Optimum Pitch Design Through the BF Method

In this test, the effectiveness of variable pitch tools for chatter suppression is demonstrated, as well as the improvement achieved with the BF criterion with respect to the BS criterion for topology 1A (Figure 4a). For this purpose, the machining of a heat resistant alloy, Jethete M-152, was considered. The technological details of the performed milling process are listed in Table 2.

Table 2. Process conditions.

f_Z (mm/tooth)	v_c (m/min)	n (rpm)	a_e (%)	Cutting Direction	L (mm)	Workpiece	Cutting Coefficient (N/mm^2)
0.20	100	636	80	X- up milling	400	Jethete M152	$K_{c,t} = 1843$ $K_{c,r} = 625$ $K_{c,a} = 467$

If the material to cut allowed a wide range of cutting speed variation, the spindle speed could be tuned to the optimum (n = 1000–1100 rpm in Figure 5b) and the variable pitch tool would be of no advantage. However, considering the target process of Jethete M152 machining (see Table 2), a particular cutting speed is regarded as optimum. If the spindle speed is shifted from this value, the tool life will decrease dramatically. Moreover, suppliers are not in a position to change any of the main technological parameters, since deviation from the designed one risks the surface integrity of a usually extremely expensive part. The specifically recommended cutting speed for this insert and workpiece material combination was v_c = 100 m/min, which in this case, corresponded to n = 636 rpm. The selected (see Figure 5a,b) and manufactured tools with rounded pitch angles using 1A topology (Figure 4a) for the different tuning methods are shown in Table 3.

Table 3. Tuned pitch solutions for spindle speed n = 636 rpm according to Figure 5a,b.

Tuning Methodology	Pitch Angles, $\varphi_{p,i}$ (deg)	Picture
none	(90, 90, 90, 90)	
BS	(77, 103, 77, 103)	
BF	(72, 108, 72, 108)	

Figure 6 shows the predicted ordinary stability diagrams (SLD's) for the selected angle distributions on a wide frequency range. The depth of cut, a, of the process can be adjusted according to the theoretical simulation for the required spindle speed of n = 636 rpm. It can be noted that the depth of cut values for the selected spindle speed in the SLD (Figure 6b) were the same as in the SYD in Figure 5b).

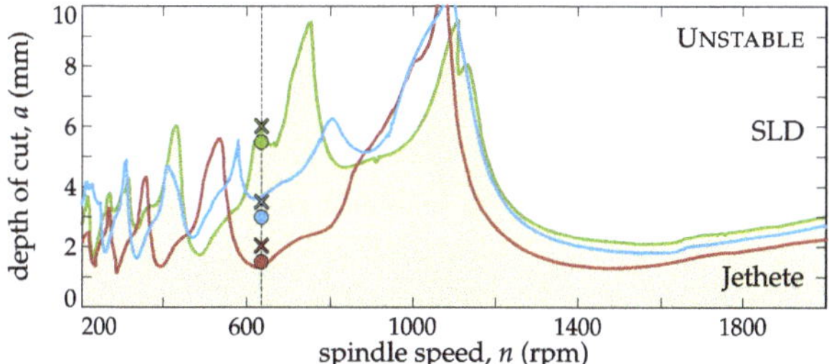

Figure 6. SLD's and experimental tests for tuned tools (Table 3) for 636 rpm according to Figure 5a,b. The workpiece material is Jethete, restricting the cutting speed to v_c = 100 m/min (green: BF, red: regular, blue: BS; circle: stable, cross: unstable tests).

The variable pitch tools designed by the presented methods were experimentally tested to assess their cutting performance. The stability of the process was monitored by using industrial accelerometers located close to the cutting point on the three directions as shown in Figure 8b. The experimental cutting results are also plotted on top of Figure 6b, with circles and crosses regarding the stable and unstable tests, respectively.

The stability obtained with the BF tuned tool was three times higher than the constant pitch tool´s stability and more than two times with respect to the tool predicted by the BS method. This tendency was also true for MRR keeping the prescribed maximum chip load per tooth, namely, the BS tuned tool performed with double MRR, while the BF tuned tool three times MRR compared to a conventional inserted cutter (see Figure 5d). With this tool, the maximal removal rate for this spindle speed was obtained, maintaining the recommended cutting speed and a proper surface finish.

4.3. Objective 2: Optimum pitch variation with spindle speed

The different methodologies for the variable pitch tuning determine the best pitch for a particular cutting frequency. However, when these non-constant pitch tools are used at a spindle speed different from the tuned one, a severe misbehaviour may occur. To show this stability variation, cutting tests at different speeds were carried out. In this case, a workpiece of C45 steel was used to cover a wide spindle speed range. The target spindle speeds for the stability lobes calculation as shown in Table 4 were n = 486, 636, 950, and 1600 rpm.

Table 4. Process conditions for steel C45 machining.

f_Z (mm/tooth)	N (rpm)	a_e (%)	Cutting Direction	L (mm)	Workpiece	Cutting Coefficient (N/mm^2)
0.20	486 636 950 1600	80	-X Up milling	400	Steel C45	$K_{c,t}$ = 1836 $K_{c,r}$ = 734 $K_{c,a}$ = 387

In Figure 7, it is demonstrated that the best tool configuration for a particular spindle speed may produce even lower stability than a constant pitch cutter at a different speed from the target of optimization (see n = 950 rpm case in Figure 7).

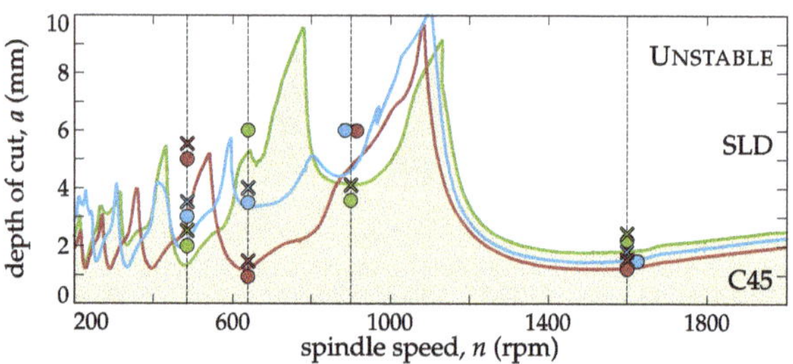

Figure 7. SLD's of different tuning methodologies for C45 steel workpiece (green: BF, red: regular, blue: BS; circle: stable, cross: unstable tests).

This means that the classic approach of random selection of variable pitch cutters can even worsen the cutting capability achieved by their regular pitch counterpart.

4.4. Objective 3: Experimental Evidence of New Flip Family

For interrupted cutting processes, the apparition of a flip chatter type with constant pitch tools has already been demonstrated [25]. However, the flip instability related to non-constant pitch tools has not been experimentally reported yet. In this test, the existence of a new family of double period or flip bifurcation chatter related to the pitch pattern was proven. In this case, a different position of the machining centre with a longer overhang (L = 610 mm) was used for the experimental testing.

To validate the existence of this new family of double period lobes, an interrupted cutting process was designed, where the tooth passing harmonics were stronger, resulting in dominant and observable flip lobes in the stability diagrams. The process conditions are shown in Table 5, using 10% radial immersion and locating the ram in its more flexible position, L = 610 mm, to facilitate the flip domination. With these parameters, new diagrams were calculated and chatter frequencies (Figure 9a) and stability lobes (Figure 9b) were predicted.

Table 5. Process conditions for new flip lobe family observation.

f_z (mm/tooth)	N (rpm)	a_e (%)	Cutting Direction	L (mm)	Workpiece	Cutting Coefficient (N/mm^2)
0.20	460 470 480 490 500	10	-X up milling	610	Jethete M152	$K_{c,t}$ = 1843 $K_{c,r}$ = 625 $K_{c,a}$ = 467

In Table 6, the three most important modes are listed, among which the second completely dominated the dynamics (Figure 8a). The fitting was performed for each spatial direction separately, due to the clear dominance of the mode in the x direction.

Table 6. Modal parameters at L = 610 mm ram overhang.

k	$\omega_{n,k}$ (Hz)	ζ_k (%)	Q_k (μm/N/s)	k_k (N/μm)	U_k/$|U_k|$ (1)
1	49.8	2.73	0.27–0.59j	264.78	$[1\ 0\ 0]^T$
2	56.1	0.58	0.66–6.26j	28.19	$[1\ 0\ 0]^T$
3	53.0	9.90	−0.47–5.24j	31.94	$[0\ 0\ 1]^T$

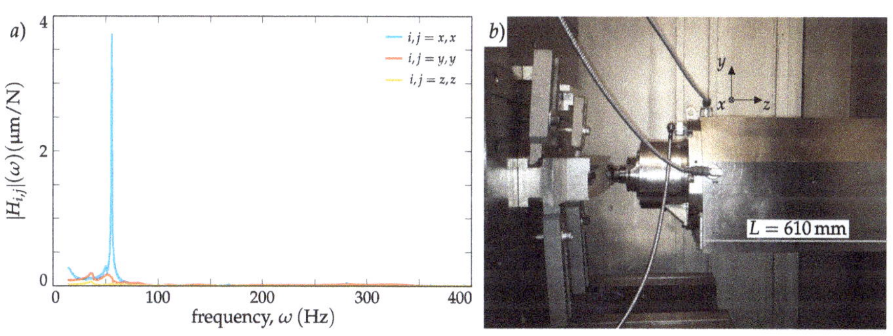

Figure 8. (a) shows the direct FRFs with L = 610 mm ram overhang presented in (b).

The SLD's (Figure 9b) are presented along with the chatter frequency diagram in Figure 9a, where different flip lobe regions can be observed. The classical flip lobes from a regular cutter corresponding to Equation (20) are lying on the black continuous line in Figure 9a. The new kind of flip lobes corresponding to Equation (18) are lying on the dashed red lines, while the saddle-node type regarding (19) is the continuous red lines shown in Figure 9a.

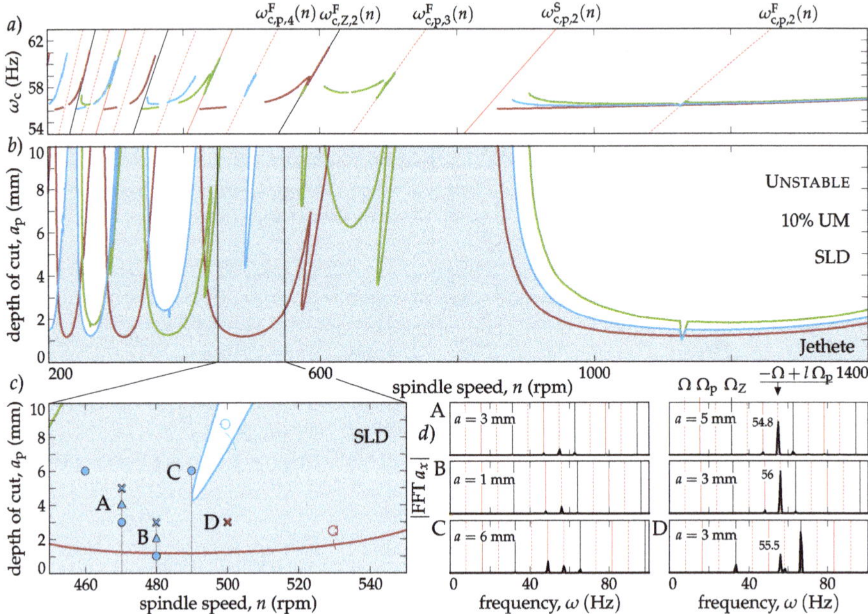

Figure 9. (**a**) presents the chatter frequency diagram, (**b**) shows the SLD's and the measurements panel (green: BF, red: regular, blue: BS.), while (**c**) is the magnified area where the tests were performed (circle: stable, cross: unstable tests, triangle: marginal). The essential spectra are presented in panel (**d**); a new kind of flip (A, B, C) and Hopf (D) stability losses can be recognized.

Cutting tests were carried out with the three tools selected in Table 3 and the results are shown in Figure 9 for spindle speeds of $n = 460, 470, 480, 490$, and 500 rpm. The existence of the flip lobe related to the variable pitch pattern was shown.

In Figure 9c, the limiting stable and chatter measurement pairs are presented. For the chatter cases, the chatter frequency can be determined by calculating the spectra. These spectra are presented in Figure 9d for the variable pitch BF case in A, B, and C points, and for the regular cutter case, D point. While, in the D point, the dominant chatter frequency was not a harmonic of a spindle speed, referring to a Hopf kind of stability loss; in A and B cases, the dominant chatter frequency was connected to $l = 4$ modulation of $\omega_{c,b} = \arg(-1)/T_p = -\pi/T_p = -2\pi/T = -\Omega$ with $\Omega_p = N\,\Omega$ ($N = 2$ for 1A, (2)) according to Equation (17). This resulted in $\omega_c = -\Omega + l\,\Omega_p = 54.8$ Hz for point A and $\omega_c = -\Omega + l\,\Omega_p = 56$ Hz for point B. C was a stable point, which means that the special flip lobe was actually slightly shifted to the lower spindle speed zone. In this manner, the chatter frequency followed the $l = 4$ modulation of the principal frequency, Ω_p, which is a clear indication of a flip kind of stability loss following the $\omega_{c,p,l}^{F}(n)$ lines defined in Equation (18).

5. Conclusions

The variable pitch cutters can be applied for vibration reduction when a difficult-to-cut material is machined in heavy duty operations limited by structural chatter. The applicability depends on the relative location in the stability lobe. If the chatter problem is lying in the first lobe ($k < 1$), the optimal pitch variations can be too big to assure a proper chip evacuation and can require a reduction in the feed to avoid overloading of some inserts. On the other hand, if it lies in a high number of lobes ($l > 10$), the required pitch variation is so subtle that it can be difficult to practically achieve an optimized performance. The optimal pitch tool chart, stabilizability diagram, and MRR diagrams for variable pitch tools were proposed to consider these limitations and optimize the application of these irregular tools. It was also shown that variable pitch cutters must be used with a moderate feed compared to conventional tools, however, the gain in stability in the axial depth of cut ensured a large improvement in the material removal rate (MRR).

A new variable pitch milling tool design method based on the semidiscretization (SD) and the brute force (BF) method was proposed. An analysis of the different variable pitch patterns for a four insert tool was carried out, concluding that the $\varphi_{p,i} = (\varphi_p, \pi - \varphi_p, \varphi_p, \pi - \varphi_p)$ topology offers a good ratio of performance over the computation time.

The BF method was theoretically and experimentally compared with current state-of-the-art analytical methods, resulting in a clear improvement with respect to those. The BS tuned tool performed with double MRR, while the BF tuned tool three times MRR compared to a conventional inserted cutter

The optimum variable pitch estimation is only valid for a given spindle speed. It was theoretically and experimentally demonstrated that the stability of the variable pitch cutter can be even worse than the one of the regular pitched tool if a different speed from the target of optimization was chosen. Therefore, the tool pitch design procedure developed allows the selection of the right tool, which will help to maximize the MRR for a specific cutting process.

Finally, a new family of double period related stability lobes, which had not been identified in any previous research, was theoretically simulated and experimentally verified. These lobes were produced under the same physical effect as the standard flip lobes. However, in this case, the main harmonics exciting the chatter were those of the variable pitch pattern passing frequency, which originates two more different families of lobes related to saddle and flip type stability losses.

Author Contributions: Conceptualization, J.M. and G.S.; Investigation, A.I. and Z.D.; Validation, G.G.

Funding: This research was partially supported by the EU H2020 MMTECH project (633776/H2020-MG-2014), European Research Council (FP7/2007-2013/ERC Advanced Grant Agreement n. 340889) and Hungarian NKFI FK 124361.

Acknowledgments: The authors would like to thank K. Scott Smith from the University of North Carolina for his useful comments to refocus the variable pitch tool cutting capability estimation procedure.

Conflicts of Interest: The authors declare no conflict of interest.

Nomenclature

a (m)	axial depth of cut,
a_e (m)	radial engagement,
$\mathbf{A}_i$ (1)	ith periodic directional coefficient matrices in (x, y, z) system,
g (1)	screen function of radial immersion,
$\mathbf{G}$ (N)	only time dependent periodic part of the cutting force,
h_i (m)	local momentary chip thickness on the ith edge,
$\mathbf{H}$ (m/N)	frequency response function (FRF),
f (m/rev)	complete feed per revolution,
f_i (m)	feed motion during ith delay τ_i,
$\mathbf{f}_i$ (m)	feed vector describing f_i motion in x direction,
f_Z (m/tooth)	feed per tooth,
$f_{Z,\max}$ (m)	maximum allowed feed per tooth allowed on an insert,
$\mathbf{F}$ (N)	momentary resultant cutting force,
$K_{c,t}$ (N/m^2)	tangential cutting coefficient,
$\mathbf{K}_e$ (N/m)	edge coefficient vector in (t,r,a) system,
k_k (N/m)	approximated modal stiffness in proportionally damped sense,
L (m)	overhang of the RAM of the machine,
n (rpm)	spindle speed,
$\mathbf{n}$ (1)	normal vector to the local edge,

$\mathbf{q}$ (m (kg/s)$^{1/2}$)	modal displacement vector,
Q_k (m/N/s)	modal scaling factor,
$\mathbf{q}_\mathrm{p}$ (m (kg/s)$^{1/2}$)	time periodic stationary solution in modal space,
r (m/N)	magnitude of the FRF,
t (s)	process time,
(t,r,a)	local edge system,
$\mathbf{T}$ (1)	transformation matrix from (t,r,a) to (x, y, z) system,
$\mathbf{u}$ (m (kg/s)$^{1/2}$)	perturbation in modal space for asymptotic analysis,
$\mathbf{U}$ ((s/kg)$^{1/2}$)	mass normalized modal transformation matrix,
$\mathbf{U}_k$ ((s/kg)$^{1/2}$)	kth mass normalized mode shape vector,
(x, y, z)	spatial system,
$\mathbf{x}$ (m)	spatial displacement vector in spatial directions (x, y, z),
$\mathbf{x}_\mathrm{p}$ (m)	spatial time periodic stationary solution,
$\mathbf{y}$ (m)	spatial perturbation for asymptotic analysis,
v_c (m/s)	cutting speed,
v_f (m/s)	feed (secondary motion) speed,
MRR (m^3/s)	material removal rate,
N (1)	integer divisor of tool period T to the principle period T_p,
N_τ (1)	number of delays,
T (s)	tool period,
T_p (s)	principle period,
Z (1)	number of inserts (teeth),
$\mathbf{z}$ (m (kg/s)$^{1/2}$)	discrete state vector for semidiscretization,
γ_MRR (1)	relative gain on MRR,
δ_f (1)	relative drop on the complete feed f,
ε (rad)	regenerative phase,
ζ_k (1)	kth damping ratio,
κ (rad)	lead angle,
$\boldsymbol{\kappa}_\mathrm{c}$ (1)	nominal cutting coefficient vector in (t,r,a) system,
λ (rad/s)	characteristic exponent,
λ_k (rad/s)	kth pole,
μ (1)	Floquet multiplier,
τ_i (s)	ith delay corresponding to the ith pitch angle $\varphi_{\mathrm{p},i}$,
τ_max (s)	maximum delay,
φ_i (rad)	position angle of the ith edge,
$\varphi_{\mathrm{p},i}$ (rad)	pitch angle between the ith and $(i+1)$th edges,
ψ (rad)	phase of the FRF
ω (rad/s, Hz)	vibration frequency,
$\omega_\mathrm{c,b}$ (rad/s)	base critical (chatter) frequency,
$\omega_{\mathrm{c,p},l}^\mathrm{F}$ (rad/s, Hz)	lth critical frequency of flip stability loss corresponding to the principal period,
$\omega_{\mathrm{c,p},l}^\mathrm{S}$ (rad/s, Hz)	lth critical frequency of the saddle-node stability loss on principal period,
$\omega_{\mathrm{c},Z,l}^\mathrm{F}$ (rad/s, Hz)	lth critical frequency of flip stability loss corresponding to tooth passing period,
$\omega_{\mathrm{n},k}$ (rad/s)	kth natural angular frequency,
Δ (rad)	regenerative phase difference,
Δt (s)	time in semidiscretization,
Λ (N/m)	eigenvalue in zeroth order solution,
$\boldsymbol{\Phi}$ (1)	Floquet transition matrix compiled by semidiscretization,
Ω (rad/s)	angular velocity of the tool,
Ω_p (rad/s)	principle angular frequency.

References

1. Munoa, J.; Beudaert, X.; Dombovari, Z.; Altintas, Y.; Budak, E.; Brecher, C.; Stepan, G. Chatter suppression techniques in metal cutting. *CIRP Ann.* **2016**, *65*, 785–808. [CrossRef]
2. Burtscher, J.; Fleischer, J. Adaptive tuned mass damper with variable mass for chatter avoidance. *CIRP Ann.* **2017**, *66*, 397–401. [CrossRef]
3. Munoa, J.; Iglesias, A.; Olarra, A.; Dombovari, Z.; Zatarain, M.; Stepan, G. Design of self-tuneable mass damper for modular fixturing systems. *CIRP Ann.* **2016**, *65*, 389–392. [CrossRef]
4. Zaeh, M.F.; Kleinwort, R.; Fagerer, P.; Altintas, Y. Automatic tuning of active vibration control systems using inertial actuators. *CIRP Ann.* **2017**, *66*, 365–369. [CrossRef]
5. Munoa, J.; Beudaert, X.; Erkorkmaz, K.; Iglesias, A.; Barrios, A.; Zatarain, M. Active suppression of structural chatter vibrations using machine drives and accelerometers. *CIRP Ann.* **2015**, *64*, 385–388. [CrossRef]
6. Seguy, S.; Insperger, T.; Arnaud, L.; Dessein, G.; Peigné, G. On the stability of high-speed milling with spindle speed variation. *Int. J. Adv. Manuf. Technol.* **2010**, *48*, 883–895. [CrossRef]
7. Bediaga, I.; Egaña, I.; Munoa, J.; Zatarain, M.; De Lacalle, L.L. Chatter avoidance method for milling process based on sinusoidal spindle speed variation method: Simulation and experimental results. In Proceedings of the 10th CIRP International Workshop on Modeling of Machining Operations, Reggio Calabria, Italy, 27–28 August 2007.
8. Bediaga, I.; Munoa, J.; Hernández, J.; De Lacalle, L.L. An automatic spindle speed selection strategy to obtain stability in high-speed milling. *Int. J. Mach. Tools Manuf.* **2009**, *49*, 384–394. [CrossRef]
9. Ezugwu, E.O. High speed machining of aero-engine alloys. *J. Braz. Soc. Mech. Sci. Eng.* **2004**, *26*, 1–11. [CrossRef]
10. Iglesias, A.; Munoa, J.; Ciurana, J. Optimisation of face milling operations with structural chatter using a stability model based process planning methodology. *Int. J. Adv. Manuf. Technol.* **2014**, *70*, 559–571. [CrossRef]
11. Budak, E.; Kops, L. Improving productivity and part quality in milling of titanium based impellers by chatter suppression and force control. *CIRP Ann. -Manuf. Technol.* **2000**, *49*, 31–36. [CrossRef]
12. Hahn, R.S. Metal-Cutting Chatter & Its Elimination. *Trans. ASME* **1952**, *74*, 1073–1080.
13. Altıntas, Y.; Engin, S.; Budak, E. Analytical stability prediction and design of variable pitch cutters. *J. Manuf. Sci. Eng.* **1999**, *121*, 173–178. [CrossRef]
14. Slavicek, J. The effect of irregular tooth pitch on stability of milling. In Proceedings of the 6th MTDR Conference, Manchester, UK, 13–15 September 1965.
15. Vanherck, P. Increasing milling machine productivity by use of cutters with non-constant cutting edge pitch. In *Proceedings of the 8th MTDR Conference*; University of Manchester: Manchester, UK, 1967; pp. 947–960.
16. Varterasian, J.H. White noise: A deterrent to milling cutter chatter. *Manuf. Eng. Manag.* **1971**, *67*, 26.
17. Tlusty, J.; Ismail, F.; Zaton, W. Use of special milling cutters against chatter. In *NAMRC*; University of Wisconsin: Madison, WI, USA, 1983; Volume 11, pp. 408–415.
18. Olgac, N.; Sipah, R. Dynamics & stability of variable-pitch milling. *J. Vib. Control* **2007**, *13*, 1031–1043.
19. Munoa, J.; Dombovari, Z.; Mancisidor, I.; Yang, Y.; Zatarain, M. Interaction between multiple modes in milling processes. *Mach. Sci. Technol.* **2013**, *17*, 165–180. [CrossRef]
20. Dombovari, Z.; Altintas, Y.; Stepan, G. The effect of serration on mechanics and stability of milling cutters. *Int. J. Mach. Tools Manuf.* **2010**, *50*, 511–520. [CrossRef]
21. Stepan, G.; Hajdu, D.; Iglesias, A.; Takacs, D.; Dombovari, Z. Ultimate capability of variable pitch milling cutters. *CIRP Ann.* **2018**, *67*, 373–376. [CrossRef]
22. Wojciechowski, S.; Maruda, R.W.; Barrans, S.; Nieslony, P.; Krolczyk, G.M. Optimisation of machining parameters during ball end milling of hardened steel with various surface inclinations. *Measurement* **2017**, *111*, 18–28. [CrossRef]
23. Wojciechowski, S.; Maruda, R.W.; Krolczyk, G.M.; Nieslony, P. Application of signal to noise ratio and grey relational analysis to minimize forces and vibrations during precise ball end milling. *Precis. Eng.* **2018**, *51*, 582–596. [CrossRef]
24. Altintas, Y. *Manufacturing Automation: Metal Cutting Mechanics, Machine Tool Vibrations, and CNC Design*; Cambridge University Press: Cambridge, UK, 2012.

25. Iglesias, A.; Munoa, J.; Ciurana, J.; Dombovari, Z.; Stepan, G. Analytical expressions for chatter analysis in milling operations with one dominant mode. *J. Sound Vib.* **2016**, *375*, 403–421. [CrossRef]

26. Dombovari, Z.; Munoa, J.; Stepan, G. General Milling Stability Model for Cylindrical Tools. *Procedia CIRP* **2012**, *4*, 90–97. [CrossRef]

27. Insperger, T.; Stepan, G. Semidiscretization method for delayed systems. *Int. J. Numer. Methods Eng.* **2002**, *55*, 503–518. [CrossRef]

28. Compean, F.I.; Olvera, D.; Campa, F.J.; López De Lacalle, L.N.; Elias-Zuniga, A.; Rodriguez, C.A. Characterization and stability analysis of a multivariable milling tool by the enhanced multistage homotopy perturbation method. *Int. J. Mach. Tools Manuf.* **2012**, *57*, 27–33. [CrossRef]

29. Bachrathy, D.; Stepan, G. Improved Prediction of Stability Lobes with Extended Multi Frequency Solution. *CIRP Ann.* **2013**, *62*, 411–414. [CrossRef]

30. Budak, E.; Altintas, Y. Analytical Prediction of Chatter Stability in Milling—Part I: General Formulation. *J. Dyn. Syst. Meas. Control.* **1998**, *120*, 22–30. [CrossRef]

31. Ewins, D. *Modal Testing: Theory, Practice, and Applications*; Research Studies Press: Letchworth, UK, 2000.

32. Budak, E. An analytical design method for milling cutters with nonconstant pitch to increase stability-Part I: Theory. *ASME J. Manuf. Sci. Eng.* **2003**, *125*, 29–34. [CrossRef]

33. Altintas, Y.; Budak, E. Analytical prediction of stability lobes in milling. *CIRP Ann. -Manuf. Technol.* **1995**, *44*, 357–362. [CrossRef]

34. Szalai, R.; Stepan, G.; Hogan, J. Global dynamics of low immersion high-speed milling. *Chaos* **2004**, *14*, 1069–1077. [CrossRef]

35. Farkas, M. *Periodic Motions*; Springer: Berlin, Germany; New York, NY, USA, 1994.

36. Dombovari, Z.; Iglesias, A.; Stepan, G. Study of the tuning of variable pitch milling cutters. In Proceedings of the 39th MATADOR Conference on Advanced Manufacturing, Manchester, UK, 5–7 July 2017.

37. Dombovari, Z.; Iglesias, A.; Zatarain, M.; Insperger, T. Prediction of multiple dominant chatter frequencies in milling processes. *Int. J. Mach. Tools Manuf.* **2011**, *51*, 457–464. [CrossRef]

38. Bachrathy, D.; Stepan, G. Bisection method in higher dimensions and the efficiency number. *Period. Polytech. Mech. Eng.* **2012**, *56*, 81–87. [CrossRef]

39. Comak, A.; Budak, E. Modeling dynamics and stability of variable pitch and helix milling tools for development of a design method to maximize chatter stability. *Precis. Eng.* **2017**, *47*, 459–468. [CrossRef]

40. Dombovari, Z. Dominant modal decomposition method. *J. Sound Vib.* **2017**, *392*, 56–69. [CrossRef]

Article

Research on High Performance Milling of Engineering Ceramics from the Perspective of Cutting Variables Setting

Rong Bian [1,2,*], Wenzheng Ding [1], Shuqing Liu [1] and Ning He [2]

[1] Industrial Center, Nanjing Institute of Technology; Nanjing 211167, China; dwz198151@njit.edu.cn (W.D.); liushuqing@126.com (S.L.)

[2] Jiangsu Key Laboratory of Precision and Micro-Manufacturing Technology, Nanjing University of Aeronautics & Astronautics; Nanjing 210016, China; drnhe@nuaa.edu.cn

* Correspondence: bianrong@njit.edu.cn; Tel.: +25-8611-8560

Received: 10 November 2018; Accepted: 26 December 2018; Published: 2 January 2019

Abstract: The setting of cutting variables for precision milling of ceramics is important to both the machined surface quality and material removal rate (MRR). This work specifically aims at the performance of corner radius PCD (polycrystalline diamond) end mill in precision milling of zirconia ceramics with relatively big cutting parameters. The characteristics of the cutting zone in precision milling ceramics with corner radius end mill are analyzed. The relationships between the maximum uncut chip thickness (h_{max}) and the milling parameters including feed per tooth (f_z), axial depth of cut (a_p) and tool corner radius (r_ε) are discussed. Precision milling experiments with exploratory milling parameters that cause uncut chip thickness larger than the critical value were carried out. The material removal mechanism was also analyzed. According to the results, it is advisable to increase f_z appropriately during precision milling ZrO_2 ceramics with corner radius end mill. There is still a chance to obtain ductile processed surface, as long as the brittle failure area is controlled within a certain range. The appropriate increasing of a_p, not only can prevent the brittle damage from affecting the machined surface, but also could increase the MRR. The milling force increases with increasing MRR, but the surface roughness can still be stabilized within a certain range.

Keywords: milling; ceramics; ductile machining; PCD; corner radius; material removal rate

1. Introduction

1.1. Machining of Ceramics

Thanks to the favourable combination of outstanding mechanical, thermal and chemical properties, advanced engineering ceramics, such as oxides, carbides, nitrides, and borides, have received growing attention in modern industries. It has broad application prospects in mechanical, aerospace, automotive, electrical, chemical and biological engineering etc. For example, the transplantation of artificial joints requires that the alternative materials have stable chemical properties in the living body, good biocompatibility, no component element dissolution and no stimulation to the body, and zirconia and alumina ceramics are ideal choices. The Power-MEMS (MEMS based power sources) project developed by researchers has a turbine diameter of only 20 mm, a working speed of at least 500,000 rpm, and a high temperature of about 1200 K. This requires a small material density, high temperature resistance, and high strength. Comprehensive assessment, advanced ceramic composite (SiN-TiN) can meet the demand [1–4]. As an example, zirconia ceramics (ZrO_2) have the highest fracture toughness (K_{IC} between 4 and 12 MPa/m$^{1/2}$) and flexural strength, combined with ionic conductivity, excellent thermal insulating properties and biocompatibility properties. Unique properties make it very suitable for advanced applications and some precision components, such as pump impellers for turbomachinery, diesel injection micro-nozzles, micro-fluidic devices, micro-molds, dental and orthopedic implants. It is also used for electronic product housing material. The case of smart phones and watches is also increasingly using zirconia ceramics with a smooth surface and good wear resistance [3].

Like some typical difficult-to-cut materials such as titanium alloys and high-temperature alloys, keeping high machined surface quality and improving machining efficiency has always been a key issue. Research works on tool wear, surface integrity and vibration are very necessary for understanding and optimizing the machining process [5–9]. Cutting force models are also used for estimation of the cutting forces to avoid unfavourable cutting parameters that lead to wrong use and premature failure of the tools [10]. At present, the machining of engineering ceramics also faces similar problems, especially high surface quality is required. Damage-free machining of ceramics has been studied by several researchers in recent years. It is reported that ceramic materials can be machined in ductile regime by decreasing the undeformed chip thickness to a sufficiently small value called "critical depth of cut" or "critical chip thickness", d_c, to get a very smooth machined surface without brittle damage [11–13]. Most of the research works are focused on precision or ultra-precision grinding of engineering ceramics in ductile or partial ductile mode [14,15]. However, precision grinding is time consuming, and it has certain limitations when it comes to complex three-dimensional structures.

To meet the demand of fabricating complex structures, alternative processes to machine ceramics have been also studied in the last decades. For example, Ferraris et al. [16,17] investigated the machining behavior of many electrically conductive ceramics, including Al_2O_3-, ZrO_2-, and Si_3Ni_4-based ceramic composites via electrical discharge machining (EDM). The results demonstrated the feasible manufacturing of complex shapes and micro features. However, the approach suitability is mostly limited to some electrically conductive ceramics, such as carbides and composites. Furthermore, it is the reported the machined surface presents thermal-induced micro-cracks, which may cause adverse effects on the operational performance of some precision components. With the development of 3D printing technology, research on the production of ceramic parts through additive manufacturing (AM) technology has emerged [18–20]. The AM processes for ceramics use ceramics in powder form, and AM techniques can directly be applied to a ceramic slurry and provide in-situ sintering. By this way, three-dimensional complex parts can be easily obtained. However, depending on the final required tolerance and surface quality of the part, especially for the precision part, final machining still might be needed [3].

1.2. Ductile Mode Milling with Ultra-Hard Tools

Micro milling has many advantages, such as high machining flexibility, removing materials quantitatively, and machining components with complex three-dimensional structures [21–23]. With the development of ultra-hard materials, different kinds of ultra-hard cutting tools such as diamond coated tools, polycrystalline diamond (PCD) tools, and cubic boron nitride (CBN) tools have been used. This makes it possible that materials with high hardness or brittleness, such as engineering ceramics [24,25], tungsten carbide [26,27], silicon [28] and pure tungsten [29] etc., can be directly machined in ductile mode by hard milling. Unlike pure cutting, in which the relative tool sharpness, known as RTS (the ratio of the uncut chip thickness to the cutting edge radius) should be ≥ 10. For some forming gear tools, RTS can be accepted ≥ 6. In the case of ductile mode milling, RTS is always below the critical value, which promotes the formation of large compressive stress into the chip formation zone, and enhances the plastic deformation of the undergoing material. Successful attempts on ductile mode milling have been conducted recently. Matsumura and Ono [30] reported that grooves with axial depth of cut in the range of 15~20 μm can be machined in glass if the CBN ball end mill is tilted at a certain angle in the feed direction. Bian [24] investigated the feasibility of ductile mode milling of ZrO_2 ceramic with diamond coated end mills, and presented a mirror-like machined surface and a three-demensional structure. Cheng [31] studied the process of micro milling tungsten carbide with PCD tools, achieved nanometric surface finish and micro-rib with a width-depth ratio of 1:10. Wu [27] also studied the tool wear of self-developed PCD tool in micro milling of tungsten carbide, and discussed the effect of tool wear on cutting force, surface quality and machined groove shape.

Thus, ductile mode milling seems a feasible way to achieve complex shape and crack free surface for hard and brittle materials, such as engineering ceramics. It is also an effective complement to the following finishing of EDM ceramics and 3D printed ceramics precision parts. However, there is still a noticeable lack of experience in this specific topic, especially in precision milling of ceramics in hard state. For instance, serious tool wear despite using ultra-hard tools and very low materials removal rate due to the micro level cutting parameters [24,32–34].

This work specifically aims at the performance of PCD end mills with corner radius in precision milling of zirconia ceramics in the hard state with Vickers hardness about HV 1180. The relationship between cutting parameters and tool corner radius are analyzed. High performance milling experiments with exploratory cutting parameters that cause uncut chip thickness larger than the critical value have been carried out to help understand the characteristics of machined surface formation. Based on that, the author put forward the strategy to improve the material removal rate while ensuring a high machined surface quality when precision milling ZrO_2 ceramics.

2. Analysis on Variables Setting in Precision Milling of Ceramics

2.1. Characteristics of Parameters in Precision Milling Ceramics

In the precision milling process, cutting parameters such as feed per tooth (f_z), axial depth of cut (a_p), are normally set to a very small value. When the parameters are reduced to a certain extent, the geometric characteristics of the cutting zone will change. The actual uncut chip thickness h is not the same as that in conventional milling.

As shown in Figure 1, when groove milling with a flat-bottom end mill, the radial uncut chip thickness f_c at any position during the single cutting process of the cutting edge changes with the tool rotation angle φ, ($f_c = f_z \sin\varphi$). In the feed direction (A-A section, $\varphi = 90°$), f_c reaches the maximum value, which is equal to f_z. In the conventional milling process, as shown in Figure 1c, the cutting section is approximately rectangular due to the relatively large axial depth of cut, a_p. The uncut chip thickness is mainly determined by the feed per tooth f_z. While in the precision milling process, the characteristics of the milling area are different from that in conventional milling. As shown in Figure 1d, when the milling parameter especially a_p is reduced to the micron level, which may be smaller than the tool corner radius r_ε, the part of the cutting edge involved in cutting process is mainly located at the bottom of the tool nose. The cutting cross-section has an irregular shape whose thickness is gradually reduced from the surface to be machined to the surface that has been machined. Thus, there is a maximum uncut chip thickness, h_{max}, as shown in Figure 2a. As shown in Figure 2b–d, the actual uncut chip thickness is different at different locations such as B1, B2, and B3 in the cutting zone.

According to the assumption of Befano [12], to get high machined surface quality, the uncut chip thickness h should be smaller than the critical chip thickness of materials d_c, the material will be removed in ductile mode, and without brittle failure. A conservative parameter setting strategy is to keep the h_{max} less than the critical value d_c of the material [24]. Therefore, it could ensure the material will be removed in completely ductile mode without brittle failure.

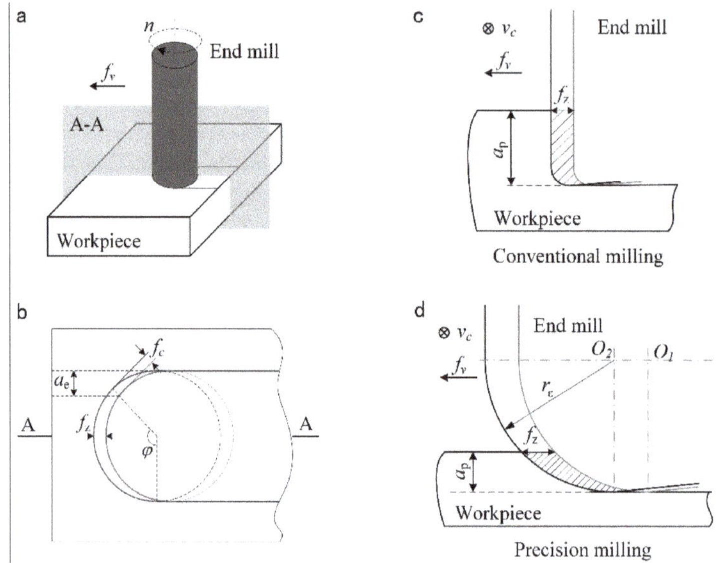

Figure 1. Schematic of groove milling from (**a**) side view, (**b**) top view and zoom in of cutting zone in (**c**) conventional milling and (**d**) micro precision milling.

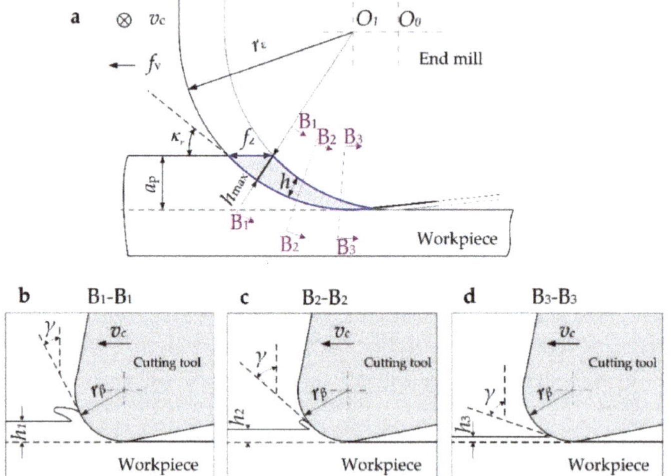

Figure 2. Schematic of (**a**) cutting zone in precision milling and the cross-section view of the uncut chip thickness at different position of (**b**) B1, (**c**) B2 and (**d**) B3.

For the specific selected f_z and a_p, and end mill with tool corner radius r_ε, the maximum uncut chip thickness, $h_{\max}$, can be approximately calculated by the Equation (1), when $\sqrt{2Ra_p - a_p^2} > f_z$ [35].

$$h_{\max} = r_\varepsilon - \sqrt{r_\varepsilon^2 + f_z^2 - 2f_z\sqrt{2r_\varepsilon a_p - a_p^2}}. \tag{1}$$

It can be seen from the Equation (1) that the maximum uncut chip thickness $h_{\max}$ in the cutting zone is affected by the feed per tooth f_z, axial depth of cut a_p and the corner radius of the tool nose r_ε.

Understanding the relationship between h_{max} and each cutting parameter helps guide the selection of cutting parameters.

Figure 3 shows the maximum uncut chip thickness h_{max} changes with f_z and a_p. For a specific end mill with tool corner radius r_ε about 50 µm, when the f_z changes from 1 to 10 µm, the a_p changes from 2 to 20 µm, the calculated h_{max} under each set of parameters show growing trend with the increasing of f_z and a_p. The change of f_z has a greater impact on h_{max} than that of a_p. It means for a specific cutting tool, smaller f_z will bring smaller h_{max}.

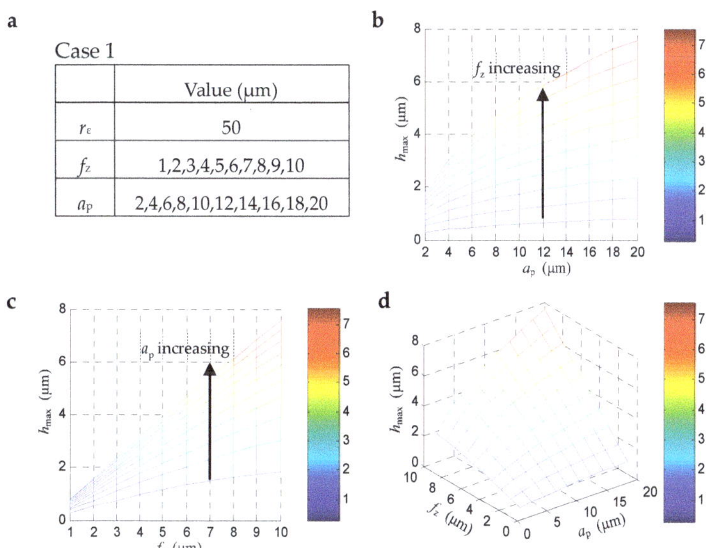

Figure 3. The maximum uncut chip thickness h_{max} changes with f_z and a_p: (**a**) parameters setting, (**b**) 2D-view from the perspective of a_p, (**c**) 2D-view from the perspective of f_z, (**d**) 3D-view from both f_z and a_p.

Figure 4 shows the maximum uncut chip thickness h_{max} changes with r_ε. and f_z. In this case, when the axial depth of cut a_p is a fixed on 10 µm, the f_z changes from 1 to 10 µm, and the tool corner radius r_ε changes from 20 to 100 µm, the h_{max} show different trends. It can be seen that the h_{max} increased with the increasing of f_z as usual, while decreased with the increasing of tool corner radius. It means that when machining with the given f_z and a_p, cutting tools with a bigger corner radius will bring a smaller h_{max}.

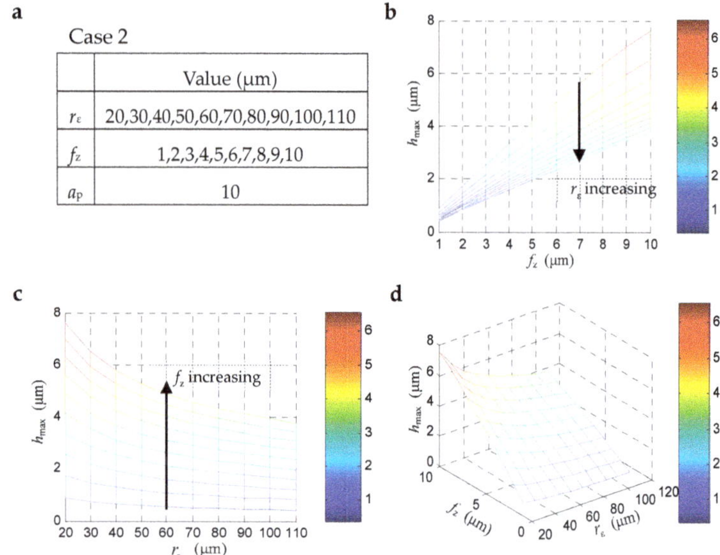

Figure 4. The maximum uncut chip thickness $h_{\max}$ changes with f_z and r_ε.: (**a**) parameters setting, (**b**) 2D-view from the perspective of f_z, (**c**) 2D-view from the perspective of r_ε, (**d**) 3D-view from both f_z and r_ε.

2.2. Parameters Selection Criteria in Precision Milling Ceramics

Generally, when the milling parameters are selected, as long as the maximum cutting thickness is limited to the critical thickness (d_c) of the material brittle-plastic transition ($h_{\max} < d_c$), any position of the cutting arc region can be ensured that the uncut chip thickness is less than d_c, and the machined surface for ductility removal can be obtained.

Figure 5 is a schematic sketch of the machined surface on the groove edge. During each cutting process, the material was removed by the cutting edge, and the machined surface area A was formed due to the tool corner radius. As the cutting process continues in the feed direction, the continuously generated new machined surface area A overlaps each other closely in accordance with the pitch of feed per tooth, f_z, to form the machined surface area B (Figure 5a). Due to the existence of the tool corner radius, the closer to the machined surface B is, the smaller the uncut chip thickness is. Therefore, it is generally always machined in ductile mode near the machined surface area B.

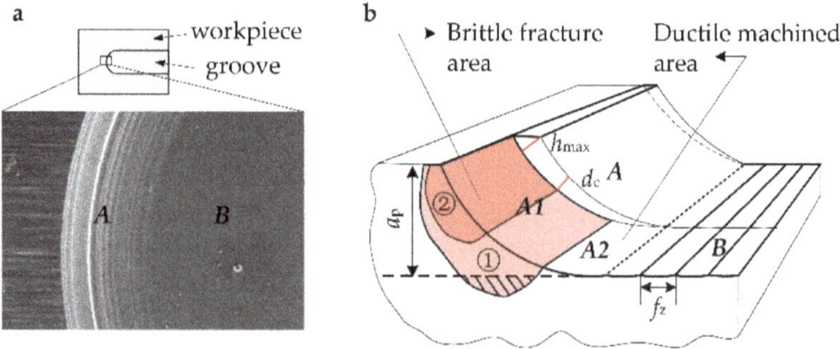

Figure 5. Schematic sketch of machined surface on groove edge: (**a**) different surface morphology of area A and B; (**b**) illustration of the effect of the brittle failure zone on the machined surface.

According to the assumption of Befano [12], it can be deduced that when the h_{max} is bigger than the d_c of material, the material on corresponding position will occur brittle damage (marked with *A1*), as shown in Figure 5b. Additionally, the area *A2* where near the machined surface still show a smooth surface. If the brittle fracture area is large enough to extend below the machined surface, as the area marked with ①, the damaged area will not be removed completely during the subsequent cutting process. Part of the damaged area will finally remain on the machined surface area *B* and present a brittle fracture surface morphology. However, if the brittle fracture area is controlled in a small range, and does not extend to the machined surface, as the area marked with ②, these damaged areas will be removed during the following cutting process completely. Therefore, a smooth surface still can be obtained finally. This reveals that when h_{max} is appropriately larger than d_c, the ductile surface can be still theoretically obtained. Therefore, when precision milling hard and brittle materials, the strategy of setting the milling parameters can be improved by appropriately increasing the h_{max}, actually the f_z and a_p, which will increase the material removal rate, and still have a chance to get asmooth machined surface.

3. Experimental Setups

3.1. Workpiece Material

As described above, ZrO_2 ceramics is currently used in a broad range of industrial applications. The workpiece material employed in this work was an yttria-stabilized tetragonal zirconia (Y-TZP) with a 2 wt% of Al_2O_3. The starting powder mixture (TM2 grade) contained yttria-free monoclinic ZrO_2 powder mixed in such a ratio that the overall content of Y_2O_3 stabilizer was 2 mol%. The material was prepared by hot pressing at a pressure of 28 MPa and at a temperature of 1450 °C for 1 h. The material was characterized by relative high fracture toughness and modest hardness; Table 1 summarizes relevant properties of the ZrO_2 ceramic work piece used during investigation.

Table 1. Chemical composition and mechanical properties of the workpiece material.

Composition	
Y_2O_3 content (mol%)	2
Al_2O_3 content (wt%)	2
Physical and Mechanical Properties	
Density ρ (g/cm^3)	6.02
Young's modulus E (GPa)	223
Fracture Toughness K_{IC} (MPam$^{1/2}$)	11.1 ± 0.7
Hardness HV$_{10}$ (kg/mm^2)	1180 ± 13

3.2. PCD Micro End Mills

The milling tools used in this study were single flute PCD end mills with a corner radius of 0.1 mm. The tool blank was made of a tungsten carbide tool shank and the PCD compact (Element 6, CMX850, average grain size s <1 µm) which was brazed on it. Figure 6 shows the structure and tool geometry of the PCD end mill. The cutting edge was designed to be straight in shape. The tool diameter was about 4 mm. The rake angle (α) and flank angle (γ) of the cutting edge were 0° and 5° respectively; hence, providing a robust cutting edge geometry to withstand cutting loads. The geometry of cutting edge was preformed by Wire cut Electrical Discharge Machining (WEDM), and then ground by a precision tool grinder. The cutting edge radius was measured to be less than 3 µm by a Leica DVM5000 microscope. The quality of the cutting edge was checked by scanning electron microscopy (SEM). The set of tool parameter is listed in Table 2.

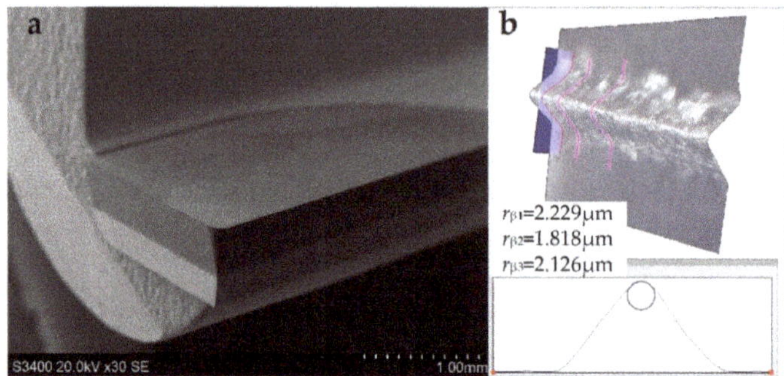

Figure 6. (**a**)SEM image of the polycrystalline diamond (PCD) end mill and (**b**) measuring of cutting edge radius.

Table 2. Tool parameters of the PCD corner radius end mills.

Tool Parameters	Value
grain size s (μm)	< 1
Tool diameter D (mm)	4
Tool corner radius r_ε (mm)	0.1
Rake angle α (°)	0
Flank angle γ (°)	5
Cutting edge radius r_β (μm)	< 3

3.3. Experimental Conditions and Procedures

In this study, two experimental campaigns were carried out. In order to clarify the characteristics of the machined surface when using corner radius end mill, the first experimental case 1 was set with fixed axial depth of cut, spindle rotating speed, but different feed per tooth, which resulted in different $h_{\max}$, as listed in case 1 in Table 3. After the experiments, the characteristics of the groove shoulder were observed by SEM to help understanding the material removal mechanism.

Table 3. Experimental parameters of the two cases.

No.	Spindle Rotating Speed n/rpm	Feed Per Tooth f_z/μm	Milling Depth a_p/μm	Maximum Uncut Chip Thickness $h_{\max}$/μm *
		Case 1		
1		3	20	1.8
2	8000	8	20	4.5
3		10	20	5.7
		Case 2		
1		5	10	2.08
2	8000	3	25	1.96
3		2.5	40	1.99
4		2	80	1.96

* Tool corner radius r_ε 0.1 mm.

According to the previous analysis, when the end mill is selected, $h_{\max}$ is mainly determined by f_z and a_p. The change of f_z has a greater influence on the value of $h_{\max}$. In the second part, case 2, by selecting different combinations of f_z and a_p, the calculated $h_{\max}$ of each group can be set to am almost similar level about 2 μm. Therefore, from test 1 to 4, f_z appears to decrease gradually from 5 to 2 μm, and a_p exhibits a multiple increase from 10 μm to a relatively bigger value about 80 μm, as listed in case 2 in Table 3. Thus, the maximum uncut chip thickness $h_{\max}$ was limited in a small value, but the material removal rate was increased. After the experiments, the cutting force and machined surface roughness were analyzed to see the performance of using exploratory milling parameters.

The experiments were conducted on a precision milling machine center (DMG Ultrasonic 20 linear, Stuttgart, Germany). A view of the experiment setups and the sketch of groove milling are shown in Figure 7. Workpieces were attached on the fixture with wax, and the surfaces of the samples were precisely ground to insure flatness and alignment. The fixture was then mounted on a dynamometer (Kistler type 9272, Kistler Group, Winterthur, Switzerland) used to measure the cutting force. The milling process was conducted in wet condition, using a water based emulsion to remove chips and debris. After the tests, the machined surface was cleaned by an ultrasonic cleaning machine and the surface roughness of the grooves were measured by a Mahr Perthomometer M1 (ISO 4287). The machined surface topography was inspected by an SEM.

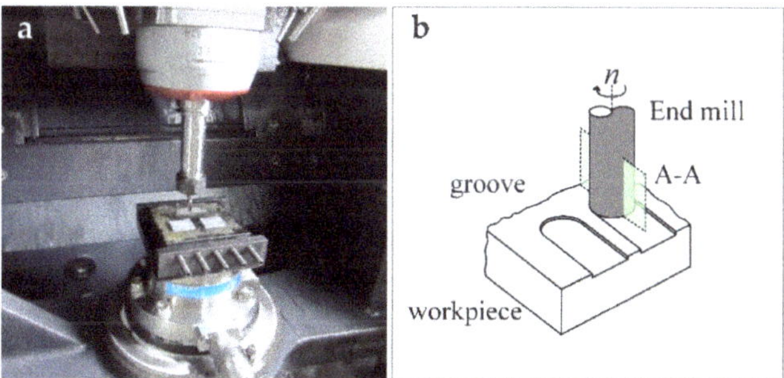

Figure 7. (**a**) Experimental setups and (**b**) a sketch of groove milling.

4. Results and Discussion

4.1. Possibility of Increasing Machining Efficiency

Figure 8 shows the SEM images of the milling shoulder in the experiment case 1. The $h_{\max}$ in the three trials is 1.8, 4.5 and 5.7 µm, respectively. According to Bifano's empirical formula, the critical uncut chip thickness of the test ZrO_2, d_c, is calculated to be about 2.6 µm by Equation (2):

$$d_c = 0.15 \left(\frac{E}{H} \right) \left(\frac{K_{IC}}{H} \right)^2,$$

(2)

where K_{IC} is the fracture toughness, H is the hardness, and E is the elastic modulus (see Table 1) [12]. Thus, in test 1, the $h_{\max}$ is about 1.8 µm, less than the critical value. The material in cutting area should be removed totally in ductile mode without brittle damage. As shown in Figure 8a, the machined area in milling shoulder presents a complete plastic texture. In test 2, due to the increase of f_z, the $h_{\max}$ is increased to about 4.5 µm that is bigger than the size of d_c. As shown in Figure 8b, the machined area near the milling shoulder where marked with $h_{\max}$, shows some brittle damage. However, the machined area away from the milling shoulder still shows the shape of ductile cutting. The same phenomenon also appears in test 3, as shown in Figure 8c, the brittle damage region is obviously larger than that in test 2. The results verify the previous analysis in part 2. When the maximum uncut chip thickness $h_{\max}$ is appropriately larger than d_c, a certain degree of brittle failure occurs near the milled shoulder. When the damaged areas are limited, the machined surface could still exhibit a ductile removal morphology.

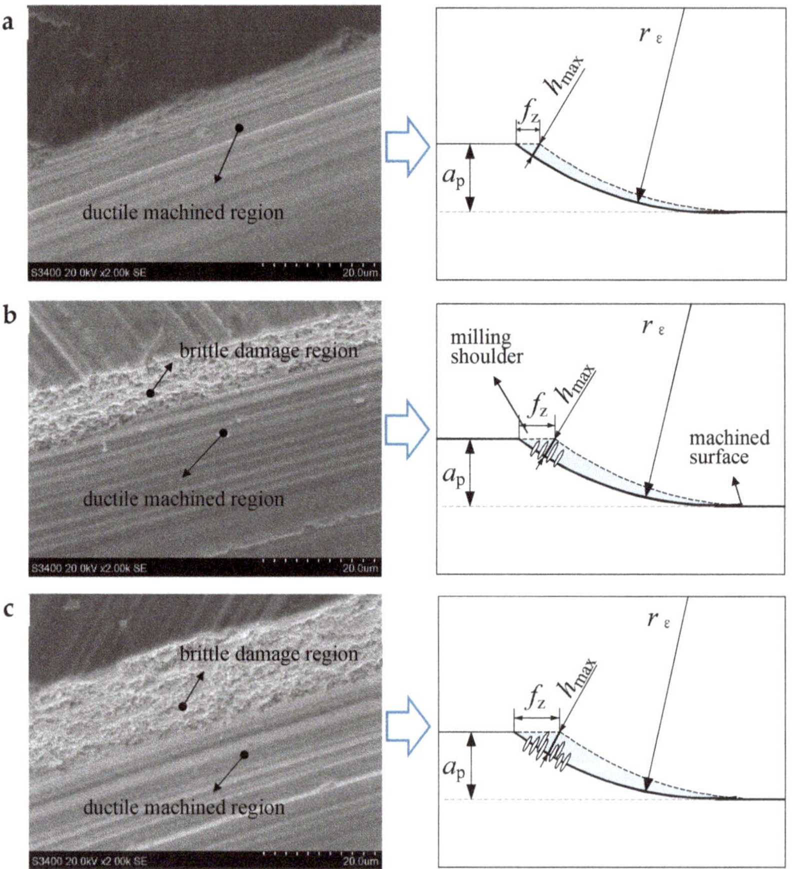

Figure 8. Characteristics of machined surface under different h_{max}, (**a**) $f_z = 3$ μm, $a_p = 20$ μm, $h_{max} = 1.8$ μm (**b**) $f_z = 8$ μm, $a_p = 20$ μm, $h_{max} = 4.5$ μm (**c**) $f_z = 10$ μm, $a_p = 20$ μm, $h_{max} = 5.7$ μm.

Figures 9 and 10 show the comparison of milling force, the machined surface roughness Ra and the relative material removal rate (take the MRR of test 1 as one unit) of each test in case 2. From test 1 to 4, the feed per tooth (f_z) was reduced by 60%, however, a significant increase of axial depth of cut (a_p) increased the material removal rate by 2.2 times. As shown in Figure 9, the milling force presents an obvious increasing trend from test 1 to 4, due to the increasing of removal material volume. In each test, due to the existence of tool corner radius, the axial milling force Fz is much bigger than Fx and Fy, which can produce large compressive stress in the chip formation zone, and it is reported benfitial for the ductile machining process [36]. It can be seen from the Figure 10, the measured surface roughness Ra of the four tests are all below 0.1 μm. Specifically, the Ra in test 3 is the largest with the average value about 0.083 μm, and in the other three tests are around 0.06 μm. It is revealed that when machining ceramics by corner radius end mill, the proper increase in the axial depth of cut can improve the material removal rate, while still possible to achieve a satisfactory surface roughness.

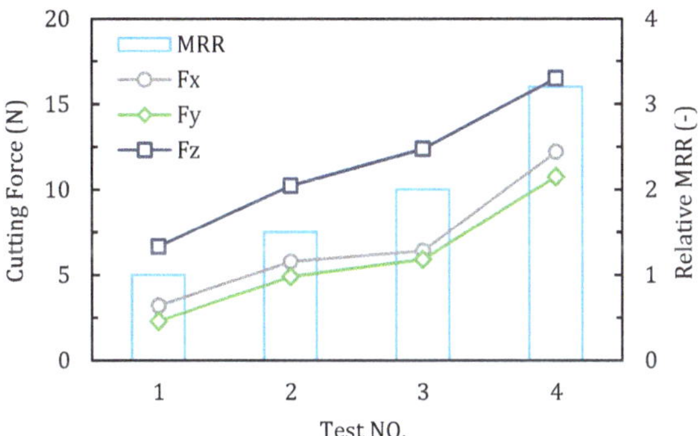

Figure 9. Cutting force and relative material removal rate (MRR) of each test in case 2.

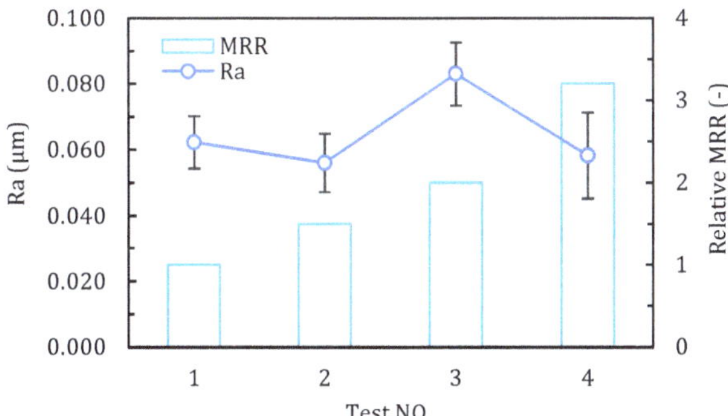

Figure 10. Surface roughness Ra and relative MRR of each test in case 2.

4.2. Material Removal Mechanism

Figure 11 shows the topography of the brittle fracture in the milled shoulder and a schematic sketch to help understand the material removal mechanism. In the area near milling shoulder, the actual uncut chip thickness is greater than the critical uncut chip thickness of the material brittle plastic transition. The material removal mode is mainly based on local brittle fracture, however, some scratches in the plastic deformation mode also can be found to be distributed in the brittle failure zone, as shown in Figure 11a. The overall morphology is somewhat different from the natural fracture surface morphology of zirconia that exhibits an uneven, rough, and highly grainy appearance (see Figure 11b). A more subtle view in Figure 11c shows that regular zirconia particles are visible in the brittle failure zone where always presented in the form of pits marked with a 1. The plastic scratched area (marked with a 2 in Figure 11c) exhibits a smooth surface.

As can be understood from Figure 11d, during the cutting process, the material to be cut is subjected to brittle fracture, and the crack is generated and expanded in the front to cause brittle fracture of the material to form an irregular fractured surface. When the cutting edge is swept, the material below the cutting surface (marked with a 1) does not contact the cutting edge and forms a lot of pits that present the original fractured surface morphology, and the material above the cutting surface are plastically deformed due to the pressing and ploughing action of the cutting edge during the cutting process. The zirconia particles are finally plastically deformed into a smooth micro plane (marked with a 2) with ductile scratches. In general, from the microscopic characteristics, the material in the brittle fracture zone is broken along the crystal, and the material in plastic scratch zone is removed with plastic deformation in the manner of transgranular failure. It was also found in the test of grinding zirconia ceramics with diamond tools by Mohammad [37].

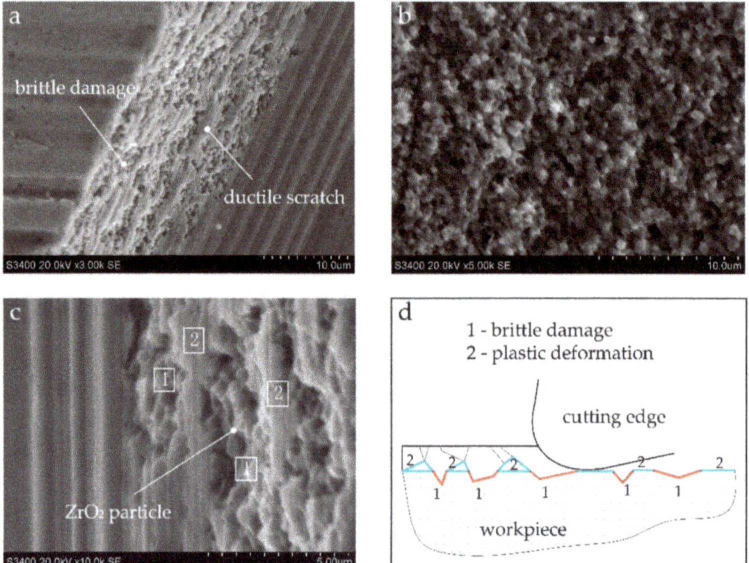

Figure 11. SEM image of (**a**) brittle fracture and plastic deformation in groove edge; (**b**) natural fracture surface of zirconia; (**c**) subtle view of micro damaged area and plastic scratched area; (**d**) schematic sketch of the material removal mechanism.

5. Conclusions

This paper presents a study on the performance of corner radius end mill in precision milling of zirconia ceramics with exploratory cutting parameters. Characteristics of precision milling ceramics and parameters selection criteria were discussed. Based on the results, the following conclusions have been drawn:

- When precision milling ZrO_2 ceramics with a specific end mill, the change in the feed per tooth f_z has a greater influence on the maximum uncut chip thickness h_{max} than that in axial depth of cut a_p. When f_z and a_p are constant, increasing the corner radius of the end mill can reduce the calculated h_{max}.

- It is advisable to increase the feed per tooth appropriately during precision milling ZrO_2 ceramics with corner radius end mills. When the calculated h_{max} is bigger than the critical value, even if the milling shoulder has brittle fractured, there is still a chance to obtain a ductile processed surface, as long as the brittle failure area is controlled within a certain range. Appropriate increase of the axial depth of cut, a_p, not only can prevent the minor brittle damage from affecting the machined surface, but also could increase the material removal rate. The milling force increases with increasing of machining efficiency, but the surface roughness can still be stabilized within a certain range.

- In the area near the milling shoulder, the actual uncut chip thickness is greater than the critical uncut chip thickness of the material brittle plastic transition. The machined surface morphology is mainly based on local brittle fractured pits and accompanied by some ductile scratches caused by plastic deformation. The material in brittle fracture zone is broken along the crystal, and the material in the plastic scratch zone is removed with plastic deformation in the manner of transgranular failure.

This study is an initial work on high performance milling of engineering ceramics. It gives a direction on how to improve the milling efficiency from the perspective of selection milling parameters. As for the extent to which the cutting parameters can be increased, further research is needed.

Author Contributions: R.B. and N.H. conceived and designed the experiments; R.B. performed the experiments; R.B., W.D. and S.L. analyzed the data; R.B. wrote the paper.

Funding: This research was funded by National Natural Science Foundation of China (Grant No. 51805242), the Natural Science Foundation of Jiangsu Province (Grant No. BK20160777), and the Scientific Foundation of Nanjing Institute of Technology (Grant No. YKJ201517). Part of the work was also funded by the Foundation of Jiangsu Key Laboratory of Precision and Micro-Manufacturing Technology.

Acknowledgments: The authors would like to sincerely thank the reviewers for their valuable comments on this work.

Conflicts of Interest: The authors declare no conflicts of interest.

References

1. Liu, K.; Reynaerts, D.; Lauwers, B. Influence of the pulse shape on the edm performance of si3n4–tin ceramic composite. *CIRP Ann.* **2009**, *58*, 217–220. [CrossRef]
2. Barry, C.C.; Grant, N.M. *Ceramic Materials/Science and Engineering*; Springer: Berlin, Germany, 2007.
3. Ferraris, E.; Vleugels, J.; Guo, Y.; Bourell, D.; Kruth, J.P.; Lauwers, B. Shaping of engineering ceramics by electro, chemical and physical processes. *CIRP Ann.* **2016**, *65*, 761–784. [CrossRef]
4. Denry, I.; Holloway, J. Ceramics for dental applications: A review. *Materials* **2010**, *3*, 351–368. [CrossRef]
5. Fernández-Valdivielso, A.; López de Lacalle, L.; Urbikain, G.; Rodriguez, A. Detecting the key geometrical features and grades of carbide inserts for the turning of nickel-based alloys concerning surface integrity. *Proc. Inst. Mech. Eng. Part C J. Mech. Eng. Sci.* **2016**, *230*, 3725–3742. [CrossRef]
6. Urbikain, G.; de Lacalle, L.N.L. Modelling of surface roughness in inclined milling operations with circle-segment end mills. *Simul. Model. Pract. Theory* **2018**, *84*, 161–176. [CrossRef]
7. Ghani, A.K.; Choudhury, I.A.; Husni. Study of tool life, surface roughness and vibration in machining nodular cast iron with ceramic tool. *J. Mater. Process. Technol.* **2002**, *127*, 17–22. [CrossRef]
8. Urbikain, G.; López de Lacalle, L.N.; Fernández, A. Regenerative vibration avoidance due to tool tangential dynamics in interrupted turning operations. *J. Sound. Vib.* **2014**, *333*, 3996–4006. [CrossRef]
9. Polvorosa, R.; Suárez, A.; de Lacalle, L.N.L.; Cerrillo, I.; Wretland, A.; Veiga, F. Tool wear on nickel alloys with different coolant pressures: Comparison of alloy 718 and waspaloy. *J. Manuf. Process.* **2017**, *26*, 44–56. [CrossRef]
10. Urbikain, G.; Artetxe, E.; López de Lacalle, L.N. Numerical simulation of milling forces with barrel-shaped tools considering runout and tool inclination angles. *Appl. Math. Model.* **2017**, *47*, 619–636. [CrossRef]
11. Shimada, S.; Ikawa, N.; Inamura, T.; Takezawa, N.; Ohmori, H.; Sata, T. Brittle-ductile transition phenomena in microindentation and micromachining. *CIRP Ann. Manuf. Technol.* **1995**, *44*, 523–526. [CrossRef]
12. Bifano, T.G.; Dow, T.A.; Scattergood, R.O. Ductile-regime grinding—A new technology for machining brittle materials. *J. Eng. Ind.* **1991**, *113*, 184–189. [CrossRef]
13. Beltrão, P.A.; Gee, A.E.; Corbett, J.; Whatmore, R.W. Ductile mode machining of commercial pzt ceramics. *CIRP Ann. Manuf. Technol.* **1999**, *48*, 437–440. [CrossRef]
14. Zhong, Z.W. Ductile or partial ductile mode machining of brittle materials. *Int. J. Adv. Manuf. Technol.* **2003**, *21*, 579–585. [CrossRef]
15. Yanyan, Y.; Bo, Z.; Junli, L. Ultraprecision surface finishing of nano-zro2 ceramics using two-dimensional ultrasonic assisted grinding. *Int. J. Adv. Manuf. Technol.* **2009**, *43*, 462–467. [CrossRef]
16. Ferraris, E.; Reynaerts, D.; Lauwers, B. Micro-edm process investigation and comparison performance of al3o2 and zro2 based ceramic composites. *CIRP Ann. Manuf. Technol.* **2011**, *60*, 235–238. [CrossRef]
17. Liu, K.; Ferraris, E.; Peirs, J.; Lauwers, B.; Reynaerts, D. Micro-edm process investigation of si3n4–tin ceramic composites for the development of micro fuel-based power units. *Int. J. Manuf. Res. (IJMR)* **2008**, *3*, 27–47. [CrossRef]
18. Shahzad, K.; Deckers, J.; Boury, S.; Neirinck, B.; Kruth, J.P.; Vleugels, J. Preparation and indirect selective laser sintering of alumina/pa micro spheres. *Ceram. Int.* **2012**, *38*, 1241–1247. [CrossRef]

19. Shahzad, K.; Deckers, J.; Zhang, Z.; Kruth, J.P.; Vleugels, J. Additive manufacturing of zirconia parts by indirect selective laser sintering. *J. Eur. Ceram. Soc.* **2014**, *34*, 81–89. [CrossRef]

20. Scheithauer, U.; Weingarten, S.; Johne, R.; Schwarzer, E.; Abel, J.; Richter, H.-J.; Moritz, T.; Michaelis, A. Ceramic-based 4d components: Additive manufacturing (am) of ceramic-based functionally graded materials (fgm) by thermoplastic 3d printing (t3dp). *Materials* **2017**, *10*, 1368. [CrossRef]

21. Ehmann, K.F.; Devor, R.E.; Kapoor, S.G. Micro/meso-scale mechanical manufacturing–opportunities and challenges. *JSME/ASME Int. Conf. Mater. Process.* **2002**, *1*, 6–13. [CrossRef]

22. Dhanorker, A.; Ozel, T. Meso/micro scale milling for micro-manufacturing. *Int. J. Mech. Manuf. Syst.* **2008**, *1*, 23–42. [CrossRef]

23. Dornfeld, D.; Min, S.; Takeuchi, Y. Recent advances in mechanical micromachining. *CIRP Ann. Manuf. Technol.* **2006**, *55*, 745–768. [CrossRef]

24. Bian, R.; Ferraris, E.; He, N.; Reynaerts, D. Process investigation on meso-scale hard milling of zro2 by diamond coated tools. *Precis. Eng.* **2014**, *38*, 82–91. [CrossRef]

25. Bian, R.; He, N.; Ding, W.; Liu, S. A study on the tool wear of pcd micro end mills in ductile milling of zro2 ceramics. *Int. J. Adv. Manuf. Technol.* **2017**, *92*, 2197–2206. [CrossRef]

26. Arif, M.; Rahman, M.; San, W.Y. A study on the effect of tool-edge radius on critical machining characteristics in ultra-precision milling of tungsten carbide. *Int. J. Adv. Manuf. Technol.* **2013**, *67*, 1257–1265. [CrossRef]

27. Wu, X.; Li, L.; He, N.; Zhao, G.; Jiang, F.; Shen, J. Study on the tool wear and its effect of pcd tool in micro milling of tungsten carbide. *Int. J. Refract. Met. Hard Mater.* **2018**, *77*, 61–67. [CrossRef]

28. Bai, J.; Bai, Q.; Tong, Z. Multiscale analyses of surface failure mechanism of single-crystal silicon during micro-milling process. *Materials* **2017**, *10*, 1424. [CrossRef] [PubMed]

29. Zhong, L.; Li, L.; Wu, X.; He, N. Micro cutting of pure tungsten using self-developed polycrystalline diamond slotting tools. *Int. J. Adv. Manuf. Technol.* **2017**, *89*, 2435–2445. [CrossRef]

30. Matsumura, T.; Ono, T. Cutting process of glass with inclined ball end mill. *J. Mater. Process. Technol.* **2008**, *200*, 356–363. [CrossRef]

31. Cheng, X.; Nakamoto, K.; Sugai, M.; Matsumoto, S.; Wang, Z.G.; Yamazaki, K. Development of ultra-precision machining system with unique wire edm tool fabrication system for micro/nano-machining. *CIRP Ann. Manuf. Technol.* **2008**, *57*, 415–420. [CrossRef]

32. Nakamoto, K.; Katahira, K.; Ohmori, H.; Yamazaki, K.; Aoyama, T. A study on the quality of micro-machined surfaces on tungsten carbide generated by pcd micro end-milling. *CIRP Ann. Manuf. Technol.* **2012**, *61*, 567–570. [CrossRef]

33. Zhan, Z.; He, N.; Li, L.; Shrestha, R.; Liu, J.; Wang, S. Precision milling of tungsten carbide with micro pcd milling tool. *Int. J. Adv. Manuf. Technol.* **2014**, *77*, 2095–2103. [CrossRef]

34. Bian, R.; Ferraris, E.; Ynag, Y.; Qian, J. Experimental investigation on ductile mode micro-milling of zro2 ceramics with diamond-coated end mills. *Micromachines* **2018**, *9*, 127. [CrossRef] [PubMed]

35. Liu, K.; Li, X.; Rahman, M.; Neo, K.; Liu, X. A study of the effect of tool cutting edge radius on ductile cutting of silicon wafers. *Int. J. Adv. Manuf. Technol.* **2007**, *32*, 631–637. [CrossRef]

36. Liu, K.; Li, X.; Liang, S. The mechanism of ductile chip formation in cutting of brittle materials. *Int. J. Adv. Manuf. Technol.* **2007**, *33*, 875–884. [CrossRef]

37. Rabiey, M.; Jochum, N.; Kuster, F. High performance grinding of zirconium oxide (zro2) using hybrid bond diamond tools. *CIRP Ann. Manuf. Technol.* **2013**, *62*, 343–346. [CrossRef]

 *materials*

Article

One-Shot Drilling Analysis of Stack CFRP/UNS A92024 Bonding by Adhesive

Fermin Bañon, Alejandro Sambruno, Sergio Fernandez-Vidal and Severo Raul Fernandez-Vidal *

Faculty of Engineering, Mechanical Engineering and Industrial Design Department, University of Cadiz,
Av. Universidad de Cadiz 10, E-11519 Puerto Real, Cadiz, Spain; fermin.banon@uca.es (F.B.);
alejandro.sambruno@uca.es (A.S.); sergio.fernandezvidal@mail.uca.es (S.F.-V.)
* Correspondence: raul.fernandez@uca.es; Tel.: +34-956-483-460

Received: 5 December 2018; Accepted: 31 December 2018; Published: 6 January 2019

Abstract: The use of adhesive layers can improve the properties and reduce the defects produced in the interfaces. This provides adherence to the structure, adapting the joining surfaces and avoiding spaces between the layers. However, the presence of the adhesive can potentiate the defects caused during drilling. In turn, a loss of adhesive in the interface can occur during machining affecting the final structure. This work has followed a conventional OSD strategy in CFRP and UNS A92024 aluminium sheet stacking with adhesive. A series of dry drilling tests have been developed with different cutting conditions and new noncoated WC-Co helical cutting tools. Analysis of Variance (ANOVA) statistical analyses and surface response models have been applied to determine the mechanical behaviour in the holes. For this purpose, the dimensional deviation, surface quality, and adhesive loss in the interface in relation to the number of holes have been considered. A combination of cutting parameters that minimizes the evaluated defects has been found. Diametric deviations and surface qualities below 2% and 3.5 μm have been measured in the materials that make up the stack with cutting speeds higher than 140 m/min and feed rates between 200 and 250 mm/min. However, the greatest adhesive losses occur at high cutting speeds.

Keywords: adhesive; machining; modelling; dry; CFRP/UNS A92024

---

1. Introduction

Today's industry is looking for a production system that approaches optimum sustainability. Industrial sectors are looking for processes that present a balanced performance in energy, economic, social, functional, and environmental aspects [1,2].

The aeronautical sector is particularly noteworthy, being a benchmark in research, development, and innovation. One of the first challenges it faces in this fourth revolution is the automation of its processes. This is especially significant in operations such relevant as the assembly ones are to this industry, which involve extensive use of manual work. For this reason, the drilling operation is considered a key process due to the high number of holes made in an aircraft [3,4] for assembly through the riveting process.

These tasks must maintain the quality of the drilled holes avoiding the later elements separation. To achieve this goal, the drilling process is opting for drilling strategies known as OSD (One Shoot Drilling) [5–7]. The aim is to make a hole in a single step, regardless of the number of plates and the type of material. All this must satisfy the quality requirements imposed by the sector and avoid the use of lubricants.

Among the varying combinations of different materials that have to be joined together to form the aircraft, one of the most common is the union of composite material such as CFRP (carbon fiber reinforced polymer) and sheets of light metals such as aluminium alloys. This structure combines the strength of the fibres and the formability of the metal alloy [8–10].

Although mechanical joints are the most commonly used in the aeronautical sector, adhesive joints have also been widely used as a complement to riveting [11]. The existence of a layer that offers continuity between the materials that constitute the stack can offer a series of advantages. These benefits can be the elimination of defects in the interfaces between them, weight reduction, increased fatigue life of the joints, and a wide adaptability to materials [12–14]. However, the effect of cutting parameters on the adhesive has not been evaluated in previous studies. The presence of a third material in the interface of the hybrid structure may increase the defects generated during machining.

This paper proposes a study of the cutting parameters in the drilling of hybrid structures composed of CFRP and alloy UNS A92024, joined by an adhesive. The current objective is to drill hybrid structures in a single step. For this reason, the study of the influence of the cutting parameters on the final quality is essential and its influence on the final quality of the adhesive [15]. Defects such as an increase in the roughness of the composite material due to metal chip evacuation [16] or diameter variation may appear [17]. In addition, the inclusion of a third material as an adhesive can generate new defects in the interface of the materials producing a nonhomogeneous hole that meets the requirements established by the aeronautical sector. For the treatment of the results, Analysis of Variance (ANOVA) analysis and response surface models have been applied, which have been widely used in various studies on this topic [18–22]. Finally, a proposal is presented for the evaluation of the defectology caused in the adhesive that joins both materials.

2. Materials and Methods

For the development of this research, a WC-Co helical model drill without coating has been used. This was selected considering the materials that make up the stacks to be drilled, their thicknesses, the required qualities, and the cutting conditions. The dimensions and main characteristics are shown in Table 1. It has a double angle point, the section closest to the center corresponds to the highest point angle (140°) and the projection of the outermost edges provides a narrower angle (118°).

Table 1. Drilling bit geometry.

DC (mm)	LU (mm)	DCON (mm)	SIG (°)	PL (mm)	OAL (mm)	LF (mm)	LCF (mm)	Helix Angle (°)	Material
7.92	25.00	8.00	118.00 140.00	1.20	80.00	78.80	30.00	29.82	Carbidre Substrate Uncoated

The material selected consisted of a hybrid structure presented in 210 mm × 210 mm sheets and composed of two different materials. It is formed by a 2 mm thick CFRP sheet and an 8 mm thick UNS A92024 alloy sheet. Specifications for both materials are shown in Tables 2 and 3.

A two-component adhesive specific for structural elements was used to obtain the hybrid structure. It was applied by means of a hydraulic press in order to apply a constant pressure and guarantee a uniform thickness of 1 mm. The properties of the adhesive used are shown in Table 4.

The adhesive was cured at room temperature in an air conditioned room. The polymerization temperature was 23 °C. After polymerization, an ultrasonic inspection was carried out to verify the correct bond between the materials. In Figure 1, the quality of the joint has been shown. It can be seen how the majority of the surface presents a continuous union. However, there are small defects in the edges of the stack, as well as air bubbles generated during the joining process.

Table 2. Properties and features of CFRP.

Orientation (°)	Reinforcement	Matrix	Tensile Strength (MPa)	ILSS (MPa)	Tg (°C)
0/90	DowAksa A42 carbon fiber 49 Vol %	Epoxy DOW Voraforce 5300 51 Vol %	871	52	121

Table 3. Properties and features of UNS A92024.

Composition	Density (g/cc)	Ultimate Tensile Strength (Mpa)	Yield Tensile Strength (Mpa)	Elongation at Break (%)	Modulus of Elasticity (GPa)
Al 90.7–94.7%, Cu 3.8–4.9%, Mg 1.2–1.8%, Mn 0.30–0.90%, Zn 0.25%, Ti 0.15%, other 0.15%	2.78	425	310	$\geq$10	73.1

Table 4. Characteristics of SAF30-LOT adhesive.

Polymerization	Operating Temperature (°C)	Tg (°C)	Elongation at Break (%)	Fixture Time (min)	Full Cure (h)	Lap Shear Strength (MPa)	Tensile Strength at Break (MPa)	Tensile Modulus (MPa)
Room Temp.	−40/150	86	50	120	24	16	8–10	180

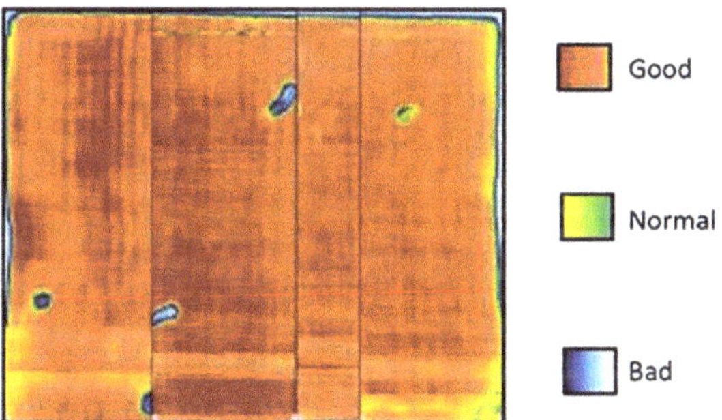

Figure 1. Ultrasound color map acquired on the hybrid stack interface for evaluating the adhesive application quality.

The tests were carried out dry on a Kondia Five 400 5-axis machining centre (Elgoibar, Guipuzcoa, Spain), controlled by a Heidenhain iTNC530 control system (Traunreut, Bavaria, Germany). Three cutting speeds and forward speeds were combined to obtain a combination of nine tests (Figure 2). The set values for the cutting parameters have been defined on the basis of other studies and real application cases, and are indicated in Table 5.

Temperature has been measured using pyrometric techniques. Macro- and microgeometric defects were evaluated after machining. Diameters were evaluated using an inside micrometer of three contacts at three different heights per material (Figure 3a). On the other hand, the arithmetic mean roughness was evaluated by means of a rugosimeter using a cutoff of 0.8 mm to establish a comparison between the results obtained. Surface quality was evaluated in 3 generatrices per drill (Figure 3b).

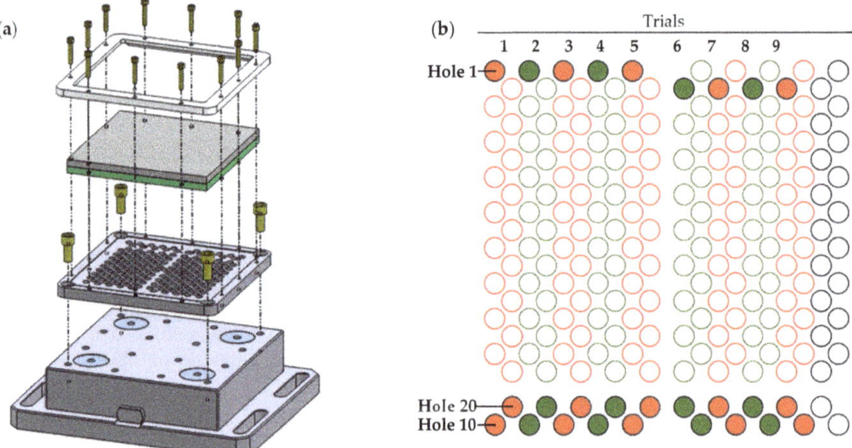

Figure 2. (**a**) Set up schematic representation. (**b**) Drill distribution for each trial.

Table 5. Combination of cutting parameters used for each test.

Trial	S (m/min)	f (mm/min)	Holes Machined [1]	Lubrication
1	85	200	20	Dry
2	85	250	20	Dry
3	85	300	20	Dry
4	115	200	20	Dry
5	115	250	20	Dry
6	115	300	20	Dry
7	145	200	20	Dry
8	145	250	20	Dry
9	145	300	20	Dry

[1] Holes 1, 10, and 20 have been analyzed.

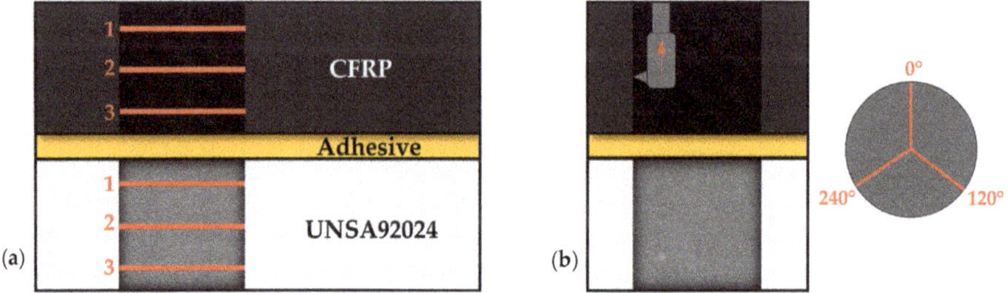

Figure 3. Evaluation methodology for (**a**) diameter and (**b**) quality surface.

Finally, through the obtained roughness profiles, the loss of the adhesive in the interface of the materials was evaluated by means of image processing software, Figure 4. For each profile obtained the maximum depth of adhesive removed has been evaluated.

In order to assess the influence of input parameters on the results obtained a statistical analysis has been carried out. An ANOVA analysis through a Response Surface Methodology (RSM) has been implemented. With this technique an empirical model will be obtained that establishes a multiple linear regression.

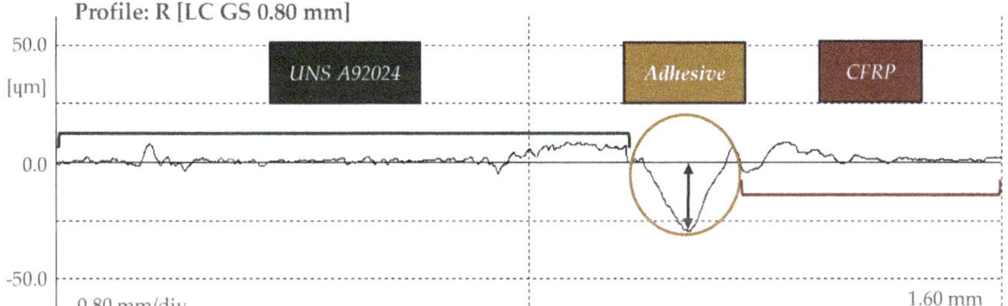

Figure 4. Evaluation of adhesive loss in the stack interface.

3. Results and Discussion

The results obtained in the evaluation of the macro and microgeometric defects are shown in Table 6. The mean values and their respective deviations are shown prior to discussion.

Table 6. Result values of each experimental parameters.

Trial	Hole	Ø [mm (±μm)]		Ra [μm (±μm)]		Depth [μm (±μm)]
		CFRP	Al	CFRP	Al	Adhesive
1	1	8.031 (±0.816)	8.042 (±0.408)	1.38 (±0.033)	1.62 (±0.065)	42.584 (±0.781)
	10	7.998 (±1.225)	8.050 (±0.816)	7.32 (±0.033)	6.23 (±0.041)	27.136 (±0.887)
	20	7.998 (±0.408)	8.060 (±0.002)	7.47 (±0.024)	12.22 (±0.139)	44.910 (±0.678)
2	1	8.026 (±0.816)	8.022 (±1.633)	1.18 (±0.065)	2.71 (±0.163)	24.730 (±0.906)
	10	8.036 (±2.041)	8.052 (±1.225)	4.43 (±0.065)	7.35 (±0.163)	33.924 (±0.746)
	20	8.057 (±0.408)	8.031 (±0.408)	2.90 (±0.098)	4.53 (±0.122)	23.937 (±0.737)
3	1	8.052 (±0.408)	8.062 (±1.225)	2.92 (±0.049)	3.20 (±0.114)	21.322 (±0.431)
	10	8.079 (±0.816)	8.066 (±1.225)	4.82 (±0.057)	5.40 (±0.220)	29.644 (±0.707)
	20	7.997 (±1.225)	8.052 (±1.225)	3.72 (±0.147)	4.97 (±0.147)	10.351 (±0.687)
4	1	8.077 (±1.225)	8.055 (±1.225)	1.20 (±0.122)	3.58 (±0.147)	53.363 (±0.296)
	10	8.091 (±0.408)	8.071 (±0.816)	5.53 (±0.008)	3.57 (±0.204)	35.714 (±0.485)
	20	8.062 (±1.225)	8.064 (±1.633)	4.06 (±0.147)	3.68 (±0.131)	47.633 (±0.892)
5	1	8.095 (±2.449)	8.082 (±1.633)	3.23 (±0.016)	4.82 (±0.057)	54.542 (±0.532)
	10	8.059 (±0.408)	8.079 (±0.816)	3.71 (±0.114)	6.06 (±0.122)	59.167 (±0.558)
	20	8.059 (±1.289)	8.081 (±0.816)	4.21 (±0.041)	7.98 (±0.090)	51.512 (±0.401)
6	1	8.100 (±1.633)	8.054 (±2.449)	1.57 (±0.024)	4.04 (±0.073)	39.640 (±0.719)
	10	8.091 (±0.816)	8.071 (±0.408)	3.96 (±0.220)	5.28 (±0.139)	66.190 (±0.441)
	20	8.064 (±0.816)	8.079 (±0.816)	5.25 (±0.106)	6.66 (±0.033)	30.546 (±0.552)
7	1	8.062 (±1.225)	8.037 (±0.408)	1.84 (±0.090)	2.28 (±0.049)	70.171 (±0.474)
	10	8.043 (±0.816)	8.041 (±0.816)	2.81 (±0.057)	2.38 (±0.163)	62.413 (±0.550)
	20	8.040 (±1.633)	8.023 (±0.408)	2.67 (±0.155)	2.58 (±0.090)	50.236 (±0.575)
8	1	7.997 (±1.225)	7.994 (±1.225)	1.50 (±0.016)	5.12 (±0.090)	78.464 (±0.495)
	10	7.992 (±1.633)	8.020 (±2.449)	4.61 (±0.024)	6.15 (±0.139)	82.222 (±0.603)
	20	7.998 (±0.408)	8.021 (±0.816)	3.61 (±0.180)	6.19 (±0.057)	72.189 (±0.612)
9	1	8.041 (±0.408)	8.006 (±1.633)	1.59 (±0.016)	6.09 (±0.171)	61.680 (±0.457)
	10	8.043 (±0.816)	8.035 (±1.225)	1.82 (±0.057)	5.76 (±0.163)	108.765 (±0.257)
	20	8.040 (±1.225)	8.030 (±0.408)	6.08 (±0.106)	6.94 (±0.057)	80.120 (±0.596)

3.1. Surface Quality

Cutting parameters are essential to obtain a correct surface quality. Figure 5 shows the values of Ra corresponding to the surface quality generated in the composite material.

It shows how the quality deteriorates as the tool increases the number of holes. This can be related to wear geometry, both abrasive and adhesive, generated in the tool. As the number of holes increases, the machining capacity decreases, resulting in a worse surface quality.

It seems that the combination of a cutting speed of 145 m/min with a feed speed of 200 mm/min produces the least variation in the geometry used. For this combination the variation of Ra in the 20 holes is the lowest.

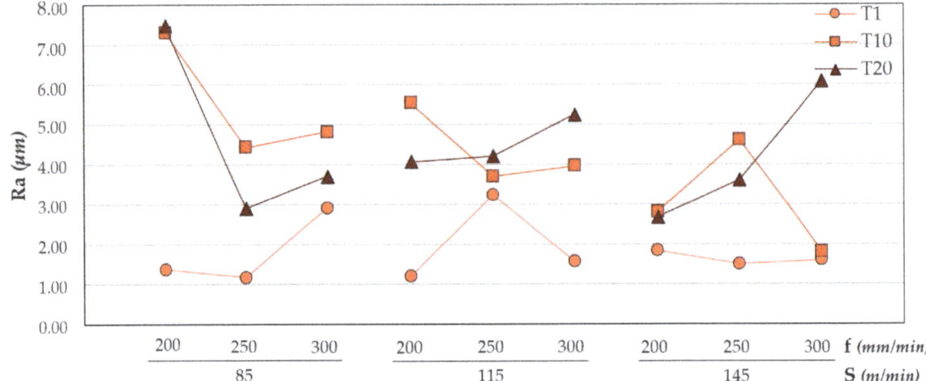

Figure 5. Ra values in CFRP for all trials carried out.

Results obtained for the aluminium alloy are shown in Figure 6. It is appreciated how the feed rate significantly influences the surface quality obtained. The use of a feed rate of 200 mm/min generates the lowest Ra values. This may be because a smaller chip is generated, as Uddin et al. [23] indicate in their results. Increases in feed rate result in poorer surface quality. These results are consistent with those obtained here and shown in [3,4,15].

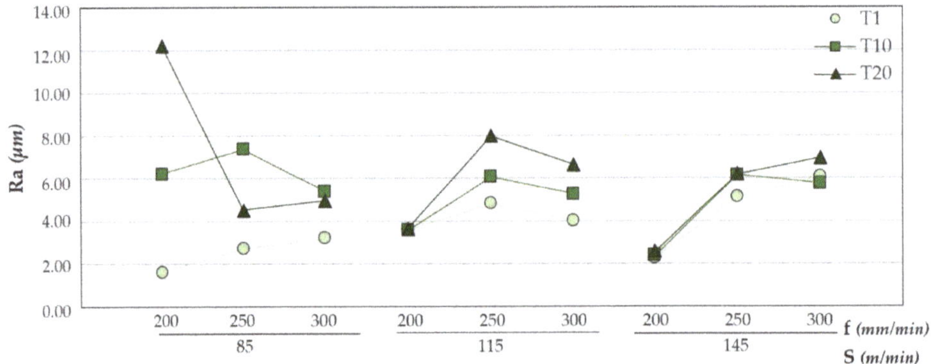

Figure 6. Ra values in UNS A92024 for all trials carried out.

There is a clear dispersion of data as the tool achieves a greater number of machined holes. However, bigger dispersion between the first and the last hole is obtained for a cutting speed of 145 m/min. As it is known, increasing the cutting speed raise the drilling temperature, which can cause greater adhesion of the adhesive on the tool (Figure 7) providing a more homogeneous surface quality.

The appreciated values of Ra in both materials are within the values established by the aeronautical sector. At the same time, the data obtained experimentally present a great dispersion in all the combinations used. This may be due to the chips generated during machining (Figure 8). The presence of the adhesive on the interface seems to exacerbate the mechanism of adhesion of the aluminium alloy on the tool. This produces long thick chips adhered to the tool that do not come off at the end of the drill. As a result, when the next hole is drilled, this chip machines the composite material and part of the aluminum alloy until it is detached. This results in an uneven surface quality and very high Ra values.

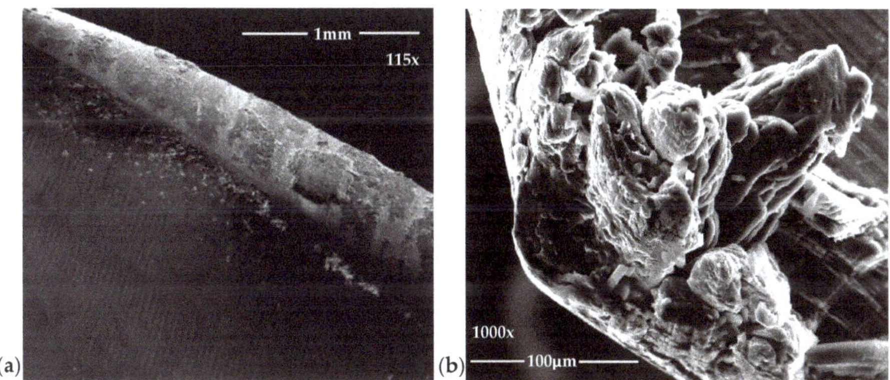

Figure 7. Adhesive wear generates in (**a**) the cutting edge and (**b**) the intersection of primary and secondary edges.

Figure 8. Adhesive and aluminium alloy chips adhered to the cutting geometry.

This could be avoided by using a different drilling strategy such as vibration-assisted drilling [24,25]. By using this technique, the tool fragments the chip, reducing its size and facilitating its evacuation. In this way, a better surface quality could be obtained.

ANOVA statistical analysis has been carried out in order to identify the most influential parameters in the drilling of both materials (Table 7).

The number of holes drilled is the most influential variable as it has a p-value of less than 0.05 in the CFRP drilling, indicating that it is a statistically significant variable.

This is in line with what it was previously said about the progressive wear generated in the tool. Nevertheless, there is no apparent influence of the cutting parameters on the quality obtained.

Similar trends are shown in the results of the drilling of UNS A92024. The number of holes drilled is the most influential variable in surface quality. There is a dispersion for the values of Ra obtained with the increase in the number of holes. On the other hand, there is no influence of the cutting parameters used in the process.

Table 7. Cutting parameters influence in Ra values by Analysis of Variance (ANOVA) analysis in both materials.

Source	DF	Adj SS	Adj MS	F-Value	P-Value
CFRP					
S (m/min)	1	5.1554	5.1554	2.65	0.122
f (mm/min)	1	0.3655	0.3655	0.19	0.670
Drills	1	30.8374	30.8374	15.88	0.001
Error	17	33.0218	1.9425		
Total	26	83.1892			
UNS A92024					
S (m/min)	1	1.416	1.4164	0.53	0.478
f (mm/min)	1	5.606	5.6057	2.08	0.167
Drills	1	27.602	27.6024	10.26	0.005
Error	17	45.741	2.6906		
Total	26	127.240			

From the results obtained, a contour diagram has been generated relating two variables. The contour diagrams are obtained from (1) and (2) with a R^2 of 60.31% and 64.05%, respectively. The average error rate for the obtained values is 34.93% for CFRP material. On the other hand, the mean error rate for the aluminium obtained values is 22.85%. This high variation is again due to the formation of long chips adhered to the tool. The chips hit the surface of both materials producing a very random surface quality. Due to this, the model obtained is not able to follow the trends obtained experimentally.

Since the number of holes is the most influential variable, both the cutting speed and the feed speed for the two materials have been confronted with it.

CFRP drilling diagrams are shown in Figure 9. For both graphs the wear is progressive, generating an increase in Ra values. However, the use of high cutting speeds, close to 145 m/min softens this trend. This may be because the amount of material removed per turn is greater, resulting in a smoother and more homogeneous surface.

$$Ra(CFRP) =$$
$$22.2 - 0.071{*}S - 0.126{*}f + 0.492{*}Holes - 0.000171{*}S^2 + 0.000161{*}f^2 - 0.01412{*}Holes^2 + \qquad (1)$$
$$0.000382{*}S{*}f - 0.00028{*}S{*}Holes - 0.000103{*}f{*}Holes,$$

$$Ra\ (UNS\ A92024) =$$
$$3.3 - 0.229{*}S + 0.077{*}f + 0.889{*}Holes + 0.000023{*}S^2 - 0.000341{*}f^2 - 0.00514{*}Holes^2 + \qquad (2)$$
$$0.001003{*}S{*}f - 0.00345{*}S{*}Holes - 0.001018{*}f{*}Holes,$$

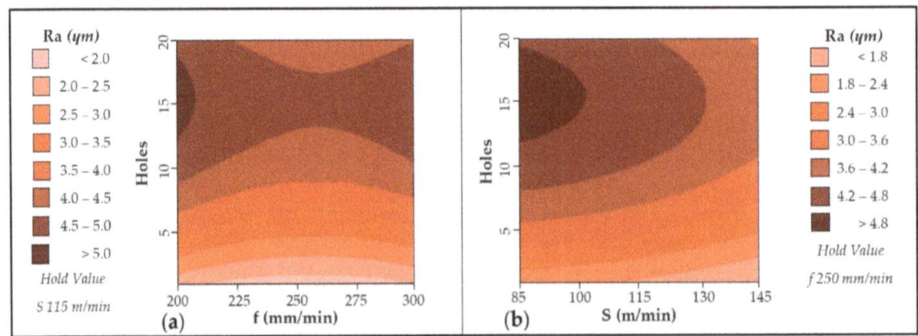

Figure 9. Contour diagram obtained by Ra values in CFRP: (**a**) Holes vs. feed rate and (**b**) holes vs. cutting speed.

In turn, by increasing the amount of material removed, the temperatures reached are lower, without damaging the matrix. This would help to obtain a smoother surface.

On the other hand, the forward speed for a fixed cutting speed does not seem to be a very influential factor. The variation of Ra is very similar in all combinations. It can be seen how an intermediate feed rate, close to 250 mm/min, generates the smoothest trend while a speed of 200 mm/min can generate the highest Ra values. Elevated temperatures are related to forward speed reductions. This can lead to deterioration in the matrix, resulting in a more irregular surface. At the same time, it can generate greater friction between tool and material, increasing abrasive wear (Figure 10) and generating greater variation in results.

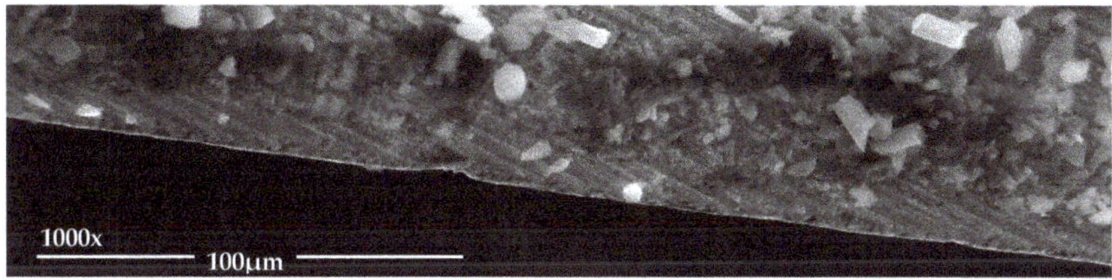

Figure 10. Abrasive wear generates in cutting edge.

Similar trends are shown in the diagrams corresponding to the drilling of the aluminium alloy (Figure 11). The selection of a high cutting speed, close to 145 m/min, together with a low feed speed, close to 200 mm/min, produces the best surface quality for both materials. This selection produces a smaller chip that is easier to evacuate and can reach lower temperature values. From a wear point of view, as the number of holes increases, there is a smoother tendency for this combination.

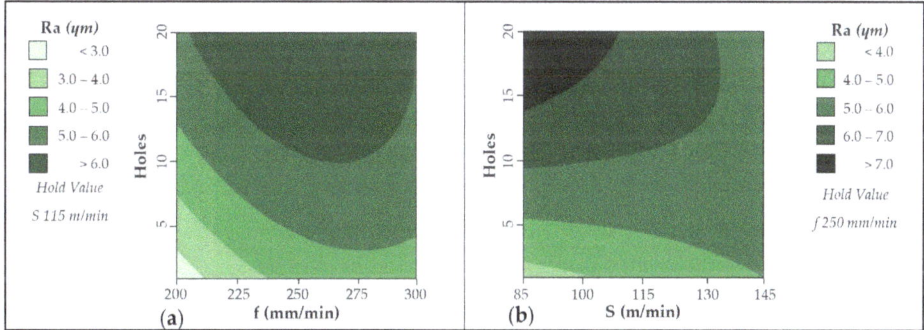

Figure 11. Contour diagram obtained by Ra values in UNS A92024: (**a**) Holes vs. feed rate and (**b**) holes vs. cutting speed.

Two contour diagrams have been overlaid to determine the cutting parameters that minimize the values of Ra (Figure 12). The cutting speed and feedrate variables have been confronted.

There is a small region generated by the combination of a cutting speed greater than 140 m/min and a feed rate close to 200 mm/min where the values of Ra are minimized. In this region it is possible to obtain values of Ra lower than 3 μm in the aluminium alloy and 3.5 μm for the composite material.

In addition, the selection of this combination is the one that offers a lower increase of Ra values by increasing the number of holes drilled as previously shown.

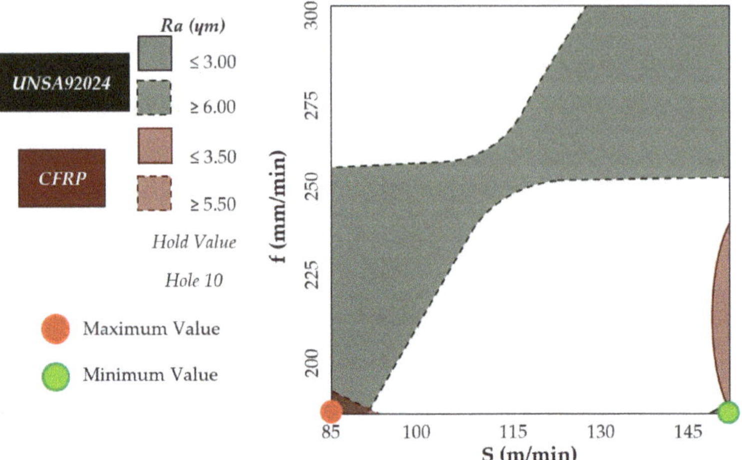

Figure 12. Overlap contour diagram by Ra values for CFRP & UNS A92024.

3.2. Diameters

Diameters evaluated in CFRP are shown in Figure 13. The number of drills performed does not seem to affect the diameters obtained in comparison with the values of Ra. The variation in results may be due to the adhesion and subsequent detachment of the aluminium alloy (Figure 14). This phenomenon modifies the geometry of the tool by varying its diameter.

However, an increase in the cutting speed results in closer values. This may be due to the increase of temperature in the cutting edges that would present a lower resistance in the matrix, facilitating its elimination. This is in line with the results obtained in the ANOVA analysis (Table 5). The cutting speed has the smallest p-value, being the most statistically influential variable.

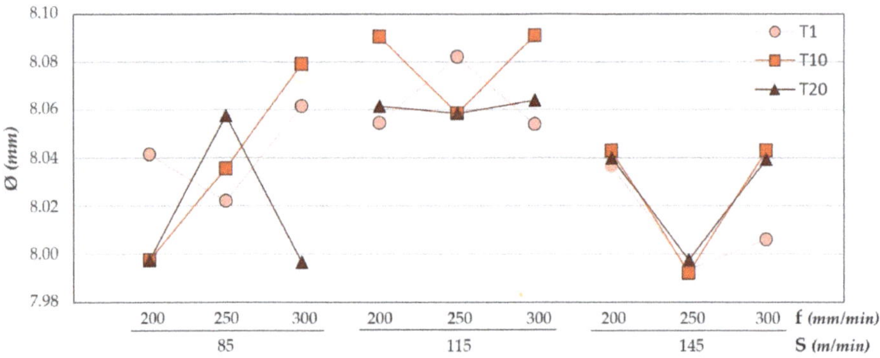

Figure 13. Diameter values in CFRP for all trials carried out.

Due to the very small thickness of the composite material, these diameter variations may also be related to the loss of adhesive in the interface. This loss could affect the measurement at a height close to the interface.

Diameters measured in aluminum alloy are shown in Figure 15. The results shown are very close to those obtained in CFRP. This may be due to the high adhesion of the material in the tool caused by process temperatures. A minimal dispersion is seen again when it comes to high cutting speeds. The increase of temperatures in the cutting zone can soften the material, facilitating its elimination and generating very uniform diameters.

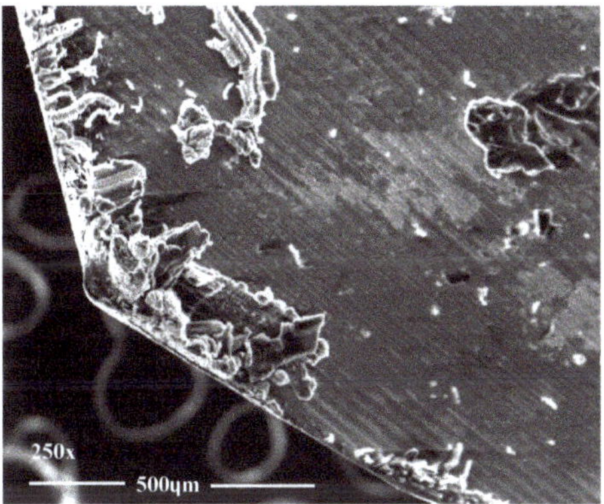

Figure 14. Adhesive wear generates in cutting edges.

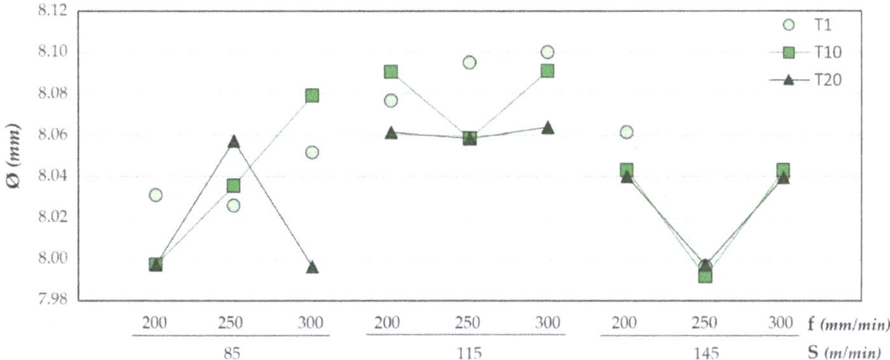

Figure 15. Diameter values in UNS A92024 for all trials carried out.

No variable appears to be statistically significant in the ANOVA analysis performed for drilling aluminium alloy (Table 8). The adhesion and subsequent detachment of the material can produce random results. This produces that no input variable presents a high influence on the diameters obtained. However, the number of holes drilled is the variable with the lowest *p*-value. This means that within the randomness described, the number of holes machined is the most influential factor in the aluminium diameters. Therefore, the diameters generated in the aluminum alloy are related to the wear mechanisms produced. Adhesive type wear is predominant.

The average diameters and their respective deviations for the nine tests performed are shown in the (Figure 16). The diameters obtained in both materials are very close to each other except for Test 1. This is reflected in the ratio obtained between the diameters measured in CFRP and those measured in the aluminium alloy. The ratios obtained are between 0.998 and 1.005. This indicates that, independently of the cutting parameters, the diameters obtained are homogeneous and constant in the drill made.

This is consistent with the results given by D'Orazio. The temperature peaks are higher in the drilling of the composite material due to the low heat transfer efficiency. This, together with the friction of the metal chip when it is evacuated, produces larger diameters very close to those obtained in the aluminum alloy.

The contour diagrams obtained are shown in Figure 17. The contour diagrams are obtained from the (3) and (4) with a R^2 of 80.97% and 66.69%, respectively. The average error rate for the obtained values is 0.104% for CFRP material. On the other hand, the mean error rate for the obtained aluminium values is 0.180%.

Table 8. Cutting parameters influence in diameter values by ANOVA analysis in both materials.

Source	DF	Adj SS	Adj MS	F-Value	P-Value
CFRP					
S (m/min)	01	0.002912	0.002912	17.56	0.001
f (mm/min)	01	0.000008	0.000008	0.05	0.834
Drills	01	0.000425	0.000425	2.57	0.128
Error	17	0.002818	0.000166		
Total	26	0.014807			
UNS A92024					
S (m/min)	01	0.000017	0.000017	0.03	0.863
f (mm/min)	01	0.000616	0.000616	1.11	0.307
Drills	01	0.001531	0.001531	2.76	0.115
Error	17	0.009441	0.000555		
Total	26	0.028340			

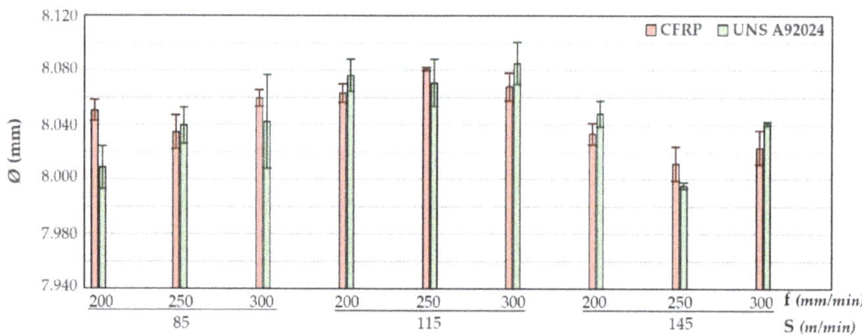

Figure 16. Diameter average values comparison for all cutting parameter combinations.

Contrary to what is stated in the literature, an increase in the number of holes drilled results in an increase in the diameters obtained in CFRP for a fixed value of cutting speed. The adhesion of the metal alloy together with the adhesive itself may be enhanced by the temperatures reached during machining. This, together with the adhesion of the chip itself, can lead to an increase in the diameters obtained.

$$\varnothing(\text{CFRP}) =$$
$$7.695 + 0.00927^*S - 0.00113^*f + 0.00112^*Holes - 0.000039^*S2 + 0.000003^*f2 - 0.000111^*Holes2 - \quad (3)$$
$$0.000003^*S^*f + 0.000006^*S^*Holes + 0.000004^*f^*Holes,$$

$$\varnothing(\text{UNS A92024}) =$$
$$7.524 + 0.01384^*S - 0.00206^*f - 0.00036^*Holes - 0.000053^*S2 + 0.000006^*f2 - 0.000037^*Holes2 - \quad (4)$$
$$0.000007^*S^*f + 0.000011^*S^*Holes - 0.000004^*f^*Holes,$$

On the other hand, it should be noted that the use of low and intermediate cutting speeds affects the diameters obtained to a greater extent. The use of a speed close to 145 m/min, on the contrary, generates a diameter with few variations.

The diameters measured in the aluminium alloy are reduced by increasing the number of holes drilled (Figure 18). For this material, it is observed that the intermediate cutting speed has a greater influence, achieving the highest values. Applying a low speed, 85 m/min, or high speed, 145 m/min, generates the smallest diameters of the stack.

The combination of cutting parameters that offer the most homogeneous diameter in both materials is shown in (Figure 19). Diameters very close to the nominal diameter are obtained by using cutting speeds higher than 140 m/min in both materials. However, the feedrate required to obtain these results must be close to 250 mm/min.

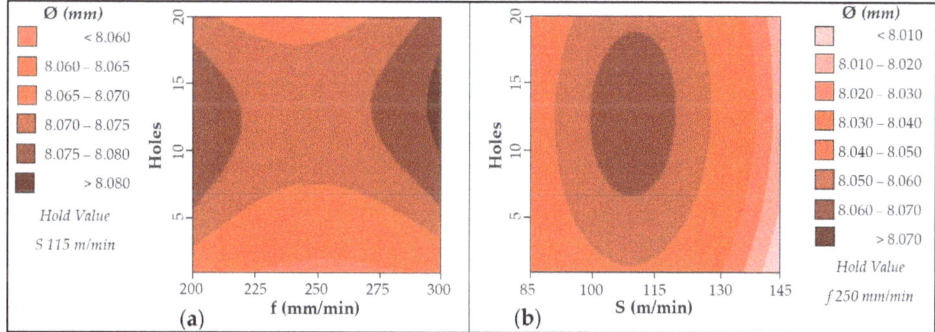

Figure 17. Contour diagram obtained by diameter values in CFRP: (**a**) Holes vs. feed rate and (**b**) holes vs. cutting speed.

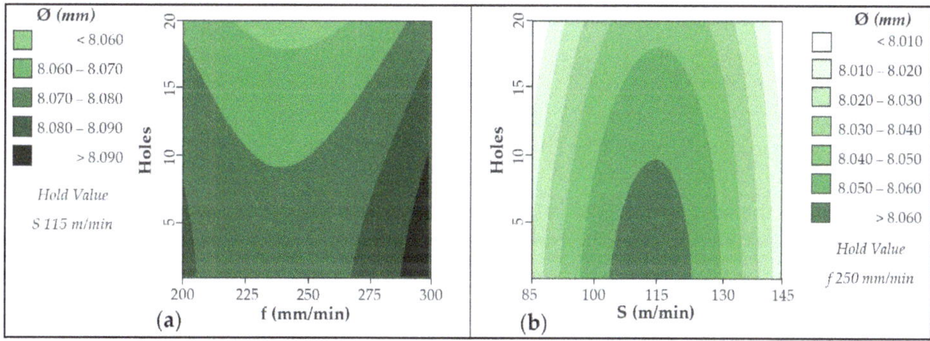

Figure 18. Contour diagram obtained by diameters values in UNS A92024: (**a**) Holes vs. feed rate and (**b**) holes vs. cutting speed.

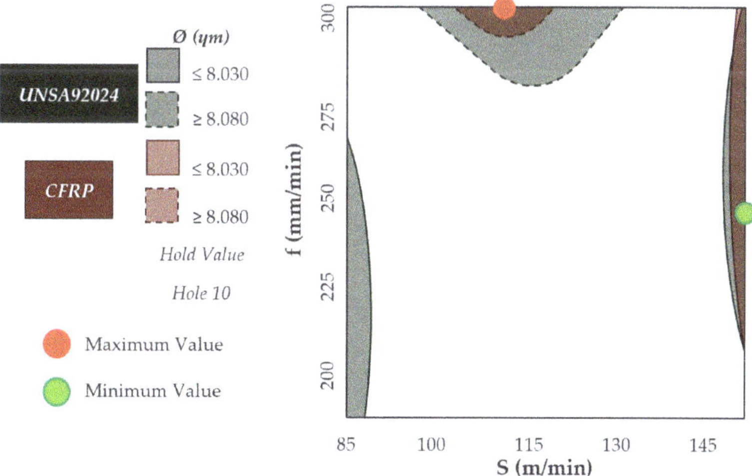

Figure 19. Overlap Contour diagram by diameter values for CFRP & UNS A92024.

3.3. Adhesive

An essential aspect is the final state of the adhesive after machining. The adhesive can be negatively affected and not perform its function. Adhesive loss in the interface can be influenced by the correct selection of cutting parameters. The maximum depth of adhesive lost after machining is shown in Figure 20.

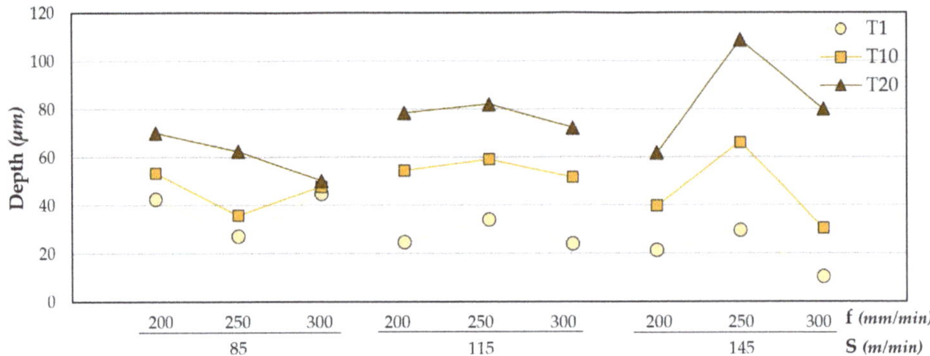

Figure 20. Adhesive loss values for all trials carried out.

This defect may be due to temperatures generated during machining (Figure 21). These may exceed the glass transition temperature of the adhesive itself. When this point is overreach, the adhesive softens and burns, facilitating its elimination by the cutting edges of the tool.

As with the results obtained in diameters and Ra, the number of holes drilled greatly affects the final quality of the adhesive: The amount of adhesive removed increases due to the progressive wear of the tool together with the adhesion of metal shavings.

The amount of adhesive removed is also related to the cutting speed. This defect increases as the cutting speed values increase for a fixed feed value. This is consistent with the results obtained by L. Sorrentino [26]. As the cutting speed increases, the friction between the cutting edges and the surface of the material becomes greater. This produces an increase in the thermal energy generated, increasing the process temperatures. In the results obtained by L. Sorrentino, temperature peaks close to 180 °C are reached when drilling CFRP.

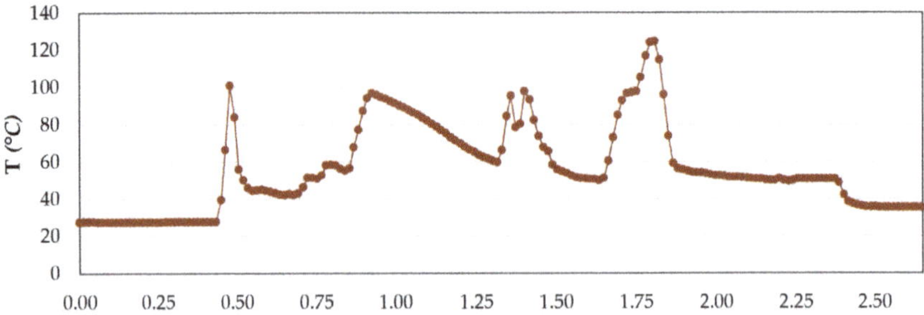

Figure 21. Temperature obtained for a single hole drilled.

These temperatures exceed the glass transition temperature of the adhesive used (86 °C). At this point the material is softened and facilitates its removal at the interface. The results obtained at a speed of 145 m/min are particularly noteworthy. The observed tendency is inverse to the diameters evaluated. This would make sense, as the diameters close to the interface would increase due to the loss of adhesive. The ANOVA analysis carried out reflects the above statement (Table 9). The most influential variables in adhesive loss are the number of holes drilled and their combination with the cutting speed used. The advance speed, on the other hand, does not seem to have an apparent influence.

Table 9. Cutting parameters influence in adhesive loss values by ANOVA analysis.

Source	DF	Adj SS	Adj MS	F-Value	P-Value
ADHESIVE					
Model	9	11168.5	1240.94	9.22	0.000
S (m/min)	1	16.6	16.6	0.12	0.730
f (mm/min	1	68.1	68.06	0.51	0.487
Drills	1	9235.4	9235.4	68.60	0.000
S (m/min) *Drills	1	1225.5	1225.47	9.10	0.008
Error	17	2288.6	134.62		
Total	26	13457.1			

The contour diagrams corresponding to the adhesive loss are shown in Figure 22. The contour diagrams are obtained from Equation (5) with a R^2 of 82.99%. The average error value obtained was 18.25%. Although the speed of advance is not a very influential factor, it is appreciated that values close to 300 mm/min generate the minimum amount of adhesive eliminated in the first holes.

$$\text{Depth (Adhesive)} =$$
$$-156 + 0.48{*}S + 1.512{*}f - 2.13{*}\text{Holes} - 0.00487{*}S^2 - 0.00339{*}f^2 + 0.0165{*}\text{Holes}^2 + \qquad (5)$$
$$0.00121{*}S{*}f + 0.0354{*}S{*}\text{Holes} + 0.00035{*}f{*}\text{Holes},$$

On the contrary, as opposed to the above for the values of diameters and Ra, a cutting speed close to 85 m/min should be selected. This combination of cutting parameters generates less adhesive loss in the interface as the number of holes increases.

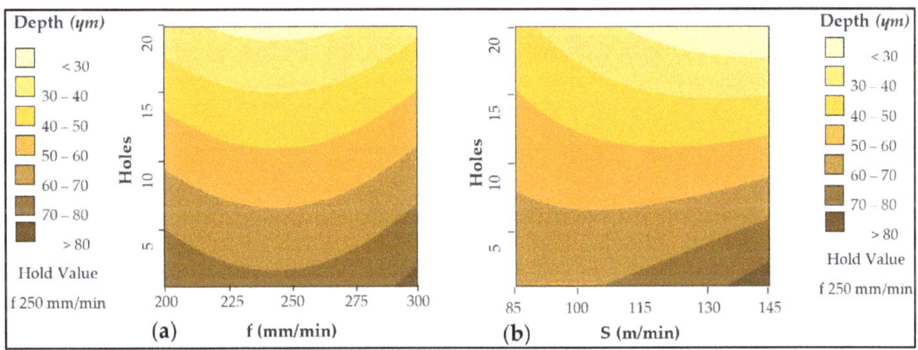

Figure 22. Contour diagram obtained by adhesive loss values: (**a**) Holes vs. feed rate and (**b**) holes vs. cutting speed.

4. Conclusions

In this work a methodology has been established for the analysis of dry drilling of adhesive bonded CRFP/UNS A92024 hybrid structures. The analysis of the results has been developed by diametric deviations, surface quality and adhesive loss through ANOVA techniques and the development of response surface models.

It has been determined that the surface quality is influenced by the cutting parameters. In addition, the loss of adhesive in the interface is related to the number of holes drilled.

The lowest values of Ra in both materials have been obtained by combining a cutting speed of 145 m/min with a feed speed of 200 mm/min. This can be produced by the smaller amount of material evacuated by the rotation of the tool, stabilizing the process temperatures.

The diametric deviations produced in both materials are very close. No variable has an influence on the diameters obtained by the ANOVA analysis. This is probably due to the randomness of the results obtained as consequence of secondary adhesion wear. This kind of wear affects the geometry of the tool with the adhesion and detachment of machined material, as a result of the stresses and temperatures generated during the process. The diametric ratios of both materials have always been near to 1. It has been seen that diameters very close to the nominal in both materials can be obtained by using cutting speeds higher than 140 m/min and feed speeds close to 250 mm/min.

Adhesive loss is directly related to increased cutting speed. This is due to the increase in temperature as consequence of the increased friction between the cutting edges and the machined materials. It has been determined that the cutting speed is the most influential factor in the process. Cutting speeds close to 85 m/min are recommended to reduce this defect. The feed rate does not have an excessive influence.

Author Contributions: F.B. and A.S. developed the drilling tests. S.F.-V and S.R.F.-V. developed the data treatment protocol. S.F.-V., F.B., A.S., and S.R.F.-V. analyzed the influence of the parameters involved. F.B. and A.S. collaborated in preparing figures and tables and F.B., A.S., S.F.-V., and S.R.F.-V. wrote the paper.

Funding: This research was funded by the Spanish Government via the Ministry of Economy, Industry and Competitiveness, the European Union (FEDER/FSE) and the Andalusian Government (PAIDI), project DIANNA.

Acknowledgments: The authors acknowledge the financial support for the work. A special mention for a passionate engineer with a tireless capacity for work, whose academic contributions have left a legacy and shown us the way forward due to his determination to continue progressing and developing not just engineering professionals, but good people overall. His great research contributions and his memory are a stimulus for us to try to be at his level, day by day, without disappointing him. Mariano Marcos-Barcena, in memoriam.

Conflicts of Interest: The authors declare no conflicts of interest.

References

1. Cirillo, P.; Marino, A.; Natale, C.; Di Marino, E.; Chiacchio, P.; De Maria, G. A low-cost and flexible solution for one-shot cooperative robotic drilling of aeronautic stack materials. *IFAC PapersOnLine* **2017**, *50*, 4602–4609. [CrossRef]
2. Gómez-Parra, A.; Álvarez-Alcón, M.; Salguero, J.; Batista, M.; Marcos, M. Analysis of the evolution of the built-up edge and built-up layer formation mechanisms in the dry turning of aeronautical aluminium alloys. *Wear* **2013**, *302*, 1209–1218. [CrossRef]
3. Wang, C.-Y.; Chen, Y.-H.; An, Q.-L.; Cai, X.-J.; Ming, W.-W.; Chen, M. Drilling temperature and hole quality in drilling of CFRP/aluminum stacks using diamond coated drill. *Int. J. Precis. Eng. Manuf.* **2015**, *16*, 1689–1697. [CrossRef]
4. Zitoune, R.; Krishnaraj, V.; Collombet, F. Study of drilling of composite material and aluminium stack. *Compos. Struct.* **2010**, *92*, 1246–1255. [CrossRef]
5. Nagaraja, R.; Rangaswamy, T. Drilling of CFRP and GFRP composite laminates using one shot solid carbide step drill K44. *AIP Conf. Proc.* **2018**, *020025*. [CrossRef]
6. Salguero, J.; Ponce, M.B.; Fernandez-Vidal, S.R.; Mayuet, P.; Rosales, E.I.; Marcos, M. Cutting Speed and Feedrate Based Analysis of Cutting Forces in the One Shot Drilling (OSD) of CFRC/Al Hybrid Stacks. In *Proceedings of the Volume 2B: Advanced Manufacturing*; ASME: Montreal, QC, Canada, 2014; Paper No. IMECE2014-38027; p. V02BT02A057. 7p, ISBN 978-0-7918-4644-5. [CrossRef]
7. Fernández-Vidal, S.R.; Mayuet, P.; Rivero, A.; Salguero, J.; Del Sola, I.; Marcos, M. Analysis of the effects of tool wear on dry helical milling of Ti6Al4V alloy. *Procedia Eng.* **2015**, *132*, 593–599. [CrossRef]
8. Lawcock, G.; Ye, L.; Mai, Y.; Sun, C. The effect of adhesive bonding between aluminum and composite prepreg on the mechanical properties of carbon-fiber- reinforced metal laminates. *Compos. Sci. Technol.* **2006**, *57*, 35–45. [CrossRef]

9. Cano, R.J.; Loos, A.C.; Jensen, B.J.; Britton, S.M.; Tuncol, G.; Long, K. Epoxy/glass and polyimide (LaRC PETI-8)/carbon fiber metal laminates made by the VARTM process. *SAMPE J.* **2011**, *47*, 50–58.

10. Huang, Z.; Sugiyama, S.; Yanagimoto, J. Adhesive-embossing hybrid joining process to fiber-reinforced thermosetting plastic and metallic thin sheets. *Procedia Eng.* **2014**, *81*, 2123–2128. [CrossRef]

11. Martinsen, K.; Hu, S.J.; Carlson, B.E. Joining of dissimilar materials. *CIRP Ann. Manuf. Technol.* **2015**, *64*, 679–699. [CrossRef]

12. Samaei, M.; Zehsaz, M.; Chakherlou, T.N. Experimental and numerical study of fatigue crack growth of aluminum alloy 2024-T3 single lap simple bolted and hybrid (adhesive/bolted) joints. *Eng. Fail. Anal.* **2016**, *59*, 253–268. [CrossRef]

13. Higgins, A. Adhesive bonding of aircraft structures. *Int. J. Adhes. Adhes.* **2000**, *20*, 367–376. [CrossRef]

14. Zhang, K.; Yang, Z.; Li, Y. A method for predicting the curing residual stress for CFRP/Al adhesive single-lap joints. *Int. J. Adhes. Adhes.* **2013**, *46*, 7–13. [CrossRef]

15. Kuo, C.L.; Soo, S.L.; Aspinwall, D.K.; Carr, C.; Bradley, S.; M'Saoubi, R.; Leahy, W. Development of single step drilling technology for multilayer metallic-composite stacks using uncoated and PVD coated carbide tools. *J. Manuf. Process.* **2018**, *31*, 286–300. [CrossRef]

16. Ashrafi, S.A.; Sharif, S.; Farid, A.A.; Yahya, M. Performance evaluation of carbide tools in drilling CFRP-Al stacks. *J. Compos. Mater.* **2013**, *48*, 2071–2084. [CrossRef]

17. D'Orazio, A.; El Mehtedi, M.; Forcellese, A.; Nardinocchi, A.; Simoncini, M. Tool wear and hole quality in drilling of CFRP/AA7075 stacks with DLC and nanocomposite TiAlN coated tools. *J. Manuf. Process.* **2017**, *30*, 582–592. [CrossRef]

18. Krishnaraj, V.; Prabukarthi, A.; Ramanathan, A.; Elanghovan, N.; Kumar, M.S.; Zitoune, R.; Davim, J.P. Optimization of machining parameters at high speed drilling of carbon fiber reinforced plastic (CFRP) laminates. *Compos. Part. B Eng.* **2012**, *43*, 1791–1799. [CrossRef]

19. Giasin, K.; Ayvar-Soberanis, S.; Hodzic, A. Evaluation of cryogenic cooling and minimum quantity lubrication effects on machining GLARE laminates using design of experiments. *J. Clean. Prod.* **2016**, *135*, 533–548. [CrossRef]

20. Kuo, C.; Li, Z.; Wang, C. Multi-objective optimisation in vibration-assisted drilling of CFRP/Al stacks. *Compos. Struct.* **2017**, *173*, 196–209. [CrossRef]

21. Wang, H.; Sun, J.; Li, J.; Lu, L.; Li, N. Evaluation of cutting force and cutting temperature in milling carbon fiber-reinforced polymer composites. *Int. J. Adv. Manuf. Technol.* **2016**, *82*, 1517–1525. [CrossRef]

22. Kyratsis, P.; Markopoulos, A.; Efkolidis, N.; Maliagkas, V.; Kakoulis, K. Prediction of thrust force and cutting torque in drilling based on the response surface methodology. *Machines* **2018**, *6*, 24. [CrossRef]

23. Uddin, M.; Basak, A.; Pramanik, A.; Singh, S. Evaluating hole quality in drilling of Al 6061 alloys. *Materials* **2018**, *11*, 2443. [CrossRef] [PubMed]

24. Bleicher, F.; Wiesinger, G.; Kumpf, C.; Finkeldei, D.; Baumann, C.; Lechner, C. Vibration assisted drilling of CFRP/metal stacks at low frequencies and high amplitudes. *Prod. Eng.* **2018**, *12*, 289–296. [CrossRef]

25. Jallageas, J.; K'Nevez, J.Y.; Chérif, M.; Cahuc, O. Modeling and optimization of vibration-assisted drilling on positive feed drilling unit. *Int. J. Adv. Manuf. Technol.* **2013**, *67*, 1205–1216. [CrossRef]

26. Sorrentino, L.; Turchetta, S.; Bellini, C. In process monitoring of cutting temperature during the drilling of FRP laminate. *Compos. Struct.* **2017**, *168*, 549–561. [CrossRef]

 *materials*

Article

A Study on the Optimal Machining Parameters of the Induction Assisted Milling with Inconel 718

Eun Jung Kim and Choon Man Lee *

Department of Mechanical Design and Manufacturing Engineering, School of Mechanical Engineering,
Changwon National University, 20, Changwondaehak-ro, Uichang-gu, Changwon-si,
Gyeongsangnam-do 51140, Korea; angel9940@hanmail.net
* Correspondence: cmlee@changwon.ac.kr; Tel.: +82-55-213-3622

Received: 9 December 2018; Accepted: 7 January 2019; Published: 11 January 2019

Abstract: This paper focuses on an analysis of tool wear and optimum machining parameter in the induction assisted milling of Inconel 718 using high heat coated carbide and uncoated carbide tools. Thermally assisted machining is an effective machining method for difficult-to-cut materials such as nickel-based superalloy, titanium alloy, etc. Thermally assisted machining is a method of softening the workpiece by preheating using a heat source, such as a laser, plasma or induction heating. Induction assisted milling is a type of thermally assisted machining; induction preheating uses eddy-currents and magnetic force. Induction assisted milling has the advantages of being eco-friendly and economical. Additionally, the preheating temperature can be easily controlled. In this study, the Taguchi method is used to obtain the major parameters for the analysis of cutting force, surface roughness and tool wear of coated and uncoated tools under various machining conditions. Before machining experiments, a finite element analysis is performed to select the effective depth of the cut. The S/N ratio and ANOVA of the cutting force, surface roughness and tool wear are analyzed, and the response optimization method is used to suggest the optimal machining parameters.

Keywords: induction assisted milling; tool wear; taguchi method; cutting tool

1. Introduction

Recently, the production of difficult-to-cut materials has increased in various engineering industries. There are difficult-to-cut materials that have properties such as a high corrosion resistance, thermal resistance, good specific strength, etc. [1,2]. However, these materials are difficult to machine due to these very higher properties. Nickel-based super-alloys are metallic materials with a high temperature strength, toughness, and resistance to deterioration in corrosive or oxidizing environments [3–5].

Thermally assisted machining is a method of machining difficult-to-cut materials such as titanium alloys, nickel-based alloys and ceramics materials. According to the heat source type, machining can be classified into categories of induction assisted milling (IAM), laser-assisted milling (LAM), electrochemical assisted machining (ECAM) and plasma assisted milling (PAM). Thermally assisted machining uses heat sources to heat the workpiece and soften it. These machining methods have some advantages, such as a decrease in the cutting force and an improvement of surface roughness and energy saving compared to the conventional machining method [6–10].

However, tool wear is an issue that causes short tool life in the area of difficult-to-cut materials. Especially when machining nickel-based alloys, various types of tool wear are generated, such as mechanical wear, adhesive wear, abrasive wear, diffusion wear and oxidation wear. In conclusion, the tool life is shortened [11–13]. In order to reduce these problems, Pimenov examined the effect of the rate flank wear teeth face mills in the processing by using the regression analysis model [14]. Zhang et al. carried out a tool life and cutting force analysis with Inconel 718 under dry and minimum quantity cooling lubrication cutting condition [15]. Pimenov et al. carried out an automatic prediction study of required surface roughness by monitoring the wear in the face milling by using artificial intelligence methods [16].

However, research on optimal parameters has been lacking in the area of induction assisted milling. The selection of optimal machining parameters is an essential parameter for any process of difficult-to-cut machining. The selection of optimal parameters has an influence on the surface quality, the required cutting force and the cutting cost. Chamarthi et al. performed an investigation analysis of plasma arc cutting parameters such as voltage, cutting speed and plasma gas flow on a 12 mm hardox-400 plate workpiece [17]. Calleja et al. made improvements to the strategies and parameters for the multi-axis laser cladding operation [18]. Kim et al. predicted the cutting force and preheating-temperature for the laser-assisted milling of Inconel 718 and AISI 1045 steel and proposed effective machining conditions for the laser-assisted milling of Inconel 718 and AISI 1045 steels [19]. Venkatesan and Ramanujam took a statistical approach by using the multi-objective optimization method for the optimization of the influencing parameters in laser-assisted machining with the Inconel alloy [20]. Abbas et al. carried out a study on the minimization of turning time for high-strength steel with a given surface roughness using the Edgeworth–Pareto optimization method [21]. In this study, the response optimization method was used to determine the optimal machining parameters. This method is useful when evaluating the effect of multiple parameters on the response.

Due to its excellent abrasion and corrosion resistance, TiN coating is widely used to increase the cutting tool life. Al_2O_3 coating as a chemically inactive material has low thermal conductivity and excellent wear resistance. Additionally, it can function as a thermal diffusion barrier to increase the plastic deformation resistance. TiAlN or AlTiN (for aluminum contents higher than 50%) coating is aimed at the high-efficiency machining of difficult-to-cut materials, rather than at attempting to induce an increase in the tool life. In order to effectively perform dry machining, a coating tool is required to have a high oxidation resistance that can withstand the high-temperatures generated during machining. Additionally, a coating tool material is required to have excellent mechanical properties such as abrasion resistance and impact resistance [22]. Aslantas et al. carried out the analysis of the wear mechanism and tool life in turning hardened alloy steel by coated and uncoated Al_2O_3/TiCN mixed ceramic tools [23]. Liu et al. carried out a tool wear analysis in the high-speed machining of titanium alloys under dry and minimum quantity lubrication conditions by nc-AlTiN/a-Si_3N_4 and nc-AlCrN/a-Si_3N_4 coated tools. The results of the experiments were to propose dry and minimum quantity lubrication conditions according to the main wear typed for the two coated carbide tools [24].

In this study, an experiment was performed to determine the optimal machining parameters and to analyze the thermal effect and machinability of Inconel 718 under various milling conditions in the IAM. To investigate the thermal effects of tools and machinability in the IAM, the experiments were performed by using an AlTiN+HH (high heat special coating) coated carbide tool and of an uncoated carbide tool. The tool can be heated by an induction heat source, and the adhesive wear of the tool can be accelerated by the coating material. Therefore, the uncoated tool and the coated tool were compared to analyze the mechanism of tool wear by high-temperature induction assisted milling. Finite element analysis (FEA) was performed to find an effective depth of cut. The influence of the machining parameters on cutting force, surface roughness and tool wear was analyzed using the Taguchi method. The cutting force was measured by dynamometer and the surface roughness was measured by the shape measuring device. The tool wear was evaluated on the flank wear by microscope in accordance with ISO 8688-2: tool life testing in milling [25]. For the determination of the optimal machining parameters, the response optimization method was performed. An efficient machining condition to increase machinability was proposed and discussed.

2. Machining Method

2.1. Induction-Assisted Milling

Induction assisted milling is a type of thermally assisted machining that uses a heat source for induction. Thermally assisted machining is an effective machining method to improve machinability. It is a method by softening the workpiece by preheating using a heat source. The advantage of induction assisted heating is its clean, rapid heating, which makes it possible to accomplish uniform heating depending on the coil size; it also has an excellent thermal conduction loss and it is cheaper than other heating methods. The preheating temperature can be controlled by the number of coils, coil size and the frequency. Generally, an eddy-current loss has more influence than hysteresis in induction heating. When a material is heated, the contribution of hysteresis is usually very small so it can be ignored. In addition, induction heating is the ability to heat only a small portion of a workpiece and it is fast and clean, therefore, induction heating is a suitable heat source. In this study, one-way slot milling was used by using an 8 mm width of cut (tool diameter) under the various machining conditions. Figure 1 shows a schematic diagram of the induction assisted milling process.

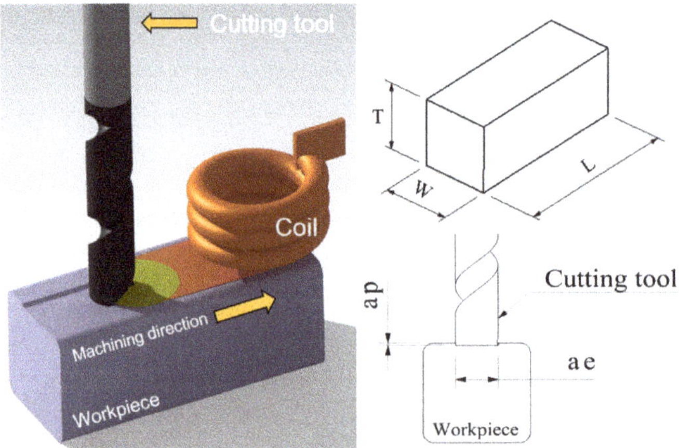

Figure 1. The schematic diagram of induction assisted milling.

2.2. Cutting Conditions

The experiments use two types of cutting tools: (WIDIN Co., Ltd., Changwon, Korea, type ZE702080 and E302080) an AlTiN-HH (high heat special coating) coated carbide tool and an uncoated carbide tool. The cutting tool has the following specifications: diameter 8 mm, length 60 mm, and 2 flutes. The distance between the tool and the coil is 5 mm. Because the distance between the tool and the induction heat source is very close, the tool can be heated by an induction heat source. The distance between the coil and the workpiece is 3 mm. Because this distance between the coil and the workpiece is long, the preheating effect is low. Figure 2 shows the distance between the tool and the coil and the workpiece. This research was carried out on a 5-axis machining center (Hyundai-WIA Inc., Changwon, Korea, Hi-560M) and a 6 kW high-frequency induction heater module (Tae-yang Induction Heater Co., Ltd., Daegu, Korea, TH-6000). The pyrometer was installed under the laser preheating module fixed to a 5-axis machining center spindle. The temperature measurement range of the IR pyrometer (Process Sensors Co., Ltd., Milford, MA, USA, PSC-CS-Laser-2MH) was 385–1600 °C, the focal length was 1100 mm and the calibration of the pyrometer was carried out by identifying the target position with dual laser aiming. The measurement of the cutting force was done using a dynamometer and an amplifier (KISTLER Co., Ltd., Winterthur, Switzerland, 9257B, 5019). The surface roughness and tool wear were measured using a shape measurement device (OPTACOM GmbH &

Co., Grettstadt, Germany, VC-10) and microscope (SOMETECH Inc., Seoul, Korea, IMS-345). Figure 3 and Table 1 show the experimental set-up and main cutting parameters.

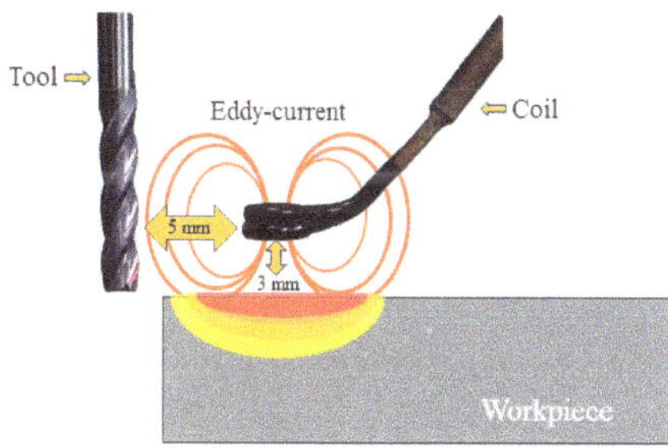

Figure 2. The distance between the tool, coil and workpiece.

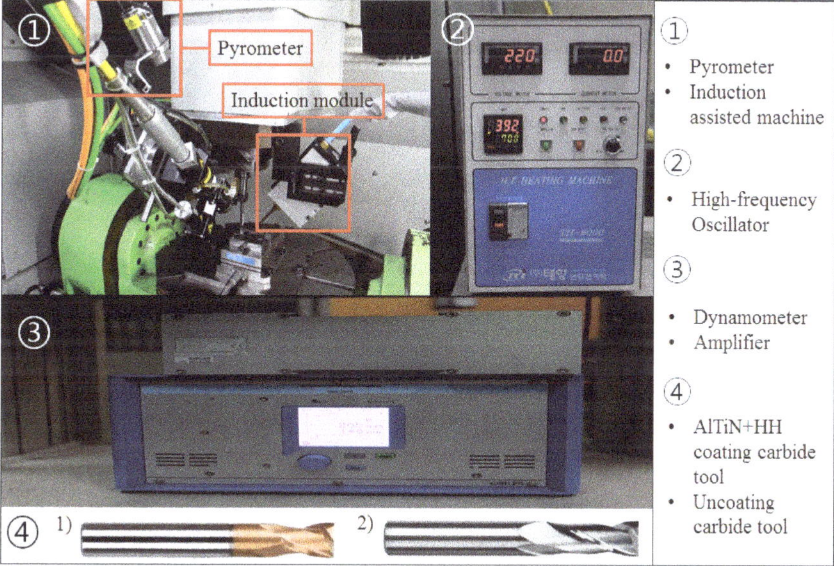

Figure 3. The experimental set-up.

Table 1. The main cutting parameters.

Material	Inconel 718
Machining method	Slot milling
Material size (T × W × L, mm)	30 × 30 × 50
Depth of cut (ap, mm)	0.15, 0.2, 0.25
Width of cut (ae, mm)	8
Spindle speed (S, rev/min)	6000, 8000, 10,000
Feed rate (f, mm/min)	100, 150, 200
Feed per tooth (fz, mm/tooth)	0.008, 0.006, 0.005, 0.012, 0.009, 0.007, 0.016, 0.012, 0.010
Coated and uncoated tool	D8 Flat end-mill, 2F, 60L
Preheating temperature (°C)	950

3. Finite Element Analysis

3.1. Finite Element Analysis

Using the multi-physics ANSYS Electromagnetics and Workbench, finite element analyses were carried out to obtain the efficient depth of cut. The governing equation of the eddy current analysis can be expressed using Maxwell's equations, as follows, Equation (1) [26].

$$\nabla \times \left(\frac{1}{\mu_0 \mu_r} \nabla \times A \right) + \sigma \frac{\partial A}{\partial t} - J_s = 0 \tag{1}$$

where A is the magnetic vector potential, J_s is the source current density, μ_0 is the vacuum magnetic permeability, μ_r is the relative magnetic permeability and σ is the electrical conductivity. The power of induction was 13 A, 380 V and the adaptive frequency was 300 kHz. These analysis conditions were determined by considering the specification of the induction oscillator. The method is an analyzing method by overlapping the heat source according to the feed rate [27]. The moving time of the heat source for the analysis was set to be 30 s by considering the lower feed rate (f = 100 mm/min) [28]. The material properties for the analysis are listed in Table 2. The chemical composition of Inconel 718 is as shown in Table 3. An electromagnetic analysis was performed first and then a temperature distribution was confirmed by thermal analysis.

Table 2. The material properties of Inconel 718.

Yield Strength (MPa)	Tensile Strength (MPa)	Elongation (%)	Hardness (HRC)
980	1280	12	42
Magnetic Permeability	Bulk conductivity (MS/m)	Specific heat (J/kg °C)	Thermal conductivity (W/mmK)
1.0011	80	435–637	11.2–25.8

Table 3. The chemical composition of Inconel 718.

Element	Ni	Cr	Nb	Mo	Ti	Al	Co	C	Mn	Si	Cu
wt%	50–55	17–21	4.75–5.5	2.8–3.3	0.65–1.15	0.2–0.8	≤1.0	≤0.08	≤0.35	≤0.35	≤0.3

3.2. Result of Analysis

When the preheating temperature is about 950 °C, the maximum depth of the cut is determined to be 0.25 mm by the finite element analysis, by considering the proper preheating effect, this could not be expected when the depth of the cut exceeds the maximum depth of the cut. The preheating temperature was chosen by considering the tensile strength of Inconel 718 according to temperature, as shown in Figure 4. Inconel 718 has its maximum elongation and minimum tensile strength at a temperature of about 950 °C. Additionally, Inconel 718 has a transformation point of its mechanical properties at about 700 °C. The temperature range of the effective depth of the cut is 700–950 °C. The depth of the cut range was determined within the maximum depth of cut of 0.25 mm by the finite element analysis. Figure 5 shows the finite element analysis results.

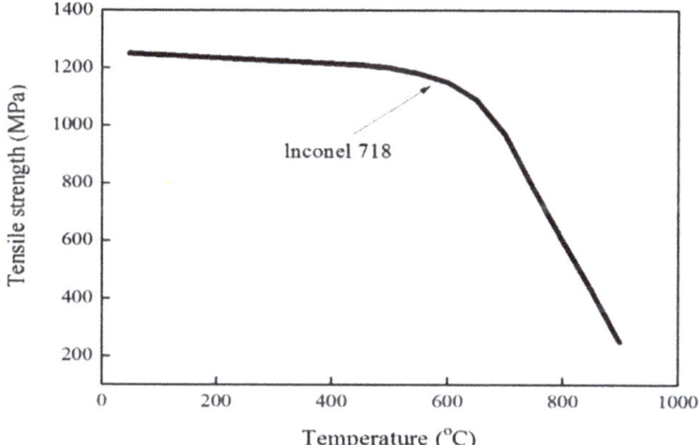

Figure 4. The tensile strength of Inconel 718 according to temperature (reproduced from [29], with permission from ELSEVIER, 2010).

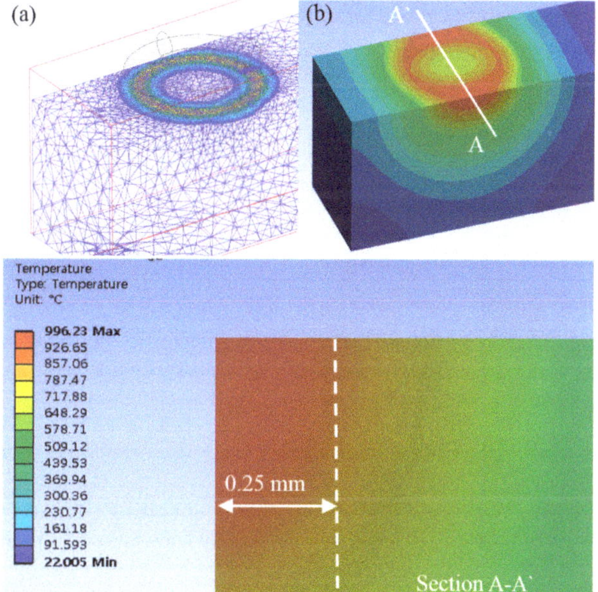

Figure 5. The results of finite element analysis; (**a**) electromagnetic, (**b**) thermal analysis.

4. Experimental Set-Up

4.1. Taguchi Method

The Taguchi method, sometimes called the robust design method, is widely used to improve the quality of machining processes. The Taguchi design optimization method has three steps: system design, parameter design and tolerance design [30]. System design is the conceptual design level. Parameter design is the detailed design level, during which the nominal values of the design parameters and various dimensions need to be set. The advantage of the Taguchi method is that the exact choices of the values required are left undecided by the performance requirements of the system. This allows the parameters to be chosen so as to minimize the effects on the performance that arise from variations in the environment [31]. Many industrials or academic researchers have used the Taguchi method for optimization.

4.2. Experimental Design

Machining experiments were carried out using an Inconel 718 square workpiece. All experiments were repeated 3 times. The cutting parameters for the induction assisted milling operation are spindle speed (S), feed rate (F) and depth of cut (D). The spindle speed range (6000, 8000 and 10,000 rev/min) and the feed rate range (100, 150 and 200 mm/min) were selected after considering previous studies [9,32]. The depth of the cut range (0.15, 0.2 and 0.25 mm) with sufficient preheating effect was selected based on the finite element analysis result. The experimental design used an orthogonal array of the Taguchi method. The machining experiments were carried out according to the 3 levels and 3-factor L_9 orthogonal array in Table 4.

Table 4. The experimental layout using an L_9 orthogonal array.

Experimental No	S	f	ap
1	6000	100	0.15
2	6000	150	0.20
3	6000	200	0.25
4	8000	100	0.20
5	8000	150	0.25
6	8000	200	0.15
7	10,000	100	0.25
8	10,000	150	0.15
9	10,000	200	0.20

5. Experimental Results and Discussion

The experimental results for cutting force, surface roughness, tool wear and the signal to noise (S/N) ratio are shown in Tables 5–7, for the coated tool and the uncoated tool. Figures 6 and 7 show the main effect of S/N ratio of cutting force, surface roughness and tool wear. All the measurements were performed after machining for the same machining length. The cutting force was measured by the dynamometer and amplifier. The dynamometer was installed under the workpiece and fixture. The measured cutting force of the dynamometer is given by Equation (2),

$$F = \sqrt{F_x{}^2 + F_y{}^2} \tag{2}$$

where Fx is the tangential force and Fy is the radial force. A piezoelectric sensor was used for the dynamometer. The initial load was set to be zero. The surface roughness was measured by the shape measuring device. The surface roughness was measured as the arithmetic mean deviation of the profile (Ra) and 0.8 mm of the cut-off value, 4.5 mm of the evaluation length and a stylus tip radius of 2 mm. The evaluation length was analyzed repeatedly 5 times and the average value was selected as the surface roughness. The tool wear was measured by the flank wear using a microscope (SOMETECH Inc., Seoul, Korea, IMS-345). The evaluation of the tool wear was measured as well as the evaluated in the worst case in both flutes. The quality characteristics are classified into three types: nominal-is-best characteristics, larger-the-better characteristics and smaller-the-better characteristics. In this study, the smaller-the-better characteristics were used. The smaller-the-better characteristics are shown in Equation (3).

$$\text{S/N ratio} = -10 \log\left(\Sigma\left(Y^2\right)/n\right) \tag{3}$$

where n is the number of observations and Y is the observed data.

5.1. S/N Ratio Analysis

According to the cutting force measurement results, the cutting force of the coated tool was approximately 1000% lower than that of the uncoated tool. The cutting force and the S/N ratio of the cutting force are shown in Table 5; the cutting force of the coated tool had the highest value of 28.62 N and the lowest value of 11.24 N. The uncoated tool had the highest value of 191.04 N and the lowest value of 138.86 N. Both the coated tool and the uncoated tool, for identical machining conditions, had highest values of $S_1F_3D_3$ and lowest values of $S_2F_1D_2$. For the coated and uncoated tools, the cutting force of the S/N ratio had the highest values of -21.73 dB and -43.45 dB. In the surface roughness measurement results, the surface roughness of the coated tool was found to be approximately 260% lower than that of the uncoated tool. The S/N ratio of the surface roughness is shown in Table 6; the surface roughness of the uncoated tool had its highest value at $S_1F_2D_2$ (0.310 μm) and its lowest value at $S_3F_1D_3$ (0.226 μm). The coated tool had its highest value at $S_3F_1D_3$ (0.212 μm) and its lowest value at $S_3F_1D_3$ (0.115 μm). The coated and uncoated tools had surface roughness S/N ratios that exhibited the highest values of 18.52 dB and 12.69 dB, respectively. For the tool wear measurement results, the tool wear of the coated tool was up to approximately 390% lower than that of the uncoated tool. The tool wear and S/N ratio of the tool wear are shown in Table 7. The tool wear of the coated tool had its highest value at $S_2F_3D_1$ (0.340 mm) and its lowest value at $S_1F_3D_3$ (1.075 mm). When the machining condition was $S_2F_1D_2$, the coated and uncoated tool to S/N ratio of tool wear had the highest values of 12.327 dB and 1.774 dB, respectively.

Table 5. The experimental results for cutting force and S/N ratio.

Experiment No	S	f	ap	Cutting Force (N)		S/N Ratio (dB)	
				Coated Tool	Uncoated Tool	Coated Tool	Uncoated Tool
1	1	1	1	15.16	164.88	−25.01	−44.72
2	1	2	2	13.23	161.90	−24.59	−44.06
3	1	3	3	28.10	191.04	−23.24	−43.73
4	2	1	2	11.24	138.86	−22.24	−43.45
5	2	2	3	15.14	167.41	−23.35	−44.16
6	2	3	1	12.72	141.66	−21.73	−43.50
7	3	1	3	28.62	177.76	−24.97	−44.10
8	3	2	1	15.88	155.40	−24.27	−44.04
9	3	3	2	12.23	149.20	−27.24	−45.03

Table 6. The experimental results for surface roughness and S/N ratio.

Experiment No	S	f	ap	Surface Roughness Ra (μm)		S/N Ratio (dB)	
				Coated Tool	Uncoated Tool	Coated Tool	Uncoated Tool
1	1	1	1	0.124	0.282	18.52	11.18
2	1	2	2	0.117	0.310	15.12	11.84
3	1	3	3	0.115	0.265	17.20	11.85
4	2	1	2	0.205	0.263	16.66	12.03
5	2	2	3	0.130	0.229	17.12	11.94
6	2	3	1	0.119	0.260	15.77	11.34
7	3	1	3	0.212	0.226	14.45	12.67
8	3	2	1	0.178	0.228	17.39	12.10
9	3	3	2	0.180	0.244	16.66	12.69

Table 7. The experimental results for tool wear and S/N ratio.

Experiment No	S	f	ap	Tool Wear (mm)		S/N Ratio (dB)	
				Coated Tool	Uncoated Tool	Coated Tool	Uncoated Tool
1	1	1	1	0.340	1.069	9.790	−0.272
2	1	2	2	0.304	0.956	10.718	1.492
3	1	3	3	0.329	1.075	11.206	−0.171
4	2	1	2	0.276	0.694	12.327	1.774
5	2	2	3	0.230	0.845	11.440	0.411
6	2	3	1	0.340	0.924	10.779	1.294
7	3	1	3	0.263	0.805	11.229	0.527
8	3	2	1	0.275	1.074	11.188	0.125
9	3	3	2	0.286	0.964	11.341	0.906

5.2. Main Effect Plot

We used the main effects plot to examine the differences between level means for one or more parameters. There is a main effect when different levels of a parameter affect the response differently. The larger the slope of the main effect plot is, the greater the magnitude of the main effect. Therefore, Figures 6 and 7 show the main effects. The Figure 6a,b show main effects of S/N ratio in the cutting force and the surface roughness of coated tool. The Figure 6c,d show main effects of S/N ratio in the cutting force and the surface roughness of uncoated tool. The Figure 7a,b show main effects of S/N ratio in the tool wear of coated and uncoated tool. It has been determined from Figure 6a that the optimum levels were S_2 (S:8000), F_2 (f:150) and D_2 (ap:0.2) and Figure 6b shows that the optimum levels were S_3 (S:10,000), F_3 (f:200) and D_3 (ap:0.25). Figure 6c shows that the optimum levels were S_2 (S:8000), F_3 (f:200) and D_2 (ap:0.2). Figure 6d shows that the optimum levels were S_1 (S:6000), F_3 (f:250) and D_1 (ap:0.15). When comparing the coated tool with the uncoated tool, the cutting force shows a similar pattern but the surface roughness shows a different pattern. In the case of the tool wear, Figure 7a shows that the optimum level is $S_3F_2D_3$, i.e., spindle speed is 10,000 rpm, the feed rate is 150 mm/min and the depth of cut is 0.25 mm. Figure 7b shows that the optimum level is $S_2F_1D_2$, i.e., the spindle speed is 8000 rpm, the feed rate is 100 mm/min and the depth of cut is 0.2 mm.

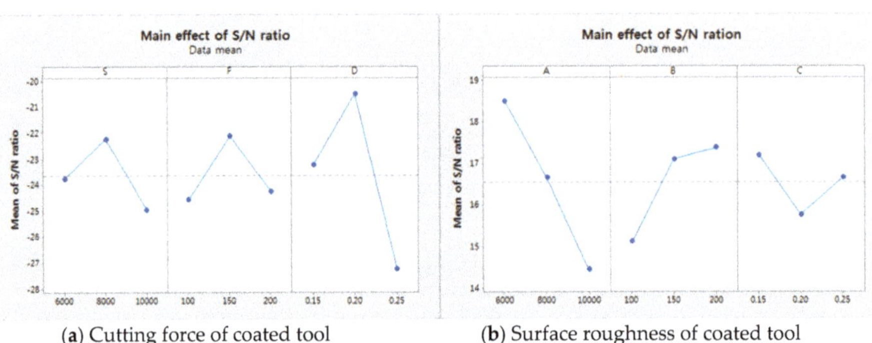

(**a**) Cutting force of coated tool   (**b**) Surface roughness of coated tool

Figure 6. *Cont.*

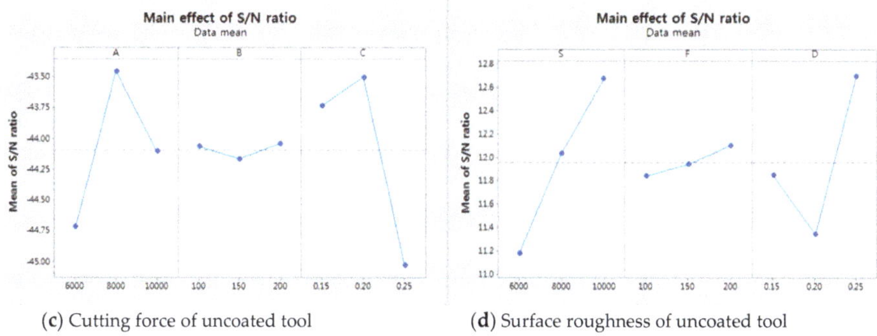

(**c**) Cutting force of uncoated tool   (**d**) Surface roughness of uncoated tool

Figure 6. The main effect plot of the coated and uncoated tools on the cutting force and the surface roughness. (**a**) Cutting force of coated tool; (**b**) Surface roughness of coated tool; (**c**) Cutting force of uncoated tool; (**d**) Surface roughness of uncoated tool.

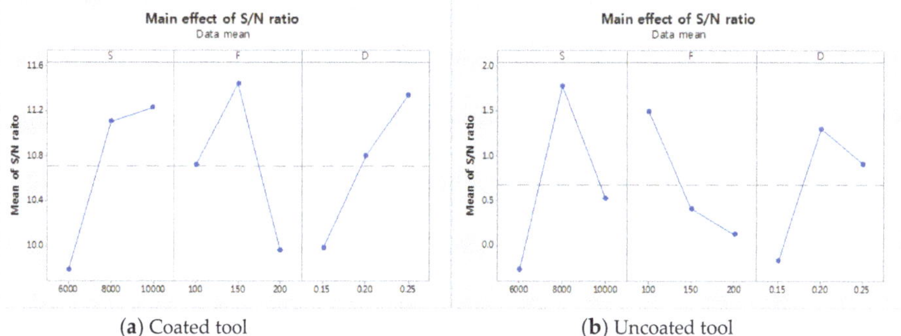

(**a**) Coated tool   (**b**) Uncoated tool

Figure 7. The main effect plot of the coated tool and uncoated tool on the tool wear. (**a**) Coated tool; (**b**) Uncoated tool.

5.3. Analysis of Variance

Analysis of Variance (ANOVA) is used to test the hypothesis that the means of two or more populations are equal. ANOVA assesses the importance of one or more parameters by comparing the response variable means at different parameter levels. ANOVA was performed to study the relative nature of the parameters. An ANOVA with 95% confidence was used. Table 8 shows the ANOVA results for cutting force, surface roughness and tool wear on the coated tool. It was observed that the depth of cut is the main contributing factor to the cutting force. The contributions of the cutting force were as follows: spindle speed at 19%, feed rate at 6% and depth of cut at 65%. It was observed that the spindle speed is the main contributing factor to the surface roughness. The contributions of the surface roughness were as follows: spindle speed at 59%, feed rate at 25% and depth of cut at 8%. It was observed that the spindle speed is the main contributing factor to tool wear. The contributions of tool wear were as follows: spindle speed at 37%, feed rate at 31% and depth of cut at 26%.

Table 9 shows the ANOVA results of cutting force, surface roughness and tool wear on the uncoated tool. It was observed that the depth of cut is the main contributing factor to the cutting force. The contributions to the cutting force were as follows: spindle speed at 35%, feed rate at 0.1% and depth of cut at 63%. It was observed that the spindle speed is the main contributing factor to the surface roughness. The contributions to the surface roughness were as follows: spindle speed at 48%, feed rate at 2% and depth of cut at 38%. It was observed that the spindle speed is the main contributing factor to the tool wear. The contributions to the tool wear were as follows: spindle speed at 48%, feed rate at 20% and depth of cut at 26%. In the contribution of the cutting force, the depth of cut has the most influence due to the decrease of the tool vibration according to the depth of heat affected zone. In the contributions of the surface roughness and tool wear, the coating blocked the heat transfer to the coated tool. So, the influence of the feed rate is increased in the coated tool. The F-test accurate only for normally distributed data. However, the three factors with three levels used in this study were not normally distributed. Therefore, the F-test was not suitable. We have performed the test for equation variance. It is possible for the null hypothesis of the variances to be equal. Table 10 shows the p-value of the test for equal variance.

Table 8. The results of the ANOVA for cutting force, surface roughness, and tool wear on the coated tool.

Cutting Parameter	DOF	SS	MS	F	P	%C
Cutting force						
S	2	68.14	34.07	2.00	0.33	19.21
f	2	21.92	10.96	0.64	0.608	6.18
ap	2	230.63	115.32	6.77	0.129	65.01
Error	2	34.05	17.02			9.60
Total	8	354.74				100.00
Surface roughness						
S	2	0.007	0.003	10.14	0.09	59.75
f	2	0.003	0.001	4.37	0.186	25.78
ap	2	0.001	5.4E−0.4	1.45	0.407	8.58
Error	2	7.5E−0.4	3.7E−0.4			5.89
Total	8	0.012				100.00
Tool wear						
S	2	0.004	0.002	9.94	0.091	37.96
f	2	0.003	0.001	8.19	0.109	31.29
ap	2	0.003	0.001	7.05	0.124	26.94
Error	2	4.0E−4	2.1E−4			3.82
Total	8	0.011				100.00

Table 9. The results of the ANOVA for cutting force, surface roughness and tool wear on the uncoated tool.

Cutting Parameter	DOF	SS	MS	F	P	%C
Cutting force						
S	2	814.16	407.08	241.15	0.004	35.79
f	2	2.04	1.02	0.60	0.623	0.09
ap	2	1455.40	727.70	431.08	0.002	63.97
Error	2	3.38	1.688			0.15
Total	8	2274.98				100.00
Surface roughness						
S	2	0.003	0.001	4.39	0.077	48.17
f	2	1.0E−4	6.5E−5	0.19	0.99	2.04
ap	2	0.002	0.001	3.54	0.189	38.82
Error	2	7.0E−4				10.97
Total	8	0.006				100.00
Tool wear						
S	2	0.068	0.034	9.71	0.093	48.44
f	2	0.028	0.014	4.07	0.197	20.28
ap	2	0.037	0.018	5.27	0.159	26.29
Error	2	0.007	0.003			4.99
Total	8	0.141				100.00

Table 10. The test for the equal equation.

Parameter	Levene's Test p-Value
Coated tool	0.009
Uncoated tool	0.001

5.4. Tool Wear and Machined Surface

For the machining of Inconel 718 using carbide tools, diffusion loss occurs due to heat. Figure 8 shows the tool wear as found in the microscope measurement results. Figure 8 shows the measured parameters for level 7, i.e., the spindle speed is 10,000 rpm, the feed rate is 100 mm/min and the depth of cut is 0.25 mm both for the coated tool and the uncoated tool. The Figure 8a shows tool wear of coated tool and 8b shows tool wear of uncoated tool. Generally, the heat generated during machining is mainly transferred to the chip and cutting tool, and less than 10% of the generated heat is transferred to the workpiece. According to Figure 8a, the wear of the coated tool is insignificant. However, the uncoated tool is negatively influenced by the friction heat and radiant heat of the induction heat source. Figure 8b shows the worn tool after machining.

Figure 9a,b show the machined surfaces and scanning electron microscope of the coated and uncoated tools. When an AlTiN+HH coated tool is used in machining, the machined surface has no heat affected zone, as can be seen in Figure 9a. However, when the uncoated tool is used in machining, the machined surface has a heat affected zone, as can be seen in Figure 9b. The heat affected zone shows chemical and structural modifications. So, this phenomenon is very important in the area of thermally assisted machining. However, the uncoated tool has a rougher surface roughness than that of the coated tool, causing the uncoated tool to have a surface build up in a built-up-edge (BUE). When the uncoated tool was used, the heat generated during IAM was transferred to the workpiece more than to the coated tool. The analysis results of the machined surface showed that the coating was effectively blocked the heat generated during IAM.

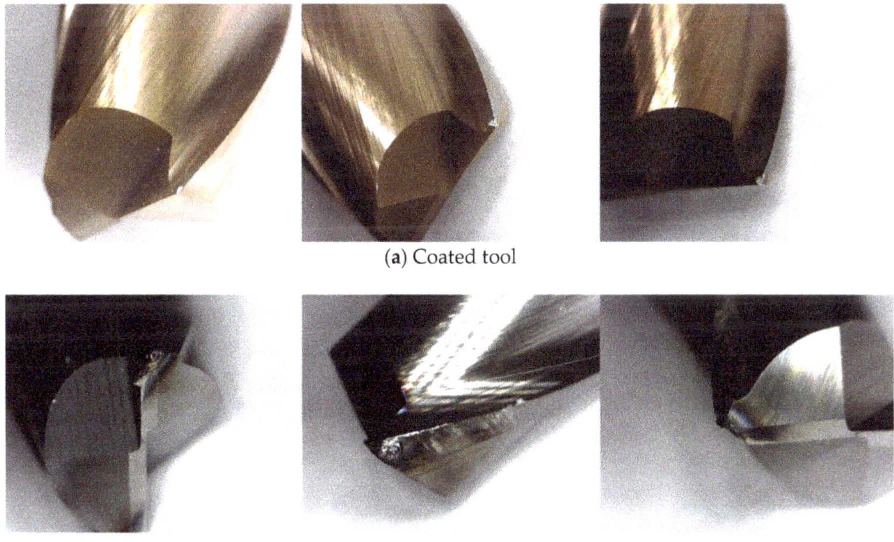

(**a**) Coated tool

(**b**) Uncoated tool

Figure 8. The tool wear of the coated tool and uncoated tool. (**a**) Coated tool; (**b**) Uncoated tool.

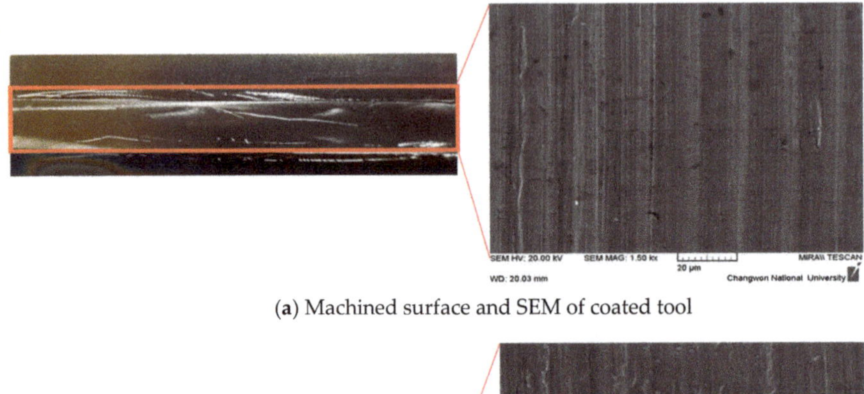

(**a**) Machined surface and SEM of coated tool

(**b**) Machined surface and SEM of uncoated tool

Figure 9. The machined surface and SEM of the coated tool and uncoated tool. (**a**) Machined surface and SEM of coated tool; (**b**) Machined surface and SEM of uncoated tool.

5.5. Response Optimization

Based on the design of the experimental results, we carried out a response optimization. The goal was to achieve a minimum and weights are shown in Table 11. The result of the response optimization is depicted in Table 12. The parameters of the coated tool $S_2F_2D_2$ were observed, i.e., a spindle speed if 8000 rpm, a feed rate of 150 mm/min, a depth of cut of 0.2 mm and a composite desirability of 0.78. Desirability assessed how well a combination of variables satisfies the goals defined for the response. The uncoated tool was $S_2F_1D_2$, i.e., a spindle speed of 8000 rpm, a feed rate of 100 mm/min, a depth of cut of 0.2 mm and a composite desirability of 0.79. The coated tool and uncoated tool both have the same spindle speed and depth of cut. The coated tool had a higher feed rate than the uncoated tool. The machining time can be decreased by using the coated tool in IAM.

Table 11. The response optimization.

Parameter	Goal	Target	Upper	Weight	Importance
Coated tool					
Cutting force	Minimum	11.240	28.62	1	1
Tool wear	Minimum	0.230	0.310	1	1
Surface roughness	Minimum	0.115	0.212	1	1
Uncoated tool					
Cutting force	Minimum	138.860	191.04	1	1
Tool wear	Minimum	0.694	1.075	1	1
Surface roughness	Minimum	0.226	0.310	1	1

Table 12. The response optimization results.

S	F	D	Cutting Force Optimization Plot	Tool Wear Optimization Plot	Surface Roughness Optimization Plot	Desirability
Coated tool 8000	150	0.2	6.16	0.25	0.15	0.78
Uncoated tool 8000	100	0.2	137.99	0.68	0.26	0.79

6. Conclusions

In this study, induction assisted milling (IAM) was carried out on Inconel 718 with an AlTiN coated carbide tool and uncoated carbide tool. The efficient depth of cut was selected using finite element analysis. The cutting force, surface roughness and tool wear were investigated using the Taguchi method for the determination of the optimal machining parameters. The findings of this study lead to the following conclusions:

1. Before the machining experiments, for an effective depth of cut determination, a multi-physics analysis by electromagnetic and thermal analyses was performed. When the preheating temperature is about 950 °C, the maximum depth of cut is determined to be 0.25 mm by the finite element analysis; the proper preheating effect could not be expected when the depth of cut exceeds the maximum depth of cut.

2. For the design of the experiments, a spindle speed of 8000 rpm, a feed rate of 150 mm/min and depth of cut 0.2 mm are the cutting parameters for the minimum cutting force for a coated tool and a spindle speed of 8000 rpm, a feed rate of 200 mm/min and depth of cut 0.2 mm are the cutting parameters for the minimum cutting force for an uncoated tool. The cutting force of the coated tool was lower than the uncoated tool. The surface roughness and tool wear, as the spindle speed increases, indicates the better quality on the coated tool.

3. Based on the ANOVA results, the main contributing factor for the cutting force of both the coated and uncoated tools were 65 and 63% depths of cut, respectively. The main contributing factor for the surface roughness of both of coated and uncoated tool was the spindle speed, 59 and 48%, respectively. The main contributing factor for tool wear of both the coated tool and uncoated tool was the spindle speed, which was approximately 37 and 48%, respectively. The uncoated tools have a lower machinability level due to the higher thermal effects by the heat source than coated tools.

4. The results of response optimization: the feed rate is different at the same conditions. A feed rate of 150 mm/min in the coated tool and a feed rate of 100 mm/min in the uncoated tool, a spindle speed of 8000 rpm and a depth of cut of 0.2 mm. An efficient preheating effect cannot be expected at a feed rate over the optimal feed rates in both tools. The coated tool has a higher feed rate due to the reduced thermal effect by the coating. The increase of feed rate can decrease the machining time and increase productivity. When the depth of cut of the analysis and the optimal parameter was compared, the FEA result showed a deeper response optimization due to the decrease of the preheating effect by the cooling and the moving.

5. Based on the machining experiment results, the uncoated tool was not recommended by induction assisted milling. The uncoated tool was affected by the thermal effect more than the coated tool, which can accelerate tool failure. The coated tool improved the tool wear up to a maximum of 390%. In the uncoated tool, the defects of the tools such as the fusion of chips, fracture and wear occurred due to thermal diffusion by the induction heat sources. During machining, the AlTiN+HH coating effectively reduces the heat generated by friction between the tool and the workpiece. The coated tool was demonstrated to be superior in efficiency for IAM compared to the uncoated tool.

Author Contributions: E.J.K. designed and performed the experiments, analyzed the experiment results, writing the paper. C.M.L. supervised all of the work carried out this research.

Funding: This research was funded by National Research Foundation of Korea (NRF), grant No. 2016R1A2A1A 05005492 and The APC was funded by National Research Foundation of Korea.

Acknowledgments: This research was supported by the Basic Science Research Program through the National Research Foundation of Korea (NRF) funded by the Ministry of Science, ICT & Future Planning (No. 2016R1A2A1A05005492).

Conflicts of Interest: The authors declare no conflict of interest. The founding sponsors had no role in the design of the study; in the collection analyses, or interpretation of data; in the writing of the manuscript; and in the decision to publish the results.

Nomenclature

IAM	Induction assisted machining		S/N	Signal to noise
S	Spindle speed		OA	Orthogonal array
f	Feed rate		DOE	Design of experiments
ap	Depth of cut		DOF	Degree of freedom
ae	Width of cut		SS	Sum of squares
Fx	Tangential force		P	Probability value
Fy	Radial force		F	Fisher ration
T	Thickness		%C	Percentage contribution
W	Width		SS	Sum of squares
L	Length		MS	Mean of squares
fz	Feed per tooth		ANOVA	Analysis of variance
Ra	Arithmetic average roughness		dB	Decibel
SEM	Scanning electron microscope			

References

1. Beranoagirre, A.; Olvera, D.; Lopez de Lacalle, L.N. Milling of gamma titanium-aluminum alloys. *Int. J. Adv. Manuf. Technol.* **2012**, *62*, 83–88. [CrossRef]
2. Pereira, O.; Rodriguez, A.; Fernandez-Abia, A.I.; Barreiro, J.; Lopez de Lacalle, L.N. Cryogenic and minimum quantity lubrication for an eco-efficiency turning of AISI 304. *J. Clean. Prod.* **2016**, *139*, 440–449. [CrossRef]
3. Singh, T.; Dureja, J.S.; Dogra, M.; Bhatti, M. Enviroment Friendly Machining of Inconel 625 under Nano-Fluid Minimum Quantity Lubrication (NMQL). *Int. J. Precis. Eng. Manuf.* **2018**, *19*, 1689–1697. [CrossRef]
4. Lian, Y.S.; Liao, C.H.; Lin, H.M. Study of Oil-Water Ratio and Flow Rate of MQL Fluid in High Speed Milling of Inconel 718. *Int. J. Precis. Eng. Manuf.* **2017**, *18*, 257–262.
5. Huang, S.; Lv, T.; Wang, M.; Xu, X. Effects of machining and oil mist parameters on electrostatic minimum quantity lubrication-EMQL Turning process. *Int. J. Precis. Eng. Manuf.-Green Technol.* **2018**, *5*, 317–326. [CrossRef]
6. Grabowski, M.; Skoczypiec, S.; Wyszynski, D. A Study on Microturning with Electrochemical Assistance of the Cutting Process. *Micromachines* **2018**, *9*, 357. [CrossRef] [PubMed]
7. Oh, N.S.; Woo, W.S.; Lee, C.M. A study on the machining characteristics and energy efficiency of Ti-6Al-4V in laser-assisted trochoidal milling. *Int. J. Precis. Eng. Manuf.-Green Technol.* **2018**, *5*, 37–45. [CrossRef]
8. Woo, W.S.; Lee, C.M. A study of the machining characterisitcs of AISI 1045 steel and Inconel 718 ith a cylindrical shape in laser-assisted milling. *Appl. Therm. Eng.* **2015**, *91*, 33–42. [CrossRef]
9. Lopez de Lacalle, L.N.; Sanchez, J.A.; Lamikiz, A.; Celaya, A. Plasma Assisted Milling of Heat-Resistant Superalloys. *J. Manuf. Sci. Eng.* **2004**, *126*, 274–285. [CrossRef]
10. Li, N.; Chen, Y.J.; Kong, D.D. Wear Mechanism Analysis and Its Effects on the Cutting performance of PCBN Inserts during Turning of Hardened 42CrMo. *Int. J. Precis. Eng. Manuf.* **2018**, *19*, 1355–1368. [CrossRef]
11. Suarez, A.; Veiga, F.; Lopez de Lacalle, L.N.; Polvorosa, R.; Lutze, S.; Wretland, A. Effecs of Ultrasonics-Assisted Face Milling on Surface Integrity and Fatigue Life of Ni-Alloy 718. *J. Mater. Eng. Perform.* **2016**, *25*, 5076–5086. [CrossRef]
12. Zhu, D.; Zhang, X.; Ding, H. Tool Wear Characteristics in Machining of Nickel-based superalloys. *Int. J. Mac. Tools Manuf.* **2013**, *64*, 60–77. [CrossRef]

13. Kang, Z.Y.; Fu, Y.H.; Chen, Y.; Ji, J.H.; Fu, H.; Wang, S.; Li, R. Experimental Investigation of Concave and Convex Micro-Textures for Improving Anti-Adhesion property of Cutting Tool in Dry Finish Cutting. *Int. J. Precis. Eng. Manuf.-Green Technol.* **2018**, *5*, 583–591. [CrossRef]

14. Pimenov, D.Y. The effect of the rate flank wear teeth face mills on the processing. *J. Frict. Wear* **2013**, *34*, 156–159. [CrossRef]

15. Zhang, S.; Li, J.F.; Wang, Y.W. Tool life and cutting forces in end milling Inconel 718 under dry and minimum quantity cooling lubrication cutting condition. *J. Clean. Prod.* **2012**, *32*, 81–87. [CrossRef]

16. Pimenov, D.Y.; Bustillo, A.; Mikolajczyk, T. Artificial intelligence for automatic prediction of required surface roughness by monitoring wear on face mill teeth. *J. Intell. Manuf.* **2018**, *29*, 1045–1061. [CrossRef]

17. Chamarthi, S.; Reddy, N.S.; Elipey, M.K.; Reddy, D.V.R. Investigation Analysis of Plasma Arc Cutting Parameters on the Unevenness Surface of Hardox-400. *Procedia Eng.* **2013**, *64*, 854–861. [CrossRef]

18. Calleja, A.; Tabernero, I.; Fernandez, A.; Celaya, A.; Lamikiz, A.; Lopez de Lacalle, L.N. Improvement of strategies and parameters for multi-axis laser cladding operations. *Opt. Laser Eng.* **2014**, *56*, 113–120. [CrossRef]

19. Kim, D.H.; Lee, C.M. A study of cutting force and preheating-temperature prediction for laser-assisted milling of Inconel 718 and AISI 1045 steel. *Int. J. Heat Mass Transf.* **2014**, *71*, 264–274. [CrossRef]

20. Venkatesan, K.; Ramanujam, R. Statstical approach for optimization of influencing parameters in laser assisted machining of Inconel alloy. *Measurement* **2016**, *89*, 97–108. [CrossRef]

21. Abbas, A.T.; Pimenov, D.Y.; Erdakov, I.N.; Mikolajczyk, E.A.; Danaf, E.; Taha, M.A. Minimization of turning time for high-stregth steel with a given surface roughness using the Edgeworth-Pareto optimization method. *Int. J. Adv. Manuf. Technol.* **2017**, *93*, 2375–2392. [CrossRef]

22. Aihua, L.; Jianxin, D.; Haibing, C.; Yangyang, C.; Jun, Z. Friction and wear properties of TiN, TiAlN, AlTiN and CrAlN PVD nitride coatings. *Int J. Refra. Metal Hard Mater.* **2012**, *31*, 82–88. [CrossRef]

23. Aslantas, K.; Ucun, I.; Cicek, A. Tool life and wear mechanism of coated and uncoated Al_2O_3/TiCN mixed ceramic tools in turning hardened alloy steel. *Wear* **2012**, *274*, 442–451. [CrossRef]

24. Liu, Z.; An, Q.; Xu, J.; Chen, M.; Han, S. Wear performance of (nc-AlTiN)/(a-Si_3N_4) coating and (nc-AlCrN)/(a-Si_3N_4) coating in high-speed machining of titanium alloys under dry and minimum quantity lubrication (MQL) conditions. *Wear* **2013**, *305*, 249–259. [CrossRef]

25. *ISO 8688-2: Tool Life Testing in Milling—Part 2: End Milling*; International Standard Organization: Geneva, Switzerland, 1989.

26. Jing, Y.P.; Wang, D.; Zhang, Q.H.; Chen, J.Q. Electromagnetic-Thermal Coupling Simulation by ANSYS Multiphysics of Induction Heater. *Appl. Mech. Mater.* **2015**, *701*, 820–825. [CrossRef]

27. Baek, J.T.; Woo, W.S.; Lee, C.M. A study on the machining characteristics of induction and laser-induction assisted machining of AISI 1045 steel and Inconel 718. *J. Manuf. Process.* **2018**, *34*, 513–522. [CrossRef]

28. Ahn, S.H.; Lee, C.M. A Study on Large-area Laser Processing Analysis in Consideration of the Moving Heat Source. *Int. J. Precis. Eng. Manuf.* **2011**, *12*, 285–292. [CrossRef]

29. Sun, S.; Brandt, M.; Dargusch, M.S. Thermally enhanced machining of hard-to-machine materials—A review. *Int. J. Mach. Tools Manuf.* **2010**, *50*, 663–680. [CrossRef]

30. Selvaraj, D.P.; Chandramohan, P.; Mohanraj, M. Optimization of surface roughness, cutting force and tool wear of nitrogen alloyed duplex stainless steel in a dry truning process using Taguchi method. *Measurement* **2014**, *49*, 205–215. [CrossRef]

31. Mandal, N.; Doloi, B.; Mondal, B.; Das, R. Optimization of flank wear using Zirconia Toughened Alumina (ZTA) cutting tool: Taguchi method and regression analysis. *Measurement* **2011**, *44*, 2149–2155. [CrossRef]

32. Kim, I.W.; Lee, C.M. A study on the machining characteristics of specimens with spherical shape using laser-assisted machining. *Appl. Therm. Eng.* **2016**, *100*, 636–645. [CrossRef]

 *materials*

Article

Predictive Modeling of Machining Temperatures with Force–Temperature Correlation Using Cutting Mechanics and Constitutive Relation

Jinqiang Ning * and Steven Y. Liang *

George W. Woodruff School of Mechanical Engineering, Georgia Institute of Technology, 801 Ferst Drive, Atlanta, GA 30332-0405, USA

* Correspondence: jinqiangning@gatech.edu (J.N.); steven.liang@me.gatech.edu (S.L.)

Received: 12 December 2018; Accepted: 11 January 2019; Published: 16 January 2019

Abstract: Elevated temperature in the machining process is detrimental to cutting tools—a result of the effect of thermal softening and material diffusion. Material diffusion also deteriorates the quality of the machined part. Measuring or predicting machining temperatures is important for the optimization of the machining process, but experimental temperature measurement is difficult and inconvenient because of the complex contact phenomena between tools and workpieces, and because of restricted accessibility during the machining process. This paper presents an original analytical model for fast prediction of machining temperatures at two deformation zones in orthogonal cutting, namely the primary shear zone and the tool–chip interface. Temperatures were predicted based on a correlation between force and temperature using the mechanics of the cutting process and material constitutive relation. Minimization of the differences between calculated material flow stresses using a mechanics model and a constitutive model yielded an estimate of machining temperatures. Experimental forces, cutting condition parameters, and constitutive model constants were inputs, while machining forces were easily measurable by a piezoelectric dynamometer. Machining temperatures of AISI 1045 steel were predicted under various cutting conditions to demonstrate the predictive capability of each presented model. Close agreements were observed by verifying them against documented values in the literature. The influence of model inputs and computational efficiency were further investigated. The presented model has high computational efficiency that allows real-time prediction and low experimental complexity, considering the easily measurable input variables.

Keywords: machining temperatures at two deformation zones; force–temperature correlation through analytical modeling; high computational efficiency; real-time prediction

---

1. Introduction

Machining is one of the most widely used manufacturing processes because of its fast speed and applicability to a broad class of materials. Karpat et al. studied the machining process of steel and aluminum alloys [1]. Danish et al. studied the machining process of magnesium alloy under dry and cryogenic cutting conditions [2]. Ning et al. studied the machining process of ultra-fine-grained titanium [3]. Machining temperature has a significant influence on tool performance and the quality of a machining part as a result of the softening effect and diffusion. Coolant [4], laser power [5], and magnetic flux [6] have been utilized to effectively control temperature in the machining process. The capabilities of temperature measurement and prediction are critical for optimizing the machining process. Predicted temperatures can be utilized to further explore machining forces, tool wear, material diffusion, etc.

Different experimental methods have been used to measure machining temperatures. Embedded thermocouples [7], tool–work thermocouples [8], infrared photography and pyrometers [9], a metallographic technique based on microstructure and hardness [10], and a metallographic technique using powders with known melting temperature [11] have been utilized in the past to measure machining temperatures. Unfortunately, experimental measurement is difficult and inconvenient to record due to the complex contact phenomena between cutting tools and workpieces, and because of restricted accessibility during the machining process [12].

Numerical methods using finite element (FE) analysis and analytical methods were developed to predict temperatures, and numerical methods using FE models have made considerable progress in their ability to predict machining processes, including machining forces, temperatures, residual stress, and chip morphology. Umbrello et al. developed an FE model to predict conventional high-speed machining processes [13]. Liu et al. developed another FE model to predict sequential machining processes [14]. Yen et al. investigated the influence of tool geometry on machining prediction using an FE model to optimize tool edge design [15]. Özel et al. investigated the influence of tool coating on machining prediction using an FE model to demonstrate the advantages of coated tool design [16]. Umbrello et al. demonstrated that machining prediction using an FE model was very sensitive to the materials constitutive model constants [17]. Arrazola et al. demonstrated that consideration of the friction coefficient at the tool–chip-work interface results in improved accuracy of machining prediction using an FE model [18]. Unfortunately, the high computational cost and low computational efficiency of numerical methods have been major limitations preventing real-time prediction and optimization with process-parameter planning.

To overcome these limitations, analytical methods were developed that could predict machining processes with comparable accuracy along with considerably high computational efficiency [19,20]. The chip formation model was modified primarily to predict machining forces in orthogonal cutting, in which the Johnson–Cook constitutive model (J–C model) is employed to calculate material flow stress. Temperatures in the J–C model, specifically at the primary shear zone (PSZ) and tool–chip interface (alternatively named secondary shear zone or SSZ), are calculated using heat partition equations as intermediate variables for force prediction [21]. Temperatures at the PSZ can also be explicitly determined by observing the energy balance between plastic works caused by shear deformation and generated heat [22]. Komanduri et al. developed a temperature model that used two heat sources at the PSZ and SSZ to predict temperature distribution at the chip formation zone [23]. The heat source caused by shear deformation at the PSZ was observed using a moving heat source solution with boundary conditions defined by appropriate image sources. The heat source caused by the friction between the tool and chip at the SSZ was observed by comparing the equivalence between two heat source solutions, namely a moving heat source in the chip and a stationary heat source in the tool. This model was further developed by considering the thermal properties of tools and tool-wear under oblique cutting conditions [24,25]. Improved prediction accuracy was reported after results were validated against experimental measurements. Shalaby et al. developed a temperature model to predict machining temperatures by considering shear deformation and friction at two precision-turning deformations zones [26]. However, these developed analytical models need temperature-dependent material properties of the workpiece that must be obtained from extensive material property tests, which are inconvenient. The shear angle and strain rate constants in the chip formation model are determined iteratively with complex mathematical calculations, which limits optimal computational efficiency, and thus restricts real-time temperature prediction.

In this work, the machining temperatures at two deformation zones were predicted by an original temperature model using the correlation between machining forces and temperatures. Machining forces can be easily and reliably measured using a piezoelectric dynamometer as reported in the literature [27]. The temperatures were correlated to forces using a constitutive model and a mechanics model with stress calculations at the PSZ and SSZ, respectively. AISI 1045 steel was chosen to test the presented models under various cutting conditions. The predicted temperatures were validated against documented values in the literature [21,28]. For comparison, the analytical model reported in the previous work used the chip-thickness and constant-material-flow-rate assumption that prevents real-time temperature prediction and optimized prediction accuracy [29]. More details of the previous model and its predictive capability can be found in reference [30]. The experimental techniques and developed models used to investigate the machining process are summarized in Table 1. In addition, sensitivity analyses were conducted to investigate the influence of input forces and J–C model constants on prediction accuracy.

Table 1. Summary of experimental and modeling methods in the investigation of the machining process.

Methods	Experimental Techniques	Numerical Methods	Analytical Methods
	Embedded thermocouple [7], tool–work thermocouple [8], infrared technique [9], graphic techniques [10,11]	FEA for machining forces, temperature distribution, residual stress, and chip morphology [13,14]	Chip formation model [21], Komanduri's model [23], Shalaby's model [26], Ning's model [29]
Major advantage	Sufficient accuracy for in-situ/post-processing measurement	Sufficient prediction capability	High computational efficiency
Major disadvantage	High experimental complexity	High computational cost	Complex input requirement; high mathematical complexity

2. Methodology

Machining temperatures were predicted at the PSZ and SSZ in orthogonal cutting with machining forces used as inputs. Machining temperatures and forces were correlated using the mechanics of the cutting process and material constitutive relation. An orthogonal cutting configuration is illustrated as in Figure 1, where F_c and F_t are the cutting force and thrust force, respectively, that can be measured using a piezoelectric dynamometer. α, β, and ϕ are the tool–rake angle, friction angle, and shear angle, respectively. V_c, V_s, and V are the chip velocity, shear velocity, and cutting velocity, respectively. w is the cutting width that is not shown. Steady-state condition and plane-strain condition were enforced in the temperature prediction.

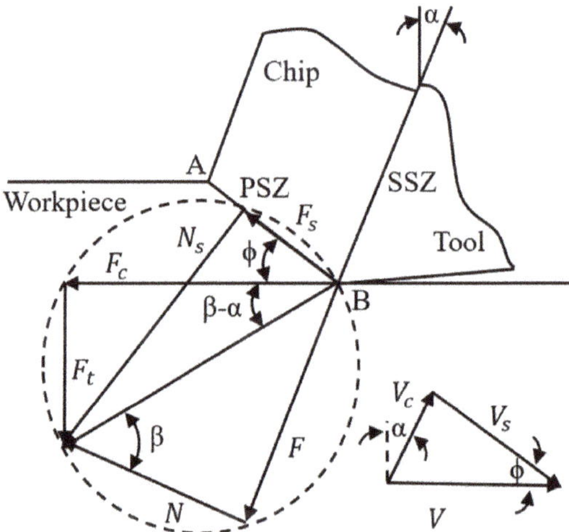

Figure 1. Schematic drawing of the orthogonal cutting process using a force circle. PSZ and SSZ denote the primary shear zone and secondary shear zone, respectively.

The stresses at two shear zones were calculated using the mechanics of the orthogonal cutting process with the given cutting force and thrust force, (see the Appendix A, Table A1, Table A2). The shear stresses at the PSZ (k_{AB}) and SSZ (τ_{int}) can be expressed using a mechanics model as

$$k_{AB} = \frac{F_s}{l_{AB}w} \tag{1}$$

$$\tau_{int} = \frac{F}{hw} \tag{2}$$

where F_s and F can be calculated with determined by the rake angle (α), friction angle (β), shear angle (ϕ), and experimental forces (F_c, F_t). l_{AB} and h are the length of the PSZ and SSZ, respectively.

The friction angle (β) can be calculated from the force circle as

$$\beta - \alpha = \mathrm{atan}\left(\frac{F_t}{F_c}\right) \tag{3}$$

The shear angle was determined by minimizing the cutting work according to the shear angle solution presented by Ernst and Merchant [31]. The cutting work was proportional to the cutting force, which can be expressed as

$$F_c = \tau \frac{wt_1}{\sin\phi} \frac{\cos(\beta - \alpha)}{\cos(\phi + \beta - \alpha)} \tag{4}$$

where τ is the shear stress, w is the width of cutting, and t_1 is the depth of cutting.

The shear angle (ϕ) can then be expressed by differentiating the above equation as

$$\phi = \frac{\pi}{4} - \frac{\beta}{2} + \frac{\alpha}{2} \tag{5}$$

The angle between shear force and resultant force (θ) can be calculated from the force circle as

$$\theta = \phi + \beta + \alpha \tag{6}$$

The lengths of the PSZ (l_{AB}) and SSZ (h) can be expressed as

$$l_{AB} = \frac{t_1}{sin\phi} \tag{7}$$

$$h = \frac{t_1 sin\theta}{cos\lambda sin\phi}\left(1 + \frac{C_0 n_{eq}}{3(1 + 2(\frac{\pi}{4} - \phi) - C_0 n_{eq})}\right) \tag{8}$$

The stresses at the two shear zones can also be calculated using the constitutive relation. The Johnson–Cook constitutive model (J–C model) was chosen for the calculation with consideration of the strain hardening effect, the strain-rate hardening effect, and the thermal softening effect. The J–C model can be expressed as

$$\sigma = (A + B\varepsilon^n)\left[1 + C\ln\left(\frac{\dot{\varepsilon}}{\dot{\varepsilon_0}}\right)\right]\left[1 - \left(\frac{T - T_r}{T_m - T_r}\right)^m\right] \tag{9}$$

where A, B, C, m, and n are five material constants that can be determined by various approaches such as Split–Hopkinson Pressure Bar (SHPB) tests [32], numerical methods [33], and analytical methods [34]. The analytical methods have less experimental complexity and high computational efficiency compared to the experimental tests and numerical methods, respectively, as discussed in the literature [35,36].

The shear stresses at the PSZ $\left(k'_{AB}\right)$ and SSZ (k_{int}) can be calculated using the J–C model with the von Mises yield criterion as

$$k'_{AB} = \frac{\sigma_{AB}}{\sqrt{3}} = \frac{1}{\sqrt{3}}(A + B\varepsilon^n_{AB})\left(1 + Cln\frac{\dot{\varepsilon}_{AB}}{\dot{\varepsilon}_0}\right)\left(1 - \left(\frac{T_{AB} - T_r}{T_m - T_r}\right)^m\right) \tag{10}$$

$$k_{int} = \frac{1}{\sqrt{3}}(A + B\varepsilon^n_{int})\left(1 + C\ln\frac{\dot{\varepsilon}_{int}}{\dot{\varepsilon}_0}\right)\left(1 - \left(\frac{T_{int} - T_r}{T_m - T_r}\right)^m\right) \tag{11}$$

where strains and strain rates are calculated as

$$\varepsilon_{AB} = \frac{\gamma_{AB}}{\sqrt{3}} = \frac{\cos\alpha}{2\sqrt{3}\sin\phi\cos(\phi - \alpha)} \tag{12}$$

$$\dot{\varepsilon}_{AB} = \frac{\dot{\gamma}_{AB}}{\sqrt{3}} = C_0\frac{V_s}{\sqrt{3}l_{AB}} \tag{13}$$

$$\varepsilon_{int} = \frac{\gamma_{int}}{\sqrt{3}} = 2\varepsilon_{AB} + \frac{h}{2\sqrt{3}\delta t_2} \tag{14}$$

$$\dot{\varepsilon}_{int} = \frac{\dot{\gamma}_{int}}{\sqrt{3}} = \frac{V_c}{\sqrt{3}\delta t_2} \tag{15}$$

The temperatures were determined by minimizing the difference between the stress calculated using the mechanics model and the same stress calculated using the J–C model at each shear zone as illustrated in Figure 2. Iterations with a defined temperature range, specifically a range between room temperature (T_r) and material melting temperature (T_m), were used in the minimization for temperature prediction. The cutting condition parameters, J–C model constants, and experimental forces were given as the inputs, and the average temperatures at the PSZ and SSZ were calculated as the outputs.

The machining temperature was predicted in the presented model using the correlation between the forces and temperature with the given forces as inputs, permitting it less mathematical complexity and thus higher computational efficiency compared to the chip formation model. The presented model also had less experimental complexity because of the following: (1) Experimental forces were reliable and easily measurable using a three-axial piezoelectric dynamometer. (2) Temperature-sensitive material properties such as thermal conductivity and specific heat (which require extensive material property tests to be obtained) were not needed in the presented model. In addition, the high computational efficiency allowed real-time temperature prediction with real-time force data. However, there were some limitations of the presented model: (1) The presented model only predicted the average temperatures at the PSZ and SSZ. (2) Prediction accuracy relied on accurate model inputs, such as forces and J–C model constants.

To further investigate the advantages and disadvantages of the presented model, machining temperatures were predicted in the orthogonal cutting of AISI 1045 steel under various cutting conditions. The following tasks were performed: (1) An investigation of prediction accuracy was conducted by validating against documented values in the literature. (2) An investigation was conducted into computational efficiency in terms of computational time. (3) An investigation was conducted on the influence of input machining forces on prediction accuracy. (4) An investigation was conducted on the influence of multiple sets of available J–C model constants on prediction accuracy. (5) A discussion was carried out on the usefulness of the predicted temperature data and future works.

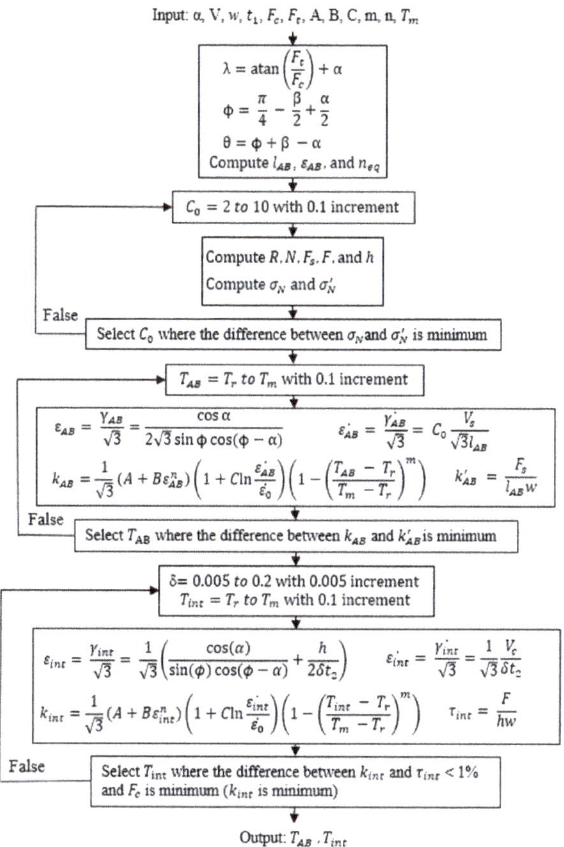

Figure 2. The algorithm of temperature predictions in the presented model.

3. Results and Discussion

In this work, machining temperatures were predicted in the orthogonal cutting of AISI 1045 steel under various cutting conditions. The model inputs of cutting condition parameters and machining forces were adopted from the literature [21,28] as presented in Table 2. The documented values in tests 1–4 were calculated using an improved chip formation model, in which machining temperatures were calculated with heat partition equations at two shear zones. The documented values in tests 5–8 were calculated using an extended chip formation model, in which machining temperatures were calculated based on two heat sources at the PSZ and SSZ. The J–C model constants of AISI 1045 steel were adopted from the literature [32], in which SHPB tests were conducted.

Table 2. Cutting condition parameters in the orthogonal machining of AISI 1045 steel ($w = 2$ mm, $\alpha = -7°$, $T_0 = 25$ °C) [21,28].

Test	V (m/min)	t_1 (mm)	F_{cR}(N)	F_{tR}(N)	T_{ABR} (°C)	T_{intR} (°C)
1	200	0.15	625.42	439.86	407.39	895.07
2	200	0.3	1077.7	637.19	383.1	992.44
3	300	0.15	574.55	364.74	393.31	947.81
4	300	0.3	1003.6	531.84	374.64	1049.8
5	200	0.15	576	500	385	942
6	200	0.3	1007	740	367	1042

Table 2. *Cont.*

Test	V (m/min)	t_1 (mm)	F_{cR}(N)	F_{tR}(N)	T_{ABR} (°C)	T_{intR} (°C)
7	300	0.15	533	478	374	1017
8	300	0.3	1041	628	387	1025

Note: Temperature and force values in Tests 1–4 were adopted from the literature [28] using an improved chip formation model, and Tests 6–8 were adopted from the literature [21] using an extended chip formation model. Subscript R denotes documented values.

In the presented model, the temperature at the PSZ (T_{AB}) was determined by minimizing the difference between the stress (k_{AB}) and the stress (k'_{AB}), while the temperature at the SSZ (T_{int}) was determined by minimizing the difference between the stress (τ_{int}) and the stress (k_{int}). The predicted temperatures were validated by the documented values in the literature as presented in Table 3. The documented values were validated through force comparison against experimental measurements using a three-axial piezoelectric dynamometer in orthogonal cutting tests [27] (the temperatures are intermediate variables in calculating machining forces). Good agreements were observed upon force validation. Other calculated variables of the shear angle and stresses are shown in Table 4. The temperature prediction was carried out using a MATLAB program on a personal computer running at 2.8 GHz. To investigate the computational efficiency, the computational time for each prediction was recorded as shown in Table 3. The average computational time was 0.27 s, which allowed real-time temperature prediction during the machining process and cutting-parameter planning with a trial-and-error method.

Table 3. Temperature prediction and validation in the orthogonal machining of AISI 1045 steel.

Test	T_{AB} (°C)	T_{int} (°C)	T_{AB} Deviation (%)	T_{int} Deviation (%)	t (s)
1	402.81	981.33	1.12	9.64	0.389
2	446.86	982.63	16.64	0.99	0.252
3	434.20	834.04	10.40	12.00	0.258
4	467.29	1089.66	24.73	3.80	0.243
5	424.04	1091.96	10.14	15.92	0.293
6	458.62	962.50	24.96	7.63	0.239
7	448.55	974.66	19.93	4.16	0.240
8	428.82	1094.65	10.81	6.79	0.242

Note: T_{AB} and T_{int} denote the average temperatures at the PSZ and SSZ respectively.

Table 4. Calculated shear angle and stresses at the PSZ and SSZ.

Test	ϕ (degs)	k_{AB} (MPa)	k'_{AB} (MPa)	τ_{int} (MPa)	k_{int} (MPa)
1	27.44	541.25	541.37	455.90	455.90
2	29.70	496.39	496.50	400.62	400.62
3	28.80	512.32	512.43	358.80	358.80
4	31.04	480.25	480.39	330.97	330.97
5	28.85	528.96	529.10	397.45	397.45
6	31.05	490.90	491.01	361.49	361.49
7	30.23	505.30	505.42	311.87	311.86
8	31.08	478.63	478.75	296.56	296.56

Good agreements were observed between the predicted temperatures and documented values as shown in Figure 3. The predicted temperatures at the PSZ and SSZ were generally larger than the documented values because of the assumption of a perfectly sharp cutting edge in the chip formation model; underestimated machining forces and temperatures were reported with this assumption in the literature, which affected the literature's predictions and experimental measurements [21,37]. The deviations between predicted temperatures and documented values might also have been affected by the deviation of input machining forces due to vibrations, which were frequently observed in heavy-duty operations [38] and in machining difficult-to-cut materials [39]. A study of the influence of input forces on the accuracy of temperature prediction is needed.

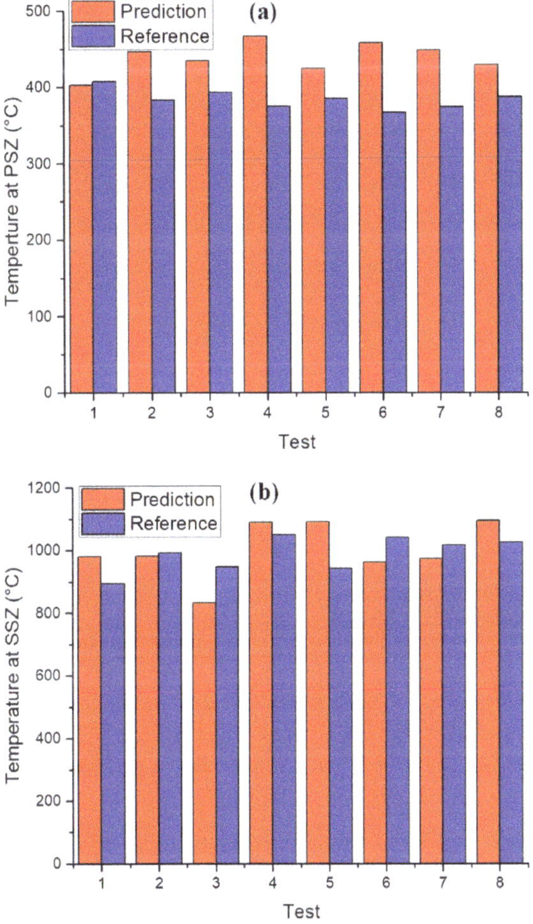

Figure 3. Temperature validation against documented values in the orthogonal cutting of AISI 1045 steel [21,28]. (**a**) Validation of temperature prediction at the primary shear zone. (**b**) Validation of temperature prediction at the secondary shear zone.

To investigate the influence of input experimental forces on the predicted temperatures, the input cutting force and thrust force were deliberately changed (separately) up to ±20% from their original values under the test 1 cutting condition. The prediction error was calculated by comparing to the documented values from the literature [28], as shown in Figure 4. For the temperature at the PSZ, the prediction error using input forces was found at the global minima. For the temperature at the SSZ, the prediction error using input forces was found near the local minima, with relatively larger values. The temperature prediction at the PSZ was more sensitive than the temperature at the SSZ in the orthogonal machining of AISI 1045 steel. The temperature deviations at the PSZ were much larger than that at the SSZ with the same amount of input-force deviations. In addition, the predicted temperatures at the PSZ were more sensitive to input-cutting-force deviations.

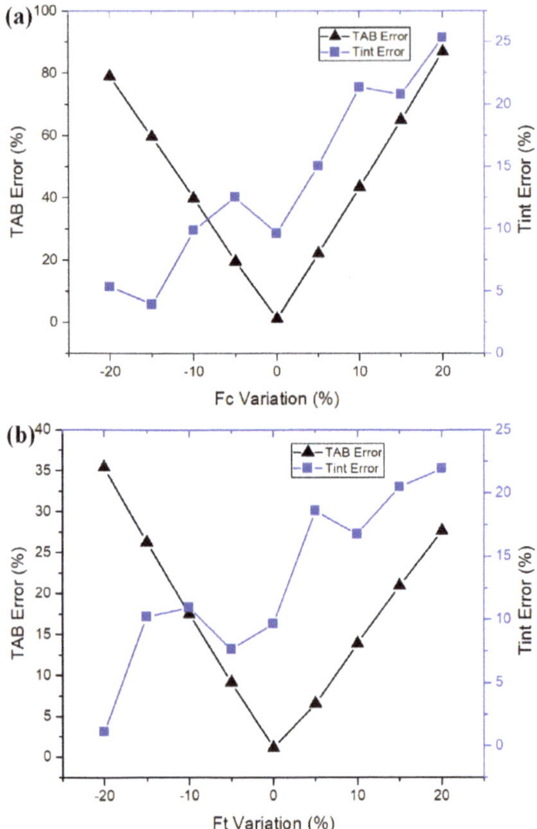

Figure 4. Sensitivity analyses of (**a**) cutting force and (**b**) thrust force on the temperature prediction.

Multiple sets of J–C constants are available for AISI 1045 steel (shown in Table 5). They were determined using different methods. To investigate the influence of J–C model constants on the accuracy of temperature prediction, different sets of J–C constants were used for prediction as shown in Table 5. The predicted temperatures were validated against documented values under the test 7 cutting condition as illustrated in Figure 5. Acceptable agreements were observed upon temperature validation.

Table 5. Johnson–Cook constitutive model constants of AISI 1045 steel ($T_m = 1460\ ^\circ$C; $\dot{\varepsilon}_0 = 1$).

Set	Method	A (MPa)	B (MPa)	C	m	n
1	SHPB [32]	553.1	600.8	0.0134	1	0.234
2	FEA [40]	546	487	0.03	0.672	0.25
3	Analytical Modeling [41]	451.6	819.5	0.0000009	1.0955	0.1736
4	PSO [42]	646.19	517.7	0.0102	0.94054	0.24597
5	PSO-c [42]	731.63	518.7	0.00571	0.94054	0.3241
6	CPSO [42]	546.83	609.35	0.01376	0.94053	0.2127

Note: SHPB: Split–Hopkinson pressure bar; PSO: particle swarm optimization algorithm; CPSO: corporative particle swarm optimization algorithm.

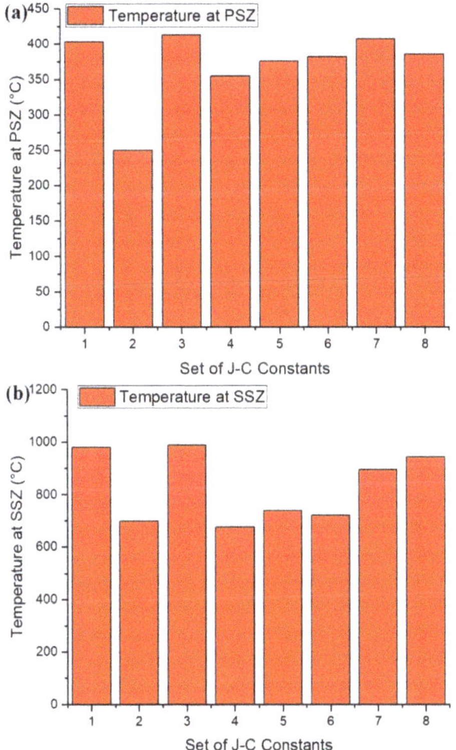

Figure 5. Sensitivity analysis of Johnson–Cook model constants on temperature predictions at (**a**) the primary shear zone; and (**b**) the secondary shear zone. The predicted temperatures under sets 1–6 were predicted using adopted J–C constants that were determined using different methodologies. The temperatures under sets 7 and 8 were adopted from the literature [21,28].

The predicted temperatures at the PSZ and SSZ are sufficient for the further investigation of machining forces [1], tool wear [43], and material diffusion [44], as reported in the literature. High computational efficiency allows process-parameter planning with a trial-and-error calculation to determine the desired temperature conditions. AISI 1045 steel was chosen for this study because of the ready availability of machining data and J–C model constants. Other metal materials should be investigated to extend the applicability of the presented model in future works.

4. Conclusions

This work presents an original analytical model for temperature prediction in the machining process. Machining temperatures and forces were correlated with the mechanics of the cutting process and constitutive relation. The stresses at two shear zones were calculated with the input cutting force and thrust force using a mechanics model. The same stresses were also calculated with unknown temperatures using the J–C model. The minimization between calculated stresses yielded an estimation of unknown temperatures at the PSZ and SSZ. Good agreements were observed based upon validation against the documented temperatures in the literature. The presented model improved an understanding of the force–temperature relationship in the machining process using mathematical calculation. The influence of input-force deviations and J–C model constants on the accuracy of temperature prediction was investigated with sensitivity analyses. Temperature prediction at the PSZ was more susceptible to input-force deviations than temperature prediction at the SSZ.

Acceptable predicted accuracy was achieved with multiple sets of available J–C constants. In addition, the average computational time for temperature prediction using the presented model was about 0.27 s, which allowed for real-time temperature prediction and process-parameter planning using trial-and-error calculations. Having achieved a high level of prediction accuracy, high computational efficiency, and low experimental complexity, this presented temperature model can be employed in the future for investigating temperature in the machining process. The applicability of the presented temperature model can further be used for prediction in machining different metals.

Author Contributions: Formal analysis, J.N.; Investigation, J.N.; Validation, J.N.; Writing, J.N. Steven Y. Liang provided general guidance.

Funding: No funding was received for this work.

Conflicts of Interest: The authors declare no conflict of interest.

Nomenclature

PSZ = primary shear zone (with subscript AB)
SSZ = secondary shear zone or tool–chip interface (with subscript int)
$A; B; C; m; n$ = yield strength; strength coefficient; strain rate coefficient; thermal softening coefficient; and strain hardening coefficient in the J–C model
$T_m; T_r; T$ = melting temperature; room temperature; temperature
$V; V_c; V_s$ = cutting velocity; chip velocity; shear velocity
$\alpha; \phi; \lambda; \theta$ = rake angle; shear angle; friction angle at the SSZ; angle between resultant force R and the PSZ
$w; t_1; t_2$ = width of cut; depth of cut; and chip thickness
$l_{AB}; h$ = length of the PSZ; the length of the SSZ (tool–chip contact length)
$\varepsilon_{AB}; \dot{\varepsilon}_{AB}; \varepsilon_{int}; \dot{\varepsilon}_{int}$ = strains and strain rates at the PSZ and SSZ
C_0 = Oxley constants (ratio of the shear plane length to the thickness of the PSZ)
δ = strain rate constant (ratio of the thickness of the SSZ to chip thickness)
n_{eq} = strain hardening constant
k_{AB} = calculated shear stress at the PSZ using the J–C model
k'_{AB} = calculated shear stress at the PSZ using a mechanics model
k_{int} = calculated shear stress at the SSZ using the J–C model
τ_{int} = calculated shear stress at the SSZ using a mechanics model
σ_N = calculated normal stress at the SSZ using a mechanics model
σ'_N = calculated normal stress at the SSZ using the J–C model
F_c = cutting force
F_t = thrust force
F_s = shear force at the PSZ
N_s = normal force at the PSZ
F = shear force at the SSZ
N = normal force at the SSZ
R = resultant force

Appendix A

Table A1. Variables in sensitivity analysis of cutting force under test 1 cutting condition.

F_c Variation (%)	T_{AB} (°C)	T_{int} (°C)	k_{AB} (MPa)	k'_{AB} (MPa)	τ_{int} (MPa)	k_{int} (MPa)
20	54.00	668.30	716.74	716.85	496.92	496.92
15	142.80	708.86	672.02	672.14	489.05	489.05
10	230.80	703.98	627.83	627.96	477.84	477.84
5	317.40	760.19	584.22	584.37	467.56	467.56
−5	402.81	981.33	541.25	541.37	455.90	455.90
−10	486.92	782.91	498.98	499.12	440.89	440.89
−15	569.44	983.53	457.50	457.64	424.49	424.49
−20	650.07	859.45	416.88	417.02	408.30	408.30

Table A2. Variables in sensitivity analysis of thrust force under test 1 cutting condition.

F_t Variation (%)	T_{AB} (°C)	T_{int} (°C)	k_{AB} (MPa)	k'_{AB} (MPa)	τ_{int} (MPa)	k_{int}(MPa)
20	520.11	698.60	484.26	484.37	481.07	481.07
15	492.60	711.75	497.56	497.70	477.07	477.07
10	463.90	745.24	511.48	511.62	471.94	471.94
5	434.10	728.80	526.03	526.17	463.62	463.62
−5	402.81	981.33	541.25	541.37	455.90	455.90
−10	370.22	826.86	557.18	557.32	444.73	444.73
−15	336.24	797.16	573.84	573.97	431.95	431.95
−20	300.46	803.64	591.29	591.42	419.36	419.36

References

1. Karpat, Y.; Özel, T. Predictive analytical and thermal modeling of orthogonal cutting process—part I: Predictions of tool forces, stresses, and temperature distributions. *J. Manuf. Sci. Eng.* **2006**, *128*, 435–444. [CrossRef]
2. Danish, M.; Ginta, T.L.; Habib, K.; Carou, D.; Rani AM, A.; Saha, B.B. Thermal analysis during turning of AZ31 magnesium alloy under dry and cryogenic conditions. *Int. J. Adv. Manuf. Technol.* **2017**, *91*, 2855–2868. [CrossRef]
3. Ning, J.; Nguyen, V.; Liang, S.Y. Analytical modeling of machining forces of ultra-fine-grained titanium. *Int. J. Adv. Manuf. Technol.* **2018**, 1–10. [CrossRef]
4. Danish, M.; Ginta, T.L.; Habib, K.; Abdul Rani, A.M.; Saha, B.B. Effect of cryogenic cooling on the heat transfer during turning of AZ31C magnesium alloy. *Heat Transfer. Eng.* **2018**, 1–10. [CrossRef]
5. Feng, Y.; Lu, Y.T.; Lin, Y.F.; Hung, T.P.; Hsu, F.C.; Lin, C.F.; Lu, Y.C.; Liang, S.Y. Inverse analysis of the cutting force in laser-assisted milling on Inconel 718. *Int. J. Adv. Manuf. Technol.* **2018**, *96*, 905–914. [CrossRef]
6. Li, F.; Li, X.; Qin, X.; Rong, Y.K. Study on the plane induction heating process strengthened by magnetic flux concentrator based on response surface methodology. *J. Mech. Sci. Technol.* **2018**, *32*, 2347–2356. [CrossRef]
7. O'sullivan, D.; Cotterell, M. Temperature measurement in single point turning. *J. Mater Process. Technol.* **2001**, *118*, 301–308. [CrossRef]
8. Leshock, C.E.; Shin, Y.C. Investigation on cutting temperature in turning by a tool-work thermocouple technique. *J. Manuf. Sci. Eng.* **1997**, *119*, 502–508. [CrossRef]
9. Sutter, G.; Faure, L.; Molinari, A.; Ranc, N.; Pina, V. An experimental technique for the measurement of temperature fields for the orthogonal cutting in high speed machining. *Int. J. Mach. Tool Manuf.* **2003**, *43*, 671–678. [CrossRef]
10. Wright, P.K. Correlation of tempering effects with temperature distribution in steel cutting tools. *J. Eng. Ind.* **1978**, *100*, 131–136. [CrossRef]
11. Kato, S.; Yamaguchi, K.; Watanabe, Y.; Hiraiwa, Y. Measurement of temperature distribution within tool using powders of constant melting point. *J. Eng. Ind.* **1976**, *98*, 607–613. [CrossRef]
12. Longbottom, J.M.; Lanham, J.D. Cutting temperature measurement while machining–A review. *Aircr. Eng. Aerosp. Technol.* **2005**, *77*, 122–130. [CrossRef]
13. Umbrello, D. Finite element simulation of conventional and high speed machining of Ti6Al4V alloy. *J. Mater. Process Technol.* **2008**, *196*, 79–87. [CrossRef]
14. Liu, C.R.; Guo, Y.B. Finite element analysis of the effect of sequential cuts and tool–chip friction on residual stresses in a machined layer. *Int. J. Mech. Sci.* **2000**, *42*, 1069–1086. [CrossRef]
15. Yen, Y.C.; Jain, A.; Altan, T. A finite element analysis of orthogonal machining using different tool edge geometries. *J. Mater. Process Technol.* **2004**, *146*, 72–81. [CrossRef]
16. Özel, T.; Sima, M.; Srivastava, A.K.; Kaftanoglu, B. Investigations on the effects of multi-layered coated inserts in machining Ti–6Al–4V alloy with experiments and finite element simulations. *CIRP Anal.* **2010**, *59*, 77–82. [CrossRef]
17. Umbrello, D.; M'saoubi, R.; Outeiro, J.C. The influence of Johnson–Cook material constants on finite element simulation of machining of AISI 316L steel. *Int. J. Mach. Tool. Manuf.* **2007**, *47*, 462–470. [CrossRef]
18. Arrazola, P.J. Investigations on the effects of friction modeling in finite element simulation of machining. *Int. J. Mech. Sci.* **2010**, *52*, 31–42. [CrossRef]

19. Lamikiz, A.; López de Lacalle, L.N.; Sanchez, J.A.; Bravo, U. Calculation of the specific cutting coefficients and geometrical aspects in sculptured surface machining. *Mach. Sci. Technol.* **2005**, *9*, 411–436. [CrossRef]

20. Calleja, A.; Alonso, M.A.; Fernández, A.; Tabernero, I.; Ayesta, I.; Lamikiz, A.; López de Lacalle, L.N. Flank milling model for tool path programming of turbine blisks and compressors. *Int. J. Prod. Res.* **2015**, *53*, 3354–3369. [CrossRef]

21. Lalwani, D.I.; Mehta, N.K.; Jain, P.K. Extension of Oxley's predictive machining theory for Johnson and Cook flow stress model. *J. Mater. Process Technol.* **2009**, *209*, 5305–5312. [CrossRef]

22. Adibi-Sedeh, A.H.; Madhavan, V.; Bahr, B. Extension of Oxley's analysis of machining to use different material models. *J. Manuf. Sci. Eng.* **2003**, *125*, 656–666. [CrossRef]

23. Komanduri, R.; Hou, Z.B. Thermal modeling of the metal cutting process—Part III: Temperature rise distribution due to the combined effects of shear plane heat source and the tool–chip interface frictional heat source. *Int. J. Mech. Sci.* **2001**, *43*, 89–107. [CrossRef]

24. Huang, Y.; Liang, S.Y. Cutting forces modeling considering the effect of tool thermal property—application to CBN hard turning. *Int. J. Mach. Tool. Manuf.* **2003**, *43*, 307–315. [CrossRef]

25. Li, K.M.; Liang, S.Y. Modeling of cutting forces in near dry machining under tool wear effect. *Int. J. Mach. Tool Manuf.* **2007**, *47*, 1292–1301. [CrossRef]

26. Shalaby, M.A.; El Hakim, M.A.; Veldhuis, S.C. A thermal model for hard precision turning. *Int. J. Adv. Manuf. Technol.* **2018**, *98*, 2401–2413. [CrossRef]

27. Ivester, R.W.; Kennedy, M.; Davies, M.; Stevenson, R.; Thiele, J.; Furness, R.; Athavale, S. Assessment of machining models: Progress report. *Mach. Sci. Eng.* **2000**, *4*, 511–538. [CrossRef]

28. Aydın, M. Cutting temperature analysis considering the improved Oxley's predictive machining theory. *J. Braz. Soc. Mech. Sci. Eng.* **2016**, *38*, 2435–2448. [CrossRef]

29. Ning, J.; Liang, S. Prediction of Temperature Distribution in Orthogonal Machining Based on the Mechanics of the Cutting Process Using a Constitutive Model. *J. Manuf. Mater. Process.* **2018**, *2*, 37. [CrossRef]

30. Ning, J.; Liang, S. Evaluation of an Analytical Model in the Prediction of Machining Temperature of AISI 1045 Steel and AISI 4340 Steel. *J. Manuf. Mater. Process.* **2018**, *2*, 74. [CrossRef]

31. Liang, S.; Shih, A.J. *Analysis of Machining and Machine Tools*; Chapter 7.4; Springer: New York, NY, USA; ISBN 978-1-4899-7643-7.

32. Kolsky, H. An investigation of the mechanical properties of materials at very high rates of loading. *Proc. Phys. Soc. Sect. B* **1949**, *62*, 676. [CrossRef]

33. Agmell, M.; Ahadi, A.; Ståhl, J.E. Identification of plasticity constants from orthogonal cutting and inverse analysis. *Mech. Mater.* **2014**, *77*, 43–51. [CrossRef]

34. Ning, J.; Liang, S.Y. Model-driven determination of Johnson-Cook material constants using temperature and force measurements. *Int. J. Adv. Manuf. Technol.* **2018**, *97*, 1053–1060. [CrossRef]

35. Ning, J.; Nguyen, V.; Huang, Y.; Hartwig, K.T.; Liang, S.Y. Inverse determination of Johnson–Cook model constants of ultra-fine-grained titanium based on chip formation model and iterative gradient search. *Int. J. Adv. Manuf. Technol.* **2018**, *99*, 1131–1140. [CrossRef]

36. Ning, J.; Liang, S.Y. Inverse identification of Johnson-Cook material constants based on modified chip formation model and iterative gradient search using temperature and force measurements. *Int. J. Adv. Manuf. Technol.* **2019**, 1–12. [CrossRef]

37. M'saoubi, R.; Chandrasekaran, H. Investigation of the effects of tool micro-geometry and coating on tool temperature during orthogonal turning of quenched and tempered steel. *Int. J. Mach. Tool Manuf.* **2004**, *44*, 213–224. [CrossRef]

38. Urbikain, G.; Campa, F.J.; Zulaika, J.J.; De Lacalle LN, L.; Alonso, M.A.; Collado, V. Preventing chatter vibrations in heavy-duty turning operations in large horizontal lathes. *J. Sound Vib.* **2015**, *340*, 317–330. [CrossRef]

39. Urbicain, G.; Palacios, J.A.; Fernández, A.; Rodríguez, A.; de Lacalle, L.L.; Elías-Zúñiga, A. Stability prediction maps in turning of difficult-to-cut materials. *Procedia Eng.* **2013**, *63*, 514–522. [CrossRef]

40. Klocke, F.; Lung, D.; Buchkremer, S. Inverse identification of the constitutive equation of Inconel 718 and AISI 1045 from FE machining simulations. *Procedia CIRP* **2013**, *8*, 212–217. [CrossRef]

41. Özel, T.; Zeren, E. A methodology to determine work material flow stress and tool-chip interfacial friction properties by using analysis of machining. *J. Manuf. Sci. Eng.* **2006**, *128*, 119–129. [CrossRef]

42. Özel, T.; Karpat, Y. Identification of constitutive material model parameters for high-strain rate metal cutting conditions using evolutionary computational algorithms. *Mater. Manuf. Process* **2007**, *22*, 659–667. [CrossRef]
43. Karpat, Y.; Özel, T. Predictive analytical and thermal modeling of orthogonal cutting process—part II: Effect of tool flank wear on tool forces, stresses, and temperature distributions. *J. Manuf. Sci. Eng.* **2006**, *128*, 445–453. [CrossRef]
44. Attanasio, A.; Ceretti, E.; Rizzuti, S.; Umbrello, D.; Micari, F. 3D finite element analysis of tool wear in machining. *CIRP Ann. Manuf. Technol.* **2008**, *57*, 61–64. [CrossRef]

 *materials*

Article

Hybrid Composite-Metal Stack Drilling with Different Minimum Quantity Lubrication Levels

J. Fernández-Pérez, J. L. Cantero *, J. Díaz-Álvarez and M. H. Miguélez

Department of Mechanical Engineering, Universidad Carlos III de Madrid, Avda. Universidad 30, Leganés, 28911 Madrid, Spain; jfperez@pa.uc3m.es (J.F.-P.); jodiaz@ing.uc3m.es (J.D.-Á.); mhmiguel@ing.uc3m.es (M.H.M.)
* Correspondence: jcantero@ing.uc3m.es; Tel.: +34-916-245-860

Received: 27 December 2018; Accepted: 29 January 2019; Published: 1 February 2019

Abstract: Hybrid stack drilling is a very common operation used in the assembly of high-added-value components, which combines the use of composite materials and metallic alloys. This process entails the complexity of machining very dissimilar materials, simultaneously, on account of the interactions that are produced between them, during machining. This study analyzed the influence of Minimum Quantity Lubrication (MQL) on the performance of diamond-coated carbide tools when drilling Ti/carbon fiber reinforced plastics (CFRP)/Ti stacks. The main wear mechanism observed was diamond-coating detachment, followed by fragile breaks in the main cutting-edge. The tests done with the lower lubrication levels have shown an important adhesion of titanium (mainly on the secondary cutting-edge) and a higher friction between the tool and the workpiece, producing higher temperatures on the cutting region and a thermal softening effect on the workpiece. These phenomena affect the evolution of cutting power consumption with tool wear in the titanium layer. Regarding the quality of the test specimen, no significant differences were observed between the lubrication levels tested.

Keywords: hybrid stacks drilling; minimum quantity lubrication; hole quality; tool wear

1. Introduction

The use of hybrid stacks for the design of high-added-value components has skyrocketed in the last decade, especially in the aeronautical industry. These stacks are composed of different materials, mainly composites and metals. The most common of these are titanium alloys (Ti) and carbon fiber reinforced plastics (CFRP) [1]. These materials stand out because of their excellent properties, and combining them provides the component with their individual characteristics. In contrast, interactions between these materials may produce a galvanic corrosion, so the use of sealants is required [2].

More than the 50% of the structural weight of some commercial aircrafts—such as the Boeing 787 and A350—is made of composite materials. Most of these are CFRP, and about 15% is made of Ti alloys [3]. To ensure the proper assembly of two components in the aerospace industry, the use of mechanical joints is still unavoidable. Drilling operations have a remarkable impact on the overall manufacturing cost, so the optimization and control of this process may have a big economic impact.

Hybrid stack drilling is a complex procedure due to the disparate nature and machinability of the materials involved. Composite fiber reinforced plastics are anisotropic, made of two faces—the fibers or reinforcements (which have a fragile behavior and low thermal conductivity) and the matrix or binder (which is more ductile). Its machining is characterized by the intermittent fracture of the fibers [4]. On the other hand, titanium alloys have a very high toughness, low elastic modulus, and resistance at high temperatures. Furthermore, they present poor machinability due to their high strength, very low thermal conductivity, and high chemical affinity with most of the cutting tools materials [5].

The study carried out by Wang et al. [6] analyzed the different wear mechanisms produced by each individual layer, when drilling hybrid stacks. They found that flank and chipping were the main mechanisms acting on Ti and edge-rounding in CFRP, which complicates the characterization of the process and the establishment of the tool life. Ramulu et al. [7] highlighted that carbides are the most appropriate materials for these processes, based on the life of the tools and the quality of the component. Park et al. [8] observed that the main wear mechanisms of uncoated carbide tools in composite/titanium stacks is flank-wear, due to the hardness of the carbon fibers and Ti adhesion. Fernandez–Vidal et al. [9] also studied the behavior of uncoated carbide tools, but on composite and aluminum stacks. They have shown that, in this case, bond-wear dominates over the abrasion of the tool surface. Rezende et al. [10] tested several cutting geometries in an aluminum/composite sandwich stack and found that the brad and spur geometry produced lower thrust forces and shorter burrs. Other authors described the influence of the tool wear in the modification of the cutting geometry and the associated machining-induced damage on the composite layers [11].

Along with the possible defects produced during the manufacturing process [12], drilling operations might cause damage on the composite layers in the form of delamination, thermal degradation, and fuzzing. When dealing with hybrid stacks, the interaction with metallic layers can further affect the quality of the hole [13]. The high temperatures produced during Ti drilling can induce thermal damage on the hole, by burning the matrix. Additionally, the metallic chips or the burrs can affect the surface quality. Furthermore, the different elastic modulus of the materials can produce variable diameters in each layer, affecting the structural integrity of the mechanical joint [14]. A completed analysis of the quality requirement in hybrid stack components was made by Shyha et al. [15].

The use of cutting fluids can help the machining process by dissipating the heat produced at the cutting region, and its use is a widespread practice [16]. The environmental impact of cutting fluids, its influence on the working environment, and its cost, justify the interest in reducing its utilization [17]. Furthermore, in hybrid composite-metal stack drilling processes, the use of conventional fluids to avoid the contamination of the composite layers is not allowed. For these operations, the Minimum Quantity Lubrication technique (MQL) is used. It is characterized by a high pressure flow of air with a small quantity of pulverized oil, which is applied directly to the cutting region, through the lubrication channels inside the tool [18], only when the tool drills through the metallic layers. To the extent possible, it is interesting to reduce the amount of oil used, to avoid cleaning operations after drilling. Several studies [19] have shown the influence of applying MQL on hole quality, in different machining processes. Giasin et al. [20] analyzed the impact of MQL lubrication in a glass fiber and aluminum stack drilling process and they observed a reduction of the burr height, at the exit of the stack, in dry conditions. Bhowmick et al. [21] studied the process of MQL drilling of a cast magnesium alloy, in dry conditions, and they found a reduction in the built-up edge formation and a reduction in the torque and thrust force. Furthermore, a softening effect of the workpiece—owing to high temperatures—was produced during dry drilling but not with MQL.

The high cost of the tools used and the number of mechanical joints required in the assembly of the high value components in the aeronautical industry makes it very difficult to complete the operations with adequate levels of competition. Furthermore, the restrictive tolerances require periodical quality controls to avoid the rejection of the component.

Hence, the objective of this study was to analyze the influence of the MQL level—through the interior of the tool—on the performance of the hybrid stack drilling process, with diamond-coated carbide drill bits. The quality of the component, the tool wear, and the evolution of the power consumption during the process were evaluated.

2. Materials and Methods

The experimental equipment used to perform the test and analyze the quality of the coupons is presented below.

2.1. Workpiece Materials and Cutting Tools

The cutting tools tested were made of a carbide substrate K10, with a diamond coating from HAM Präzision. Figure 1 shows one of the drill bits tested. These tools were specially designed following the Airbus Getafe specifications. They had two cutting edges, with a point angle of 140°, a diameter of 7.6 mm, and a split-point geometry. Two lubrication channels were located at the frontal relief surface of each edge.

(a)

(b)

Figure 1. The diamond-coated carbide drill bit with a 7.6 mm diameter that were tested. (**a**) Lateral view. (**b**) Frontal view.

The specimen tested was a hybrid stack used in the aeronautical industry, with a Ti/CFRP/Ti configuration. On one hand, the Ti alloy used on this coupon was Ti6AlV. On the other hand, the composite material was a carbon-fiber-reinforced polymer, composed of multiple unidirectional layers with different orientations. These were covered with an epoxy pre-impregnated copper foil, in the upper part, and with a prepreg made from a glass fiber fabric pre-impregnated with epoxy, in the lower part.

2.2. Machining Conditions

The drilling tests were done with the machines, tools, materials, and cutting conditions used for the assembly process of aeronautical components. Hence, the results presented can be directly applied to industrial production conditions.

The machine used was a computer numerical control (CNC) gantry drilling machine. The spindle head was a MFW 1406/24/2 from Fischer (Switzerland), with a nominal power of 15.200 W and a feed movement motor 1FK7 from Siemens (Berlin, Germany), with a nominal power of 940 W. The MQL lubrication system used was a Lubrix V5, and the lubricant oil used was Accu-Lube LB5000 (Maulbronn, Germany).

Two MQL lubrication levels, corresponding to the nominal conditions of the equipment, were tested. One was a low level, with an oil flow of 3 mL/h, and another was a high level, with a flow of 15 mL/h. For each condition, 40 holes were made with a new tool.

Table 1 shows the main cutting parameters used and the thickness of each layer of the stack. Different cutting conditions were used for the composite and metallic layer, due to their disparate machinability properties. For Ti drilling, a smaller cutting speed, MQL lubrication, and peck drilling were required to mitigate the formation of long and continuous chips, which can damage the composite layer and promote tool wear. On the CFRP layer, the lubrication flow had to be interrupted, since it was not needed for this material, and the excess oil could have affected the behavior of the composite materials. The possible influence of the residual lubrication of the CFRP layer is minimal, since the amount of oil used on the metallic layers is very low. The staking of the coupon was made by applying pressure with clamps.

Table 1. The stack configuration and cutting parameters tested.

Material	Thickness (mm)	Cutting Speed (m/min)	Feed (mm/rev)	Minimum Quantity Lubrication (MQL) Lubrication
Ti	9.3	15	0.05	Yes
CFRP	8.4	70	0.05	No
Ti	9.3	15	0.05	Yes

The CNC machine used was equipped with sensors to independently measure the power consumptions of the spindle and the feed motors, which were closely related to the cutting torque and the thrust force, respectively. Being able to measure and analyze these variables provided important information on the process [22].

2.3. Hole Quality and Tool Wear Evaluation

Controlling the quality of the holes is mandatory for fulfilling the requirements which ensures a proper behavior of the manufacturing process and the mechanical joint. The tolerances of the different quality requirements depended on the applications of the process. In this case, the application was the assembly of the aeronautical components through mechanical joints. Hence, the tolerances required were considered to be "low clearance fit assemblies in high-load transfer applications".

- Diameter: The measurements were done with the digital bore gauge XT3 from Bowers (Bradford, England). Calibrated setting rings were used to check the repeatability and accuracy of the measurements and variations of ± 2 µm were observed. To define each diameter, three measurements were done at the entrance and exit of each layer. The average value was taken.
- Hole surface quality: This was measured with a contact profilometer MARWIN XCR20 from Mahr (Esslingen, Germany), using a 5 µm styli, with a measurement speed of 0.5 mm/s. The roughness parameter used in this case was the average roughness parameter (Ra). Three consecutives measurements of the inner surface of the hole were done, taking the highest value obtained as the result was based on Standard DIN 4774, to ensure the fulfillment of the dimensional tolerance required in all cases. Additionally, the skewness parameter (Rsk) was also calculated. Figure 2 shows the contact profilometer equipment used to analyze the hole surface quality.
- Machining induced damage on the composite layer: Holes were visually inspected to look for damage in the form of delamination and scratching produced by metallic chips, fuzzing, or thermal degradation. No damages were observed in any of the holes analyzed in this study.
- Burr: This phenomenon is characteristic of the metallic layers and it is produced at the entrance and exit of the hole. A Mitotuyo 2046S dial gauge (Kawasaki, Japan) was used. Several measurements were done on the same hole, to determine the maximum value. Differences in the values obtained for each hole were in the order of 0.05 mm.

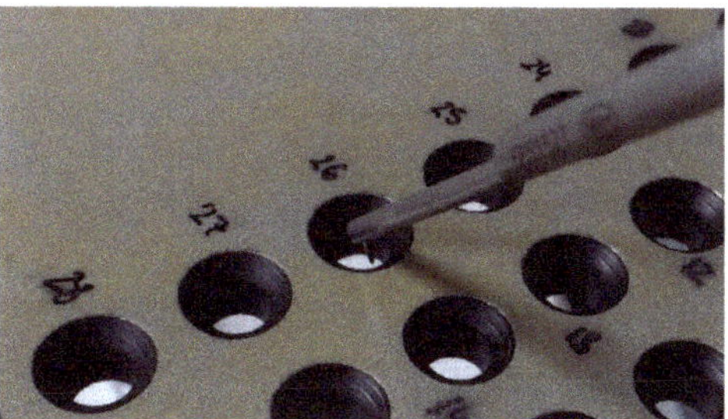

Figure 2. Hole profile and roughness analysis of a carbon fiber reinforced plastics (CFRP) coupon with a contact profilometer.

Tool wear was monitored and characterized using a scanning electron microscope (SEM) Philips XL-30 (Amsterdam, Netherlands), with an EDSDX4i system.

3. Results and Discussion

The influence of the lubrication level on the performance of the hybrid stack drilling processes was analyzed in terms of tool wear, hole quality, and cutting and feed movement power consumption.

3.1. Tool Wear Characterization

Figures 3 and 4 show the condition of the tools after 40 holes. These were tested with a lubrication level of 3 and 15 mL/h, respectively. The main wear mechanisms produced in both tools were the loss of the diamond coating by delamination and fragile breakages of the main cutting edges. Furthermore, Ti adherence was found in both conditions, but a stronger built-up edge was produced on the secondary cutting-edge of the tool tested with the lower lubrication level.

The loss of the diamond layer affected the whole length of the main cutting-edges and the secondary cutting-edge. It was produced by the high temperatures generated during the Ti layer drilling, which induced thermal stresses on the coating from the differences between the coefficients of thermal expansion of the diamond layer and the carbide substrate. Furthermore, the fluctuations of the cutting forces, during the CFRP layer machining aggravated the tool wear mechanism [23]. Therefore, the hardness of the cutting-edge regions where the coating was lost was reduced and the substrate was exposed to the abrasiveness of the carbon fibers [24].

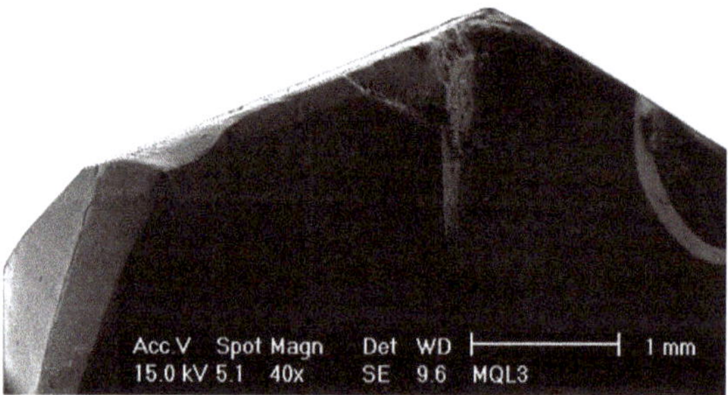

Figure 3. Wear of the tool tested on a Ti/CFRP/Ti stack with the lower lubrication level (3 mL/h), after 40 holes.

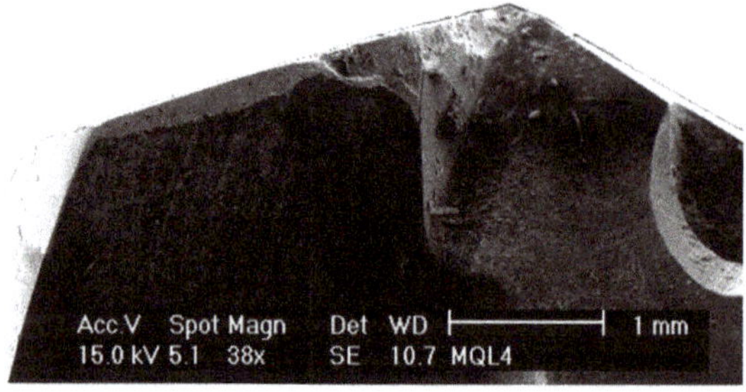

Figure 4. Wear of the tool tested on a Ti/CFRP/Ti stack with the higher lubrication level (15 mL/h), after 40 holes.

Regarding the fragile breaks, it can be seen in Figures 3 and 4 that they were produced from the middle region of the cutting-edge to the tip of the tool. The morphology of the damage was very similar for both lubrication levels, but its extension was larger at the lower lubrication level condition.

Furthermore, the tool tested with the lower lubrication level also suffered damage on the tip of the cutting-edge (Figure 3).

The description of the wear suffered on the tools was focused on the rake surface of the main cutting-edges. Figure 5 shows a detailed view of the relief surface where a loss of diamond coating by delamination can also be seen, but not any other additional types of tool wear.

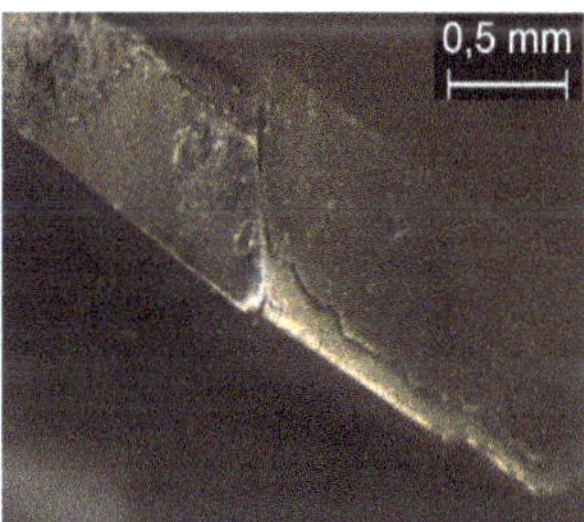

Figure 5. Detailed view of the relief surface of the tool with the lower lubrication level, after 40 holes.

Ti has a strong affinity with the substrate material of the cutting tools, especially at high temperatures. Figure 6 shows a composition analysis of the cutting-edges of the tools tested with a lubrication level of 3 mL/h (Figure 6a) and 15 mL/h (Figure 6b). The black color corresponds to carbon components, the grey tones are the titanium alloy, and the white color indicates the tungsten of the carbide substrate. The tools tested with the lower lubrication condition suffered a greater Ti-adhesion on the rake surface of the main cutting-edge and a strong built-up edge formation on the secondary cutting-edge. The higher lubrication level reduces the cutting temperatures [21] and the coefficient of friction between the tool and the chip [25].

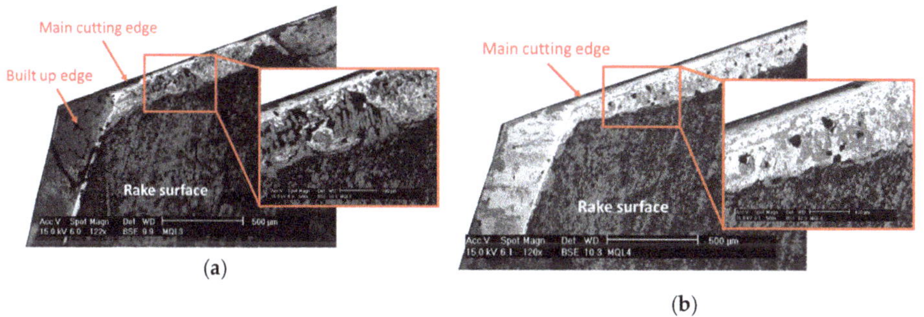

Figure 6. Composition analysis of the rake surface, after 40 holes, in a Ti/CFRP/Ti coupon. The black color corresponds to the carbon components, the grey tones are titanium alloy, and the white indicates the tungsten of the carbide substrate. (**a**) A 3 mL/h lubrication level. (**b**) A 15 mL/h lubrication level.

3.2. Hole Quality Analysis

Controlling the dimensional quality of the holes is important in guaranteeing the suitable behavior of the mechanical joint. Furthermore, this analysis will determine whether the hole is valid for the application or a subsequent reaming operation is required.

3.2.1. Diameter Analysis

Figure 7 shows the diameter measurements. The values obtained from the upper and lower Ti layers were named Ti1 and Ti2, respectively. In this case, no differences were observed among the entrance and exit side of each Ti layer, so just one value was represented for each of them. On the other hand, in the composite layer, a different behavior was observed among the entrance and exit side of the layer, so one value for each side was represented—CFRP in and CFRP out, respectively.

The diameter of the holes in the Ti layers showed a very similar behavior, in both lubrication levels. For the lower lubrication level, a higher dispersion of the diameters of the holes on the CFRP layer was observed at the beginning of the tool's life. This could be explained by the higher temperatures produced on the Ti layers [13], which generated longer and continuous chips. This chip morphology was not desirable since it over-sized the holes of the CFRP layer, during the chip evacuation. Once the number of holes had increased, the tool wear modified the cutting geometry towards a more negative one, producing shorter and discontinuous chips [15]. Beyond the 30th hole, the diameter of the holes obtained on the CFRP and the Ti layers became similar.

For higher lubrication levels, the diameters of the holes on the CFRP layer were similar to those in the Ti layer, and they remained nearly constant, throughout the tested lifespan of the tool.

The exit side of the hole of the CFRP layer had a greater diameter, compared to the entrance side, due to the larger influence of the chips produced in the lower Ti layer drilling, during its evacuation [26].

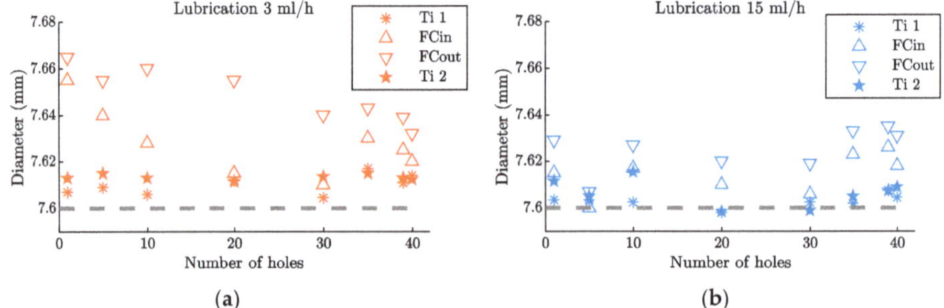

Figure 7. Hole diameter evolution, after 40 holes, in a hybrid Ti/CFRP/Ti stack. The dashed line represents the nominal diameter of the tool with (**a**) a lubrication level of 3 mL/h and (**b**) a lubrication level of 15 mL/h.

3.2.2. Burr Height Analysis

Burr height values, represented in Figure 8, showed no variation as the tested lubrication levels changed. With the fresh tool, the height values for both tools were around 0.05 mm at the entrance of the stack, and 0.1 mm at the exit. As the tool wear evolved, the burr height at the entrance increased up to 0.15 mm and, at the exit side of the stack, it reached values close to 0.4 mm. In both cases, they were acceptable values [27] although, for assembly purposes, a deburring process would be required [28].

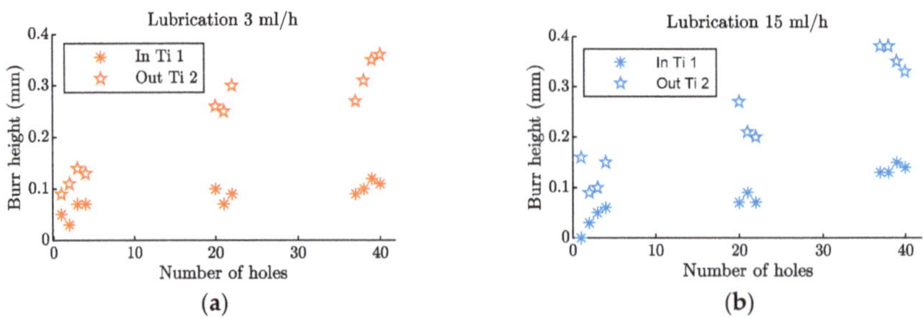

Figure 8. Burr height evolution, after 40 holes, in a hybrid Ti/CFRP/Ti stack with a (**a**) lubrication level of 3 mL/h and a (**b**) lubrication level of 15 mL/h.

3.2.3. Hole Surface Quality Analysis

The study on roughness shows important differences in the machining behavior between the composite material and the metal, which can be seen in Figure 9.

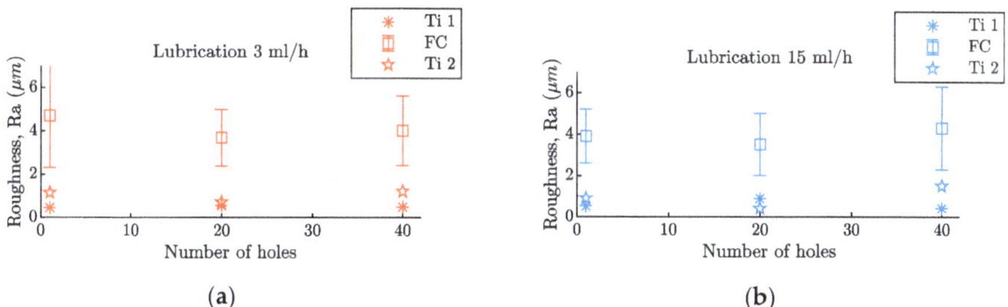

Figure 9. Roughness evolution, after 40 holes, in a hybrid Ti/CFRP/Ti stack with a (**a**) lubrication level of 3 mL/h and a (**b**) lubrication level of 15 mL/h.

The chip formation mode on composite materials, which depends on the relative orientation of the cutting-edge with respect to the fibers, has a big influence on surface quality [29]. Owing to this phenomenon, the roughness obtained during the drilling processes of these materials is variable. Furthermore, interactions with the metallic chips can further deteriorate the obtained roughness [30].

More stable values of the surface quality of the Ti layers were obtained. Roughness values (Ra) of around 1 μm were found from the fresh tool, up to the 40th hole. No influence from the lubrication level or the tool wear, was observed.

The skewness analysis also showed differences between the hole surface quality of the metallic layers and the composite layers. In Figure 10, it can be seen that the parameter Rsk was always negative, for the CFRP layer, while it was around zero in the Ti layers, with negative and positive values.

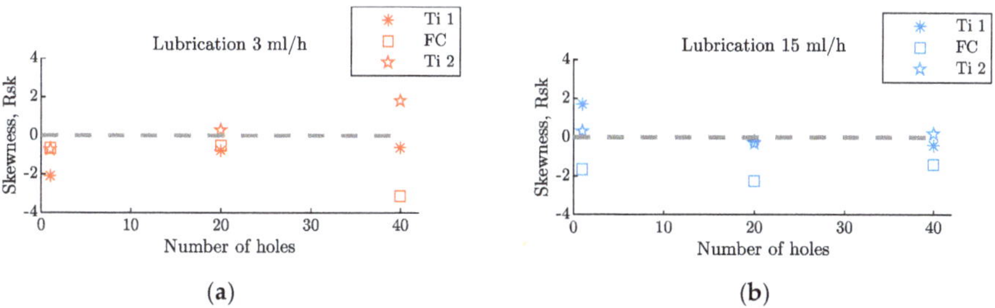

Figure 10. Skewness evolution, after 40 holes, in a hybrid Ti/CFRP/Ti stack with a (**a**) lubrication level of 3 mL/h and a (**b**) lubrication level of 15 mL/h.

3.3. Power Consumption Analysis

The spindle and feed motor power consumptions were recorded, individually, and these measurements were expressed as a percentage of their nominal power consumptions. Along with the net power associated to the drilling process, the signal included friction, noise, or perturbations. Thus, the values obtained were analyzed qualitatively.

Figure 11 shows the spindle power consumption raw signal obtained during the Ti/CFRP/Ti stack drilling process of the 40th hole. Noticeable fluctuations could be seen while the tool was spinning in the air, before the tool penetrated the coupon (first 10 s of the signal), which needed to be filtered out. For this reason, a low-pass filter was applied to the signal; this is represented in Figure 11 as a solid black line.

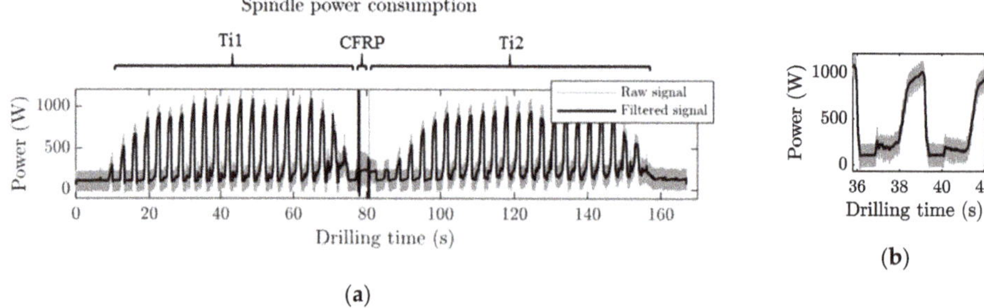

Figure 11. Spindle power consumption signal of hole 40 with the higher lubrication level (15 mL/h) recorded during Ti/CFRP/Ti stack drilling. (**a**) The complete signal of the drilling process. (**b**) A detailed view of a peck drilling cycle on the first Ti layer.

During the Ti drilling, oscillations on the power consumption signal were produced by the peck drilling cycle. The change of the cutting parameters between the Ti and CFRP layers could be observed as a sharp increase in consumption, in order to adapt the spindle speed. For the CFRP layer, the consumption was constant, since no peck drilling was done. However, there were big oscillations in the signal, owing to the fluctuation of the cutting forces produced during the fiber-reinforced composite machining [31]. After the CFRP layer, a new sharp increase of the power consumption was produced, again, to match the Ti cutting parameters.

The Ti power consumption was approximately 10% of the nominal power of the spindle motor, with very similar values on both layers, Ti1 and Ti2. Thus, hereafter, the average value of both layers was considered as the characteristic value for the Ti layers' consumptions. At the CFRP layer, the power consumption was an order of magnitude smaller (by 1%) than the nominal power consumption.

Figure 12 shows the feed motor power consumption obtained during a Ti/CFRP/Ti stack drilling process. It had a similar trend to the spindle power consumption. However, in each peck cycle, a sharp increase of the power was produced.

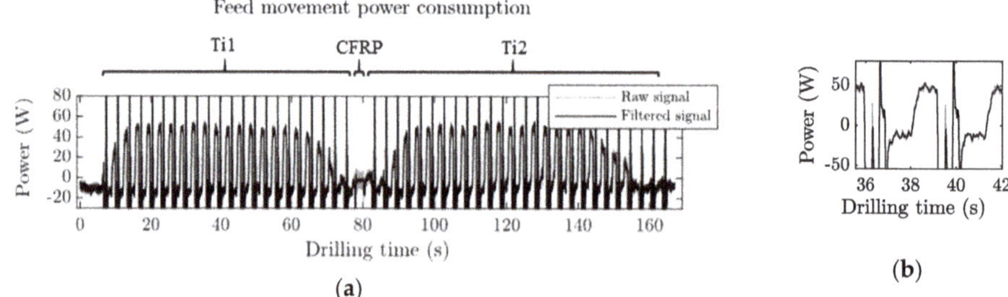

Figure 12. Feed movement power consumption signal of the 40th hole with the higher lubrication level (15 mL/h) recorded during the Ti/CFRP/Ti drilling. (**a**) The complete signal of the drilling process. (**b**) A detailed view of a peck drilling cycle on the first Ti layer.

It can be seen in Figures 11b and 12b that the shape of power consumption in each individual peck drilling cycle followed the same behavior as that of a conventional continuous drilling process [30]. In all cases, the amplitude of the signal measured, while the tool was approaching the material, was eliminated.

The influence of the lubrication level on the evolution of the spindle power consumption with the tool wear can be seen in Figure 13a, for the Ti layers, and Figure 13b, for the composite layer. With the fresh tool, the power consumption for the Ti layer with the higher lubrication level (blue asterisks) was 8% smaller, due to the reduction of the friction coefficient between the tool and the workpiece [32].

However, this behavior was not maintained with the evolution of the tool wear. The gradient of the relationship between the spindle power consumption and the number of holes was steeper for the higher lubrication level (blue asterisks) and flatter for the lower lubrication level (red triangles). This could be related to the different Ti adhesions observed on the rake surface of the tools. At lower lubrication levels, the higher coefficients of friction between the tool and the workpiece, combined with larger temperatures, produced a modification of the chip formation mode and led to a greater influence of the thermal softening effect [33]. Therefore, the gradient of the spindle power consumption with the number of holes was less steep.

Hence, the spindle power consumption increment on the Ti layer, after 40 holes, with respect to the fresh tool, was 38% for the lower lubrication level and 76% for the higher lubrication level.

On the CFRP layer drilling process, no lubrication was used and the thermal softening effect had less influence. Thus, the increment of the spindle power consumption with the number of holes was higher for the lower lubrication level, owing to the higher tool wear suffered from the metallic layers (fragile breakages). Furthermore, no differences were observed with the fresh tool, which indicated that the residual oil did not have an influence on the CFRP layer drilling. In this case, the increment of the spindle power consumption, after 40 holes, with respect to the fresh tool, was 88% for the lower lubrication level and a 56% for the higher lubrication level.

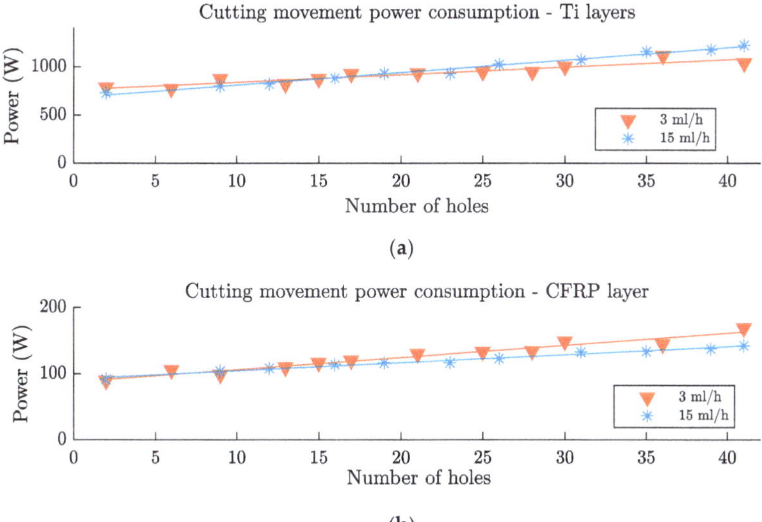

Figure 13. Spindle power consumption, after 40 holes, for a Ti/CFRP/Ti stack on (**a**) the Ti layers and (**b**) on the CFRP layer.

Feed movement power consumption can be seen in Figure 14. The behavior was very similar to the cutting movement power consumption. With the fresh tool, the power of the higher lubrication level was 6% smaller due to the reduction of the friction coefficient. This tendency was reversed, and the evolution of the feed movement power consumption with the number of holes was faster for the higher lubrication level (blue asterisks) and more gradual for the lower lubrication level (red triangles). This was also explained by the phenomenon related to the modification of the chip formation mode, and the thermal softening effect. In this case, after 40 holes, the feed movement power consumption increased by 24% and 61% for the 3 mL/h and the 15 mL/h lubrication levels, respectively.

Regarding the feed movement power consumption on the CFRP layer, no analysis was carried out since the magnitude of the signal was too small. There was not enough sensibility in order to draw solid conclusions.

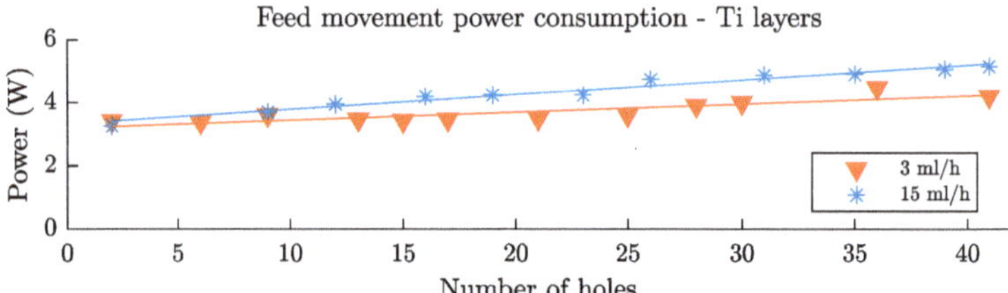

Figure 14. Feed movement power consumption on the Ti layer, after 40 holes, for a Ti/CFRP/Ti stack drilling process.

4. Conclusions

The use of the MQL lubrication system during a Ti/CFRP/Ti hybrid stack drilling process was analyzed. With a minimum amount of lubrication, the behavior of the process during the metallic layers drilling was significantly affected without having an influence on the performance of the composite layer drilling.

It was found that the main wear mechanisms suffered by the diamond-coated carbide drill bits, during this process, were the loss of the coating by delamination and fragile breaks on the middle region of the main cutting-edge. Additionally, the level of lubrication used affected the Ti adhesion to the cutting tool. For the lower lubrication level, a strong Ti adhesion was found on the rake surface of the main cutting-edge and the secondary cutting-edge, due to the higher friction coefficient and temperatures produced. In contrast, this was not produced for the higher MQL level.

The differences induced by the level of lubrication also affected the evolution of the cutting and feed movement power consumption on the Ti layers. With the fresh tool, less consumption was found for the lower lubrication level but, as the tool wear increased, this tendency was reversed due to the differences in the Ti adhesion suffered by the tools and the influence of the thermal softening effect. In contrast, the level of lubrication did not directly affect the drilling process during the CFRP layer. However, in this material, the gradient of the relationship between the spindle power consumption and the number of holes was steeper for the lower lubrication level, due to the higher wear suffered by the Ti layers.

No significant differences were observed on the quality of the components between the lubrication levels analyzed. However, a higher dispersion of the diameter of the holes in the composite layer for the lower lubrication condition was produced.

Author Contributions: J.F.-P., J.D.-Á., and J.L.C. performed the experiment design and drilling tests. J.F.-P. and J.L.C. analyzed the quality of the coupons and characterized the tool wear mechanisms. All authors collaborated on the discussion and analysis of the results. Finally, J.F.-P. and J.L.C. wrote the paper, and M.H.M. reviewed and edited it.

Funding: The authors acknowledge the financial support of Airbus Defense and Space, through the project Drilling Processes Improvement for Multi Material CFRP-Al-Ti Stacks, and the Ministry of Economy and Competitiveness of Spain through the grant PTA2015-10741-I and the project DPI2017-89197-C2-1-R.

Acknowledgments: The authors acknowledge the Airbus for the availability of the drilling machines used to perform the experiments and the provision of the materials and tools tested.

Conflicts of Interest: The authors declare no conflict of interest.

References

1. Zweben, C. Advanced composites for aerospace applications: A review of current status and future prospects. *Composites* **1981**, *12*, 235–240. [CrossRef]

2. Pan, L.; Ding, W.; Ma, W.; Hu, J.; Pang, X.; Wang, F.; Tao, J. Galvanic corrosion protection and durability of polyaniline-reinforced epoxy adhesive for bond-riveted joints in AA5083/Cf/Epoxy laminates. *Mater. Des.* **2018**, *160*, 1106–1116. [CrossRef]

3. Kuo, C.L.; Soo, S.L.; Aspinwall, D.K.; Carr, C.; Bradley, S.; M'Saoubi, R.; Leahy, W. Development of single step drilling technology for multilayer metallic-composite stacks using uncoated and PVD coated carbide tools. *J. Manuf. Process.* **2018**, *31*, 286–300. [CrossRef]

4. Alonso-Pinillos, U.; Girot-Mata, F.A.; Polvorosa-Teijeiro, R.; López-De-Lacalle-Marcaide, L.N. Taladrado de materiales compuestos: Problemas, prácticas recomendadas y técnicas avanzadas. *DYNA* **2017**, *92*, 188–194. [CrossRef]

5. Cantero, J.L.; Tardío, M.M.; Canteli, J.A.; Marcos, M.; Miguélez, M.H. Dry drilling of alloy Ti-6Al-4V. *Int. J. Mach. Tools Manuf.* **2005**, *45*, 1246–1255. [CrossRef]

6. Wang, X.; Kwon, P.Y.; Sturtevant, C.; Kim, D.; Lantrip, J. Comparative tool wear study based on drilling experiments on CFRp/Ti stack and its individual layers. *Wear* **2014**, *317*, 265–276. [CrossRef]

7. Ramulu, M.; Branson, T.; Kim, D. A study on the drilling of composite and titanium stacks. *Compos. Struct.* **2001**, *54*, 67–77. [CrossRef]

8. Park, K.-H.; Beal, A.; Kim, D.; Kwon, P.; Lantrip, J. Tool wear in drilling of composite/titanium stacks using carbide and polycrystalline diamond tools. *Wear* **2011**, *271*, 2826–2835. [CrossRef]

9. Fernandez-Vidal, S.R.; Fernandez-Vidal, S.; Batista, M.; Salguero, J. Tool wear mechanism in cutting of stack CFRP/UNS A97075. *Materials* **2018**, *11*, 1276. [CrossRef]

10. Rezende, B.A.; Silveira, M.L.; Vieira, L.M.G.; Abrão, A.M.; de Faria, P.E.; Rubio, J.C.C. Investigation on the effect of drill geometry and pilot holes on thrust force and burr height when drilling an aluminium/PE sandwich material. *Materials* **2016**, *9*, 774. [CrossRef]

11. Fernández-Pérez, J.; Cantero, J.L.; Díaz-Álvarez, J.; Miguélez, M.H. Influence of cutting parameters on tool wear and hole quality in composite aerospace components drilling. *Compos. Struct.* **2017**, *178*. [CrossRef]

12. Anyfantis, K.; Stavropoulos, P.; Chryssolouris, G. Fracture mechanics based assessment of manufacturing defects laying at the edge of CFRP-metal bondlines. *Prod. Eng.* **2018**, *12*, 173–183. [CrossRef]

13. D'Orazio, A.; El Mehtedi, M.; Forcellese, A.; Nardinocchi, A.; Simoncini, M. Tool wear and hole quality in drilling of CFRP/AA7075 stacks with DLC and nanocomposite TiAlN coated tools. *J. Manuf. Process.* **2017**, *30*, 582–592. [CrossRef]

14. Kim, D.; Beal, A.; Kang, K.; Kim, S.-Y. Hole Quality Assessment of Drilled CFRP and CFRP-TI Stacks Holes Using Polycrystalline Diamond (PCD) Tools. *Carbon Lett.* **2017**, *23*, 1–8.

15. Shyha, I.S.; Soo, S.L.; Aspinwall, D.K.; Bradley, S.; Perry, R.; Harden, P.; Dawson, S. Hole quality assessment following drilling of metallic-composite stacks. *Int. J. Mach. Tools Manuf.* **2011**, *51*, 569–578. [CrossRef]

16. Pereira, O.; Martín-Alfonso, J.E.; Rodríguez, A.; Calleja, A.; Fernández-Valdivielso, A.; López de Lacalle, L.N. Sustainability analysis of lubricant oils for minimum quantity lubrication based on their tribo-rheological performance. *J. Clean. Prod.* **2017**, *164*, 1419–1429. [CrossRef]

17. Weinert, K.; Inasaki, I.; Sutherland, J.W.; Wakabayashi, T. Dry Machining and Minimum Quantity Lubrication. *CIRP Ann. Manuf. Technol.* **2004**, *53*, 511–537. [CrossRef]

18. Sharma, A.K.; Tiwari, A.K.; Dixit, A.R. Effects of Minimum Quantity Lubrication (MQL) in machining processes using conventional and nanofluid based cutting fluids: A comprehensive review. *J. Clean. Prod.* **2016**, *127*, 1–18. [CrossRef]

19. Sidik, N.A.C.; Samion, S.; Ghaderian, J.; Yazid, M.N.A.W.M. Recent progress on the application of nanofluids in minimum quantity lubrication machining: A review. *Int. J. Heat Mass Transf.* **2017**, *108*, 79–89. [CrossRef]

20. Giasin, K.; Ayvar-Soberanis, S.; Hodzic, A. The effects of minimum quantity lubrication and cryogenic liquid nitrogen cooling on drilled hole quality in GLARE fibre metal laminates. *Mater. Des.* **2016**, *89*, 996–1006. [CrossRef]

21. Bhowmick, S.; Lukitsch, M.J.; Alpas, A.T. Dry and minimum quantity lubrication drilling of cast magnesium alloy (AM60). *Int. J. Mach. Tools Manuf.* **2010**, *50*, 444–457. [CrossRef]

22. Stavropoulos, P.; Papacharalampopoulos, A.; Vasiliadis, E.; Chryssolouris, G. Tool wear predictability estimation in milling based on multi-sensorial data. *Int. J. Adv. Manuf. Technol.* **2016**, *82*, 509–521. [CrossRef]

23. Kuo, C.; Wang, C.; Ko, S. Wear behaviour of CVD diamond-coated tools in the drilling of woven CFRP composites. *Wear* **2018**, *398–399*, 1–12. [CrossRef]

24. Iliescu, D.; Gehin, D.; Gutierrez, M.E.; Girot, F. Modeling and tool wear in drilling of CFRP. *Int. J. Mach. Tools Manuf.* **2010**, *50*, 204–213. [CrossRef]

25. Giasin, K.; Ayvar-Soberanis, S.; Hodzic, A. Evaluation of cryogenic cooling and minimum quantity lubrication effects on machining GLARE laminates using design of experiments. *J. Clean. Prod.* **2016**, *135*, 533–548. [CrossRef]

26. Wang, C.Y.; Chen, Y.H.; An, Q.L.; Cai, X.J.; Ming, W.W.; Chen, M. Drilling temperature and hole quality in drilling of CFRP/aluminum stacks using diamond coated drill. *Int. J. Precis. Eng. Manuf.* **2015**. [CrossRef]

27. Abdelhafeez, A.M.; Soo, S.L.; Aspinwall, D.K.; Dowson, A.; Arnold, D. Burr formation and hole quality when drilling titanium and aluminium alloys. *Procedia CIRP* **2015**, *37*, 230–235. [CrossRef]

28. Ton, T.P.; Park, H.Y.; Ko, S.L. Experimental analysis of deburring process on inclined exit surface by new deburring tool. *CIRP Ann. Manuf. Technol.* **2011**, *60*, 129–132. [CrossRef]

29. König, W.; Wulf, C.; Graß, P.; Willerscheid, H. Machining of Fibre Reinforced Plastics. *CIRP Ann.* **1985**, *34*, 537–548. [CrossRef]

30. Xu, J.; El Mansori, M. Experimental study on drilling mechanisms and strategies of hybrid CFRP/Ti stacks. *Compos. Struct.* **2016**, *157*, 461–482. [CrossRef]

31. DiPaolo, G.; Kapoor, S.G.; DeVor, R.E. An Experimental Investigation of the Crack Growth Phenomenon for Drilling of Fiber-Reinforced Composite Materials. *J. Eng. Ind.* **1996**, *118*, 104–110. [CrossRef]

32. Jun, S. Lubrication Effect of Liquid Nitrogen in Cryogenic Machining Friction on the Tool-chip Interface. *J. Mech. Sci. Technol.* **2005**, *19*, 936–946. [CrossRef]

33. Parida, A.K. Simulation and experimental investigation of drilling of Ti-6Al-4V alloy. *Int. J. Lightweight Mater. Manuf.* **2018**, *1*, 197–205. [CrossRef]

Article

Modeling and Optimization of Fractal Dimension in Wire Electrical Discharge Machining of EN 31 Steel Using the ANN-GA Approach

Arkadeb Mukhopadhyay [1], Tapan Kumar Barman [1], Prasanta Sahoo [1,* and J. Paulo Davim [2]

[1] Department of Mechanical Engineering, Jadavpur University, Kolkata 700032, India; arkadebjume@gmail.com (A.M.); tkbarman@gmail.com (T.K.B.)

[2] Department of Mechanical Engineering, University of Aveiro, 3810-193 Aveiro, Portugal; pdavim@ua.pt

* Correspondence: psjume@gmail.com; Tel.: +91-33-2457-2660

Received: 17 December 2018; Accepted: 29 January 2019; Published: 1 February 2019

Abstract: To achieve enhanced surface characteristics in wire electrical discharge machining (WEDM), the present work reports the use of an artificial neural network (ANN) combined with a genetic algorithm (GA) for the correlation and optimization of WEDM process parameters. The parameters considered are the discharge current, voltage, pulse-on time, and pulse-off time, while the response is fractal dimension. The usefulness of fractal dimension to characterize a machined surface lies in the fact that it is independent of the resolution of the instrument or length scales. Experiments were carried out based on a rotatable central composite design. A feed-forward ANN architecture trained using the Levenberg-Marquardt (L-M) back-propagation algorithm has been used to model the complex relationship between WEDM process parameters and fractal dimension. After several trials, 4-3-3-1 neural network architecture has been found to predict the fractal dimension with reasonable accuracy, having an overall R-value of 0.97. Furthermore, the genetic algorithm (GA) has been used to predict the optimal combination of machining parameters to achieve a higher fractal dimension. The predicted optimal condition is seen to be in close agreement with experimental results. Scanning electron micrography of the machined surface reveals that the combined ANN-GA method can significantly improve the surface texture produced from WEDM by reducing the formation of re-solidified globules.

Keywords: WEDM; EN 31 steel; surface roughness; fractal dimension; ANN; GA

1. Introduction

A very popular non-conventional machining process which is capable of machining parts with intricate shapes and sharp edges is wire electrical discharge machining (WEDM). Though this process is widely used in tool and die making industries, its usage has been further extended to making micro-scale parts with high dimensional accuracy and surface finish [1,2]. The WEDM process has gained immense popularity for machining wear resistant and hard materials such as ceramics, nano-structured hardfacing alloys metal matrix composites, etc., with high machining accuracy [3–5]. WEDM utilizes a series of discrete sparks between the workpiece and tool electrode resulting in material erosion. The melted debris is flushed away by the dielectric medium (generally deionized water) [6]. The tool electrode in WEDM is a wire with a small diameter ranging between 0.05–0.25 mm. Since this is a non-contact type process, vibrations and chatter are prevented which introduce inaccuracies to machined parts [7]. However, wire breakage and bending are challenges that tend to limit the capabilities of this process, and a significant amount of research has been carried out to address this issue [8]. A schematic diagram of WEDM is shown in Figure 1.

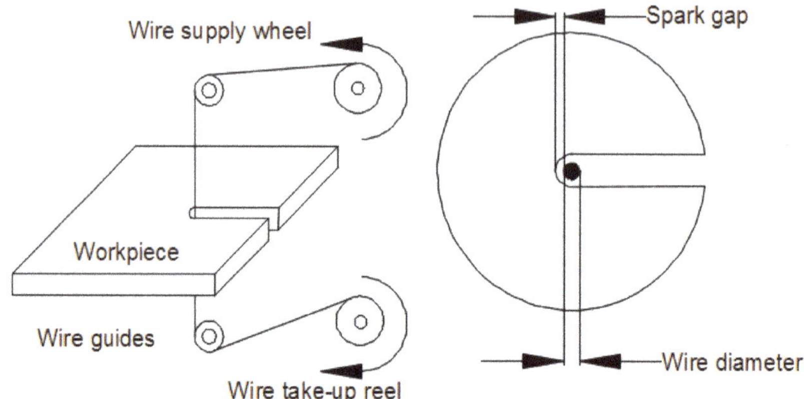

Figure 1. Schematic representation of the wire electrical discharge machining (WEDM) process.

From the tribological point of view, the quality of a machined surface, i.e., surface roughness, is an important parameter that affects the proper functioning of mating components in machines. Therefore, improvement of surface roughness and studies of the effect of controllable parameters on surface roughness have always been in focus. Such controllable parameters for WEDM include discharge current, open circuit voltage, wire speed, wire diameter, dielectric flushing pressure, pulse-on time, pulse-off time, spark gap, etc. [1,6,8]. The effect of WEDM process parameters and their significance in controlling the surface roughness has been widely studied through the use of design of experiments (DOE) and statistical tools [9–12]. The effect of pulse duration, open circuit voltage, wire speed and dielectric flushing pressure on the surface roughness of WEDM workpieces was investigated by Tosun et al. [9]. It was seen that surface roughness increases with an increase in pulse duration, open circuit voltage, or wire speed. Further, surface roughness improved with an increase in dielectric fluid pressure. The potential of WEDM for machining stainless clad steel in terms of surface roughness as a quality characteristic was investigated by Ishfaq et al. [7] using Taguchi's orthogonal array (OA) and grey relational technique. Several parameters, such as the orientation of the workpiece (mild steel or stainless steel on top), the thickness of layers, wire diameter, pressure ratio, servo voltage, pulse-on time and wire feed rate were varied, and the surface roughness was observed. The highest significance in controlling the roughness was concluded for wire diameter. A larger wire diameter resulted in enhanced surface finish owing to a lower cutting speed and appropriate flushing of melted debris [7]. With the assistance of Taguchi's quality design technique, a surface roughness of the order of 0.22 μm could be achieved by controlling process parameters such as machining voltage, current limiting resistance, type of pulse generating circuit, and capacitance [10]. In the case of a newly developed DC 53 die steel, pulse on time and pulse peak current were observed to have a significant effect on the surface roughness [11]. Pulse-on time, pulse-off time, and spark voltage have a major influence on the surface roughness and material removal rate (MRR) while machining high strength armor steel [12].

The influence of machining parameters on surface roughness and integrity has been also reported for a wide variety of materials of industrial importance, such as commercially pure titanium and its alloys, Inconel, multiwall carbon nanotube (MWCNT)-alumina composites, aluminum metal matrix composites, etc. [13–17]. The surface damage of WEDMed Ti-6Al-4V and Inconel 718 could be reduced by employing ultra-high frequency/short duration pulses and multiple trim passes [14]. This resulted in improved surface roughness and integrity with a low thickness recast layer which may be removed easily by etching. To improve the surface integrity of Nimonic 80A, Goswami and Kumar [15] suggested the setting of high pulse-off duration and low pulse-on duration. This again resulted in a lower recast layer thickness. Recast layer thickness and porosity in WEDM of MWCNT filled alumina composites may be reduced by adopting a multi-pass technique [16]. With each pass, the surface was smoothened by the removal of already-formed debris and cracks rather than removal of fresh material. The surface roughness and integrity may be also improved by carrying out post-processing such as grinding and etching-grinding to remove the recast layer [18]. This process was found to effectively reduce the surface roughness of Nimonic C 263 super alloy post-machining in WEDM [18].

Analysis of the surface and sub-surface layers formed after WEDM of Ti-6Al-4V alloy with and without heat treatment was reported by Mouralova et al. [19]. In both of the cases, a layer of molten metal stuck to the surface was observed. But for the heat-treated samples, this thickness was only 5–10 μm instead of 10–20 μm for the non-heat-treated samples. In fact, Goyal [20] reported the use of a cryogenically-treated zinc coated wire electrode to improve the machining performance and surface roughness for WEDM of Inconel 625 super alloy. Thus, even surface pre-treatments and wire electrode treatments also influence surface integrity.

Due to the stochastic nature of WEDM and too many adjustable machining parameters, it becomes rather difficult to utilize the machine optimally. Hence, in order to achieve efficient machining, modeling between input parameters and response variables is necessary. This may be carried out by adopting theoretical/empirical or artificial intelligence (AI) techniques. An efficient mathematical and statistical technique that has been employed in several research works is response surface methodology (RSM) [21]. Modeling and analysis of micro-WEDM of a titanium alloy were carried out by Sivaprakasam et al. [22]. A relationship between voltage, capacitance, and feed rate with MRR, kerf width, and surface roughness could be established with reasonable accuracy (coefficient of determination, $R^2 > 0.95$). Finally, the genetic algorithm (GA) was employed for multi-objective optimization. A non-linear regression model [23], as well as a mathematical model using Buckingham's pi theorem [24], were seen to have good modeling capabilities for surface roughness and WEDM process parameters.

Owing to simplified and unavoidable assumptions in mathematical models, AI-based techniques were seen to have better prediction capabilities. Such AI-based models include artificial neural networks (ANN), fuzzy logics, adaptive neuro-fuzzy inference systems (ANFIS), etc. Tzeng et al. [25] used RSM and a back-propagation neural network (BPNN) for analyzing the dependence of MRR and the surface roughness of pure tungsten on WEDM process parameters. There it was seen that the average prediction error was lower for the BPNN model in comparison with RSM based on regression models. Similarly, Saha et al. [26] observed overall mean prediction errors of 3.29% and 6.02% in BPNN and RSM approaches, respectively, for prediction of cutting speed and surface roughness in WEDM of WC-Co composites. Combining the modeling function of fuzzy inference and the learning ability of ANN led to the formation of an ANFIS model of WEDM with pulse duration, open circuit voltage, dielectric flushing pressure, and wire feed rate as input parameters [27]. The output parameters, namely white layer thickness and average surface roughness, could be predicted with reasonable accuracy using ANFIS architecture. Majumder and Maity [28] developed a general regression neural network (GRNN) architecture to model the surface roughness of nitinol with five critical WEDM parameters, namely pulse-on time, discharge current, wire feed, wire tension, and flushing pressure. The developed GRNN model had prediction capabilities with ±10% error.

From the existing literature, it may be observed that WEDM has gained immense importance due to its numerous advantages. Due to this, it has found its application in machining hard, fragile, and difficult to process materials in tool and die industries, as well as for generating complex surface geometries in mold walls [27,29]. WEDM also finds usage in aerospace, automobile, and nuclear industries since it provides an effective solution for machining materials with high hardness properties [25]. Several studies have been carried out to investigate the surface roughness and integrity of materials that find wide usage in industries (especially ceramics, titanium alloys, metal matrix composites, etc.) post-machining using WEDM [15–17]. The effect of WEDM process parameters and post-processing on surface integrity has been thoroughly investigated. Moreover, due to the stochastic nature of WEDM, mathematical as well as AI techniques have been employed to predict optimal machining parameters and to improve the surface roughness. However, the fractal dimension has been seldom used as a parameter to characterize surface roughness in WEDM [30–32]. The center line average surface roughness (R_a) is dependent on parameters such as the resolution of the instrument used and the sampling length. To overcome this problem, the fractal dimension has been used in the present work to characterize surface roughness in WEDM. The effect of process parameters such as discharge current, voltage, pulse-on time, and pulse-off time on fractal dimension has been investigated. Furthermore, ANN has been used to model the complex relationship between fractal dimension and considered process parameters. Finally, the genetic algorithm (GA) is used to predict the optimal parametric combination using the developed ANN model.

2. Materials and Methods

The material on which WEDM was carried out is EN 31 tool steel. It is a high carbon steel with high hardness, compressive strength, and abrasion resistance [1]. Blocks with dimensions of 20 mm × 20 mm × 15 mm were selected as workpiece materials. The composition of EN 31 steel is 1.07% C, 0.57% Mn, 0.32% Si, 0.04% P, 0.03% S, 1.13% Cr, and 96.84% Fe. The modulus of elasticity, yield strength, ultimate tensile strength, and Poisson's ratio of EN 31 steel are 197.37 GPa, 528.97 MPa, 615.40 MPa, and 0.294, respectively. Experiments were carried out on a five-axis CNC WEDM (Elektra, Maxicut 434, Kolkata, India). The wire electrode was zinc coated copper with a 0.25 mm diameter. The workpiece and electrode were separated by deionized water dielectric medium. The dielectric flow pressure was kept at 1.30 bar. The wire was fed at a rate of 8 m/min, and wire tension was 1000 gf. Discharge current, voltage, pulse-on time and pulse-off time were considered as the controllable/process parameters. They were varied at five levels, as shown in Table 1. A rotatable central composite design (CCD) was selected to reduce the number of experiments. Accordingly, 31 experiments were carried out with 16 factorial points, 8 axial points, and 7 center points. The roughness profiles of the machined workpieces were analyzed using a stylus-type profilometer (Talysurf, Taylor Hobson, Leicester, UK). A cutoff length, traverse speed, and traverse length of 0.8 mm, 1 mm/s and 4 mm were respectively selected. The obtained profile was then processed using Talyprofile software, and the fractal dimension was evaluated based on the structure-function method [33]. An average of 4 readings was considered. The sequence of experimental runs and the corresponding fractal dimension is outlined in Table 2. A scanning electron microscope (SEM) was used to observe the morphology of the machined surface at a near-optimum combination of parameters (JEOL, JSM 6360, Tokyo, Japan).

Table 1. WEDM process parameters and their levels.

Controllable Factors	Unit	Levels				
		1	2	3	4	5
Discharge current	Amp	2	4	6	8	10
Voltage	Volt	40	45	50	55	60
Pulse-on time	μs	1	2	3	4	5
Pulse-off time	μs	1	2	3	4	5

Table 2. Combination of process parameters and experimental results.

Sl. No.	Discharge Current (Amp)	Voltage (V)	Pulse-On Time (μs)	Pulse-Off Time (μs)	Fractal Dimension
1	6	50	3	3	1.428
2	6	50	3	3	1.428
3	6	50	3	3	1.428
4	6	40	3	3	1.415
5	4	45	2	4	1.408
6	8	55	2	4	1.36

Table 2. *Cont.*

Sl. No.	Discharge Current (Amp)	Voltage (V)	Pulse-On Time (µs)	Pulse-Off Time (µs)	Fractal Dimension
7	8	55	4	4	1.403
8	4	45	4	2	1.363
9	6	50	3	3	1.428
10	6	50	3	3	1.428
11	6	50	3	3	1.428
12	8	55	2	2	1.39
13	6	50	5	3	1.27
14	8	45	4	2	1.383
15	4	55	4	4	1.373
16	6	60	3	3	1.44
17	6	50	3	1	1.403
18	4	55	4	2	1.263
19	4	55	2	4	1.398
20	6	50	3	5	1.383
21	6	50	3	3	1.428
22	8	55	4	2	1.325
23	8	45	2	2	1.428
24	4	45	2	2	1.353
25	8	45	2	4	1.043
26	6	50	1	3	1.423
27	10	50	3	3	1.393
28	8	45	4	4	1.32
29	4	55	2	2	1.383
30	4	45	4	4	1.388
31	2	50	3	3	1.425

3. WEDM Process Modeling and Optimization Methodology Using ANN Integrated with GA

Since WEDM is a random and stochastic process, it is very difficult to map the input and output parameters. Neural networks have the flexibility of modeling input process parameters with output response variables without having prior knowledge of their relationship [26]. ANN is designed based on copying the human brain artificially [34]. ANN consists of a number of neurons organized in different layers. The neurons in different layers are connected by weights. The neural network architecture can be trained by adjusting the weights and other parameters. Several types of neural network architectures may be found in the literature [34]. In the present work, a feed-forward network with the Levenberg-Marquardt (L-M) back-propagation algorithm is used to train the network. It combines the Gauss-Newton algorithm with the steepest descent method to minimize the mean square error (MSE) of the output of the network. The feed-forward network consists of neurons that are grouped into the input, hidden, and output layers interconnected by weights. These weights are adjusted during the training stage of the learning process. The output O_j from a jth neuron at any layer may be calculated as:

$$O_j = f \sum_{i=1}^{n} w_{ij} x_i + b_j \tag{1}$$

where, f is the activation function (linear, logsigmoidal, tansigmoidal, etc.), n is the number of inputs to the jth neuron or rather the number of neurons in the previous layer, w_{ij} is the corresponding weight, x_i is the output from i^{th} neuron, and b_j is the corresponding bias.

In the present work, 2 hidden layers are considered with 3 neurons while 4 and 1 neurons are considered in the input and output layers respectively. Thus, this leads to a feed forward back-propagation network with 4-3-3-1 architecture. The activation function used between the input and hidden layers is "tansig" whereas a "linear" activation function is considered between the hidden and output layers. The aforesaid neural network architecture and the number of hidden neurons is selected carefully after several trial and error runs based on minimum MSE and higher correlation coefficient (R) of the model. Input parameters are normalized between −1 and 1. The results laid down in Table 2 are used to train, test, and validate the model. To train the network, 20 data is selected, while for testing, 6 data is chosen. The remaining 5 are used as validation sets. The ANN model may be represented as shown in Figure 2. MATLAB R2014b is used to develop the ANN model of WEDM.

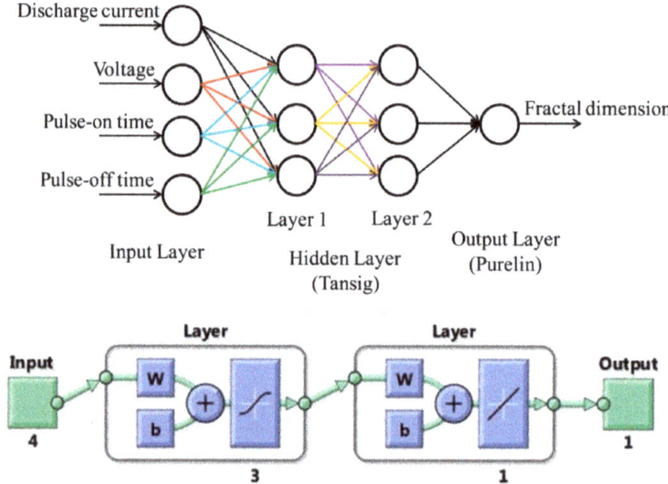

Figure 2. A 4-3-3-1 artificial neural network (ANN) architecture for the prediction of the fractal dimension.

On the other hand, GA has the capability to solve linear as well as non-linear stochastic problems by mimicking the principles of biological evolution. The essence of GA lies in the principle of "survival of the fittest", wherein a population continuously evolves with iterations to achieve a better solution. Initially, the process begins with a set of potential solutions (in the form of bit strings) known as chromosomes. Each chromosome represents a set of process parameters. A population of possible solutions is created randomly within the search space from the chromosomes. The fitness of each of the chromosomes is determined from an objective function. After the fitness of the chromosomes is evaluated, then they evolve through biologically inspired processes in a generation or an iteration. In a generation, some individuals are selected based on their fitness value, which undergoes crossover and mutation. A pair of chromosomes exchanges genetic material in the crossover operation, while in mutation, changes are made to an individual chromosome. This enables the exploration of a broader search space. Thus, a new population is created and they again undergo the process of selection, reproduction, and evaluation in successive generations. This continues unless the global optimum solution is reached or the stopping criteria are met.

In the present work, the ANN model of fractal dimension is used as an objective function for optimization using GA. The objective is to find out the optimal combination of discharge current, voltage, pulse-on time and pulse-off time to maximize the fractal dimension. A higher fractal dimension signifies a denser profile and a smoother topography [1]. This may be represented as:

Maximize: Fractal dimension (D)

Subject to: 2 Amp $\leq$ Discharge current $\leq$ 10 Amp

40 V $\leq$ Voltage $\leq$ 60 V

1 μs $\leq$ Pulse-on time $\leq$ 5 μs

1 μs $\leq$ Pulse-off time $\leq$ 5 μs

The neural network model of the process parameters and fractal dimension obtained is used as an input to the Genetic Algorithm Toolbox in MATLAB. Values of the fractal dimension are predicted by the ANN model for a given population size. In GA, there are no specific set of parameters mentioned in the literature that may yield the best results. But in the present work, the major parameters under consideration were population size, number of generations, crossover rate, and mutation rate. The population size considered is 50, and fitness scaling is done using a rank function. A uniform stochastic selection function is chosen for the determination of the parents for reproduction in the successive generation. The crossover function chosen is "scattered", while 80% of the population size is selected for crossover. Lastly, mutation is carried out, and for this, an adaptive feasible mutation function is selected. This process continues for 500 generations unless the change in fitness function value is observed to be 10^{-12}. The aforesaid GA parameters are considered based on the literature review [25] as well as the trial and error method. A schematic diagram of the combined ANN-GA approach is shown in Figure 3.

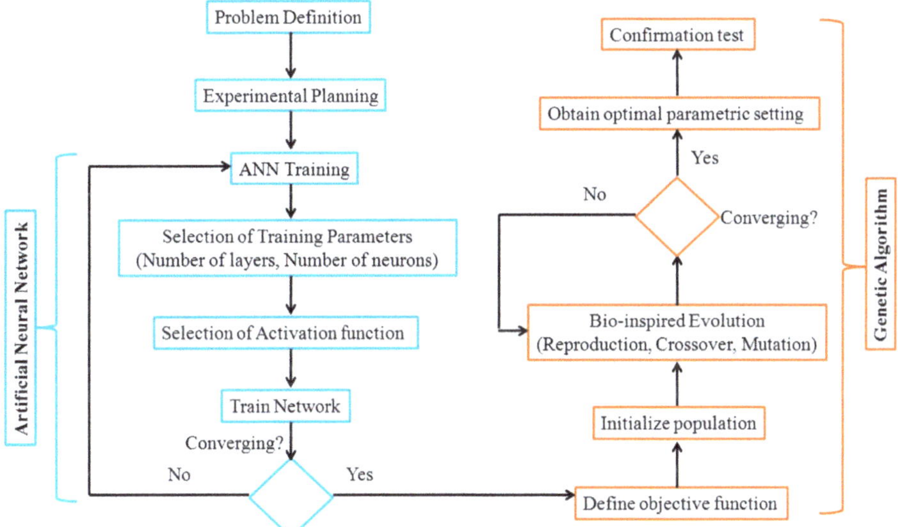

Figure 3. The integrated ANN-genetic algorithm (GA) approach for the modeling and prediction of fractal dimension in WEDM.

4. Results and Discussion

The ANN model for the prediction of fractal dimension in WEDM of EN 31 is developed using MATLAB. A feed-forward network with an L-M training algorithm and 4-3-3-1 architecture with "tansig" activation function in hidden layers and "purelin" in the output layer gives the best correlation of WEDM process parameters with fractal dimension. The best architecture is also defined by the lowest MSE of the validation set. The best validation was observed at epoch 7 with an MSE of 0.0040197, as may be seen in Figure 4. The performance of the trained network may be also observed in Figure 5. The R-value of training is seen to be 0.99, which is significantly high and very near to 1. Moreover, the R-values of testing and validation are ~0.9 and ~0.99 respectively. For the overall prediction model, the R-value comes out to be ~0.97. Finally, the MSE values for training and testing patterns are 0.0001 and 0.0017, respectively. Recently, Yusoff et al. [35] employed a cascade forward back-propagation neural network (CFNN) for modelling of the multi-performances WEDM on Inconel 718. A 5-14-4 CFNN architecture could efficiently correlate machining parameters, namely pulse-on time, pulse-off time, peak current, servo voltage, and flushing pressure with material removal rate, R_a, cutting speed, and sparking gap. An average error of 5.16% was generated with good agreement between predicted and experimental results. The results of the present work also indicate a high efficiency of ANN in modeling a non-linear and stochastic process like WEDM of EN 31 steel. The results obtained are also in accordance with the literature [26,35].

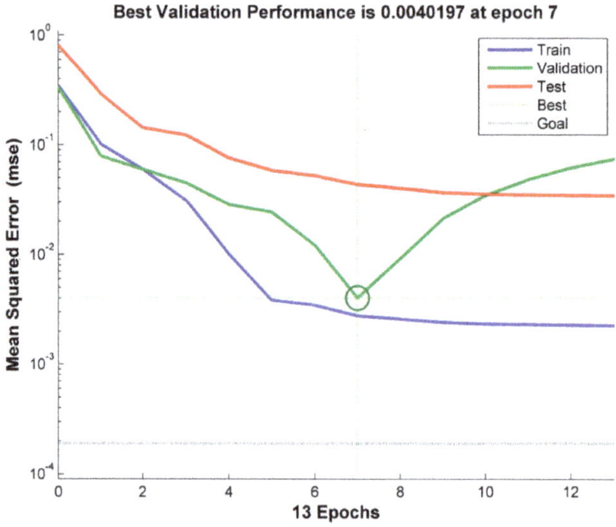

Figure 4. The lowest mean squared error for validation in 4-3-3-1 neural network architecture for the prediction of fractal dimension in WEDM.

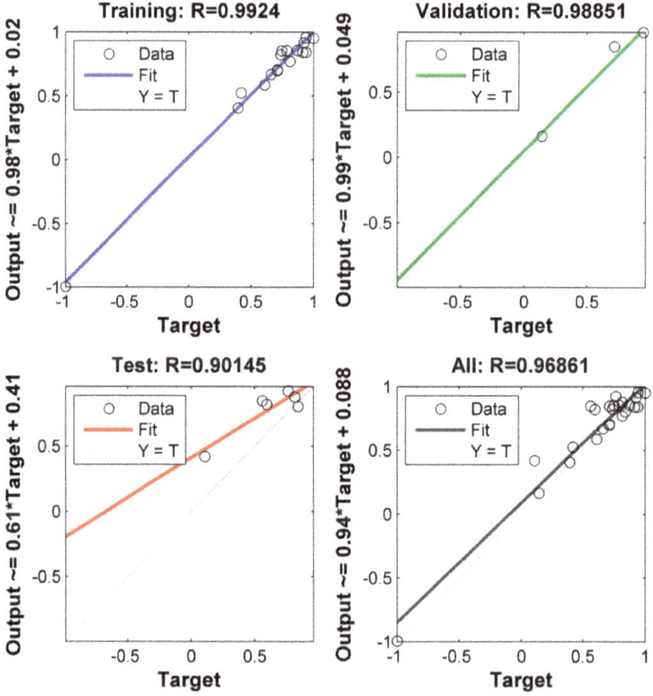

Figure 5. Training, testing, and validation performances of the 4-3-3-1 neural network architecture for the prediction of fractal dimension in WEDM.

The trained ANN model with high efficiency is used as the objective function for optimization using GA. The objective is to maximize the fractal dimension. Variation of the fitness value with the number of generations may be observed in Figure 6. After ~50 generations, a significant change in best fitness or mean fitness is not observed. The algorithm is hence terminated after 176 generations due to insignificant changes in the results in successive generations. Thus, the selection of GA parameters for optimization is justified. The optimal combination of machining parameters predicted by GA is 7.064 Amp discharge current, 60 V voltage, 3.215 μs pulse-on time, and 5 μs pulse-off time. The predicted value of the fractal dimension by the ANN-GA model is ~1.395. Exact values of the predicted process parameters could not be used for carrying out a confirmation test due to limitations with the WEDM setup. However, from Table 2, a combination of 6 Amp discharge current, 50 V voltage, 3 μs pulse-on time, and 5 μs pulse-off time results in a fractal dimension of 1.383 (Experiment 20) which is in close agreement with the predicted results. A higher value of fractal dimension was observed by Sahoo and Barman [1] with an increase in current and voltage for WEDM of EN 31 steel. Thus, the optimization results are in good agreement with their study. On the other hand, Das et al. [36] reported the use of an artificial bee colony (ABC) algorithm for the optimization of WEDM parameters considering the average surface roughness as the response. A parametric setting of 2 Amp discharge current, 60 V voltage, 5 μs pulse-on, time and 1 μs pulse-off time was predicted to yield a lower surface roughness of 2.244 μm. Thus, a clear difference in the predicted results may be observed for a scale-dependent roughness parameter and a scale-independent roughness parameter. From GA predicted results, it may be also seen that a lower pulse-on time and a higher pulse-off time is suggested for better surface finish. A higher pulse-on time causes a more powerful explosion and an increase in discharge energy and has a detrimental effect on surface roughness [25]. Similar observations have been also made for WEDM of pure tungsten, as well as armor materials [24,25]. Kumar et al. [17] observed an increase in R_a of aluminum metal matrix composites with an increase in pulse-on time, and a decrease in R_a was observed for an increase in pulse-off time. A higher pulse-off time results in an increased time gap between two successive sparks and hence it allows better solidifying and molten metal to be washed away from the cutting zone [37].

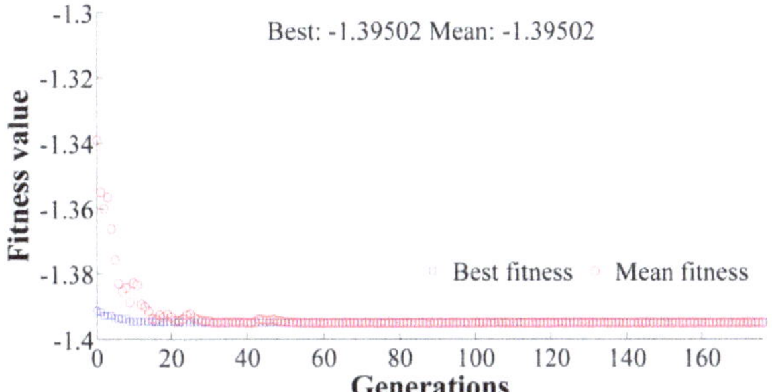

Figure 6. Variation of fractal dimension (fitness values) with generations.

The SEM micrograph of a machined surface at 8 Amp current, 45 V voltage, 2 μs pulse-on time, and 4 μs pulse-off time is shown in Figure 7. This corresponds to the lowest value of fractal dimension (1.043) as can be observed in Table 2 (experiment 25). Pits are visible on the machined surface along with re-solidified particles of molten metal sticking onto each other. Consequently, the surface exhibits a high roughness. A high temperature is obtained in WEDM which causes the formation of molten metal during machining. Since it cannot flush away or vaporize all the molten metal, it re-solidifies to form lumps [28]. This re-solidified of molten metal thus forms a recast layer. In a recent study by Mandal et al. [18], grinding and etching have been proposed for the achievement of a surface finish of the order of 0.024 μm and complete removal of the recast layer. Such processes may be further adopted to improve surface integrity of EN 31 steel post-machining in WEDM. Also, higher current leads to higher discharge energy. This results in higher melting. Inadequate flushing/evaporation of molten metal may lead to a rough surface, as can be observed in Figure 7. Furthermore, trapped air bubbles may lead to the formation of micro-pits [28].

Figure 7. The machined surface corresponding to experiment number 25.

SEM micrography of a machined surface at a discharge current of 6 Amp, voltage of 50 V, pulse-on time of 3 µs, and pulse-off time of 5 µs is shown in Figure 8. This is the machined surface at a combination of parameters at near optimal conditions (experiment number 20 in Table 2) and here the surface appears to be quite smooth in comparison with Figure 7. Any cracks or craters may not be observed on the surface. The surface is also characterized by lower re-solidified globules due to the lower discharge current in comparison with Figure 7. This again establishes the fact that the relationship between WEDM process parameters and roughness has a complex nature, and GA has the capability of predicting optimal combinations with reasonable accuracy. In a recent study by Singh et al. [16], a multi-pass WEDM was seen to be capable of reducing the surface roughness drastically for MWCNT alumina composites in comparison with a single pass. Muthuramalingam et al. [38] observed that a diffused type of brass electrode produces better surface morphology in comparison with conventional brass wire or zinc coated electrodes. Reolon et al. [39] concluded that zinc coated copper wire produces better surface integrity for WEDMed Inconel alloy IN718 in comparison with uncoated brass wire. Thus, from the present analysis, the surface integrity of EN 31 could be improved without any post-processing or complexity while machining using WEDM. Nevertheless, such modifications may be included in future research works and prove to be beneficial in improving in fractal characteristics of EN 31 steel post-machining in WEDM.

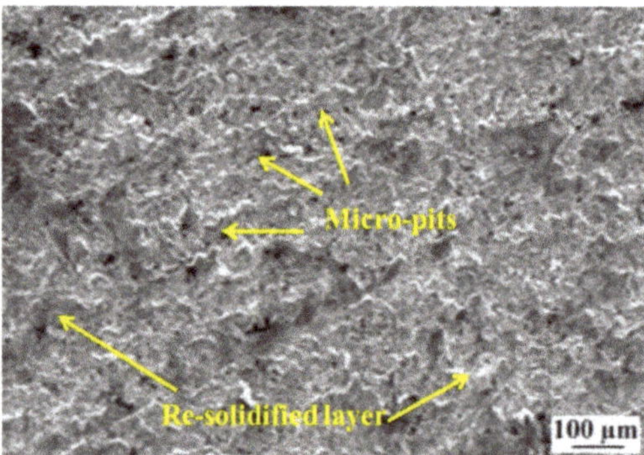

Figure 8. The machined surface corresponding to experiment number 20.

5. Conclusions

In the present work, the fractal dimension is used to characterize the machined surface produced by WEDM on EN 31 steel. Four process parameters, namely discharge current, voltage, pulse-on time and pulse-off time, are varied at five equally-spaced levels. To reduce the number of experiments, a rotatable CCD experimental design methodology is adopted. The complex relationship between considered WEDM parameters and fractal dimension is modeled using ANN. After extensive modeling of the process, the following results may be directly inferred:

- A feed-forward network with an L-M training algorithm and 4-3-3-1 architecture is seen to have better prediction capability.
- The R-values during training, testing, and validation are 0.99, 0.9, and 0.99, respectively. The overall R-value of the model is 0.97.
- The MSE for training is 0.0001, and for the testing pattern it is 0.0017. This indicates the high efficiency of the formed ANN architecture in predicting fractal dimension with high accuracy. A good correlation between predicted and experimental results is therefore concluded.
- The predicted optimal combination of parameters from the combined ANN-GA is 7.064 Amp, 60 V, 3.215 μs, and 5 μs of discharge current, voltage, pulse-on time, and pulse-off time respectively. The predicted fractal dimension is 1.395.
- The SEM micrograph of the machined surface for the lowest value of the fractal dimension is characterized by re-solidified layers or recast layer and micro-pits.
- The SEM micrograph for a combination of parameters at a near-optimal combination is seen to be devoid of any pits or cracks. The re-solidified globules are lesser, indicating better vaporization and flushing. Thus, the integrated ANN-GA methodology is effective in optimizing the quality of a machined surface using WEDM.
- From the present model, a significant improvement in surface quality for WEDM of EN 31 steel could be achieved without any post-processing or pre-treatment, thereby optimizing time and cost of machining as well.

Author Contributions: Conceptualization, P.S. and J.P.D.; investigation, A.M. and T.K.B.; writing—original draft preparation, A.M.; writing—review & editing, T.K.B., P.S. and J.P.D.

Funding: This research received no external funding.

Conflicts of Interest: The authors declare no conflict of interest.

References

1. Sahoo, P.; Barman, T.K. Response surface modelling of fractal dimension in WEDM. In *Design of Experiments in Production Engineering*; Davim, J.P., Ed.; Springer International Publishing: Cham, Switzerland, 2016; pp. 135–149.
2. Ho, K.H.; Newman, S.T.; Rahimifard, S.; Allen, R.D. State of the art in wire electrical discharge machining (WEDM). *Int. J. Mach. Tools Manuf.* **2004**, *44*, 1247–1259. [CrossRef]
3. Sanchez, J.A.; Cabanes, I.; López de Lacalle, L.N.; Lamikiz, A. Development of optimum electrodischarge machining technology for advanced ceramics. *Int. J. Adv. Manuf. Technol.* **2001**, *18*, 897–905. [CrossRef]
4. Saha, A.; Mondal, S.C. Experimental investigation and modelling of WEDM process for machining nano-structured hardfacing material. *J. Braz. Soc. Mech. Sci. Eng.* **2017**, *39*, 3439–3455. [CrossRef]
5. Chen, Z.; Zhang, G.; Yan, H. A high-precision constant wire tension control system for improving workpiece surface quality and geometric accuracy in WEDM. *Precis. Eng.* **2018**, *54*, 51–59. [CrossRef]
6. Kumar, A.; Kumar, V.; Kumar, J. Surface crack density and recast layer thickness analysis in WEDM process through response surface methodology. *Mach. Sci. Technol.* **2016**, *20*, 201–230. [CrossRef]
7. Ishfaq, K.; Mufti, N.A.; Ahmed, N.; Mughal, M.P.; Saleem, M.Q. An investigation of surface roughness and parametric optimization during wire electric discharge machining of cladded material. *Int. J. Adv. Manuf. Technol.* **2018**, *97*, 4065–4079. [CrossRef]

8. Dauw, D.F.; Albert, L. About the evolution of wire tool performance in wire EDM. *CIRP Ann.-Manuf. Technol.* **1992**, *41*, 221–225. [CrossRef]

9. Tosun, N.; Cogun, C.; Inan, A. The effect of cutting parameters on workpiece surface roughness in wire EDM. *Mach. Sci. Technol.* **2003**, *7*, 209–219. [CrossRef]

10. Liao, Y.S.; Huang, J.T.; Chen, Y.H. A study to achieve a fine surface finish in Wire-EDM. *J. Mater. Process. Technol.* **2004**, *149*, 165–171. [CrossRef]

11. Kanlayasiri, K.; Boonmung, S. An investigation on effects of wire-EDM machining parameters on surface roughness of newly developed DC53 die steel. *J. Mater. Process. Technol.* **2007**, *187–188*, 26–29. [CrossRef]

12. Bobbili, R.; Madhu, V.; Gogia, A.K. Effect of Wire-EDM machining parameters on surface roughness and material removal rate of high strength armor steel. *Mater. Manuf. Process.* **2013**, *28*, 364–368. [CrossRef]

13. Chalisgaonkar, R.; Kumar, J. Multi-response optimization and modeling of trim cut WEDM operation of commercially pure titanium (CPTi) considering multiple user's preferences. *Eng. Sci. Technol. Int. J.* **2015**, *18*, 125–134. [CrossRef]

14. Aspinwall, D.K.; Soo, S.L.; Berrisford, A.E.; Walder, G. Workpiece surface roughness and integrity after WEDM of Ti–6Al–4V and Inconel 718 using minimum damage generator technology. *CIRP Ann.-Manuf. Technol.* **2008**, *57*, 187–190. [CrossRef]

15. Goswami, A.; Kumar, J. Investigation of surface integrity, material removal rate and wire wear ratio for WEDM of Nimonic 80A alloy using GRA and Taguchi method. *Eng. Sci. Technol. Int. J.* **2014**, *17*, 173–184. [CrossRef]

16. Singh, M.A.; Sarma, D.K.; Hanzel, O.; Sedláček, J.; Šajgalík, P. Surface characteristics enhancement of MWCNT alumina composites using multi-pass WEDM process. *J. Eur. Ceram. Soc.* **2018**, *38*, 4035–4042. [CrossRef]

17. Kumar, H.; Manna, A.; Kumar, R. Modeling of Process Parameters for Surface Roughness and Analysis of Machined Surface in WEDM of Al/SiC-MMC. *Trans. Indian Inst. Met.* **2018**, *71*, 231–244. [CrossRef]

18. Mandal, A.; Dixit, A.R.; Chattopadhyaya, S.; Paramanik, A.; Hloch, S.; Królczyk, G. Improvement of surface integrity of Nimonic C 263 super alloy produced by WEDM through various post-processing techniques. *Int. J. Adv. Manuf. Technol.* **2017**, *93*, 433–443. [CrossRef]

19. Mouralova, K.; Kovar, J.; Klakurkova, L.; Blazik, P.; Kalivoda, M.; Kousal, P. Analysis of surface and subsurface layers after WEDM for Ti-6Al-4V with heat treatment. *Measurement* **2018**, *116*, 556–564. [CrossRef]

20. Goyal, A. Investigation of material removal rate and surface roughness during wire electrical discharge machining (WEDM) of Inconel 625 super alloy by cryogenic treated tool electrode. *J. King Saud Univ.-Sci.* **2017**, *29*, 528–535. [CrossRef]

21. Spedding, T.A.; Wang, Z.Q. Study on modeling of wire EDM process. *J. Mater. Process Technol.* **1997**, *69*, 18–28. [CrossRef]

22. Sivaprakasam, P.; Hariharan, P.; Gowri, S. Modeling and analysis of micro-WEDM process of titanium alloy (Ti-6Al-4V) using response surface approach. *Eng. Sci. Technol. Int. J.* **2014**, *17*, 227–235. [CrossRef]

23. Nain, S.S.; Garg, D.; Kumar, S. Modeling and optimization of process variables of wire-cut electric discharge machining of super alloy Udimet-L605. *Eng. Sci. Technol. Int. J.* **2017**, *20*, 247–264. [CrossRef]

24. Bobbili, R.; Madhu, V.; Gogia, A.K. Modelling and analysis of material removal rate and surface roughness in wire-cut EDM of armour materials. *Eng. Sci. Technol. Int. J.* **2015**, *18*, 664–668. [CrossRef]

25. Tzeng, C.-J.; Yang, Y.-K.; Hsieh, M.-H.; Jeng, M.-C. Optimization of wire electrical discharge machining of pure tungsten using neural network and response surface methodology. *Proc. IMechE Part B J. Eng. Manuf.* **2010**, *225*, 841–852. [CrossRef]

26. Saha, P.; Singha, A.; Pal, S.K.; Saha, P. Soft computing models based prediction of cutting speed and surface roughness in wire electro-discharge machining of tungsten carbide cobalt composite. *Int. J. Adv. Manuf. Technol.* **2008**, *39*, 74–84. [CrossRef]

27. Çaydaş, U.; Hasçalýk, A.; Ekici, S. An adaptive neuro-fuzzy inference system (ANFIS) model for wire-EDM. *Expert Syst. Appl.* **2009**, *36*, 6135–6139. [CrossRef]

28. Majumder, H.; Maity, K. Prediction and optimization of surface roughness and micro-hardness using grnn and MOORA-fuzzy-a MCDM approach for nitinol in WEDM. *Measurement* **2018**, *118*, 1–13. [CrossRef]

29. López de Lacalle, L.N.; Lamikiz, A.; Salgado, M.A.; Herranz, S.; Rivero, A. Process planning for reliable high-speed machining of moulds. *Int. J. Prod. Res.* **2002**, *40*, 2789–2809. [CrossRef]

30. Ding, H.; Guo, L.; Zhang, Z.; Cui, H. Study on fractal characteristic of surface micro-topography of micro-WEDM. *Appl. Mech. Mater.* **2009**, *16–19*, 1273–1277. [CrossRef]
31. Geng, L.; Zhong, H. Evaluation of WEDM surface quality. *Adv. Mater. Res.* **2010**, *97–101*, 4080–4083. [CrossRef]
32. Luo, G.; Ming, W.; Zhnag, Z.; Liu, M.; Li, H.; Li, Y.; Yin, L. Investigating the effect of WEDM process parameters on 3D micron-scale surface topography related to fractal dimension. *Int. J. Adv. Manuf. Technol.* **2014**, *75*, 1773–1786. [CrossRef]
33. Sahoo, P.; Barman, T.K.; Davim, J.P. *Fractal Analysis in Machining*; Springer: Heidelberg, Germany; Dordrecht, The Netherlands; London, UK; New York, NY, USA, 2011.
34. Pratihar, D.K. *Soft Computing Fundamentals and Application*; Narosa Publishing House: New Delhi, India, 2015.
35. Yusoff, Y.; Zain, A.M.; Sharif, S.; Sallehuddin, R.; Ngadiman, M.S. Potential ANN prediction model for multiperformances WEDM on Inconel 718. *Neural Comput. Appl.* **2018**, *30*, 2113–2127. [CrossRef]
36. Das, M.K.; Barman, T.K.; Kumar, K.; Sahoo, P. Optimization of Surface Roughness in WEDM Process Using Artificial Bee Colony Algorithm. *Int. J. Appl. Eng. Res.* **2014**, *9*, 8748–8751.
37. Vijayabhaskar, S.; Rajmohan, T. Experimental investigation and optimization of machining parameters in WEDM of nano-SiC particles reinforced magnesium matrix composites. *Silicon* **2018**. [CrossRef]
38. Muthuramalingam, T.; Ramamurthy, A.; Sridharan, K.; Ashwin, S. Analysis of surface performance measures on WEDM processed titanium alloy with coated electrodes. *Mater. Res. Express* **2018**, *5*, 126503. [CrossRef]
39. Reolon, L.W.; Laurindo, C.A.H.; Torres, R.D.; Amorim, F.L. WEDM performance and surface integrity of Inconel alloy IN718 with coated and uncoated wires. *Int. J. Adv. Manuf. Technol.* **2018**, 1–11. [CrossRef]

Article

A New Cutting Tool Design for Cryogenic Machining of Ti–6Al–4V Titanium Alloy

Alborz Shokrani [1,*] **and Stephen T Newman** [1]

Department of Mechanical Engineering, University of Bath, BA2 7AY Bath, UK; s.t.newman@bath.ac.uk
* Correspondence: a.shokrani@bath.ac.uk; Tel.: +44-1225-38-6588

Received: 3 January 2019; Accepted: 31 January 2019; Published: 4 February 2019

Abstract: Titanium alloys are extensively used in aerospace and medical industries. About 15% of modern civil aircrafts are made from titanium alloys. Ti–6Al–4V, the most used titanium alloy, is widely considered a difficult-to-machine material due to short tool life, poor surface integrity, and low productivity during machining. Cryogenic machining using liquid nitrogen (LN_2) has shown promising advantages in increasing tool life and material removal rate whilst improving surface integrity. However, to date, there is no study on cutting tool geometry and its performance relationship in cryogenic machining. This paper presents the first investigation on various cutting tool geometries for cryogenic end milling of Ti–6Al–4V alloy. The investigations revealed that a 14° rake angle and a 10° primary clearance angle are the most suitable geometries for cryogenic machining. The effect of cutting speed on tool life was also studied. The analysis indicated that 110 m/min cutting speed yields the longest tool life of 91 min whilst allowing for up to 83% increased productivity when machining Ti–6Al–4V. Overall the research shows significant impact in machining performance of Ti–6Al–4V with much higher material removal rate.

Keywords: cryogenic machining; cutting tool; cutting geometry; titanium

1. Introduction

Titanium is one of the most desirable materials in engineering applications where weight is a major concern. It has the highest strength-to-weight ratio amongst all structural materials and can withstand high temperatures. Grade 5 titanium, Ti–6Al–4V, with α-β microstructure is one of the most used titanium alloys in industry, forming more than 50% of the global production [1]. Due to its inherent mechanical properties, titanium alloys are extensively used in aerospace industries. It is reported that 14% of the Airbus A350-900 [2] and 15% of Boeing 787 [3] aeroframes are made of titanium alloys. Manufactured from wrought material, investment casting or even additive manufacturing, all parts require machining processes for finishing in order to achieve the required surface finish, engineering tolerances, and mechanical properties such as predictable fatigue life.

The material properties that make titanium alloys an attractive material for many engineering applications are also responsible for making them difficult-to-machine materials [4]. Machining titanium is often associated with short tool life, poor surface integrity, and low productivity [5]. Due to the high material strength and poor thermal conductivity, high temperatures are generated during machining titanium alloys [6]. High temperatures at the cutting zone can result in thermal softening of the cutting tool material and increase chemical reaction between the cutting tool and workpiece materials leading to diffusion and adhesion wear. In recent years, significant research has been conducted to improve the machinability of titanium alloys [5–10]. One of the methods which has attracted many researchers is cryogenic machining using liquefied gases such as liquid nitrogen (LN_2) and liquid carbon dioxide (CO_2) [11–13]. In this method, a liquefied gas at very low temperature is sprayed over the cutting zone to cool the workpiece, the cutting tool and the cutting zone. Super cold liquefied gases can favorably alter the material properties of the workpiece and/or cutting tool and enhance heat removal from the cutting zone [14,15].

Investigations by various researchers have shown that cryogenic cooling can enhance the machinability of difficult-to-machine materials by extending tool life and improving surface finish [16–19]. Lee et al. [20] reported that using cryogenic cooling with LN_2 and a preheated Ti–6Al–4V workpiece can result in up to 90% increased tool life. Similarly, Park et al. [21] reported that using LN_2 coolant resulted in reduced tool wear when face milling Ti–6Al–4V alloy. Pereira et al. [12] compared the impact of different machining environments in turning AISI 304 stainless steel. In their study, the effect of LN_2 and CO_2 as well as combined cryogenic cooling and minimum quantity lubrication (MQL) on tool life, cutting forces and surface integrity were investigated. They noted that using CO_2 as a cryogen performs better than LN_2 coolant when turning AISI 304. The addition of MQL lubrication further improved machinability irrespective of the cryogen used [12]. In a similar study [22], the impact of various cooling and lubrication scenarios were tested in turning Ti–6Al–4V. Iqbal et al. [22] reported that the lowest tool wear was associated with combined LN_2 cryogenic cooling and MQL followed by LN_2 cryogenic machining and combined CO_2 and MQL. This indicated that the effect of cryogenic cooling and MQL depends on the workpiece and cutting tool material pair and setup as identified by Zhao and Hong [23].

Rotella et al. [24] investigated the effect of various cooling conditions on surface finish of Ti–6Al–4V alloy in turning operations. They noted that cryogenic cooling significantly improved the surface finish of the machined parts in terms of surface roughness and surface hardness. Bordin et al. [25,26] studied the effect of cryogenic cooling in machining additively manufactured Ti–6Al–4V alloy. The investigations indicated that cryogenic cooling can inhibit adhesion wear and reduce the tool-chip contact length and concluded that increasing feed rate and cutting speed results in higher tool wear. The investigations also indicated that improved surface roughness can be achieved using cryogenic cooling. Pereira et al. [27] proposed a nozzle design based on the Coanda effect for combined cryogenic cooling with CO_2 and MQL. The nozzle was tested in turning Inconel 718 alloy and compared with flood cooling. Using only CO_2 cryogenic cooling, the tool life was only 68% that of flood cooling. The addition of MQL to the cryogenic cooling increased the tool life to 93% of the tool life achieved with flood cooling.

A number of researchers studied the effect of cryogenic delivery methods on the machining performance by developing new cutting tool designs or modifying existing tools [28]. Hong et al. [29], Dhananchezian and Kumar [30], and Bellin et al. [31] proposed drilling holes on flank and/or rake faces of turning inserts for delivery of liquid gases in cryogenic turning operations. Wang and Rajurkar [32], Ahmed et al. [33], Venugopal and Chattopadhyay [34], and Wang et al. [35] suggested developing caplets which mount on top of the cutting inserts or modifying cutting tool holder for delivery of liquefied gases in turning operations. Lu et al. [15,36] proposed through the spindle and cutting tool delivery of LN_2 in milling operations. In their design, LN_2 is delivered through the spindle into the coolant channels inside a solid-end mill tool. Similarly, Georgiou and Azzopardi [37] patented a through spindle cryogenic cooling tool for boring applications. Various tool designs for delivering cryogens are discussed by Astakhov and Godlevskiy [28].

Reviews of the literature [38–41] indicate that cryogenic cooling using LN_2 can improve the machinability of Ti–6Al–4V titanium alloy used in aerospace and medical industries. The improvements are attributed to the increased heat removal during machining and reduced fracture toughness and ductility [42]. Various cutting tool designs were found in the literature for cryogenic machining. These designs concentrated on the delivery of liquid gases such as LN_2 and CO_2 into the cutting zone. Currently there is no study on the effect of cutting tool geometries and their relationship specifically for cryogenic machining.

In this paper, the cutting tool geometries of an end mill cutting tool for cryogenic milling of Ti–6Al–4V are investigated. In particular, the rake angle and primary clearance angle are considered for the cutting tool design. The selected cutting tool geometry is further analyzed by changing the cutting speed in an attempt to identify the effect of cutting speed on tool life in cryogenic machining when a specially designed cutting tool is used. A systematic methodology has been developed for machining experiments and the collected data is analyzed and thoroughly discussed.

2. Materials and Methods

Titanium alloy, Ti–6Al–4V, is commonly machined using flood cooling. In cryogenic machining, LN_2 at $-197\,^{\circ}C$ is sprayed into the cutting zone. This would change the temperature of both cemented tungsten carbide cutting tool and the Ti–6Al–4V workpiece material. It is well known that mechanical properties such as material strength, hardness, and toughness are temperature-dependent parameters and they change as the temperature changes [23,43]. Whilst the design of a cutting tool depends on the application, machining operation, and workpiece, the cutting geometries are selected based on material properties of the workpiece and cutting tool. For this research, three rake angles of $10°$, $12°$, and $14°$ are selected to be tested using an LN_2 cryogenic machining environment. The levels were based on preliminary studies and cutting tool manufacturer's recommendation for machining titanium. The clearance angle affects the friction between the cutting tool and a newly machined surface. Since the cutting edge is the hypothetical point where the rake face and flank face meet, the clearance angle also affects the sharpness and robustness of the cutting edge, and therefore, a correct balance between robustness and sharpness is required when selecting the clearance angle. For this study, the primary clearance angle was tested at two levels of $8°$ and $10°$. The cutting tools were specifically manufactured for this investigation from Extramet EMT 210 solid tungsten carbide with average grain size of 0.8 µm. The tools were coated with average 3 µm TiSiN–TiN Hardcut physical vapor deposition (PVD) coating from IonBond recommended for machining titanium and nickel alloys. Apart from the rake and primary clearance angles, the remaining geometries of the cutting tools were kept constant for this study. The cutting tools had a 12-mm diameter with 3 flutes, $23°$ second clearance angle, and 250-µm chamfer at $45°$ for protecting the tool nose. The core diameter of the tools was 7.8 mm with $34°/35°$ variable helix angle to prevent chatter. In addition, the tools had a 100-µm wiper edge at the minor cutting face to enhance the surface finish of the bottom machined face.

The machining experiments for this investigation was an end milling operation along the length of a Ti–6Al–4V workpiece material with 50 mm x 50 mm x 150 mm dimensions. A new block of workpiece material was used for each machining experiment using a climb milling strategy. The cutting parameters used for experiments are provided in Table 1 based on preliminary studies. The tool overhang was kept constant at 50 mm for all experiments. The tools were balanced at 20,000 rpm. In order to encourage rapid tool wear and limit the experimental time, a high cutting speed of 200 m/min was selected for testing the cutting tool geometries based on previous studies [42,44]. This cutting speed is at the boundary of transition region and high-speed machining for titanium alloys as defined by Schulz and Moriwaki [45].

Table 1. Cutting parameters used for testing cutting tool rake angle and primary clearance angle.

Cutting speed	Chip load	Axial depth of cut	Radial depth of cut
200 m/min	0.03 mm/tooth	1 mm	4 mm

All machining experiments were conducted on a VMC XP10 Bridgeport CNC milling center equipped with a retrofittable external cryogenic cooling nozzle. The cryogenic cooling nozzle used for the experiments is shown in Figure 1. The nozzle was placed around the cutting tool covering the periphery of the tool with 1 bar of LN_2 at 20 kg/h flowrate. There was a 0.5 mm clearance between the nozzle and the cutting tool's outer diameter. The slow-motion time lapse of the LN_2 delivered into the cutting zone through the flute of the tool is shown in Figure 2 with the circle highlighting the LN_2. This figure clearly demonstrates that the LN_2 was only delivered into the cutting zone without flooding the workpiece. Therefore, the cooling was limited to the cutting zone without significantly affecting the temperature of the workpiece. The tight (0.5 mm) clearance between the nozzle and the cutting tool ensured that LN_2 can only flow through the cutting tool's flute into the cutting zone.

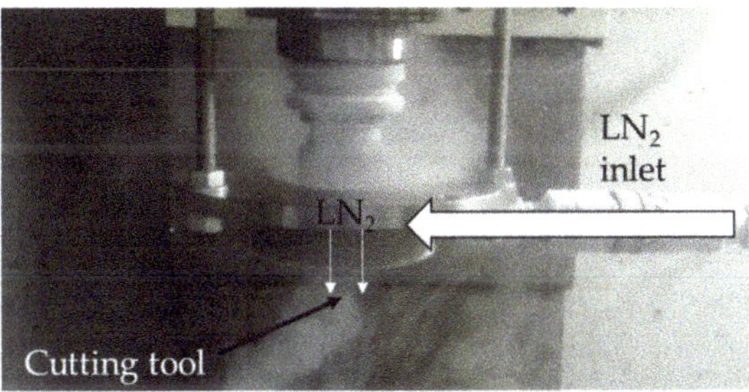

Figure 1. Cryogenic machining setup.

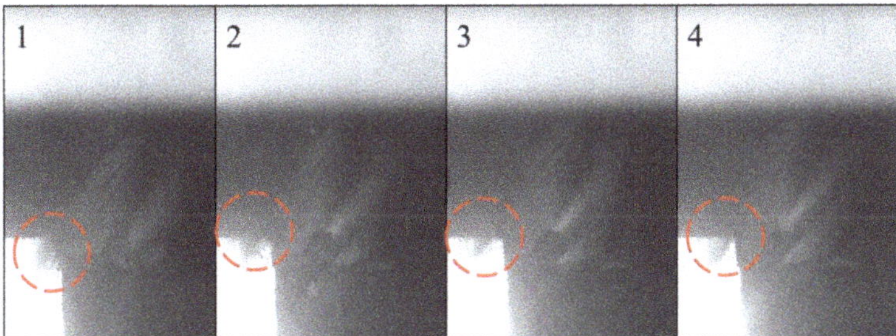

Figure 2. Slow-motion time laps of cryogenic machining showing the delivery of LN_2 into the cutting zone through the tool's flute.

In order to measure the tool life for each experiment, the ISO 8688-2 [46] was followed. A maximum flank wear of 300 μm was defined as the end of tool life criterion and other tool wear modes such as notching and chipping were treated as flank wear to identify tool life. The machining experiments were interrupted regularly to measure the tool wear during the experiments. Once the tool life criterion of 300 μm was reached, the tool life was recorded, and the experiment was stopped. The tool wear was measured using a digital optical microscope. In order to ensure the repeatability of the results, the experiments were repeated at least three times. If chipping or notching caused tool failure, the experiment was repeated an extra two times to ensure that the failure was not caused by manufacturing defects.

The surface roughness of the machined samples was measured at the start of each experiment to minimize the effect of tool wear on surface finish. The surface roughness was measured at three points namely, start, middle, and end of the machining path along the length of the workpiece and each measurement was repeated three times.

After investigating the effect of rake and primary clearance angles on tool life and surface roughness, the tool which performed best in terms of tool life and surface finish was selected to analyze the effect of cutting speed on tool life. The cutting speed was varied from 90 m/min to 200 m/min with 10 m/min intervals at 12 levels following a similar procedure as explained above.

3. Results

In this section, the results are presented and analyzed. Firstly, the effect of cutting tool geometry on tool life and surface roughness are presented. Based on the results, the optimum cutting geometry is selected for further investigation to study the effect of cutting speed in cryogenic end milling of Ti–6Al–4V titanium alloy.

3.1. Cutting Tool Geometry

The effect of rake and primary clearance angle on tool life, tool wear, and surface roughness are shown in this section.

3.1.1. Tool Life

A series of machining experiments were conducted by varying rake and primary clearance angles in cryogenic end milling of Ti–6Al–4V as described in Section 2. The tool life for each experiment was recorded and is presented in Figure 3. This shows that increasing the rake angle from 10° to 14° results in increased tool life. The longest tool life of 9.7 min was achieved using the tool with a 14° rake angle and 10° primary clearance angle. In all experiments, the tool with 10° primary clearance angle performed superior to their counterpart with a 8° primary clearance angle. However, the impact of primary clearance angle on tool life was minimal at lower rake angles of 10° and 12°. It is known that increasing positive rake angle can lead to reductions in cutting forces, power consumption, and cutting temperatures [47–50]. However, after reaching a certain limit, increasing the rake angle will weaken the cutting edge and results in rapid tool wear [48]. In this study, increasing the rake angle from 10° to 14° favorably resulted in increased tool life. Kaymakci et al. [51] reported that increasing clearance angle to an optimum angle results in improved tool life by minimizing flank wear. Further increasing of the clearance angle above the optimum angle, weakens the cutting edge leading to chipping and tool failure.

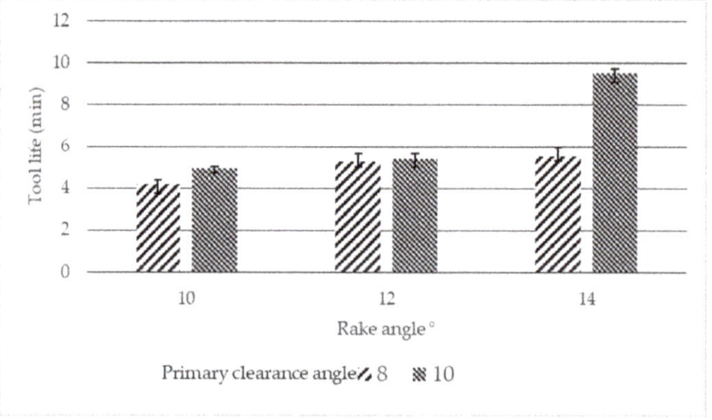

Figure 3. Tool life graph for various rake and primary clearance angles.

Analysis of means was also performed on the data for tool life to investigate the effect of rake and primary clearance angle on tool life in cryogenic machining of Ti–6Al–4V as illustrated in Figure 4. The analysis indicated that within the investigation range, increasing the rake angle from 10° to 14° directly results in enhanced tool life irrespective of the primary clearance angle. However, the impact of primary clearance angle is more significant at higher rake angle of 14°. Similarly, as shown in Figure 4, the analysis indicated that 10° primary clearance angle is more suitable when machining Ti–6Al–4V alloy using LN_2 coolant. Analysis of variance was also performed on the data to identify the significance of rake and primary clearance angles on tool life as provided in Table 2. The analysis indicated that both rake and primary clearance angles have significant effect on tool life with rake angle being the most significant factor.

Since a steady increase in tool life was achieved by increasing the rake angle, a further experiment was conducted using a 16° rake angle. The cutting edge chipped within the first few minutes of the machining experiment and the tool failed catastrophically. As explained above, increasing the rake angle beyond a certain point will result in weakening of the cutting edge and rapid tool wear as reported by Sabberwal and Fleischer [47].

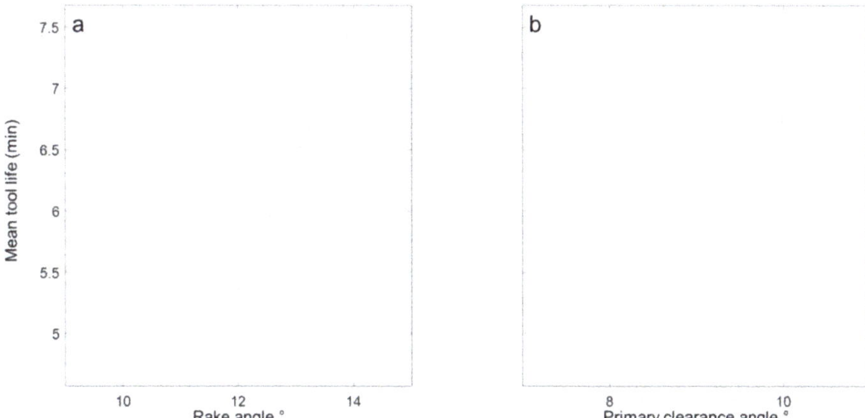

Figure 4. Main effect plots showing the effect of (**a**) rake angle and (**b**) primary clearance angle on tool life.

Table 2. ANOVA of the results for tool life based on rake and primary clearance angles.

Source	Sum of squares	Degree of freedom	Mean square	P
Rake angle	28.72	2	14.36	0.0004
Primary clearance angle	11.60	1	11.60	0.0045
Error	14.26	14	1.02	
Total	54.58	17		

3.1.2. Tool Wear

The machining experiments were interrupted when 300 µm tool wear was reached as per ISO 8688-2 [46]. The cutting tools were analyzed after each machining experiment and the micrographs of the tools were generated using an optical microscope. The micrographs of the tools for 10°, 12°, and 14° rake angles are shown in Figures 5–7, respectively. All cutting tools, irrespective of cutting tool geometry, suffered from flank wear and crater wear. Adhesion and abrasion wear were dominant on all cutting tools. Adhesion was in the form of built-up edge and welded chips on the rake face of the tools and smearing on the flank face. The tool wear was initiated by notch wear at the depth of cut followed by adhesion and abrasion on the flank wear leading to tool failure. For all cutting tools, the flank wear was concentrated at the depth of cut as shown in Figures 5–7. Figure 7 compares the tool wear of the cutting tools with a 14° rake angle and 8° and 10° primary clearance angles. Whilst there are minimal differences in the abrasive tool wear between the tools with identical rake angles, there is a clear difference in the adhesion wear mechanism particularly on the flank face. The flank face of the tool with a 8° primary clearance angle is fully covered by the Ti–6Al–4V workpiece material beyond the abrasion wear region whilst the adhesion was limited to where the substrate was exposed on the tool with a 10° primary clearance angle. This phenomenon was also observed in tools with 10° and 12° rake angles illustrated in Figures 5 and 6. This is attributed to the fact that at shallow primary clearance angles, the friction between the machined surface and tool flank face is higher. As shown in Figures 6b and 7b, chipping of the cutting edge was identified in machining with a 8° primary clearance angle. In addition, crater wear was noticed on the tools adjacent to the cutting edge on the rake faces of the tool as shown in Figures 6 and 7. The cutting chips adhered onto the rake face of the cutting tool forming a built-up edge (BUE). The BUE was removed by the flow of the chips taking some parts of the coating and cutting tool leaving a crater on the rake face of the tools. The formation of crater weakens the cutting edge leading to chipping and notch wear. This exposed the cutting tool substrate at the cutting edge resulting in accelerated flank wear. Flank wear was the dominant tool wear pattern in all machining experiments.

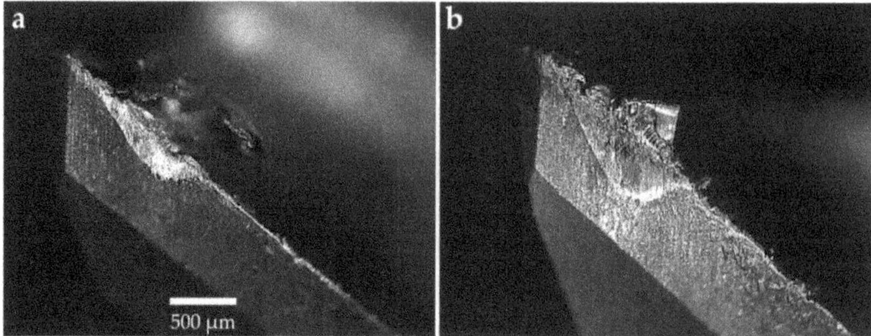

Figure 5. Micrographs of the tool with a 10° rake angle: (**a**) 10° primary clearance angle and (**b**) 8° primary clearance angle.

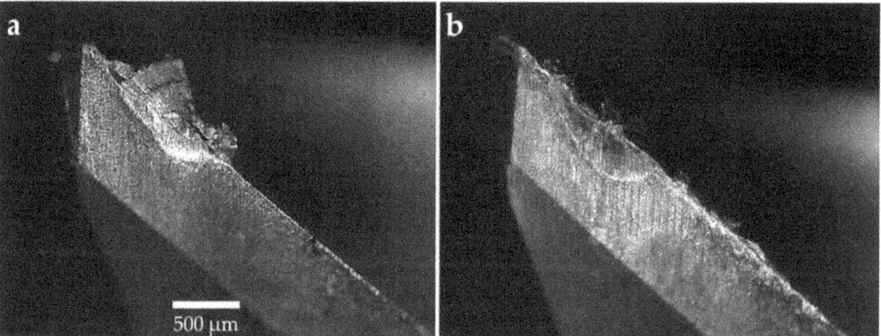

Figure 6. Micrographs of the tool with a 12° rake angle: (**a**) 10° primary clearance angle and (**b**) 8° primary clearance angle.

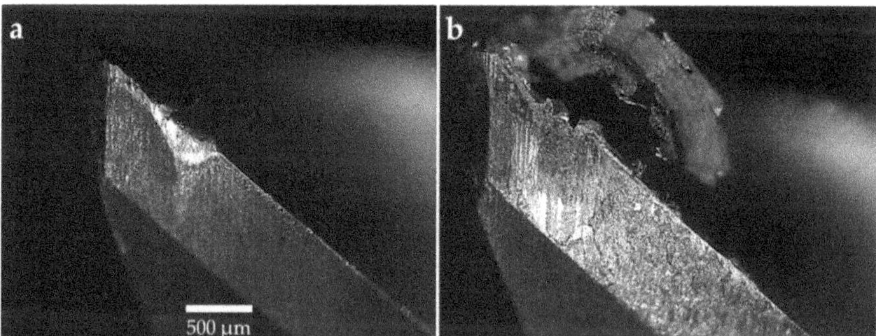

Figure 7. Micrographs of the tool with a 14° rake angle: (**a**) 10° primary clearance angle and (**b**) 8° primary clearance angle.

3.1.3. Surface Roughness

The surface roughness of the workpieces from each machining experiment was measured. The surface roughness measurement was performed at the start of the experiments to minimize the effect of tool wear on surface roughness. The surface roughness was measured at three points namely, start, middle, and end of the machining path and each measurement was repeated three times to ensure repeatability. The average surface roughness Ra from each machining experiment is presented in Figure 8. As shown in the figure, the surface roughness was reduced by increasing both rake angle and primary clearance angle. The lowest surface roughness Ra of 0.27 μm was measured for the tool with a 14° rake angle irrespective of the primary clearance angle. Increasing the primary clearance angle from 8° to 10° in tools with a rake angle of 10° and 12° resulted in improved surface roughness. The highest surface roughness of 0.9 μm was recorded at the end of the machining path for the tool with 12° rake and 8° primary clearance angles. The average surface roughness was 0.6 μm for this experiment. In general, the surface roughness was higher at the exit of the tool for all machining experiments where the tools experiences instability. Analysis of variance indicated that the rake angle was the most significant factor affecting surface roughness Ra with a *p*-value of 0.01.

In machining, the flank face rubs against the newly machined surface when the cutting material causing plastic deformation of the surface. Increasing the primary clearance angle reduces the contact length between the cutting tool and the newly machined surface. Therefore, less plastic deformation takes place on the machined surface leading to improved surface roughness.

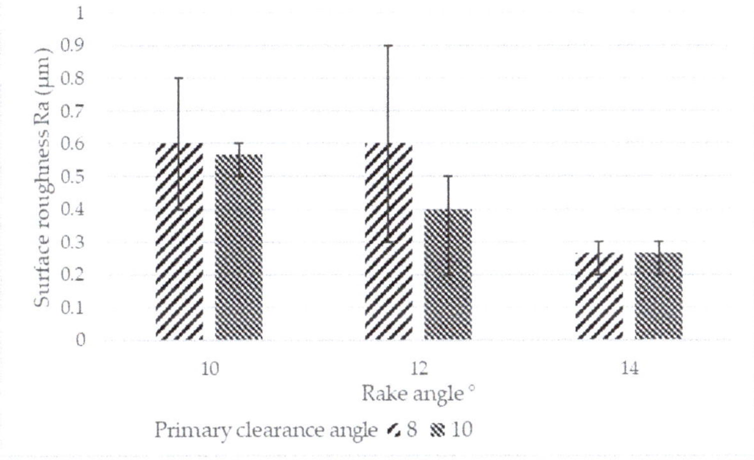

Figure 8. Average surface roughness Ra for tools with different rake and primary clearance angles.

3.2. Cutting Speed

Based on the analysis in Section 3.1, the tool with a 14° rake angle and a 10° primary clearance angle was selected as the best option within the investigated range for further analysis. One of the major advantages of cryogenic cooling using LN_2 is that it allows for using higher cutting speeds than flood cooling. Cutting speed is the major contributor to heat generation during machining and is the most dominant parameter affecting the tool life [52]. Effective use of LN_2 coolant would allow using higher cutting speeds without sacrificing tool life when compared to conventional flood cooling. Therefore, the effect of cutting speed on tool life was investigated as part of this research.

Cutting speeds of 90 m/min to 200 m/min with intervals of 10 m/min (12 levels) were used for machining experiments. The remaining cutting parameters were kept constant at 0.03 mm/tooth feed rate, 1 mm axial depth of cut, and 4 mm radial depth of cut. A series of cutting tools with a 14° rake angle and a 10° primary clearance angle were manufactured from EMT 210 solid tungsten carbide and coated with TiSiN–TiN Hardcut coating. The setup explained in Section 2 was used for the experiments.

The tool life was measured at each cutting speed ranging from 90 m/min to 200 m/min. Following the recommendation from ISO 8688, the results were presented in minutes for tool life as shown in

Figure 9. However, as the cutting speed changes, the feed rate also changes. As a result, higher volume of material can be machined at higher speeds within the same time period. Therefore, the second tool life graph based on machined volume was developed as illustrated in Figure 10. Comparing the graphs in Figures 9 and 10, it is clearly shown that whilst tool life in minutes at 90 min/min cutting speed is higher than that of 100 m/min, more material is machined within the tool life of the experiment at 100 m/min cutting speed. Similar comparisons can be made for the tool life results at 100 m/min and 120 m/min cutting speeds.

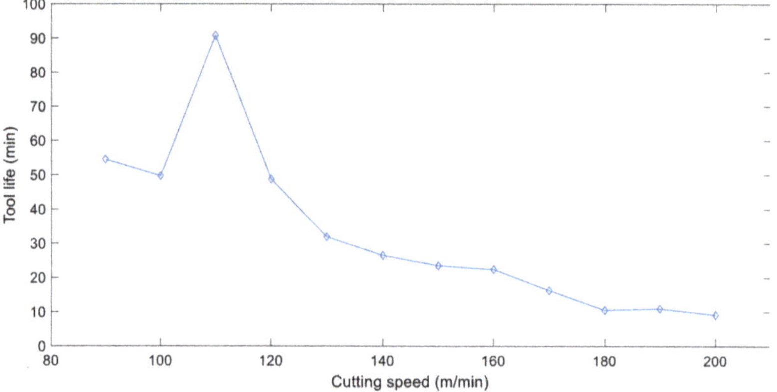

Figure 9. Graph of tool life (min) versus cutting speed.

As shown in Figures 9 and 10, the tool life increases sharply when increasing cutting speed from 100 m/min to 110 m/min. The tool life gradually decreases by increasing cutting speed from 110 m/min to 200 m/min. The longest tool life was recorded to be 91 min equivalent to machining 95400 mm^3 of Ti–6Al–4V workpiece material achieved at 110 m/min. This constitutes that 10% increase in cutting speed from 100 m/min resulted in 83% increased tool life in minutes and 101% increased material removal.

As shown in Figure 11, the cutting tools suffered mechanical wear such as chipping of the nose and notching at the depth of cut at lower cutting speeds. The tool wear mechanism transformed into thermal wear coupled with adhesion and abrasion at higher cutting speeds. Welding of the chips onto the rake face of the tools was evident in all machining experiments above 100 m/min cutting speed. The tool wear was initiated by removal of the coating and exposing the tungsten carbide substrate. This has resulted in accelerated abrasion and adhesion wear. It is known that titanium is chemically reactive to all known cutting tool materials and the chemical reaction is enhanced at higher temperatures [53]. This is evident by the higher level of adhesion of the workpiece material on both the rake and flank face of the cutting tools from increasing the cutting speed. Shalaby and Veldhuis [54] reported that in machining Ti–6Al–4V with tungsten carbide tools, carbides and oxides of titanium and vanadium workpiece material can form at the cutting zone. These hard materials can act as thermal barriers and solid lubricants at the cutting zone enhancing machining performance at certain cutting speeds. Further investigation is required to confirm this occurrence in cryogenic machining at various cutting speeds.

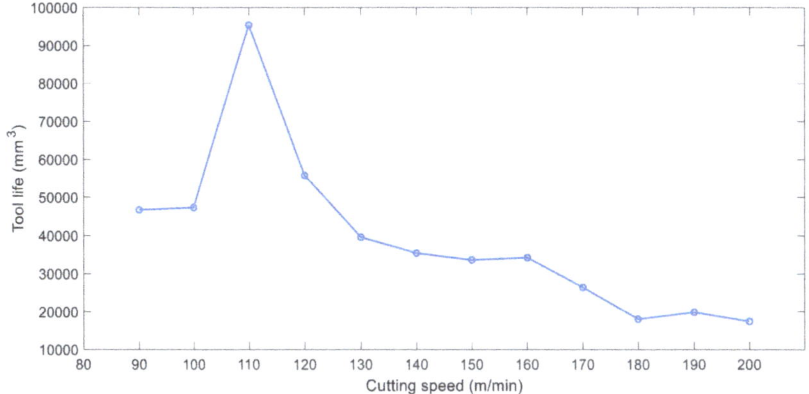

Figure 10. Graph of tool life (mm^3) versus cutting speed.

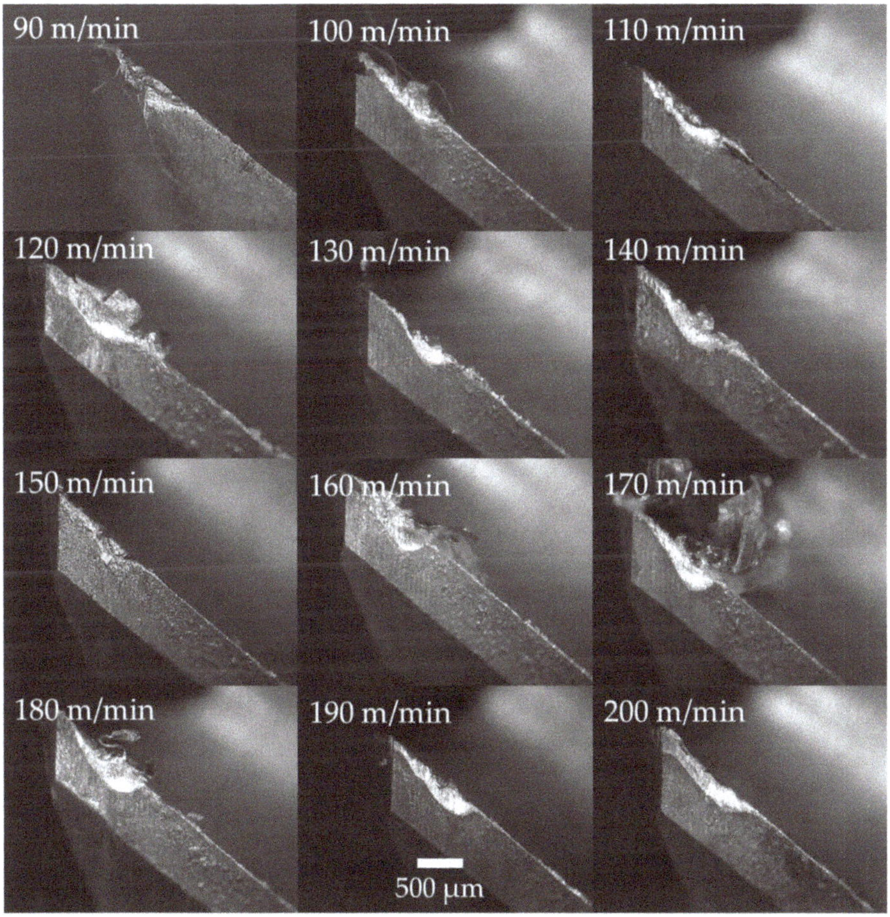

Figure 11. Micrographs of cutting tool wear after machining experiments showing the flank face of the cutting tool.

4. Discussion

Exposing the cutting tool and workpiece material to extremely low temperatures ($-197\,°C$) in cryogenic machining impacts on the material properties of the tool and workpiece. The material hardness of both tungsten carbide tool and Ti–6Al–4V increases by reducing the temperature. In addition, the ductility and fracture toughness of Ti–6Al–4V reduces below 100 K [42]. As a result, tungsten carbide cutting tools with geometries optimized for machining Ti–6Al–4V at ambient temperatures does not necessarily perform well in machining at cryogenic conditions.

In this study, three levels of rake angles and two levels of primary clearance angles were investigated. The analysis indicates that increasing the rake angle from 10° to 14° results in a significant improvement in tool life. However, increasing the rake angle from 14° to 16° proved to weaken the cutting edge leading to catastrophic failure of the cutting edge. Nouari and Makich [55] compared the machinability of Ti–6Al–4V with Ti-55531 alloy by varying cutting speed and tool rake angle. In their study, Nouari and Makich [55] used 0° and 20° rake angles tested at 20 m/min, 35 m/min, and 65 m/min. They noted that cutting forces were lower using 20° rake angle whilst the friction coefficient was higher.

All tools suffered notch wear at the depth of cut followed by flank wear concentrated at the depth of cut. A similar observation is reported by Sadik et al. [5] in cryogenic milling where the authors identified notch wear as the dominant tool wear mechanism. Strain hardening of the workpiece material at the depth of cut results in increased workpiece hardness resulting in increased tool wear at the depth of cut. In addition, the cryogenic cooling setup used in this study sprayed LN_2 along the cutting tool into the cutting zone as well as the top surface of the workpiece. This can result in increased material hardness on the surface of the workpiece leading to increased tool wear at the depth of cut knowing that the hardness of Ti–6Al–4V increases by reducing temperature.

This study showed that using higher rake and primary clearance angles results in improved surface finish from 0.6 μm using the tool with 8° primary clearance and 10° rake angles to 0.26 μm for the tool with 10° primary clearance and 14° rake angles. Higher rake and clearance angles mean sharper cutting edges leading to improved surface finish which is also reported in conventional machining of titanium alloys [55]. In this study, the interactions between the rake and primary clearance angles were not investigated. This will be further investigated in the future.

Cutting speed was identified as the most significant factor affecting the tool life [52,56]. Therefore, cutting speed at 12 levels was investigated using a cutting tool with 14° rake and 10° primary clearance angle. The analysis indicated that the tool life increases by increasing cutting speed from 100 m/min to 110 m/min. The tool life takes a downward slope when cutting speed is further increased as demonstrated in Figure 10. According to Astakhov [57], in machining operations, there is an optimum cutting temperature which yields to minimum tool wear and cutting forces and high-quality surface finish. "This temperature is invariant to the way it has been achieved" (p. 641, [57]). In this research, the combination of heat generation through increased cutting speed and heat removal through cryogenic cooling has resulted in the longest tool life of 91 min at 110 m/min cutting speed. The measurement of cutting temperature was beyond this investigation which will be further studied. Ti–6Al–4V alloy is commonly machined at cutting speeds ranging from 60 m/min to 80 m/min with solid carbide tools using conventional flood cooling [55,58,59]. Based on this, up to 83% increased cutting speed and therefore material removal rate is achieved by using cryogenic cooling and the new proposed cutting tool geometry.

5. Conclusions

A comprehensive review of literature in cryogenic machining revealed that there is a significant need for investigating the effect of various cutting tool geometries in cryogenic end milling of Ti–6Al–4V titanium alloy. This study identified that:

The material properties of workpiece and cutting tool are affected by cryogenic cooling. A new cutting tool with improved geometries was generated to improve tool life and surface roughness in cryogenic machining of Ti–6Al–4V. The new cutting tool can withstand machining at 200 m/min cutting speed for over 9 min.

This new cutting tool was used for machining at various cutting speeds. The investigations indicated that the balance between heat generation and heat dissipation at the cutting zone are most favorable at 110 m/min cutting speed yielding to 91 min tool life equivalent to 95,400 mm^3 of machined material volume.

This research has shown that the application of cryogenic cooling using LN_2 together with the proposed cutting tool has a significant effect in the finish machining of Ti–6Al–4V titanium alloy used in aerospace and medical industries, resulting in up to 83% increased material removal rate and improved productivity.

Author Contributions: Conceptualization, AS and STN; Experiments and analysis, AS; Writing, AS; Review and editing, AS and STN; Management, AS and STN.

Funding: This research was funded by the UK Engineering and Physical Sciences Research Council, grant number EP/K503654/1 and Innovate UK, grant number KTP010108.

Acknowledgments: The authors would like to acknowledge Scorpion Tooling UK Limited for supporting this research.

Conflicts of Interest: The authors declare no conflict of interest.

References

1. Lütjering, G.; Williams, J.C. *Titanium*, 2nd ed.; Springer: New York, NY, USA, 2007.
2. Criou, O. *A350 XWB Family & Technologies*; Airbus S.A.S: Leiden, The Netherlands, 2007.
3. Scot, A. Boeing looks at pricey titanium in bid to stem 787 losses. *Reuters.* 2015. Available online: https://www.reuters.com/article/us-boeing-787-titanium-insight/boeing-looks-at-pricey-titanium-in-bid-to-stem-787-losses-idUSKCN0PY1PL20150724 (accessed on 2 February 2019).
4. Ezugwu, E.; Bonney, J.; Yamane, Y. An overview of the machinability of aeroengine alloys. *J. Mater. Process. Technol.* **2003**, *134*, 233–253. [CrossRef]
5. Sadik, M.I.; Isakson, S.; Malakizadi, A.; Nyborg, L. Influence of coolant flow rate on tool life and wear development in cryogenic and wet milling of Ti-6Al-4V. *Procedia CIRP* **2016**, *46*, 91–94. [CrossRef]
6. Revuru, R.S.; Posinasetti, N.R.; VSN, V.R.; M, A. Application of cutting fluids in machining of titanium alloys—A review. *Int. J. Adv. Manuf. Technol.* **2017**, *91*, 2477–2498. [CrossRef]
7. Li, G.; Yi, S.; Wen, C.; Ding, S. Wear mechanism and modeling of tribological behavior of polycrystalline diamond tools when cutting Ti6Al4V. *J. Manuf. Sci. Eng.* **2018**, *140*, 121011. [CrossRef]
8. Liu, J.; Jiang, X.; Han, X.; Gao, Z.; Zhang, D. Effects of rotary ultrasonic elliptical machining for side milling on the surface integrity of Ti-6Al-4V. *Int. J. Adv. Manuf. Technol.* **2018**. [CrossRef]
9. Machai, C.; Biermann, D. Machining of β-titanium-alloy Ti–10V–2Fe–3Al under cryogenic conditions: Cooling with carbon dioxide snow. *J. Mater. Process. Technol.* **2011**, *211*, 1175–1183. [CrossRef]
10. Biermann, D.; Abrahams, H.; Metzger, M. Experimental investigation of tool wear and chip formation in cryogenic machining of titanium alloys. *Adv. Manuf.* **2015**, *3*, 292–299. [CrossRef]
11. Jawahir, I.S.; Attia, H.; Biermann, D.; Duflou, J.; Klocke, F.; Meyer, D.; Newman, S.T.; Pusavec, F.; Putz, M.; Rech, J.; et al. Cryogenic manufacturing processes. *CIRP Ann.* **2016**, *65*, 713–736. [CrossRef]
12. Pereira, O.; Rodríguez, A.; Fernández-Abia, A.I.; Barreiro, J.; López de Lacalle, L.N. Cryogenic and minimum quantity lubrication for an eco-efficiency turning of aisi 304. *J. Clean. Prod.* **2016**, *139*, 440–449. [CrossRef]
13. Pereira, O.; Català, P.; Rodríguez, A.; Ostra, T.; Vivancos, J.; Rivero, A.; López-de-Lacalle, L.N. The use of hybrid CO_2+MQL in machining operations. *Procedia Eng.* **2015**, *132*, 492–499. [CrossRef]
14. Yildiz, Y.; Nalbant, M. A review of cryogenic cooling in machining processes. *Int. J. Mach. Tools Manuf.* **2008**, *48*, 947–964. [CrossRef]
15. Lu, T.; Kudaravalli, R.; Georgiou, G. Cryogenic machining through the spindle and tool for improved machining process performance and sustainability: Pt. I, system design. *Procedia Manuf.* **2018**, *21*, 266–272. [CrossRef]
16. Jawahir, I.; Brinksmeier, E.; M'Saoubi, R.; Aspinwall, D.; Outeiro, J.; Meyer, D.; Umbrello, D.; Jayal, A. Surface integrity in material removal processes: Recent advances. *CIRP Ann.* **2011**, *60*, 603–626. [CrossRef]

17. Kaynak, Y.; Karaca, H.E.; Noebe, R.D.; Jawahir, I.S. Tool-wear analysis in cryogenic machining of niti shape memory alloys: A comparison of tool-wear performance with dry and MQL machining. *Wear* **2013**, *306*, 51–63. [CrossRef]

18. Pu, Z.; Outeiro, J.C.; Batista, A.C.; Dillon, O.W. Jr.; Puleo, D.A.; Jawahir, I.S. Enhanced surface integrity of AZ31B mg alloy by cryogenic machining towards improved functional performance of machined components. *Int. J. Mach. Tools Manuf.* **2012**, *56*, 17–27. [CrossRef]

19. Mia, M.; Dhar, N.R. Influence of single and dual cryogenic jets on machinability characteristics in turning of Ti-6Al-4V. *Proc. Inst. Mech. Eng. Part B J. Eng. Manuf.* **2019**, *233*, 711–726. [CrossRef]

20. Lee, I.; Bajpai, V.; Moon, S.; Byun, J.; Lee, Y.; Park, H.W. Tool life improvement in cryogenic cooled milling of the preheated Ti–6Al–4V. *Int. J. Adv. Manuf. Technol.* **2015**, *79*, 665–673. [CrossRef]

21. Park, K.-H.; Yang, G.-D.; Lee, M.-G.; Jeong, H.; Lee, S.-W.; Lee, D.Y. Eco-friendly face milling of titanium alloy. *Int. J. Precis. Eng. Manuf.* **2014**, *15*, 1159–1164. [CrossRef]

22. Iqbal, A.; Zhao, W.; Zaini, J.; He, N.; Nauman, M.M.; Suhaimi, H. Comparative analyses of multi-pass face-turning of a titanium alloy under various cryogenic cooling and micro-lubrication conditions. *Int. J. Lightweight Mater. Manuf.* **2018**. [CrossRef]

23. Zhao, Z.; Hong, S. Cooling strategies for cryogenic machining from a materials viewpoint. *J. Mater. Eng. Perform.* **1992**, *1*, 669–678. [CrossRef]

24. Rotella, G.; Dillon, O.W., Jr.; Umbrello, D.; Settineri, L.; Jawahir, I.S. The effects of cooling conditions on surface integrity in machining of Ti6Al4V alloy. *Int. J. Adv. Manuf. Technol.* **2014**, *71*, 47–55. [CrossRef]

25. Bordin, A.; Bruschi, S.; Ghiotti, A.; Bariani, P.F. Analysis of tool wear in cryogenic machining of additive manufactured Ti6Al4V alloy. *Wear* **2015**, *328–329*, 89–99. [CrossRef]

26. Bordin, A.; Sartori, S.; Bruschi, S.; Ghiotti, A. Experimental investigation on the feasibility of dry and cryogenic machining as sustainable strategies when turning Ti6Al4V produced by additive manufacturing. *J. Clean. Prod.* **2017**, *142*, 4142–4151. [CrossRef]

27. Pereira, O.; Rodríguez, A.; Barreiro, J.; Fernández-Abia, A.I.; de Lacalle, L.N.L. Nozzle design for combined use of MQL and cryogenic gas in machining. *Int. J. Precis. Eng. Manuf.-Green Technol.* **2017**, *4*, 87–95. [CrossRef]

28. Astakhov, V.P.; Godlevskiy, V. 3—Delivery of metalworking fluids in the machining zone. In *Metalworking Fluids (MWFS) for Cutting and Grinding*; Astakhov, V.P., Joksch, S., Eds.; Woodhead Publishing: Sawston, UK, 2012; pp. 79–134.

29. Hong, S.Y.; Ding, Y. Cooling approaches and cutting temperatures in cryogenic machining of Ti-6Al-4V. *Int. J. Mach. Tools Manuf.* **2001**, *41*, 1417–1437. [CrossRef]

30. Dhananchezian, M.; Kumar, M.P. Cryogenic turning of the Ti–6Al–4V alloy with modified cutting tool inserts. *Cryogenics* **2011**, *51*, 34–40. [CrossRef]

31. Bellin, M.; Sartori, S.; Ghiotti, A.; Bruschi, S. New tool holder design for cryogenic machining of Ti6Al4V. *AIP Conf. Proc.* **2017**, *1896*, 090001.

32. Wang, Z.; Rajurkar, K. Cryogenic machining of hard-to-cut materials. *Wear* **2000**, *239*, 168–175. [CrossRef]

33. Ahmed, M.; Ismail, A.; Abakr, Y.; Amin, A. Effectiveness of cryogenic machining with modified tool holder. *J. Mater. Process. Technol.* **2007**, *185*, 91–96. [CrossRef]

34. Venugopal, K.; Paul, S.; Chattopadhyay, A. Growth of tool wear in turning of Ti-6Al-4V alloy under cryogenic cooling. *Wear* **2007**, *262*, 1071–1078. [CrossRef]

35. Wang, Z.Y.; Rajurkar, K.; Fan, J.; Lei, S.; Shin, Y.C.; Petrescu, G. Hybrid machining of inconel 718. *Int. J. Mach. Tools Manuf.* **2003**, *43*, 1391–1396. [CrossRef]

36. Lu, T.; Kudaravalli, R.; Georgiou, G. Cryogenic machining through the spindle and tool for improved machining process performance and sustainability: Pt. II, sustainability performance study. *Procedia Manuf.* **2018**, *21*, 273–280. [CrossRef]

37. Georgiou, G.; Azzopardi, D. Temperature Management for a Cryogenically Cooled Boring Tool. US 2018 / 0345387 A1, 06 December 2018.

38. Hong, S.Y. Economical and ecological cryogenic machining. *J. Manuf. Sci. Eng.* **2001**, *123*, 331. [CrossRef]

39. Bermingham, M.; Palanisamy, S.; Kent, D.; Dargusch, M. A comparison of cryogenic and high pressure emulsion cooling technologies on tool life and chip morphology in Ti-6Al-4V cutting. *J. Mater. Process. Technol.* **2011**, *212*, 752–765. [CrossRef]

40. Cordes, S.; Hübner, F.; Schaarschmidt, T. Next generation high performance cutting by use of carbon dioxide as cryogenics. *Procedia CIRP* **2014**, *14*, 401–405. [CrossRef]
41. Kaynak, Y.; Lu, T.; Jawahir, I. Cryogenic machining-induced surface integrity: A review and comparison with dry, mql, and flood-cooled machining. *Mach. Sci. Technol.* **2014**, *18*, 149–198. [CrossRef]
42. Shokrani, A.; Dhokia, V.; Newman, S.T. Energy conscious cryogenic machining of Ti-6Al-4V titanium alloy. *Proc. Inst. Mech. Eng. Part B J. Eng. Manuf.* **2018**, *232*, 1690–1706. [CrossRef]
43. Zhao, Z.; Hong, S. Cryogenic properties of some cutting tool materials. *J. Mater. Eng. Perform.* **1992**, *1*, 705–714. [CrossRef]
44. Shokrani, A.; Dhokia, V.; Newman, S.T. Comparative investigation on using cryogenic machining in cnc milling of Ti-6Al-4V titanium alloy. *Mach. Sci. Technol.* **2016**, *20*, 475–494. [CrossRef]
45. Schulz, H.; Moriwaki, T. High-speed machining. *CIRP Ann.* **1992**, *41*, 637–643. [CrossRef]
46. International-Organization-for-Standardization. *Tool Life Testing in Milling—Part 2: End Milling*; ISO-8688-2; International-Organization-for-Standardization: Geneva, Switzerland, 1989.
47. Sabberwal, A.J.P.; Fleischer, P. The effect of material and geometry on the wear characteristics of cutting tools during face milling. *Int. J. Mach. Tool Des. Res.* **1964**, *4*, 47–71. [CrossRef]
48. Krain, H.; Sharman, A.; Ridgway, K. Optimisation of tool life and productivity when end milling inconel 718tm. *J. Mater. Process. Technol.* **2007**, *189*, 153–161. [CrossRef]
49. Ezugwu, E.O.; Pashby, I.R. High speed milling of nickel-based superalloys. *J. Mater. Process. Technol.* **1992**, *33*, 429–437. [CrossRef]
50. Saglam, H.; Unsacar, F.; Yaldiz, S. Investigation of the effect of rake angle and approaching angle on main cutting force and tool tip temperature. *Int. J. Mach. Tools Manuf.* **2006**, *46*, 132–141. [CrossRef]
51. Kaymakci, M.; Kilic, Z.M.; Altintas, Y. Unified cutting force model for turning, boring, drilling and milling operations. *Int. J. Mach. Tools Manuf.* **2012**, *54-55*, 34–45. [CrossRef]
52. Hong, S.Y.; Markus, I.; Jeong, W. New cooling approach and tool life improvement in cryogenic machining of titanium alloy Ti-6Al-4V. *Int. J. Mach. Tools Manuf.* **2001**, *41*, 2245–2260. [CrossRef]
53. Abele, E.; Fröhlich, B. High speed milling of titanium alloys. *Adv. Prod. Eng. Manag.* **2008**, *3*, 131–140.
54. Shalaby, M.A.; Veldhuis, S.C. Some observations on flood and dry finish turning of the Ti-6Al-4V aerospace alloy with carbide and pcd tools. *Int. J. Adv. Manuf. Technol.* **2018**, *99*, 2939–2957. [CrossRef]
55. Nouari, M.; Makich, H. Analysis of physical cutting mechanisms and their effects on the tool wear and chip formation process when machining aeronautical titanium alloys: Ti-6Al-4V and Ti-55531. In *Machining of Titanium Alloys*; Davim, J.P., Ed.; Springer: Berlin/Heidelberg, Germany, 2014; pp. 79–111.
56. Bermingham, M.J.; Kirsch, J.; Sun, S.; Palanisamy, S.; Dargusch, M.S. New observations on tool life, cutting forces and chip morphology in cryogenic machining Ti-6Al-4V. *Int. J. Mach. Tools Manuf.* **2011**, *51*, 500–511. [CrossRef]
57. Astakhov, V.P. The assessment of cutting tool wear. *Int. J. Mach. Tools Manuf.* **2004**, *44*, 637–647. [CrossRef]
58. Rahman, M.; Wong, Y.S.; Zareena, A.R. Machinability of titanium alloys. *JSME Int. J. Ser. C* **2003**, *46*, 107–115. [CrossRef]
59. Zoya, Z.A.; Krishnamurthy, R. The performance of cbn tools in the machining of titanium alloys. *J. Mater. Process. Technol.* **2000**, *100*, 80–86. [CrossRef]

 *materials*

Article

Model Based on an Effective Material-Removal Rate to Evaluate Specific Energy Consumption in Grinding

Amelia Nápoles Alberro [1], Hernán A. González Rojas [1], Antonio J. Sánchez Egea [2,*], Saqib Hameed [1] and Reyna M. Peña Aguilar [3]

[1] Department of Mechanical Engineering (EPSEVG), Universidad Politécnica de Cataluña,
Av. de Víctor Balaguer, 1, Vilanova i la Geltrú, 08800 Barcelona, Spain; amelia.napoles@upc.edu (A.N.A.);
hernan.gonzalez@upc.edu (H.A.G.R.); hameeds@tcd.ie (S.H.)

[2] Department of Mechanical Engineering (EEBE), Universidad Politécnica de Cataluña, Av. Eduard Maristany,
16, 08019 Barcelona, Spain

[3] Department of Fluid Mechanics (EEBE), Universidad Politécnica de Cataluña, Av. Eduard Maristany, 16,
08019 Barcelona, Spain; reyna.mercedes.pena@upc.edu

* Correspondence: antonio.egea@upc.edu

Received: 21 February 2019; Accepted: 17 March 2019; Published: 21 March 2019

Abstract: Grinding energy efficiency depends on the appropriate selection of cutting conditions, grinding wheel, and workpiece material. Additionally, the estimation of specific energy consumption is a good indicator to control the consumed energy during the grinding process. Consequently, this study develops a model of material-removal rate to estimate specific energy consumption based on the measurement of active power consumed in a plane surface grinding of C45K with different thermal treatments and AISI 304. This model identifies and evaluates the dissipated power by sliding, ploughing, and chip formation in an industrial-scale grinding process. Furthermore, the instantaneous positions of abrasive grains during cutting are described to study the material-removal rate. The estimation of specific chip-formation energy is similar to that described by other authors on a laboratory scale, which allows to validate the model and experiments. Finally, the results show that the energy consumed by sliding is the main mechanism of energy dissipation in an industrial-scale grinding process, where it is denoted that sliding energy by volume unity decreases as the depth of cut and the speed of the workpiece increase.

Keywords: power consumption; material-removal rate; specific energy consumption; grain density; modeling

---

1. Introduction

Efficiency in machining processes requires more attention due to the high cost of energy, in which the manufacturing cost represents a significant proportion of the total cost of the final product [1]. The Industry 4.0 philosophy presents a global vision of virtualization for manufacturing high-quality parts [2,3]. These models and simulations help to optimize the conditions to execute the work cycle and desired results in manufacturing parts [4,5]. Hence, it is deduced that both energy efficiency and virtualization require a model to analyze the behavior of the different manufacturing processes with respect to operating conditions.

The models of specific energy are divided into two main groups: the models that evaluate specific cutting energy SCE, and the models that calculate specific energy consumption SEC. The first group is based on experimental measurements of cutting forces during machining by using a piezoelectric dynamometer located on the table of the grinding machine [6] or in the spindle where the grinding wheel is attached [7]. In this case, the recorded values of forces are multiplied with the peripheral speed of the grinding wheel to define cutting-power consumption. Other authors estimated SCE by developing a function that related the active power of motor with the mechanical power developed by the spindle during turning [8]. The same strategy was also used by González et al. [9] in drilling to investigate the influence of different cutting conditions.

However, the second group evaluated *SEC* during the process by measuring the active power of the motor, as was done by Diaz et al. [10] during milling, or Sánchez Egea et al. [11] during turning operations. Moreover, there are two ways to obtain material-removal rate Q_w. The first is defined as the product of the cutting cross section and workpiece speed. The other is defined as the product of the effective section of cutting grains and the cutting speed of the grinding wheel [12]. Generally, the authors used the first model, in which they considered the cutting section by the depth of cut and width of the grinded zone [13]. Conversely, in the second model, the authors considered the effective cross section of the cutting chip and number of grains corresponding to the contact area [14]. To make the second model applicable, the researchers used an equation for the maximum thickness of the undeformed chip [15]. Based on the geometrical characteristics of chip formation, this thickness is defined as a function of cutting speed, workpiece speed, depth of cut, and the diameter of the grinding wheel. This equation also included the normalized density of static grains C_g and a constant that indicated the average grain geometry. There are cases in which C_g is defined as a function of angle of attack of the grain [16]. Normally, the authors calculated chip thickness by using the empirical data of C_g [17]. Due to the complexity of the cutting edges of the grinding wheel, it is well known that C_g significantly influenced grinding behavior. Therefore, several authors measured the topography of a grinding wheel through an electron microscope [18]. Unlike the classic methods of estimating Q_w, Nadolny [19] proposed a new index SI_Q that defines the material-removal rate of a single abrasive grain, which is based on the number of active kinematic cutting grains. So far, *SEC* models are characterized by the macro level during iteration between workpiece and grinding wheel to predict the average value of chip thickness. A recent work developed the model of normal and tangential forces by considering the microinteraction between workpiece and grinding tool [20].

In the present work, *SEC* is obtained by measuring the active power consumed by the motor that drives the grinding wheel. A model was developed to calculate the Q_w under different cutting configurations and taking into account the interaction between grains and workpiece. Active power consumption is measured by a power analyzer connected to the three-phase electric motor. An equation of deformed-chip thickness and effective cutting section is also proposed to accurately define the material-removal rate. Finally, the chip-thickness equation is defined as a function of the radial position of each grain, cutting parameters, and actual grain density of the grinding wheel. Additionally, a laser distance sensor was used to measure the topography of the abrasive wheel and, ultimately, to calculate grain density.

2. Model of Specific Energy Consumption

In grinding, three mechanisms occurred between grinding wheel and workpiece. First, the friction between wheel and workpiece, characterized by negligible small Q_w. When the force of grains increased on the workpiece, elastic and plastic deformation occurred, which produced a scratch with crests on the sides. The material was removed by increasing the force to produce chip formation [21]. In this work, it was considered that the consumed power in the grinding is due to the power dissipated by different mechanisms involved in the process. These mechanisms were the friction between wheel and workpiece (sliding), plastic deformation without breakage (ploughing), and chip removal by shearing (chip formation) [22]. The power consumed by the sliding, ploughing, and chip-formation mechanisms are P_{sl}, P_{pl} and P_{ch}, respectively. Then, total consumed power P during the process is equal to the sum of power consumed by each of the above-mentioned mechanisms.

$$P = P_{sl} + P_{pl} + P_{ch} \tag{1}$$

In the chip-removal process, SEC is directly proportional to the relation between consumed power and Q_w [23]. If Equation (1) is divided by Q_w and reorganized, then Equation (2) is obtained as follows:

$$\frac{P - P_{sl}}{Q_w} = \frac{P_{pl}}{Q_w} + SEC_{ch} \tag{2}$$

where SEC_{ch} is the specific energy consumed by chip formation.

During grinding, two types of cutting operations were defined due to the alternative movement of the table on which the workpiece was placed. If the movement was in the opposite direction to peripheral speed v_c of the grinding wheel, then this operation is called up-grinding. There are three associated mechanisms here: sliding, ploughing, and chip formation. If the movement is in the same direction as the v_c of the wheel, then the operation is called down-grinding. In this particular case, there is only one mechanism associated, which is sliding. In this work, grinding was conceived in the following way: the depth of cut was applied to the workpiece when it started its movement in up-grinding. This step should not be repeated until down-grinding is completed. Therefore, during the up-grinding movement, sliding, ploughing, and chip formation existed simultaneously. On the other hand, during down-grinding, only sliding was expected. Accordingly, the power consumed during up-grinding and the power consumed during down-grinding are known, so then the mechanisms of ploughing and chip formation can be isolated. Hence, the difference between the power consumed in up- and down-grinding was due to ploughing and chip formation, which were the mechanisms that characterized cutting [24]. In this study, the power of the two trajectories was measured by a power analyzer during grinding in a dry condition with different cutting conditions and metallic alloys. The power consumed by the motor was also measured during an idle condition, i.e., when the wheel was not in contact with the workpiece. Therefore, active power consumption can be calculated by subtracting the power measured without cutting (idle) from the power in up- and down-grinding.

Model of Effective Material-Removal Rate in Grinding

In grinding, it is difficult to define the geometry of the cutting tool, as the grinding wheel has different cutting grains, distributed irregularly on the working surface and, at the same time, grains have different cutting edges. The material-removal rate is obtained by considering the geometric intersections between grinding wheel and workpiece, as well as the multiple grains involved in cutting. To define the model of material-removal rate Q_w, the equation of chip thickness and the section cut by a grain A_{cg} was first obtained. Subsequently, the Q_w by all cutting grains is simultaneously calculated. Figure 1 represents the section of the removed material during up-grinding. It defined the radius of grinding wheel R_M, angular position of grain θ, contact length between grinding wheel and workpiece l_c, and cutting parameters such as speed of grinding wheel v_c, speed of workpiece v_w, and depth of cut a. Undeformed chip thickness h was measured in the XY plane, and A_{cg} was evaluated in the ZY plane, which is perpendicular to the plane of the grinding wheel and is represented by the A-A section.

In this section, the evolution of chip thickness as a function of angular position of grain θ was analyzed. To define chip thickness, it was assumed that the grains of the grinding wheel were equally spaced, like the teeth of a milling cutter. Accordingly, Figure 1 shows the trajectories G_1 and G_2 of two abrasive grains that were consecutively cut. Trajectory G_2 has a center displaced at a distance OO', equivalent to feed rate f that depends on the distance between grains l_g, and speeds of workpiece v_w and grinding wheel v_c. The zone of interest was defined by points $BEE'F$, where the l_g between grinding wheel and workpiece was defined by arc BE, and maximum thickness by points $E'B'$. To obtain the coordinates of the intersection of line $E''B''$ with curves G_1 and G_2, equations were developed to define the circumferential arcs of G_1 and G_2 and line $O'B''$. Then, point E'' was defined by the intersection of curve G_1 and line $O'E''$ as a function of θ. Therefore, the equations can be defined as a function of dimensionless angular position θ^* defined as the ratio of θ and θ_{max}.

$$\theta^* = \frac{\theta}{\sqrt{2 \cdot a / R_M}} \tag{3}$$

Chip thickness was defined as:

$$e = 2\theta^* \cdot \left(l_g \cdot \frac{v_w}{v_c} \right) \cdot \left(\frac{a}{D_M} \right)^{1/2} \tag{4}$$

where D_M is the diameter of the grinding wheel.

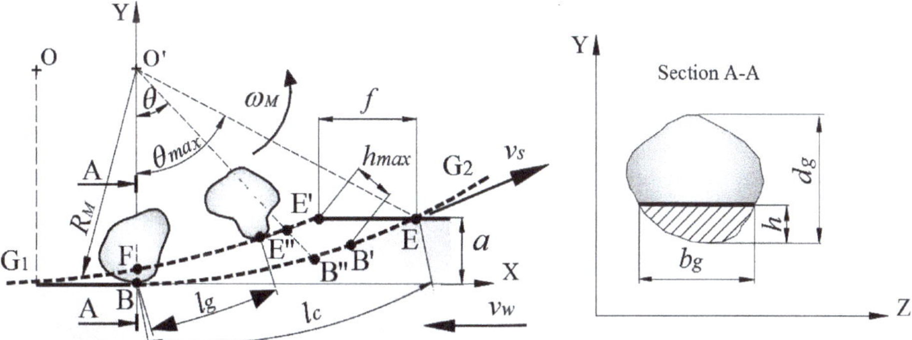

Figure 1. Characteristics of interaction between grinding wheel and workpiece.

By considering the static density of the grain constant, it was estimated that the distance between grains is constant throughout the perimeter of the grinding wheel. Then length between grains l_g can be deduced as:

$$l_g = \frac{1}{C_g \cdot b_g} \tag{5}$$

where b_g is the width of grain as a function of undeformed chip thickness h and diameter of grain d_g.

$$b_g = 2 \cdot \sqrt{d_g \cdot h} \tag{6}$$

Replacing Equations (5) and (6) in Equation (4) gave a useful expression for h, as follows:

$$h = \left(\frac{\theta^* \cdot v_w}{C_g \cdot v_c} \right)^{2/3} \cdot \left(\frac{a}{d_g \cdot D_M} \right)^{1/3} \tag{7}$$

The area of the material removed by grain A_{cg} corresponded to the effective section of cutting by grain. To estimate A_{cg}, it was assumed that the geometric shape of grain can be approximated to a sphere, and only a part of the grain cut the material [15]. For a sphere, the effective cutting area is a function of h and the radius of grain R_g:

$$A_{cg} = \arccos(1 - h/R_g) \cdot R_g^2 - (R_g - h) \cdot R_g \cdot sen(\arccos(1 - h/R_g)) \tag{8}$$

A_{cg} is different for each relative position of the grain as chip thickness e increases with the increase of θ^*. The total area of cutting depends on the number of grains and is equal to the sum of instantaneous areas of each grain present along the contact length between wheel and workpiece.

Finally, considering number of grains N_g cut in the grinding width, material-removal rate Q_w by all grains in the ZY plane is calculated as:

$$Q_w = \sum_{i=1}^{N_g} (\overrightarrow{A_{cg}}(h_i) \times \overrightarrow{v_c}) \tag{9}$$

3. Experiment Setup

In this work, two types of experiments were performed. The topography of the grinding wheel was evaluated and the power consumed by motor was measured during grinding test. Two types of metallic alloys were selected, ductile and brittle. This helps to understand the effect of material hardness and cutting parameters on the SEC.

3.1. Estimation of Grain Number Per Unit Area in Grinding Wheel

The distance between two adjacent grains depends on the structure of grinding wheel. In grinding tests, the grinding wheel of aluminium oxide A36H5V was used, which has an outside diameter of 250 mm, a 76 mm mounting hole, 40 mm width, and grain-size number 36 according to the manufacturer's certificate. According to the FEPA standard [25], the characteristics of this grinding wheel are d_g = 0.337 mm and l_g = 0.67 mm. The topography of the wheel was measured to confirm the information provided by the manufacturer. The wheel was mounted on a divider head located on the table of a vertical milling machine. Measurements were made by using a laser (LDS-Laser distance sensor, model: LDS90/40, LMI Technologies Inc., Burnaby, BC, Canada) located on the spindle of the machine. Surface roughness was measured with an accuracy of 0.001 mm according to the data-acquisition equipment (HBM, model: Spider-8, Hottinger Baldwin Messtechnik GmbH, Darmstadt, Germany). In total, eight profiles of surface roughness, with an evaluation length of 5 mm each, were measured across the width of the grinding wheel. For statistical analysis, the average length value between grains was calculated. The Anderson Darling test was applied to the specimen, and a probability of 0.570 was found. Consequently, it could be assumed that its distribution had normal behavior, as the p-value was greater than 0.05. Figure 2a shows the surface-roughness profile of the grinding wheel in which the distance between grains was identified. The average grain number per unit area can be calculated by evaluating the number of peaks in the specimen. The average distance between grains was 0.775 mm, with a confidence interval of 0.650–0.900 mm, and the averaged C_g was 3.05 grains/mm^2, with a confidence interval of 3.030–3.676 grains/mm^2. The value of distance between grains was greater than the theoretical value indicated by the manufacturer and, therefore, the C_g was slightly smaller. A single-tip diamond test was also performed with a maximum depth of cut of 0.05 mm and an axial table feed speed of 1.6 mm/s. A total of six tests were performed with a 1.5 carat single-tip diamond cut to collect detached grains and analyze grain size. The main length of grains was measured by using optical magnifiers (Leica, model: M165C, Leica Microsistemas S.L.U., Wetzlar, Germany) shown in Figure 2b. Then, the equivalent diameter was calculated by assuming the grain geometry as a sphere. The Anderson Darling test was applied to the diameters, and a p-value of 0.65 was obtained. Consequently, it could be assumed that the equivalent diameter had normal behavior, with an average value of 0.347 mm and confidence interval of 0.300–0.394 mm.

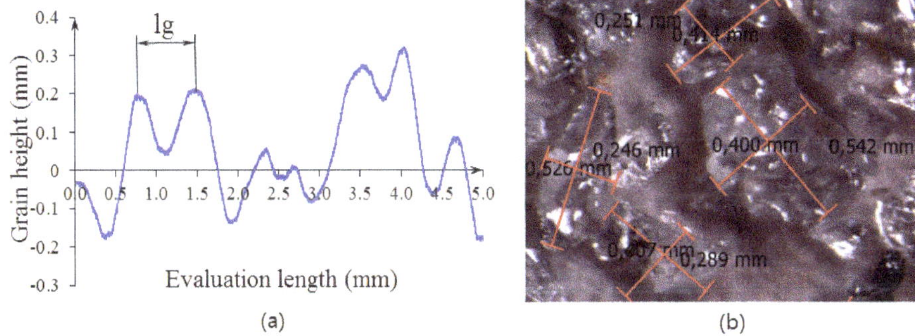

Figure 2. Characteristics of grinding wheel A36H5V: (**a**) surface-roughness profile and (**b**) size of detached grains.

3.2. Power-Consumption Measurement

The plane dry grinding experiments were carried out in a grinding machine (KAIR, model: T650, KEHREN, Hennef, Germany) with a nominal power of 2.24 kW. The cutting conditions used during the experimental tests are listed in Table 1. These conditions are similar to the range of values used by Singh et al. [26]. Five grinding passes were made for each test specimen with dimensions of 30 mm × 10 mm × 130 mm. Material hardness was measured with a durometer (Wolpert, model: Testor HT, Buehler, Esslingen am Neckar, Germany). Table 1 shows the average error dispersion, with an interval of 95% confidence, of the material hardness of each metallic alloy. In total, five hardness values were recorded for each alloy. Then, an Anderson Darling test was applied to verify a normal distribution. The confidence interval was estimated by using a t-student test in the material's hardness measurements.

Table 1. Cutting conditions used during the experimental tests.

Cutting Parameter	Magnitude of Values		
Depth of cut (mm)	0.010	0.015	0.020
Peripheral cutting speed (m/s)	22.9	22.9	22.9
Speed of workpiece (mm/s)	57	101	150

To evaluate the power consumption, the active power of an electric motor was recorded in three conditions: idle, up-grinding, and down-grinding. Power was measured by an energy analyzer (HBM, model: Genesis eDrive Testing, Hottinger Baldwin Messtechnik GmbH, Darmstadt, Germany), where the current intensity, voltage, and the power consumed by the motor were recorded [27]. Since the grinding machine was a three-phase machine, a wattmeter recorded the measurements of the three phases. Then, these measurements were saved on files in ASCI format to be postprocessed. These results allowed to identify the tie periods and power consumption during cutting in up- and down-grinding, and an idle condition. Figure 3 shows the signals of consumed power while grinding C45K steel with a depth of cut of 0.020 mm and workpiece speed of 101 mm/s. In this figure, the path of up- and down-grinding and the idle condition were identified, with average active power consumption of 259, 240, and 54 W, respectively.

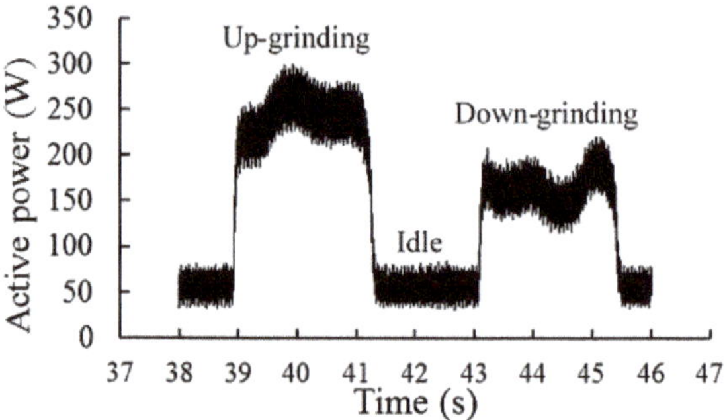

Figure 3. Signal of active power consumption by the electric motor.

4. Results and Discussion

In Equation (2), if terms P, P_{sl}, and Q_w are known, then it is possible to find P_{pl} and SEC_{ch} by performing regression. These regression curves are estimated from experimental data obtained with different cutting conditions, such as depths of cut of 0.010, 0.015, and 0.020 mm; average workpiece speed of 57, 101, and 150 mm/s; and a constant peripheral cutting speed of 22.9 m/s. These cutting conditions are similar to those frequently used by other authors [26]. Figure 4 shows the regression curves of each material from the experimental data. In Equation (2), the specific energy consumed during grinding SEC is defined as the ratio between P and Q_w, while the specific energy consumed in sliding SEC_{sl} is the ratio between the P_{sl} and Q_w.

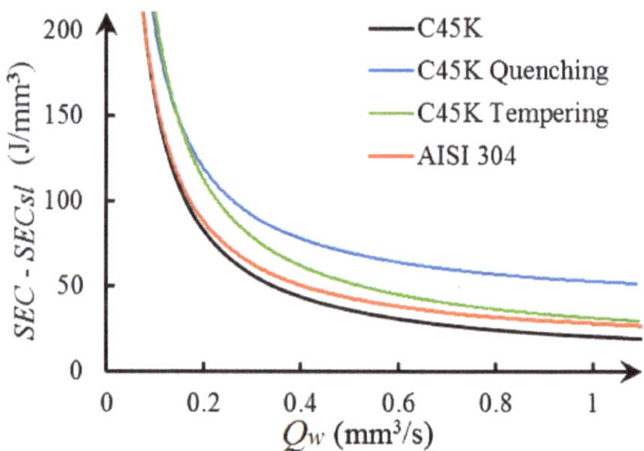

Figure 4. Specific energy consumption and material-removal rate.

The regression applied to the experimental data of materials C45K, C45K quenching, C45K tempering, and AISI 304, have adjustment quality R^2 of 0.82, 0.84, 0.76, and 0.8, respectively. Figure 4 exhibits that, when Q_w increases, SEC gradually decreases. From the graph, it is also noted that, if Q_w is very small, then SEC is higher, which is defined as a size effect [15]. The quality of the adjustment allows to validate the hypothesis that SEC has asymptotic behavior defined by the Equation (2). This behavior is similar to the model developed by Diaz et al. [10] and by Zhong et al. [28] for milling and turning operations, respectively. Moreover, Table 2 shows tha material hardness of each metalic alloy and Table 3 exhibits the results of SEC associated with the mechanism involved in grinding, SEC_{sl}, SEC_{pl}, and SEC_{ch}, where the specific energy consumed by ploughing SEC_{pl} is the ratio between P_{pl} and Q_w. The average energy consumed by sliding SEC_{sl} is 92%, 85%, 57%, and 94% of the total energy consumed by C45K, C45K quenching, C45K tempering, and AISI 304, respectively. This work characterizes the industrial-scale grinding process, where SEC_{sl} is an order of magnitude greater than SEC_{ch}, as compared to other authors who studied grinding at the laboratory scale by using a single grain grinding wheel [6,26], which found small SEC_{sl} values. In addition, the SEC_{ch} values for C45K steel and C45K quenching reported in this work are similar in magnitude to the SEC_{ch} values reported by Marinescu et al. [29]. Furthermore, SEC_{sl}, SEC_{pl}, and SEC_{ch} of the C45K quenching steel present greater values than the other metallic alloys. This is due to the fact that the greater the material hardness of the workpiece is, the higher the required SEC for chip cutting is [10].

Table 2. Material hardness of metallic alloys.

Hardness Material	C45K	C45K Quenching	C45K Tempering	AISI 304
(HRC)	17.35 ± 1.38	56.16 ± 0.52	25.72 ± 0.72	19.85 ± 0.68

Table 3. Average specific energy consumption of different indices in plane dry grinding.

Metallic Alloy	SEC (J/mm^3)	SEC_{sl} (J/mm^3)	SEC_{pl} (J/mm^3)	SEC_{ch} (J/mm^3)
C45K	655	602	30	8
C45K quenching	1805	1541	132	36
C45K tempering	351	201	113	11
AISI 304	958	901	36	13

Figure 5a shows the results of SEC_{sl} for different depths of cut, types of alloys, and thermal treatments. It is shown that the greater the depth of cut is, the lower the contribution of SEC_{sl} is, which is similar to the behavior reported by Ghosh et al. [30]. This is due to the presence of a large number of cutting grains and, subsequently, the area subjected to friction is smaller [12]. Figure 5b exhibits the results of SEC_{sl} for an average depth of 0.015 mm at different workpiece speeds and different materials. In general, SEC_{sl} decreases as the speed of the workpiece increases. A similar behavior was reported by Bakkal et al. [23], who described that, in grinding, the ratio between tangential and normal forces increased as the speed of the workpiece increased.

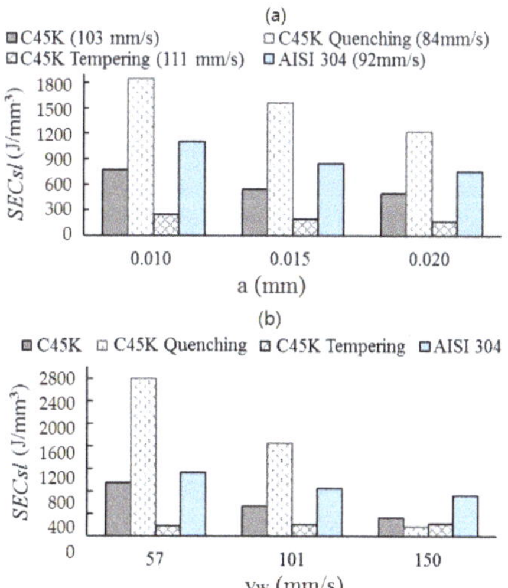

Figure 5. Relationship between specific energy consumption and (**a**) depth of cut, and (**b**) workpiece speed.

The results show that high energy consumption is found for lower depths of cut and workpiece speeds, except for in the C45K tempering material, which shows constant values of energy consumption when these two operational parameters are increased. In particular, quenching requires more energy for low and medium speeds of the workpiece and depths of cut, whereas tempering presents similar low values of energy consumption for higher depths of cut and workpiece speeds. This can be due to differences in hardness at the surface of the materials and their elastoplastic behavior. Finally, both the C45K and AISI 304 materials exhibited the same trend of decreasing energy consumption by increasing depth of cut and workpiece speed. Thus, thermal treatments had a noticeable influence on energy consumption, but temperature in grinding is also crucial and depends on the selection of the operational parameters [31]. The model of material-removal rate Q_w developed in the present study is different from other models, as the thickness of chip (7) and the section of cutting grain (8) are the function of the angular position of the grain. This is different from Zhenzhen et al. [16], who considered the maximum value of chip thickness to estimate material-removal rate Q_w. On the other hand, chip thickness (7) has the same variables and structure as defined by Malkin et al. [15]. The only difference is in the exponent that affects C_g and the speed of the grinding wheel and workpiece. Other models calculated the material-removal rate as a product of depth of cut, grinding width, and workpiece speed [32]. This last model did not incorporate the speed of the grinding wheel in the definition of the material-removal rate as compared to the model presented in this work. Furthermore, this model stated the relationship between the three mechanisms in grinding. The energy consumed by ploughing depends on the material-removal rate that is, ultimately, associated with the selected cutting conditions. Thus, an increase of Q_w produces a decrease of $SEC-SEC_{sl}$ and, also, a decrease of P_{pl}. This is not evident due to the nonlinear behavior of $SEC-SEC_{sl}$, as shown in Figure 4. On the other hand, SEC_{ch} mainly depends on the material and is not sensitive to cutting conditions. Accordingly, SECch is constant and defined by the limit value of the asymptote. Furthermore, SECsl presents a linear trend, with an R^2 greater than 0.999 for cutting conditions a and v_w, as shown in Figure 5a,b. In particular, if a and v_w increase, then, Q_w and P_{sl} also increase due to the linear behavior of SEC_{sl} with respect to Q_w. Finally, sliding is the main mechanism of energy consumption in industrial-scale grinding. Therefore, an increase of Q_w produces an increase in energy consumption in this process.

5. Conclusions

The present paper proposed a model to calculate material-removal rate Q_w and specific energy consumption in grinding, where depth of cut, workpiece speed, effective cutting section, grain density, and material hardness play a crucial role. Accordingly, the main conclusions can be summarized as follows:

- A model was successfully developed to evaluate the dissipated energy by the sliding, ploughing, and chip-formation mechanisms in an industrial-scale grinding process. In general, sliding energy governs the process of energy dissipation in grinding.
- The dissipated energy by the sliding mechanism decreases when the depth of cut and workpiece speed increase, allowing to reduce energy consumption and manufacturing cost during grinding. The sliding mechanism represents, on average, 90% of the total energy consumed for the following materials: C45K, C45K quenching, and AISI 304.
- The model also allows to find the specific energy consumed by chip formation, which is the limit value defined by the asymptotic behavior experienced by SEC. This validates the hypothesis that, during down-grinding, the energy calculated by the analyzer corresponds to the energy dissipated by sliding.

For future work, we propose to study the relationship of the three mechanisms of sliding, ploughing, and chip formation when performing up- or down-grinding in industrial-scale grinding. Additionally, this study helps to optimize this process with the aim of reducing the energy consumption during up- or down-grinding operations. Accordingly, it is necessary to use a wider range of operational parameters v_w and a to investigate SEC behavior and its local minimum.

Author Contributions: Conceptualization: H.A.G.R. and A.N.A. Data curation: A.N.A., H.A.G.R., and A.J.S.E. Formal analysis: H.A.G.R., A.J.S.E., and S.H. Funding acquisition: A.J.S.E. and H.A.G.R. Methodology: A.N.A., R.M.P.A., H.A.G.R., and S.H. Software: A.J.S.E., R.M.P.A., and A.N.A. Supervision: H.A.G.R. and A.J.S.E. Validation: H.A.G.R. and R.M.P.A. Writing—original draft: A.N.A. and S.H. Writing—review and editing: H.A.G.R. and A.J.S.E.

Funding: The authors acknowledge the funding from the Serra Húnter program (Generalitat de Catalunya) with a reference number UPC-LE-304.

Acknowledgments: H.B.M. Ibérica is acknowledged for letting us use the device for measuring power consumption in this work. Thanks also to Juan Alsina from HBM for the valuable support with the energy analyzer (Genesis eDrive Testing).

Conflicts of Interest: All authors who sign this manuscript do not have any conflict of interest to declare. Furthermore, the corresponding author certifies that this work has not been submitted to or published in any other journal.

Nomenclature

θ	Angular position (°)
θ^*	Dimensionless angular position
a	Depth of cut (mm)
A_{cg}	Section cut by a grain (mm^2)
b_g	Width of grain during cutting (mm)
C_g	Grain number per unit area (mm^{-2})
d_g	Grain diameter (mm)
D_M	Diameter of the grinding wheel (mm)
f	Feed per grain (mm)
h	Undeformed chip thickness (mm)
k	Constant of proportionality
l_c	Contact length between wheel and workpiece (mm)
l_g	Distance between grains (mm)
N_g	Number of grains
P	Total power consumption (W)
P_{ch}	Power consumption by chip formation (W)
P_{pl}	Power consumption by ploughing (W)
P_{sl}	Power consumption by sliding (W)
P_v	Idle power consumption (W)
Q_w	Material removal rate (mm^3/s)
R_g	Grain radius (mm)
R_M	Radius of the grinding wheel (mm)
SCE	Specific cutting energy (J/mm^3)
SEC	Specific energy consumption (J/mm^3)
SEC_{ch}	Specific energy consumed by chip formation (J/mm^3)
SEC_{pl}	Specific energy consumed by ploughing (J/mm^3)
SEC_{sl}	Specific energy consumed by sliding (J/mm^3)
v_c	Peripheral cutting speed (m/s)
v_w	Speed of workpiece (m/s)

References

1. Merchant, M.E. An interpretive look at 20th century research on modeling of machining. *Mach. Sci. Technol.* **1998**, *2*, 157–163. [CrossRef]

2. Zheng, P.; Sang, Z.; Zhong, R.Y.; Liu, Y.; Liu, C.; Mubarok, K.; Yu, S.; Xu, X. Smart manufacturing systems for Industry 4.0: Conceptual framework, scenarios, and future perspectives. *Front. Mech. Eng.* **2018**, *13*, 137–150. [CrossRef]

3. Sánchez Egea, A.J.; López de Lacalle, L.N.L. Máquinas, procesos, personas y datos, las claves para la revolución 4.0. *DYNA Ing. Ind.* **2018**, *93*, 576–577. [CrossRef]

4. Aurich, J.C.; Linke, B.; Hauschild, M.; Carrella, M.; Kirsch, B. Sustainability of abrasive processes. *CIRP Ann. Manuf. Technol.* **2013**, *62*, 653–672. [CrossRef]

5. Calleja, A.; Tabernero, I.; Fernández, A.; Celaya, A.; Lamikiz, A.; López de Lacalle, L.N. Improvement of strategies and parameters for multi-axis laser cladding operations. *Opt. Lasers Eng.* **2014**, *56*, 113–120. [CrossRef]

6. Azizi, A.; Mohamadyari, M. Modeling and analysis of grinding forces based on the single grit scratch. *Int. J. Adv. Manuf. Technol.* **2015**, *78*, 1223–1231. [CrossRef]

7. Li, L.J.; Yan, J.; Xing, Z. Energy requirements evaluation of milling machines based on thermal equilibrium and empirical modelling. *J. Clean. Prod.* **2013**, *52*, 113–121. [CrossRef]

8. Hameed, A.; Rojas, H.A.G.; Benavides, J.I.P.; Alberro, A.N.; Egea, A.J.S. Influence of the regime of electropulsing-assisted machining on the plastic deformation of the layer being cut. *Materials* **2018**, *11*, 886. [CrossRef]

9. González Rojas, H.A.; Nápoles Alberro, A.; Sánchez Egea, A.J. Machinability estimation by drilling monitoring. *DYNA Ing. Ind.* **2018**, *93*, 663–667. [CrossRef]

10. Diaz, N.; Redelsheimer, E.; Dornfeld, D. Energy Consumption Characterization and Reduction Strategies for Milling Machine Tool Use. In *Glocalized Solutions for Sustainability in Manufacturing*; Hesselbach, J., Herrmann, C., Eds.; Springer: Berlin/Heidelberg, Germany, 2011; pp. 263–267. [CrossRef]

11. Sánchez Egea, A.J.; González Rojas, H.A.; Montilla Montaña, C.A.; Kallewaard Echeverri, V. Effect of electroplastic cutting on the manufacturing process andsurface properties. *J. Mater. Process. Technol.* **2015**, *222*, 327–334. [CrossRef]

12. Hecker, R.L.; Liang, S.Y.; Wu, X.J.; Xia, P.; Jin, D.G.W. Grinding force and power modeling based on chip thickness analysis. *Int. J. Adv. Manuf. Technol.* **2007**, *33*, 449–459. [CrossRef]

13. Anderson, D.; Warkentin, A.; Bauer, R. Experimental validation of numerical thermal models for dry grinding. *J. Mater. Process. Technol.* **2008**, *204*, 269–278. [CrossRef]

14. Agarwal, S.; Rao, P.V. Predictive modeling of force and power based on a new analytical undeformed chip thickness model in ceramic grinding. *Int. J. Mach. Tools Manuf.* **2013**, *65*, 68–78. [CrossRef]

15. Malkin, S.; Guo, C. *Grinding Technology—Theory and Applications of Machining with Abrasives*, 2nd ed.; Industrial Press: New York, NY, USA, 2008; ISBN 9780831132477.

16. Zhenzhen, C.; Jiuhua, X.; Wenfeng, D.; Changyu, M. Grinding performance evaluation of porous composite-bonded CBN wheels for Inconel 718. *Chin. J. Aeronaut.* **2014**, *27*, 1022–1029. [CrossRef]

17. Lee, Y.M.; Jang, S.G.; Jang, E.S. Grinding characteristics of polycrystalline silicon. *Rev. Adv. Mater. Sci.* **2013**, *33*, 287–290.

18. Nguyen, A.T.; Butler, D.L. Correlation of grinding wheel topography and grinding performance: A study from a viewpoint of three-dimensional surface characterisation. *J. Mater. Process. Technol.* **2008**, *208*, 14–23. [CrossRef]

19. Nadolny, K. Estimation of the active grains load in different kinematic variations of the internal cylindrical grinding process. *Int. J. Adv. Manuf. Technol.* **2017**, *89*, 3337–3348. [CrossRef]

20. Li, H.N.; Yu, T.B.; Wang, Z.X.; Zhu, L.D.; Wang, W.S. Detailed modeling of cutting forces in grinding process considering variable stages of grain-workpiece micro interactions. *Int. J. Mech. Sci.* **2017**, *126*, 319–339. [CrossRef]

21. Rowe, W.B. *Principles of Modern Grinding Technology*; William Andrew: Burlington, VT, USA, 2014; ISBN 978-0-323-24271-4.

22. Malkin, S.; Guo C. Thermal Analysis of Grinding. *CIRP Ann. Manuf. Technol.* **2007**, *56*, 760–782. [CrossRef]

23. Bakkal, M.; Serbest, E.; Karipçin, I.; Kuzu, A.T.; Karagüzel, U.; Derin, B. An experimental study on grinding of Zr-based bulk metallic glass. *Adv. Manuf.* **2015**, *3*, 282–291. [CrossRef]

24. Kumar, V.; Salonitis, K. Empirical estimation of grinding specific forces and energy based on a modified Werner grinding model. *Procedia CIRP* **2013**, *8*, 287–292. [CrossRef]

25. *Grinding Wheels and Abrasives*; ISO/TC 29/SC 5; International Organization for Standardization: Berlin, Germany, 2006. Available online: https://www.iso.org/committee/47502.html (accessed on 14 March 2018).

26. Singh, V.; Venkateswara Rao, P.; Ghosh, S. Development of specific grinding energy model. *Int. J. Mach. Tools Manuf.* **2012**, *60*, 1–13. [CrossRef]

27. Gontarz, A.M.; Weiss, L.; Wegener, K. Energy Consumption Measurement with a Multichannel Measurement System on a machine tool. *Proc. Int. Conf. Innov. Technol.* **2010**, 499–502. [CrossRef]

28. Zhong, Q.; Tang, R.; Lv, J.; Jia, S.; Jin, M. Evaluation on models of calculating energy consumption in metal cutting processes: A case of external turning process. *Int. J. Adv. Manuf. Technol.* **2015**, *82*, 2087–2099. [CrossRef]

29. Marinescu, I.; Hitchiner, M.M.; Uhlmnn, E.E. *Brian Rowe W. Handbook of Machining with Grinding Wheels*, 1st ed.; CRC Press: Boca Raton, FL, USA, 2007; ISBN 978-1-57444-671-5.

30. Ghosh, S.; Chattopadhyay, A.B.; Paul, S. Modelling of specific energy requirement during high-efficiency deep grinding. *Int. J. Mach. Tools Manuf.* **2008**, *48*, 1242–1253. [CrossRef]

31. Urgoiti, L.; Barrenetxea, D.; Sánchez, J.A.; Pombo, I.; Álvarez, J. On the Influence of Infra-Red Sensor in the Accurate Estimation of Grinding Temperatures. *Sensors* **2018**, *18*, 4134. [CrossRef]

32. Hacksteiner, M.; Peherstorfer, H.; Bleicher, F. Energy efficiency of state-of-the-art grinding processes. *Procedia Manuf.* **2018**, *21*, 717–724. [CrossRef]

Article

Multi-Response Optimization of WEDM Process Parameters for Machining of Superelastic Nitinol Shape-Memory Alloy Using a Heat-Transfer Search Algorithm

Rakesh Chaudhari [1], Jay J. Vora [1,*], S. S. Mani Prabu [2], I. A. Palani [2,3], Vivek K. Patel [1], D. M. Parikh [4] and Luis Norberto López de Lacalle [5]

[1] Department of Mechanical engineering, School of Technology, Pandit Deendayal Petroleum University, Raisan, Gandhinagar 382007, India; chaudharirakesh5@gmail.com (R.C.); viveksaparia@gmail.com (V.K.P.)
[2] Metallurgy Engineering and Materials Science, Indian Institute of Technology Indore, Indore 453552, India; maniprabhu.sakthivel@gmail.com (S.S.M.P.); palaniia@iiti.ac.in (I.A.P.)
[3] Discipline of Mechanical Engineering, Indian Institute of Technology Indore, Indore 453552, India
[4] Department of Industrial engineering, School of Technology, Pandit Deendayal Petroleum University, Raisan, Gandhinagar 382007, India; dm.parikh@sot.pdpu.ac.in
[5] Department of Mechanical Engineering. University of the Basque Country, Escuela Superior de Ingenieros Alameda de Urquijo s/n., 48013 Bilbao, Spain; norberto.lzlacalle@ehu.eus
*   Correspondence: vorajaykumar@gmail.com; Tel.: +91-9099962298

Received: 24 March 2019; Accepted: 16 April 2019; Published: 18 April 2019

Abstract: Nitinol, a shape-memory alloy (SMA), is gaining popularity for use in various applications. Machining of these SMAs poses a challenge during conventional machining. Henceforth, in the current study, the wire-electric discharge process has been attempted to machine nickel-titanium (Ni55.8Ti) super-elastic SMA. Furthermore, to render the process viable for industry, a systematic approach comprising response surface methodology (RSM) and a heat-transfer search (HTS) algorithm has been strategized for optimization of process parameters. Pulse-on time, pulse-off time and current were considered as input process parameters, whereas material removal rate (MRR), surface roughness, and micro-hardness were considered as output responses. Residual plots were generated to check the robustness of analysis of variance (ANOVA) results and generated mathematical models. A multi-objective HTS algorithm was executed for generating 2-D and 3-D Pareto optimal points indicating the non-dominant feasible solutions. The proposed combined approach proved to be highly effective in predicting and optimizing the wire electrical discharge machining (WEDM) process parameters. Validation trials were carried out and the error between measured and predicted values was negligible. To ensure the existence of a shape-memory effect even after machining, a differential scanning calorimetry (DSC) test was carried out. The optimized parameters were found to machine the alloy appropriately with the intact shape memory effect.

Keywords: shape memory alloy; superelastic nitinol; WEDM; heat transfer search algorithm; DSC test; shape memory effect

---

1. Introduction

Nitinol is a nickel-titanium alloy which exhibits outstanding properties such as shape memory, superelasticity, and biocompatibility. A shape-memory alloys exhibit in numerous forms. One of the shape memory alloys which possesses superelasticity and biocompatibility is nickel–titanium alloy. Nickel-titanium alloys are also commonly referred to as nitinol in honour of its discovery at the Naval Ordnance Laboratory (NOL) and are a preferred material specifically for biomedical applications due to their high corrosion and wear resistance, pseudoelasticity and biocompatibility [1–3]. Shape-memory alloys (SMAs) when heated above the transition temperature recover their previous deformed shape. Recoverable elastic deformation of superelastic nitinol which is considered to be a new generation smart material is significantly larger than the conventional materials [1]. Phase transformations of SMAs are exhibited in the austenite and martensite phase. Austenite is stable at a higher temperature and has a body center cubic structure whereas martensite is stable at a lower temperature and has a monoclinic crystal structure. SMAs exhibit two effects, super-elastic effect and shape-memory effect. If the austenite finish temperature is below room temperature then particular material is said to produce a super-elastic effect whereas if austenite finish temperature is above room temperature then the material exhibits a shape-memory effect [1]. Due to large deformation recovery and higher wok density, NiTi alloy also finds its application in various areas like aerospace, sensors, robotics, actuators, automotive, structural elements etc. [4,5]. However, high ductility, typical stress-strain behavior, and superelasticity make nitinol difficult to cut using conventional machining processes due to short tool life, a reduced quality of the workpiece, and burr formation [6–9]. Past studies of SMAs with conventional machining reports poor chip breaking, high tool wear, poor surface quality, low-dimensional accuracy and most importantly retaining the shape memory effect after machining [9–11]. Thus, SMAs and most preferably nitinol are best machined through non-conventional machining techniques.

Wire electrical discharge machining (WEDM) is one of the non-conventional machining processes which is a non-contact type process, where the material is removed with the help of high frequency sparks generated between the tool and workpiece [12]. Due to the absence of physical contact of tool and workpiece, this process can be used for any type of material regardless of their hardness, provided the material is electrically conductive [13,14]. This process is used to obtain intricate and complex shape geometries with close tolerances [15]. However, WEDM process involves a high number of input process parameters which needs to be set to their optimal level for achieving the required geometry with close co-relationship between multiple output responses along with the metallurgical and mechanical properties. In accordance with this, most of the past research studies are pivoted to parametric optimization of the WEDM process using different optimization techniques.

Bisaria and Shandilya [16] used the WEDM process for the machining of Ni55.95Ti44.05 SMA. The influence of input parameters such as pulse on time (T_{on}), pulse off time (T_{off}), spark gap voltage (SV), wire tension (WT), wire feed rate (WF) has been studied on material removal rate (MRR), surface roughness (SR) and surface characteristics of material. They found that MRR and SR values increase significantly with the increase in T_{on} whereas MRR and SR values decrease with an increase in SV and T_{off}. Micro-cracks, craters, and debris were observed on the machined surface. The defect could be eliminated by more precise controlling of WEDM process parameters. Sharma et al. [17] conducted the parametric optimization of Ni40Ti60 alloy using the WEDM process. MRR, SR and the dimensional shift have been considered as an output response variables under the influence of input variables such as T_{on}, T_{off}, peak current (IP) and servo voltage (SV). Output responses were optimized using desirability approach for SMA alloy and close correlation between predicted and experimented values were obtained. T_{on} of 124 µs, T_{off} of 25 µs, SV of 30 V, and IP of 110 mu was the obtained optimal parameter setting for multi-response optimization with desirability value of 0.708. In another study by Soni et al. [18], WEDM machining of Ti50Ni40CO10 SMA has been explored. The final result revealed formation of microcracks can be avoided and recast layer thickness can be reduced by setting pulse on time lower than 125 µs and servo voltage larger than 20 V. Majumder and Maity [19] conducted a similar study wherein microhardness (MH) and SR were considered as output response variables and they are optimized with the help of a fuzzy technique for the SMA alloy Ni55Ti45. T_{on} was identified as the main significant input process parameter as compared to other input variables. Manjaiah et al. [20] used a L_{27} orthogonal array to perform the experiment and optimized output responses of MRR and SR during the machining of SMA. The study highlighted the significant effect of T_{on} and T_{off} and SV for the maximization of MRR and minimization of SR under the influence of brass wire and zinc-coated brass wire. B. Jabbaripour et al. [15] states that the electrical discharge machining (EDM) process is suitable for machining titanium alloys. They used Ti6Al4V to investigate various output performance characteristics. Increase in T_{on} resulted in increase in MRR. In another study by Ramamurthy et al. [21], machining of Ti6Al4V alloy was conducted using the EDM process by using Taguchi L9 orthogonal array. They observed that T_{off} has more influent nature on the output performances of machining characteristics because the T_{off} influences the discharge energy in WEDM.

From this preliminary survey, it can be summarized that the optimization of process parameters of machining of SMA alloys has been carried out primarily for MRR, SR, and MH. However, the effect of these optimized parameters has not been explored on the shape memory effect of the machined surface. In addition to this, the majority of the studies optimized an individual response rather than simultaneous optimization.

WEDM is a multi-input multi-output process encompassing complex dependencies on individual parameters as well as their interactions [21,22]. Generally, wherever multiple objectives are considered, several conflicting situations arise wherein there is a requirement is to settle at a tradeoff between these multiple responses. This tradeoff is best presented by optimal Pareto points generated using advanced evolutionary optimization techniques. This method gives the advantage by finding the solution near global optimum with reduced time and computational efforts by generating optimal Pareto points. Different algorithms such as particle swarm optimization, genetic algorithms, ant colony optimization etc. which are nature-based optimization ideologies have been widely experimented with for different optimization problems including WEDM [23–26]. However, these evolutionary algorithms function under a set of assigned values which is algorithm-specific. Precise control of these parameters will dictate the performance of those algorithms for optimization. To overcome this challenge, new algorithms were developed by researchers wherein tuning of those algorithm-specific parameters was not required. The heat-transfer search (HTS) algorithm is one such technique with the major advantage being proper balancing between exploration and exploitation. The proper balancing is incorporated by introducing six different search mechanisms in algorithm. The different search mechanism is generated by number of generations of the algorithm. In addition to that HTS is easy to implement and can find the global optimal solution for complex problems. These noticeable advantages of the HTS algorithm are observed at the cost of computation time. HTS has been successfully applied to different benchmarking problems pertaining to different fields [27–31]. However, to the best of the authors' knowledge, no study has been reported on the application of HTS for manufacturing problems.

Pursuant to a detailed review of an available research article, it can be recognized that T_{on}, T_{off} and the current are the three most notable input process parameters while MRR, SR, and MH as the output response variables. The prime requirement after machining of SMA is its shape-memory effect. Shape-memory effect has been correlated to MH [32]. The differential scanning calorimetry (DSC) test is one of the techniques to ensure shape memory effect. Along with this, either single or multi-objective optimizations with limited consideration to actual industrial requirements are targeted in past published studies. However, to the best of the author's knowledge, generated Pareto curves are targeted only for two responses for the study of SMAs using the WEDM process [33]. Thus, the present study addresses an evident research gap by using pareto curves incorporating three simultaneous responses generated using a novel HTS algorithm.

In the current study, a detailed study on WEDM of superelastic Ni55.8Ti alloy has been carried out and the aforementioned research gap has been fulfilled. The Box-Behnken technique of response surface methodology (RSM) has been used to conduct the experiments and three important output responses were recorded. Furthermore, for each of those output responses, mathematical models were generated and tested by analysis of variance (ANOVA) and residual plot analysis was carried out to verify ANOVA results. Using an advanced parameter less evolutionary algorithm known as HTS, multiple case studies including real-time manufacturing scenario along with simultaneous optimization of the output responses have been carried out. 3D and 2D Pareto curves have been generated with the help of multi-objective HTS algorithm which displays different non-dominant optimal points. Confirmation trails have been conducted to compare the results of predicted and measured responses. To check the shape memory effect, the DSC test has been carried out on the machined surface obtained with optimized parameters during the validation test. The authors firmly believe the study would provide substantial input to the end users working of WEDM of superelastic SMAs.

2. Materials and Methods

In the present study, superelastic shape memory nitinol (Ni5.8Ti) is selected as a workpiece material. In all the experimental trials, slices of 1.5 mm were cut from the rod of 8 mm diameter with the WEDM process. A schematic of the WEDM process for the present study is shown in Figure 1. Chemical composition (wt.%) of Ni55.8Ti is as shown in Table 1. As received SMA has a tensile strength of 750 MPa with 11% elongation. Experiments were conducted on a Concord WEDM machine DK7732 (Concord Limited, Bangalore, India) with dielectric fluid. Table 2 shows the selected range of machining parameters such as T_{on}, T_{off}, and current along with 3 different levels which have been selected on the basis of existing literature, machining capability and their influence on selected output response parameters. Ni55.8Ti was used as a workpiece (anode) and 0.18 mm diameter molybdenum wire was used as a tool electrode (cathode). Experiments (3 number of trials at each parameter setting) have been conducted following the Box-Behnken (BBD) technique of RSM as shown in Table 3. Response surface methodology has been used to minimize the number of trials which reduces the cost of material as well as reduces the machining time.

Table 1. Chemical composition (wt.%) of Nitinol.

Element	Ti	Ni	Co	Cu	Cr	Fe	Nb	C	H	O	N
Content (wt.%)	Balance	55.78	0.005	0.005	0.005	0.012	0.005	0.04	0.001	0.035	0.001

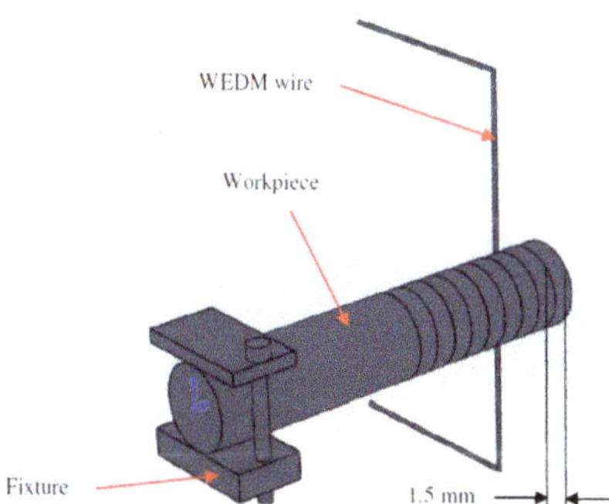

Figure 1. Schematic representation of wire electrical discharge machining (WEDM) process.

Table 2. Process parameters and their levels.

Factors	Process Parameters	Level 1	Level 2	Level 3
A	Pulse on time (T_{on}), μs	35	45	55
B	Pulse off time (T_{off}), μs	10	15	20
C	Discharge current, Ampere	2	3	4

Table 3. Process parameters and their levels.

Run	Pulse on Time (µs)	Pulse off Time (µs)	Current (Ampere)	MRR (mm³/s)	SR (µm)	Microhardness (HV)
1	35	10	3	1.230948122	5.637	330.2
2	55	10	3	1.065245598	5.986	342.3
3	35	20	3	0.714888337	6.453	275.2
4	55	20	3	0.756738988	6.82	342.8
5	35	15	2	0.675983558	5.322	308.6
6	55	15	2	0.666859456	6.58	301
7	35	15	4	1.066357739	6.595	346.5
8	55	15	4	1.103461538	5.577	351.3
9	45	10	2	0.845274725	4.944	341.7
10	45	20	2	0.541463415	5.925	354.2
11	45	10	4	1.23034188	5.638	383.9
12	45	20	4	0.874333587	5.762	374.6
13	45	15	3	0.92275641	6.053	326.3
14	45	15	3	0.925	5.484	309.5
15	45	15	3	0.921634615	6.098	316.3

The material removal rate is calculated by the difference in weight of the sample before and after machining carried out per second. Equation (1) shows the method to calculate MRR in mm³/s.

$$MRR = \frac{\Delta W \times 1000}{\rho \times t} \tag{1}$$

where, ΔW = weight loss from the workpiece,

t = duration of the machining process in second,
ρ = 6.5 g/cm³ the density of the workpiece

The Mitutoyo Surftest SJ-410 model (Mitutoyo ltd., New-Delhi, India) surface roughness tester was used to measure the surface roughness of the machined sample. The SR of each sample was measured at four different locations and the average value was taken as the output response. The cutoff length (λ_c) was selected as 0.8 mm with the evaluation length of 7 mm. Ra values were recorded and analyzed to indicate the surface quality of the cut. A mirror finish was developed on the machined sample for the micro-hardness testing. A Vickers microhardness tester (MVH-S Auto Omintech, Pune, India) was used to calculate microhardness of the surface at 500 gf load at 10 s. The measured values of MRR, SR, and MH are analyzed and shown in Table 3 and the mathematical correlation was developed for each response. Further, the optimization route was followed as given in Figure 2. Validation of the shape-memory effect was conducted by the DSC technique. The DSC test was used to study the phase transformation behavior for both machined/unmachined surfaces and the results were compared. Using a Netzsch DSC 214 polyma machine (Netzsch, Selb, Germany), the DSC test was performed with a sample weight of 20 g at heating/cooling rate of 10 °C/min and a constant flow of nitrogen. The sample was place in a pan and a small spear hole was drilled on the top of the pan. This pan was kept within the machine for testing.

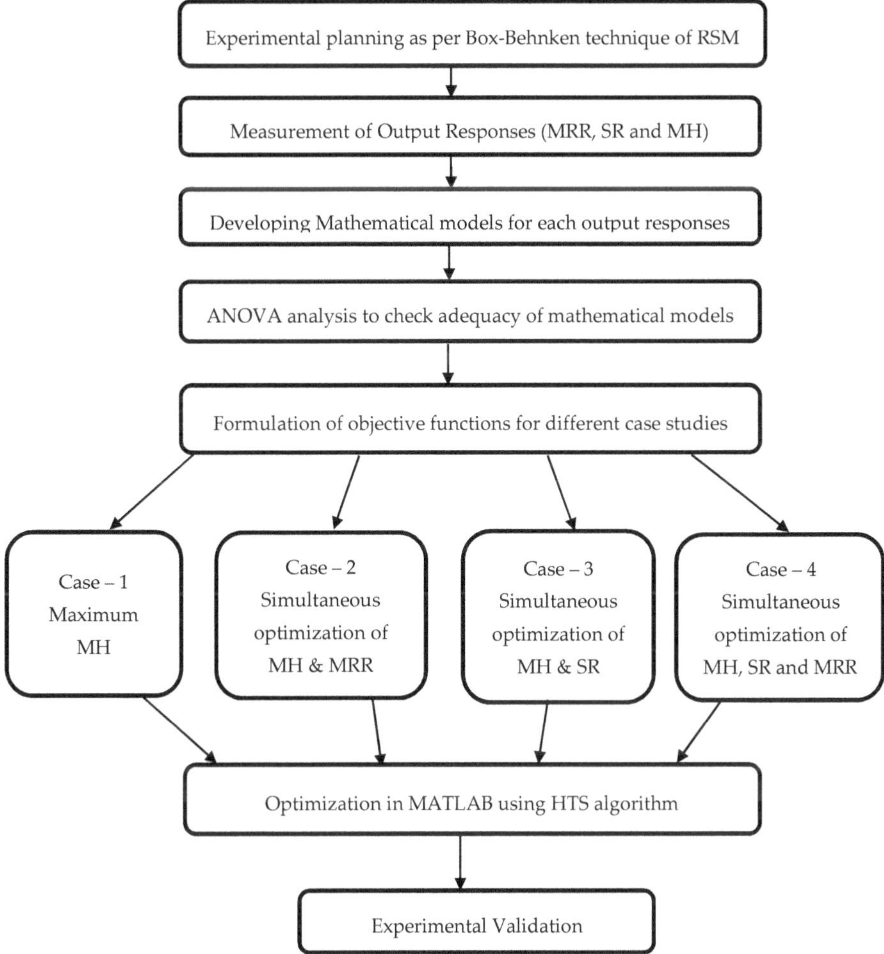

Figure 2. Proposed optimization route.

3. Results and Discussion

3.1. Generation of Mathematical Model

Table 3 shows the results obtained for MRR, SR, MH by using the design matrix of BBD. The maximum and minimum MRR values of 1.2309 mm^3/s to 0.5414 mm^3/s respectively were obtained. SR value ranging from 4.944 μs to 6.82 μs and MH ranging from 275.2 HV to 383.9 HV were achieved. It can be observed from Table 3 that a different set of output responses was obtained at a different input set of parameters. The output response values mentioned in Table 3 were analyzed using Minitab software (Version 17) to generate in terms of input variables as shown in Equations (2)–(4):

$$MRR = 1.1400 - 0.0168 \cdot A - 0.0838 \cdot B + 0.5430 \cdot C - 0.0548 \cdot C^2 + 0.0010 \cdot A \cdot B \tag{2}$$

$$SR = 0.2300 - 0.18000 \cdot A + 0.1974 \cdot B + 4.8290 \cdot C + 0.0040 \cdot A^2 - 0.2540 \cdot C^2 - 0.0569 \cdot A \cdot C - 0.0429 \cdot B \cdot C \tag{3}$$

$$MH = 521 + 10.9700 \cdot A - 39 \cdot B - 132.5000 \cdot C - 0.1575 \cdot A^2 + 0.8400 \cdot B^2 + 25.2300 \cdot C^2 + 0.2770 \cdot A \cdot B \tag{4}$$

where A is pulse on time, B is pulse off time and C is current.

Equations (2)–(4): are regression equations which primarily give the information on the effect of independent parameters (T_{on}, T_{off} and Current) and their interactions on the dependent quality parameter (MRR, SR and MH). The ANOVA technique is used to investigate the significant and non-significant process parameters. The significance of input process parameter on the output response is indicated by F value and P value. The significance of process parameter on output response can be known from either higher F value or lower P value. The value of P must be lower than 0.05 to keep the particular process parameter significant for the 95% confidence level. The significance of T_{on} and T_{off} and current is shown in Table 4 on MRR, SR, and MH. T_{off} and current are found to be significant for the output response of MRR while T_{off} for SR and current for MH is considered to be the significant process parameters. Lack of fit was observed to be insignificant for all the responses which mean that the model is adequate for predicting the output responses under any combination of the process parameters considered in the range [34]. The 20% difference between R-squared and Adjusted (Adj) R-squared values means that the model is the best fit for selected responses [17]. For all the output responses considered in this study a difference of less than 20% was achieved.

Figure 3 shows the residual plot for MRR which includes normality plot, fitted versus predicted plot, histogram, and residual versus observation order plot. The normality plot shows that all the residuals are on the straight line which indicates the fitness of the proposed model. The random allocation of all the residuals on both the sides of the reference line can be seen from residual versus fitted plot. In a histogram plot, the bell-shaped curve is observed which support the normality of data. Furthermore, the residual versus observation order plot does not follow any pattern which is a mandatory case for significant ANOVA. This verifies that all the four tests confirms a satisfactory future outcome for the proposed model. In a similar way, residual plots as shown in Figures 4 and 5 for SR and MH respectively verify ANOVA results.

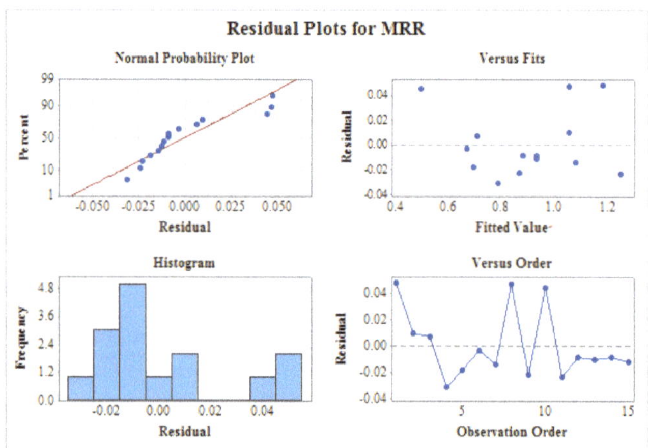

Figure 3. Residual plot for material removal rate (MRR).

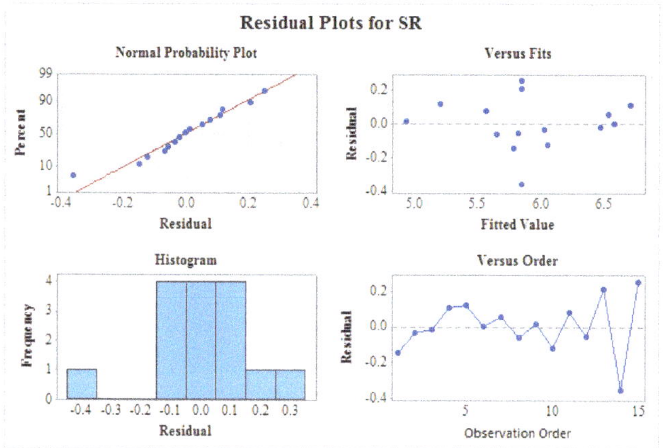

Figure 4. Residual plot for surface roughness (SR).

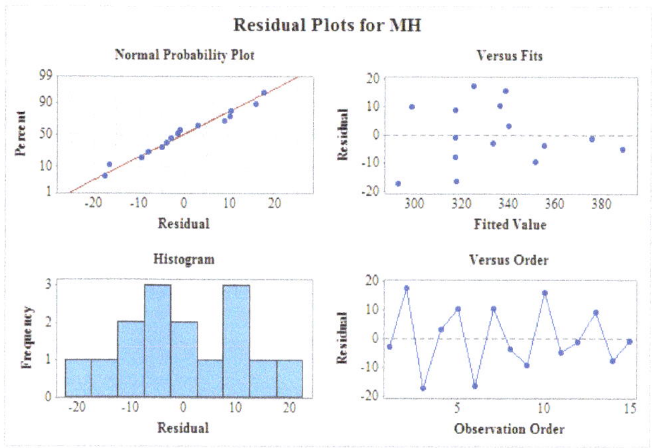

Figure 5. Residual plot for microhardness (MH).

Table 4. Analysis of variance (ANOVA) for MRR, SR and microhardness.

ANOVA for MRR					
Source	**SS**	**MS**	**F**	**P**	**Significance**
T_{on}	0.001149	0.001149	1.07	0.328	-
T_{off}	0.275425	0.275425	256.99	0.000	Significant
Current	0.298345	0.298345	278.37	0.000	Significant
R–Sq = 98.41 %, R–Sq (Adj) = 97.53%					
ANOVA for SR					
Source	**SS**	**MS**	**F**	**P**	**Significance**
T_{on}	0.11424	0.11424	2.51	0.157	-
T_{off}	0.94875	0.94875	20.85	0.003	Significant
Current	0.08020	0.08020	1.76	0.226	-
R–Sq = 91.71 %, R–Sq (Adj) = 83.41%					
ANOVA for MH					
Source	**SS**	**MS**	**F**	**P**	**Significance**
T_{on}	739.2	739.2	3.14	0.120	-
T_{off}	329.0	329.0	1.40	0.276	-
Current	2842.6	2842.6	12.07	0.010	Significant
R–Sq = 85.52 %, R–Sq (Adj) = 71.03%					

The prediction of two simultaneous factors (when the third value is held constant) on a single output response is shown in Figure 6 for MRR. A higher value of MRR (>1.05 mm^3/s) can be obtained when T_{off} value is near to 10 µs and for any value of T_{on} (between 35 µs to 55 µs). The effects of T_{on} and T_{off} also predict that MRR value increases with a decrease in T_{off}. Similarly, MRR increases with the decrease in T_{off} value and increase in current for the effect of contour for current and T_{off}. This is due to the fact that the number of spark decreases with an increase in pulse off time which thereby lowers MRR and discharge energy increase with an increase in current which thereby increases MRR [35,36]. The effect of T_{on} and current predicts the higher value of MRR when the current is in between 3.2 to 4 A and for any value of T_{on} (between 35 µs to 55 µs).Figure 7 shows the contour plots for SR with the variation of two alternative input variables keeping the third variable as constant. Effect of T_{on} and T_{off} predicts the lower value of SR when T_{on} value is in between 35 µs to 50 µs and T_{off} value is near to 10 µs. A lower value of SR is observed at a lower value of T_{on} and current and maximum SR is obtained at a higher value of T_{on} and current. The discharge of higher pulse energy penetrates into the surface by forming a deep crater and leads to higher SR [20]. Furthermore, a contour plot for current and T_{off} shows the least value of SR when T_{off} varies from 10 to 12 µs and for the current value of 2 A. Figure 8 shows the contour plots for MH. Higher MH is obtained at the three different conditions viz. of T_{on} 35 µs and T_{off} 16 to 19 µs, current near to 4 A and T_{on} 40 µs to 55 µs, and again current 4 A and T_{off} near to 10 µs.

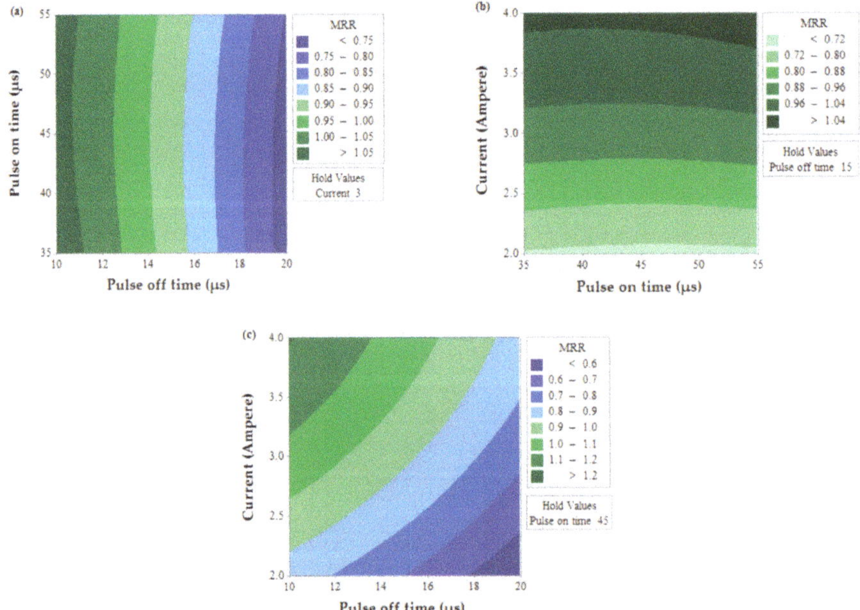

Figure 6. Contour plot for MRR.

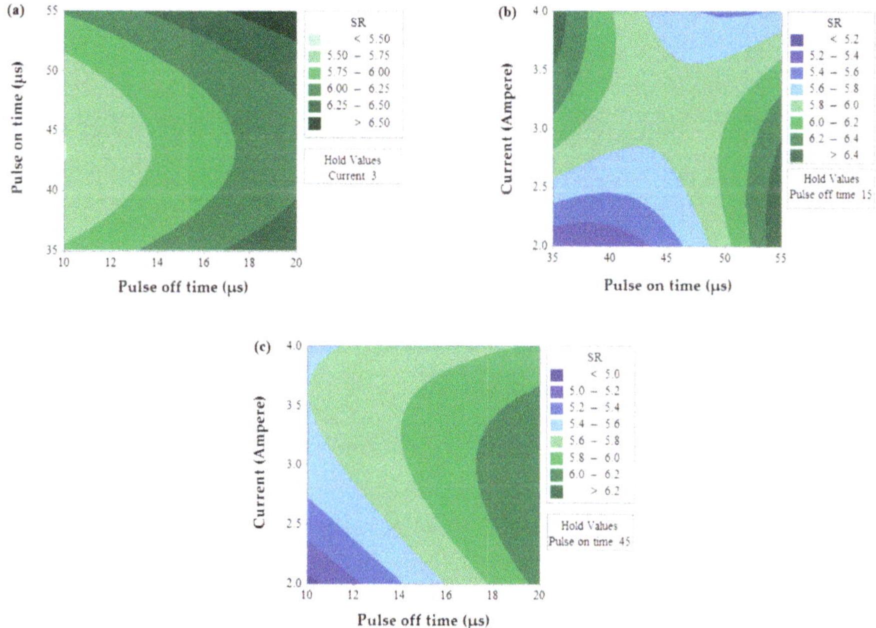

Figure 7. Contour plot for SR.

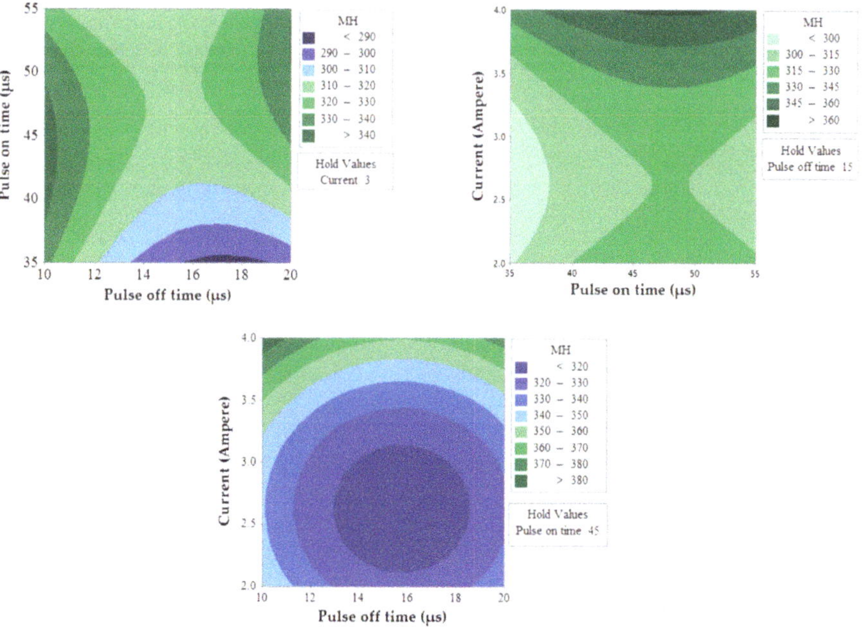

Figure 8. Contour plot for MH.

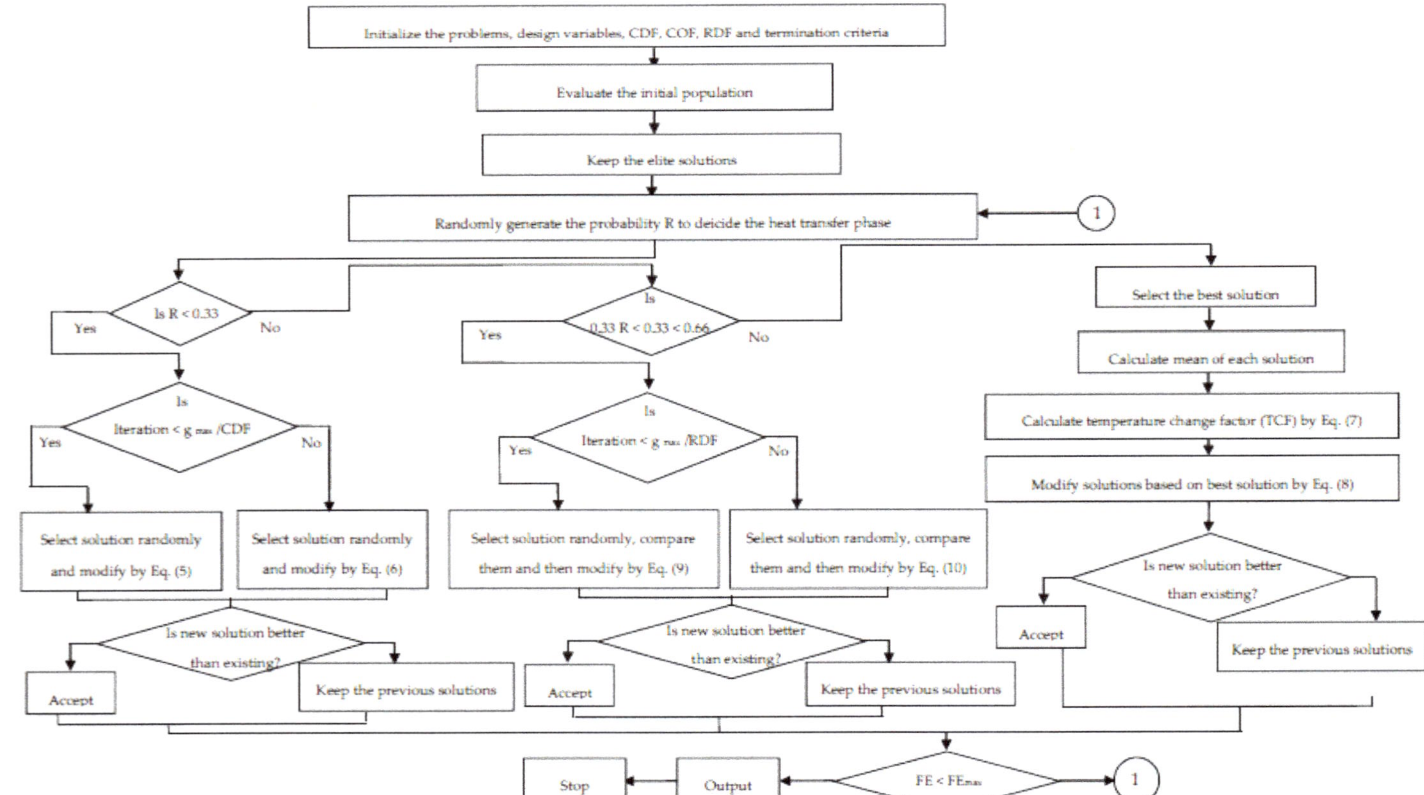

Figure 9. Flow chart of heat-transfer search (HTS) algorithm. Adapted from [31], with permission from © 2017 Elsevier.

3.2. Optimization of Case Studies and Its Validation

The HTS algorithm, proposed by Patel and Savsani [37] works on heat transfer due to the interaction between the system molecules as well as with the surroundings to reach thermal equilibrium. A thermodynamically imbalanced system always tries to achieve thermal equilibrium by heat transfer between the system and its surroundings. The modes of heat transfer are conduction, convection, and radiation plays a major role in setting thermal equilibrium. Thus, the HTS algorithm considers 'the conduction phase,' 'the convection phase, ' and 'the radiation phase' to reach an equilibrium state. In the HTS algorithm, all three modes of heat transfer have an equal chance to transfer heat, and one of the heat transfer modes is decided randomly for each generation. The HTS algorithm initiates with a randomly generated population, where the system has 'n' number of molecules (i.e., population size) and the temperature level (i.e., design variables). In the next stage, the population is updated in each generation by one of the randomly selected heat transfer modes. Moreover, the updated solution in the HTS algorithm is accepted if it has a better functional value. Subsequently, the worst solutions of the population are replaced by elite solutions. The entire working process of the HTS algorithm is presented in the form of flow chart as shown in Figure 9.

3.2.1. Conduction Phase

The solutions are updated in the conduction phase as per the below Equations (5) and (6),

$$X'_{j,i} = \begin{cases} X_{k,i} + \left(-R^2 X_{k,i}\right), & \text{iff}\left(X_j\right) > f(X_k) \\ X_{j,i} + \left(-R^2 X_{j,i}\right), & \text{iff}\left(X_j\right) < f(X_k) \end{cases} ; \text{If } g \leq \frac{g_{max}}{CDF} \tag{5}$$

$$X'_{j,i} = \begin{cases} X_{k,i} + \left(-r_i X_{k,i}\right), & \text{iff}\left(X_j\right) > f(X_k) \\ X_{j,i} + \left(-r_i X_{j,i}\right), & \text{iff}\left(X_j\right) < f(X_k) \end{cases} ; \text{If } g > \frac{g_{max}}{CDF} \tag{6}$$

where, $X'_{j,i}$ is the updated solution; $j = 1,2, \ldots ,n$; k is a randomly selected solution; $j \neq k$; $k \in (1,2, \ldots ,n)$; i is a randomly selected design variable; $i \in (1,2, \ldots ,m)$; g_{max} is the maximum number of generation specified; CDF is the conduction factor; R is the probability variable; $R \in \{0, 0.3333\}$; $r_i \in \{0, 1\}$ is a uniformly distributed random number.

3.2.2. Convection Phase

The solutions are updated in the convection phase as per the below Equations (7) and (8),

$$X'_{j,i} = X_{j,i} + R \times (X_s - X_{ms} \times TCF) \tag{7}$$

$$TCF = \begin{cases} abs(R - r_i), & \text{If } g \leq \frac{g_{max}}{COF} \\ round(1 + r_i), & \text{If } g > \frac{g_{max}}{COF} \end{cases} \tag{8}$$

where, $X'_{j,i}$ is the updated solution; $j = 1,2, \ldots ,n$; $i = 1,2, \ldots ,m$.COF is the convection factor; R is the probability variable; $R \in \{0.6666, 1\}$; $r_i \in \{0, 1\}$ is a uniformly distributed random number; X_s be the temperature of the surrounding and X_{ms} be the mean temperature of the system; TCF is a temperature change factor.

3.2.3. Radiation Phase

The solutions are updated in the radiation phase as per the below Equations (9) and (10),

$$X'_{j,i} = \begin{cases} X_{j,i} + R \times \left(X_{k,i} - X_{j,i}\right), & \text{iff}\left(X_j\right) > f(X_k) \\ X_{j,i} + R \times \left(X_{j,i} - X_{k,i}\right), & \text{iff}\left(X_j\right) < f(X_k) \end{cases} ; \text{If } g \leq \frac{g_{max}}{RDF} \tag{9}$$

$$X'_{j,i} = \begin{cases} X_{j,i} + r_i \times \left(X_{k,i} - X_{j,i}\right), \ \text{iff}\left(X_j\right) > f(X_k) \\ X_{j,i} + r_i \times \left(X_{j,i} - X_{k,i}\right), \ \text{iff}\left(X_j\right) < f(X_k) \end{cases} ; \text{If } g > \frac{g_{max}}{RDF} \tag{10}$$

where, $X'_{j,i}$ is the updated solution; $j = 1,2,\dots,n$; $i = 1,2,\dots,m$; $j \neq k$; $k \in (1,2,\dots,n)$ and k is a randomly selected molecules; RDF is the radiation factor; R is the probability variable; $R \in \{0.3333, 0.6666\}$; $r_i \in \{0, 1\}$ is a uniformly distributed random number.

For each of the objective function considered in different case studies, the HTS algorithm was executed with 10000 function evaluation.

The machining range considered in the HTS algorithm for all case studies is as follows:

For Pulse on time: 1 µs ≤ Pulse on time ≥ 110 µs

For Pulse off time: 1 µs ≤ Pulse off time ≥ 32 µs

For Current: 1 A ≤ Current ≥ 6 A

Case: I Optimization of Microhardness (MH)

Machining of shape memory alloys could vastly extend their applications without the deterioration of shape memory effect. MH value is one of the available measures to check the shape memory effect. Lotfi Neyestanak and Daneshmand [32] reported that a larger value of MH shall indicate the continuation of shape memory effect even after machining. In the present case, Equation (11) shows a single objective function, for which optimization was carried out using the HTS algorithm.

$$\text{Obj} (v_1) = (MH) \tag{11}$$

The maximum attainable value was found to be 870.86 HV from the optimization results. For this maximum value of MH, corresponding input parameters of a pulse on time of 63 µs, pulse off time of 32 µs and current of 6 A was observed. The other output responses were also predicted for the present optimal condition and the validation test was also carried out at predicted input parameters. As the objective of the present case was to obtain maximum MH, other output response variables were not at their optimal values. The experimental validation result of 861.7 HV was observed for MH. The developed model along with HTS algorithm was found to be capable of predicting and optimizing the process parameters as a close relation can be seen between the predicted and measured value. Furthermore, wherever shape memory effect is of prime most requirements with lesser importance to MRR and SR, this present case could become highly useful for industrial applications.

Case: II Optimization of MH and Material Removal Rate (MRR)

MRR plays an important role in increasing productivity and thereby making the machining process cost-effective for industries. Without losing the shape memory effect, higher MRR is always preferable for any kind of machining process. Hence in the present case, MH and MRR are combined with different weights by formulating the objective function (shown in Equation (12)) and the HTS algorithm was used to analyze this multi-objective optimization problem.

$$\text{Obj} (v_2) = w_1 \cdot (MH) + w_2 \cdot (MRR) \tag{12}$$

As already discussed, higher MH could maintain the shape memory effect, slightly higher weight of 0.6 was assigned to MH and 0.4 for MRR to achieve a higher value of MRR. The optimal value of the objective function for the input variables wasa pulse on time of 28 µs, pulse off time of 5 µs and current of 6 A observed. A confirmation trial was conducted on the obtained set of input parameters (pulse on time of 28 µs, pulse off time of 5 µs and current of 6 A) and are shown in Table 5. The developed model along with HTS algorithm was found to be capable of predicting and optimizing the process parameters as the close relation can be seen between the predicted and measured value as shown in Table 5.

Table 5. Validation results for case study II.

Condition	MRR (mm^3/s)	SR (μm)	Microhardness (HV)
Predicted by HTS Algorithm	1.6938	8.62	679.05
Experimentally measured values	1.5921	8.89	654.32
% ERROR	6	3.13	3.64

Case: III Optimization of MH and Surface Roughness (SR)

Apart from MH, SR of the machined surface becomes an important parameter which denotes the machining quality for certain application requiring aesthetic features or the mating of parts. In the present case, multi-objective optimization was conducted for simultaneous optimization of MH and SR by considering a single objective function wherein suitable weights were given to both responses. MH is considered as higher-the-better and SR as lower-the-better, which are typically a measure of a higher quality of machined surface. Equation (13) shows the objective function which was optimized using the HTS algorithm.

$$\text{Obj } (v_3) = w_1 \cdot (\text{MH}) + w_2 \cdot (\text{SR}) \tag{13}$$

Equal weights of 0.5 were assigned to both the output parameters. Corresponding input parameters of a pulse on time of 65 μs, pulse off time of 32 μs and current of 6 A were observed to give the optimal value of the present objective function. Output responses were mentioned in Table 6 which shows a negligible error between the predicted values and the measured value. This shows that the HTS algorithm performs fairly well in getting required results for multi-objective optimization of the WEDM process.

Table 6. Validation results for case studies III and IV.

Condition	MRR (mm^3/s)	SR (μm)	Microhardness (HV)
Predicted by HTS Algorithm	0.81324	1.28	870.21
Experimentally measured values	0.77271	1.35	855.55
% ERROR	4.98	5.46	1.68

Case: IV Simultaneous Optimization of MH, SR, and MRR

In the present case, a single objective function was formulated by combining all the output responses of MH, SR, and MRR by weights of 0.5, 0.3 and 0.2 respectively. Weights were assigned to output responses by considering their importance for shape memory alloys. In the present objective function (as shown in Equation (14)), MH and MRR are considered as higher-the better whereas SR as lower-the better performance characteristics. Further, the HTS algorithm was used to conduct the multi-objective optimization.

$$\text{Obj } (v_4) = w_1 \cdot (\text{MH}) + w_2 \cdot (\text{SR}) + w_3 \cdot (\text{MRR}) \tag{14}$$

Corresponding input parameters of pulse on time of 65 μs, pulse off time of 32 μs and current of 6 A were observed to give the optimal value of the present objective function. Input process parameters obtained for the present case was found to be the same as that of case III. Hence, the similar validation results were used for the present case and were mentioned in Table 6 which shows a negligible error between the predicted values and the measured value. This shows that the HTS algorithm performs fairly well in obtaining the required results for multi-objective optimization of the WEDM process. This shows that the developed models and the combined approach can be useful to predict the accurate and robust results.

3.3. Differential Scanning Calorimetry (DSC) Test

A DSC test was carried out to ensure the presence of crucial shape memory effect after machining. This test typically consists of heating and cooling the sample under a controlled environment wherein a plot indicating relationship between temperate and phase change is determined. The machined surface for validation trial of case IV was selected for the DSC test. The austenite start temperature (As) denotes the start of the austenite transformation whereas the austenite finish temperature (Af) represents the end of the transformation. The same type of notation applies for martensite transformation temperatures too (Ms, Mf). Hysteresis is the measure of the difference in transformation temperatures between heating and cooling i.e., |As − Mf|and |Ms − Af| [38–40]. The hysteresis and the phase transformation temperatures are mentioned in Table 7. Figure 10a,b shows the DSC curve for the unmachined and machined sample (at final optimized parameters). To determine the transformation temperature, the intersection of a baseline and the tangent to each peak has been considered. DSC curves of the unmachined and machined samples depict the exothermic peak on cooling for the forward transformation from austenite to martensite (B2 to B19′). Furthermore, both samples show endothermic peaks on heating during the reverse transition from martensite to austenite (B19′ to B2). The transformation temperature was observed to be very close to each other for unmachined and machined samples. This shows that the shape memory effect was retained even after machining using the WEDM process. Acceptable change in hysteresis has been observed between unmachined and machined sample values.

Table 7. Phase transformation temperatures and hysteresis.

Nitinol Sample	A_s (°C)	A_f (°C)	M_s (°C)	M_f (°C)	Hysteresis, $|A_s - M_f|$ (°C)
Unmachined	−58	−39.7	−88.5	−110.4	52.4
Machined	−61.3	−40.1	−98.9	−118.3	57

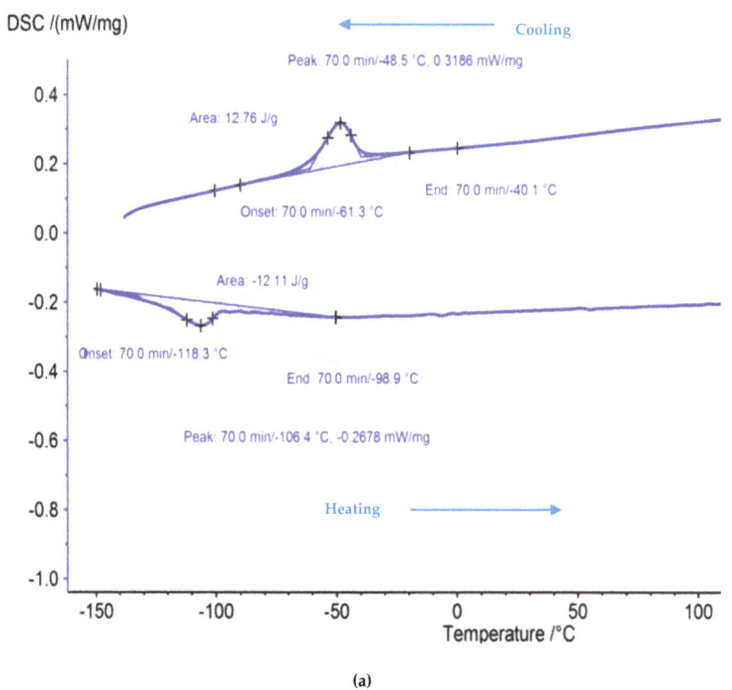

(a)

Figure 10. *Cont.*

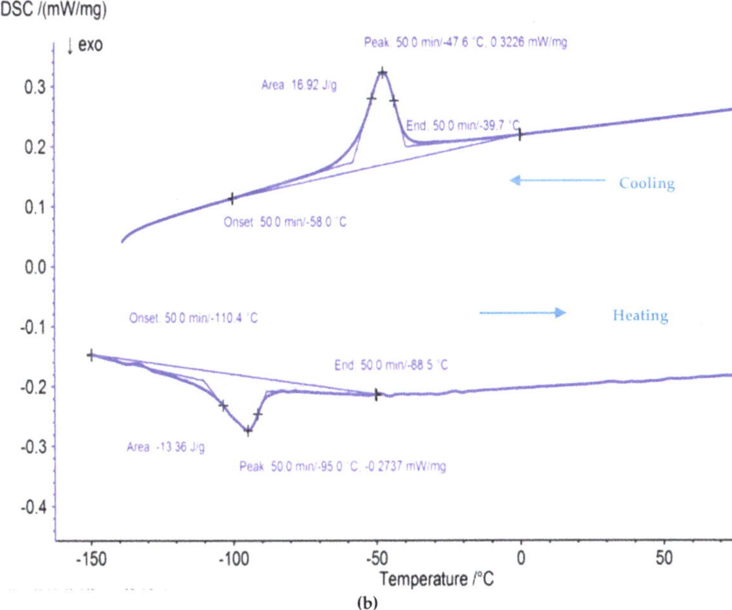

Figure 10. Differential scanning calorimetry (DSC) curve of NiTi alloy for (**a**) unmachined sample (**b**) machined sample at optimized parameter.

3.4. Generation of Pareto Optimal Set

Table 8 shows the individual single objective optimization using HTS algorithm for each of the objective (MRR, SR, and MH). It can be observed that when one of the objectives is at its optimal value, the remaining two objectives are away from its maximum/minimum values. For example, when maximization of MRR was carried out, the corresponding values of MH and SR obtained were 423.05HV and 13.70 μm, respectively. However, those values of MH and SR are not clearly at their optimal level. The same can be seen for the remaining two objective function results. Another important conclusion that can be drawn from the table is that, for obtaining the optimum value of each of the objectives individually, the value of input variables are different. These results indicate the conflicting effect of input parameters on the output responses. Such problems can be efficiently tackled by obtaining Pareto optimal points which are basically tradeoffs between output responses.

Table 8. Results of single objective optimization.

Objective Function	Design Variables			Objective Function Value		
	Pulse on Time	Pulse off Time	Current	MRR (mm^3/s)	MH (HV)	SR (μm)
Maximum MRR	10	5	5	**1.9503**	423.05	13.70
Maximum MH	63	32	6	0.7803	**870.86**	1.29
Minimum SR	65	32	6	0.8132	870.21	**1.28**

Simultaneous optimization of two or more than two objectives can be achieved with the help of multi-objective heat transfer search (MOHTS) algorithm which is a multi-objective version of the heat transfer search algorithm [41–43]. The non-dominated solutions generated by the MOHTS algorithm are stored in the external archive. Furthermore, ε-dominance based updating method is used to check the domination of solutions kept in the external archive. The Pareto front in MOHTS algorithm was obtained with the help of these non-dominated solutions. The archiving process in the MOHTS algorithm employed a grid-based approach with fixed size archive. The archive stored the best solutions obtained during the execution of the HTS algorithm. Furthermore, the archive is updated in every generation during the execution of the HTS algorithm by adapting the ε-dominance method. The ε-dominance method adopted in the MOHTS algorithm presumes a space having dimensions equal to the number of objectives of the optimization problem. This space further converted into the boxes of ε to ε size by slicing each dimension. Each box holds the solutions generated during the course of optimization. After that, the boxes which are dominated by the other boxes are removed first. Thus, the solutions in those boxes are removed. Afterward, in the remaining box if more than one solution exists then the dominated ones are removed from that box. Therefore, only one solution remains in the box which is non-dominated in nature. Thus, only non-dominated solutions are retained in the archive.

For the simultaneous optimization of all the output responses viz. MRR, SR and MH, the MOHTS algorithm was implemented to obtain the non-dominant Pareto optimal points. Figure 11 shows the Pareto optimal points plotted 3D space. In the 3D plot, MRR is represented on the X-axis, SR on Y-axis and MH on Z-axis. These Pareto optimal points have been obtained at the end of 10,000 evolution functions. Figure 11 shows 100 feasible Pareto points, however, the number of Pareto points can be obtained depending upon the requirement. Each of the point represented on the Pareto curve gives a unique solution and has a corresponding input process parameter. The operator can select any Pareto point and its corresponding input process parameter for machining based on the requirement. In several methods of generating Pareto optimal points, the non-dominant Pareto points have to be identified from the resultant Pareto points. However one of the inherent benefits of using the MOHTS algorithm is the fact that the resultant Pareto points are non-dominant solutions and are obtained in a single step.

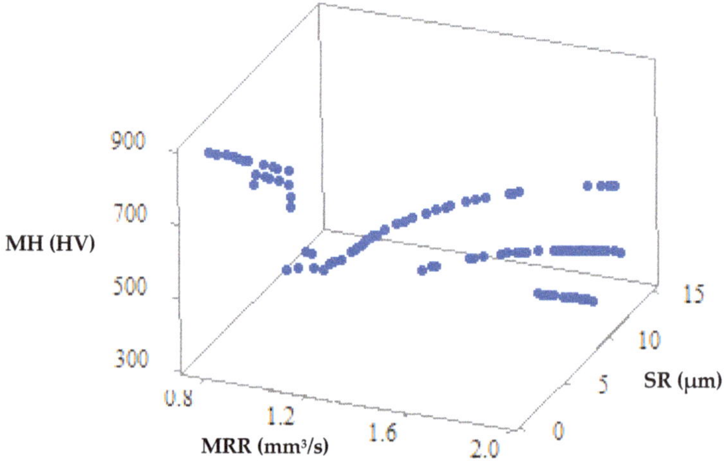

Figure 11. 3D Pareto curve of MRR vs. SR vs. MH.

Figures 12–14 show the 2-D views of the generated solutions for the better understanding of 3-D Pareto points. The effect of third response parameter is included in all the 2D views as shown in Figures 12–14. Over the entire search space, a discrete distribution of Pareto points has been obtained. These points give a clear picture for better understanding along with the inclusion of the effect from the third response parameter. Complex dependencies and highly conflicting effects can be evidently seen from 2D views of input variables on output responses. In existing literature, simultaneous optimization of two output responses has been reported completely neglecting the third output response [33], incorporated in the present study. This gives a better idea to designers and manufacturers on the selection of input parameters for the WEDM process for SMA. Figure 12 shows the maximum and values of MRR and SR are 1.95 mm³/s and 1.28 µm (as shown by red points) respectively. This shows that when maximum MRR needs to be achieved for higher production rate, the SR values are also at higher side showing the conflicting occurrence. Hence a Pareto point with its corresponding input parameters will be selected which would be a trade-off between these two values. A similar conclusion can be drawn from Figure 13 for the 2D view of MRR vs. MH which also shows that when maximum MRR needs to be achieved for higher production rate, the MH values are at lower side showing the conflicting occurrence. Whereas when maximum MH is considered as an output, MRR value is at a minimum. The maximum and minimum values of MRR and MH are 1.95 mm³/s and 870.86 HV (as shown by red points), respectively. Figure 14 gives the maximum and minimum values of SR and MH as 1.28 µm and 870.86 HV (as shown by red and yellow points) respectively for the 2D view of SR vs. MH. For some Pareto points, when SR is at its peak value, corresponding values of MH are not at its peak. This might be due to the effect of the third response parameter evident on the generated 2-D graphs. The selection of input and output variable values in the actual WEDM process is very complex, which requires true dependencies for the decision of input process parameter selection. In the current study, MH value becomes more important considering the fact that the possession of a shape memory effect is a must after machining.

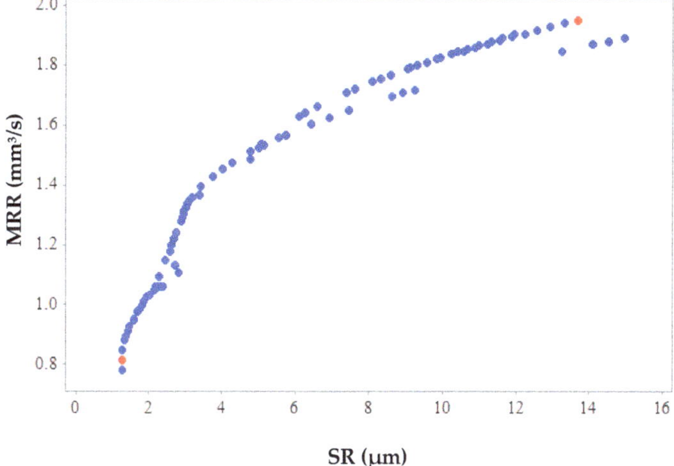

Figure 12. 2-D Pareto optimal points for MRR vs SR.

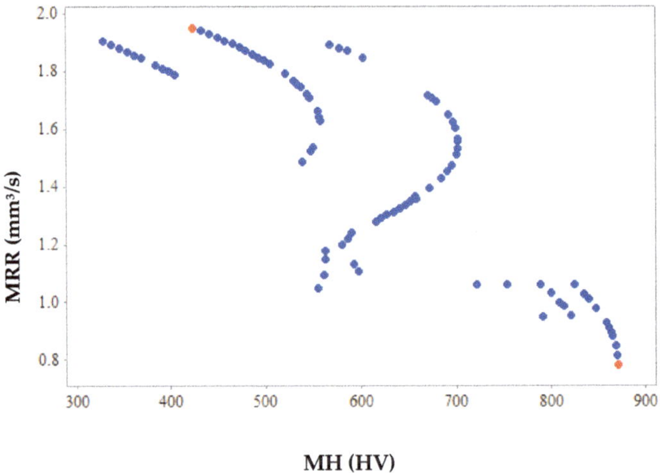

Figure 13. 2-D Pareto optimal points for MRR vs. MH.

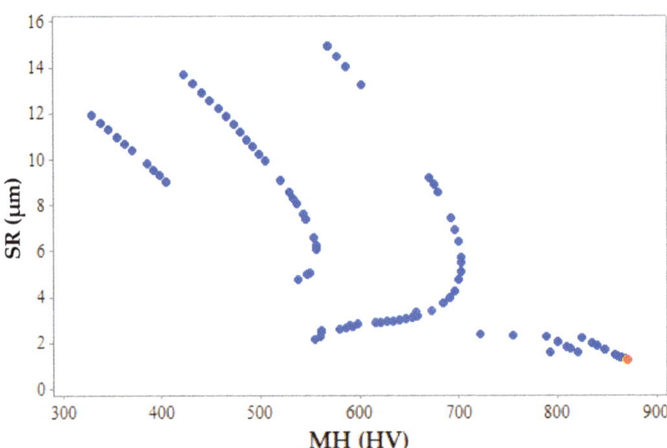

Figure 14. 2-D Pareto optimal points for SR vs. MH.

4. Conclusions

In the current study, desirable results for the parametric optimization of superelastic SMAs during the WEDM process have been achieved. The following important conclusions can be drawn from the study:

- The regression models generated for the selected output response variables were found to be robust, verified using ANOVA. Residual plot analysis confirmed the prediction capability of the generated models of MRR, SR and MH.
- T_{off} and current were found to be most significant parameters influencing SR and MH respectively, where as T_{off} and current were significantly influencing MRR. Contour plots analyses were used to identify the significance of input variables on the individual output responses.
- The heat-transfer search (HTS) algorithm was found effective in predicting and optimizing the input values for four different case studies under consideration. The same was confirmed using validation tests. A close correlation between predicted and achieved values was obtained.

- DSC tests carried out for case IV indicated negligible difference between A_f temperature for the machined and unmachined surface which confirmed the retention of shape memory effect even after WEDM. This can be considered as one of the most notable outcomes of the study.
- 3-D Pareto curves were generated which effectively presented the solution for the simultaneous optimization of three output responses such as MRR, SR, and MH. A multi-objective version of HTS termed the MOHTS algorithm was successfully implemented for this.
- The complex relationship and conflicting nature of between input parameters, their interactions and output responses was evident from the scattered nature of the 2-D Pareto fronts.

Author Contributions: R.C. designed and performed the experiments, analyzed the experiment results, writing the paper. J.J.V. and D.M.P. supervised all of the work carried out this research and reviewed the manuscript. V.K.P. analyzed the parameters through optimization technique. S.S.M.P. and I.A.P. performed and analyzed DSC testing work. L.N.L.d.L. reviewed and analyzed the optimization work.

Funding: This research received no external funding.

Acknowledgments: The authors would like to thank Faculties and especially Head of the department, Metallurgy department of GEC, Gandhinagar for helping in carrying out the Micro-Hardness testing. Bhavesh Patel, research scholar, Indian institute of technology, Indore is also acknowledged for proving support in sample preparation for the DSC testing. The authors would also like to thank Messer Salutation MituToyo for proving the facilities for surface roughness measurements. Lastly, the editors and reviewers of this research paper are greatly acknowledged for enhancing the quality of the manuscript.

Conflicts of Interest: The authors declare no conflict of interest.

Nomenclature

A	Ampere
A_f	Austenitic start
A_f	Austenitic finish
ANOVA	Analysis of variance
BBD	Box-Behnken design
DSC	Differential scanning calorimetry
HTS	Heat transfer search
IP	Peak current (Ampere)
M_f	Martensitic start
M_s	Martensitic finish
MH	Microhardness (HV)
MOHTS	Multi-objective heat transfer search
MRR	Material removal rate (mm^3/s)
RSM	Response surface methodology
SMA	Shape memory alloy
SMAs	Shape memory alloys
SR	Surface roughness (μm)
SV	Spark gap voltage
T_{on}	Pulse on time (μs)
T_{off}	Pulse off time (μs)
WEDM	Wire electric discharge machine

References

1. Jani, J.M.; Leary, M.; Subic, A.; Gibson, M.A. A review of shape memory alloy research, applications and opportunities. *Mater. Des.* **2014**, *56*, 1078–1113. [CrossRef]
2. Henderson, E.; Buis, A. Nitinol for prosthetic and orthotic applications. *J. Mater. Eng. Perform.* **2011**, *20*, 663–665. [CrossRef]
3. Elahinia, M.H.; Hashemi, M.; Tabesh, M.; Bhaduri, S.B. Manufacturing and processing of NiTi implants: A review. *Prog. Mater. Sci.* **2012**, *57*, 911–946. [CrossRef]

4. Kannan, T.D.; Pegada, R.; Sathiya, P.; Ramesh, T. A comparison of the effect of different heat treatment processes on laser-welded NiTinol sheets. *J. Braz. Soc. Mech. Sci. Eng.* **2018**, *40*, 562. [CrossRef]

5. Marattukalam, J.J.; Balla, V.K.; Das, M.; Bontha, S.; Kalpathy, S.K. Effect of heat treatment on microstructure, corrosion, and shape memory characteristics of laser deposited NiTi alloy. *J. Alloys Compd.* **2018**, *744*, 337–346. [CrossRef]

6. Elahinia, M.H.; Ahmadian, M. An enhanced SMA phenomenological model: II. The experimental study. *Smart Mater. Struct.* **2005**, *14*, 1309. [CrossRef]

7. Yuan, B.; Chung, C.Y.; Zhang, X.P.; Zeng, M.Q.; Zhu, M. Control of porosity and superelasticity of porous NiTi shape memory alloys prepared by hot isostatic pressing. *Smart Mater. Struct.* **2005**, *14*, S201. [CrossRef]

8. Markopoulos, A.; Pressas, I.; Manolakos, D. Manufacturing processes of shape memory alloys. In *Materials Forming and Machining*; Elsevier: Amsterdam, The Netherlands, 2016; pp. 155–180.

9. Velmurugan, C.; Senthilkumar, V.; Dinesh, S.; Arulkirubakaran, D. Machining of NiTi-shape memory alloys—A review. *Mach. Sci. Technol.* **2018**, *22*, 355–401. [CrossRef]

10. Prabu, S.M.; Madhu, H.C.; Perugu, C.S.; Akash, K.; Kumar, P.A.; Kailas, S.V.; Anbarasu, M.; Palani, I.A. Microstructure, mechanical properties and shape memory behaviour of friction stir welded nitinol. *Mater. Sci. Eng. A* **2017**, *693*, 233–236. [CrossRef]

11. Weinert, K.; Petzoldt, V. Machining of NiTi based shape memory alloys. *Mater. Sci. Eng. A* **2004**, *378*, 180–184. [CrossRef]

12. Tripathy, S.; Tripathy, D. Multi-response optimization of machining process parameters for powder mixed electro-discharge machining of H-11 die steel using grey relational analysis and topsis. *Mach. Sci. Technol.* **2017**, *21*, 362–384. [CrossRef]

13. Dave, S.; Vora, J.J.; Thakkar, N.; Singh, A.; Srivastava, S.; Gadhvi, B.; Pate, V.; Kumar, A. Optimization of EDM drilling parameters for Aluminium 2024 alloy using Response Surface Methodology and Genetic Algorithm. In *Key Engineering Materials*; Transtech Publication: Stafa-Zurich, Switzerland, 2016; pp. 3–8.

14. Kumar, P.; Parkash, R. Experimental investigation and optimization of EDM process parameters for machining of aluminum boron carbide (Al–B$_4$C) composite. *Mach. Sci. Technol.* **2016**, *20*, 330–348. [CrossRef]

15. Jabbaripour, B.; Sadeghi, M.H.; Faridvand, S.; Shabgard, M.R. Investigating the effects of EDM parameters on surface integrity, MRR and TWR in machining of Ti–6Al–4V. *Mach. Sci. Technol.* **2012**, *16*, 419–444. [CrossRef]

16. Bisaria, H.; Shandilya, P. Experimental studies on electrical discharge wire cutting of Ni-rich NiTi shape memory alloy. *Mater. Manuf. Process.* **2018**, *33*, 977–985. [CrossRef]

17. Sharma, N.; Raj, T.; Jangra, K.K. Parameter optimization and experimental study on wire electrical discharge machining of porous Ni$_{40}$Ti$_{60}$ alloy. *Proc. Inst. Mech. Eng. Part B J. Eng. Manuf.* **2017**, *231*, 956–970. [CrossRef]

18. Soni, H.; Sannayellappa, N.; Rangarasaiah, R.M. An experimental study of influence of wire electro discharge machining parameters on surface integrity of TiNiCo shape memory alloy. *J. Mater. Res.* **2017**, *32*, 3100–3108. [CrossRef]

19. Majumder, H.; Maity, K. Application of GRNN and multivariate hybrid approach to predict and optimize WEDM responses for Ni-Ti shape memory alloy. *Appl. Soft Comput.* **2018**, *70*, 665–679. [CrossRef]

20. Manjaiah, M.; Narendranath, S.; Basavarajappa, S.; Gaitonde, V.N. Effect of electrode material in wire electro discharge machining characteristics of Ti$_{50}$Ni$_{50-x}$Cux shape memory alloy. *Precis. Eng.* **2015**, *41*, 68–77. [CrossRef]

21. Ramamurthy, A.; Sivaramakrishnan, R.; Muthuramalingam, T.; Venugopal, S. Performance analysis of wire electrodes on machining Ti-6Al-4V alloy using electrical discharge machining process. *Mach. Sci. Technol.* **2015**, *19*, 577–592. [CrossRef]

22. Shabgard, M.; Farzaneh, S.; Gholipoor, A. Investigation of the surface integrity characteristics in wire electrical discharge machining of Inconel 617. *J. Braz. Soc. Mech. Sci. Eng.* **2017**, *39*, 857–864. [CrossRef]

23. Kuruvila, N. Parametric influence and optimization of wire EDM of hot die steel. *Mach. Sci. Technol.* **2011**, *15*, 47–75. [CrossRef]

24. Dang, X.-P. Constrained multi-objective optimization of EDM process parameters using kriging model and particle swarm algorithm. *Mater. Manuf. Process.* **2018**, *33*, 397–404. [CrossRef]

25. Somashekhar, K.; Ramachandran, N.; Mathew, J. Optimization of material removal rate in micro-EDM using artificial neural network and genetic algorithms. *Mater. Manuf. Process.* **2010**, *25*, 467–475. [CrossRef]

26. Prabhu, S.; Vinayagam, B. Optimization of carbon nanotube based electrical discharge machining parameters using full factorial design and genetic algorithm. *Aust. J. Mech. Eng.* **2016**, *14*, 161–173. [CrossRef]

27. Patel, V.K.; Raja, B.D. A comparative performance evaluation of the reversed Brayton cycle operated heat pump based on thermo-ecological criteria through many and multi objective approaches. *Energy Convers. Manag.* **2019**, *183*, 252–265. [CrossRef]

28. Raja, B.D.; Jhala, R.; Patel, V. Multiobjective thermo-economic and thermodynamics optimization of a plate–fin heat exchanger. *Heat Transf.—Asian Res.* **2018**, *47*, 253–270. [CrossRef]

29. Patel, V.K. An efficient optimization and comparative analysis of ammonia and methanol heat pipe for satellite application. *Energy Convers. Manag.* **2018**, *165*, 382–395. [CrossRef]

30. Tawhid, M.A.; Savsani, V. ϵ-constraint heat transfer search (ϵ-HTS) algorithm for solving multi-objective engineering design problems. *J. Comput. Des. Eng.* **2018**, *5*, 104–119. [CrossRef]

31. Raja, B.; Patel, V.; Jhala, R. Thermal design and optimization of fin-and-tube heat exchanger using heat transfer search algorithm. *Therm. Sci. Eng. Prog.* **2017**, *4*, 45–57. [CrossRef]

32. LotfiNeyestanak, A.A.; Daneshmand, S. The effect of operational cutting parameters on Nitinol-60 in wire electrodischarge machining. *Adv. Mater. Sci. Eng.* **2013**, *2013*, 1–6. [CrossRef]

33. Magabe, R.; Sharma, N.; Gupta, K.; Davim, J.P. Modeling and optimization of Wire-EDM parameters for machining of $Ni_{55.8}Ti$ shape memory alloy using hybrid approach of Taguchi and NSGA-II. *Int. J. Adv. Manuf. Technol.* **2019**, 1–15. [CrossRef]

34. Zhao, D.; Wang, Y.; Wang, X.; Wang, X.; Chen, F.; Liang, D. Process analysis and optimization for failure energy of spot welded titanium alloy. *Mater. Des.* **2014**, *60*, 479–489. [CrossRef]

35. Mahanta, S.; Chandrasekaran, M.; Samanta, S.; Arunachalam, R.M. EDM investigation of Al 7075 alloy reinforced with B_4C and fly ash nanoparticles and parametric optimization for sustainable production. *J. Braz. Soc. Mech. Sci. Eng.* **2018**, *40*, 1–17. [CrossRef]

36. Tonday, H.; Tigga, A. Analysis of effects of cutting parameters of wire electrical discharge machining on material removal rate and surface integrity. *IOP Conf. Ser. Mater. Sci. Eng.* **2016**, *115*, 012013. [CrossRef]

37. Patel, V.K.; Savsani, V.J. Heat transfer search (HTS): A novel optimization algorithm. *Inf. Sci.* **2015**, *324*, 217–246. [CrossRef]

38. Buehler, W.J.; Gilfrich, J.; Wiley, R. Effect of low-temperature phase changes on the mechanical properties of alloys near composition TiNi. *J. Appl. Phys.* **1963**, *34*, 1475–1477. [CrossRef]

39. La Roca, P.; Isola, L.; Vermaut, P.; Malarría, J. Relationship between grain size and thermal hysteresis of martensitic transformations in Cu-based shape memory alloys. *Scr. Mater.* **2017**, *135*, 5–9. [CrossRef]

40. Kato, H.; Yasuda, Y.; Sasaki, K. Thermodynamic assessment of the stabilization effect in deformed shape memory alloy martensite. *Acta Mater.* **2011**, *59*, 3955–3964. [CrossRef]

41. Raja, B.D.; Jhala, R.; Patel, V. Thermal-hydraulic optimization of plate heat exchanger: A multi-objective approach. *Int. J. Therm. Sci.* **2018**, *124*, 522–535. [CrossRef]

42. Patel, V.; Savsani, V.; Mudgal, A. Many-objective thermodynamic optimization of Stirling heat engine. *Energy* **2017**, *125*, 629–642. [CrossRef]

43. Patel, V.; Savsani, V.; Mudgal, A. Efficiency, thrust, and fuel consumption optimization of a subsonic/sonic turbojet engine. *Energy* **2018**, *144*, 992–1002. [CrossRef]

Article

Parametric Optimization of Trochoidal Step on Surface Roughness and Dish Angle in End Milling of AISID3 Steel Using Precise Measurements

Santhakumar J and Mohammed Iqbal U *

Department of Mechanical Engineering, Faculty of Engineering and Technology, SRM Institute of Science and Technology, Kattankulathur, Tamil Nadu 603203, India; jsanthakumar1987@gmail.com
* Correspondence: umiqbal@gmail.com or mohammediqbal.u@ktr.srmuniv.ac.in

Received: 1 April 2019; Accepted: 23 April 2019; Published: 24 April 2019

Abstract: Tool steel play a vital role in modern manufacturing industries due to its excellent properties. AISI D3 is a cold work tool steel which possess high strength, more hardenability and good wear resistance properties. It has a wide variety of applications in automobile and tool and die making industries such as blanking and forming tools, high stressed cutting, deep drawing and press tools. The novel ways of machining these steels and finding out the optimum process parameters to yield good output is of practical importance in the field of research. This research work explores an attempt to identify the optimized process parameter combinations in end milling of AISI D3 steel to yield low surface roughness and maximum dish angle using trochoidal milling tool path, which is considered as a novelty in this study. 20 experimental trials based on face centered central composite design (CCD) of response surface methodology (RSM) were executed by varying the input process factors such as cutting speed, feed rate and trochoidal step. Analysis of variance (ANOVA) was adopted to study the significance of selected process parameters and its relative interactions on the performance measures. Desirability-based multiple objective optimization was performed and the mathematical models were developed for prediction purposes. The developed mathematical model was statistically significant with optimum conditions of cutting speed of 41m/min, feed rate of 120 mm/min and trochoidal step of 0.9 mm. It was also found that the deviation between the experimental and predicted values is 6.10% for surface roughness and 1.33% for dish angle, respectively.

Keywords: dish angle; trochoidal step; response surface methodology; surface roughness; desirability approach

---

1. Introduction

The overall machining economy, performance and efficiency are largely affected by the cutting tool and its geometry. Cutting tools generally fail due to rapid plastic deformation or mechanical breakage at the cutting edges and often because of gradual wear [1,2]. Failures due to plastic deformation and sudden breakage have considerable adverse effect. Generally, the helical end mills having two or more flutes are widely utilized for slotting, facing, profiling and grooving of narrow surfaces. The end mill cutter edges accomplish the foremost cutting work and generate the flow surface profile at the time of machining process. Dish angle is one of most significant parameters of cutting tool edge which ensures that the flat surface is produced by the cutter during machining. Angle formed between end cutting edge and perpendicular to plane of the cutter axis is known as dish angle as shown in Figure 1. During milling operation, the cutting tool is subjected to wear; as a result, the dish angle value decreases, which influence the more engagement length between workpiece and tool during machining. Thus, the height of residual ridge is increased, which leads to poor surface finish. Therefore, a cutting tool with dish angle is required to achieve a moderate surface quality and tool life [3,4].

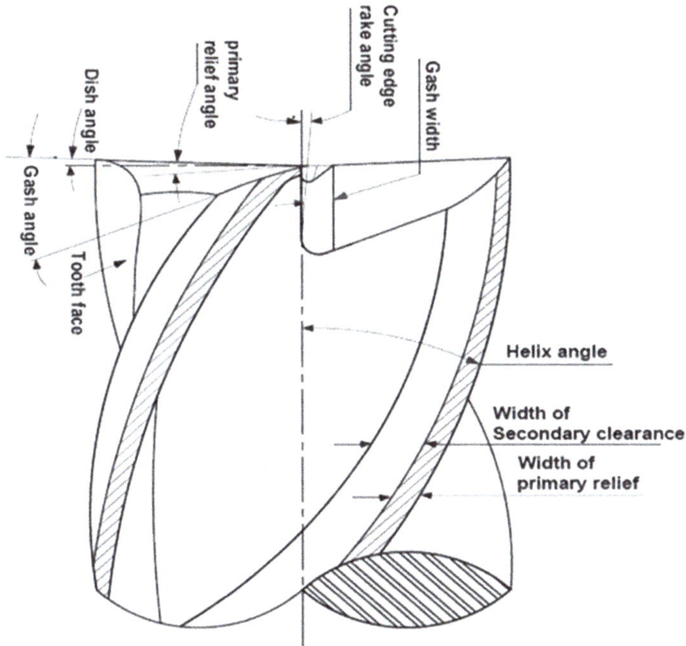

Figure 1. Geometrical parameters of an end mill.

The quality of the machined surface is an essential parameter to assess the productivity of machine tools and the machined components. Therefore, obtaining the required surface quality is vital to the functional behavior aspects of mechanical parts [5]. Hence, it is required to assess the important parameter that affects the quality characteristic of the machined surface. Not only process parameter, but also geometric parameters such as relief angle, gash angle, gash rake angle, helix angle, pitch angle and fluting rake angle difference influence the surface roughness on machined surfaces [6,7].

In many manufacturing sectors, end milling operation is most utilizing processes for making the variety of components such as die and mold [8]. Milling of these components is performed by employing different tool paths such as linear and non-linear tool path strategies [9]. The implementation of traditional tool paths during high speed machining has induced effects such as cutting forces, machining vibrations, damage to the tool and precision loss. To eliminate these effects, many researches have started focusing on the trochoidal milling, which is an amalgamation of circular and linear motions with simultaneous execution. Trochoidal milling is a promising tool path, which can reduce the sudden change of the dynamic cutting force due to the continuous change of radial depth of cut [10]. This is one of the most utilizing tool paths for machining narrow slot and other complex cavities. The major concern in this tool path is the large trochoidal tool path step parameter which leads to increase the cutting force and tool wear resulting in poor surface quality, while the conservative trochoidal step restricts the efficiency of machining [11,12]. Therefore, it is required to analyze and predict the effect of tool path parameter in order to yield the better machining performance in trochoidal milling. In recent years, certain commercial computer aided manufacturing (CAM) software has inbuilt function of trochoidal machining methods. Many researchers have also propelled corresponding studies on trochoidal machining.

Deng et al. [13] have proposed a method to choose optimal trochoidal tool path parameters to minimize the machining time and have analyzed the relationship between MRR and cutting forces. Lopez de lacalle et al. [14] developed modeling of the end milling process to estimate the real tool-path and the tool deflection on cutting force. Modifications were made in the model for minimum chip thickness influence and its relationship with the large tool deflection and the effect of tool cutting edge radius with respect to coefficients of cutting force. It was reported that consistency of simulated cutting force and measured cutting force are in good agreement during end milling process. Patil et al. [15] implemented the trochoidal tool path strategy on Ti6Al4V to study its impact on quality and productivity, and concluded that the tool path generates better surface roughness and tool life. Shixiong [16] studied the trochoidal machining on high speed pocket milling operation and it was reported that path geometry has to be optimized to reduce the trochoidal tool path length feed rate and depth of cut, and the other parameters have to be increased within a bearing capability of tool which enhances the machining efficiency.

Generally, the quality of the end milled components depend upon the other machining factors such as depth of cut, feed rate, cutting speed etc. [17]. Therefore, there is a need of optimizing the end mill parameters in addition to trochoidal tool path parameter required for the better machining performance characteristics. Conventionally, many trials were performed for the selection of end milling process parameters but it was time consuming and costlier. Hence, it is required to develop a multiple objective optimization methodology that can predict the output responses for reducing the surface roughness and maximize the dish angle of end milling process.

Many researchers have reported on multi-response optimization in the end milling process. Sivaraosa et al. [18] stated that the Response Surface Methodology (RSM) approach is an effective technique by its precision towards experimental validation and mathematical modeling. The prominence of an interactive term and square term of factors are accurately predicted by RSM. Two-dimensional (2D) contour and three-dimensional (3D) surface plot developed by RSM helped in identifying the interactive influence of factors on output responses within the range of specified limits. Mia [19] adopted the RSM technique for developing mathematical modeling, in which the optimization was attained by the composite desirability function on the cutting force, surface roughness and specific cutting energy in slot milling operation using AISI 1060 steel performed through-tool cryogenic cooling condition. Similarly, Abou-El-Hossein et al. [20] developed a 1st and 2nd order model for cutting force that has been generated during the end milling process of modified AISI P20 tool steel using RSM approach. The effect of cutting speed, axial depth of cut, feed rate and stepover on cutting force was stated. It was observed that there is a strong interaction among feed rate and axial depth and a quite significant interaction of feed rate and stepover. Calıskan et al. [21] studied the effect of various coating tools on surface roughness and cutting force using RSM technique in the face milling operation of hardened steel. It was concluded that an interaction of different coating type of tools and depth of cut influence the surface roughness, whereas cutting force was not affected by hard coating tool.

From the literature studies, it was analyzed that the researchers have investigated the effects of trochoidal machining extensively without considering the influence of trochoidal step parameter on surface roughness and dish angle. The purpose of the present investigation was to employ experimental and statistical methods to examine the role of process parameters on the surface roughness and dish angle in AISI D3 cold work tool steels using trochoidal tool path. Novel parameters such as trochoidal step and dish angle are taken as an input and output responses for this study. Predictive modeling for the input variable such as cutting speed, feed rate and trochoidal step combinations in end milling is another aspect that has not been explored by researchers. To investigate the effect of output responses with respect to input variable, the RSM technique is well suited. Hence, RSM-based desirability multi-objective optimization approach was employed to arrive the optimal solution during end milling process.

The layout of this article is composed of various sections in which the first section outlines the gaps identified in the extensive literature studies, need and objectives of the proposed research. Section two deals with the materials and methodology of this research. It elaborates the importance of AISI D3 applications and its properties, selection of process parameter and simulation of trochoidal step increment followed by cutting tool calibration measurement technique on dish angle and surface roughness. In the third section, the design of the experiments' orientation plan was formed with the input factors based on central composite design (CCD) for formulating the estimation models with respect to dependent factors and evaluating the adequacy of the model based on analysis of variance (ANOVA) was discussed. The fourth section elaborates the effect of each output responses, using 3D RSM surface plots and determining the optimum process parameters using desirability based multi objective optimization. The last section describes the conclusion made from the investigation.

2. Materials and Methods

The material utilized for this study is AISI D3 cold work tool steel (Tradewell Ferromet Private Limited, Mumbai, India), of which the chemical composition in wt% and its mechanical properties are shown in Tables 1 and 2 respectively. It is widely utilized in the production of cold forming dies and molds due to its excellent wear resistance from gliding contact with distinct metals and deep-hardening characteristics for automobile and aerospace applications. End milling operation were executed on a BFW Gaurav BMV 35 T12 (3-AxisVMC, Bharat Fritz Werner Limited, Bengaluru, India) in built with Siemens Sinumerik 828D Basic controller. The traverse xyz axis is $450 \times 350 \times 350$ mm, respectively. The maximum spindle speed and feed rate is 8000 rpm and 10000 mm/min, respectively with spindle motor has a power of 3.7 kW. The machine positioning accuracy and repeatability; accuracy is ±0.005 mm and ±0.003 mm, respectively. All the experiments were performed under dry conditions. The slot size was cut into 35×10 mm cross section and 75 mm in length to conduct the experiments and overall 22,500 mm^3 volume of materials was removed at each experiment. Table 3 illustrates the list of process parameters and its levels used in this study for machining AISID3 steel. The cutting parameters are selected based on the pilot experiments conducted above and below the selected ranges of input parameter from Table 3, cutting speed of 60 m/min, feed rate 450 mm/min and loop spacing 2.5 mm. After conducting the pilot experiments, the machined surface and cutting tool was inspected using surface roughness testing machine and Zoller presetter. It was found that surface roughness value increased above 1 μm due to high feed marks presented on the surfaces and also tooltip got damaged. There was no considerable effect of wear noticed with the experiments conducted below the lower level of the parameters as shown in Table 3. Thus, feasible working limits of cutting speed was safely selected in the range of 15–45 mm/min, feed rate in the range of 120 to 360mm/min and loop spacing 0.6 mm to 1.8 mm for end milling operation. 20 end milling experiment trials based on the CCD of RSM as shown in Table 3 were conducted individually using fresh uncoated solid tungsten carbide cutting tools (Make: Addison & Company Limited, Chennai, India) for each experiment. The tool has an overall length of 50 mm, diameter of 4 mm with two flutes and a helix angle of 30°.

In trochoidal milling, cutting tool is subjected to gradual milling by consecutive continuous circles and simultaneous forward moments. The trochoidal machining module in (NX.10, Siemens, Plano, TX, USA) CAM software was used to perform the trochoidal trajectory simulation as shown in Figure 2. In this study trochoidal step (S_{tr}) parameter is a function of diameter of the tool. Therefore, for tool safety purposes, step values were taken less than 50% of tool diameter which is in the range 0.6 to 1.8 mm and loop diameter is automatically adjusted by CAM software based on the cavity. Trochoidal trajectory simulation was applied to machine the narrow slot cavity (75 mm × 30 mm) cross section. Trochoidal steps were taken as 0.6 mm, 1.2 mm and 1.8 mm. The milling operation was carried out in the direction of width of the workpiece. The experimental setup details are shown in Figure 3.

Table 1. Chemical composition in wt% of AISID3 cold work tool steel.

Element	Carbon (C)	Silicon (Si)	Chromium (Cr)	Manganese (Mn)	Nickel (Ni)	Vanadium (V)	Iron (Fe)
Content (wt%)	2.1	0.3	11.5	0.4	0.31	0.25	Balance

Table 2. AISI D3 cold work tool steel mechanical properties.

Workpiece Materials	Mechanical Properties of D3 Cold Work Tool Steel				
	Hardness, (HRC)	Density, (kg/cm^3)	Tensile Strength, (N/mm^2)	Yield Strength, (N/mm^2)	Heat Conductivity, (W/mK)
AISI D3	30	7.7	970	850	20

Table 3. Process factors and their coded values.

Factors	Symbol	Units	Coded Values		
			(−1)	(0)	(+1)
Cutting speed	v_c	m/min	15	30	45
Feed rate	v_f	mm/min	120	240	360
Trochoidal step	s_{tr}	mm	0.6	1.2	1.8

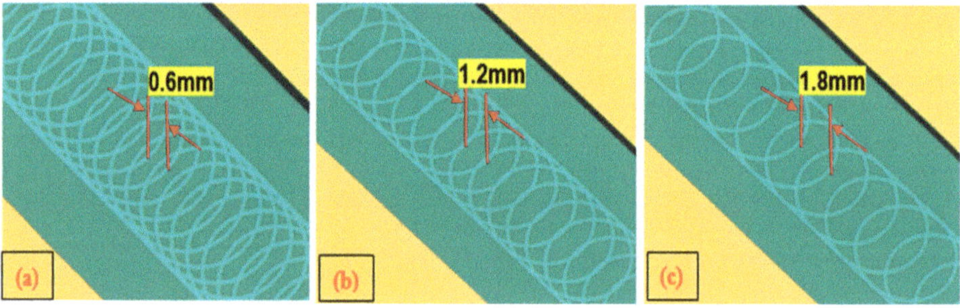

Figure 2. Trochoidal tool path step simulation at (**a**) s_{tr} = 0.6 mm (**b**) s_{tr} = 1.2 mm (**c**) s_{tr} = 1.8 mm.

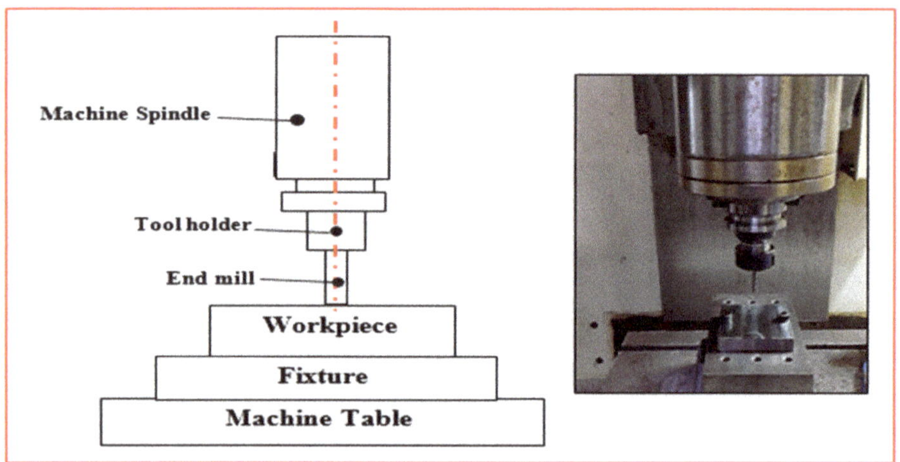

Figure 3. Experimental setup details.

Measurement of Output Responses

The surface roughness for each specimen was measured in the bottom of machined surface using a surface roughness testing machine (Model: Surfcom 1400 G, ACCRETECH, SIEMITSU, Tokyo, Japan) with 0.8 mm of sampling length and 4 mm of cut-off length. Specification of the roughness measuring instrument used are as follows: type: contact stylus instrument, stylus arm length of 60 mm and stylus radius of 2 µmR (60° conical diamond tip size), stylus measuring force of 0.75 mN and scanning rate of 1.5 mm/s. The detector has maximum stroke of 800 µm and the resolution is 0.1 nm in vertical range. Tracing driver column up/down speed is 10 mm/s. The accuracy of machine is 3 nm resolution at 0.2 mm and 15 nm resolution at 1mm in vertical range. An example of the surface roughness measurement graph obtained from "Accretech" software during the measurement is shown in Figure 4. The measurements were consequently repeated thrice at different locations along the feed direction of end milled surface and the arithmetic average value of surface roughness (R_a), root mean square roughness (R_q) and ten-point average height (R_z) was noted as shown in Table 4. The initial cutting tool geometrical characteristics were determined using 2D and 3D measurements. The tool wear was examined by Vision measuring system (VMS) after performing milling operations. Surface roughness image and tool wear image were captured using Optive Lite Model (OLM) 3020 Vision with a color of CCD camera of 1/3 inches high resolution capacity with least count of 1µm integrated with LED stage light and ring light with field view of 30X to 180X magnification. The images were processed using VMS 3.1 software (Hexagon Manufacturing Intelligence, Noida, India). The Zoller Junior Plus (Zoller Inc., Deutschland, Germany) tool pre-setter as depicted in Figure 5 was utilized to calibrate the dish angle deviation. The measuring range of the pre setter is 420 mm in Z-axis and 210 mm in X-axis. The positioning accuracy of horizontal and vertical axis are ±0.003 mm and ±0.005 mm, respectively. The ZOLLER SK50 (steep taper 50, Zoller Inc., Deutschland, Germany) high-precision spindle with its concentricity of 0.002 mm used for indexing for picking up attachment holders for tool calibration. The maximum safe load on the table is 50 kg and maximum tool length and diameter is 320 mm and 620 mm, respectively. The chip set camera type (charge couple device monochrome model) and lighting system is 7 × 6 mm with 12 LEDs of red color for cutting edge calibration.

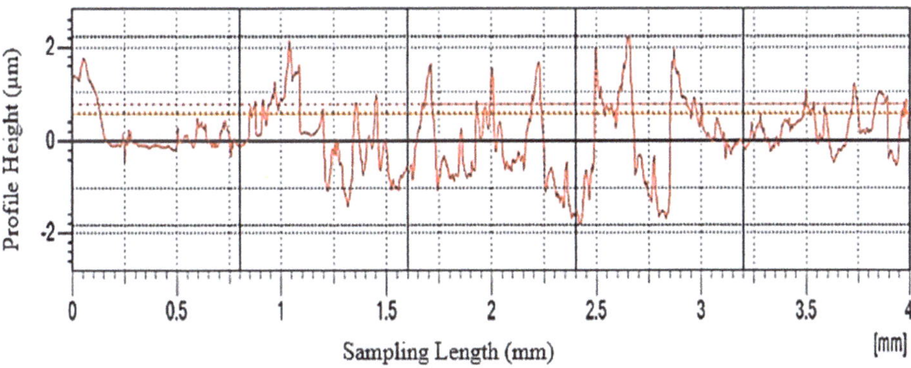

Figure 4. Surface roughness measurement graph.

The dish angle value for the given set of runs were calibrated by Zoller tool pre-setter "Pilot 2mT" software system before and after machining. The initial dish angle (X_1) is 1.55° for all the tools. The dish angle deviation was calculated using Equation (1):

$$\text{Dish angle deviation } (\%) = \left(\frac{X_1 - X_2}{X_2}\right) \times 100 \tag{1}$$

where, X_1 is the dish angle before machining of the tool and X_2 is the dish angle after machining of the tool. The experimental design procedure for the performance of tool determination is shown in Figure 6.

3. Response Surface Methodology (RSM) Experimental Design Matrix

RSM is a highly established technique for formulating the estimation models that rely on experimental observations or physical experiments. RSM is extensively used for optimization and development of mathematical models, which define the interdisciplinary relations of the process parameters and responses. The procedure for RSM follows six steps. The first step is to define the dependent responses and independent parameter. Next, the design of experiments orientation plan was formed with the independent factors based on CCD. Then, the appropriate multiple regression analysis was carried out [22]. The identification of significance factors and interactions using statistical analysis (ANOVA) follows. Finally, a confirmation test was performed to justify the developed model, after which the decision for the model's acceptance or rejection was taken. Here, the dependent parameters taken were cutting speed, feed rate and trochoidal step, which are numeric, while the independent responses considered were surface roughness and dish angle. The measured output values are shown in Table 4.

Equations (2) and (3) represent the 1st and 2nd order developed mathematical correlation among the data sets, respectively [23,24], which was used for developing the empirical models relationships of the data that implies the best possible accuracy towards prediction:

$$X_i = d_0 + d_1 x_1 + d_2 x_2 + \ldots\ldots + d_n x_n \tag{2}$$

$$X_i = d_0 + \sum_{i=1}^{k} d_i x_i + \sum_{i=1}^{k} d_i x_i^2 + \sum_{i=1}^{k} \sum_{j=1_{i<j}}^{k} d_{ij} x_i x_j \pm \epsilon \tag{3}$$

where X_i represents output responses, i.e., surface roughness(R_a) and Dish angle; d_0 is a constant term, $d_1, d_2 \ldots\ldots .d_n$ in Equation (3) represents the coefficients of linear terms, while d_i, d_{ii}, d_{ij} in Equation (4) denote the coefficients of linear terms, square terms and interaction terms, respectively; x_i represents the input parameter i.e, Cutting speed (v_c), federate (v_f) and Trochoidal step (s_{tr}).

Figure 5. Zoller Junior Plus tool pre-setter used for dish angle measurement.

3.1. Developing Mathematical Relationships and Regression Analysis

The 2nd order polynomial quadratic model also known as regression model, describes the system behavior. Nonlinearity in Equation (3) is changed into its linear form using logarithmic transformation in order to generate the regression models. Design expert software version 11.0 (Stat-Ease, Inc., Minneapolis, MN, USA) was employed to determine the coefficients of response surface regression model in an empirical form. All the main parameters and its interaction parameters may not lead to vital consequences on the machining performance. In order to determine the significance of parameter ANOVA was utilized.

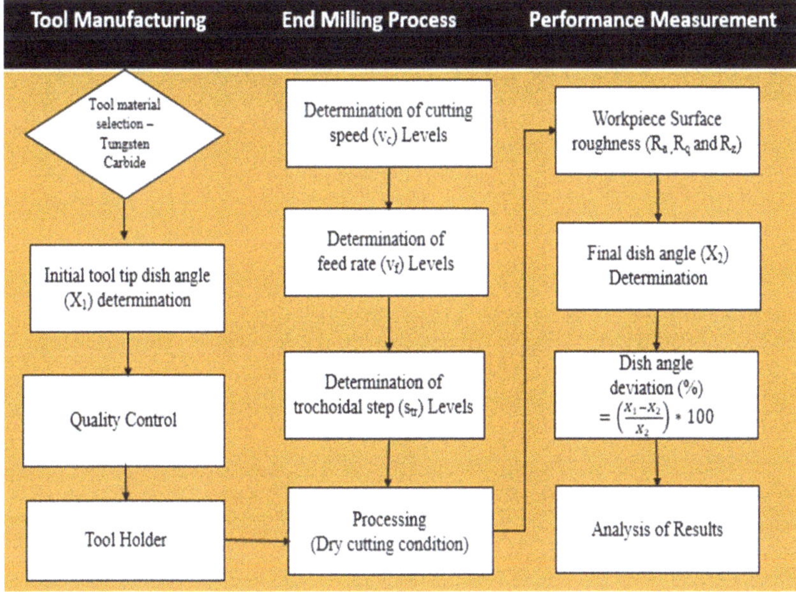

Figure 6. The experimental design procedure for cutting tool performance determination.

Table 4. Experimental results for the end milling operation.

Run	Input Parameters			Output Responses				
	v_c (m/min)	v_f (mm/min)	s_{tr} (mm)	Ra (µm)	R_q (µm)	R_z (µm)	Dish Angle (degree)	Dish Angle Deviation (%)
1	15	120	0.6	0.2968	0.3135	2.0018	1.42	8.38
2	45	120	0.6	0.3292	0.3592	2.4050	1.51	2.5
3	15	360	0.6	0.4759	0.5070	3.3420	1.25	19.35
4	45	360	0.6	0.4983	0.5288	3.6948	1.36	12.25
5	15	120	1.8	0.3783	0.4130	2.8704	1.45	6.45
6	45	120	1.8	0.4006	0.4309	3.5797	1.53	1.29
7	15	360	1.8	0.5574	0.6017	4.9839	1.34	13.54
8	45	360	1.8	0.6297	0.7478	5.6560	1.47	5.16
9	15	240	1.2	0.4271	0.5324	3.7702	1.38	10.96
10	45	240	1.2	0.4494	0.4812	2.8073	1.52	1.93
11	30	120	1.2	0.3387	0.3963	2.1260	1.49	3.87
12	30	360	1.2	0.5178	0.6138	4.2041	1.37	11.61
13	30	240	0.6	0.3875	0.4334	2.6660	1.40	9.67
14	30	240	1.8	0.4690	0.5012	3.1440	1.48	4.51
15	30	240	1.2	0.3949	0.4237	2.4030	1.45	6.45
16	30	240	1.2	0.3824	0.4131	2.2032	1.44	7.09
17	30	240	1.2	0.4032	0.4631	2.4601	1.46	5.8
18	30	240	1.2	0.3786	0.4284	2.3031	1.45	6.45
19	30	240	1.2	0.3678	0.4057	2.4216	1.46	5.8
20	30	240	1.2	0.3956	0.4317	2.6856	1.44	7.09

Based on the ANOVA results, the most significant process parameters were observed, and these parameters were incorporated on the final mathematical model relationships. Thus, obtained mathematical model relationships are listed below in Equations (4) and (5):

$$Ra = 0.3980 + 0.0172 \times v_c + 0.0936 \times v_f + 0.0447 \times s_{tr} + 0.0050 \times v_c \times v_f + 0.0050 \times v_c \times s_{tr} + 0.0075 \times v_f \times s_{tr} + 0.0240 \times v_c^2 + 0.0140 \times v_f^2 + 0.0140 \times s_{tr}^2 \tag{4}$$

$$Dish\ Angle = 1.45 + 0.0530 \times v_c - 0.0630 \times v_f + 0.0310 \times s_{tr} + 0.0113 \times v_c \times v_f + 0.0038 \times v_c \times s_{tr} + 0.0213 \times f_z \times s_{tr} - 0.0005 \times v_c^2 - 0.0205 \times v_f^2 - 0.0105 \times s_{tr}^2 \tag{5}$$

3.2. Evaluating the Correctness of the Empirical Relationship

The potential of the obtained empirical model is evaluated by ANOVA. Tables 4 and 5 represent the ANOVA results for surface roughness and dish angle, respectively. The value of F indicates the significance of model. In Tables 5 and 6, the value of p is greater than F and less than 0.0001, which implies that the developed models are vital [25]. Similarly, the effect of individual input terms ($v_f \times s_{tr}$) found to be significant for Ra and ($v_c \times v_f \times s_{tr}$) found to be significant for dish angle and its interaction terms and 2nd order terms were found to be not significant for all the two output responses. Lack of fit value is smaller, and hence, it is not significant as desired.

The obtained models possess high value of coefficient determination (R^2) and adequate precision (AP). The obtained values are: $R^2 = 0.9687$ and AP = 22.203 for surface roughness; and $R^2 = 0.9844$ and AP = 35.64 for dish angle, which implies the goodness of fit of the models for the prediction of experimental results. The value of R^2 adjusted is 0.9405 and 0.9705 for surface roughness and dish angle, respectively, which are higher, and denotes that higher importance of the developed model. R^2 (predicted) and R^2 (adjusted) are also in best agreement with each other. The low value of coefficient of variation (Cv) is 4.67 and 0.8132 for surface roughness and dish angle, respectively, which indicates the conducted experiments are reliable with high precision. Figure 7 shows the relationship between the experimental and predicted value. Each observed values of output responses of the AISI D3 tool steel samples were compared with an actual and predicted value obtained from empirical model, and its corresponding correlation plots. The 'R^2' values for the obtained empirical relationship models seem to be within the range which implies that higher correlation persists between the predicted values and the actual values.

Table 5. Analysis of variance (ANOVA) table for surface roughness.

Source	SS	d.f	MS	F-Value	p-Value Prob > F	Remarks
Model	0.1213	9	0.0255	34.38	<0.0001	significant
v_c	0.0029	1	0.2165	7.52	0.0208	
v_f	0.0875	1	0.0013	223.18	<0.0001	
s_{tr}	0.0200	1	0.0001	51.02	<0.0001	
$v_c \times v_f$	0.0002	1	0.0050	0.5100	0.4915	
$v_c \times s_{tr}$	0.0002	1	0.0002	0.5049	0.4936	
$v_f \times s_{tr}$	0.0005	1	0.0039	1.15	0.3092	
v_c^2	0.0016	1	0.0008	4.03	0.0726	
v_f^2	0.0005	1	0.0018	1.37	0.2694	
s_{tr}^2	0.0005	1	0.0000	1.37	0.2694	
Residual	0.0039	10	0.0001			
Lack of fit	0.0031	5		3.56	0.0947	not significant
Pure Error	0.0009	5				
Total	0.1252	19				
Standard Deviation	0.0198			R^2	0.9687	
Mean	0.4239			Adjusted R^2	0.9405	
Cv %	4.67			Predicted R^2	0.8152	
				Adeq Precision	22.203	

Table 6. ANOVA table for Dish angle.

Source	SS	d.f	MS	F-Value	p-Value Prob > F	Remarks
Model	0.0861	9	0.0096	70.34	<0.0001	significant
v_c	0.0281	1	0.0281	206.44	<0.0001	
v_f	0.0397	1	0.0397	291.69	<0.0001	
s_{tr}	0.0096	1	0.0096	70.63	<0.0001	
$v_c \times v_f$	0.0010	1	0.0010	7.44	0.0213	
$v_c \times s_{tr}$	0.0001	1	0.0001	0.8268	0.3846	
$v_f \times s_{tr}$	0.0036	1	0.0036	26.55	0.0004	
v_c^2	5.682×10^{-7}	1	5.682×10^{-7}	0.0042	0.9498	
V_f^2	0.0012	1	0.0012	8.46	0.0156	
s_{tr}^2	0.0003	1	0.0003	2.21	0.1680	
Residual	0.0014	10	0.0001			
Lack of fit	0.0010	5		2.40	0.1792	not significant
Pure Error	0.0004	5				
Total	0.0875	19				

Standard Deviation	0.0117		R^2	0.9844	
Mean	1.43		Adjusted R^2	0.9705	
Cv. %	0.8132		Predicted R^2	0.9067	
			Adeq Precision	35.6438	

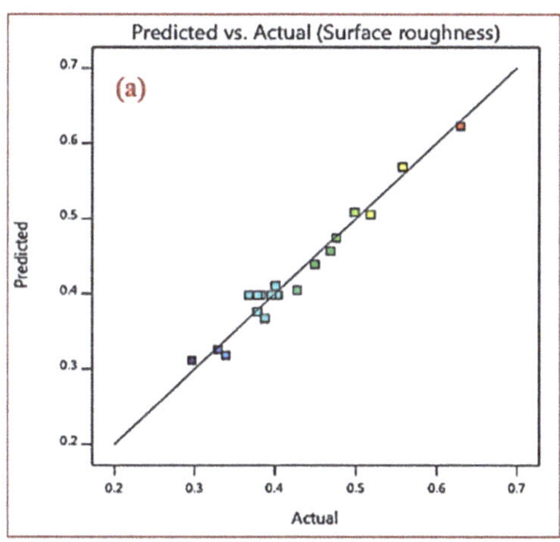

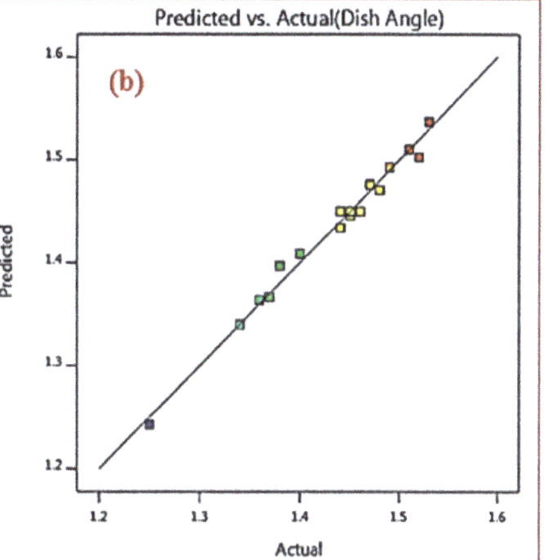

Figure 7. Prediction versus Actual correlation graphs (**a**) surface roughness (**b**) dish angle.

4. Results and Discussion

4.1. Effect of Process Parameter on Surface Roughness

Surface irregularities cannot be eliminated in any kind of machined surface. These irregularities can be examined by means of 2D roughness parameters (R_a, R_q, R_z, R_{sm}, R_t, R_{sk} and R_{ku} etc.). In this current investigation, surface roughness was measured based on the value of Ra. Ra remains useful as a general guideline of surface texture ensuring uniformity in measurement of multiple surface. In addition, R_q and R_z was taken into consideration for assessing the influence of other amplitude parameters on surface texture of the machined surface. R_q denotes the root mean square average of the profile heights over the evaluation length and R_z is the average value of the absolute values of the heights of five highest profile peaks and the depths of five deepest valleys within the evaluation length. The rest of amplitude parameters, R_{sk}, R_{ku}, R_{sm} and R_t, were not measured in the current study.

The 3D surface plots act as an effective tool for investigating the behavior of responses with respect to two factors. In these plots, the dependent response is generally assigned to Z-axis where the independent factors are assigned to X and Y-axis. Figure 8a shows the surface plots of the mean Ra with respect to v_c, v_f and s_{tr}. From Figure 8a it is evident from the plot that mean Ra increases when increasing trochoidal step but decreases with increase in cutting speed. This can be interpreted by the fact that as the cutting speed increases, friction between cutting tool and work piece has been reduced that leads to suppression of built up edge (BUE) formation. On the other side increasing trochoidal step load on the tooltip is high which leads to increasing cutting force, thus causing poor surface roughness [10]. Figure 8b shows that mean Ra is increased with increase in feed rate and trochoidal step. This phenomenon may be explained by the fact that physical impression on the machined surface is forming of laces.

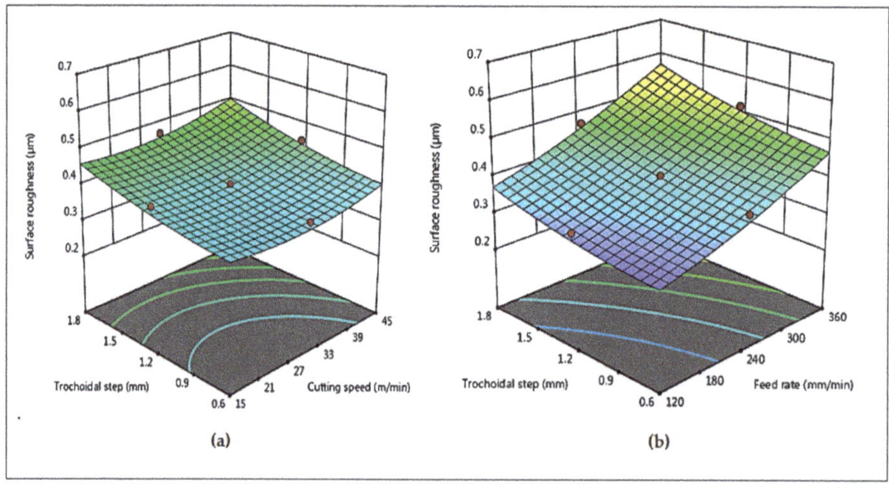

Figure 8. Combined effects of (**a**) v_c and s_{tr} on R_a (**b**) v_f and s_{tr} on R_a.

The intensity of the appearance over the machined surface keep on increasing as the feed rate and trochoidal step is increased as shown in Figure 9. Additionally, the geometry of the tool influences the surface roughness. During milling operation with time, the cutting tool undergoes wear, which tends to change the tool profile during the operation. Similar results on surface roughness with respect to feed rate in end milling process of AISID2 steel was reported [26]. In addition, the effect of R_q and R_z were taken in to account. From Table 4 (Run 1 and Run 5), it can be observed that the trochoidal step has a predominant influence on the R_q and R_z parameter. The R_q value increases distinctly from 0.3135 μm to 0.4130 μm and R_z value increased from 2.0018 μm to 2.8704 μm when trochoidal step varying from 0.6 mm to 1.8 mm with respect to constant cutting speed of 15 m/min and feed rate of 120 mm/min. Similarly, for run 5 and run 8, the same behavior was observed for both R_q and R_z parameter.

4.2. Effect of Process Parameter on Dish Angle

Figure 10 reveals the 3D surface interaction plots for the dish angle in terms of v_c, v_f and s_{tr}, respectively. Figure 10a describes that the dish angle increases when cutting speed and trochoidal step increases. In trochoidal milling, the cutting tool is not always engaged with the workpiece for a period of time, as the material is subjected to gradual milling by consecutive continuous circles. Hence, the load or stress on the cutting tool is highly reduced due to the increase in trochoidal step with the increase of cutting speeds. These can be interpreted by the fact that due to increase in trochoidal step, the tool exhibits good heat dissipation. Therefore, it leads to a lower tool wear and an increased tool life [16]. From Figure 10b it is observed that the dish angle decreases with increase in feed rate and trochoidal step. This phenomenon may be explained by the fact that increasing trochoidal step built up edge is formed on the tool as shown in Figure 11. This may be the reason for the decrease in sharpness of cutting tool that tends to increase the tool wear resulting in the decrease of dish angle. Meanwhile, increase in feed rate is influenced with effect of high strain-hardening due to the plastic deformation and simultaneous increase in tool wear [26].

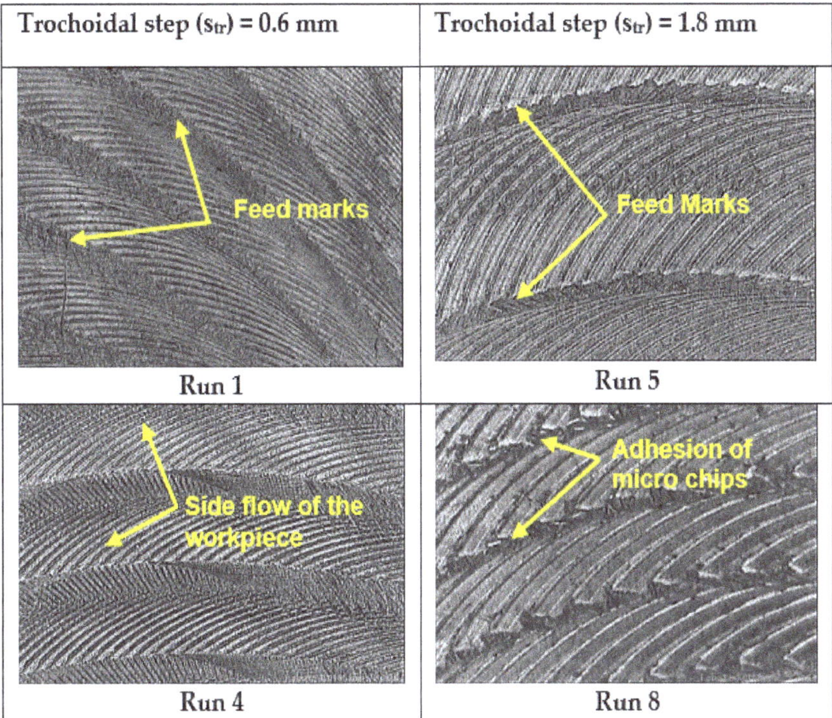

Figure 9. Effect of trochoidal step on surface roughness using Vision measuring system (corresponding to run numbers in Table 4).

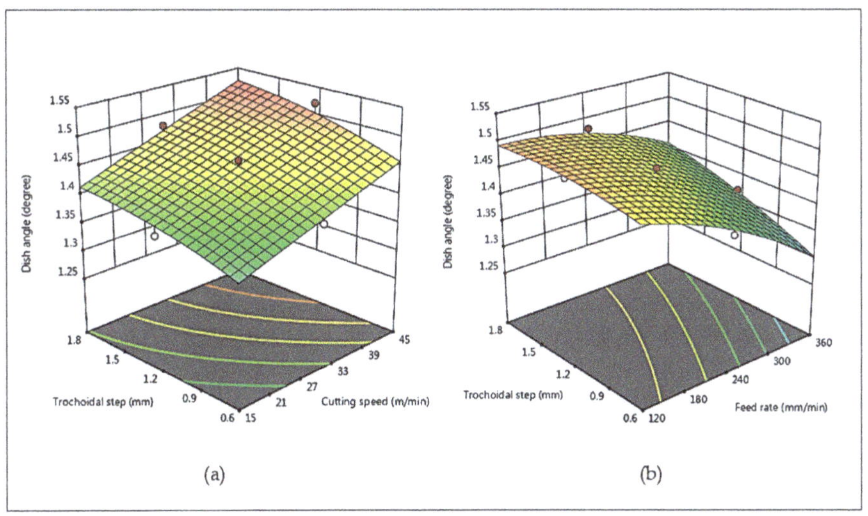

Figure 10. Combined effects of (**a**) v_c and s_{tr} on Dish angle (**b**) v_f and s_{tr} on Dish angle.

4.3. Multi Objective Optimization

The desirability approach is an optimization tool, which is a widely utilized technique in the industries for the multi-objective problem to determine the optimum values for the input process parameters and was proposed by Derringer and Suich [27]. The value of desirability varies between 0 and 1, which depends on the closeness of the output performance. To evaluate the desirability value of this experiment, Design Expert software version 11.0 package was used. Surface roughness and dish angle were optimized by the use of a set of values obtained from RSM. The ramp of numerical optimization graph and 2D composite desirability histogram plot of desirability are shown in Figures 12 and 13 respectively. The ramp function shows the desirability values for each variable and each output response, as well as the composite desirability. Input parameter set or output quality prediction on a specific quality characteristic is quantified as per the dots on ramps. Each dot's height signifies the desirability of the output quality response [27]. The nearest optimum region has an overall composite desirability value of 0.932, designating the closeness of the target value. The multiple regression model has been developed for the output responses (Ra, Dish angle) and was verified with the obtained experimental values, and these values were correlated with the results of confirmation experiments. The confirmation experiments were carried out thrice with optimum input process parameters of 41 m/min of v_c, 210 mm/min of v_f and 0.9 mm of s_{tr}. The confirmatory test and the predicted and obtained values of the output responses based on the optimization approach are shown in Table 7.

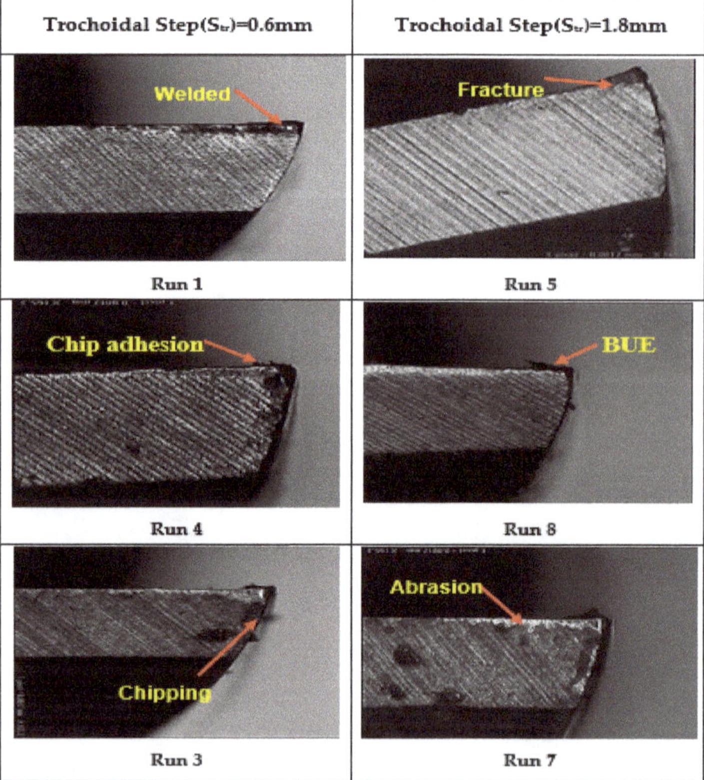

Figure 11. Effect of trochoidal step on dish angle (Tool wear) using Vision measuring system (corresponding to run numbers in Table 4).

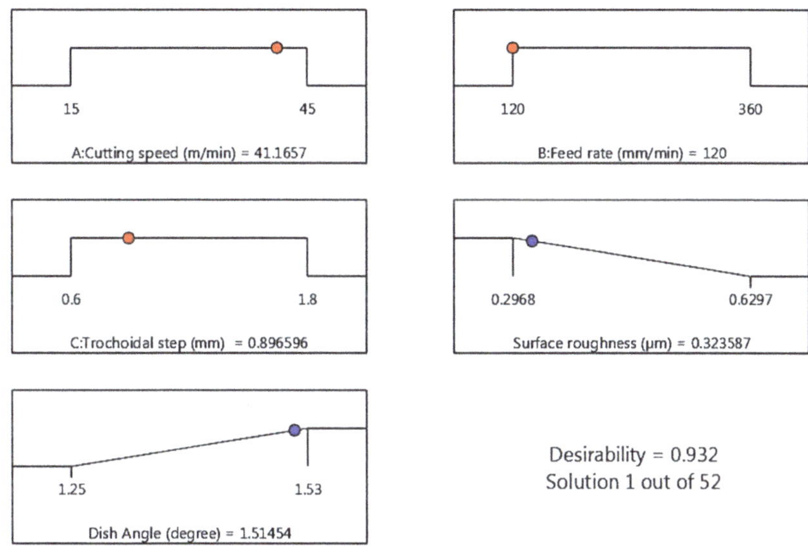

Figure 12. Ramps of numerical optimization.

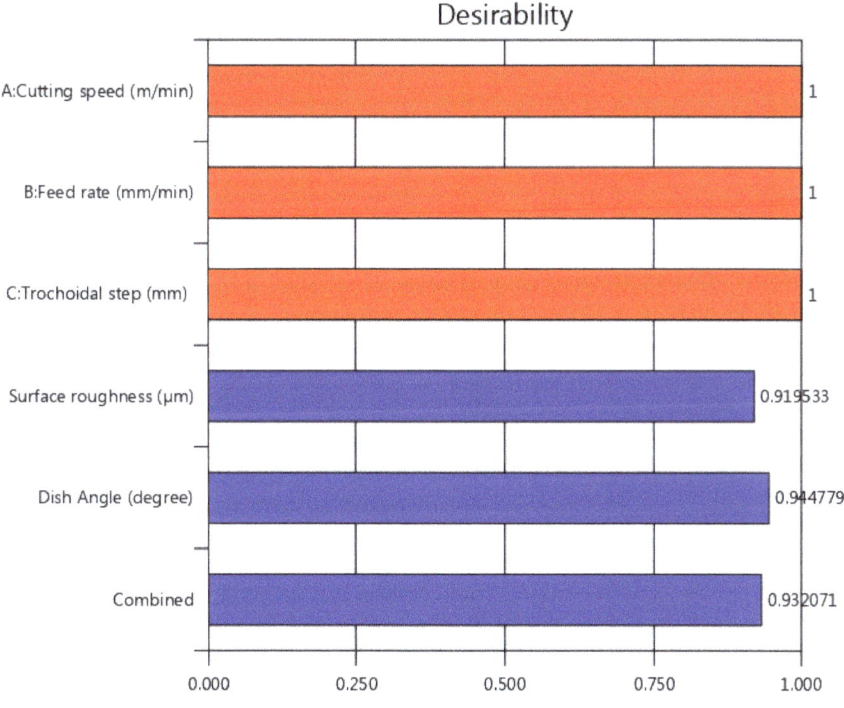

Figure 13. Two-dimensional (2D) composite desirability histogram plot.

Table 7. Optimum parameter level and output response.

Optimum Run	Input Parameters			Ra (µm)		Dish Angle (°)	
	v_c	v_f	s_{tr}	Value	%[Error]	Values	%[Error]
Optimum (RSM)	41	120	0.9	0.323	6.10	1.51	1.34
Optimum (Actual)	41	120	0.9	0.344	-	1.49	-

It is understood from Table 7 that, the average of two output responses, mean surface roughness (0.344 µm) and dish angle (1.49°) of the experimental results seems to be very closer with RSM predicted values and its deviation is much smaller, which authenticate the predicted models is vital within the range of input parameters levels.

5. Conclusions

In this research work, experimental investigation and optimization studies in end milling operation on AISID3 cold work tool steel were performed using a trochoidal tool path to determine the effects of the cutting speed, feed rate and trochoidal step on the output responses. 20 trials for three input variables and three levels were successfully performed using face centered CCD of RSM. The 2nd order polynomial models were generated to predict the Ra and dish angle using RSM. The following inferences have been derived from the proposed work:

1. The developed mathematical modeling relationships of surface roughness and dish angle agreed well with the experimental results, since the error between the experimental and predicted value is within 6.10%, and 1.33%, respectively. The high rating of determination coefficients (R^2) value prove their credibility.
2. The ANOVA studies substantiate that the most influence process parameter affecting the surface roughness is feed rate and trochoidal step, and the parameter dish angle is influenced by all the three-input parameters based on F-ratio value.
3. The formation of laces and adhesion of microchips on the machined surface leads to decrease the surface finish.
4. From the tool wear studies determined by the vision measurement system, it was concluded that the built-up-edge, chipping, abrasion and fracture leads to reduction in the dish angle. It was also concluded that the higher dish angle deviation was observed at a lower cutting speed, lower trochoidal step and higher feed rate.
5. Desirability based multi-objective optimization approach revealed that an optimum process parameter setting of 41m/min of v_c, 210 mm/min of v_f and 0.9mm of s_{tr}. It is summarized that the decrease in feed rate with increase in cutting speed and trochoidal step improve the output quality characteristics.

The results and conclusions obtained from this research will be more beneficial and advantageous for the researchers and machine tool industries for choosing optimum parameter setting for attaining desired surface finish and dish angle using trochoidal tool path, which is considered a novel parameter in end milling process within the range of limit considered for this study.

Moreover, the obtained optimum level of process parameter will improve the quality of machined parts, thereby leading to substantial savings in the tool cost and improve the productivity.

Author Contributions: Conceptualization, S.K.J.; Software, S.K.J.; Investigation, S.K.J. and M.I.U.; Writing-Original Draft Preparation, S.K.J.; Writing-Review & Editing, S.K.J. and M.I.U.; Supervision, M.I.U.

Funding: This research received no external funding.

Acknowledgments: The authors are grateful to the department of mechanical engineering of SRM institute of Science and Technology for providing the facilities to carry out the research.

Conflicts of Interest: The authors declare no conflict of interest.

Abbreviations

Notation

AISI	American Iron & Steel Institute
ANOVA	Analysis of variance
HRC	Hardness measured with the Rockwell test for hard materials
CAM	Computer aided manufacture
RSM	Response surface methodology
BUE	Built up edge
CCD	Central composite design
MRR	Material removal rate
SS	Sum of Squares
MS	Mean Square Symbols
VMS	Vision measuring system
d.f	Degree of freedom

Symbol

s_{tr}	The trochoidal step means the distance between adjacent centers of revolution path (mm)
v_c	Cutting speed (m/min)
v_f	Feed rate(mm/min)
R_a	Surface roughness (μm)
R_z	Ten point average height (μm)
R_q	Root-mean-square roughness (μm)
R_{sk}	skewness
R_{ku}	kurtosis
R_{sm}	Mean spacing between profile peaks at the mean line
X_i	Estimated output response
x_i	Represents the input parameter
d_0	Free term of regression equation
d_i	Coefficients of linear terms
d_{ij}	Coefficients of square terms
$\in$	Experimental error
$d_1, d_2 \dots d_n$	coefficients of linear terms
X_1	Before machining of the tool
X_2	After machining of the tool

References

1. Kuczmaszewski, J.; Zaleski, K.; Matuszak, J.; Palka, T.; Madry, J. Studies on the effect of mill microstructure upon tool life during slot milling of Ti6Al4V alloy parts. *Maintenance Reliab.* **2017**, *19*, 590–596. [CrossRef]
2. Polvorosa, R.; Suarez, A.; Lopez de Lacalle, L.N.; Cerrillo, I.; Wretland, A.; Veiga, F. Tool wear on nickel alloys with different coolant pressures: Comparison of Alloy 718 and Waspaloy. *J. Manuf. Process.* **2017**, *26*, 44–56. [CrossRef]
3. Chen, W.J.; Hsu, C.C.; Yang, Y.L. Improving roughness quality of end milling Al 7075-T6 alloy with Taguchi based multi objective quantum behaved particle swarm optimisation algorithm. *Mater. Res. Innov.* **2014**, *18*, 647–653.
4. Wang, M.-Y.; Chang, H.-Y. Experimental study of surface roughness in slot end milling AL2014-T6. *Int. J. Mach. Tools Manuf.* **2004**, *44*, 51–57. [CrossRef]
5. Mahesh, G.; Muthu, S.; Devadasan, S.R. Prediction of surface roughness of end milling operation using genetic algorithm. *Int. J. Adv. Manuf. Technol.* **2015**, *77*, 369–381. [CrossRef]
6. Ren, J.X.; Zhou, J.H.; Wei, J.W. Optimization of Cutter Geometric Parameters in End Milling of Titanium Alloy Using the Grey-Taguchi Method. *Adv. Mech. Eng.* **2015**, *7*, 1–10. [CrossRef]
7. Wang, Y.-C.; Chen, C.-H.; Lee, B.-Y. Analysis model of parameters affecting cutting performance in high-speed machining. *Int. J. Adv. Manuf. Technol.* **2014**, *72*, 521–530. [CrossRef]
8. Topal, E. The role of stepover ratio in prediction of surface roughness in flat end milling. *Int. J. Mech. Sci.* **2009**, *51*, 782–789. [CrossRef]

9. Gologlu, C.; Sakarya, N. The effects of cutter path strategies on surface roughness of pocket milling of 1.2738 steel based on Taguchi method. *J. Mater. Process. Technol.* **2008**, *206*, 7–15. [CrossRef]

10. Li, H.; Peng, F.Y.; Tang, X.W.; Xu, J.W.; Zeng, H.H. Stability prediction and step optimization of Trochoidal Milling. *J. Manuf. Sci. Eng.* **2017**, *139*. [CrossRef]

11. Akhavan Niaki, F.; Pleta, A.; Mears, L. Trochoidal milling: investigation of a new approach on uncut chip thickness modeling and cutting force simulation in an alternative path planning strategy. *Int. J. Adv. Manuf. Technol.* **2018**, *97*, 641–656. [CrossRef]

12. Ibaraki, S.; Yamaji, I.; Matsubara, A. On the Removal of Critical Cutting Regions by Trochoidal Grooving. *Precis. Eng.* **2010**, *34*, 467–473. [CrossRef]

13. Deng, Q.; Mo, R.; Chen, Z.C.; Chang, Z.Y. A new approach to generating trochoidal tool paths for effective corner machining. *Int. J. Adv. Manuf. Technol.* **2017**, *95*, 3001–3012. [CrossRef]

14. Uriarte, L.; Azcarate, S.; Herrero, A.; Lopez de Lacalle, L.N.; Lamikiz, A. Mechanistic modelling of the micro end milling operation. *Proc. Inst. Mech. Eng. B J. Eng. Manuf.* **2008**, *222*, 23–33. [CrossRef]

15. Patil, P.; Polishetty, A.; Golberg, M.; Littlefair, G.; Nomani, J. Slot Machining of TI6AL4V with Trochoidal Milling Technique. *J. Mach. Eng.* **2014**, *14*, 42–54.

16. Wu, S.X.; Ma, W.; Li, B.; Wang, C.Y. Trochoidal machining for high speed milling of pockets. *J. Mater. Process. Technol.* **2016**, *233*, 29–43.

17. Palanisamy, P.; Rajendran, I.; Shanmugasundaram, S. Prediction of tool wear using regression and ANN models in end-milling operation. *Int. J. Adv. Manuf. Technol.* **2008**, *37*, 29–41. [CrossRef]

18. Sivaraosa, K.R.; Milkey, A.R.; Samsudin, A.K.; Dubey Kidd, P. Comparison between Taguchi method and response surface methodology (RSM) in modelling CO2 laser machining. *Jordan J. Mech. Ind. Eng.* **2014**, *8*, 35–42.

19. Mia, M. Multi-response optimization of end milling parameters under through-tool cryogenic cooling condition. *Measurement* **2017**, *111*, 134–145. [CrossRef]

20. Abou-El-Hossein, K.A.; Kadirgamaa, K.; Hamdi, M.; Benyounis, K.Y. Prediction of cutting force in end-milling operation of modified AISI P20 tool steel. *J. Mater. Process. Technol.* **2007**, *182*, 241–247. [CrossRef]

21. Calıskan, H.; Kurbanoglu, C.; Panjan, P.; Kramar, D. Investigation of the performance of carbide cutting tools with hard coatings in hard milling based on the response surface methodology. *Int. J. Adv. Manuf. Technol.* **2013**, *66*, 883–893. [CrossRef]

22. Rajeswari, B.; Amirthagadeswaran, K.S. Experimental investigation of machinability characteristics and multi response optimization of end milling in aluminium composites using RSM based grey relational analysis. *Measurement* **2017**, *105*, 78–86. [CrossRef]

23. Mohammed Iqbal, U.; Senthil Kumar, V.S.; Gopala Kannan, S. Application of Response Surface Methodology in optimizing the process parameters of Twist Extrusion process for AA6061-T6 aluminum alloy. *Measurement* **2016**, *94*, 126–138. [CrossRef]

24. Chauhan, S.R.; Dass, K. Optimization of Machining Parameters in Turning of Titanium (Grade-5) Alloy Using Response Surface Methodology. *Mater. Manuf. Process.* **2012**, *27*, 531–537. [CrossRef]

25. Ravi Kumar, K.; Sreebalaji, V.S. Desirability based multi objective optimisation of abrasive wear and frictional behaviour of aluminium (Al/3.25Cu/8.5Si)/fly ash composites. *Tribol. Mater. Surf. Interfaces* **2015**, *9*, 128–136. [CrossRef]

26. Koshy, P.; Dewes, R.C.; Aspinwall, D.K. High speed end milling of hardened AISI D2 tool steel (58 HRC). *J. Mater. Process. Technol.* **2002**, *127*, 266–273. [CrossRef]

27. Mohammed Iqbal, U.; Senthil Kumar, V.S. Modeling of twist extrusion process Parameters of AA6082-T6 alloy by response surface approach. *Proc. Inst. Mech. Eng. B J. Eng. Manuf.* **2014**, *228*, 1458–1468. [CrossRef]

Article

Investigation of Tool Wear and Chip Morphology in Dry Trochoidal Milling of Titanium Alloy Ti–6Al–4V

Dongsheng Liu, Ying Zhang, Ming Luo * and Dinghua Zhang

Key Laboratory of Contemporary Design and Integrated Manufacturing Technology of Ministry of Education, Northwestern Polytechnical University, Xi'an 710072, China; 2014200870@mail.nwpu.edu.cn (D.L.); zhangyingcdim@nwpu.edu.cn (Y.Z.); dhzhang@nwpu.edu.cn (D.Z.)
* Correspondence: luoming@nwpu.edu.cn; Tel.: +86-29-8849-3232 (ext. 409)

Received: 21 May 2019; Accepted: 14 June 2019; Published: 16 June 2019

Abstract: Titanium alloys are widely used in the manufacture of aircraft and aeroengine components. However, tool wear is a serious concern in milling titanium alloys, which are known as hard-to-cut materials. Trochoidal milling is a promising technology for the high-efficiency machining of hard-to-cut materials. Aiming to realize green machining titanium alloy, this paper investigates the effects of undeformed chip thickness on tool wear and chip morphology in the dry trochoidal milling of titanium alloy Ti–6Al–4V. A tool wear model related to the radial depth of cut based on the volume of material removed (VMR) is established for trochoidal milling, and optimized cutting parameters in terms of cutting speed and axial depth of cut are selected to improve machining efficiency through reduced tool wear. The investigation enables the environmentally clean rough machining of Ti–6Al–4V.

Keywords: tool wear; trochoidal milling; titanium alloy; chip morphology

---

1. Introduction

The application of titanium alloys has dramatically increased in many industries over the past half century, particularly in the aviation and aerospace industries, where about 80% of titanium production is used [1]. Such extensive application of titanium alloys is due to their excellent properties, such as high strength-to-weight ratio, good anticorrosion performance, and strong fracture resistance. Ti–6Al–4V accounts for the majority of the total consumption of all kinds of titanium alloys.

For most titanium alloy parts, especially aviation parts, traditional cutting methods such as turning, milling, drilling, and grinding are still the main means of processing and manufacturing [2,3]. However, titanium alloy is known as a kind of hard-to-cut material because of its several inherent properties, such as high chemical reactivity, low thermal conductivity, and low modulus of elasticity [4]. High chemical reactivity makes chip easily adhere to the tool-cutting edge. Low thermal conductivity increases the temperature at the tool cutting edge [5], which has an adverse effect on tool condition. Hence, tool wear is severe in the machining process of titanium alloy, and it will lead to serious machining vibration [6,7] or damage the machined surface quality [8]. In order to improve the tool life in the machining of titanium alloy, many research efforts have been made to assist in choosing suitable machining conditions. Effective cooling methods can significantly improve tool life by lowering the temperature of the tool cutting edge. Sun et al. [9] adopted the cryogenic compressed air cooling technique to cool the tool edge during the turning of Ti–6Al–4V. Pittalà [10] used CO_2 cryogenic coolant in the end milling of Ti–6Al–4V. Bermingham et al. [11] made a comparison between cryogenic and high-pressure emulsion cooling technologies during the turning of Ti–6Al–4V, and found that high-pressure emulsion cooling achieved a slightly better tool life. Recently, minimum quantity lubrication (MQL) combined with cryogenic gas [12–15], such as carbon dioxide (CO_2) and liquid nitrogen (LN2), has drawn great attention due to its excellent performance in improving tool life during machining hard-to-cut materials. Machining parameters such as the cutting speed, cutting depth, and feed rate have critical effects on tool life. Hou et al. [16] investigated the influence of cutting speed on tool wear in the end milling of Ti–6Al–4V, and the results showed the mean flank temperature and cutting force increased significantly as the cutting speed increased, which accelerated the tool wear. Jaffery et al. [17] indicated that the depth of cut had a significant effect on wear performance in machining Ti–6Al–4V alloy. To improve the machinability of Ti–6Al–4V, other approaches, such as laser-assisted machining [18], hybrid machining [19], and ultrasonic machining [20] were adopted by many researchers.

The improvement of tool life by the above-mentioned methods is at the expense of increased cost and energy consumption, even environmental pollution. To achieve the goals of green machining as well as the reduction of production cost, dry machining is considered as a promising and satisfactory solution in the future [21]. In the case of dry machining, the cleaning process of the workpiece can be spared, and the chips can be disposed of immediately without other treatment. Dry machining also has some positive effects, such as a reduction in thermal shock and no intrusion introduced by cooling equipment. However, the dry machining of titanium alloys is detrimental to the tool condition, because there will be more friction and adhesion between the tool and the workpiece without coolant, which will accelerate the tool wear [22]. Li et al. [23] investigated the effect of high cutting speed on tool wear in the dry milling of Ti–6Al–4V, and found that the increase in the cutting speed and feed rate accelerated the tool wear and drastically decreased the tool lifetime. Deng et al. [24] studied tool wear in the dry milling of Ti–6Al–4V, and found element diffusion from the Ti–6Al–4V titanium alloy to WC/Co carbide tools (and vice versa) at temperatures up to 400 °C. Li et al. [25] indicated that the main wear mechanisms of coated cemented carbide tools were the complicated interactions of several kinds of wear patterns—such as abrasive wear, coating delamination, adhesive wear, oxidation wear, and diffusion wear—at high cutting speed in the dry face milling of Ti–6Al–4V.

Trochoidal milling is a promising technology for the high-efficiency machining of hard-to-cut materials. This method is a combination of circular milling and slicing [26], which possesses a small engagement region between the tool and workpiece, and enables a relatively high feed rate and axial depth of cut to increase the machining efficiency. In trochoidal milling, the small engagement angle between the tool and workpiece generates low cutting forces, and a long period of non-engagement between the tool and workpiece ensures the effective cooling of the tool. These advantages enable a long tool life in trochoidal milling. Some research studies about trochoidal milling showed good practical results. Uhlmann et al. [27] indicated that trochoidal milling enabled lower energy consumption and process time in the machining of Ti–6Al–4V, compared with the conventional milling strategy. Wu et al. [28] investigated trochoidal milling in nickel-based superalloy, and the results showed that trochoidal milling significantly reduced tool flank wear. Patil et al. [29] adopted a trochoidal milling technique in the slot machining of Ti–6Al–4V, and found that it was better for surface finish and chip evacuation compared with the traditional slot milling method. In view of the merits of trochoidal milling, it is an effective method to reduce tool wear, as well as improve the machining efficiency and quality.

As mentioned above, dry trochoidal milling can be a potential high-efficiency green machining strategy. However, on reviewing the current research for the green milling of Titanium alloys, there appears to be limited reports on the machinability of the dry trochoidal milling of titanium alloy. To fill this research gap, this paper investigates the tool wear as well as the chip morphology in the dry trochoidal milling of titanium alloy Ti–6Al–4V based on a set of cutting experiments. Firstly, the effect that undeformed chip thickness had on tool wear and chip morphology was investigated in the dry milling of Ti–6Al–4V. The effects of radial cutting depth as well as cutting speed on tool wear were studied, and a tool wear model based on the volume of material removed (VMR) was then established based on the flank milling test. Finally, a set of optimized cutting parameters for the dry trochoidal milling of Ti–6Al–4V was selected and validated by cutting experiments.

2. Experimental Procedures

The trochoidal model enables continuity in both feed rate and acceleration due to its continuity in tangency and curvature [30], which is considered better concerning the kinematical behavior of the machine tool. The trochoidal model is shown in Figure 1. The tool center position O_t (x_c, y_c) parameterized by angle θ is expressed as follows:

$$x_c = c\frac{\theta}{2\pi} + R_c \sin\theta$$
$$y_c = R_c \cos\theta \tag{1}$$

where c is the trochoidal stepover, and R_c is the trochoidal radius. Point S is on the tool path envelope, and can be calculated using:

$$\vec{OS} = \vec{OO_t} + R_t \cdot \frac{\vec{O_tS}}{\|\vec{O_tS}\|} \tag{2}$$

where R_t is the tool radius, and vector $\vec{O_tS}$ is normal to the trochoidal tool path. The tool contour (x_t, y_t) is determined by following expression:

$$\left(x_t - c\frac{\theta}{2\pi} - R_c \sin\theta\right)^2 + (y_t - R_c \cos\theta)^2 = R_t^2 \tag{3}$$

Point P is the intersection point of the tool contour and tool path envelope, and can be solved by Equations (2) and (3). After calculating P, the instantaneous radial depth of cut can be obtained using:

$$a_e = \frac{\vec{PS} \cdot \vec{O_tS}}{\|\vec{O_tS}\|} \tag{4}$$

Point A and point B are the tool entry and exit points in one trochoidal cycle. Between A and B, the instantaneous radial depth of cut increases continuously until reaching its maximal value when point P reaches the cusp [31], and then decreases to zero. The trochoidal model generates a non-symmetric tool path that is smooth toward its cusp, which is not located on x axis.

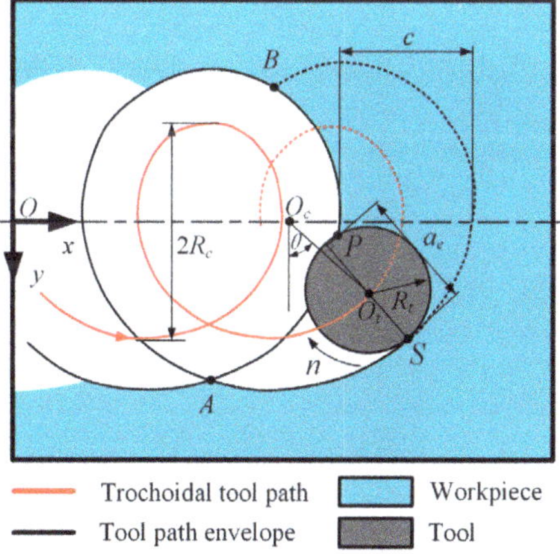

Figure 1. Geometric model of trochoidal milling. Here, O_t is the tool center position; O_c the current revolution center of the tool center; θ is the revolution angle; c is the trochoidal stepover; R_c is the trochoidal radius; R_t is the tool radius; and a_e is the current radial depth of cut.

To improve machining efficiency, trochoidal milling usually adopts a large axial depth of cut and a small radial depth of cut. In combination with suitable cutting parameters and machining equipment, an axial depth of cut up to two times the tool diameter can be reached [27]. Compared with the axial depth of cut, the radial depth of cut related to chip thickness has a greater effect on the tool wear, since it affects the maximum undeformed chip thickness, which determines the maximum cutting force applied on the tool cutting edge. Additionally, cutting speed also has a great effect on tool life in milling titanium alloy [16]. Therefore, before performing the dry trochoidal milling test, the effects of radial cutting depth and cutting speed on the dry milling of Ti–6Al–4V are investigated by flank milling tests first.

2.1. Experimental Setup

The workpieces used in this study were annealed Ti–6Al–4V blocks with a size of 230 mm × 230 mm × 27 mm. The nominal chemical components and physical properties of the titanium alloys are listed in Table 1. The specifications of the used cutting tools are presented in Table 2. All the milling tests were conducted on an YHV850 three-axis machine tool (Yonghua Machinery Co., Ltd., Yanzhou, China) shown in Figure 2. Down milling was adopted in all the tests.

Table 1. Nominal chemical components and physical properties of the titanium alloys.

Element	Al	V	Fe	O	C	N	H	Ti
wt.%	5.5–6.8	3.5–4.5	<0.5	<0.2	<0.1	<0.05	<0.015	balance

Physical Properties	Density (kg/m^3)	Elastic Modulus (GPa)	Yield Strength (MPa)	Thermal Conductivity (W/(m·K))	Hardness (HRC)	Melting Point (°C)
value	4430	113.8	880	6.7	36	1604–1660

Table 2. Specifications of the helical milling cutter.

Diameter (mm)	Number of Flutes	Helix Angle (°)	Corner Radius (mm)	Coating	Rake Angle (°)	Clearance Angle (°)	Second Clearance Angle (°)	Hardness (HRC)
12	4	35	0.5	TiAlN (3 μm)	4	8	20	65

Figure 2. Experimental setup.

2.2. Experimental Design

The experimental design is focused on two main cutting parameters: the cutting speed and the radial depth of cut. The axial depth of cut is fixed at 2 mm, and the feed rate is fixed at 0.1 mm per tooth. Three levels of cutting speeds are selected, i.e., Vc = 60 m/min, 130 m/min, and 200 m/min. For each cutting speed, there are three levels of radial depths of cut, i.e., a_e = 0.2 mm, 0.4 mm, and 0.6 mm. Detailed cutting parameters for all nine groups of tests are listed in Table 3.

Table 3. Cutting parameters used in the experiments (f_z = 0.1 mm/tooth, a_p = 2 mm).

Test No.	Cutting Speed (m/min)	Radial Depth of Cut a_e (mm)	Spindle Speed (rpm)	Feed Rate (mm/min)
1		0.2		
2	60	0.4	1592	637
3		0.6		
4		0.2		
5	130	0.4	3448	1379
6		0.6		
7		0.2		
8	200	0.4	5305	2122
9		0.6		

For each test, the tool rejection or failure is judged according to the following criteria:

(1) Average flank wear VB = 0.2 mm;
(2) Maximum flank wear VB_max = 0.3 mm;
(3) Excessive chipping/flaking or catastrophic failure.

The test will be stopped when any of the above criteria is reached. For the third criterion, the test result is invalid, and the test will be repeated until meeting one of the first two criteria. Then, the VMR will be calculated according to the cutting length.

3. Results and Analysis

3.1. Tool Wear Model Related with Radial Depth of Cut

Tool life results in terms of VMR are shown in Table 4, where the third column "effective cutting time" denotes the total time when the tool is engaged with the workpiece, and "MRR" represents the material removal rate. Suppose that the relationship between the radial depth of cut and VMR possesses a quadratic form as follows:

$$VMR = A_2 a_e^2 + A_1 a_e + A_0 \tag{5}$$

where A_2, A_1, and A_0 are unknown coefficients.

According to the results in Table 4, the relationships between radial depth of cut and VMR for three levels of cutting speeds can be obtained by quadratic fitting (see Figure 3):

$$\begin{aligned}
VMR_1 &= 1.2713 \times 10^6 a_e^2 - 0.9974 \times 10^6 a_e + 0.2259 \times 10^5 \\
VMR_2 &= 5.5603 \times 10^5 a_e - 5.3556 \times 10^5 a_e + 1.2995 \times 10^5 \\
VMR_3 &= 1.4433 \times 10^5 a_e^2 - 1.4502 \times 10^5 a_e + 0.3657 \times 10^5
\end{aligned} \tag{6}$$

where VMR_1, VMR_2, and VMR_3 correspond to cutting speeds of 60 m/min, 130 m/min, and 200 m/min, respectively. In Figure 3, it is observed that VMR decreases with the increase of cutting speed in the cutting speed range of 60 to 200 m/min. For each cutting speed, there is a radial depth of cut corresponding to the lowest VMR, which should be avoided in practical milling. More specifically, VMR at 0.4-mm radial depth of cut is lower than that at 0.2 mm and 0.6 mm radial depths of cut. This could be explained by the cutting force load increasing with the increase of radial depth of cut, thus accelerating the tool wear. However, when the radial depth of cut increased to 0.6 mm, a thicker chip carried away more cutting heat, which kept the tool in a better cutting condition despite a larger force load.

Table 4. Test results. VMR: volume of material removed.

Test No.	VMR (mm^3)	Effective Cutting Time (s)	MRR (mm^3/min)
1	77,280	18,198	254.8
2	30,360	3575	509.6
3	85,146	6683	764.4
4	45,080	4904	551.6
5	4692	255	1103.2
6	8786	319	1654.8
7	13,340	943	848.8
8	1472	52	1697.6
9	1518	36	2546.4

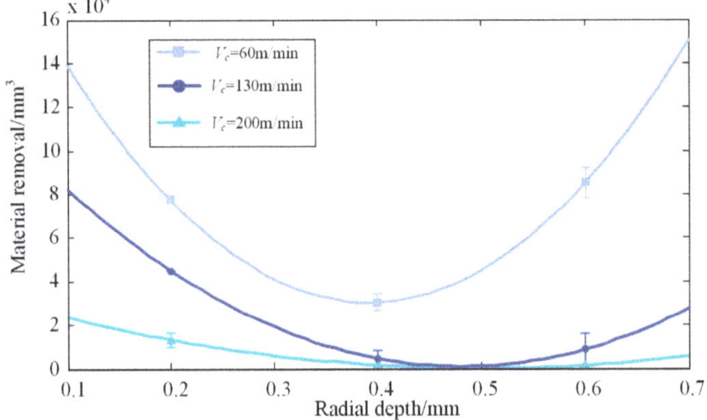

Figure 3. Relationship between radial depth of cut and VMR.

3.2. Tool Wear and Chip Morphology

Tool edges of worn tools were scanned using Alicona Infinitefocus G4 (Graz, Austria) with a magnification of 50×, as shown in Figure 4. For three groups of tests at 60 m/min cutting speed (test no. 1, 2, and 3), the tool wear pattern is uniform flank wear. As the cutting speed increases to 130 m/min and 200 m/min, the tool wear pattern changes to notch wear, and the maximum wear position locates on the upper part of the tool edge. At the same radial depth of cut, the maximum flank wear increases as the cutting speed increases. At the same cutting speed, the higher the radial depth, the more serious the tool wear. For all the tests, tool wear on the tool tip is slight. As reported in [9], in milling titanium alloy, the cutting temperature increases dramatically with the increase in the cutting speed, and the tool wear progresses rapidly due to the high temperature and strong adhesion between the tool and workpiece material.

Chips during tests were collected and analyzed. The macro images of chips in test no. 1 are shown in Figure 5, where chip morphologies at different VMRs are presented. At the initial stage of tool life, the chips are uniform, regular, curled, and separated in shape due to intermittent milling and good tool condition [32] (see Figure 5a). Each chip can be divided into two sections: the corner section and the major section (see Figure 6a). The corner section was formed by the tool nose edge, and the major section was formed by the side cutting edge [33]. The back surface of the chip came into contact with the rake face of the tool under high contact pressure, and the friction is smooth and shiny, while the free surface of the chip is rough (see Figure 6a). When the VMR reaches 9200 mm^3, chip separation has failed, which results in several chips sticking to each other (see Figure 5b). With the VMR increased to 27,000 mm^3, chips with tight spiral shapes appear (see Figure 5c). As the VMR increases, the spiral radius increases (see Figure 5c,e). When the VMR is 46,000 mm^3, the chip is twisted and irregular in shape. Besides, obvious scratches related to tool wear are observed on the back surface of the chip (see Figure 6b). At the end stage of tool life, serious distortion and scratches occur on the chip (see Figure 6c). It can be inferred that the worse the tool wear, the more severe the chip distortion.

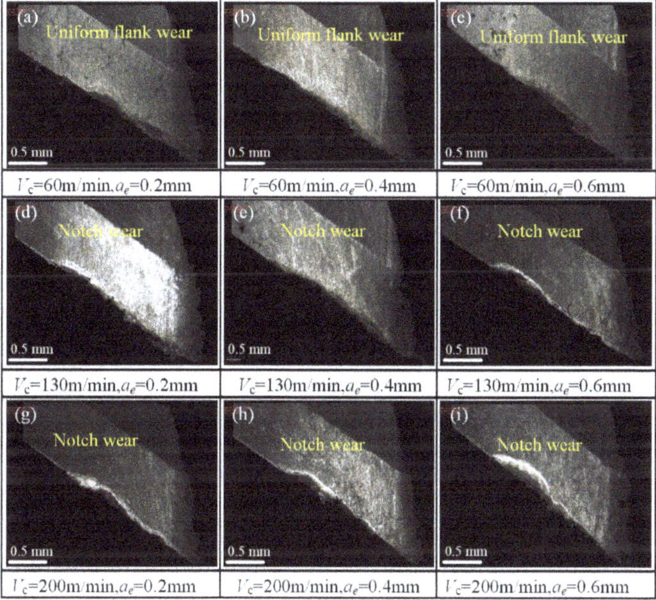

Figure 4. Photos taken by Alicona Infinitefocus G4 for worn tools with cutting parameters (**a**) V_c = 60 m/min, a_e = 0.2 mm, (**b**) V_c = 60 m/min, a_e = 0.4 mm, (**c**) V_c = 60 m/min, a_e = 0.6 mm, (**d**) V_c = 130 m/min, a_e = 0.2 mm, (**e**) V_c = 130 m/min, a_e = 0.4 mm, (**f**) V_c = 130 m/min, a_e = 0.6 mm, (**g**) V_c = 200 m/min, a_e = 0.2 mm, (**h**) V_c = 200 m/min, a_e = 0.4 mm, and (**i**) V_c = 200 m/min, a_e = 0.6 mm.

Chip morphologies in tests no. 5 and no. 9 are shown in Figures 7 and 8, respectively. Comparisons of chip morphology between test no. 1, no. 5, and no. 9 are shown in Figure 9. The evolutions of chip morphology are similar for each test. At the initial stage of tool life, the chips are separated and regular in shape (see Figures 7a, 8a and 9a). At the middle stage, the chips begin to stick together (see Figures 7b, 8b and 9b). At the end stage, continuous chips with long spiral shapes appear (see Figures 7c, 8c and 9c), while the chips in test no. 1 are broken into pieces. Such a difference of chip morphologies between test no. 1 and test no. 5 or no. 9 is caused by the chip thickness. To be more specific, the radial depth of cut in test no. 1 is smaller than those in test no. 5 and no. 9, which causes the chip in test no. 1 to be thinner and easier to break. Chip morphologies in other tests were also observed, and it was found that the evolutions of chip morphology at the same radial depth were similar. The change of chip morphology is mainly caused by the tool wear and the formation of a built-up edge [34].

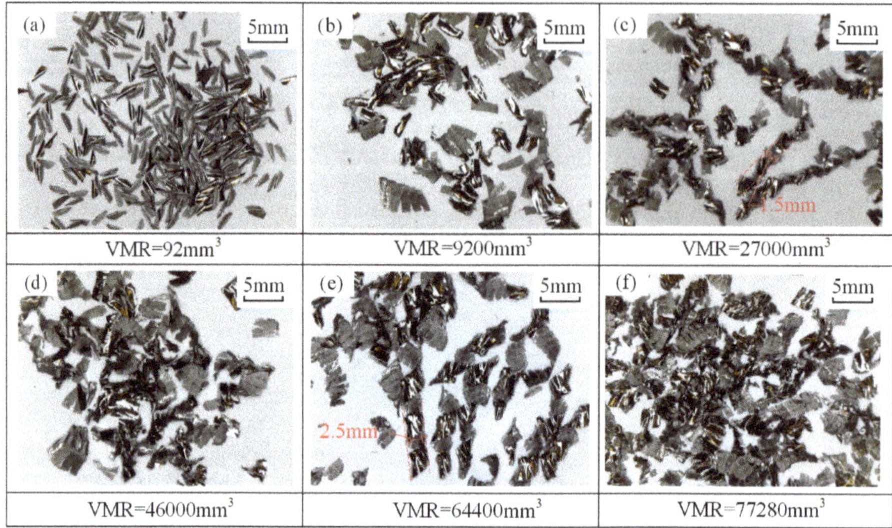

Figure 5. Chip morphology (test no. 1: $V_c = 60$ m/min, $a_p = 2$ mm, $a_e = 0.2$ mm, $f_z = 0.1$ mm/tooth) under VMR of (**a**) 92 mm³, (**b**) 9200 mm³, (**c**) 27,000 mm³, (**d**) 46,000 mm³, (**e**) 64,400 mm³, and (**f**) 77,280 mm³.

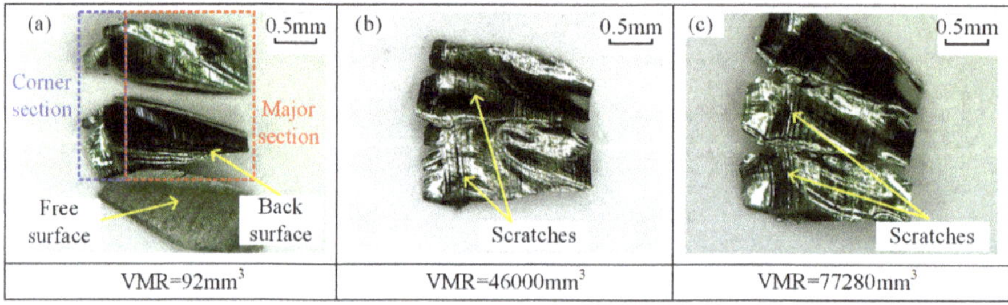

Figure 6. Chip morphology in enlarged view (test no. 1: $V_c = 60$ m/min, $a_p = 2$ mm, $a_e = 0.2$ mm, $f_z = 0.1$ mm/tooth) with VMR of (**a**) 92 mm³, (**b**) 46,000 mm³, and (**c**) 77,280 mm³.

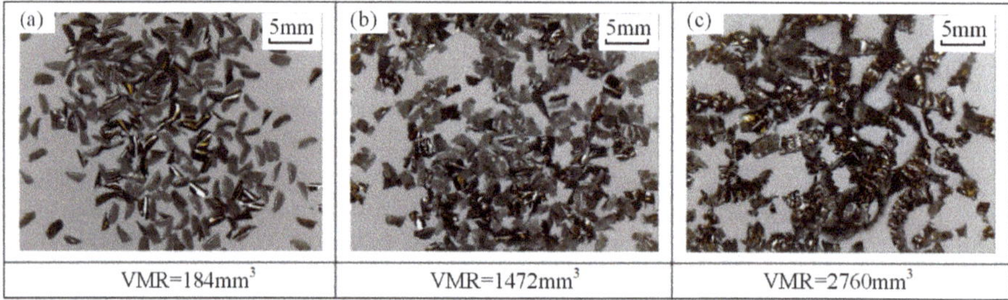

Figure 7. Chips morphology (test no. 5: $V_c = 130$ m/min, $a_p = 2$ mm, $a_e = 0.4$ mm, $f_z = 0.1$ mm/tooth) under VMR of (**a**) 184 mm³, (**b**) 1472 mm³, and (**c**) 2760 mm³.

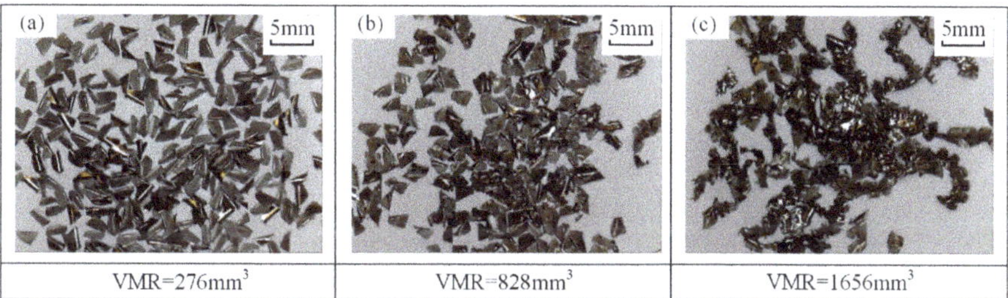

VMR=276mm^3	VMR=828mm^3	VMR=1656mm^3

Figure 8. Chips morphology (test no. 9: V_c = 200 m/min, a_p = 2 mm, ae = 0.6 mm, f_z = 0.1 mm/tooth) under VMR of (**a**) 276 mm^3, (**b**) 828 mm^3, and (**c**) 1656 mm^3.

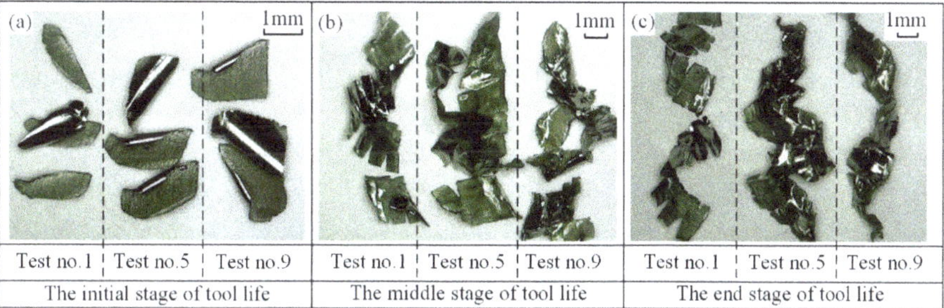

Test no.1	Test no.5	Test no.9	Test no.1	Test no.5	Test no.9	Test no.1	Test no.5	Test no.9
The initial stage of tool life			The middle stage of tool life			The end stage of tool life		

Figure 9. Chip morphology at (**a**) initial stage (the VMR values for test no. 1, no. 5, and no. 9 are 92 mm^3, 184 mm^3, and 276 mm^3), (**b**) middle stage (the VMR values at the middle stage for test no. 1, no. 5, and no. 9 are 46,000 mm^3, 1472 mm^3, and 828 mm^3) and (**c**) end stage (the VMR values at the end stage for test no. 1, no. 5, and no. 9 are 77,280 mm^3, 2760 mm^3, and 1656 mm^3) of tool life.

The results show that the cutting speed and the radial depth of cut have great influence on the VMR. To achieve a maximum VMR, the suggested cutting parameters are V_c = 60 m/min and a_e = 0.6 mm when a_p = 2 mm and f_z = 0.1 mm/tooth in drying milling Ti–6Al–4V.

3.3. Trochoidal Milling Test

The test setup for trochoidal milling is shown in Figure 10a. The tool has the same specifications as those used in the above tests. Figure 10b shows the trochoidal toolpath, where c denotes the trochoidal stepover. One trochoid cycle is composed of a toolpath of retraction denoted by the red dot line and a toolpath of engagement denoted by the black dot line (see Figure 10b). According to the results of the flank milling tests, a combination of a 60 m/min cutting speed and 0.6-mm radial depth of cut achieved a maximum VMR. Thus, a set of cutting parameters—i.e., V_c = 60 m/min, f_z = 0.1 mm/tooth, and c = 0.6 mm—was adopted in the trochoidal milling test, where the stepover corresponded to the radial depth of cut. Unlike the above tests, the axial depth of cut was set to 10 mm in this test to achieve a high material removal rate.

This test achieved a material removal rate of 1165.9 mm^3/min and a total volume of 206,940 mm^3 of material removed. The total cutting time, including the tool engaged and retracted time, was 177.5 min. The general view of the tool after the test is shown in Figure 11a, where lots of chip debris adhered to the rake face of the tool are observed, and the shapes of chips are diverse along the cutting edge. In addition, the TiAlN coating (dark color) on the engaged part of the tool had been worn away and the substrate material (bright color) of the tool was exposed. Figure 11b,c show enlarged views of the middle and tip position of the engaged cutting edge, respectively. In the figures, the materials removed from the workpiece were adhered to the flank face by cold welding. Moreover, small chippings, which perform as chip breakers, can be observed on the cutting edge [35].

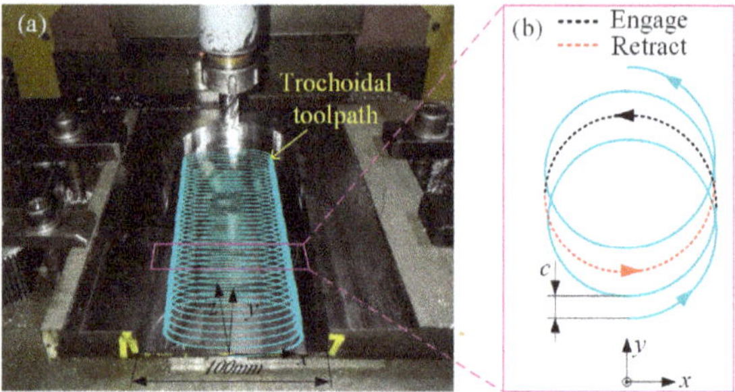

Figure 10. (**a**) Dry trochoidal milling test; (**b**) Trochoidal toolpath.

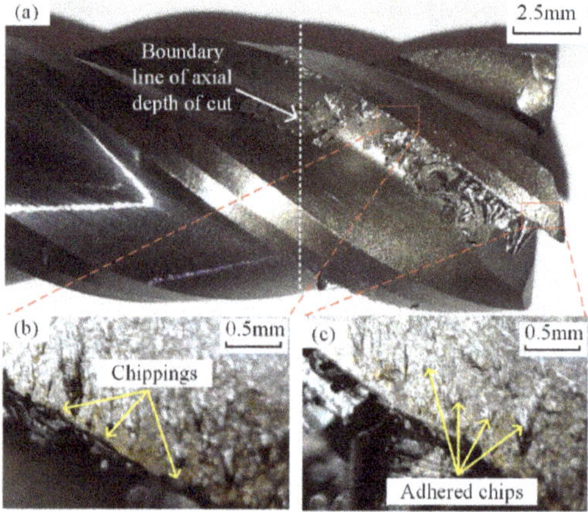

Figure 11. Tool condition after dry trochoidal milling test. (**a**) General view of the tool; (**b**) Top position; (**c**) Middle position; (**d**) Bottom positon.

The chips at different stages of tool life were collected in the trochoidal milling tests. Chip morphology images are shown in Figure 12. It is observed that each chip had an intact curled needle shape with serrated boundaries at an early stage of 10 min (see Figure 12a). At 80 min, distorted and obvious scratches could be seen on the back surface of the chip (see Figure 12b). At 177.5 min, several chips were twisted into a lump, and intact chips had broken into small pieces (see Figure 12c). It can be concluded that chip morphology is closely related to the tool wear condition and provides an important indication for tool condition monitoring.

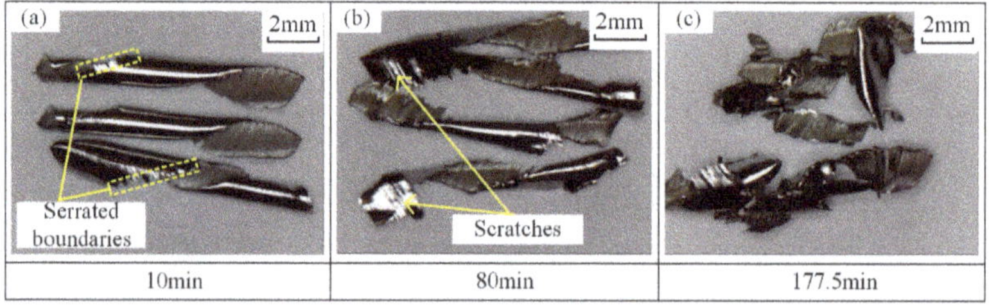

Figure 12. Chip morphology in trochoidal milling tests at cutting stage of (**a**) 10 min, (**b**) 80 min and (**c**) 177.5 min.

4. Conclusions

This paper investigated the tool wear and chip morphology in the dry milling of titanium alloy Ti–6Al–4V for the aim of green machining titanium alloy. The effect of the radial depth of cut on tool wear was studied considering cutting speed, and a corresponding tool wear model related with radial cutting depth was established. According to the test results, cutting parameters in terms of radial depth of cut were optimized to improve the machining efficiency and reduce the tool wear. Then, the optimized cutting parameters were adopted by following the trochoidal milling test. Meanwhile, chips over the tool life cycle were collected and observed. Based on the results and analysis, the following conclusions can be drawn:

(1) For each cutting speed, $V_c = 60$, 130, and 200 m/min, there is a radial depth of cut corresponding to the lowest VMR, which should be avoided in practical milling. In this research, a 0.4-mm radial depth of cut should be avoided.

(2) In order to improve cutting efficiency in material removal, an appropriate radial depth of cut is necessary. In this paper, a 0.6-mm radial depth of cut combined with a 60 m/min cutting speed are suggested when $a_p = 2$ mm and $f_z = 0.1$ mm/tooth in drying milling titanium alloy Ti–6Al–4V.

(3) Trochoidal milling is a promising method for green machining titanium alloy. By using a 0.6-mm radial depth of cut and 60 m/min cutting speed, the material removal rate is 1165.93 mm^3/min, and the tool life reaches 177.5 min in dry trochoidal milling.

(4) Chip morphology is closely associated with tool condition, and chip morphology can be adopted to monitor the tool condition.

Author Contributions: Conceptualization, M.L.; Investigation, D.L. and Y.Z.; Methodology, D.L. and M.L.; Supervision, M.L. and D.Z.; Validation, D.L. and Y.Z.; Writing—original draft, D.L. and Y.Z.; Writing—review & editing, M.L. and D.Z.

Funding: This work was supported by the National Science and Technology Major Project (Grant No. 2017ZX04013001).

References

1. Cui, C.X.; Hu, B.M.; Zhao, L.C.; Liu, S.J. Titanium alloy production technology, market prospects and industry development. *Mater. Des.* **2011**, *32*, 1684–1691. [CrossRef]
2. Yao, Q.; Luo, M.; Zhang, D.H.; Wu, B. Identification of cutting force coefficients in machining process considering cutter vibration. *Mech. Syst. Sig. Process.* **2018**, *103*, 39–59. [CrossRef]
3. Han, C.; Luo, M.; Zhang, D.H.; Wu, B. Iterative Learning Method for Drilling Depth Optimization in Peck Deep-Hole Drilling. *J. Manuf. Sci. Eng.-Trans. ASME.* **2018**, *140*, 121009. [CrossRef]
4. Komanduri, R.; Von Turkovich, B.F. New observations on the mechanism of chip formation when machining titanium alloys. *Wear* **1981**, *69*, 179–188. [CrossRef]
5. Che-Haron, C.H.; Jawaid, A. The effect of machining on surface integrity of titanium alloy Ti–6% Al–4% V. *J. Mater. Process. Technol.* **2005**, *166*, 188–192. [CrossRef]
6. Luo, M.; Yao, Q. Vibrations of Flat-End Cutter Entering Workpiece Process: Modeling, Simulations, and Experiments. *Shock Vib.* **2018**, *2018*, 8419013. [CrossRef]
7. Yuan, H.; Wan, M.; Yang, Y. Design of a tunable mass damper for mitigating vibrations in milling of cylindrical parts. *Chin. J. Aeronaut.* **2019**, *32*, 748–758. [CrossRef]
8. M'Saoubi, R.; Axinte, D.; Soo, S.L.; Nobel, C.; Attia, H.; Kappmeyer, G.; Engin, S.; Sim, W.-M. High performance cutting of advanced aerospace alloys and composite materials. *CIRP Ann-Manuf. Technol.* **2015**, *64*, 557–580. [CrossRef]
9. Sun, S.; Brandt, M.; Dargusch, M.S. Machining Ti–6Al–4V alloy with cryogenic compressed air cooling. *Int. J. Mach. Tools Manuf.* **2010**, *50*, 933–942. [CrossRef]
10. Pittalà, G.M. A study of the effect of CO_2 cryogenic coolant in end milling of Ti-6Al-4V. *Procedia CIRP.* **2018**, *77*, 445–448. [CrossRef]

11. Bermingham, M.J.; Palanisamy, S.; Kent, D.; Dargusch, M.S. A comparison of cryogenic and high pressure emulsion cooling technologies on tool life and chip morphology in Ti–6Al–4V cutting. *J. Mater. Process. Technol.* **2012**, *212*, 752–765. [CrossRef]

12. Pereira, O.; Rodríguez, A.; Fernández-Abia, A.I.; Barreiro, J.; López de Lacalle, L.N. Cryogenic and minimum quantity lubrication for an eco-efficiency turning of AISI 304. *J. Cleaner Prod.* **2016**, *139*, 440–449. [CrossRef]

13. Pereira, O.; Rodríguez, A.; Barreiro, J.; Fernández-Abia, A.I.; de Lacalle, L.N.L. Nozzle design for combined use of MQL and cryogenic gas in machining. *Int. J. Precis. Eng. Manuf.-Green Technol.* **2017**, *4*, 87–95. [CrossRef]

14. Pereira, O.; Martín-Alfonso, J.E.; Rodríguez, A.; Calleja, A.; Fernández-Valdivielso, A.; López de Lacalle, L.N. Sustainability analysis of lubricant oils for minimum quantity lubrication based on their tribo-rheological performance. *J. Cleaner Prod.* **2017**, *164*, 1419–1429. [CrossRef]

15. Pereira, O.; Rodríguez, A.; Ayesta, I.; García, J.B.; Fernández-Abia, A.I.; De Lacalle, L.N.L. A cryo lubri-coolant approach for finish milling of aeronautical hard-to-cut materials. *Int. J. Mechatron. Manuf. Syst.* **2016**, *9*, 370–384. [CrossRef]

16. Hou, J.Z.; Zhou, W.; Duan, H.J.; Yang, G.; Xu, H.W.; Zhao, N. Influence of cutting speed on cutting force, flank temperature, and tool wear in end milling of Ti-6Al-4V alloy. *Int. J. Adv. Manuf. Technol.* **2014**, *70*, 1835–1845. [CrossRef]

17. Jaffery, S.I.; Mativenga, P.T. Assessment of the machinability of Ti-6Al-4V alloy using the wear map approach. *Int. J. Adv. Manuf. Technol.* **2009**, *40*, 687–696. [CrossRef]

18. Bermingham, M.J.; Palanisamy, S.; Dargusch, M.S. Understanding the tool wear mechanism during thermally assisted machining Ti-6Al-4V. *Int. J. Mach. Tools Manuf.* **2012**, *62*, 76–87. [CrossRef]

19. Dandekar, C.R.; Shin, Y.C.; Barnes, J. Machinability improvement of titanium alloy (Ti–6Al–4V) via LAM and hybrid machining. *Int. J. Mach. Tools Manuf.* **2010**, *50*, 174–182. [CrossRef]

20. Pujana, J.; Rivero, A.; Celaya, A.; López de Lacalle, L.N. Analysis of ultrasonic-assisted drilling of Ti6Al4V. *Int. J. Mach. Tools Manuf.* **2009**, *49*, 500–508. [CrossRef]

21. Sreejith, P.S.; Ngoi, B.K.A. Dry machining: Machining of the future. *J. Mater. Process. Technol.* **2000**, *101*, 287–291. [CrossRef]

22. Suárez, A.; Veiga, F.; Lópezde Lacalle, L.N.; Polvorosa, R.; Wretland, A. An investigation of cutting forces and tool wear in turning of Haynes 282. *J. Manuf. Process.* **2019**, *37*, 529–540. [CrossRef]

23. Li, A.H.; Zhao, J.; Luo, H.B.; Pei, Z.Q.; Wang, Z.M. Progressive tool failure in high-speed dry milling of Ti-6Al-4V alloy with coated carbide tools. *Int. J. Adv. Manuf. Technol.* **2012**, *58*, 465–478. [CrossRef]

24. Deng, J.X.; Li, Y.S.; Song, W.L. Diffusion wear in dry cutting of Ti–6Al–4V with WC/Co carbide tools. *Wear* **2008**, *265*, 1776–1783.

25. Li, A.H.; Zhao, J.; Hou, G.M. Effect of cutting speed on chip formation and wear mechanisms of coated carbide tools when ultra-high-speed face milling titanium alloy Ti-6Al-4V. *Adv. Mech. Eng.* **2017**, *9*, 1–13. [CrossRef]

26. Otkur, M.; Lazoglu, I. Trochoidal milling. *Int. J. Mach. Tools Manuf.* **2007**, *47*, 1324–1332. [CrossRef]

27. Uhlmann, E.; Fürstmann, P.; Rosenau, B.; Gebhard, S.; Gerstenberger, R.; Müller, G. The Potential of Reducing the Energy Consumption for Machining TiAl6V4 by Using Innovative Metal Cutting Processes. In Proceedings of the Global Conference on Sustainable Manufacturing (GCSM), Berlin, Germany, 23–25 September 2013.

28. Wu, B.H.; Zheng, C.Y.; Luo, M.; He, X.D. Investigation of trochoidal milling nickel-based superalloy. *Mater. Sci. Forum.* **2012**, *723*, 332–336. [CrossRef]

29. Patil, P.; Polishetty, A.; Goldberg, M.; Littlefair, G.; Nomani, J. Slot machining of Ti6Al4V with trochoidal milling technique. *Mach. Eng.* **2014**, *14*, 42.

30. Luo, M.; Han, C.; Hafeez, H.M. Four-axis trochoidal toolpath planning for rough milling of aero-engine blisks. *Chin. J. Aeronaut.* **2018**. [CrossRef]

31. Rauch, M.; Duc, E.; Hascoet, J.-Y. Improving trochoidal tool paths generation and implementation using process constraints modelling. *Int. J. Mach. Tools Manuf.* **2009**, *49*, 375–383. [CrossRef]

32. Luo, M.; Luo, H.; Axinte, D.; Liu, D.S.; Mei, J.W.; Liao, Z.R. A wireless instrumented milling cutter system with embedded PVDF sensors. *Mech. Syst. Sig. Process.* **2018**, *110*, 556–568. [CrossRef]

33. Li, A.H.; Zhao, J.; Zhou, Y.H.; Chen, X.X.; Wang, D. Experimental investigation on chip morphologies in high-speed dry milling of titanium alloy Ti-6Al-4V. *Int. J. Adv. Manuf. Technol.* **2012**, *62*, 933–942. [CrossRef]

34. Álvarez, M.; Gómez, A.; Salguero, J.; Batista, M.; Huerta, M.M.; Marcos Bárcena, M. SOM-SEM-EDS Identification of Tool Wear Mechanisms in the Dry-Machining of Aerospace Titanium Alloys. *Adv. Mater. Res.* **2010**, *107*, 77–82. [CrossRef]
35. Fernández-Vidal, S.R.; Mayuet, P.; Rivero, A.; Salguero, J.; del Sol, I.; Marcos, M. Analysis of the Effects of Tool Wear on Dry Helical Milling of Ti6Al4V Alloy. *Procedia Eng.* **2015**, *132*, 593–599. [CrossRef]

 *materials*

Article

Analysis of Secondary Adhesion Wear Mechanism on Hard Machining of Titanium Aerospace Alloy

Moises Batista Ponce [1,*], Juan Manuel Vazquez-Martinez [1], Joao Paulo Davim [2] and Jorge Salguero Gomez [1,*]

[1] Mechanical & Industrial Design Deptartment, University of Cadiz, Avda. de la Universidad de Cádiz 10, E11519 Puerto Real-Cádiz, Spain; juanmanuel.vazquez@uca.es
[2] Department of Mechanical Engineering, University of Aveiro, Campus Santiago, 3810-193 Aveiro, Portugal; pdavim@ua.pt
* Correspondence: moises.batista@uca.es (M.B.P.); jorge.salguero@uca.es (J.S.G.); Tel.: +34-956483200 (M.B.P.)

Received: 27 May 2019; Accepted: 21 June 2019; Published: 23 June 2019

Abstract: Titanium alloys are widely used in important manufacturing sectors such as the aerospace industry, internal components of motor or biomechanical components, for the development of functional prostheses. The relationship between mechanical properties and weight and its excellent biocompatibility have positioned this material among the most demanded for specific applications. However, it is necessary to consider the low machinability as a disadvantage in the titanium alloys features. This fact is especially due to the low thermal conductivity, producing significant increases in the temperature of the contact area during the machining process. In this aspect, one of the main objectives of strategic industries is focused on the improvement of the efficiency and the increase of the service life of the elements involved in the machining of this alloy. With the aim to understand the most relevant effects in the machinability of the Ti6Al4V alloy, an analysis is required of different variables of the machining process like tool wear evolution, based on secondary adhesion mechanisms, and the relation between surface roughness of the work-pieces with the cutting parameters. In this research work, a study on the machinability of Ti6Al4V titanium alloy has been performed. For that purpose, in a horizontal turning process, the influence of cutting tool wear effects has been evaluated on the surface finish of the machined element. As a result, parametric behavior models for average roughness (Ra) have been determined as a function of the machining parameters used.

Keywords: titanium alloys; machining; turning; machinability; tool wear

---

1. Introduction

Traditionally, the machinability concept involves the ability of materials for the development of removal processes to obtain a manufactured part [1], or the ability to be machined by using machine tools [2,3]. However, the initial concept is incomplete mainly due to the lack of definition of the variables of influence in the machinability and the experimental tests for a correct evaluation of the machinability properties. This problem was already described by Trent [4] studying the difficulty in the understanding of materials behavior during machining and the most relevant variables in the cutting processes. In order to obtain a more uniform machinability concept, Trent described five different criteria, the tool life, the material remove rate, the cutting forces, the surface finish and the chip shape. However, a general rate is not established for each specific criterion, therefore, it is not possible to make comparisons between different conditions of machinability, on the other hand, this consideration is specially based on economic-energy considerations, and currently the industry requires new criteria to evaluate the processes. This fact favors that the initially proposed considerations can be updated and extended to environmental and functional aspects [5]. In this sense, the study of the environmental impact of the process and the machined work piece features evaluated through the evaluation of the surface integrity is shown as an important research line [6]. Therefore, a relevant growth in the interest for dry machining was found by several authors in the last years [7,8].

Currently, using different methods of monitoring the machining process, the influence of different factors in this process can be accurately known. These elements can be considered as specific machinability criteria for each material and process. Special consideration is required in the case of materials with low machinability capacity such as titanium alloys, where the selection of the control parameters should allow knowing the range of parameters where the material can be machined under proper process conditions.

Another criterion to consider in the study of tool wear is the functional performance, related to the work behavior of the involved elements. Titanium alloys are widely used in the development of structural and strategic components of the aerospace industry [9]. The titanium alloys, and especially Ti6Al4V is considered a strategic material for the global market of the aerospace industry. In recent years, an important industrial interest in the use of this type of alloy has favored a significant increase in global consumption [9,10]. Due to the tolerance requirements of the aerospace industry, and the large volume of developed components with this type of materials, machining processes require accurate assembly conditions favoring the achievement of optimal dimensional and geometrical characteristics and excellent biocompatibility [11] for their use in work conditions. As a result, the machinability properties may be considered of strategic interest for the development of the manufacturing processes in the aerospace industry, especially due to the high influence of the process conditions on the surface integrity of the machined work-pieces [12,13].

Additionally, in order to reduce finishing procedures, maintaining micro-geometrical requirements of the surface, the study of the range of the parameters where the behavior of the tool reduces the development of a specific roughness is shown as a relevant aspect in the improvement of the manufacturing process performance [5].

However, when a titanium alloy is machined, the tool wear may imply an important problem and different ways are proposed to solve it. Lack of thermal conductivity of the substrate has been observed for titanium alloys, in many cases interfering with the machining process development [14–17] and being replaced in some cases by coatings such as PCD to improve the tool life [14]. With the aim to improve the economic performance and the lack of uniformity of a large volume of coated tools types, a generic uncoated tool was used to carry out the machining tests. In this way, the identification and characterization of the wear mechanisms that appears during the machining process is described [18–24].

The tool wear is related to the cutting speed. For titanium alloy, a cutting speed lower than 60 m/min is commonly used. Above of this cutting speed, in titanium, may be considered as high speed machining. In addition, cutting speeds higher than 150 m/min involves a quick high wear of the tool. This condition is considered ultra-high speed machining for titanium alloy, studied by Rashid et al. [18] and corroborate for Liang et al. [19] and Balaji et al. [25] for orthogonal cutting process. In an industrial process this consideration may be critical to improve the performance.

In this research, a study of the machinability of titanium alloy Ti6Al4V near the industrial environment is proposed based on the relationship between tool wear effects and micro-geometrical characteristics of machined parts in low, high and ultrahigh speed machining. With the aim of simplifying the study by reducing the control variables involved in the process and reduce the environmental impact, machining tests have been carried out by horizontal turning process in dry conditions.

2. Materials and Methods

Cylindrical bars of 150 mm length of UNS R56400 titanium alloy (Ti6Al4V) with an initial composition (wt%) shown in Table 1, were used as work-pieces to perform horizontal turning tests by means of a Kingsbury MHP 50 CNC Lathe Center (Kingsbury, Hampshire, UK). Cutting tools used were WC-Co uncoated inserts described by ISO ref DCMT 11T308, Figure 1.

Table 1. Composition (wt%) of Ti6Al4V titanium alloy.

Al	V	Fe	C	O	N	H	Ti
6.29	4.04	0.16	0.008	>0.05	>0.05	>0.05	Rest.

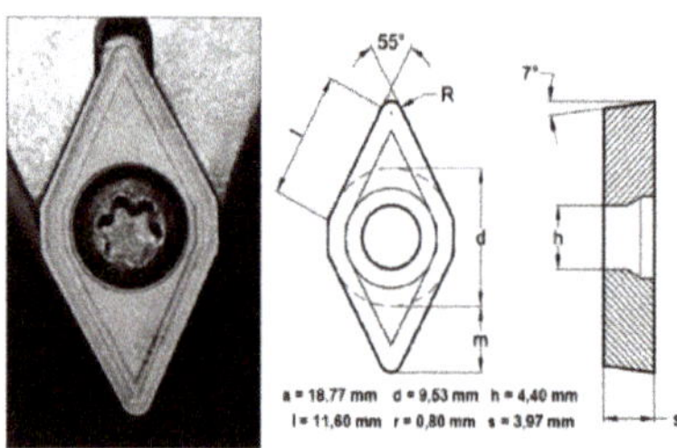

Figure 1. Cutting tool features.

Three different types of dry turning tests have been carried out by varying machining conditions. A first set of the test has been performed to evaluate the wear effects on the tool surface and to analyze the appearance and evolution of adhesion wear mechanism, taking into account the analysis of the oxidation phenomena that takes place near the contact area, Table 2.

Table 2. Machining conditions for the evaluation of adhesion wear evolution.

Cutting Speed (m/min)			Feed Rate (mm/rev)		Cutting Depth (mm)	
25	50	100	0.05	0.1	1	2

Additional tests were performed taking as reference the results of the main influence parameters with the aim to obtain a deeper understanding of the effects of the feed rate increase on the maximum speed studied (100 m/min). For this study, a cutting depth of 1 mm was kept and two higher different values of feed rate were selected (0.2 and 0.4 mm/rev).

Finally, taking the starting point the evaluations of the main influence parameters of the adhesive wear, the study of maximum tool life time for different machining process, including hard conditions were performed, from commonly used cutting speed ranges to ultrahigh speed. For this analysis, the machining conditions of Table 3 were used.

Table 3. Machining parameters for the evaluation of tool life time under standard and hard conditions.

Cutting Speed (m/min)				Feed Rate (mm/rev)	Cutting Depth (mm)
50	75	100	200	0.2	1

For the characterization of wear effects on the cutting tool an analysis of the main wear mechanisms that takes place in the process was carried out. Based on this consideration, the formation of adhered layers on the tool surface have been studied by means of scanning electron microscopy (SEM) (FEI Quanta 200, ThermoFisher Scientific, Hillsboro, OR, USA) and stereoscopic optical microscopy (SOM) techniques (SMZ-800, Nikon, Tokyo, Japan), taking in to account the oxidation processes on these layers. Oxidation phenomena mainly caused by high temperature of the contact area during machining stage have been identified using energy density spectroscopy (EDS) (Phoenix, EDAX, Mahwah, NJ, USA).

The use of a wide range of the machining parameters results in some different performances of the cutting effects. In this context, chip development can also be affected, giving rise to variations in the shape, length and thickness values.

Finally, as a consequence of geometrical variations of the cutting tool, mainly due to wear effects, and a wide range of chip growth behavior, the surface finish over machined work-pieces may be affected.

Due to this fact, the evaluation of roughness of the parts surface was evaluated by a Hommel-Etamic T1000 Perthometer roughness measurement system (Jenoptik Group, Thuringia, Germany). It was analyzed in three different machining sections, initial, medium and final, and in each zone, a total of four roughness measurement was carried out in four different generatrices.

In Table 4, a brief description of the experimental methodology is shown.

Table 4. Experimental methodology followed in this research.

Study	Method	Objective
Tool wear mechanism characterization	Single test SEM/EDS analysis Cross-section of the tool evaluation	Characterization of tool wear mechanism for Ti6Al4V
Wear evolution by secondary adhesion mechanisms	Parametric test: f(S,f,d) SOM analysis	Evolution of the tool wear as a function of cutting parameters
Influence of secondary adhesion wear on the service tool life of the cutting tool	Tool life test: f(S) SOM analysis Tool wear measurement	Evolution of the tool wear regarding cutting time and establishment the tool life.
Tool wear effects on the surface finish of machined parts	Parametric test: f(S,f,d) Surface roughness analysis	Relation between cutting parameters and roughness as a function of tool wear.

3. Results and Discussion

The analysis of wear effects on the cutting tools and machined parts caused by hard machining implies the study, identification and description of the most relevant mechanisms that takes place in the process. In this case, with adhesion being the main wear mechanism, the study of this mechanism needs a greater depth evaluation of the oxidation effects on the adhered layers. Additionally, the variation of the initial geometry and mechanical properties of the tool surface in the contact area during the machining stage may be the cause of the development of a wide variety of chip sections, with different features as length, width or thickness. Finally, the combination of tool modification by wear and the friction with different types of chip may affect the surface finish, in terms of roughness, of the machined parts, being relevant for specific applications.

3.1. Tool Wear Mechanism Characterization

Wear by adhesive phenomena are shown as the main cause of modification in the initial geometry of the cutting area and shortening of the service life of the tools. This fact makes necessary a deeper analysis of the adhered layer development and their implications in the modification of the initial properties of the surface. Adhered material on the flank and rake face of the tool during the machining stage shows a direct influence of the aggressiveness of the cutting conditions, based on the feed rate, cutting speed and depth parameters [26].

The evolution of the adhesion phenomenon starts with the growth of thin extruded layers of adhered material from the titanium alloy over the rake face-giving rise to primary and secondary built-up layers (BUL) and similar trends may be detected on the rake face, Figure 2. The thermo-mechanical effect is the main responsible of the adhesion phenomena on the tool surface, promoting the wear mechanism.

High temperature reached during the dry machining process near the contact area between the tool and work-piece, and the non-protective atmosphere where the test was performed, caused the oxidation of the extruded Ti6Al4V, Figure 3.

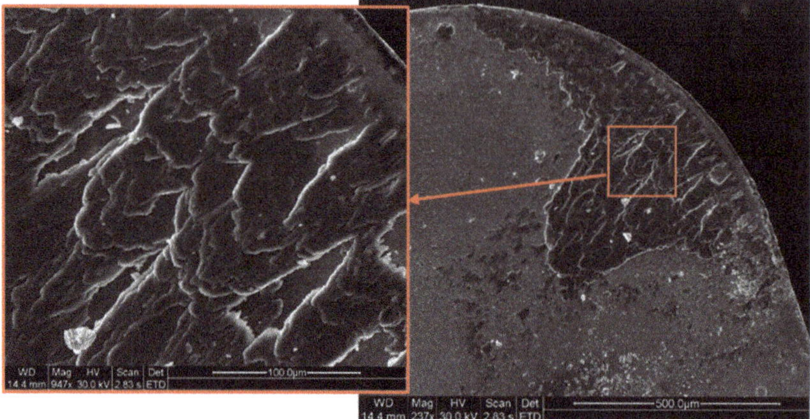

Figure 2. Built-up stratified layers on the rake face.

This fact results in significant changes on the contact behavior in the chip-tool interface, involving the development of a secondary BUL with a decrease of the oxidation effects from the tool surface to outer layers. Under this consideration, an influence may be confirmed of the titanium oxidized layers on the multilayer formation of the BUL and the built-up edge (BUE), Figure 4a.

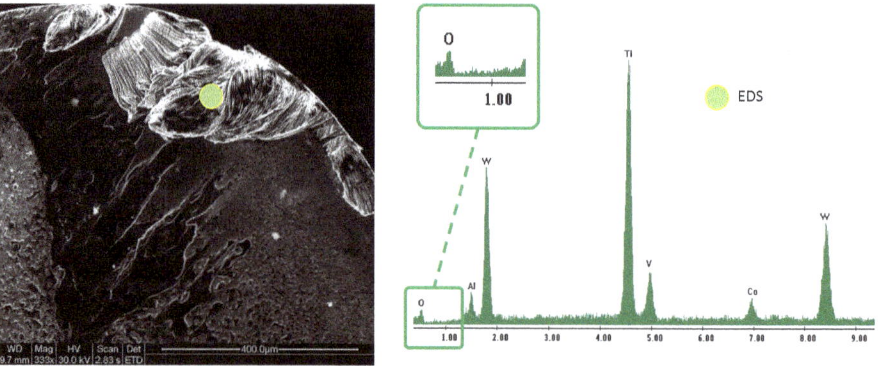

Figure 3. Oxidation phenomena on adhered material over rake face.

The development of BUE implies the increase of adhered multilayer secondary BUL to rake face being controlled by machining conditions. The increase of the cutting speed parameter results in the extrusion of the adhered material over farthest areas from the edge of the tool [27]. On the cross-section of the cutting tool two main phenomena can be detected. On the one hand, stratified layers of BUL extruded over the rake face allows to identify the flow lines of the material. On the other hand, the sliding process of the chip over the rake face implies friction phenomena and the corresponding temperature rise that promotes the adhesive wear mechanism. This fact is also associated with the development of wear effects on the clearance, as can be observed in Figure 4b, and according to [26].

Morphological changes mainly due to adhesion wear may cause a significant variation on the rake angle, giving rise to an increase of friction in the interface and conditioning the service life time of the tools. These geometrical variations also show critical implications on the process development, being relevant the damage consequences over the surface finish on the machined work-pieces and the shape and length of the chip.

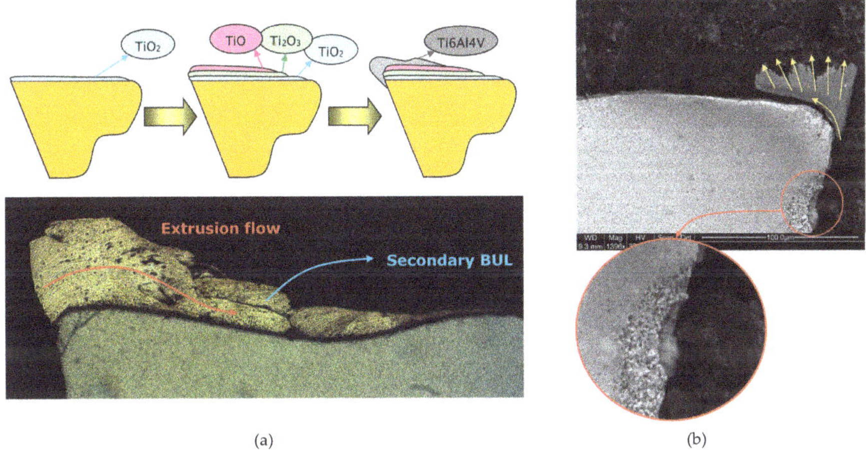

(a)   (b)

Figure 4. (**a**) Multilayer built-up layer (BUL) development process, (**b**) wear effects on clearance.

3.2. Wear Evolution by Secondary Adhesion Mechanisms

The evaluation of wear effects on the uncoated carbide tools shows two different and representative regimes in the wear phenomena behavior based on the cutting speed parameter. For lower studied speeds (25–50 m/min) non-relevant changes can be detected for different values of the feed in mm/rev. Additionally, the increase on the cutting depth from 1 to 2 mm have not caused significant damages over the tool surface. However, the use of higher speed (100 m/min) involves variations in the wear behavior, giving rise to increase in the aggressiveness of the machining and highlighting the influence of feed in the cutting process, Figure 5.

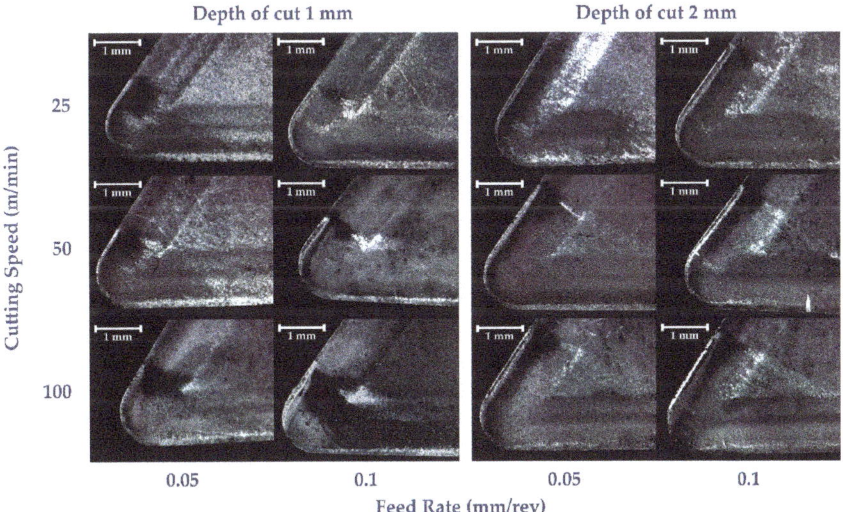

Figure 5. Wear evolution over the rake face.

Under these considerations, abrasion wear marks are manifested by a dark area on the surface of the tool in Figure 5. This wear effect can be related to the friction between the chip and the rake face of the tool. Due to this abrasion mechanism, a drastic increase of the temperature in the surface is caused.

As a consequence of high temperature and friction in the contact area, burn cases may occur giving rise to the darkening process of this area of the tool. Moreover, this increase in temperature can lead the diffusion of Ti in the tool matrix [24,28].

Due to the described abrasion mechanism, an important growth of the surface temperature occurs, being the cause in some cases of a burning process of the chip and darkening this area of the tool, according to [17]. This increase in temperature can lead Ti diffusion processes in the tool substrate [24,28]. As can be seen, this effect seems to increase as a function of the cutting speed and decrease as a function of feed parameter.

This described behavior can also be observed over the flank face where the use of 100 m/min speed results in the development of color tone modification over the tool surface increases in the feed values, Figure 6. This fact confirms a more important influence of the feed rate for higher cutting speed according to [21,29]. In addition, color tone variation is related to the appearance of titanium oxide layers on the surface of the tool [30], induced by the high temperatures reached in the machining. Color tone variation is related to the secondary adhesion wear mechanism [14,21] and is manifested by the appearance of an oxide multi-layer on the tool that favors adhesion of the alloy material on the rake face and tool edge, developing the BUL-BUE morphologies [14,21,26]. The development of oxidation layers favors the modification of the surface properties of the cutting tool in the contact area.

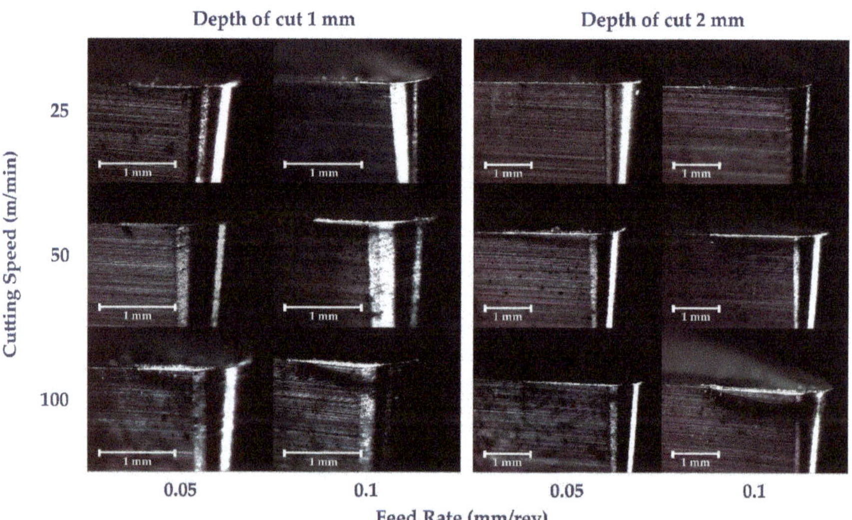

Figure 6. Wear evolution over the flank face.

As described above, feed rate variations carried out on the higher studied cutting speed (100 m/min) show an important influence in the wear evolution of the uncoated tool. In this sense, with the aim to obtain a better understanding of this situation, a deeper study was performed on this behavior from the most aggressive machining conditions setting initially. This study has evaluated the wear effects in 1 mm depth of cutting and a range of feed rate values of 0.2 and 0.4 mm/rev by using a cutting speed of 100 m/min. Under these conditions, a severe growth of the wear effects have been detected, and adhesion mechanisms are shown as the most adverse phenomena against the tool life time, Figure 7a.

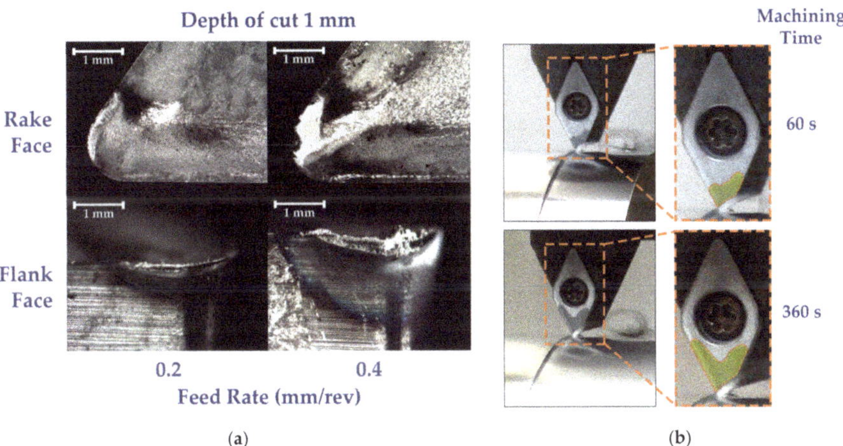

Figure 7. (**a**) Wear effects on flank and rake face of the turning tool for different feed rate using 100 m/min cutting speed; (**b**) oxidative layer growth over tool surface during machining process.

The oxidative phenomena that takes place under these machining conditions results in a quick darkening of surface color tone of the uncoated tool. The thermal oxidative layer growth along the tool surface as a function of the machining time, mainly due to high temperatures reached in air atmosphere (without protective environment), can be observed in Figure 7b.

3.3. Influence of Secondary Adhesion Wear on the Service Tool Life of the Cutting Tool

Adhesive wear on the cutting tool surfaces may be the main cause to reduce the life time under working conditions. The use of aggressive machining conditions as well as the lack of lubricating or cooling fluids accelerates the wear effects on the tool, and results in the most severe conditions that imply critical failures. With the aim to understand the influence of wear on the tool life time, turning tests were conducted using cutting speed as reference parameter, due to their influence in adhesive wear mechanisms.

Flank wear (VB) was determined on a range of cutting speed from 50 to 100 m/min reaching a maximum duration of the machining process previous to the failure of the tool. As can be seen in Figure 6, the test for 50 m/min cutting speed did not show significant wear damages, being interrupted after 30 min. Additionally, the ultra-high speed test (200 m/min) was carried out to evaluate the system behavior on maximum aggressive conditions and ensure a direct influence of the cutting speed in the life time of the cutting tool. As can be expected, the life time of the tool, based on the VB analysis, is strongly affected by cutting speed, reaching a minimum time of 29 s from the start of the machining process to the tool failure for a cutting speed of 200 m/min, Figure 8 according to [31].

The VB evaluation on different cutting speed have allowed to estimate the coefficients for the life time of the tool as a function of the speed through the Taylor tool life equation:

$$T = C \cdot v^k \tag{1}$$

where T is the time life of the tool in minutes and v is the cutting speed (m/min).

By means of the Taylor equation on the experimental data, a new expression was obtained where the k coefficient allows corroborating the results with other authors' researches [10,18,25,28] on carbide tools and according to the ISO 3685 standard.

$$T = 58.12 \cdot 10^6 \cdot v^{-3.58} \tag{2}$$

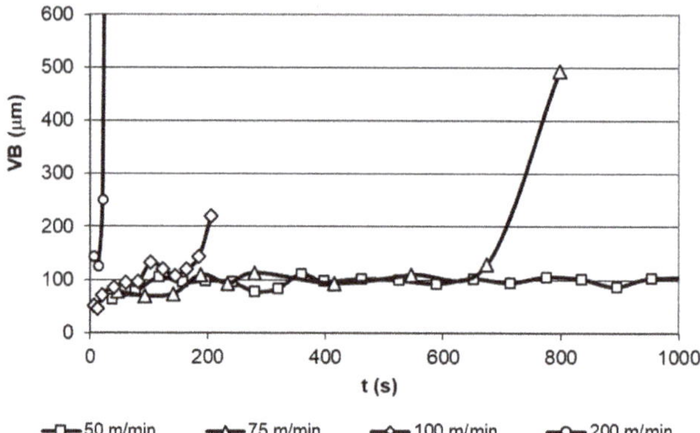

Figure 8. Life time of the tool as a function of the cutting speed.

As described above, the tool life end in previous experimental tests results in a critical failure that could be the cause of significant damages on the different elements involved in the machining process. This critical failure implies a drastic variation of the cutting edge geometry, giving rise to the modification of the surface finish, in terms of roughness, of the work-piece. Additionally, this edge variation in combination with adhesive wear effects, causes relevant changes in the rake angle, changing the initial thickness, shape and length of the chip.

On the final stage of the machining, near critical failure, some specific phenomena were detected on the development of the tests, corresponding to visual and profilometry analysis of the machined surface, Figure 9.

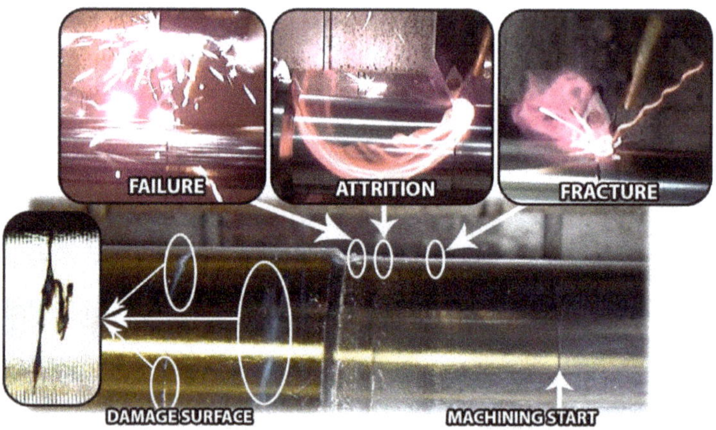

Figure 9. Tool wear evolution in the last stage of the life time.

After the stationary stage of the machining process, the critical failure phenomenon starts with the increase of the wear damage over the clearance, promoting a growth in the contact interface between the cutting tool and work-piece. This effect is generally associated with the fracture/attrition of small fragments of the tool and an important rise of the friction between contact elements. In most cases, an increase of friction implies an increase of the temperature, reaching burning situations of the chip. These aggressive conditions involve severe adhesion effects and the fracture of the tool through the detachment of the adhered material and removing particles from the carbide surface.

3.4. Tool Wear Effects on the Surface Finish of Machined Parts

In order to evaluate the impact of the tool wear effects on the surface finish of the machined work-pieces, roughness measurements have been performed over the turned sections. Figure 10 shows the evolution of the roughness as a function of the feed for the three cutting speed studied compared with the theoretical value obtained by the classical expression exposed in Equation (3). For the measurement of surface finish, roughness average (Ra) has been selected as a control parameter to quantify the variations caused by the wear effects.

$$Ra_t = \frac{a^2}{32 \cdot R} \cdot 1000 \tag{3}$$

An important dependence was detected on the surface finish, in terms of Ra, with respect to the feed. This fact is mainly due to the formation of specific microgeometry relative to the displacement of the tool on the machined surface. Furthermore, lower dependence of Ra on the cutting speed is observed, which increases for higher values of feed. This dependence decreases for 2 mm depth of the cut.

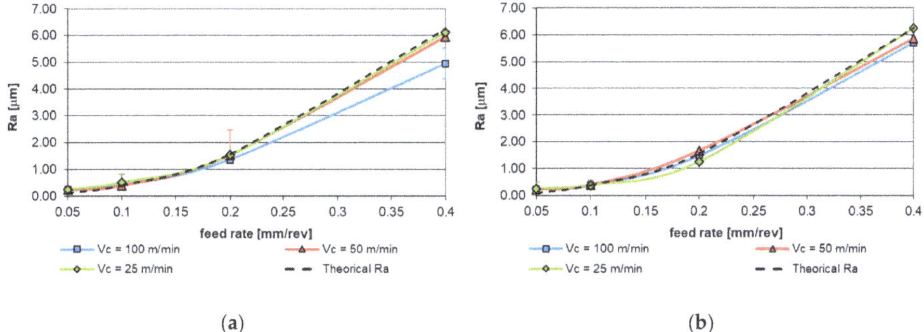

(a)   (b)

Figure 10. Evolution of the roughness average (Ra): (**a**) Cutting depth 1 mm; (**b**) cutting depth 2 mm.

In order to clarify this effect, parametric models have been established for each value of the cutting depth used. Equation (4) corresponds to the depth of the cut of 1 mm and Equation (5) is related to 2 mm of depth.

$$Ra_{p1} = 30.649 \cdot f^{1.554} \cdot v^{-0.109} \tag{4}$$

$$Ra_{p2} = 19.668 \cdot f^{1.566} \cdot v^{0.011} \tag{5}$$

As can be observed in previous equations, the exponent of the feed (f) is close to a constant value, ensuring a similar behavior for 1 mm and 2 mm depth, however, in the case of the cutting speed (v) relevant variations in the exponent have been detected between values of depth.

When the depth of the cut is included in the surface finish behavior expression, the feed exponent maintaining similar values as previous equations show the consistency of the proposed model.

$$Ra_p = 23.662 \cdot f^{1.561} \cdot v^{-0.041} \cdot p^{0.012} \tag{6}$$

However, the tool wear may affect the surface quality of the machined parts. In this way, it is known that the wear behavior is not linear, showing a steadily increase until reaching a rapid growth point. For this reason, the proposed models described in Equations (4) to (6) have been determined for an intermediate machined section where the wear behavior could be considered in a stationary stage. Furthermore, previous studies have determined an effect that reduces the surface quality of the machined parts in a stable wear section [32,33], as in the aluminum case [34–37]. This effect is related to the accumulation of adhered material into the tool, which causes a reduction in the values of the roughness, according to [36,37].

4. Conclusions

A direct relationship has been detected, in constant machining length conditions, between the tool wear and cutting parameters. This behavior is mainly related to thermo-mechanical effects during machining process.

The feed is shown as a control parameter for the adhesion mechanisms and cutting speed for the abrasion wear phenomena. By increasing these parameters, the wear mechanism effects also increase, maintaining this trend for all the studied depth of the cut.

Regarding the surface finish of the machined parts, the roughness, in terms of average roughness parameter, of the work-pieces show a direct dependence with the feed parameter. This fact is mainly due to the micro-geometrical effects of the cutting tool displacement, therefore, the experimental Ra obtained is relatively closed to the theoretical Ra based on the geometric model.

An increase in the cutting speed results in variations between theoretical and experimental values of the roughness of the machined parts. The lack of concordance between theoretical and experimental models can be associated with the appearance of the tool wear in the flank face for high values of cutting speed.

Through increases in 50 m/min of cutting speed, a significant reduction in the life time of the tool have been detected. This effect may be produced especially by a rapid growth of the flank wear of the tool. This behavior has as a result that, in the case of higher cutting speeds, the tool life is greatly reduced and critical failures can be caused.

A near exponential behavior has been detected between the increase of the cutting speed and the toll life, showing an approximately value of 75%–90% reduction for the tool life under high speed conditions (>50 m/min).

Author Contributions: M.B.P. and J.P.D. conceived and designed the experiments; M.B.P. performed the experiments; M.B.P. and J.S.G. analyzed the data; M.B.P. and J.M.V.-M. wrote the paper.

Funding: This work has received financial support by the Spanish Government with the Ministry of Science and Innovation (Project DPI2008-06771-C04-01), the European Union (FEDER/FSE) and the Andalusian Government (PAIDI).

Acknowledgments: Authors want also to thank the experimental support of the Dimensional Metrology Lab of the Industrial Metrology Center of the University of Cadiz and the Machining & Tribology Research Group of the Aveiro University. A special acknowledge to Mariano Marcos Bárcena, a great scientific and engineer, and personally our father in research works. In memoriam.

References

1. Black, J.T. Introduction to Machining Processes. In *Metals Handbook*, 9th ed.; Pitler, R.K., Zwilsky, K.M., Wood, W.G., Halverstadt, R.D., Eds.; ASM International: Geauga County, OH, USA, 1989; ISBN 0-87170-007-7.
2. Arriola, I.; Whitenton, E.; Heigel, J.; Arrazola, P.J. Relationship between machinability index and in-process parameters during orthogonal cutting of steels. *CIRP Ann. Manuf. Technol.* **2011**, *60*, 93–96. [CrossRef]
3. Rajshekhar, L.; Krishna, P.; Mohankumar, G.C. An Experimental Investigation on Machinability Studies of Steels by Face Turning. *Procedia Mater. Sci.* **2014**, *6*, 1386–1395. [CrossRef]
4. Wright, P.K.; Trent, E. *Metals Cutting*; Butterworth-Heinemann: Oxford, UK, 2000; ISBN 9780750670692.
5. Salguero, J.; Batista, M.; Sanchez-Carrilero, M.; Alvarez, M.; Marcos, M. Sustainable manufacturing in aerospace industry. Analysis of the viability of intermediate stages elimination in sheet processing. *Adv. Mater. Res.* **2010**, *107*, 9–14. [CrossRef]
6. Davim, J.P. *Surface Integrity in Machining*; Springer: London, UK, 2010; ISBN 978-1-84882-874-2.
7. Shokrani, A.; Dhokia, V.; Newman, S.T. Environmentally conscious machining of difficult-to-machine materials with regard to cutting fluids. *Int. J. Mach. Tools Manuf.* **2012**, *57*, 83–101. [CrossRef]

8. Goindi, G.; Sarkar, P. Dry machining: A step towards sustainable machining-challenges and future directions. *J. Clean. Prod.* **2017**, *165*, 1557–1571. [CrossRef]

9. Batista, M.; Gerez, J.M.; Gomez-Parra, A.; Salguero, J.; Marcos, M. Micro and Macrogeometrical based Study of the influence of the Tool Coating in the Dry Turning of UNS R56400 Ti alloy. In Proceedings of the Coatings Science International, Noordwijk, The Netherlands, 25–29 June 2011.

10. Salguero, J.; Gerez, J.; Batista, M.; Gómez, A.; Mayuet, P.F.; Marcos Bárcena, M. Roughness based Analysis of the Influence of Tool Coating in the Dry Turning of UNS R56400 Ti Alloy. *Appl. Mech. Mater.* **2012**, *152*, 647–652. [CrossRef]

11. Kyzioł, K.; Kaczmarek, Ł.; Brzezinka, G.; Kyzioł, A. Structure, characterization and cytotoxicity study on plasma surface modified Ti-6Al-4V and γ-TiAl alloys. *Chem. Eng. J.* **2014**, *240*, 516–526. [CrossRef]

12. Moussaoui, K.; Mousseigne, M.; Senatore, J.; Chieragatti, R.; Lamesle, P. Influence of Milling on the Fatigue Lifetime of a Ti6Al4V Titanium Alloy. *Metals* **2015**, *5*, 1148–1162. [CrossRef]

13. Beranoagirre, A.; Lopez de Lacalle, L.N. Optimising the milling of titanium aluminide alloys. *Int. J. Mechatron. Manuf. Syst.* **2010**, *3*, 425–436. [CrossRef]

14. Gerez, J.M.; Sanchez-Carrilero, M.; Salguero, J.; Batista, M.; Marcos, M. A SEM and EDS based Study of the Microstructural Modifications of Turning Inserts in the Dry Machining of Ti6A14V Alloy. *AIP Conf. Proc.* **2009**, *1181*, 567–574. [CrossRef]

15. Hatt, O.; Crawforth, P.; Jackson, M. On the mechanism of tool crater wear during titanium alloy machining. *Wear* **2017**, *374*, 15–20. [CrossRef]

16. Ramirez, C.; Ismail, A.I.; Gendarme, C.; Dehmas, M.; Aeby-Gautier, E.; Poulachon, G.; Rossi, F. Understanding the diffusion wear mechanisms of WC-10%Co carbide tools during dry machining of titanium alloys. *Wear* **2017**, *390*, 61–70. [CrossRef]

17. Xie, J.; Luo, M.J.; Wu, K.K.; Yang, L.F.; Li, D.H. Experimental study on cutting temperature and cutting force in dry turning of titanium alloy using a non-coated micro-grooved tool. *Int. J. Mach. Tools Manuf.* **2013**, *74*, 25–36. [CrossRef]

18. Rashid, R.R.; Palanisamy, S.; Sun, S.; Dargusch, M.S. Tool wear mechanisms involved in crater formation on uncoated carbide tool when machining Ti6Al4V alloy. *Int. J. Adv. Manuf. Technol.* **2016**, *83*, 1457–1465. [CrossRef]

19. Liang, X.; Liu, Z. Tool wear behaviors and corresponding machined surface topography during high-speed machining of Ti-6Al-4V with fine grain tools. *Tribol. Int.* **2018**, *121*, 321–332. [CrossRef]

20. Álvarez, M.; Gómez, A.; Salguero, J.; Batista, M.; Huerta, M.; Marcos, M. SOM-SEM-EDS Identification of Tool Wear Mechanisms in the Dry-Machining of Aerospace Titanium Alloys. *Adv. Mater. Res.* **2010**, *107*, 77–82. [CrossRef]

21. Nouari, M.; Makich, H. On the Physics of Machining Titanium Alloys: Interactions between Cutting Parameters, Microstructure and Tool Wear. *Metals* **2014**, *4*, 335–358. [CrossRef]

22. Armendia, M.; Garay, A.; Iriarte, L.M.; Arrazola, P.J. Comparison of the machinabilities of Ti6Al4V and TIMETAL®54M using uncoated WC–Co tools. *J. Mater. Process. Technol.* **2010**, *210*, 197–203. [CrossRef]

23. Arrazola, P.J.; Garay, A.; Iriarte, L.M.; Armendia, M.; Marya, S.; Le Maître, F. Machinability of titanium alloys (Ti6Al4V and Ti555.3). *J. Mater. Process. Technol.* **2009**, *209*, 2223–2230. [CrossRef]

24. Bermingham, M.J.; Palanisamy, S.; Dargusch, M.S. Understanding the tool wear mechanism during thermally assisted machining Ti-6Al-4V. *Int. J. Mach. Tools Manuf.* **2012**, *62*, 76–87. [CrossRef]

25. Balaji, J.H.; Krishnaraj, V.; Yogeswaraj, S. Investigation on High speed turning of titanium alloys. *Procedia Eng.* **2013**, *64*, 926–935. [CrossRef]

26. Niknam, S.A.; Kouam, J.; Songmeme, V.; Balazinski, M. Dry and Semi-Dry turning of titanium metal matrix composites (Ti-MMCs). *Procedia CIRP* **2018**, *77*, 62–65. [CrossRef]

27. Olander, P.; Heinrichs, J. Initiation and propagation of tool wear in turning of titanium alloys – Evaluated in successive sliding wear test. *Wear* **2019**, *426*, 1658–1666. [CrossRef]

28. Perez del Pino, A.; Fernandez-Pradas, J.M.; Serra, P.; Morenza, J.L. Coloring of titanium through laser oxidation: Comparative study with anodizing. *Surf. Coat. Technol.* **2004**, *187*, 106–112. [CrossRef]

29. Boujelbene, M. Investigation and modelling of the tangential cutting force of the Titanium alloy Ti-6Al-4V in the orthogonal turning process. *Procedia Manuf.* **2018**, *20*, 571–577. [CrossRef]

30. Sánchez-Sola, J.M. Parametric Analysis of Machining Aluminum Alloys. Relation to the Topography of the Machined Samples. Ph.D. Thesis, National Distance Education University (UNED), Madrid, Spain, July 2004.

31. Caggiano, A. Tool Wear Prediction in Ti-6Al-4V Machining through Multiple Sensor Monitoring and PCA Features Pattern Recognition. *Sensors* **2018**, *18*, 823. [CrossRef] [PubMed]

32. Rubio, E.M.; Bericua, A.; de Agustina, B.; Marín, M.M. Analysis of the Surface roughness of titanium pieces obtained by turning using different cooling systems. *Procedia CIRP* **2019**, *79*, 79–84. [CrossRef]

33. Ramesh, S.; Karunamoorthy, L.; Palanikumar, K. Measurement and analysis of surface roughness in turning of aerospace titanium alloy (gr5). *Measurement* **2012**, *45*, 1266–1276. [CrossRef]

34. Rubio, E.M.; Camacho, A.M.; Sánchez-Sola, J.M.; Marcos, M. Surface roughness of AA7050 alloy turned bars Analysis of the influence of the length of machining. *J. Mater. Process. Technol.* **2005**, *162*, 682–689. [CrossRef]

35. Gómez-Parra, A.; Álvarez-Alcón, M.; Salguero, J.; Batista, M.; Marcos, M. Analysis of the evolution of the Built-Up Edge and Built-Up Layer formation mechanisms in the dry turning of aeronautical aluminium alloys. *Wear* **2013**, *302*, 1209–1218. [CrossRef]

36. Pérez, J.; Llorente, J.I.; Sanchez, J.A. Advanced cutting conditions for the milling of aeronautical alloys. *J. Mater. Process. Technol.* **2000**, *100*, 1–11. [CrossRef]

37. Puga, H.; Grilo, J.; Carneiro, V.H. Ultrasonic Assisted Turning of Al alloys: Influence of Material Processing to Improve Surface Roughness. *Surfaces* **2019**, *2*, 326–335. [CrossRef]

Article

A Study on the Laser-Assisted Machining of Carbon Fiber Reinforced Silicon Carbide

Khulan Erdenechimeg [1], Ho-In Jeong [1] and Choon-Man Lee [2,*]

[1]   Mechanical Design and Manufacturing, School of Mechatronics Engineering, Changwon National University, 20, Changwondaehak-ro, Uichang-gu, Changwon-si, Gyeongsangnam-do 51140, Korea
[2]   Departement of Mechanical Engineering, College of Mechatronics, Changwon National University, 20, Changwondaehak-ro, Uichang-gu, Changwon-si, Gyeongsangnam-do 51140, Korea
*   Correspondence: cmlee@changwon.ac.kr; Tel.: +82-55-213-3622

Received: 31 May 2019; Accepted: 26 June 2019; Published: 27 June 2019

Abstract: In recent years, as replacements for traditional manufacturing materials, monolithic ceramics and carbon fiber reinforced silicon carbide (C/SiC) ceramic matrix composites have seen significantly increased usage due to their superior characteristics of relatively low density, lightweight, and good high temperature mechanical properties. Demand for difficult-to-cut materials is increasing in a variety of area such as the automotive and aerospace industries, but these materials are inherently difficult to process because of their high hardness and brittleness. When difficult-to-cut materials are processed by conventional machining, tool life and quality are reduced due to the high cutting force and temperatures. Laser-assisted machining (LAM) is a method of cutting a workpiece by preheating with a laser heat source and lowering the strength of the material. LAM has been studied by many researchers, but studies on LAM of carbon–ceramic composites have been carried out by only a few researchers. This paper focuses on deducing the optimal machining parameters in the LAM of C/SiC. In this study, the Taguchi method is used to obtain the major parameters for the analysis of cutting force and surface roughness under various machining conditions. Before machining experiments, finite element analysis is performed to determine the effective depth of the cut. The cutting parameters for the LAM operation are the depth of cut, preheating temperature, feed rate, and spindle speed. The signal to noise (S/N) ratio and variance analysis (ANOVA) of the cutting force and surface roughness are analyzed, and the response optimization method is used to suggest the optimal machining parameters.

Keywords: laser-assisted machining; Taguchi method; optimal machining conditions; machiningcharacteristic

---

1. Introduction

In recent years, due to their superior properties, carbon matrix ceramic composites have been increasingly utilized instead of monolithic ceramics and traditional manufacturing materials. Carbon matrix ceramic composites have low density, are lightweight, and have good high temperature strength. The demand for ceramic composite materials is growing in many fields such as, the aerospace and automobile industry, because of their high mechanical properties. However, the carbon matrix ceramic composites have high hardness, brittleness, are inhomogeneous and anisotropic in nature because its structure is composed of a brittle matrix and reinforcing fibers. The impact behaviors of the fibers, such as pullout and delamination, cause low surface quality after machining. The particles such as SiC and Al2O3 cause tool wear during machining [1–5]. Therefore, many researchers have studied the advanced machining technologies to machine the composite material effectively. The thermally assisted machining (TAM) is a method which is an effective process of machining difficult-to-cut materials [6]. TAM uses heat sources such as laser, induction, and plasma to locally preheat the workpiece and soften it. In particular, as laser technology has improved, many researchers have studied laser-assisted machining (LAM), a type of TAM [7,8]. The LAM is an eco-friendly machining method that uses the high-density laser beam to soften workpieces and removes the material with cutting tools along the machining path. In LAM, the thermal conductivity and specific heat of ceramic matrix composite is increased as the temperature is increased. When the C/SiC is preheated above 400 °C with a laser, the composite is oxidized. The material behavior is changed from brittle to ductile in the range of glass transition temperature (1050–1250 °C). When the oxidation rate is high, the coefficient of friction is decreased. When the coefficient of friction is decreased and material behavior is changed from brittle to ductile, material removal rate (MRR) and machinability are improved [9–13]. Figure 1 shows a schematic diagram of the LAM process.

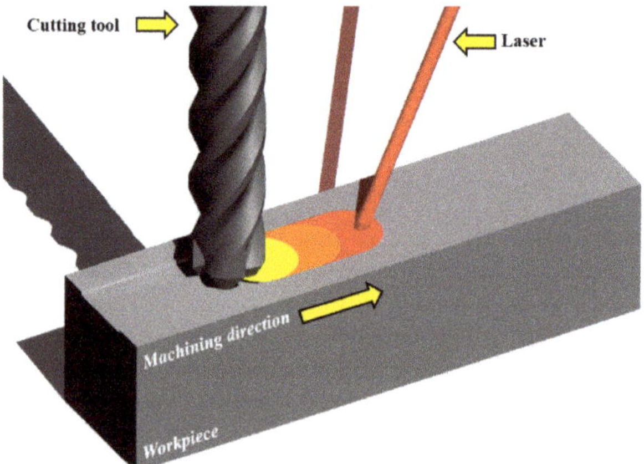

Figure 1. The schematic diagram of laser-assisted machining (LAM).

Many researchers are still investigating efficient methods for processing ceramic composites materials. Many researchers have been actively studying the carbon fiber reinforced silicon carbide (C/SiC) composites and also have considered the structural integrity and reliability of high temperature structures such as exhaust valves, automobile parts, aircraft parts, and nozzle necks by composites [14–16]. Hui et al. investigated the changes in the tensile strength of C/SiC composites according to the changes of specimen cross-section and heat treatment conditions [17]. It was found that the fracture work and residual strain increased as the cross-section of the specimen decreased. Tao et al. conducted static and dynamic compression tests in various temperature ranges and studied the resulting changes in compressive strength [18]. Research has found that compressive strength increases with increasing temperature, but decreases with decreasing temperature. Fattahi et al. focused on the analytical prediction of delamination during drilling composite laminates [19]. Chi et al. investigated the effects of cylindrical specimen size for IG-110 and NBG-18 on the compressive strength and Weibull modulus [20]. Chinmaya et al. compared experimental and simulated results for cutting forces for machining of A359/SiC/20p composites [21]. The effects of operating conditions under LAM were determined for the shear zone stress of silicon nitride by experimentation. Shuting Lei et al. and Damian Przestackid focused on improving the Al/SiC composite machinability by LAM, and compared the results with those obtained when using the conventional turning process [22]. Except for the above-mentioned studies on the machining of composite materials using LAM, most studies have been performed with titanium and nickel alloys, and with various steels [23–26]. To date, there has been no research on using laser heat sources for preheating and machining of C/SiC composite materials. In this study, the response optimization method is used to determine the optimal machining parameters in LAM of C/SiC composite materials. This method is useful when evaluating the effect of multiple parameters on the response.

In this study, the experiments are performed to determine the optimal machining parameters and to analyze the thermal effect and machinability of the C/SiC composite under various machining conditions in the LAM process. The effective depth of cut for LAM of C/SiC composite is obtained by finite element analysis according to preheating temperature. Then, LAM experiments are performed for flat shaped C/SiC. The influence of the machining parameters such as depth of cut, preheating temperature, feed rate, and spindle speed on cutting force, surface roughness, and tool wear is analyzed using the Taguchi method. The cutting force is measured by a dynamometer and the surface roughness is measured by the shape measuring device. For the determination of the optimal machining parameters, the response optimization method is performed. An efficient machining condition to increase machinability is proposed and discussed.

2. Finite Element Analysis

2.1. Finite Element Analysis

The finite element analysis method is performed to study the overlapping in the laser heat source according to the feed rate. The moving time of the heat source for the analysis was set to be in the range of 10–30 sec by considering the feed rate (f = 100–300 mm/min). The material properties for the analysis are listed in Table 1. The main components of C/SiC composite are reported in Table 2. The preheating temperature was chosen by considering the tensile strength of C/SiC composite material according to the temperature, as shown in Figure 2. The tensile strength of the C/SiC composite decreases at the temperature ranges from the room temperature to 1300 °C [27]. Additionally, C/SiC composite material has its maximum elongation and minimum tensile strength at a temperature of about 1300 °C. The temperature range of the effective depth of the cut is 1100–1300 °C. To increase the accuracy of the analysis, the mesh elements are divided into squares of 1 mm and the preheated sections are localized at a mesh size of 0.5 mm. The mesh consists of 66,267 nodes and 16,425 elements. Figure 3 shows the analysis model of the specimen and shows the shape of the generated mesh.

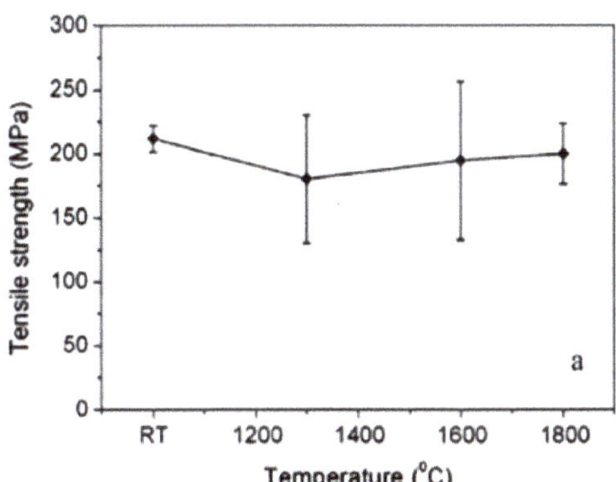

Figure 2. The tensile stress value according to temperature of C/SiC composite.

Table 1. Mechanical properties of C/SiC composite.

Density (g/cm^3)	Young Modulus (GPa)	Thermal Conductivity (W/mm-K)	Specific Heating (J/kg-K)	Flexural Strength (MPa)
2.1	35	40	1200	67

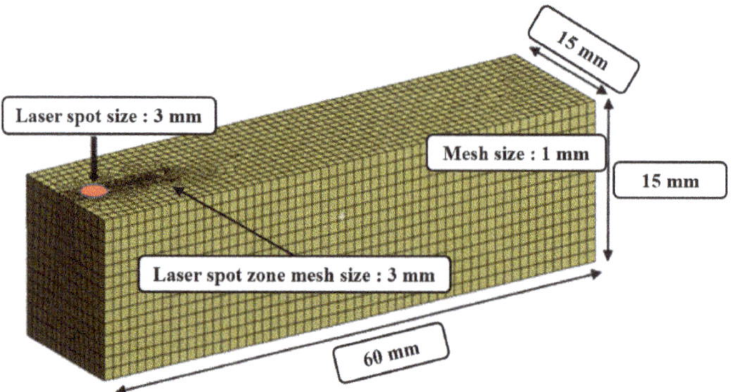

Figure 3. The finite element analysis model.

Table 2. The main components of C/SiC composite.

The Composition by X-ray Analysis (%)			Open Porosity (qv) (%)
C	SiC	Residual Si	
50.47	44.81	4.72	5.40

2.2. Result of Analyis

The finite element analysis result, the temperature distribution for the section view of the workpiece was used to determine the effective depth of cut according to preheating temperature. The effective depth of the cut is selected to be the depth at the temperature range of 1100–1300 °C where the strength of the material decreases. The analysis result, the effective depth of the cut was determined to be in the region of 0.2–0.4 mm at a preheating temperature range of 1100–1300 °C. Figure 4 shows the finite element analysis results, and effective depth of cut according to the preheating temperature.

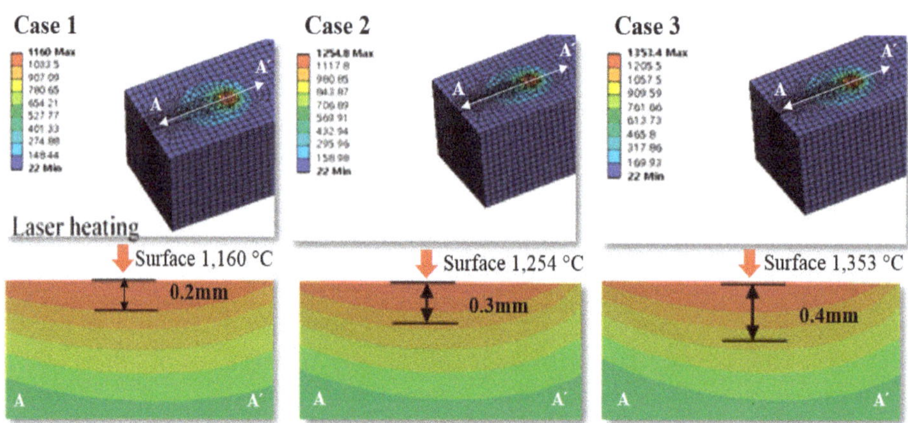

Figure 4. The finite element analysis result and the effective depth of the cut according to the preheating temperature.

3. Laser-Assisted Machining

3.1. Procedure

The design of the experiment was performed to determine optimal machining conditions for the LAM of C/SiC composites. The object functions are selected as cutting force and surface roughness. The parameters such as depth of cut, preheating temperature, spindle speed, and feed rate are selected. The depth of cut is determined by the results of finite element analysis. Figure 5 shows the flow chart of the design of experiments for the LAM of C/SiC composite.

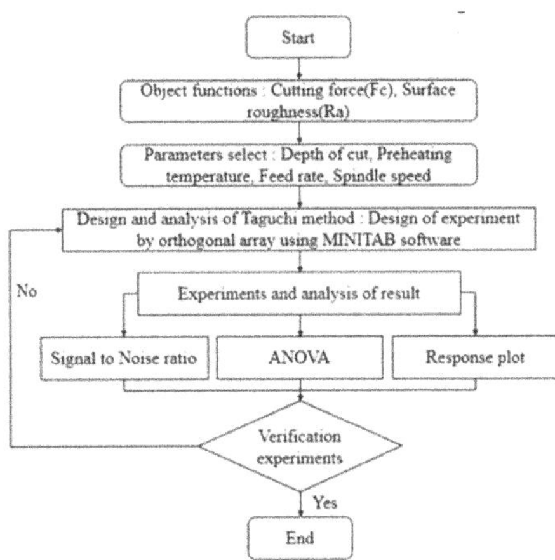

Figure 5. Flow chart of the design of experiments for the laser-assisted machining (LAM) of the C/SiC composite.

3.2. Machining Conditions

The LAM process was performed on the 5-axis machining center (Hyundai WIA., Type Hi-V560M) with laser module. The laser module is a high-power diode laser (HPDL) with a wavelength range of 940–980 nm (Laser line Inc., Type LDM 1000-100). To measure the preheating temperature, a pyrometer (Dr. Mergemthaler GmbH & Co. KG, LPC03) with a range of 400–3000 °C was used. The dynamometer (Kistler Inc., 9257B) attached to the indexing table was used to measure the cutting forces. A dynamometer measures the three orthogonal components of a force using the quartz three-component measurement. The measurement range is −5 kN to 5 kN, and the rigidity is 1 kN/μm to 2 kN/μm. The surface roughness measurement device (Kosaka Inc., SE-3500K) and field emission scanning electron microscope (ZEISS Inc., MERLIN) were used to measure the surface roughness. The surface roughness measurement device was a probe type with a resolution of 32,000 points/16 bite. The surface roughness measurement device used the Gaussian profile filter to separate the long and short wave of a surface profile, and a cut-off value of 0.25 mm was used in this study. Figure 6 shows the experimental set-up and Table 3 shows machining conditions.

Table 3. The machining conditions.

Material	C/SiC Composite
Material size (T × W × L, mm)	15 × 15 × 60
Machining method	Slot milling
Cutting tool	D8 CBN flat end-mill, 2F, 70L

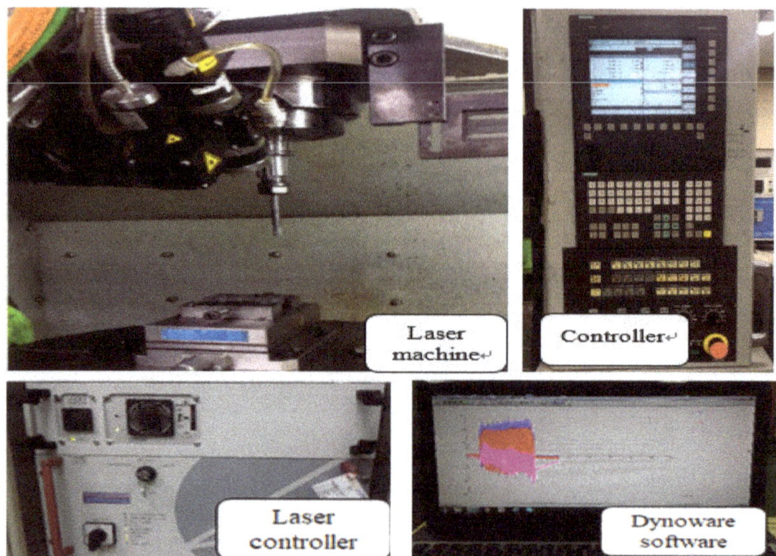

Figure 6. The experimental setup.

3.3. Experimental Design

In LAM, machining characteristics of C/SiC composite are affected by various factors. To analyze the influence of the factors on the machining characteristics, the experiments should be performed after considering all the combinations of factors. However, in this case, the number of experiments increases and it is costly and time-consuming. Therefore, the design of experiment should be performed to reduce the cost and time. The experimental design proposed by the Taguchi method uses orthogonal arrays to organize the parameters that affect the process and varies the levels of those parameters. The Taguchi approach values the importance of the logic of the parameters and has a strong effect compared to actual experiments [28–32]. The Taguchi method used signal to noise (S/N) as a quality characteristic. The S/N ratio characteristics can be divided into three types: nominal-is-best characteristics, larger-the-better characteristics, and smaller-the-better characteristics. In this study, the smaller-the-better characteristics were used. The smaller-the-better characteristics are shown in Equation (1).

$$\mathrm{SN} = -10\log\left[\frac{1}{n}\sum_{i=1}^{n} y_i^2\right] \tag{1}$$

where, y_i is the average of the observed data, n is the number of observations, and y represents the observed data or each type of characteristic; with the above S/N ratio transformation, the smaller the S/N ratio is, the better are the results for cutting force and surface roughness.

The cutting parameters for the LAM are the depth of cut (A), preheating temperature (B), feed rate (C), and spindle speed (D). The depth of the cut range (0.2–0.4 mm) was selected based on the finite element analysis result. The preheating temperature range (1100–1300 °C) was selected based on the tensile strength of C/SiC composite material according to the temperature. The feed rate range (100–300 mm/min) and the spindle speed range (2000–8000 rpm) were selected after considering previous studies. The conventional machining (CM) was performed to verify the efficiency of the LAM. Table 4 shows the factors and levels used in the experiments and Table 5 shows the experimental layout using an L₉ orthogonal array [33,34].

Table 4. The factors and levels used in the experiments.

Symbol	Factor	Level 1	Level 2	Level 3
A	Depth of cut (mm)	0.2	0.3	0.4
B	Preheating temperature (°C)	1100	1200	1300
C	Feed rate (mm/min)	100	200	300
D	Spindle speed (rpm)	2000	5000	8000

Table 5. The experimental layout using an L_9 orthogonal array.

Experiment No.	Depth of Cut (mm)	Preheating Temperature (°C)	Feed Rate (mm/min)	Spindle Speed (rpm)
CM	0.2	1100	100	2000
1	0.2	1100	100	2000
2	0.2	1200	200	5000
3	0.2	1300	300	8000
4	0.3	1100	200	8000
5	0.3	1200	300	2000
6	0.3	1300	100	5000
7	0.4	1100	300	5000
8	0.4	1200	100	8000
9	0.4	1300	200	2000

3.4. Experimental Results on LAM

The cutting forces were measured using a tool dynamometer during the machining of C/SiC composite material. The surface integrity was analyzed by the surface roughness measurement device and a field emission scanning electron microscope (FE-SEM). All experiments were repeated three times. The cutting force was calculated by the average value of each experiment and the surface roughness was used as the lowest value of each experiment. Figure 7 shows the microphotographs of the machined surfaces of C/SiC composite material in all experiments. Table 6 shows the measured cutting force and surface roughness according to the four factors of the machining conditions. In LAM, the cutting force was decreased by about 40.7% and the surface roughness was decreased by about 33.8%, compare to the CM. The lowest cutting force value at $A_2B_3C_1D_2$ was 42.25 N, and the surface roughness of the S/N ratio had the highest value of −32.5165 dB. Figure 8a shows that the optimal levels were found to be A_2 (depth of cut: 0.3 mm), B_3 (preheating temperature: 1300 °C), C_1 (feed rate: 100 mm/min), and D_2 (spindle speed: 5000 rpm). The lowest surface roughness value at $A_2B_3C_1D_2$ was 1.26 μm, and the surface roughness of the S/N ratio had the highest value of −2.0074 dB. Figure 8b shows the optimal levels were found to be A_2 (depth of cut: 0.3 mm), B_3 (preheating temperature: 1300 °C), C_1 (feed rate: 100 mm/min), and D_2 (spindle speed: 5000 rpm). Tables 7 and 8 shows the response table mean S/N ratio for the cutting force and surface roughness according to the machining conditions.

Table 6. The experimental data value of cutting force and surface roughness.

No.	Depth of Cut (mm)	Preheating Temperature (°C)	Feed Rate (mm/min)	Spindle Speed (rpm)	Surface Roughness (μm)	Cutting Force (N)
CM	0.2	1100	100	2000	5.95	105.90
1	0.2	1100	100	2000	3.94	62.80
2	0.2	1200	200	5000	3.20	87.77
3	0.2	1300	300	8000	6.80	50.79
4	0.3	1100	200	8000	2.54	55.60
5	0.3	1200	300	2000	4.65	129.50
6	0.3	1300	100	5000	1.26	42.25
7	0.4	1100	300	5000	1.95	72.58
8	0.4	1200	100	8000	1.85	159.36
9	0.4	1300	200	2000	4.30	90.63

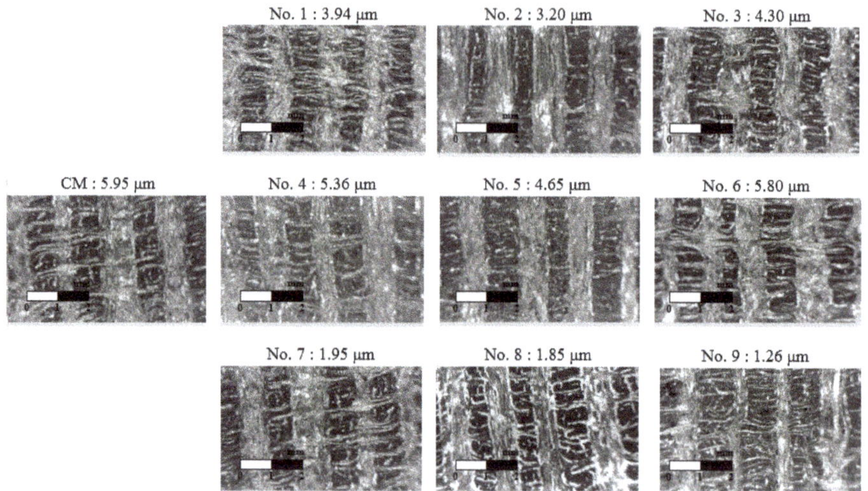

Figure 7. The microphotographs of machined surfaces of C/SiC composite material in LAM.

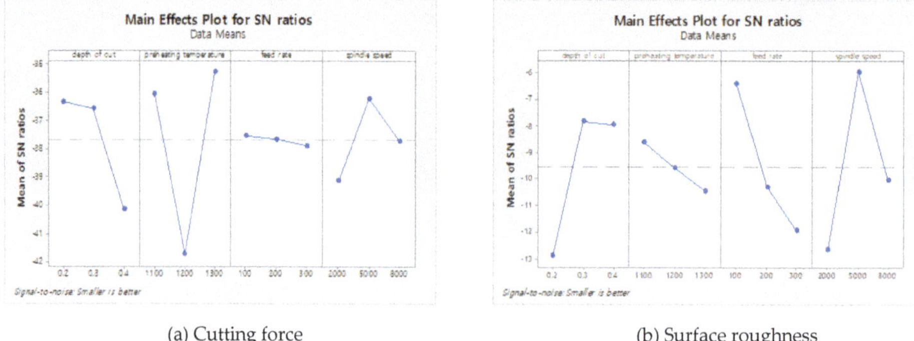

(a) Cutting force

(b) Surface roughness

Figure 8. The main effect plot of C/SiC composite on the cutting force and the surface roughness. (**a**) Cutting force; (**b**) Surface roughness.

Table 7. The response table mean signal to noise (S/N) ratio for the cutting force according to the machining conditions.

Level	Depth of Cut (A)	Preheating Temperature (B)	Feed Rate (C)	Spindle Speed (D)
1	−36.31	−36.03	−37.51	−39.12
2	−36.55	−41.72	−37.64	−36.20
3	−40.14	−35.26	−37.86	−37.69
Delta	3.82	6.46	0.35	2.92
Rank	2	1	4	3

Table 8. The response table mean S/N ratio for the surface roughness according to the machining conditions.

Level	Depth of Cut (A)	Preheating Temperature (B)	Feed Rate (C)	Spindle Speed (D)
1	−12.888	−8.602	−6.420	−12.643
2	−7.818	−9.598	−10.290	−5.970
3	−7.938	−10.442	−11.933	−10.030
Delta	5.070	1.840	5.513	6.672
Rank	3	4	2	1

3.5. Variance Analysis

Variance analysis (ANOVA) was applied to the S/N ratios to determine the relations between machining parameters relating to surface roughness and cutting force. The calculated S/N ratio for the four factors of the surface roughness and cutting force in the machining of C/SiC composites are shown in Tables 9 and 10. In these results, the most significant influences on the cutting force value for each factor were the percentage contributions of the factors of preheating temperature, depth of cut, spindle speed, and feed rate; these percentages were 66.23%, 22.55%, 9.91%, and 1.31%, respectively. Also, the factors that contributed to the surface roughness were determined; the most important factor was 31.24% for the spindle speed, the second factor was 29.69% for the depth of cut, the third factor was 27. 93% for the preheating temperature, and the fourth factor was 11.13% for the feed rate.

Table 9. The analysis results of variance for cutting force.

Factors	Degree of Freedom	Sum of Squares	Mean of Squares	Contribution (%)
Feed rate	2	157.1	78.56	1.31%
Spindle speed	2	1192.9	596.44	9.91%
Depth of cut	2	2714.9	1357.46	22.55%
Preheating temperature	2	7972.5	3986.27	66.23%
Error	0	*	*	*
Total	8	12037.5	-	100

Table 10. The analysis results of variance for surface roughness.

Factors	Degree of Freedom	Sum of Squares	Mean of Squares	Contribution (%)
Feed rate	2	6.728	3.364	27.93%
Spindle speed	2	7.525	3.763	31.24%
Depth of cut	2	7.152	3.576	29.69%
Preheating temperature	2	2.681	1.341	11.13%
Error	0	*	*	*
Total	8	24.087	-	100

4. Experimental Results and Discussion

4.1. Signal to Noise (S/N) Ratio of Analysis

S/N ratio is a very important measurement in the Taguchi method for experimental data analysis. According to the Taguchi approach, optimal machining condition values should lead to a maximum S/N ratio. Parameter values are important factors for evaluating the surface roughness and cutting force. Other characteristics contribute slightly to the cutting force and surface roughness evaluation. Results of machining experiments have been studied using the S/N ratio. Based on the predictions and response results of the ANOVA analyses, optimal machining parameters for cutting force and surface roughness were obtained and verified. The correlation test between the cutting force and surface roughness was performed by analyzing the correlation coefficient (r). The result of the correlation test, the correlation between the two parameters was a positive correlation and r was 0.223. The correlation coefficient ranged from −1 to 1 and describes the parametric value of linear relationship.

4.2. Response Optimization

The objective of this experiment is to optimize the machining parameters and to develop better (i.e., low value) surface roughness and cutting force values; the "smaller the better" characteristic was used. The optimal machining conditions, which were the depth of cut of 0.3 mm, preheating temperature of 1100 °C, the feed rate of 200 mm/min, and a spindle speed of 5000 rpm were obtained for the best cutting force and surface roughness values. The desirability is confirmed at 1. According to the Taguchi design results obtained for the cutting force and surface roughness, the response optimization results are given in Tables 11 and 12.

Table 11. The response optimization.

Parameter	Goal	Target	Upper	Weight	Importance
Cutting force	Minimum	42.25	159.36	1	1
Surface roughness	Minimum	1.26	6.80	1	1

Table 12. Response optimization results.

Depth of Cut (mm)	0.3
Preheat temperature (°C)	1100
Feed rate (mm/min)	200
Spindle speed (rpm)	5000
Cutting force optimization plot (N)	34.55
Surface roughness optimization plot (μm)	0.946667
Desirability	1

4.3. Prediction Equations and Confirmation Experiments of the Optimal Condition

Confirmation experiments were conducted to calculate the suitability of the analysis results. The prediction equations for the cutting force and the surface roughness are shown in Equations (2) and (3).

$$Fc = 83.48 + 10.83 \, (S2000) - 15.94 \, (S5000) + 5.108 \, (S8000) + 4.661 \, (F100)$$
$$- 5.476 \, (F200) + 0.8144 \, (F300) - 19.82 \, (P1100) + 42.07 \, (P1200) - 22.25 \, (P1300) \tag{2}$$
$$- 16.36 \, (D0.2) - 7.692 \, (D0.3) + 24.05 \, (0.4)$$

$$Ra = 3.388 + 0.9089 \, (S2000) - 1.251 \, (S5000) + 0.3422 \, (S8000) - 1.038 \, (F100)$$
$$- 0.04111 \, (F200) + 1.079 \, (F300) - 0.5778 \, (P1100) - 0.1544 \, (P1200) + 0.7322 \, (P1300) \tag{3}$$
$$+ 1.259 \, (D0.2) - 0.5711 \, (D0.3) - 0.6878 \, (D0.4)$$

where, Fc represents the cutting force, and Ra represents the surface roughness. Table 13 shows the machining conditions of the confirmation experiment. The three experiments were conducted by randomly adding four machining conditions, including optimal machining conditions (Exp. No. 1) and the main effect of surface roughness (Exp. No. 2), are deducted in Table 13. All experiments were repeated three times. Figure 9 shows the comparison of the results of the prediction equation and the confirmation experiments for the cutting force. As a result of the comparison, the maximum error rate was confirmed to be approximately 7.55%. Figure 10 shows the comparison of the results of the prediction equation and the confirmation experiments for the surface roughness. As a result of the comparison, the maximum error rate was confirmed to be approximately 8.76%. Confirmation experiments were conducted to verify the optimal machining parameters.

Table 13. The machining conditions for confirmation experiments.

Exp. No.	Depth of Cut (mm)	Preheating Temperature (°C)	Feed Rate (mm/min)	Spindle Speed (rpm)
1	0.2	1200	100	2000
2	0.3	1200	200	5000
3	0.4	1200	300	2000

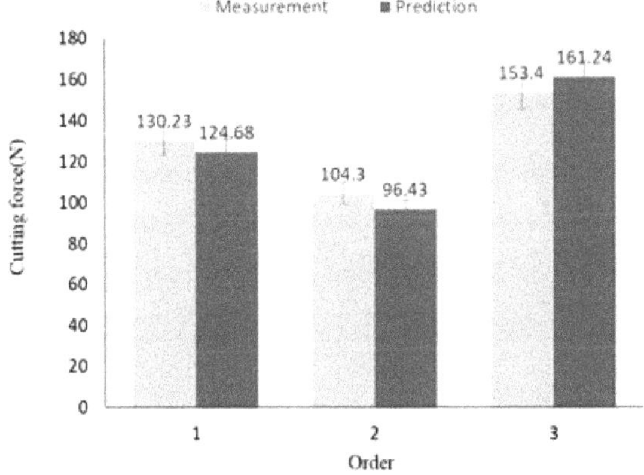

Figure 9. The comparison of the cutting force between the prediction equation results and the confirmation experiment results.

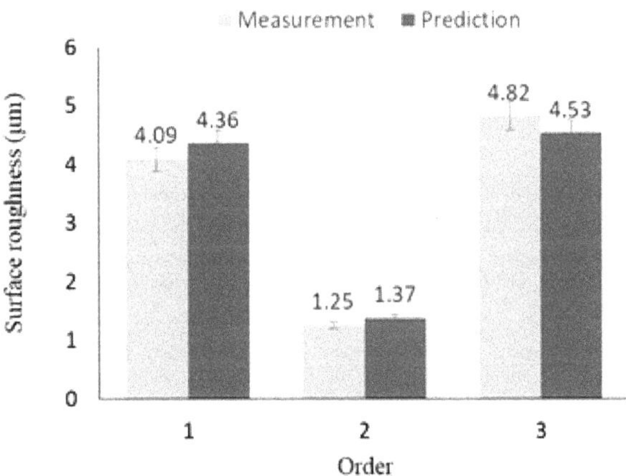

Figure 10. The comparison of the surface roughness between the prediction equation results and the confirmation experiment results.

5. Conclusions

In this study, LAM was carried out on the C/SiC composite material. The effective depth of cut was selected using the finite element analysis. The optimal machining conditions were obtained using the Taguchi method, which uses cutting force and surface roughness as objective function. The conclusions obtained from this study are as follows.

(1). The finite element analysis was performed to determine the preheating temperature and the depth of cut depending on the tensile strength of the C/SiC composite material. When the preheating temperature is in the tensile strength decreasing range (1100–1300 °C), the effective depth of cut is determined to be in the range of 0.2–0.4 mm.

(2). According to the Taguchi standard design concept in this experiment, at three levels with four factors of each one, nine experiments must be performed, and fractional design was selected in a standard L9 orthogonal array. The maximum value was found using the S/N ratio equation of "the smaller-the better"; the maximum S/N ratio yielded the optimal machining parameters.

(3). In same case of the machining conditions, the cutting force was decreased by about 40.7% compared to CM in LAM of the C/SiC composite material, and the surface roughness was decreased by about 33.8% compared to CM in LAM of the C/SiC composite material.

(4). Variance analysis (ANOVA) was applied to the S/N ratio to discover the interactions between the parameters relating to surface roughness (Ra) and cutting force (Fc). Based on the ANOVA results, the main contributing factor for the cutting force was 66.23% preheating temperature. The main contributing factor for the surface roughness was 31.24% spindle speed.

(5). The verification experiment was performed to construct the predictive equation and to ensure the reliability of the predictive equation. The verification experiment confirmed that the maximum error was 7.55% between the prediction equation for cutting force and measurement experiment value. The maximum error was 8.76% between the prediction equation for surface roughness and measurement experiment value. The prediction equation demonstrated the reliability of low error.

The results of response optimization, the optimal machining conditions for LAM of the C/SiC composite material were obtained as spindle speed (5000 rpm), feed rate (200 mm/min), preheating temperature (1100 °C), and DOC (0.3 mm). When the experiment was performed by optimal machining conditions, the cutting force was measured to be 34.55 N and surface roughness was measured to be 0.95 μm.

Author Contributions: Conceptualization, K.E.; Data curation, H.J.; Formal analysis, H.J.; Funding acquisition, C.L.; Supervision, C.L.; Writing—original draft, K.E.; Writing—review & editing, H.J.

Funding: This work was supported by the National Research Foundation of Korea (NRF) grant funded by the Korea government (MSIT) (No. 2019R1A2B5B03070206).

Conflicts of Interest: The authors declare no conflict of interest.

References

1. M'Saoubi, R.; Axinte, D.; Soo, S.L.; Nobel, C.; Attia, H.; Kappmeyer, G.; Engin, S.; Sim, W.M. High performance cutting of advanced aerospace alloys and composite materials. *CIRP Ann.* **2015**, *64*, 557–580. [CrossRef]

2. Mei, H.; Li, H.; Bai, Q.; Zhang, Q.; Cheng, L. Increasing the strength and toughness of a carbon fiber/silicon carbide composite by heat treatment. *Carbon* **2013**, *54*, 42–47. [CrossRef]

3. Appiah, K.A.; Wang, Z.L.; Lackey, W.J. Characterization of interfaces in C fiber-reinforced laminated C-SiC matrix composites. *Carbon* **2000**, *38*, 831–838. [CrossRef]

4. Mei, H.; Cheng, L.; Zhang, L.; Xu, Y. Modeling the effects of thermal and mechanical load cycling on a C/SiC composite in oxygen/argon mixtures. *Carbon* **2007**, *45*, 2195–2204. [CrossRef]

5. Shirvanimoghaddam, K.; Hamim, S.U.; Karbalaei, A.M.; Fakhrhoseini, S.M.; Khayyam, H.; Pakseresht, A.H.; Ghasali, E.; Zabet, M.; Munir, K.S.; Jia, S.; et al. Carbon fiber reinforced metal matrix composites: Fabrication processes and properties. *Composites Part A* **2017**, *92*, 70–96. [CrossRef]

6. Lo´pez de Lacalle, L.N.; Sa´nchez, J.A.; Lamikiz, A.; Celaya, A. Plasma Assisted Milling of Heat-Resistant Superalloys. *J. Manu. Sci. Eng.* **2004**, *126*, 274–285. [CrossRef]

7. Chen, S.H.; Tsai, K.T. The study of plasma-assisted machining to Inconel-718. *Adv. Mech. Eng.* **2017**, *9*, 1–7. [CrossRef]

8. Ló´pez de Lacalle, L.N.; Lamikiz, A.; Celaya, A. Simulation of Plasma Assisted Milling of Heat Resistant Alloys. *Int. J. Simul. Modll.* **2002**, *1*, 5–15.

9. Jeon, Y.; Lee, C.M. Current research trend on laser assisted machining. *Int. J. Precis. Eng. Manuf.* **2012**, *13*, 311–317. [CrossRef]

10. Kim, T.W.; Lee, C.M. Determination of the machining parameters of nickel-based alloys by High-Power diode laser. *Int. J. Precis. Eng. Manuf.* **2015**, *16*, 309–314. [CrossRef]

11. Sun, S.; Brandt, M.; Dargusch, M.S. Thermally enhanced machining of hard-to-machine materialsA review. *Int. J. Mach. Tools Manuf.* **2010**, *50*, 663–680. [CrossRef]

12. Bermingham, M.J.; Palanisamy, S.; Dargusch, M.S. Understanding the tool wear mechanism during thermally assisted machining Ti-6Al-4V. *Int. J. Mach. Tools Manuf.* **2012**, *62*, 76–87. [CrossRef]

13. Woo, W.S.; Lee, C.M. A study of the machining characteristics of AISI 1045 steel and Inconel 718 with a cylindrical shape in laser-assisted milling. *Appl. Therm. Eng.* **2015**, *91*, 33–42. [CrossRef]

14. Mei, H. Measurement and calculation of thermal residual stress in fiber reinforced ceramic matrix composites. *Compos. Sci. Technol.* **2008**, *68*, 15–16. [CrossRef]

15. Krenkel, W.; Berndt, F. C/C-SiC composites for space applications and advanced friction systems. *Mater. Sci. Eng., A* **2005**, *412*, 177–181. [CrossRef]

16. Leatherbarrow, A.; Wu, H. Mechanical behaviour of the constituents inside carbon-fibre/carbon-silicon carbide composites characterised by nano-indentation. *J. Eur. Ceram. Soc.* **2012**, *32*, 579–588. [CrossRef]

17. Mei, H.; Xiao, S.; Bai, Q.; Wang, H.; Li, H.; Cheng, L. The effect of specimen cross-sectional area on the strength and toughness of two-dimensional C/SiC composites. *Ceram. Int.* **2015**, *41*, 2963–2967. [CrossRef]

18. Suo, T.; Fan, X.; Hu, G.; Li, Y.; Tang, Z.; Xue, P. Compressive behavior of C/SiC composites over a wide range of strain rates and temperatures. *Carbon* **2013**, *62*, 481–492. [CrossRef]

19. Fattahi, Z.; Hegab, H.; Kishawy, H.A. Analytical Prediction of Delamination during Drilling Composite Laminates. *Procedia Manuf.* **2018**, *26*, 237–244. [CrossRef]

20. Chi, S.H. Specimen size effects on the compressive strength and Weibull modulus of nuclear graphite of different coke particle size: IG-110 and NBG-18. *J. Nucl. Mater.* **2013**, *436*, 185–190. [CrossRef]

21. Przestacki, D. Conventional and laser assisted machining of composite A359/20SiCp Procedia. *Procedia CIRP* **2014**, *14*, 229–233. [CrossRef]

22. Dandekar, C.R.; Shin, Y.C. Modeling of machining of composite materials: A review. *Int. J. Mach. Tools Manuf.* **2012**, *57*, 102–121. [CrossRef]

23. Brecher, C.; Emonts, M.; Rosen, C.J.; Hermani, J.P. Laser-assisted milling of advanced materials. *Physics Procedia* **2011**, *12*, 599–606. [CrossRef]

24. Dandekar, C.R.; Shin, Y.C.; Barnes, J. Machinability improvement of titanium alloy (Ti-6Al-4V) via LAM and hybrid machining. *Int. J. Mach. Tools Manuf.* **2010**, *50*, 174–182. [CrossRef]

25. Kang, D.W.; Lee, C.M. A study on the development of the laser-assisted milling process and a related constitutive equation for silicon nitride. *CIRP Ann.* **2014**, *63*, 109–112. [CrossRef]

26. Kim, D.H.; Lee, C.M. A study of cutting force and preheating-temperature prediction for laser-assisted milling of Inconel 718 and AISI 1045 steel. *Int. J. Heat Mass Transfer* **2014**, *71*, 264–274. [CrossRef]

27. Li, Z.; Xiao, P.; Xiong, X.; Huang, B.Y. Manufacture and properties of carbon fibre-reinforced C/SiC dual matrix composites. *New Carbon Mater.* **2010**, *25*, 225–231. [CrossRef]

28. Nalbant, M.; Gökkaya, H.; Sur, G. Application of Taguchi method in the optimization of cutting parameters for surface roughness in turning. *Mater. Des.* **2007**, *28*, 1379–1385. [CrossRef]

29. Somashekara, H.M.; Lakshmana, S.N. Optimizing Surface Roughness and MRR in turning operation using Taguchi's design of experiments approach. *Int. J. Appl. Eng. Res.* **2012**, *7*, 887–895.

30. Fratila, D.; Caizar, C. Application of Taguchi method to selection of optimal lubrication and cutting conditions in face milling of AlMg3. *J. Cleaner Prod.* **2011**, *19*, 640–645. [CrossRef]

31. Ghani, J.A.; Choudhury, I.A.; Hassan, H.H. Application of Taguchi method in the optimization of end milling parameters. *J. Mater. Process. Technol.* **2004**, *145*, 84–92. [CrossRef]

32. Kim, E.J.; Lee, C.M. A Study on the Optimal Machining Parameters of the Induction Assisted Milling with Inconel 718. *Materials* **2019**, *12*, 233. [CrossRef] [PubMed]

33. Erdenechimeg, K. A study on the machining characteristics of composite material using Laser-assisted machining. M.S. dissertation, Changwon National University, Changwon, Gyeongsangnam-do, Korea. 2019. Available online: http://lib.changwon.ac.kr/search/DetailView.External.ax?edp1=edsker&edp2=edsker.000004669582&sid=1&widget_id=56&linkID=1&q= (accessed on 10 December 2018).
34. Research report of the National Research Foundation of Korea (NRF) (2019) A new conceptual 3-dimensional laser assisted machining system 2016R1A2A1A05005492. Available online: http://ernd.nrf.re.kr/ (accessed on June 2019).

Article

Trochoidal Milling and Neural Networks Simulation of Magnesium Alloys

Ireneusz Zagórski [1,*], Monika Kulisz [2], Mariusz Kłonica [1] and Jakub Matuszak [1]

[1] Department of Production Engineering, Mechanical Engineering Faculty, Lublin University of Technology, 20-618 Lublin, Poland

[2] Department of Enterprise Organisation, Management Faculty, Lublin University of Technology, 20-618 Lublin, Poland

* Correspondence: i.zagorski@pollub.pl; Tel.: +48-815-384-240

Received: 30 May 2019; Accepted: 24 June 2019; Published: 27 June 2019

Abstract: This paper set out to investigate the effect of cutting speed v_c and trochoidal step s_{tr} modification on selected machinability parameters (the cutting force components and vibration). In addition, for a more detailed analysis, selected surface roughness parameters were investigated. The research was carried out for two grades of magnesium alloys—AZ91D and AZ31—and aimed to determine stable machining parameters and to investigate the dynamics of the milling process, i.e., the resulting change in the cutting force components and in vibration. The tests were performed for the specified range of cutting parameters: v_c = 400–1200 m/min and s_{tr} = 5–30%. The results demonstrate a significant effect of cutting data modification on the parameter under scrutiny—the increase in v_c resulted in the reduction of the cutting force components and the displacement and level of vibration recorded in tests. Selected cutting parameters were modelled by means of Statistica Artificial Neural Networks (Radial Basis Function and Multilayered Perceptron), which, furthermore, confirmed the suitability of neural networks as a tool for prediction of the cutting force and vibration in milling of magnesium alloys.

Keywords: milling; the cutting force components; vibrations; magnesium alloys; artificial neural networks

---

1. Introduction—State-Of-The-Art

The aircraft and automotive industries demand innovative modern components and parts of uncompromising quality, which would be unattainable without employing innovative materials and high-efficiency machiningemploying advanced tools and machining centres. In order to fulfil the challenges of the modern market, manufacturers often face substantial production costs; therefore, engineers are relentless in their striving for the optimisation of production costs while maintaining high quality. In order to enhance the stability and effectiveness of subtractive machining, a thorough analysis of real cutting forces and computer simulation are indispensable. Artificial neural network simulation continues to confirm its applicability as an effective cost-optimisation solution, primarily by limiting the time and amount of testing required to specify and optimise machining parameters, which consequently leads to minimising errors in machining.

In the search for optimisation in manufacturing, it is the selection of production materials that may eventually prove to play the leading role. The use of magnesium alloys significantly contributes to reducing the design weight and boosts savings in production and maintenance costs. On the other hand, the use of Mg alloys also has a beneficial effect on an inherent component of machining—effective removal of machining allowances—which is often carried out by means of milling. Forces occurring in machining are critical to the performance and quality of the process itself, which is why they are typically considered among the key "functional" machinability parameters. With respect to "physical" and additional absolute machinability parameters, the following may be distinguished; damping characteristics of materials, the amplitude and frequency of vibrations. In unfavourable conditions, different the cutting force components acting on the workpiece may cause uncontrollable deformation of the workpiece during machining, e.g., leading to the reduction of the undeformed chip thickness, which subsequently results in the increase in the cutting force. Then again, as a result of the increased shear energy per unit of volume, both the amount of subtracted workpiece material and the value of cutting forces increase [1]. Adhesion and build-up may similarly cause fluctuation in the components of the cutting force and cause insufficient surface finish quality or shape and dimensional accuracy. Compared to the machinability of other materials, magnesium alloys are characteristically quick and effective to cut, which enables executing the process at large depths of cut and high feed.

1.1. The Cutting Force Components

The analysis of the components of cutting forces is virtually indispensable in subtractive machining for medical and aircraft purposes—the components implemented in these industries must meet particularly severe quality requirements. This is even more difficult given the frequently manufactured thin-walled workpieces of complex geometry [2–4]. Considerable cutting forces could potentially decrease the surface finish quality, as a consequence of excessive chatter in the milling machine-milling cutter–workpiece–fixture system [5]. Moreover, adhesion and build-up are similarly considered to be detrimental to the total cutting force. They are a source of major fluctuations of the cutting force components during cutting, and, as a result, deteriorate the condition of the surface finish of the final piece as well as its dimensional and shape accuracy [6].

Tool geometry is found to strongly affect the total cutting force and the amplitudes of its components. A study including Kordell geometry tools showed that the F_x and F_y components as well as their amplitudes rise in response to growth in feed per tooth f_z. In addition, in the case of tools of classic geometry, the cutting forces and their amplitudes tended to respond more to the modification of feed per tooth f_z than to the cutting speed v_c; the results from the study showed that the cutting force components peaked in machining with PCD tools and AZ91HP alloy. Finally, the cutting force components were shown to drop with the cutting speed was raised to $v_c = 1200$ m/min when employing traditional-geometry tools [7]. Research studies in the field of the cutting force components and their amplitude tend to focus on their response to particular technological milling parameters, such as v_c, f_z and a_p. Kuczmaszewski and Pieśko's study investigated the milling process performed with carbide cutters of variable tooth geometry ($\gamma = 5°$ and $30°$). It showed that the cutting force components and their amplitudes—both important indicators of cutting stability—were shown to decrease when cutting with the tool $\gamma = 30°$. Furthermore, the cutting forces in milling exhibited great dependence on the cutting tool coating material (here: TiB2 and TiAlCN). The lowest values of the cutting forces (F_x, F_y) in milling Al6082 aluminium alloy were recorded when cutting with TiB2-coated tools. What is more, a typical transition point was reflected in the measurement data when v_c reached the high-speed cutting levels ($v_{cgr} = 450 \div 600$ m/min). In addition, the cutting force component amplitudes—relevant from the viewpoint of the cutting process dynamics—have been reported to attain the highest values for indexable tools, which are due proper consideration when selecting tools for particular applications [8].

In the study by Monies et al. [9], extremely deep pockets were cut in MRI301F (Mg-Nd-Y-Zr-Zn) workpiece, by means of two methods: plunge milling and BotTCF ramping at successive depth levels. The cutting force components were derived from test data. The maximum values of F_t, F_r and F_a were as follows; in zone 1 (the start and end of each trajectoryone cutting tooth submerging in the workpiece): $F_t = 1136$ N, $F_r = 450$ N and $F_a = 253$ N—cutting force of 1248 N; in zone 2 (two teeth cutting): $F_t = 572$ N, $F_r = 330$ N and $F_a = 468$ N—cutting force $F_{Rmax} = 827$ N. Maladjusted cutting forces are the source of plastic strain in workpiece and generally increase the thickness of the undeformed chip, which, consequently, affects the shear angle and results in higher temperature in the area of cut. On the other hand, it was found that the lower the undeformed chip thickness, the higher the cutting forces [1].

Interesting insights into the behaviour of the cutting force components are found in works of Oczoś, who investigated the process of AZ91HP alloy milling with fluid by means of PCD cutters. It was shown that the cutting force components remain low and grow steadily with feed, which is of considerable significance to tool wear. Lower cutting forces curb heat, owing to reduced friction in the cutting zone. Furthermore, especially with respect to layers subtracted at a small radial depth of cut, the temperature in the area of cut is reportedly smaller [6]. PCD-cutter tools have been widely reported to produce better quality of surface finish, generate lower cutting force values and prevent tool overheating by decreasing friction at tool–workpiece interface [10]; consequently, the temperatures in the area of cut are significantly lower [11–13]. The cutting tool was the subject of analysis in the study by Zagórski et al. [14,15], which investigated the AZ91HP magnesium alloy machining with "serrated", i.e., "wavy-edged", cutting tools. In the study, the cutting forces and amplitudes were considerably higher compared with classic-geometry tools, whose amplitudes and components moreover exhibited higher sensitivity to modification of feed per tooth f_z rather than the cutting speed v_c. While the primary objective of the study by Shi et al. [16] was to assess how tool wear affects cutting performance and to investiagte the wear mechanism of carbide inserts (in dry milling at speeds ranging from 1600 to 2000 m/min); it also provided data on the behaviour of the cutting forces. The analysed mechanisms were: abrasion, adhesion wearand diffusion. It was shown that prior to tool damage at speeds amounting to 1800 and 2000 m/min, extensive flaking appeared at the speed of 1600 m/min. The conclusion from the study was that increased tool wear causes the cutting forces to rise—resulting in the rise in cutting resistance.

There is a well-evidenced tendency showing that the cutting force components largely depend on the cutting process conditions, as concluded by Sivam et al. [17], whose study involved the analysis of cutting forces and their response to selected input parameters (i.e., depth of cut, feed per tooth, spindle speed and tool diameter) recorded in ZE41 magnesium alloy dry face milling. The cutting forces were reported to increase gradually when the input parameters increased and, furthermore, they were to a greater extent affected by the change in spindle speed than in feed per tooth. Among the conclusions from the study, the relevant one from the point of view of behaviour of cutting forces was that there is a proportional relationship between the constant depth of cut and tool diameter and the total cutting force when operating at higher spindle speeds and feed per tooth respectively; moreover, it was evidenced that by monitoring cutting forces we may prolong tool life.

In conclusion, there are numerous factors that exert a direct and indirect impact on the total cutting force, such as the type of tool and the material of the workpiece, tool geometry or perhaps, most essential, cutting parameters. Machining is more efficient when executed at higher speeds, feed rates and depths of cut [18–20]. Simultaneously, due to the fact that the relationships between the said factors show a nonlinear character, there is a growing interest in the application of mathematical analytical methods [2,18,21,22] or artificial intelligence systems [15,23–25] in modelling machining processes.

The main objective when modelling nonlinear milling processes is to predict the technological process, which could subsequently lead to developing a decision-making support system for an enterprise (e.g., optimising the milling process by determination of suitable machining parameters). The cutting forces are typically modelled with artificial neural networks (ANNs), with the application of e.g., Statistica software, which offers a selection of available network solutions, such as RBF (Radial Basis Function) and MLP (Multilayered Perceptron), which were, e.g., employed in the study by Zagórski et al. [15]. The authors investigated a simple milling process model including two variable parameters, cutting speed and feed per tooth, whereas others—axial depth of cut, radial depth of cut and helix angle—were fixed. The modelled machining process tested the effect of change in v_c and f_z at the same time eliminating the need for further real-life testing. Mathematical modelling was also included in Wu et al. [26], which showed a good fit of results from tests and simulations. The model for predicting the cutting forces was developed for trochoidal milling and enabled predicting the curvature effects of tool path on chip thickness and the impact of entry and exit angles on the process itself.

This section briefly introduced the problem of the cutting force components through relevant literature. Monitoring forces in cutting is critical to obtaining desired milling effects. Numerical simulation of nonlinear machining processes exponentially aids this analysis, particularly with respect to the investigation of optimal machining parameters at limited knowledge.

1.2. Vibrations (Displacement and Acceleration)

Vibration is typical for milling, which significantly affects the quality of finish and is decisive for maintaining specified parameters of the product. Due to its great relevance for the process, it is considered as one of the additional machinability parameters in milling. This relevance is, however, a negative one, i.e., it is highly undesirable during machining, particularly due to chatter occuring at the tool–workpiece interface (due to immediate surface disproportions). Vibrations heavily contribute to poor finish, dynamic tool wear (the consequence of high friction and temperature in cutting zone) [12,13] or decrease in machining accuracy and even damage of machine tool components. Surfaces that are predominantly subject to accelerated wear are those in permanent or temporary joints or actively involved in machining [27]. Moreover, vibrations in milling deteriorate machining efficiency and may even disrupt the entire production process, not to mention the fact that they may put the machine operators' health at risk. Another reason why vibration is highly undesirable in production is the randomness of the end product characteristics (i.e., fulfilling target dimensions, shape and surface finish). Vibrations are even less acceptable in finishing [28] and milling thin-walled structures [29–31]. It is for these reasons that the stability of the process is of utmost importance. In an attempt to ensure it, engineers and researchers typically employ stability lobe diagrams (or the analysis of recurrence quantification), which effectively determines stable machining regions for workpieces of decreasing (nonuniform) wall thickness.

It seems then that the key question in milling isetermination of machining conditions imparting stability by preventing vibration [32,33], which is in fact frequently performed with the application of various specialist software. Milling parameters that are instrumental to machining stability are spindle speed n and axial depth of cut a_p. Modern sophisticated innovative cutting tools, machine tools and materials highlight the need for controlling and modelling the mill cutting process. The unending race for efficiency and cost-effectiveness fuels the development of new methods that would resolve the problem of the cutting tool and workpiece vibration [21].

Although highly undesirable, vibrations are an inherent element of various technological operations, and are thus impossible to eliminate. Their intensity is predominantly the consequence of the machining conditions, tool wear, inconsistency of environmental conditions or workload. Vibration may be successfully curbed and maintained; however, it is essential to possess information regarding the work of particular machine elements and how they interact with each other [21,28].

A range of chatter prevention methods have been established: numerical simulation and modelling the process dynamics, the use of automation devices, pulsating frequency of spindle speed or work at high rates of feed per tooth. The study by Quintana et al. [34] distinguishes two major groups of vibration prevention methods, which embrace: determination of stable cutting parameters from lobe diagrams (group one) and methods enabling the change of system behaviour and adjustment of stability lobes (group two).

In the study by Weremczuk et al. [27] the active chatter control in open and closed loop consisted in the introduction of an external force. The numerical analysis executed in MATLAB–Simulink employed the fourth-order Runge–Kutta method with variable step of integration into the simulation of a nonlinear model of two degrees of freedom, which accounted for the phenomena accompanying the process, the susceptibility of the tool for nonlinearity and harmonic motion of the workpiece in the feed direction. Stable machining conditions and the value of the external excitation of the workpiece were derived from stability lobe diagrams. It was shown that the determined parameters denoted unstable machining. In the case of the closed-loop system, the control is executed by means of a proportional integrating PD controller, and it was the combined effect of external forces, suitably adjusted machining parameters and the controller in question that produced the desired effect of chatter reduction.

An alternative approach for stability analysis was employed in Wu et al. [31], which combined phase plane method, Poincare method and spectral analysis. The researchers explored the relationship between the maximum Lyapunov exponent, and the changes in milling depth and spindle speed, which allowed them to determine stable milling conditions in contouring. The maximum Lyapunov exponent of vibration signals, established as the stability criterion, was shown to increase with the rising depth of cut. Furthermore, the maximum Lyapunov index equal to 0.61 was selected as a threshold value of milling chatter nonlinear criterion. The study has, therefore, concluded that the maximum Lyapunov exponent of the workpiece vibration signal may be employed in the processing parameters optimisation.

The most relevant factors to the stability of milling are: machining strategy, machine tool, the cutting tool and its mounting system, chucks, the workpiece, cooling and lubrication and cutting data. Due to their specific machinability characteristics, resulting in their susceptibility to chatter during milling, thin-walled aluminium alloys, such as EN AW-7075 T6 [31,35], EN AW-2024-T351 [36] and EN AW-6061-T6 [30,37], typically undergo stability analyses. Rusinek et al. [35] investigated the cutting forces and acceleration to specify stable machining conditions for workpieces withdecreasing wall thickness. The analysis was performed by means of the recurrence plots and also involved recurrence quantification analysis (RQA), and showed that it is crucial for the determination of stability diagrams are variations in the modal cutting parameters for thin-walled features.

Vibration in the area of cut can also be modelled by means of finite element method (FEM) simulation. The scientific literature in the field presents different approaches, such as the one given by Yang et al. [30]. The proposed model of tool–workpiece system accounts for its dynamic behaviour as well as the effect of tool engagement and direction of feed. The numerical and experimental results proved that chatter could be accurately predicted for peripheral milling of thin-walled workpieces. In a different study, by Yusoff [37], another improving stability analysis parameter was considered, namely the root mean square (RMS) of vibration displacement, which was employed in the investigations of the positive effect of varying angle of inclination of the helix on the modulation of regenerative effect and thus on vibration damping. FEM was also applied in simulation works conducted in milling EN AW-2024-T351 alloy, and was shown to aid the determination of tool vibration effect on chip morphology, cutting force, and surface topology. In the referred study, by Rusinek and Zaleski [35], chip morphology was predicted under dynamic cutting conditions with the application of a hybrid dynamic cutting model (HDC). The data obtained from the numerical analysis showed that cutting speed and chip thickness increase vibration. Milling process limit stability is also determined by the use of methods based on constant training algorithms. The works reported in Friedrich et al. [38] introduced a novel method for the assessment of process stability, carried out by measuring acceleration signals. The paper introduced a new criterion providing information on the prediction assessment for a given input region—the multidimensional stability lobe diagram (MSDL). The research works were conducted with the application of a support vector machine (SVM) and ANNs. Whereas in the study by Olvera et al. [39], the stability charts were obtained from enhanced multistage homotopy perturbation (EMHP) method, which accounts for the impact of the helix angle and, what is essential, of the runout and cutting speed in machining of Al 7075 T6. As a result, thus obtained stability boundaries show a significantly improved accuracy.

Another work, by Munoa et al. [40], overviewed chatter damping techniques from the perspective of machinability and stiffness of workpiece material in the machine–tool–workpiece–fixture system. With the view to ensuring process stability, the work employed stability lobe diagrams. SLD-based (stability lobe diagrams) analyses, such as in the study by Hsiao and Huang [41], are applied with the purpose of investigating the relationships between the tool diameter chatter generation in milling. The study concluded that stability lobes increase withthe cutter diameter. It was, also shown that increasing the number of cutter teeth tends to move SLDs to the right (increased range of n).

Despite the numerous advantages, there are certain drawbacks of the SLD technique. The SLD data is derived from machine tools in their downtime. As a result, there are factors that cannot be accounted for in the analysis, such as the rigidity of the spindle system, the effect of excessive heat of machine tool parts in operation and other workpiece-related factors (e.g., mass-related damping properties of the workpiece, material defects or the variable diameter of cut) [40,42]. It is important that the obtained SLD results be cross-referenced with the cutting data from dynamic milling tests.

A different approach is employed by Campa et al. [43], which presents the process of calculating the stability diagrams that establish stable spindle speeds during the machining of thin floors. The study employs the milling model of the process carried out in the tool axis direction with bull-nose end mills. The model is combined with an analysis of modal parameters variation for machining thin floor workpieces, which consequently provides data for the calculation of the stability diagrams.

Typical chatter prevention and damping methods consist in either adjustment of technological cutting parameters, or restructuring the design of the machine tool. However, their common drawback results from the fact that both these classic methods are reactive rather than predictive, and respond to vibration observed in milling and not prevent it. From the perspective of implementation cost and time, modification of cutting data is more feasible, and the parameter that directly affects the entire process is spindle speed [44].

Milling nowadays is undergoing the speed revolution. High-performance cutting, which allows workpiece material to be efficiently removed, is slowly becoming a standard in machining. In HPC milling it is passive force component F_p that is key todiagnostics. The work by Burek et al. [45] investigated HPC milling of AlZn5.5MgCu (EN AW-7075) utilising the cutting force, acoustic emission and vibration signals. Excessive cost and complicated set-up requirements of modern dynamometers cause that alternative methods for the determination of the passive force component are developed, as exemplified by RMS acoustic emission or vibration measurement. The capacity of the former method consists in the high correlation between the signals of acoustic emission and the recorded values of the passive force component and, secondly, the acoustic method is sufficiently sensitive while being resistant to background noise. The major advantage of vibration amplitude measurement with accelerometer is that the obtained data describes the process dynamics while exhibiting good fit with the force data; on the other hand though, the method is sensitive to background noise interference.

ANN modelling of machining dynamics is yet to reach its full capabilities and scope of applications [38]. It may be regarded as a booster of efficiency, as it facilitates the selection of optimal technological parameters of milling [15,23]. Computer simulation is highly productive—it replaces long-term and costly experimental testing, thus limiting the involvement of cutters and machine tools. The models determine cutting parameters that guarantee minimum chatter, which is highly desirable in manufacturing performance and economy. Thanks to neural network simulations, initial process parameters are determined, along with their predicted outcome in the form of vibration [46].

1.3. Trochoidal Milling of Light Alloys

A range of CAM software applications enable executing trochoidal milling [47]: NX (Siemens, Munich, Germany), Catia (Dassault Systèmes, Vélizy-Villacoublay, France), hyperMILL (OPEN MIND Technologies AG, Wessling, Germany) and PowerMILL (Delcam, Birmingham, UK). The spiral tool path used in this method proves highly beneficial predominantly in difficult materials; nevertheless, with light alloys, trochoidal milling reduces cutting forces, vibrations and heat in the cutting zone [48,49].

Pleta et al. [50] investigated chip thickness and force modelling in trochoidal milling of 7075-T651 aluminium alloy. The novel approach to modelling chip thickness was employed in the low-to-medium range of cutting speeds ($v_c \approx 15$–30 m/min). The immediate chip thickness was validated experimentally with the use of the semimechanistic cutting force model, exhibiting good fit of the simulated and experimental results. The cutting force coefficients for trochoidal milling were studied in another paper by the same author, [51], primarily to investigate the behaviour of machining parameters and coefficient values recorded for different tool paths. The tool feed and speed between trochoidal and slotting tool paths were found to produce large differences, which may be reduced given that the specified slotting parameters were derived from trochoidal milling chip geometry, with the closest agreement resulting from matching the maximum chip thicknesses. The 5-axis mill cutting was performed on 7075-T651 alloy, at n = 300–1450 rpm and v_f = 456–729 mm/min.

Otkur and Lazoglu [52] introduce an analytical model for defining the engagement along with the force model relative to engagement to determine machining forces. In addition, a numerical engagement model was developed. What is more, for the sake of trochoidal milling efficiency optimisation, the study investigates an alternative tool path strategy and double trochoidal milling. The resulting models were tested against the cutting results obtained from milling of Al 7039 alloy, executed at v_c = 22 m/min, v_f = 240 mm/min and a_p = 1 mm, only to show good correlation of the numerical and experimental results.

Rauch et al. [53] puts forward improvements for trochoidal tool path implementation in roughing. First, the maximum radial depth of cut according to tool path parameterisation was determined, which involved the use of two interpolation models. The second case was an improved tool path generation model for 5086 aluminium alloy pocket milling applications. As a result of the works carried out in the study, trochoidal tool paths were corrected according to objective process constraints. The following milling parameters (a 5-axis machine tool) were employed in the study; v_c = 2400 m/min, f_z = 0.35 mm/tooth, a_p = 6.66–13.50 mm and s_{tr} = 2–8 mm. The recorded maximal cutting force components obtained in the trochoidal milling tool path was equal to 3668 N, while 5670 N was recorded when following a zigzag tool path. The numerical analysis results were validated by the experimental tests. Although the models employed in the study in question appear to have overemphasised the radial depth of cut, according to the trochoidal model employed for programming the tool paths, they do, however, offer short calculation times, and are therefore suitable for quick tool path parameters selection and tool load evaluation.

Gao et al. [54] performed simulation analysis (with a 3D FEM model using Coupled Eulerian Lagrangian (CEL) to simulate milling) along with experimental validation on Al6061-T6 alloy. In the study, linear motion of the workpiece was adopted as a simplification of the trochoidal motion of the end mill. The machining was analysed for two axial depths of cut a_p = 0.3 and 0.5 mm, and three feed rate scenarios f_z = 0.04, 0.07, 0.10 mm/tooth. The cutting speed was constant for all cutting processes–n = 4500 rpm. The verification confirmed the usefulness of the model: the prediction error of the observed cutting forces can be maintained below 12%. The proposed model also shows high accuracy of predicted chip morphology. Unlike other 3D Lagrangian FEM models, the model put forward in this paper does not require a continuous remeshing algorithm to obtain high prediction accuracy. The described FEM model provides additional data, otherwise unextractable from classic cutting experiments, with respect to, e.g., contact forces and interfacial stresses.

Although magnesium alloy machining is a highly advantageous process, it involves several risks, such as the emergence of build-up, as a consequence of adhesion processes. Such conditions may contribute to the occurrence of variations in the components of the cutting force, which may consequently compromise the surface finish quality and the size and shape accuracy; secondly, they may lead to increased vibration and elevated temperature during the machining processes. Magnesium is widely known for its tendency for self-ignition, occurring when the temperature in the area of cut suddenly increases. Lastly, magnesium dust is a serious health hazard to machine tool operators and to the working of the machinery itself [6].

The cutting force components, vibration and computer simulations enhance the stability, security and efficiency of magnesium alloy machining. Mathematical models enable selecting optimal cutting parameters from the viewpoint of self-excited vibration, without engendering redundant machining errors. During milling, temporary surface irregularities may appear in response to chatter between the tool and the workpiece, and it is therefore absolutrly critical to determine stable machining conditions that will ensure a stable tool path and counteract vibration, and chatter. Simulation can be considered as a decision support tool for engineers with respect to specifying favourable technological parameters.

ANNs have the capability to predetermine the particular machining parameters and predict vibration. The validation of numerical results provides positive evidence for further use of artificial neural networks for designing nonlinear processes. Finally, it appears that the scientific literature in the field of trochoidal milling is relatively scant, and therefore there emerges a necessity to investigate how to improve effectiveness, efficiency and stability of magnesium alloy milling.

2. Materials and Methods

The presented study was performed on two types of workpiece materials—AZ91D ($MgAl_9Zn_1$) and AZ31 ($MgAl_3Zn_1$) magnesium alloys—exhibiting high mechanical properties and very good corrosion resistance, which in turn earmark them for a number of different industrial applications. Die-cast magnesium alloys (e.g., AZ91D and Mg–Li) are found in a range of parts and elements manufactured for the aerospace and automotive industries, such as gear housing (e.g., of a steering gear), clutches, gears and components of combustion engines. Another type of magnesium alloys suitable for plastic forming applications (e.g., AZ31, AZ80 and WE43) can be used in aircraft elements, such as the cockpit control panel or transmission. The chemical compositions of the alloys are presented in Table 1.

Table 1. The chemical composition (wt%) of the tested workpiece magnesium alloys.

Chemical Composition	Al	Zn	Mn	Si	Cu	Fe	Ni	Be	Mg
AZ91D	8.91	0.66	0.22	0.016	0.002	0.002	0.001	0.001	rest/other
AZ31	2.9	0.81	0.25	0.01	-	0.003	0.0004	-	rest/other

All milling operations were performed on AVIA VMC 800 HS (Warszawa, Poland) vertical machining centre with Heidenhain iTNC 530 control (power 25 kW, n_{max} up to 24,000 rev./min and $v_{f\,max}$ up to 40 m/min). The cutter tool was mounted in HSK 63A tool holder, capable of performing the cut at speeds from the high-speed machining range. With respect to cutting data, the range of changeable milling parameters was cutting speed v_c = 400–1200 m/min and trochoidal step s_{tr} = 5–30% of the cutter diameter and the fixed parameters of the process were feed per tooth f_z = 0.15 mm/tooth and axial depth of cut a_p = 6 mm.

Trochoidal milling operations were carried out with the straight shank solid carbide (VHM) end mill Fenes (Siedlce, Poland) with TiAlN coating, recommended for aluminium and magnesium milling applications. It is a 2-flute tool z = 2, diameter d = 16 mm of overall dimensions 16 mm × 25 mm × 100 mm and helix angle λ_s = 30°.

The cutting tool was mounted in SECO (Fagersta, Sweden) HSK63A SFD 16 × 120 thermal shrink-fit tool holder. Prior to mounting in the spindle, the tool holder system was balanced in CIMAT (Bydgoszcz, Poland) CMT-15V2N balancing machine, balancing class G 2.5 at 25 000 rev./min, according to ISO1940:2003, which permits a slight residual unbalance tolerance of 1 g mm/kg. In the study, the analysed milling cutter's residual unbalance was 0.77 g mm.

Figure 1 shows the schematic diagram of the tested system (Figure 1a) and the test set-up (Figure 1b) with the object of the study. The model and the machining sequence were designed by means of NX 10.0–SIEMENS (Munich, Germany) software. The measurement apparatus applied in the study was, for cutting force measurement, piezoelectric dynamometer 9257B by Kistler (Winterthur, Switzerland); for vibration acceleration measurement, PCB 352B10 accelerometer by Piezotronics (New York, NY, USA); for vibration displacement, optoNCDT LD1605-2 laser sensor by Micro-Epsilon Masstechnik (Ortenburg, Germany); and the 3D surface roughness profiles were obtained from the T8000RC120-400 profilographometer provided by Hommel–Etamic Jenooptik (Jena, Germany).

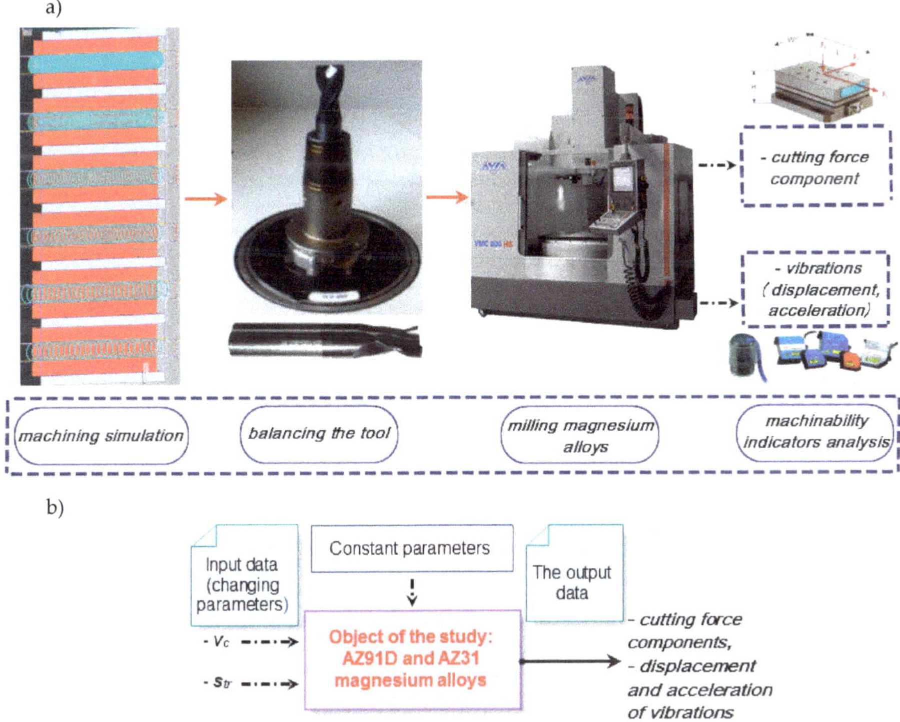

Figure 1. Schematic diagram (**a**) and test set-up (**b**) with the object of the study

In the reported tests, vibration displacement and acceleration measurements were performed in the Y-axis (along the machine tool axis). The cutting forces were measured with 9257B Kistler piezoelectric dynamometer and equipped with a charge amplifier 5017B. The techincal data for the Kistler dynamometer is presented in Table 2.

Table 2. Basic specifications of Kistler 9257B and amplifier 5017B [55]

9257B/5017B*	Measuring Range [kN]	Sensitivity [pC/N]	Operating Temp. Range [°C]	Measuring Range* [pC]	Linearity* [%]	Sampling Rate* [kHz]
Value	±5	≈−7.5	0–70	±10–999 000	≤±0.05	10

The basic specifications of the PCB 352B10 sensor are: sensitivity 1.02 mV/(m/s^2), measurement range ±4905 m/s^2 pk, frequency range 2 to 10,000 Hz, resonant frequency ≥65 kHz and broadband resolution 0.03 m/s^2 rms. The specifications of the laser optical sensor optoNCDT LD1605-2 were measuring range 2 mm, resolution from 0.5 μm and spot diameter 0.3 mm. 3D surface roughness measurements were performed at the following specifications; Lt = 4.8 mm, Lc = 0.8 mm and v_t = 0.5 mm/s. The scanned sample surface area was 4.8 × 4.8 mm and it was measured in approximately 200 scanning steps.

Modal analysis is a process typically employed in various research works with the purpose of investigating basic dynamic parameters (i.e., parameters determined in milling, e.g., the cutting force components and vibration displacement/acceleration) of given systems as well as to generate stability lobe diagrams. The SLD analysis was performed with the use of CutPro software (9.0, Manufacturing Automation Laboratories Inc., Vancouver, BC, Canada), a 352B10 PCB accelerometer and a 070A02 PCB scope input adaptor.

In addition, numerical simulation of process parameters was employed to predict the nonlinear technological process. It is rather difficult to define nonlinear processes by means of mathematical equations. Given that the milling process is a control object, then the simulated parameters, the cutting force components F_x and F_y and the vibration displacement (x) constitute its output parameters. The variable input parameters of milling were cutting speed v_c and trochoidal step s_{tr}, while the remaining parameters were fixed. Figure 2 presents the schematic model of the process, where v_c and s_{tr} denote variable input parameters and nn gives the currently simulated corresponding parameter. Since there were three machining parameters and two workpiece materials analysed in the study—AZ31 and AZ91D—the total number of six different neural networks were obtained.

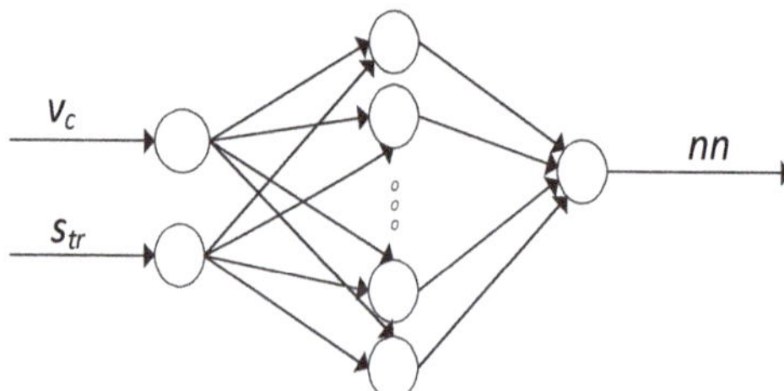

Figure 2. Schematics of the artificial neural network with the analysed process parameters

The network used in this study employed the black box model, as this solution offers good performance on the training set. Furthermore, the selected model is utilised whenever determination of mathematical equations describing the analysed process is complicated, as in the case with the process investigated in this study. It was thus resolved that the black box approach would most accurately reflect the complex character of milling. Since machining is repeatable, then the force input and cutting conditions in selected points of the tool path may be similarly considered repeatable. However, for the sake of accuracy of the model, it must be assumed that cutting is performed without tool wear and at constant cutting parameters in all consecutive machining cycles.

The cutting force components (F_x and F_y) and vibration displacement in AZ31 and AZ91D alloy milling by means of artificial neural networks were modelled with the application of Statistica Neural Networks software. The network models employed in the study were RBF (Radial Basis Function) and MLP (Multilayered Perceptron). For both alloys, all machining parameters, the component F_x of the cutting force, the component F_y of the cutting force and the displacement of vibrations x were modelled separately. The established parameters were the average of the maximum values from the 10 ranges separated from the stable machining area and constituted the output value for individual models. The MLP network was trained with the application of the BFGS (Broyden–Fletcher–Goldfarb–Shanno) method, while the RBF network was trained with the RBFT. The former modelling method employed the linear, exponential, logistic, tanh and sinusoidal activation functions. The activation function of the RBF for hidden neurons is the Gaussian distribution and for the output neurons—a linear function. Both MLP and RBF networks contained one hidden layer, which was dictated by simplicity factors. In the input layer, there are two neurons (cutting speed v_c and trochoidal step s_{tr}) and, in the output layer, the neuron represents a currently simulated parameter, i.e., F_x or F_y component of the cutting force or vibration displacement x). The number of training epochs was in the range of 150 to 250 epochs and the hidden neurons (2–10) were selected experimentally.

The key indicators of the network fitting were training quality, validation quality and training error determined with the least squares method. The learning group used 80% of measurement results and the remaining 20% of results served the purpose of validation. As postulated in the research by Zagórski et al. [46], given that the amount of available data is insufficient, the test group may be omitted, which is what occurred in the reported study.

3. Results and Discussion

SLD analysis is a tool for the determination of stable machining areas. Operating within thus established parameters, i.e., in this case rotational speed n and axial depth of cut a_p, should ensure that the system remains stable throughout milling. However, SLD analysis is carried out in static tests and fails to account for the dynamics of the machining system (resulting from the complexity of the machine tool and of the milling process itself). Figure 3 shows SLD diagrams for the tested cutting tool parameters and magnesium alloys.

a)

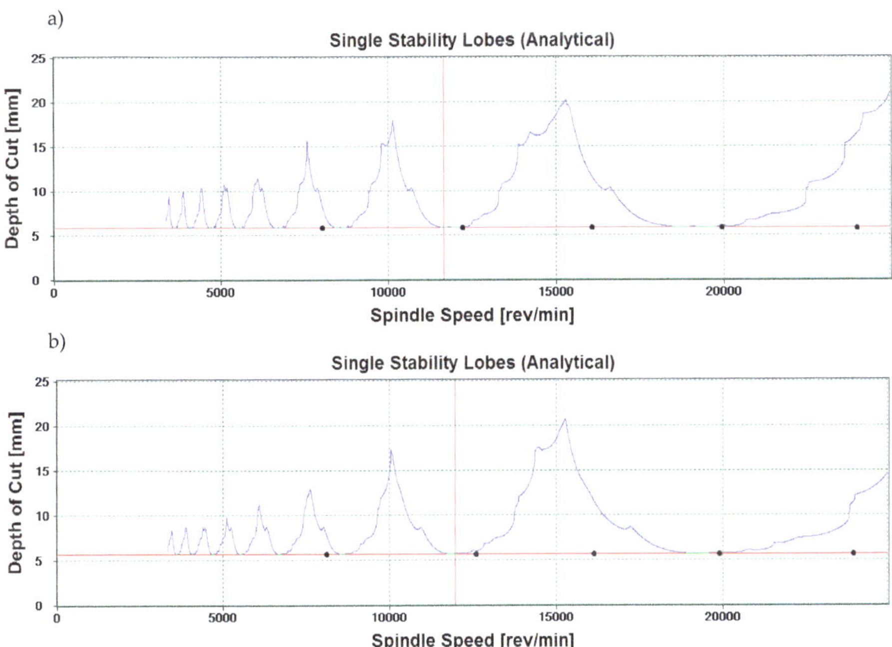

b)

Figure 3. Stability lobes diagrams for (**a**) AZ31 and (**b**) AZ91D magnesium alloys

The markings in the chart define the given machining regions as stable—in the upper part of the chart above the stability lobes (often marked as "+"), unstable—below the stability lobes (often marked as "–")—and "●" the settings selected as milling parameters.

3.1. The Cutting Force Components

The components of the cutting forces F_x, F_y and the displacement and acceleration of vibrations were recorded throughout the milling of magnesium alloys AZ91D and AZ31. The test results were presented in the form of box plot charts and bar charts showing the maximum values of the analysed machinability parameters. These charts provide the information regarding the maximum range of values of vibrations and of the total cutting force components, as well as the range, i.e., the difference between the minimum and the maximum values of the parameters in question. For certain machining settings, the charts also show outliers and/or extremes, which may be associated with the increase in the cutting force components and vibrations caused by temporary stability loss under given machining parameters. A tighter range of outliers/extreme values in box plots and lower values of standard deviation in bar charts may be indicative of improved stability of the process.

Figures 4 and 6 show box plot charts, whereas Figures 5 and 7 show the maximum values of F_x, F_y, the amplitude of these components—AF_x and AF_y—and the normalised force coefficient RMS values—F_{x_RMS} and F_{y_RMS}. In addition, the standard deviation for individual components of the cutting force and its amplitude are presented.

It may be seen from Figure 4a that the most advantageous speeds from the perspective of the cutting force components are: 600 m/min, 1000 m/min and 1200 m/min, as it is under these conditions that the most beneficial conditions occur, i.e., minimum F_x and the tightest ranges of outliers and extremes. Extreme values tend to occur at v_c 400 m/min and 800 m/min. By analysing Figure 4b, it can be seen that trochoidal step change does not significantly affect the increase in the outliers and extreme ranges for the F_x component (over the entire s_{tr} change interval). Nevertheless, the change in the s_{tr} typically leads to the higher value of the F_x component (in the majority of cases) than in the change in v_c. In conclusion, it is advantageous to carry out trochoidal milling within the scope of parameters pertaining to the HSM machining range. Similar to the use of straight tool paths [8], the components of the cutting forces drop with the increase in the cutting speed v_c.

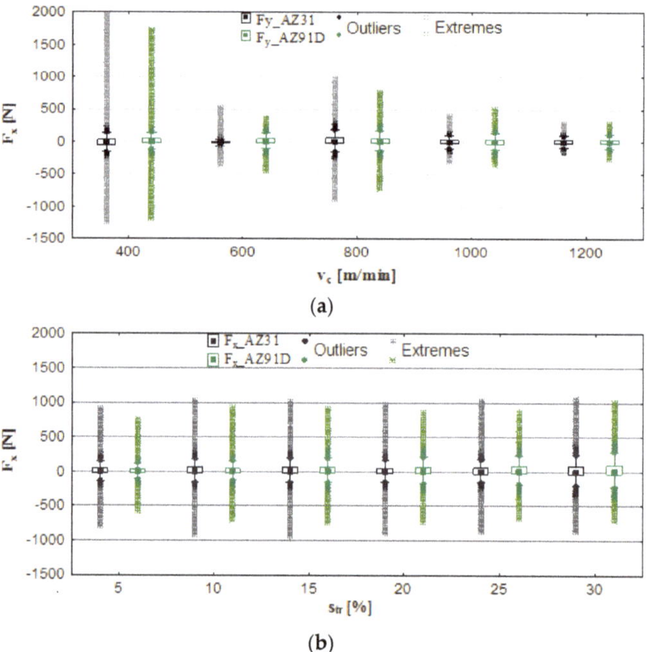

Figure 4. Effect of cutting speed and trochoidal step on cutting force component F_x in milling of magnesium alloys: (**a**) v_c (f_z = 0.15 mm/tooth, a_p = 6 mm, s_{tr} = 15%) and (**b**) s_{tr} (v_c = 800 m/min, f_z = 0.15 mm/tooth, a_p = 6 mm).

Similar tendencies as in Figure 4 are clearly visible in Figure 5, where the component F_x, its amplitude AF_x and the normalised force coefficient RMS value F_{x_RMS} generally show a tendency to decrease practically over the entire tested cutting speed range (except for v_c = 800 m/min). Maximum F_x was recorded for the AZ31 alloy amounted to F_x = 1940 N and the minimum F_x = 291 N, in the AZ91D alloy: maximum F_x = 1707 N and minimum F_x = 301 N. In s_{tr} analysis, its values fluctuated in the range of 901 to 1020 N (for AZ31) and approximately 753 to 991 N (for AZ91D). Lower values of the F_x component were generally observed in AZ91D alloys.

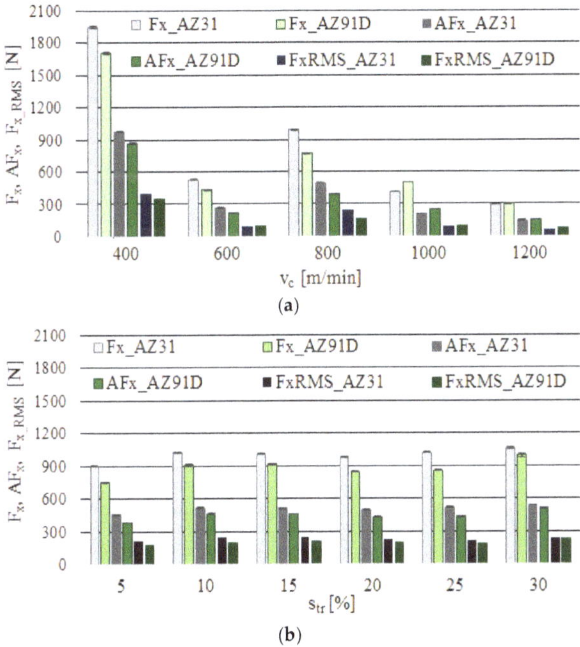

Figure 5. Maximum value of cutting force component F_x, amplitude and RMS in the milling of magnesium alloys: (**a**) effect of cutting speed v_c (f_z = 0.15 mm/tooth, a_p = 6 mm, s_{tr} = 15%) and (**b**) effect of trochoidal step s_{tr} (v_c = 800 m/min, f_z = 0.15 mm/tooth, a_p = 6 mm).

On the basis of Figure 5, it can be observed that the normalised force coefficient RMS values exhibit a close correlation with the amplitudes: in the majority of cases, they constitute up to approximately 50% of this value. Upon a close investigation of the problem and having obtained the results from measurements, it may be concluded that the highest tool damage and workpiece deformation occur at the cutting speed v_c = 400 m/min and v_c = 800 m/min, whereas the smallest at v_c = 1200 m/min. It is, therefore, this machining region that should be indicated as "safe" considering the execution of milling processes, particularly in the case of thin-walled elements and parts (less machining deformations).

To an extent, Figure 6a appears to confirm the previous findings: more favourable machining parameters (as exemplified by fewer outliers and extremes) occur in the range of v_c = 1000 to 1200 m/min. However, in the case of trochoidal step (Figure 6b), the findings differ. Here the F_y component develops the minimum values for p = 5% of the cutter diameter. This is a consequence of the increase in the s_{tr} parameter and the simultaneous increase in the diameter of the area of cut. In this case, therefore, it is preferable to use high values of v_c speed and small trochoidal steps: s_{tr} = 5–15% for AZ31 alloy and p = 5–10% for AZ91D alloy.

The analysis of the maximum values, the amplitude and the normalised force coefficient RMS value of F_y (Figure 7) show that the aforementioned exhibit a similar response to the change in v_c as in the case of the F_x. Here, the most favourable processing conditions were established as v_c = 1200 m/min, f_z = 0.15 mm/tooth, a_p = 6 mm and s_{tr} = 15%, for which the values of the analysed indicator were AZ31 alloy—F_y = 202 N, AF_y = 104 N and F_{y_RMS} = 50N, while in the AZ91D alloy—F_y = 236 N, AF_y = 120 N and F_{y_RMS} = 61N. Lower values of the F_y component of the total cutting force were typically observed: under changing v_c for AZ91D alloy, and in the case of the change in s_{tr} for AZ31 alloy.

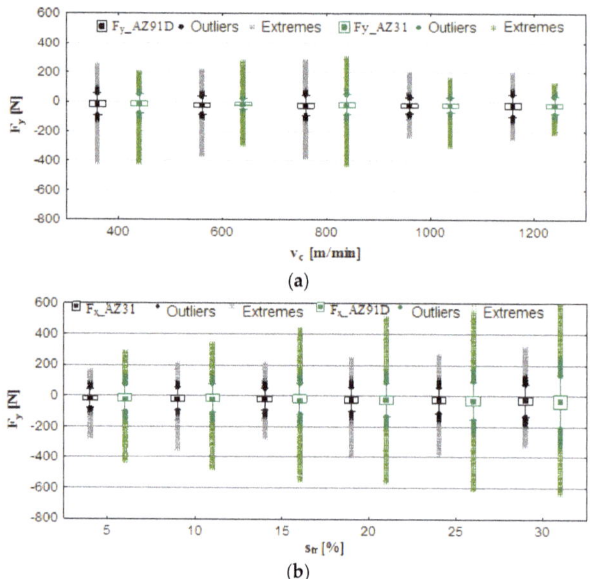

(a)

(b)

Figure 6. Effect of cutting speed and trochoidal step on cutting force component F_y in milling of magnesium alloys: (**a**) v_c (f_z = 0.15 mm/tooth, a_p = 6 mm, s_{tr} = 15%) and (**b**) s_{tr} (v_c = 800 m/min, f_z = 0.15 mm/tooth, a_p = 6 mm).

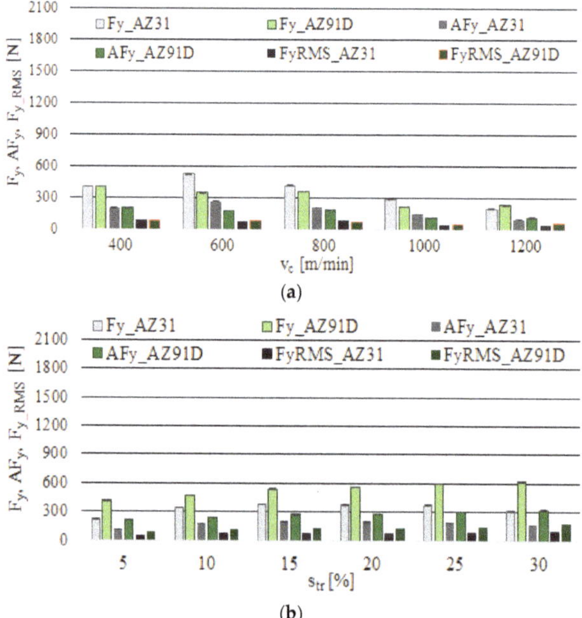

(a)

(b)

Figure 7. Maximum value of cutting force component F_y, amplitude and root mean square (RMS) in the milling of magnesium alloys: (**a**) effect of cutting speed v_c (f_z = 0.15 mm/tooth, a_p = 6 mm, s_{tr} = 15%) and (**b**) effect of trochoidal step s_{tr} (v_c = 800 m/min, f_z = 0.15 mm/tooth, a_p = 6 mm).

3.2. Vibrations (Displacement)

With respect to vibration displacement, the analysed output data were: displacement of vibrations x, their amplitude Ax and the normalised force coefficient RMS value of vibration displacement $x_{\_RMS}$. The analytical procedure for vibration displacement resembled that of the cutting force, i.e., the first stage consisted in computing the box plot charts, which subsequently provided the data for the bar graphs and next for determining the standard deviation. These relationships (x, Ax and $x_{\_RMS}$) are presented in the function of the technological cutting parameter—cutting speed v_c and the additional parameter is trochoidal step s_{tr}.

Figure 8a indicates that the most favorable conditions regarding vibration displacement occur when milling is performed at v_c = 1000–1200 m/min. Considering AZ91D alloy, one may see no outliers or extremes, which clearly suggests very good machining stability throughout the entire range of cutting speeds v_c. In the case of AZ31 alloy, although outliers/extremes are admittedly present in graphs, the box plot range is the smallest for the entire analysed range of v_c. Figure 8b shows that trochoidal step change leads to an increase in the range of recorded values and outliers/extremes for vibration displacement. What follows from these observations is that it is more advantageous to repeatedly increase the cutting speed rather than the trochoidal step, due to the fact that higher values of v_c translate to a more favorable range of x. The most unfavourable range of displacement vibrations x occurs at s_{tr} = 20–30%.

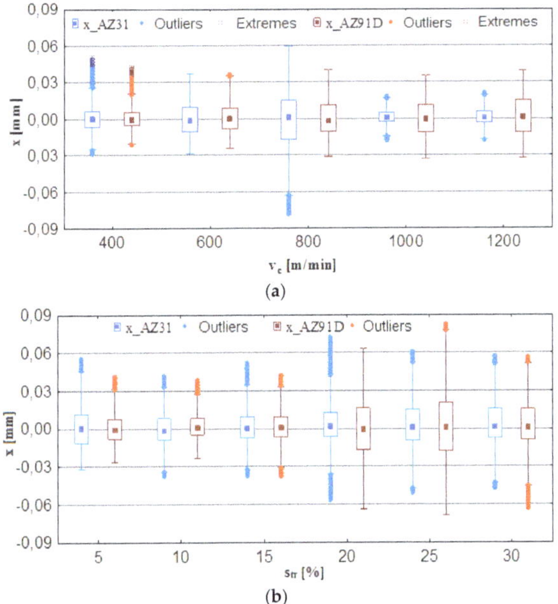

Figure 8. Effect of cutting speed and trochoidal step on vibrations displacement in milling of magnesium alloys: (**a**) v_c (f_z = 0.15 mm/tooth, a_p = 6 mm, s_{tr} = 15%) and (**b**) s_{tr} (v_c = 800 m/min, f_z = 0.15 mm/tooth, a_p = 6 mm).

As shown in Figure 9a, the most favorable cutting conditions are obtained for the speeds v_c = 600 m/min and v_c = 1000–1200 m/min. Comparable values of x, Ax and $x_{\_RMS}$ (Figure 9b) were recorded, even for s_{tr} = 15%, however, at a lower cutting speed—v_c = 800 m/min. Therefore, the data suggest that in order to achieve an increase machining efficiency while not contributing to the deterioration of the machining conditions defined by the vibration displacement, it is more convenient to increase the cutting speed v_c rather than trochoidal step s_{tr}.

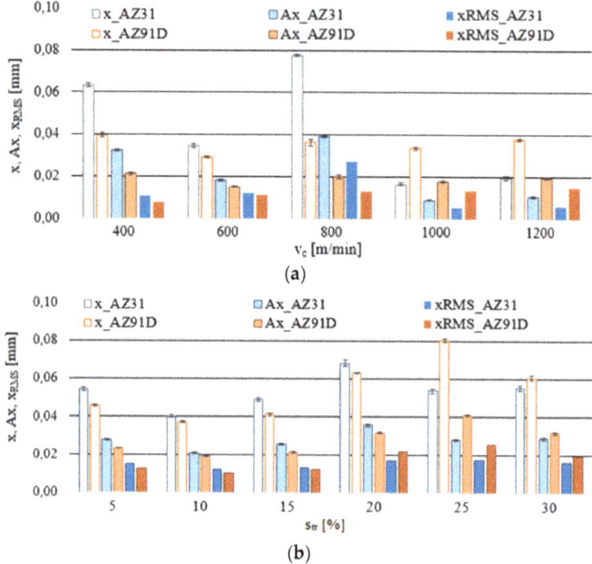

Figure 9. Maximum value of vibrations displacement, amplitude and RMS in the milling of magnesium alloys: (**a**) effect of cutting speed v_c ($f_z = 0.15$ mm/tooth, $a_p = 6$ mm and $s_{tr} = 15\%$) and (**b**) effect of trochoidal step s_{tr} ($v_c = 800$ m/min, $f_z = 0.15$ mm/tooth and $a_p = 6$ mm).

Generally, it should be emphasised that vibration displacement values that can be regarded as "favourable" milling conditions do not exceed $x = 0.04$ mm. The normalised force coefficient RMS value of vibration displacement x_{RMS} is in the range of $x_{RMS} = 0.005$-0.012 mm.

3.3. Vibrations (Acceleration)

The analysis of the output vibration acceleration data was performed according to the procedure applied in the vibration displacement analysis. The selected output parameters were: vibration acceleration a, vibration acceleration amplitude Aa and normalised force coefficient RMS value of vibration acceleration a_{RMS}. The results obtained from tests provided data for the computation of box plot charts and bar graphs with the standard deviation. Similarly to the formerly analysed parameters, the functional relations (a, Aa and a_{RMS}) are presented in the function of the technological cutting parameter—cutting speed v_c and the additional parameter is trochoidal step s_{tr}.

From Figure 10a, it may be seen that the most favourable cutting parameters with respect to vibration acceleration (and similarly to displacement analysis) are $v_c = 600$ m/min and $v_c = 1000$–1200 m/min. Under these conditions, the recorded vibration acceleration value does not exceed the range of ± 75 m/s^2. A considerably higher range of values of vibration acceleration is observed in the trochoidal step change (Figure 10b). In general, outliers/extreme occur for all presented cutting conditions; however, the intensity of their occurrence is notably lower under the conditions of changing cutting speed v_c.

Figure 11a further confirms the previous observations regarding the most favourable cutting parameters as occurring at the speed range of $v_c = 600$ m/min and $v_c = 1000$–1200 m/min. At the given levels of v_c, there are practically no instances in which the level of 50 m/s^2 would be exceeded. The trochoidal step change (Figure 11b) results in the increase in a, Aa and a_{RMS}. Therefore, it is in another instance that it becomes confirmed that machining efficiency is boosted by increasing the cutting speed v_c. In this case, higher values of vibration acceleration were observed for the foundry alloy AZ91D.

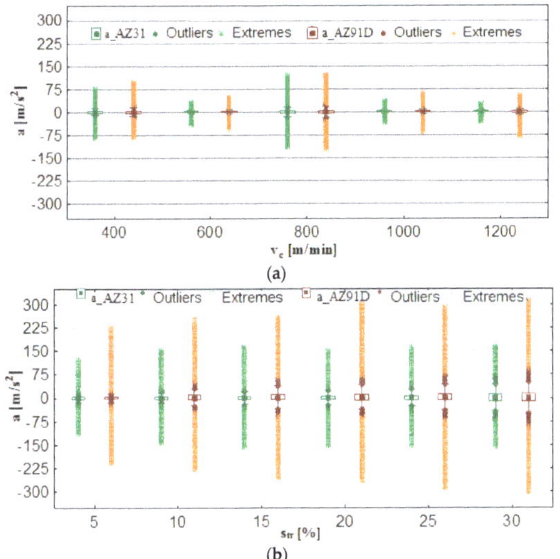

Figure 10. Effect of cutting speed and trochoidal step on vibrations acceleration in milling of magnesium alloys: (**a**) v_c (f_z = 0.15 mm/tooth, a_p = 6 mm, s_{tr} = 15%) and (**b**) s_{tr} (v_c = 800 m/min, f_z = 0.15 mm/tooth, a_p = 6 mm).

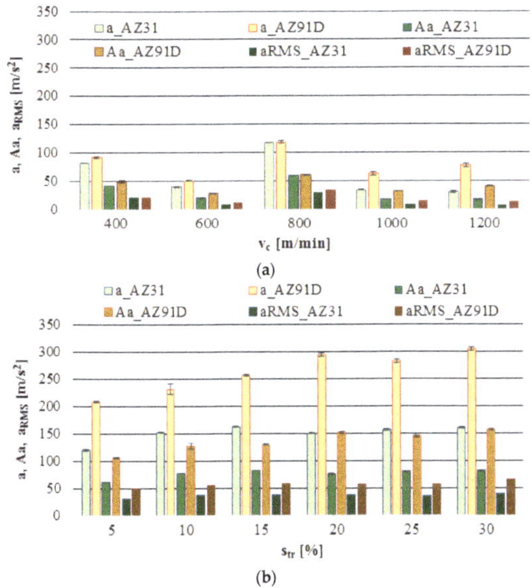

Figure 11. Maximum value of vibrations acceleration, amplitude and RMS in the milling of magnesium alloys: (**a**) effect of cutting speed v_c (f_z = 0.15 mm/tooth, a_p = 6 mm and s_{tr} = 15%) and (**b**) effect of trochoidal step s_{tr} (v_c = 800 m/min, f_z = 0.15 mm/tooth and a_p = 6 mm).

3.4. 3D Surface Roughness

With the purpose of providing a more detailed analysis of milling effects and relationships between forces occurring in the process, selected 3D models of the machined workpiece surface roughness profiles are presented in Table 3, along with more important 3D roughness surface parameters. What is instantly visible is the characteristic lay of machining marks carved in the workpiece during each pass along the trochoidal tool path.

Table 3. Selected isometric images of AZ31 and AZ91D specimen surfaces after trochoidal milling.

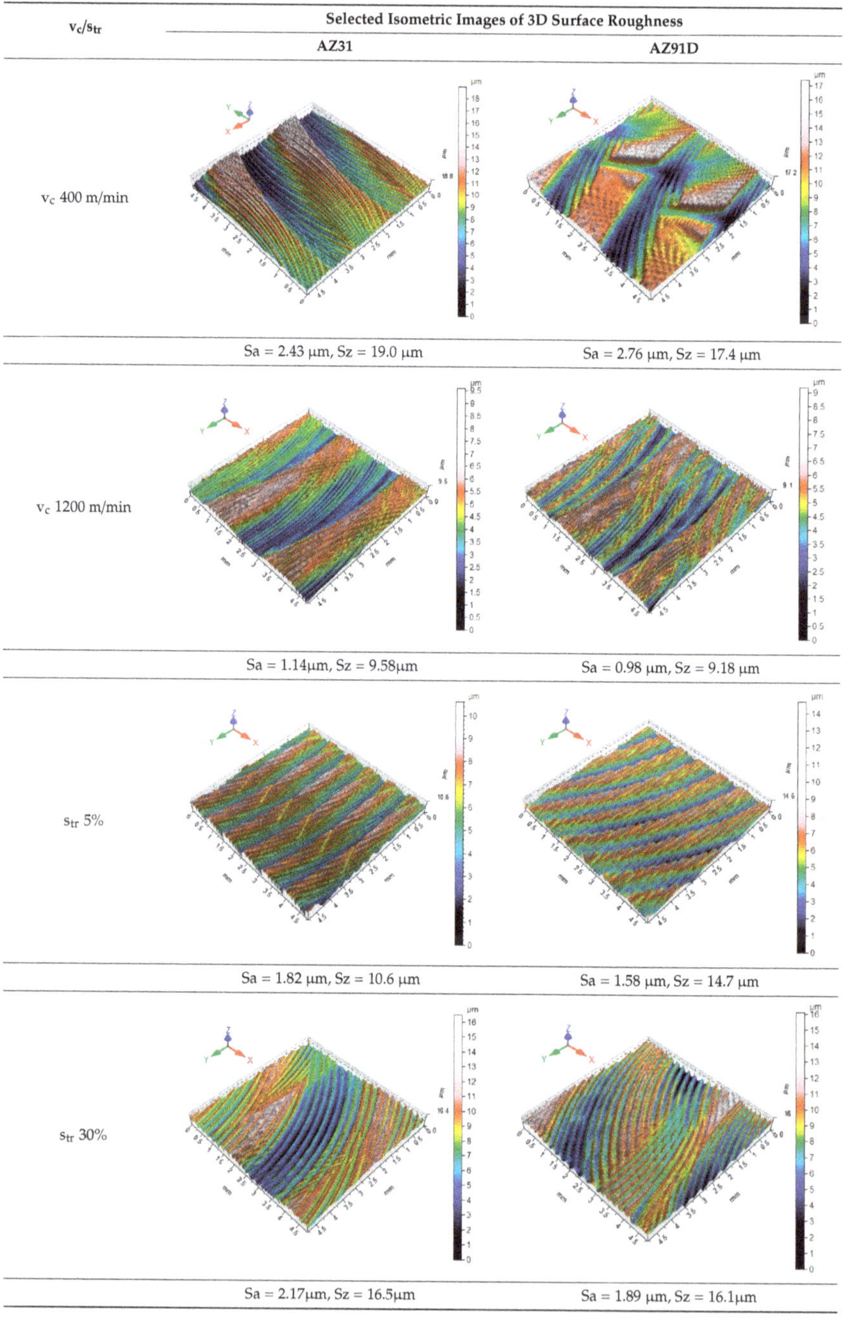

v_c/s_{tr}	Selected Isometric Images of 3D Surface Roughness	
	AZ31	AZ91D
v_c 400 m/min	Sa = 2.43 µm, Sz = 19.0 µm	Sa = 2.76 µm, Sz = 17.4 µm
v_c 1200 m/min	Sa = 1.14µm, Sz = 9.58µm	Sa = 0.98 µm, Sz = 9.18 µm
s_{tr} 5%	Sa = 1.82 µm, Sz = 10.6 µm	Sa = 1.58 µm, Sz = 14.7 µm
s_{tr} 30%	Sa = 2.17µm, Sz = 16.5µm	Sa = 1.89 µm, Sz = 16.1µm

From the perspective of improving machining conditions with respect to surface roughness, the presented visual representation of the machined surface appears to indicate that higher cutting speed provides more favourable cutting conditions than higher trochoidal step value. In the presented example (cf. Table 3), the 3D roughness parameters are approximately 50% lower when machining at v_c = 1200 m/min, than for s_{tr} = 30%. This confirms the validity of thesis postulating that the increase in the cutting speed increases the efficiency and effectiveness of milling magnesium alloys. These values are typical for the surface obtained in the finishing milling proces (for s_{tr} = 30%) and comparable with rough grinding (for v_c = 1200 m/min).

3.5. Numerical Modelling of the Cutting Force Components and Vibrations with Artificial Neural Networks

The experimental data on the components of cutting forces as well as displacement of vibrations during milling of AZ91D and AZ31 alloys were utilised as input data for simulation of these quantities by means of suitable artificial neural networks. Simulations were carried out with the Statistica Neural Networks software and involved MLP and RBF networks.

There are several key network quality indicators assessing whether a suitable network type was selected: the quality of learning, the quality of validation as well as the learning error and validation error derived by means of the least-squares method. In the study, 200 networks were experimentally determined for each simulation; subsequently, the quality of networks was evaluated with the indicators above in order to select the most suitable alternative—MLP or RBF. The created network parameters for vibration displacement and the cutting force components F_x and F_y in milling of AZ31 alloy are given in Table 4. The analysis of the tested neural network models indicates that it is the RBF network that produces better results of simulation for all modelled parameters for AZ31 alloy: for vibrations displacement x the RBF neural network structure is 2-7-1 with seven hidden neurons, considering the F_y component the RBF network structure was 2-6-1 and, for the F_x component of the cutting forces, the network included five hidden neurons (RBF 2-5-1).

Table 4. Characteristics of multilayered perceptron (MLP) and radial basis function (RBF) networks for the cutting force components Fx, Fy and vibrations displacement x for AZ31 alloy.

Network No.	Network Name	Quality (Training, %)	Quality (Validation, %)	Error (Training)	Error (Validation)	Activation (Hidden)	Activation (Output)
			Cutting force component F_x				
1	RBF 2-2-1	92.54	86.32	403.45	573.96	Gaussian	Linear
2	RBF 2-5-1	96.88	95.23	158.67	202.54	Exponential	Sinusoidal
			Cutting force component F_y				
3	MLP 2-10-1	87.34	82.74	404.32	562.43	Linear	Tanh
4	RBF 2-6-1	99.89	89.45	25.16	43.54	Gaussian	Linear
			Vibrations displacement x				
5	MLP 2-8-1	99.96	97.13	0.0001	0.0002	Tanh	Sinus
6	RBF 2-7-1	99.99	84.06	0.0001	0.0001	Gaussian	Linear

The MLP and RBG network parameters for vibration displacement are F_x and F_y and the cutting force components modelled in AZ91D alloy milling are presented in Table 5. The comparative analysis of the neural network models leads to the conclusion that for F_x component of the cutting force it was the RBF 2-5-1 network with five neurons that provided better results. However, with regard to other simulated parameters, better fit with the experimental data was obtained for MLP networks: the network with three neurons (MLP 2-3-1) for the F_y component of the cutting force, and with six (MLP 2-6-1) for vibrations displacement x.

Table 5. Characteristics of MLP and RBF networks for the cutting force components F_x, F_y and vibrations displacement x for AZ91D alloy.

Network No.	Network Name	Quality (Training, %)	Quality (Validation, %)	Error (Training)	Error (Validation)	Activation (Hidden)	Activation (Output)
			Cutting force component F_x				
1	MLP 2-7-1	95.79	89.43	387.98	422.63	Tanh	Linear
2	RBF 2-5-1	98.66	95.34	234.67	364.37	Gaussian	Linear
			Cutting force component F_y				
3	MPL 2-3-1	95.93	94.32	198.65	224.56	Exponential	Logistics
4	RBF 2-7-1	91.94	90.24	459.32	502.56	Gaussian	Linear
			Vibrations displacement x				
5	MLP 2-6-1	97.88	98.84	0.0005	0.0004	Tanh	Tanh
6	RBF 2-5-1	95.97	99.43	0.0008	0.0005	Gaussian	Linear

The numerical results of the total the cutting force components F_x and F_y and vibration displacement x according to the cutting speed v_c and trochoidal step s_{tr} for AZ31 substrate are shown in Figure 12, while for AZ91D alloy in Figure 13. Once v_c and s_{tr} are fed into Statistica, the corresponding parameters (F_x, F_y and x) are obtained.

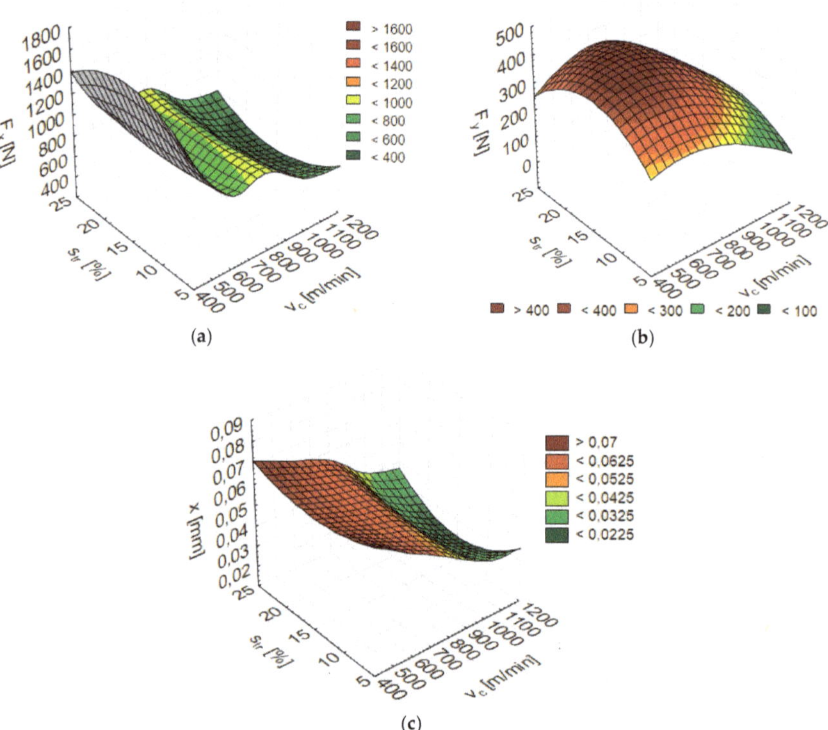

Figure 12. Numerical results of (**a**) Fx, (**b**) F_y and (**c**) x depending on the cutting speed v_c and trochoidal step s_{tr} for AZ31 alloy (F_x: RBF 2-5-1, F_y: RBF 2-6-1, x: RBF 2-7-1).

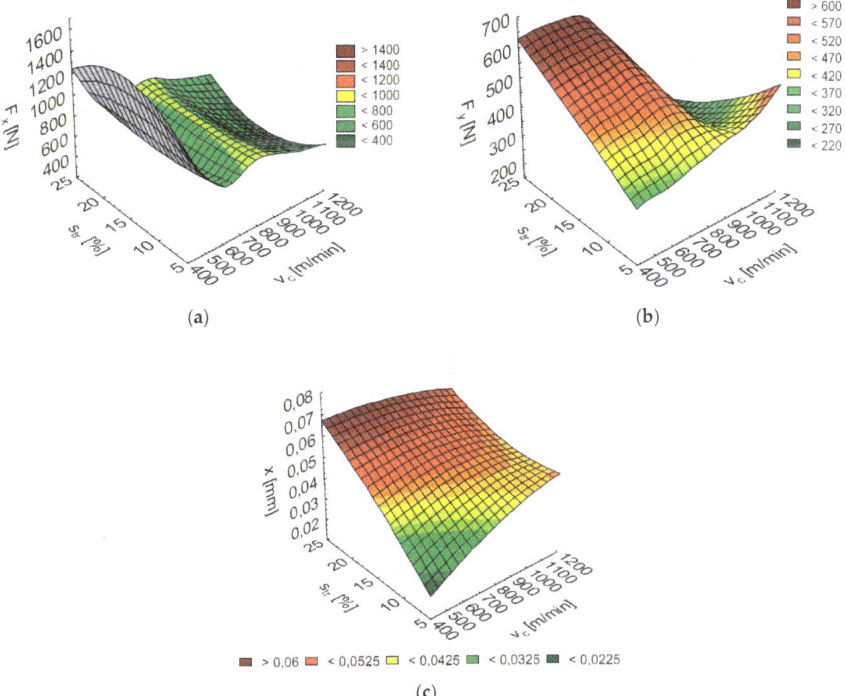

Figure 13. Numerical results of (**a**) Fx, (**b**) Fy and (**c**) x depending on the cutting speed v_c and trochoidal step s_{tr} for AZ91D alloy (F_x: RBF 2-5-1, F_y: MLP 2-3-1, x: MLP 2-6-1).

The accuracy of the modelled networks is shown in Figure 14, which provides a comparison of F_x and F_y components of the cutting force for AZ31 and AZ91D alloy material with numerical values depending on the cutting speed v_c at fixed trochoidal step s_{tr} = 15%. Figure 15 presents a comparison of modelled and experimental data for vibration displacement x for the same parameters of both alloys. Analysing simulation results with actual values of simulated parameters, the relative error did not exceed 15%.

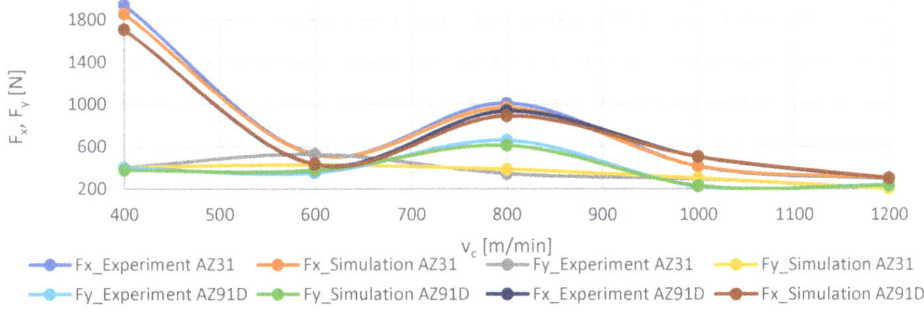

Figure 14. Comparison of experimental and numerical results depending on the cutting speed v_c for trochoidal step s_{tr} = 15% for F_x and F_y components of the cutting force in milling of AZ31 and AZ91D alloys.

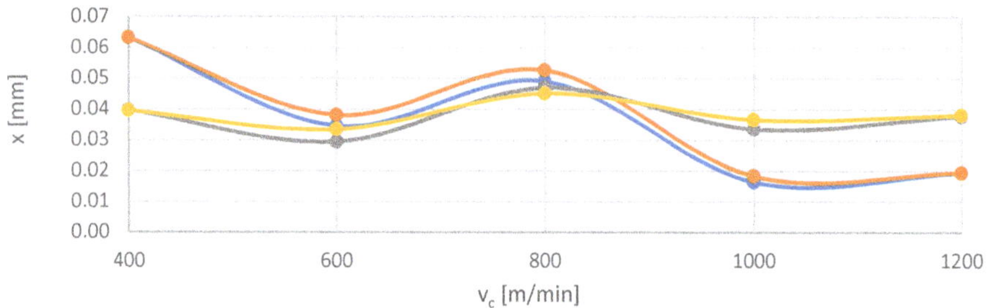

Figure 15. Comparison of experimental and numerical results depending on the cutting speed v_c for trochoidal step $s_{tr} = 15\%$ for vibrations displacement x in milling of AZ31 and AZ91D alloys.

The numerical results obtained from simulations exhibit an acceptable level of error, below 15%. On the one hand, this is a result, and on the other, the indication of artificial neural networks suitability for the use as a tool for simulating, e.g., parameters of abrasive water jet machining. Such data becomes instrumental for the elaboration of working numerical models. The capabilities of artificial neural networks earmark them as a perfect tool for numerical modelling of machining processes.

4. Conclusions

The analysis of the experimental and mathematical data produced in the presented study leads to the formulation of the following conclusions.

1. Modification of either of the tested machining parameters, v_c or s_{tr} affects the values of the cutting force components: F_x is shown to increase when $v_c = 1200$ m/min—$F_x \approx 300$ N (in both tested magnesium alloys) and $s_{tr} = 30\%$—$F_x \approx 1000$ N (in both tested magnesium alloys), compared with F_y for $v_c = 1200$ m/min—$F_y \approx 200$ N (for AZ31) and $F_y \approx 235$ N (for AZ91D), while for $s_{tr} = 30\%$—$F_y \approx 313$ N (for AZ31) and $F_y \approx 620$ N (for AZ91D).
2. More stable and, therefore, effective machining, as well as lower values of the cutting forces and vibrations were obtained for the range of cutting speeds $v_c = 1200$ m/min.
3. The character of vibrations is relative to milling parameters; stable machining conditions may be obtained by increasing the cutting speed and reducing trochoidal step.
4. Dry milling of magnesium alloys can be performed in a wide range of technological parameters of machining. With the increase in v_c, the values of the analysed machinability indicators decrease, which is an indirect confirmation that dry milling of magnesium alloys is a safe process.
5. Comparing both grades of AZ31/AZ91D magnesium alloys, it was found that

    -   in general, higher values of the cutting force components prevail in AZ31 alloy milling,
    -   vibration displacement x analysis shows that milling performed at lower parameters (v_c and s_{tr}), i.e., up to 800 m/min and 20% trochoidal step, results in higher values of x in AZ31 alloy, whereas at higher cutting parameters ranges $v_c > 800$ m/min and $s_{tr} > 20\%$—higher displacement values were recorded in AZ91D; this is potentially due to the differences in the character of the decohesion mechanism typical of the analysed workpiece material (brittle elastic or elastic–plastic) at higher v_c and s_{tr},
    -   from analysis of vibration acceleration a, it can be seen that vibration acceleration is higher in milling of AZ91D
    -   and there are no clear differences in the 3D roughness of the magnesium alloys surface AZ91D and AZ31.

6. In both tested magnesium alloys, the limit cutting speed $v_c = 800$ m/min; this limit marks the point exceeding which results in the decrease in the values of the analysed machinability parameters (the cutting force components and vibration).

7. With regard to the simulation results of the cutting forces and their correlation with actual cutting data, the analysis confirms that no discrepancies between the modelled and experimental data exceeded 15%, which is within the permissible level of error.

8. The simulation by means of artificial neural networks shows sufficient capacity and accuracy to initially determine such machining parameters as the components of the cutting force and the displacement of vibrations.

9. The relationships between nonlinear dependences between machining parameters and the values of the cutting force components or vibrations displacement x, represented in the neural structure of networks, enable investigating the process without the need for laborious, time-consuming and often cost-intensive machining tests.

10. By means of simulation, it is possible to model processes with a nonlinear course, under conditions of incomplete information about the process itself. The simulation results may be utilised in designing a tool for modelling phenomena occurring during machining, which will aid the technologist in the decision process providing them with machining parameters that ensure maintaining the stability of the process.

11. ANNs "transfer" discrete data (input data derived from tests) to continuous data (simulation data builds response surface diagrams).

12. A 50% drop in the value of 3D roughness parameters is observed at $v_c = 1200$ m/min, compared to the change as a result of modification of trochoidal step $s_{tr} = 30\%$. The results from the tests prove that it is the cutting speed that should be increased in order to boost the efficiency and effectiveness of milling magnesium alloys.

Author Contributions: Conceptualization, I.Z., M.K. (Monika Kulisz), M.K. (Mariusz Kłonica) and J.M.; Data curation, I.Z. and M.K. (Monika Kulisz); Formal analysis, M.K. (Monika Kulisz); Investigation, I.Z. and J.M.; Methodology, I.Z.; Software, M.K. (Monika Kulisz); Visualization, I.Z. and M.K. (Monika Kulisz); Writing—original draft, I.Z. and M.K. (Monika Kulisz); Writing—review & editing, I.Z. and M.K. (Mariusz Kłonica) and M.K. (Monika Kulisz), J.M.; Supervision, I.Z.; Project Administration, I.Z.; Funding Acquisition, I.Z.

Funding: The project/research was financed under the project Lublin University of Technology-Regional Excellence Initiative, funded by the Polish Ministry of Science and Higher Education (contract No. 030/RID/2018/19).

Conflicts of Interest: The authors declare no conflicts of interest.

References

1. Fang, F.Z.; Lee, L.C.; Liu, X.D. Mean Flank Temperature Measurement in High Speed Dry Cutting. *J. Mater. Process. Technol.* **2005**, *167*, 119–123. [CrossRef]

2. Danis, I.; Monies, F.; Lagarrigue, P.; Wojtowicz, N. Cutting forces and their modelling in plunge milling of magnesium-rare earth alloys. *Int. J. Adv. Manuf. Technol.* **2016**, *84*, 1801–1820. [CrossRef]

3. Monies, F.; Danis, I.; Lagarrigue, P.; Gilles, P.; Rubio, W. Balancing of the transversal cutting force for pocket milling cutters: Application for roughing a magnesium-rare earth alloy. *Int. J. Adv. Manuf. Technol.* **2016**, *89*, 45–64. [CrossRef]

4. Saptaji, K.; Gebremariam, M.A.; Azhari, M.A.B. Machining of biocompatible materials: A review. *Int. J. Adv. Manuf. Technol.* **2018**, *97*, 2255–2292. [CrossRef]

5. Józwik, J.; Mika, D.; Dziedzic, K. Vibration of thin walls during cutting process of 7075 T651 aluminium alloy. *Manuf. Technol.* **2016**, *16*, 113–120.

6. Oczoś, K.E.; Kawalec, A. *Light Metals Forming*; Wydawnictwo Naukowe PWN: Warszawa, Poland, 2012. (In Polish)

7. Zagórski, I.; Kuczmaszewski, J. The study of cutting forces and their amplitudes during high-speed dry milling magnesium alloys. *Adv. Sci. Technol. Res. J.* **2013**, *7*, 61–66. [CrossRef]

8. Kuczmaszewski, J.; Pieśko, P. Wear of milling cutters resulting from high silicon aluminium alloy cast AlSi21CuNi machining. *Maint. Reliab.* **2014**, *16*, 37–41.

9. Monies, F.; Danis, I.; Bes, C.; Cafieri, S.; Mongeau, M. A new machining strategy for roughing deep pockets of magnesium-rare earth alloys. *Int. J. Adv. Manuf. Technol.* **2017**, *92*, 3883–3901. [CrossRef]

10. Kuczmaszewski, J.; Login, W.; Piesko, P.; Zawada-Michalowska, M. Assessment of the accuracy of high-speed machining of thin-walled EN AW-2024 aluminium alloy elements using carbide milling cutter and with PCD blades. In *Advances in Manufacturing, Lecture Notes in Mechanical Engineering*; Springer: Cham, Germany, 2018; pp. 671–680.

11. Kuczmaszewski, J.; Zagórski, I.; Zgórniak, P. Thermographic study of chip temperature in high-speed dry milling magnesium alloys. *Manag. Prod. Eng. Rev.* **2016**, *7*, 86–92. [CrossRef]

12. Zgórniak, P.; Grdulska, A. Investigation of temperature distribution during milling process of AZ91HP magnesium alloys. *Mech. Mech. Eng.* **2012**, *16*, 33–40.

13. Zgórniak, P.; Stachurski, W.; Ostrowski, D. Application of thermographic measurements for the determination of the impact of selected cutting parameters on the temperature in the workpiece during milling process. *Strojniški Vestn. J. Mech. Eng.* **2016**, *62*, 657–664. [CrossRef]

14. Zagórski, I.; Kuczmaszewski, J. Temperature measurements in the cutting zone, mass, chip fragmentation and analysis of chip metallography images during AZ31 and AZ91HP magnesium alloy milling. *Aircr. Eng. Aerosp. Technol.* **2018**, *90*, 496–505. [CrossRef]

15. Zagórski, I.; Kulisz, M.; Semeniuk, A. Artificial neural network modelling of cutting force components in milling. *ITM Web Conf.* **2017**, *15*, 02001. [CrossRef]

16. Shi, K.; Ren, J.; Zhang, D.; Zhai, Z.; Huang, X. Tool wear behaviors and its effect on machinability in dry high-speed milling of magnesium alloy. *Int. J. Adv. Manuf. Technol.* **2017**, *90*, 3265–3273. [CrossRef]

17. Sivam, S.P.; Bhat, M.D.; Natarajan, S.; Chauhan, N. Analysis of residual stresses, thermal stresses, cutting forces and other output responses of face milling operation on ZE41 Magnesium alloy. *Int. J. Mod. Manuf. Technol.* **2018**, *10*, 92–101.

18. Fu, Z.T.; Yang, W.Y.; Wang, X.L.; Leopold, J. Analytical Modelling of Milling Forces for Helical End Milling Based on a Predictive Machining Theory. *Procedia CIRP* **2015**, *31*, 258–263. [CrossRef]

19. Salguero, J.; Batista, M.; Calamaz, M.; Girot, F.; Marcos, M. Cutting Forces Parametric Model for the Dry High Speed Contour Milling of Aerospace Aluminium Alloys. *Procedia Eng.* **2013**, *63*, 735–742. [CrossRef]

20. Shi, K.; Zhang, D.; Ren, J.; Yao, C.; Huang, X. Effect of cutting parameters on machinability characteristics in milling of magnesium alloy with carbide tool. *Adv. Mech. Eng.* **2016**, *8*, 1–9. [CrossRef]

21. Weremczuk, A.; Rusinek, R.; Warminski, J. Bifurcation and stability analysis of a nonlinear milling process. *AIP Conf. Proc.* **2018**, *1922*, 100008. [CrossRef]

22. Beranoagirre, A.; Urbikain, G.; Marticorena, R.; Bustillo, A.; Lopez de Lacalle, L.N. Sensitivity Analysis of ToolWear in Drilling of Titanium Aluminides. *Metals* **2019**, *9*, 297. [CrossRef]

23. Lipski, J.; Zaleski, K. Optimisation of milling parameters using neural network. *ITM Web Conf.* **2017**, *15*, 01005. [CrossRef]

24. Kilickap, E.; Yardimeden, A.; Celik, Y.H. Mathematical Modelling and Optimization of Cutting Force, Tool Wear and Surface Roughness by Using Artificial Neural Network and Response Surface Methodology in Milling of Ti-6242S. *Appl. Sci. Basel* **2017**, *7*, 1064. [CrossRef]

25. Kazemi, P.; Khalid, M.H.; Szlek, J.; Mirtic, A.; Reynolds, G.K.; Jachowicz, R.; Mendyk, A. Computational intelligence modeling of granule size distribution for oscillating milling. *Powder Technol.* **2016**, *301*, 1252–1258. [CrossRef]

26. Wu, B.H.; Yan, X.; Luo, M.; Gao, G. Cutting force prediction for circular end milling process. *Chin. J. Aeronaut.* **2013**, *26*, 1057–1063. [CrossRef]

27. Weremczuk, A.; Rudzik, M.; Rusinek, R.; Warmiński, J. The concept of active elimination of vibrations in milling process. *Procedia CIRP* **2015**, *31*, 82–87. [CrossRef]

28. Seguy, S.; Insperger, T.; Arnaud, L.; Desseiu, G.; Peigne, G. On the stability of high-speed milling with spindle speed variation. *Int. J. Adv. Manuf. Technol.* **2010**, *48*, 883–895. [CrossRef]

29. Rusinek, R. Stability criterion for aluminium alloy milling expressed by recurrence plot measures. *Proc. Inst. Mech. Eng. Part B J. Eng. Manuf.* **2012**, *226*, 1976–1985. [CrossRef]

30. Yang, Y.; Zhang, W.-H.; Ma, Y.-C.; Wan, M. Chatter prediction for the peripheral milling of thin-walled workpieces with curved surfaces. *Int. J. Mach. Tools Manuf.* **2016**, *109*, 36–48. [CrossRef]

31. Wu, S.; Li, R.; Liu, X.; Yang, L.; Zhu, M. Experimental study of thin wall milling chatter stability nonlinear criterion. *Procedia CIRP* **2016**, *56*, 422–427. [CrossRef]

32. Comak, A.; Budak, E. Modeling dynamics and stability of variable pitch and helix milling tools for development of a design method to maximize chatter stability. *Precis. Eng.* **2017**, *47*, 459–468. [CrossRef]

33. Eynian, M. Vibration frequencies instable and unstable milling. *Int. J. Mach. Tools Manuf.* **2015**, *90*, 44–49. [CrossRef]

34. Quintana, G.; Ciurana, J. Chatter in machining processes: A review. *Int. J. Mach. Tools Manuf.* **2011**, *51*, 363–376. [CrossRef]

35. Rusinek, R.; Zaleski, K. Dynamics of thin-walled element milling expressed by recurrence analysis. *Meccanica* **2015**, *51*, 1275–1286. [CrossRef]

36. Asad, M.; Mabrouki, T.; Rigal, J.-F. Finite-element-based hybrid dynamic cutting model for aluminium alloy milling. *Proc. Inst. Mech. Eng. Part B J. Eng. Manuf.* **2010**, *224*, 1–13. [CrossRef]

37. Yusoff, A.R. Identifying bifurcation behavior during machining process for an irregular milling tool geometry. *Measurement* **2016**, *93*, 57–66. [CrossRef]

38. Friedrich, J.; Hinze, C.; Renner, A.; Verl, A.; Lechler, A. Estimation of stability lobe diagrams in milling with continuous learning algorithms. *Robot. Comput. Integr. Manuf.* **2017**, *43*, 124–134. [CrossRef]

39. Olvera, D.; Urbikain, G.; Elías-Zuñiga, A.; Lopez de Lacalle, L.N. Improving Stability Prediction in Peripheral Milling of Al7075T6. *Appl. Sci.* **2018**, *8*, 1316. [CrossRef]

40. Munoa, J.; Beudaert, X.; Dombovari, Z.; Altintas, Y.; Budak, E.; Brecher, C.; Stepan, G. Chatter suppresion techniques in metal cutting. *CIRP J. Manuf. Sci. Technol.* **2016**, *65*, 785–808. [CrossRef]

41. Hsiao, T.C.; Huang, S.C. The Effect of Cutting Process Parameters on the Stability in Milling. *Adv. Mater. Res.* **2014**, *887–888*, 1200–1204. [CrossRef]

42. Le Lan, J.V.; Marty, A.; Debongnie, J.F. Providing stability maps for milling operations. *Int. J. Mach. Tools Manuf.* **2007**, *47*, 1493–1496. [CrossRef]

43. Campa, F.J.; Lopez de Lacalle, L.N.; Celaya, A. Chatter avoidance in the milling of thin floors with bull-nose end mills: Model and stability diagrams. *Int. J. Mach. Tools Manuf.* **2011**, *51*, 43–53. [CrossRef]

44. Madoliat, R.; Hayati, S.; Ghalebahman, A.G. Modeling and Analysis of Frictional Damper Effect on Chatter Suppression in a Slender Endmill Tool. *J. Adv. Mech. Des. Syst. Manuf.* **2011**, *5*, 115–128. [CrossRef]

45. Burek, J.; Babiarz, R.; Słutkowicz, P.; Sałata, M. Diagnostics of high performance milling of aluminum alloys. *Mechanik* **2016**, *11*, 1652–1653. [CrossRef]

46. Zagórski, I.; Kulisz, M.; Semeniuk, A.; Malec, A. Artificial neural network modelling of vibration in the milling of AZ91D alloy. *Adv. Sci. Technol. Res. J.* **2017**, *11*, 261–269. [CrossRef]

47. Lopez de Lacalle, L.N.; Lamikiz, A.; Muñoa, J.; Sánchez, J.A. The CAM as the centre of gravity of the five-axis high speed milling of complex parts. *Int. J. Prod. Res.* **2005**, *43*, 1983–1999. [CrossRef]

48. Santhakumar, J.; Iqbal, M.U. Parametric Optimization of Trochoidal Step on Surface Roughness and Dish Angle in End Milling of AISID3 Steel Using Precise Measurements. *Materials* **2019**, *12*, 1335. [CrossRef]

49. Celaya, A.; Bo, P.; González, H.; Bartoň, M.; Lopez de Lacalle, L.N. Highly accurate 5-axis flank CNC machining with conical tools. *Int. J. Adv. Manuf. Technol.* **2018**, *97*, 1605–1615. [CrossRef]

50. Pleta, A.; Niaki, F.A.; Mears, L. Investigation of chip thickness and force modelling of trochoidal milling. *Procedia Manuf.* **2017**, *10*, 612–621. [CrossRef]

51. Pleta, A.; Niaki, F.A.; Mears, L. A comparative study on the cutting force coefficient identification between trochoidal and slot milling. *Procedia Manuf.* **2018**, *26*, 570–579. [CrossRef]

52. Otkur, M.; Lazoglu, I. Trochoidal milling. *Int. J. Mach. Tools Manuf.* **2007**, *47*, 1324–1332. [CrossRef]

53. Rauch, M.; Duc, E.; Hascoet, J.Y. Improving trochoidal tool paths generation and implementation using process constraints modelling. *Int. J. Mach. Tools Manuf.* **2009**, *49*, 375–383. [CrossRef]

54. Gao, Y.; Ko, J.H.; Lee, H.P. 3D Eulerian Finite Element Modelling of End Milling. *Procedia CIRP* **2018**, *77*, 159–162. [CrossRef]

55. Kistler Eastern Europe s.r.o. Available online: https://www.kistler.com/ (accessed on 11 June 2019).

MDPI

St. Alban-Anlage 66

4052 Basel

Switzerland

Tel. +41 61 683 77 34

Fax +41 61 302 89 18

www.mdpi.com

Materials Editorial Office

E-mail: materials@mdpi.com

www.mdpi.com/journal/materials